Energy Sources from Agriculture and Rural Areas

Energy Sources from Agriculture and Rural Areas

Editors

Vitaliy Krupin
Roman Podolets

MDPI • Basel • Beijing • Wuhan • Barcelona • Belgrade • Manchester • Tokyo • Cluj • Tianjin

Editors
Vitaliy Krupin
Institute of Rural and Agricultural Development,
Polish Academy of Sciences (IRWiR PAN)
Warsaw
Poland

Roman Podolets
Institute of Economics and Forecasting of the
National Academy of Sciences of Ukraine
Kyiv
Ukraine

Editorial Office
MDPI
St. Alban-Anlage 66
4052 Basel, Switzerland

This is a reprint of articles from the Special Issue published online in the open access journal *Energies* (ISSN 1996-1073) (available at: https://www.mdpi.com/journal/energies/special_issues/energy_sources_agricultural_rural_areas).

For citation purposes, cite each article independently as indicated on the article page online and as indicated below:

LastName, A.A.; LastName, B.B.; LastName, C.C. Article Title. *Journal Name* **Year**, *Volume Number*, Page Range.

ISBN 978-3-0365-8214-6 (Hbk)
ISBN 978-3-0365-8215-3 (PDF)

Cover image courtesy of Vitaliy Krupin

Contents

About the Editors

Vitaliy Krupin

Dr Vitaliy Krupin holds an assistant professor position in the Department of Economic Modelling at the Institute of Rural and Agricultural Development, Polish Academy of Sciences (IRWiR PAN). With a primary major in international economics and trade, over the past years he has focused on the issues of rural development, agricultural and environmental economics, renewable energy sources and climate change. He is involved in several international research projects: BioMonitor4CAP and SoilValues in the Horizon Europe program; SURE-Farm, LIFT and TRADE4SD in the Horizon 2020 program; LIFE Climate CAKE PL and LIFE VIIEW 2050 in the EU's LIFE program; and has formerly contributed to projects financed by USAID, the Ministry of Agriculture and Rural Development of Poland, and the National Academy of Sciences of Ukraine. He is an expert evaluator at the Polish Agency for Enterprise Development (PARP) and the National Research Foundation of Ukraine (NRFU). He is the author of over 150 scientific publications, among which are 12 monographs (1 personal and 11 in co-authorship), and over 50 articles in peer-reviewed scientific journals. He has been awarded for his work on rural and agricultural development by the Parliament of Ukraine (2016) and by the Minister of Agriculture and Rural Development of Poland (2022).

Roman Podolets

Dr Roman Podolets is a Ukraine-based researcher, specializing in the country for over twenty years. He heads the Energy Department at the Institute for Economics and Forecasting of the National Academy of Sciences of Ukraine, where he conducts studies and offers government consultancy on various energy and environmental issues. His most recent studies focus on the economic and energy repercussions of Russian aggression and the rebuilding of Ukraine with a resilient, carbon-neutral energy system. Dr Podolets either leads or contributes expertise to a multitude of research and consultancy projects. He also served as the Head of the Department of Corporate Management and Modelling at the Institute of Oil and Gas Industry "Naukanaftogaz" for two years, and completed an internship at DIW-Berlin, focusing on energy modelling. Author of numerous scientific and analytic publications, including in peer-review journals with high impact factors.

Preface to "Energy Sources from Agriculture and Rural Areas"

This Special Issue is devoted to energy generation within rural areas, including the agricultural sector. Such technologies and application practices vary depending on the type of agricultural activity, local natural conditions, and external factors, thus globally creating a multitude of possible approaches and applications of technologies under particular conditions. We strive to go beyond strict technological perception and enrich it based on a multidisciplinary approach, making it possible to pursue an understanding not only of energy generation technologies, but also the conditions of their implementation and possible measures to increase their efficiency (technological, economic, environmental, and others).

Energy produced within or in addition to agricultural activities supplies the direct needs of farms, as well as other entities, through the energy market. Even though it is an integral part of so-called "renewable sources", agriculture's distinct features create the preconditions and necessity for research on technologies aimed at the generation of energy within agriculture itself. This is needed for both directly addressing such technologies, as well as for the substantiation of measures available to ensure the development of complex approaches towards ensuring sustainability in agriculture. It is also crucial considering the policies aimed particularly at the development of agriculture and rural areas (e.g., the EU's Common Agricultural Policy).

A multitude of economic, social, environmental, and institutional factors constantly modify the conditions for agricultural production, inflicting positive or negative effects upon its structure, output, and efficiency. Climate change issues are forcing agricultural entities to mitigate their negative effect upon the environment, while also creating the necessity to adapt and maintain proper efficiency and output levels. All of these influence the sustainability of rural areas and agriculture, both in terms of its primary production focus, as well as its input into the generation and use of energy.

Vitaliy Krupin and Roman Podolets
Editors

energies

MDPI

Article

Crop Residue Removal: Assessment of Future Bioenergy Generation Potential and Agro-Environmental Limitations Based on a Case Study of Ukraine

Sergii Kyryzyuk [1], Vitaliy Krupin [2,*], Olena Borodina [1] and Adam Wąs [3]

[1] Institute of Economics and Forecasting of the National Academy of Sciences of Ukraine, Panasa Myrnoho 26, 01011 Kyiv, Ukraine; kyryzyuk.ief@gmail.com (S.K.); oborodina@ief.org.ua (O.B.)
[2] Institute of Rural and Agricultural Development, Polish Academy of Sciences, Nowy Świat 72, 00-330 Warsaw, Poland
[3] Institute of Economics and Finances, Warsaw University of Life Sciences (SGGW), Nowoursynowska 166, 02-787 Warsaw, Poland; adam_was@sggw.edu.pl
* Correspondence: vkrupin@irwirpan.waw.pl

Received: 19 September 2020; Accepted: 12 October 2020; Published: 14 October 2020

Abstract: This study assesses the bioenergy generation potential of crop residues in Ukraine for the year 2030. Projections of agricultural development are made based on the Global Biosphere Management Model (GLOBIOM) and verified against available Agricultural Member State Modeling (AGMEMOD) results in regard to the six main crops cultivated in Ukraine (wheat, barley, corn, sunflower, rape and soya). Two agricultural development scenarios are assessed (traditional and innovative), facilitating the projection of future crop production volumes and yields for the selected crops. To improve precision in defining agro-environmental limitations (the share of crop residues necessary to be kept on the fields to maintain soil fertility for the continuous cultivation of crops), yield-dependent residue-to-product ratios (RPRs) were applied and the levels of available soil nutrients for regions of Ukraine (in regard to nitrogen, phosphorus, potassium and humus) were estimated. The results reveal the economically feasible future bioenergy generation potential of crop residues in Ukraine, equaling 3.6 Mtoe in the traditional agricultural development scenario and 10.7 Mtoe in the innovative development scenario. The projections show that, within the latter scenario, wheat, corn and barley combined are expected to provide up to 81.3% of the bioenergy generation potential of crop residues.

Keywords: crop residue; bioenergy; generation potential; residue-to-product ratio; soil nutrient balance; cereals; industrial crops; GLOBIOM model; Ukraine

---

1. Introduction

According to the national strategic documents from 2017 [1], renewable energy is expected to play a growing role in Ukraine, reaching 15.5 Mtoe or 17% of total energy supply by 2030, while the energy generated from biomass, biofuels and waste is projected to reach 8.8% or 8 Mtoe. This is deemed crucial for the diversification of energy sources and for increasing Ukraine's independence from foreign energy suppliers, while having a favorable impact on climate change and the environment.

Currently (2018 being the latest available data), renewable sources in Ukraine generate 4.3 Mtoe or 4.6% of the total energy supply, while biomass, biofuels and waste combined generate 3.2 Mtoe or a 3.4% share [2]. To reach the goals set, an intense structural transformation of the energy sector is needed, shifting it toward renewable energy generation. Private investors (as the key actors in this transformation) must follow the indicative development path and implement the available production technologies.

As one of the leading domestic entities in the energy sector, the Bioenergy Association of Ukraine [3] states that the majority of bioenergy generation potential in Ukraine is based on the crop residues of mainly cereals and oleaginous plants (both from field and processing) and energy crops (corn, rape, energy trees, and shrubs), which—combined—provide up to 83% of the country's total biomass potential (Figure 1).

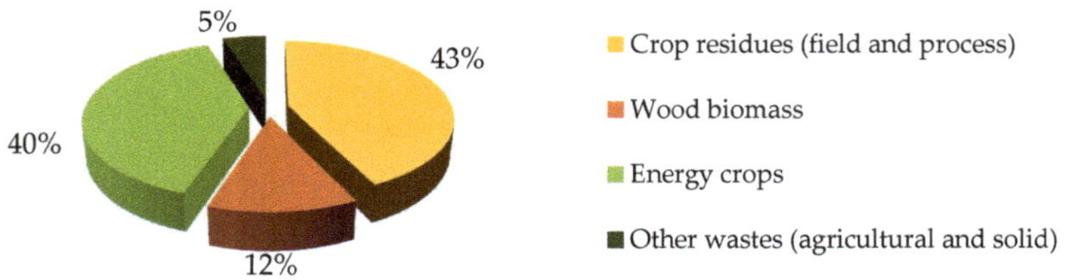

Figure 1. Economic potential of energy generation from biomass in Ukraine by inputs (based on 2017 data). Source: own aggregation based on Bioenergy Association of Ukraine estimations [4].

The current high energy generation potential of biomass in Ukraine has also been confirmed by other studies. In particular, [5] provides an assessment of the so-called technical sustainable potential of crop residues based on five key crops (wheat, barley, corn, sunflower and rape) equaling ca. 6.1 Mtoe (estimations based on 2013 data). In the study, the technical sustainable potential (i.e., one that ensures the required return of nutrients to the soil) is derived from the theoretical potential, and takes into account limitations for yield increases in residues due to technical constraints. In addition, sustainability constraints are considered, although the study stresses that sustainability is limited to environmental factors and that socioeconomic factors are not covered. Another study [6] by the Food and Agriculture Organization (FAO) provides an overview of different assessments of the economic potential of primary agricultural residues based on data from the first half of the 2010s, with the annual results varying within the 9.9–12.8 Mtoe range.

Energy generation based on biomass is becoming increasingly important as a low-carbon, widely available, renewable component of national energy matrices [7], with land-intensive bioenergy already constituting a significant part of the global energy mix [8]. In the past two decades, studies have made efforts to substantiate bioenergy development directions and functions [9–13] focusing on the diverse uses of biomass, and have thereby also tried to define the existing and future energy potential of biomass globally [14–18] or for particular countries [19–21]. As more and more of the peculiarities and limitations of biomass utilization become visible, increasing numbers of studies [8,22–24] are also seeking to go beyond simple assessments of bioenergy potential, additionally defining if and how its utilization will be feasible and relevant in the long run.

The generation potential of agricultural crop residues is among the most frequently discussed topics [5,25–27]. The utilization of this potential is especially beneficial to countries focusing on crop production and boasting vast agricultural areas. Ukraine is one example, boasting 41.5 million ha [28] of agricultural land and intensifying its crop specialization in its constantly growing agricultural production. This explains the considerable interest in researching crop residues' potential in Ukraine [4,6,29–36]. Nevertheless, the variety of approaches implemented in the cited studies exemplify the complexity of the issue and the existence of different methods to estimate both crop residues' availability and the bioenergy generation potential of the withdrawn biomass. Most agree that there is a continuous need to improve the methods and approaches utilized, taking into account more factors and assessing existing uncertainties.

Among the research uncertainties are future agricultural development trends; thus, many studies prefer to assess the present potential of crop residues based on the available statistical data.

Diversified crop yields and their expected values in the future represent other uncertainties, as climate change, extreme weather events and the limited availability of organic and mineral fertilization have substantial influence, yet are difficult to predict. Regional disparities exist in terms of soil fertility and nutrient deficits. Overall, approaches to estimating crop residues' volumes differ depending on the research's purpose. As [37] states, agricultural crop residue estimations are often limited to multiplying crop yields by a harvest index (the ratio of non-grain plant material to grain material), yet estimates can be refined by accounting for factors that might limit their quantities.

Crucially, the removal of crop residues from fields must be managed in a balanced manner in order to satisfy the sustainability of agricultural production and development. Residues play a number of critical roles within an agronomic system, and have direct and indirect impacts on physical, chemical, and biological processes in the soil. Excessive residue removal can degrade the long-term productive capacity of soil resources [27].

Thus, bioenergy generation potential must be realized while maintaining sustainable crop production through ensuring adequate levels of soil nutrients and aiming at the optimization of mineral fertilization according to agronomic and environmental norms, as well as economic viability. This means that the key criterion for crop residue removal is the optimal withdrawal level, ensuring the balanced satisfaction of both continuous crop production and bioenergy needs. Therefore, the objective of this research is to define the future bioenergy generation potential of crop residues in Ukraine while satisfying these agro-environmental limitations.

This article is divided into five sections. Following the introduction, Section 2 explains the methods used to model Ukrainian agricultural development (serving as the basis for the estimation of selected crop production volumes), the methods used to calculate the environmental limitations for the removal of crop residues, and the methods used to estimate the energy potential of crop residues, supplemented by a description of the data sources. Section 3 presents the research results as follows: (a) a brief description of Ukrainian agriculture; (b) projections for the development of selected crops' cultivation; (c) an assessment of the availability of crop residues based on agro-environmental limitations; and (d) estimations of bioenergy generation. Subsequently, Section 4 is devoted to a discussion and comparison of the results obtained with similar studies. Finally, Section 5 presents the conclusions, implications and limitations of the work.

2. Methods and Materials

Typical agricultural crop residues in Ukraine can be divided into two main groups: (1) field residues, comprising straw, stalks, stubble, leaves, seedpods, etc.; and (2) process residues, comprising husks, seeds, roots, and bagasse. The substantiations presented in this article strictly deal with the former (also referred to as crop residues).

2.1. Modeling Ukraine's Agricultural Development

Given that the research approach has been designed to estimate the future bioenergy generation potential of crop residues, 2030 is set as the target year. Thus, it is crucial to identify the current and projected trends in the development of Ukraine's agricultural sector that will determine the potential future availability of bioenergy resources. These estimations serve as the basis for understanding the economic potential of the crop residues available for energy generation. Trends in the agricultural production of selected crops are modeled using the Global Biosphere Management Model (GLOBIOM), including elements such as crop yields, gross harvest, and regional distribution. GLOBIOM is a partial equilibrium model of the global agriculture and forestry sectors, providing data for crop and livestock production at the level of simulation units. There are two sets of variables: endogenous (e.g., process, production, land uses) and exogenous (e.g., population, GDP growth, technological changes) Prices are determined at the regional level to establish market equilibrium to reconcile demand, domestic supply, and international trade. The average yield for each crop in each major region or country is taken from the Food and Agriculture Organization Corporate Statistical Database (FAOSTAT). Potential and average yields as well as yield variability dependent on climatic and weather conditions are provided in simulated units, generated based on suitability studies by the Environmental Policy Integrated Climate (EPIC) model. Land and other resources are allocated to different production and processing activities to maximize a social welfare function, which consists of the sum of the producer and consumer surplus, based on cost structure and input use in terms of simulated units. The model simulates four different crop management systems—irrigation, high-input/rainfed, low-input/rainfed, and subsistence farming—with different levels and structures of production cost and profitability. It is based on the following key determinants: on the one side, demand, and, on the other, the profitability of the various land-based production activities. Thus, the model allows the land use change between crops and management systems to maximize the social welfare level [38].

While the estimations for this research are conducted solely with the use of the GLOBIOM, the results are also compared to the Agricultural Member State Modeling (AGMEMOD) results available in the literature [39].

Two modeling scenarios have been developed for the purpose of this research with the use of the GLOBIOM model: business-as-usual (BAU) and innovative (INNO). The BAU scenario is based on the assumption that the current development trends in Ukrainian agriculture (covering 2000–2015) will be preserved. The INNO scenario assumes the increasing implementation of agricultural production technologies through intensifying investment (based on the possible increase in the availability of financial resources), as well as scientific development, the modernization of existing agricultural machinery, the application of more advanced agricultural technologies, the use of high-quality seeds, and improvements in infrastructure. Technological changes (exogenous variables in the model) depend on GDP growth, via the elasticity function. These changes determine the level of crops' yields, the production costs, and the profitability of the different land-based production activities. In general, the INNO scenario provides a better realization of crops' potential yields (at least 80% of potential yield for a high-input system) and the widespread implementation of the high-input management system.

2.2. Methods for Evaluating the Environmental Limitations to Removal of Crop Residues

In order to conduct agricultural activities ensuring sustainable development, it is important to satisfy the agrochemical Law of Return, which requires that the soil be compensated for the nutrients removed by crops. In addition, soil nutrient losses due to general agricultural activities need to be taken into account. Typically, the control is conducted by determining the balance of (a) humus, and (b) nutrients. The main sources of humus are the humification of crop residues and organic fertilizers, while its losses occur due to its mineralization and erosion. Nutrient uptake is usually measured on the basis of experimental reference values that take into account their content in organic and mineral fertilizers and crop residues, as well as their replenishment due to biological nitrogen fixation and precipitation. Soil nutrient losses are estimated through their removal by crops during harvest and due to soil erosion (via the processes of leaching and weathering).

In the following, the estimates for humus and nutrient balances within the cultivation of selected crops in Ukraine are calculated based on the methodology developed by the National Academy of Agrarian Sciences of Ukraine (NAASU) [40], which takes into account the aforementioned characteristics. It includes evaluations of the balances of the essential nutrients (N, P_2O_5, K_2O) and humus. The nutrient balance is calculated as the difference between nutrient supplies and losses. Nutrients are supplied through the application of mineral and organic fertilizers, with precipitation, seeds, and symbiotic and non-symbiotic nitrogen fixation. Nutrient losses occur due to crop harvest, weeds, erosion, denitrification, and irrigation (although only for irrigated land). Similarly, the balance of humus is calculated by taking into account all sources of its accumulation (humification of crop residues, humification of organic fertilizers) and losses (humus mineralization, losses due to erosion).

2.3. Energy Potential of Crop Residues

Although numerous investigations have sought to estimate the energy potential of crop residues (many of which are referred to in this study), most [5,6,34,36,41] use similar methods based on the assessment of theoretical potential, while the next estimation steps are referred to differently by the authors (e.g., technical potential, technical sustainable potential, economic potential). Nevertheless, the key features they share are derived from the method [30] provided by the National Academy of Sciences of Ukraine (NASU). The [33] provides further development of this method, outlining three types (levels) of potential: theoretical, technical, and economic.

Theoretical potential: the overall maximum amount of terrestrial biomass that can be considered theoretically available for bioenergy generation within fundamental biophysical limits. The theoretical potential for field residues is calculated based on the maximum crop yields within the particular climate limits.

Technical potential: the fraction of the theoretical potential achievable for energy generation under the specific techno-structural conditions with the current technological possibilities (such as harvesting and processing techniques) available. It can be limited by various factors, in particular spatial allocation, competition between land users, and ecological limits. The technical potential is calculated by multiplying the theoretical potential and the coefficient of technical availability.

Economic potential: the share of the technical potential suitable for energy generation under current market and economic conditions.

The abovementioned elements are calculated as:

(a) theoretical potential:

$$P_t = C_r * K_r * K_{ce}$$

(b) technical potential:

$$P_{te} = C_r * K_r * K_{ce} * K_t$$

(c) economic potential:

$$P_e = C_r * K_r * K_{ce} * K_t * K_e$$

where: P_t—theoretical potential; P_{te}—technical potential; P_e—economic potential; C_r—total production of crop r; K_r—residue-to-product ratio (RPR) of r crop; K_t—coefficient of technical availability of crop residue; K_e—coefficient for crop residues' utilization for energy purposes; K_{ce}—oil equivalent for crop residues (toe); r—crops.

Thus, crop residues' economic potential depends on a number of the abovementioned coefficients used for the assessment. What is important to note is that there is still uncertainty within scientific publications as to their proper levels. The first uncertainty exists in the case of the residue-to-product ratio (RPR). The recommendations used nationwide, developed by the NASU, offer fixed RPRs [30]. However, both theoretical and applied studies indicate an inverse correlation between the yields of primary and secondary crops in conditions of the increasing productivity of most primary crops [5,32,34,41]; thus, yield-dependent RPR values seem to be more suitable. In addition, the influence of crop varieties and hybrids needs to be taken into account (in particular, cover crops), the cultivation of which is becoming especially important due to climate change's impacts (lower water availability, extreme weather conditions [39]). Therefore, RPRs for both approaches are presented in Table 1 for selected crops. In the second approach, the unknown x in the equation for the crop residue evaluation is the yield, while the variable y is the crop residue yield depending on the main crop yield, both in decitons.

Table 1. RPRs (fixed and yield-dependent) for selected crops.

Crops	Approach 1: Fixed RPR (K_r)	Approach 2:		
		Main Crop Yield, Decitons per ha	Crop Residue Dependency on Yield Equation	Yield-Dependent RPR (K_r)
Wheat, including:	1.0	-	-	-
-Winter	-	10–25	y = 1.7x + 3.4	2.0–1.8
		26–40	y = 0.8x + 25.9	1.8–1.4
-Spring	-	10–20	y = 1.3x + 4.2	1.7–1.5
		21–30	y = 0.5x + 19.8	1.5–1.2
Barley	0.8	10–20	y = 0.9x + 6.5	1.5–1.2
		21–35	y = 0.9x + 7.2	1.2–1.1
Corn	1.3	10–35	y = 1.2x + 17.5	3.0–1.7
		36–110	y = 0.95x + 25.95	1.7–1.2
Sunflower	1.9	8–30	y = 1.8x + 5.3	2.5–2.0
		31–40	y = 1.1x + 24.7	1.9–1.7
Rape	-	5–15	y = 1.7x + 4.7	2.6–2.0
		16–30	y = 0.79x + 18.43	1.9–1.4
Soya	0.9	5–15	y = 1.7x + 4.7	2.6–2.0
		16–30	y = 0.8x + 18.4	2.0–1.4

Source: based on [30,32] (pp. 20–21).

In general, the RPRs estimated within the second approach based on empirical Ukrainian data for the period 2000–2015 are higher than the RPRs estimated based on the European Union's data [32] (pp. 12–13). These differences can be explained by the higher crop yields and different technologies, crop varieties and hybrids used in the EU. They could be used to understand possible future development trends in Ukraine in case of the implementation of similar production approaches.

Additional uncertainties in regard to RPRs persist due to existing or periodically arising challenges, such as seasonal and regional weather uncertainties, the availability of nitrogen, and the application of herbicides [5] (pp. 73–79).

Some uncertainties also occur when evaluating the coefficient of the technical availability of crop residues (K_t). Although it is generally characterized by a lower level of uncertainty, some variability is still present due to differences in the technologies used for harvesting (i.e., requirements for the height of the cut stem, and availability of equipment on a farm allowing the residues to be harvested). At the same time, the value of the technical potential of crop residues could be reduced by the negative effects of weather events (hail, squalls, droughts). In the aforementioned NASU recommendations [30], the value of the coefficient of technical availability is defined at the level of 0.8. Notably, the use of slightly lower values has been substantiated in some studies by Ukrainian scientists. In particular, for the Khersonska region, the coefficient of technical availability of the main grain crops based on data from 2012–2013 was estimated at 0.5, and for corn and major industrial crops at 0.7 [34] (p. 113). Another study dealing with the Sumska region defined a coefficient of technical availability of 0.5–0.7 for grains [42] (p. 113).

Concerning the coefficient for crop residues' utilization for energy purposes K_e, uncertainties depend on market and technical factors, including market competition and the economic efficiency of different uses of crop residues (for animal feed, as a fertilizer), the availability of infrastructure for transportation, and the availability and capacity of biomass processing for other purposes. Traditionally, the use of cereals' straw is directed at the needs of livestock (as roughage and bedding). However, the role of straw in animal nutrition (as a source of fiber) in the transition to intensive technology is minimized, retaining its functional purpose only as bedding for animals. With market changes in Ukraine, the profile of agricultural producers has changed significantly due to intensifying production specialization and a general reduction in livestock numbers. There are fewer and fewer agricultural enterprises with diversified crop and animal production. Currently, there are regional zones of different agricultural specializations in Ukraine, most notably a few livestock production zones around key urban agglomerations [43] (pp. 131–133). Thus, several conclusions can be drawn that are important from the point of view of competition over access to agricultural residues for livestock production purposes:

- a general declining trend in demand for straw for livestock production purposes;
- distribution inequalities in the regional concentration of specialized livestock enterprises;
- improvement of livestock production technologies, leading to the reduced use of biomass as bedding.

Decreasing competition from the livestock sector may be partially offset by increasing demand for the use of crop residues for agro-environmental purposes within land use, aimed at restoring soil fertility. Limiting the possibility of applying organic fertilizers (natural manure) increases the role of crop residues for the restoration of the organic component of the soil: 1 ton of straw equals 2.5–2.8 tons of bedding manure in equivalent humus [35] (p. 6). However, the use of straw as a fertilizer leads to additional costs, which is mainly due to increased fuel consumption to mechanically prepare the straw for further use, and the higher application rates of nitrogen fertilizers to accelerate the mineralization of straw (10–12 kg of N per 1 ton of straw) [35].

In the last decade, demand for crop residues has been increasing due to climate change's impacts (higher temperatures and water evaporation, lower rain frequency) through the implementation of adaptation measures in crop production. Under new climate conditions, the efficiency of traditional agricultural practices (primarily based on deep tillage) has been decreasing, especially in the southern and eastern regions of Ukraine. Accordingly, the issue of the selection and implementation of alternative technologies and practices aimed at minimizing the negative impacts of weather factors has gained in importance. Such technologies include strip-, mini, or no-till practices. The use of these technologies limits the possibility of using crop residues for purposes other than as a cover material [44].

However, it should be noted that currently increasing crop yields provide higher residue outputs, creating some difficulties for those producers who apply no-till technologies. Therefore, in these cases, a certain proportion of the crop residues still needs to be removed despite the conditions of highly productive agriculture. These practices have already spread beyond the country's arid and wet regions, being actively implemented on the farms of western, central, and northern Ukraine [45]. The efficiency of no-till technologies has been confirmed by the experiences of South American countries (in particular Argentina, with 80% of its main cropland being cultivated with the use of no-till technology [46]), yet they also have certain disadvantages that reduce their attractiveness for widespread implementation. The key disadvantages are [44]:

- increased costs of weed (15–100%) and pest (by two to three times) control;
- lower seed germination, requiring increased seeding rates (by 15–25%);
- higher price of equipment for direct seeding;
- higher fire hazard due to the presence of crop residues on the soil surface;
- need to control the application of additional doses of mineral fertilizers (phosphorus, nitrogen) under certain conditions.

Given the established practice of agricultural technologies in Ukraine, as well as the objective shortcomings of minimal tillage systems, a significant shift in the structure of the technologies applied by agricultural producers should not be expected in the near future. Therefore, a surplus of crop residues is expected to be available in Ukraine overall; however, due to uneven spatial distribution (regional, district, local), more comprehensive and regionally targeted approaches are required to assess the economic potential of biomass for energy generation purposes.

As for the coefficient for crop residues' utilization for energy purposes, according to the NASU recommendations, a fixed value of 0.25 is applied to all crops [30]. This value is based on the demand of livestock production and levels to maintain balanced organic matter in the soil. Under visible conditions of increasing crop productivity, it is possible to expect the achievement of a higher level of K_e coefficient (0.3 for cereals, 0.4 for industrial crops and corn) without harming sustainable land use, but with the application of the recommended rates of mineral and organic fertilizers [29] (p. 17). The use of the average value of the K_e coefficient is understandable in strategic documents at the national level, but in in-depth calculations, its values should be detailed, taking into account regional factors: spatial features of production allocation and local demand from livestock production; technologies used for crop production; types of crops; and available options to maintain the soil's fertility by biomass application (e.g., the digestate from biogas generation, the ash and sludge from biomass combustion process [47]).

2.4. Data Sources

Data on key agricultural trends (yields, area, production, nutrient supply with mineral and organic fertilizer, etc.) are provided by the State Statistics Service of Ukraine. Data on the production costs of different agricultural products are calculated based on the annual database "Main economic indicators of farm activity" are available for 2000–2015 (based on a survey with approximately 8000 farms carried out annually), which form the basis for the Ukrainian input data used by the GLOBIOM used within this study. National and regional data on soil types and their nutrient and humus contents are provided by the State Statistics Service of Ukraine. The spatial allocation of soil types is gathered from the State Service of Ukraine for Geodesy, Cartography and Cadastre [48].

3. Research Results

3.1. Background: Ukrainian Agriculture and Its Transformation

Since 1991, major transformations in Ukrainian agriculture have taken place. Overall, livestock production has gradually and substantially declined (the cattle population decreasing eightfold and pig production by 3.4 times). Consequently, there are fewer opportunities available for the application of organic fertilizers in the form of manure, as was widely used in Ukraine's pre-independence era (6208 kg/ha of agricultural land in 1990 vs. 281 kg/ha in 2018 [49]). In parallel, through the 1990s there was an initially sharp reduction in the application of mineral fertilizers in crop production due to agricultural transformation processes and overall economic instability. Thus, the average application of NPK mineral fertilizers (consisting of nitrogen, phosphorus and potassium components) dropped from 105 kg/ha of agricultural land in 1990 to its lowest record of 6.7 kg/ha in 2000, and only partially regaining in intensity in the following years, reaching 56.5 kg/ha in 2018 [49]. However, crop production after the beginning of the 2000s due to the acceleration of investments (including the inflow of foreign capital) began its steady increase. Even though (since 2010) the rate of mineral fertilization has been growing, it still does not usually satisfy agronomic requirements to maintain the proper level of soil nutrients [40] (p. 97). Farming approaches and abilities vary greatly depending on a farm's type, size, and market focus. Currently, a wide range of medium and large agricultural enterprises follow an export-oriented production strategy based on cultivating a limited number of crops: wheat, barley, corn, rape, sunflower, and soya. Due to their limited access to capital, some (mostly small and medium-sized) farms do not have the capacity to maintain the necessary equipment for cultivating a diversified number of crops. Therefore, they usually grow no more than two or three crops. Such limited crop rotation leads to soil depletion. Indeed, according to previous Ukrainian soil quality studies [50], humus content decreased by 15% between 1991 and 2005.

Both 2004 and 2005 saw new intense growth in agricultural activity after 13 years of economic instability and production decline. This growth was accompanied by corporatization and the introduction of state support for agricultural producers, seeing the area used for arable land increase by 95,000 ha (0.3% of the total, albeit excluding the area of the currently occupied Autonomous Republic of Crimea) to a total of 29.9 million ha in 2017. This is still 7% less compared to 1990, yet provides evidence of growing pressure upon agricultural land resources, including those left unattended during the 1990s and the beginning of the 2000s. The area represented by pastures diminished by over 100,000 ha from 2004 to 2017. Key changes in the sown area structure since 2004 include the increase in areas (and their share in total) being used for industrial crops, structural changes in cereals (replacement of rye and barley by wheat and corn in northern regions). These key changes are summarized in Table 2.

Table 2. Changes in agricultural land use in Ukraine from 2004 to 2017.

Area Type and Crops	Area, Thousand ha		Change	
	2004	2017	Thousand ha	%
Agricultural area, including:	35,819	34,958	−861	−2.4
Sown area, including:	25,952	27,434	1482	5.7
Cereals and leguminous, including:	**14,887**	**14,607**	**−280**	**−1.9**
Wheat	5357	6368	1011	18.9
Barley	4492	2506	−1986	−44.2
Corn	2462	4522	2061	83.7
Rye	732	168	−564	−77.0
Industrial crops, including:	**4891**	**9161**	**4270**	**87.3**
Rape	113	789	676	598.3
Sunflower	3486	5943	2457	70.5
Soya	269	1994	1725	641.3
Sugar beet	732	318	−414	−56.6
Fodder crops	**4128**	**1826**	**−2303**	**−55.8**

Note: excluding data from the Autonomous Republic of Crimea. Source: own compilation based on [51].

Crop production currently prevails in the total agricultural output (72% in 2017), having grown by 93% from 2000 to 2017, while the growth of livestock production equals just 20%. Cereals, leguminous and industrial crops make up over two thirds of the gross crop production. Aside from having relatively high shares in Ukrainian exports, these subsectors are also providing most inputs for the production of fodder necessary for the domestic livestock sector. The aforementioned crop groups cover approximately 87% of total sown areas or 68% of total agricultural land in the country, with the six main crops (wheat, barley, corn, sunflower, rape, and soya) covering 80% and 63%, respectively [51] (Figure 2).

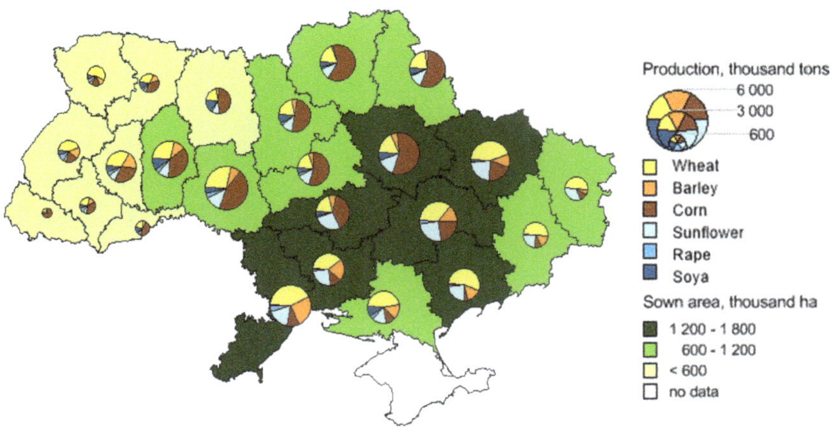

Figure 2. Sown areas and production volumes of main crops in Ukraine (based on 2017 data). Source: own compilation based on [51].

3.2. *Projection for the Development of Selected Crops' Cultivation by 2030*

Ukraine is among the world's leading countries with the largest unrealized agricultural potential [52–54]. According to the Global Yield Gap Atlas (GYGA), the current level of crop yields in Ukraine equal 30–45% of the estimated potential [54]. Therefore, current relatively lower yield levels could mean bigger opportunities compared to other countries concerning the possible increase in productivity in the future, even in the case the development is realized within the BAU development scenario. In addition, in the next decade the positive impact of climate change will still mostly prevail in the case of yields of major crops (excluding in south-eastern regions of the country), increasing average crop productivity relative to the baseline period.

The increasing yield forecasts derived within the GLOBIOM simulation are also confirmed by other modeling results. This is the case of the report [39] published in 2017, revealing the modeling results for Ukrainian agricultural development up to 2030, based on the AGMEMOD model. Thus, Figure 3 represents the results for the selected main crops (which combined cover approximately 90% of the total sown area in Ukraine), comprising both the data for two developed scenarios within this research (BAU and INNO) and the aforementioned AGMEMOD traditional development scenario results.

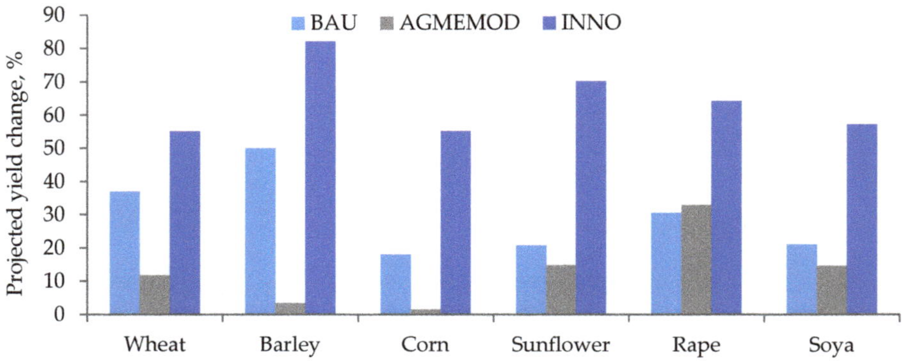

Figure 3. Projection of changes in yields for selected crops in Ukraine by 2030 (compared to average yields in 2008–2014). Source: Agricultural Member State Modeling (AGMEMOD) modeling results [39] and own calculations (business-as-usual (BAU), innovative (INNO)) based on the Global Biosphere Management Model (GLOBIOM).

The modeling results show the following yield growths by 2030: 10–20% for oil crops, and 18–50% for cereals in the BAU scenario. The AGMEMOD modeling confirms these projections, although, for the cereals, the expected growth is slightly lower. However, the forecasts for the AGMEMOD model tend to underestimate the available potential of cereals (Figure 4) when compared to the latest available statistical data. This provides additional reassurance concerning the credibility of the forecasts, including the innovative scenario results (INNO). The probability of the innovative scenario's realization is confirmed by the visible acceleration of yield levels from 2015 to 2019, which was possible due to the actual introduction of new crop varieties and plant protection systems minimizing the negative effects (temperature variations, precipitation reduction) while maximizing the positive impacts (higher temperatures) of climate change, thereby facilitating the increased adaptation abilities of domestic agricultural producers [55,56].

The projections of crops' production volumes obtained via the GLOBIOM approach are compared with the AGMEMOD results (Figure 5). The general trend of increasing production volumes according to the results of the GLOBIOM is consistent with the conclusions obtained within the AGMEMOD modeling, with the exceptions of barley and, to some extent, sunflower. This result can be a guide for domestic producers to make efforts to improve the economic efficiency of barley production, as its permanent growth in global demand and available production potential provides barley with a competitive advantage compared to the key cereal crop in Ukraine: winter wheat.

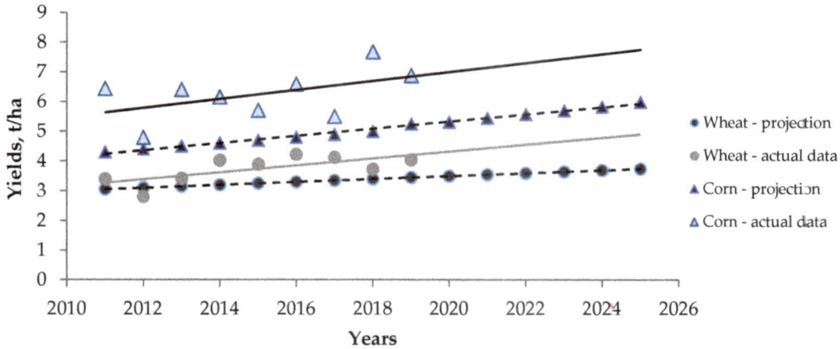

Figure 4. Wheat and corn crop yields: comparison of available statistical data and projections obtained with the AGMEMOD model. Source: own compilation based on [51], verified against [39].

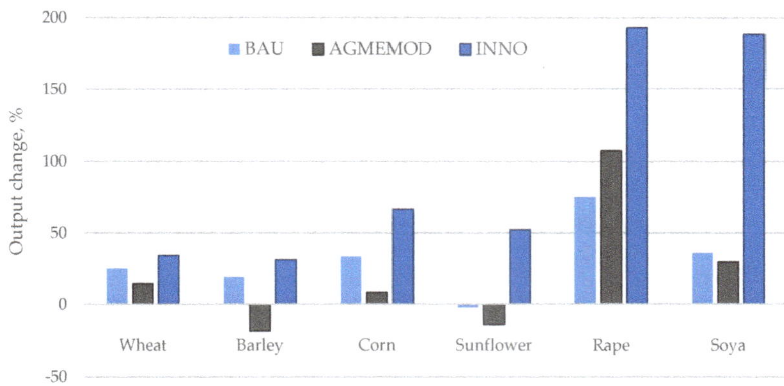

Figure 5. Projection of production (output) change for selected crops by 2030 (compared to 2008–2014 average volumes). Source: own calculations.

3.3. Assessment of Crop Residues' Availability Based on Agro-Environmental Limitations

Based on the crop production projections, the theoretical potential of crop residues for the investigated crops (Table 3) is assessed. For this purpose, both of the abovementioned approaches to the assessment of RPRs (K_r) are used: K_{r1}, provided in Table 1, and K_{r2}, estimated with the use of the EU's RPR data [41].

Table 3. Assessment of the theoretical potential of crop residues for the selected crops in 2030, ktoe.

Crops	BAU		INNO	
	K_{r1}	K_{r2}	K_{r1}	K_{r2}
Wheat	10,850	15,249	11,645	14,824
Barley	3390	3962	3736	5618
Corn	11,892	13,262	14,812	13,137
Sunflower	5214	6022	7823	7983
Rape	3039	3667	4712	5780
Soya	1061	3035	2096	5071
Total	**35,446**	**45,197**	**44,824**	**52,413**

Source: own calculations.

The assumptions of improving Ukrainian farmers' access to technologies and financing support the expectations for the growing technical availability of crop residues. However, there is a high risk of adverse weather events that may reduce the technical availability of crop residues in certain regions and particular years covered by the projection. Given these uncertainties, the coefficient of the technical availability of crop residues at the level of 0.8 (provided by the NASU [30]) is used within the estimations.

Currently, the main share of the technical potential of crop residues is being utilized for the purposes of livestock production and stabilization of soil fertility, leaving only the remainder to be potentially used for renewable energy generation purposes. To determine future development trends within the economic potential of crop residues for energy production, some general assumptions are used for the years to come:

- declining share of roughage in livestock feed;
- a reduction in the use of straw for bedding due to the increasing use of alternative materials (e.g., rubber, sawdust).

According to both forecasts obtained based on the GLOBIOM for the period up to 2030, a further reduction in cattle numbers is expected, with a slight (1–7%) increase in pigs and the further growth (25–40%) of poultry (depending on the scenario). At the same time, the share of all livestock kept in farm households will continue to decline according to the ongoing trends. Given these forecasts, the use of straw for livestock production purposes is expected to equal the baseline values or even decrease due to the expected gradual implementation of agricultural innovations in corporate livestock production. Accordingly, this will determine the trends in the amounts of manure available for various purposes, mainly for application to the soil as organic fertilizers. This, in turn, will engender certain risks for sustainable land use—specifically compliance with the agrochemical Law of Return. Under such conditions, the need for crop residues to maintain agro-environmental functions will increase, limiting the possibilities of its use for bioenergy generation purposes.

The need to return a proportion of agricultural residues to the fields in Ukraine owes to the low use of organic fertilizers, a consequence of the decline and disproportionate development of the livestock sector, this being the main source of organic matter for the soil due to manure humification. A deficit-free balance of humus manure application should be at least at the level of 8 t/ha in the southeastern regions, and up to 14 t/ha in the northwest when the average level in Ukraine is below 1 t/ha [50] (p. 49). Farmers try to ensure the return of nutrients by more intense mineral fertilization and by leaving 100% of crop residues in the fields. However, mineral fertilization is often either conducted in violation of agro-ecological norms of required nutrient balance, or the amount of mineral fertilization is insufficient for an adequate process of humification of crop residues.

The conducted assessment of the production processes of the main crops in Ukraine confirms the existence of issues preventing the satisfaction of deficit-free requirements. Thus, wheat production with a 100% return of crop residues ensures a positive balance of humus (H) and nitrogen (N) in most regions of Ukraine, but does not ensure the return of phosphorus (P) and potassium (K). The worst condition with the return of nutrients is observed with respect to the cultivation of barley and corn (Table 4).

The production of industrial crops also shows unbalanced soil content and an inability to balance the removed nutrients and humus, although due to the application of sufficient amounts of nitrogen-based mineral fertilizers, a positive balance for N can be achieved in the cultivation of sunflower and nitrogen-fixing soybean. Nevertheless, there is still a shortage of humus and potassium in the cases of the aforementioned crops. The cultivation of rape shows a slightly different picture: there is an acute deficit of nutrients, but, for humus, the balance of return is sufficient. This can be explained by the high output of crop root residues of rape, which, during its humification, provide the return of sufficient organic matter to the soil (Table 5).

Thus, it can be concluded that, in most regions of Ukraine, the currently applied technologies for the production of the main crops do not satisfy agro-ecological requirements, and limit the opportunity to remove crop residues for other purposes, including bioenergy generation.

Table 4. Soil nutrient balance for the production of cereals in Ukraine by region (calculated for 2017).

Regions	Wheat				Barley				Corn			
	N	P	K	H	N	P	K	H	N	P	K	H
Vinnytska	1.8	−18.9	−33.1	0.68	−9.4	−15.7	−30.9	0.38	−14.2	−20.6	−61.6	0.33
Volynska	24.7	−3.6	8.8	0.78	1.5	5.2	16.8	0.32	−28.9	−13.0	−29.9	0.89
Dnipropetrovska	−2.1	−10.4	−16.3	−0.07	−6.8	0.1	−9.4	−0.37	−3.8	0.1	−28.4	−1.06
Donetska	−16.5	−18.6	−26.1	−0.04	−17.2	−2.8	−15.0	−0.35	−3.8	−7.2	−34.1	−1.28
Zhytomyrska	6.2	−11.8	−7.1	0.60	7.1	−5.7	2.5	0.26	−32.0	−22.5	−51.3	0.44
Zakarpatska	42.9	−8.0	−6.0	0.37	37.2	10.1	12.9	0.15	36.4	−25.7	−50.8	0.07
Zaporizka	12.5	−4.8	−14.5	−0.01	−8.2	−1.4	−11.4	−0.25	−1.4	2.0	−23.1	−0.80
Ivano-Frankivska	22.3	−10.5	3.7	0.76	−7.7	−4.9	−4.4	0.37	−9.7	−9.4	−39.3	−0.24
Kyivska	29.0	−3.6	4.9	0.06	8.3	1.4	12.8	−0.06	4.4	−4.7	−29.7	−0.32
Kirovohradska	1.7	−9.4	−15.5	−0.25	−9.9	1.2	−7.9	−0.49	−10.4	−6.6	−34.5	−1.11
Luhanska	−23.7	−15.8	−32.7	0.06	−23.4	−7.9	−24.6	−0.36	−9.5	−9.6	−35.8	−1.20
Lvivska	18.2	−12.2	−8.7	0.62	−16.3	−18.6	−9.3	0.49	−12.7	−0.6	−46.9	−0.10
Mykolaivska	−5.3	−5.9	−16.7	0.10	−11.7	−1.4	−12.5	−0.17	−1.3	11.6	−31.1	−0.71
Odeska	5.6	−1.7	−17.2	0.37	−10.1	−1.7	−14.1	0.20	−4.2	−5.1	−34.1	−0.31
Poltavska	13.7	−3.7	−4.4	0.16	7.8	−2.6	−1.7	−0.18	1.3	−13.7	−41.2	−0.69
Rivnenska	18.1	−6.0	10.0	0.77	−11.5	−12.4	−8.7	0.50	−35.4	−21.7	−44.2	0.39
Sumska	−13.2	−12.7	−7.3	0.48	−6.3	−3.2	6.0	0.00	−23.4	−22.2	−56.0	−0.01
Ternopilska	23.7	−13.3	−4.6	0.66	−12.1	−18.2	−8.3	0.39	−13.5	−26.3	−61.8	0.25
Kharkivska	7.6	−9.2	−18.5	−0.04	2.7	−3.4	−10.9	−0.51	26.4	−2.1	−28.3	−1.78
Khersonska	−15.7	−19.1	−23.8	0.44	−12.8	−10.3	−17.8	0.16	−31.6	−24.0	−66.9	0.90
Khmelnytska	−5.0	−18.7	−12.5	0.72	−18.8	−16.1	−9.5	0.34	−34.9	−25.3	−64.5	0.06
Cherkaska	27.1	−6.3	−3.1	0.06	15.0	3.1	11.5	−0.08	1.0	−8.0	−34.3	−0.55
Chernivetska	−6.6	−13.4	−10.5	1.03	3.7	−13.3	−12.3	0.56	8.9	1.4	−24.6	0.35
Chernihivska	6.0	−7.2	−0.5	0.41	−2.4	−4.7	8.0	0.15	−14.5	−18.9	−52.3	−0.23

Note: N—nitrogen, kg a.i./ha; P—phosphorus, kg a.i./ha; K—potassium, kg a.i./ha; H—humus, tons/ha. Source: own calculations based on [40].

Table 5. Soil nutrient balance for the production of industrial crops in Ukraine by region (calculated for 2017).

Regions	Soya				Sunflower				Rape			
	N	P	K	H	N	P	K	H	N	P	K	H
Vinnytska	41.7	1.3	−28.8	−0.97	22.5	8.0	−65.1	−0.88	37.7	−1.3	−21.5	1.14
Volynska	40.4	3.2	−19.6	−0.75	12.9	18.6	−33.0	−0.86	−2.5	−12.3	−30.2	1.70
Dnipropetrovska	30.4	10.0	−20.4	−1.53	1.7	10.4	−48.2	−1.45	−23.6	−17.7	−53.8	0.36
Donetska	27.4	3.5	−8.1	−1.93	0.7	11.0	−44.9	−1.63	−40.2	−31.0	−71.0	0.37
Zhytomyrska	36.4	−6.1	−33.2	−0.94	22.4	9.0	−42.5	−1.12	−16.0	−13.2	−44.3	1.14
Zakarpatska	39.1	−19.1	−68.6	−0.33	44.5	20.6	−31.7	−0.97	12.4	−15.1	−37.8	0.65
Zaporizka	44.8	0.8	−40.0	−0.82	5.7	10.0	−39.5	−1.29	−13.8	−15.6	−50.4	0.36
Ivano-Frankivska	46.9	2.7	−29.4	−1.35	23.5	17.6	−46.6	−1.58	13.4	−13.6	−32.4	1.16
Kyivska	42.4	7.1	−10.2	−1.41	22.5	15.9	−34.6	−1.37	11.1	−12.9	−32.9	0.70
Kirovohradska	24.3	1.2	−20.4	−1.78	−1.6	11.5	−47.8	−1.68	−30.5	−17.5	−54.7	0.17
Luhanska	10.9	−10.1	−26.6	−1.63	−12.2	2.5	−54.8	−1.41	−29.2	−20.4	−62.2	0.29
Lvivska	46.5	1.9	−30.0	−1.36	25.7	7.8	−40.4	−1.59	9.3	−19.5	−54.9	1.13
Mykolaivska	32.5	−1.4	−26.2	−1.16	1.6	10.2	−43.5	−1.13	−32.0	−2.0	−55.2	0.43
Odeska	49.5	7.1	−31.6	−0.48	10.3	14.3	−50.3	−0.70	−3.7	2.1	−45.2	0.63
Poltavska	42.3	3.6	−18.2	−1.60	19.8	10.7	−48.6	−1.45	−37.3	−30.0	−65.2	1.03
Rivnenska	17.5	−11.0	−44.1	−0.80	7.5	5.3	−61.5	−0.92	−14.0	−15.2	−35.6	1.68
Sumska	44.2	4.1	−17.2	−1.42	11.5	13.5	−43.6	−1.36	−27.3	−16.0	−53.7	1.13
Ternopilska	48.3	−1.5	−32.0	−1.06	20.2	6.1	−59.1	−1.19	10.3	−13.6	−42.7	0.97
Kharkivska	51.6	3.3	−17.5	−2.39	18.6	11.4	−54.8	−2.16	51.4	−19.8	−58.8	0.13
Khersonska	26.5	−14.0	−64.3	0.10	3.6	2.0	−42.4	−0.54	−39.9	−29.5	−64.5	0.89
Khmelnytska	41.1	−6.9	−40.1	−1.30	10.9	7.5	−64.5	−1.30	0.1	−14.4	−50.8	0.90
Cherkaska	49.2	4.5	−17.5	−1.54	27.3	17.0	−38.2	−1.39	2.9	−11.2	−36.6	0.67
Chernivetska	28.3	−5.7	−30.3	−0.46	23.5	4.1	−60.5	−0.42	−10.2	−17.2	−51.7	1.13
Chernihivska	34.3	3.6	−19.6	−1.76	23.0	15.0	−35.8	−1.82	−16.5	−14.8	−49.0	1.06

Note: N—nitrogen, kg a.i./ha; P—phosphorus, kg a.i./ha; K—potassium, kg a.i./ha; H—humus, tons/ha. Source: own calculations based on [40].

Soils with low humus content are especially vulnerable to various climatic factors, which negatively affect soil composition, structure and quality. In Ukraine, 1.7 million ha (4.1% of the country's agricultural land) is subjected to wind erosion, 13.3 million ha (32%) to water erosion, and over 2 million ha (4.8%) to both of these types of erosion. In addition, 10.7 million ha (25.8% of agricultural land) is classified as acidic, 1.7 million ha (4.1%) as saline, and 1.9 million ha as waterlogged soils. Over 20% of the territory of Ukraine is contaminated with various toxic compounds. Some areas are contaminated with radioactive compounds. Negative geological phenomena are common in over 50% of the country [57] (pp. 7–8).

Within the developed scenarios, the rate of mineral fertilizer application will increase gradually by 2030. The BAU and INNO scenarios are based on the gradual growth of mineral fertilizer application until 2030, at least by 50% and 85% of the base level (2017) for the farms ranking in the top 10% for intensive production technologies for these scenarios, respectively. Organic fertilization is expected to increase as well, but—on average—not more than 2–5 t/ha in the BAU and 3–7 t/ha in the INNO scenario, depending on crops.

Thus, the main sources of nutrient return are expected to be mineral fertilizers (up to two thirds of the return amount of nutrients depending on the crop) and crop residues (15–35% accordingly) for all crops except soya (due to nitrogen fixation, the returns are up to 50–60% of used nitrogen). The organic matter of the soil (humus) in conditions of limited application of organic fertilizers can be maintained only by returning the crop residues to the soil. Given that maintaining a positive balance of humus is a decisive factor for farmers, the possibility of removing part of the crop residues for energy purposes is therefore further limited. To understand these limitations, Table 6 presents the humus balance for the investigated crops within the two developed scenarios.

Table 6. Humus balance under conditions of residues' return to the soil, t/ha.

Regions	Wheat		Barley		Corn		Soya		Sunflower		Rape	
	B	I	B	I	B	I	B	I	B	I	B	I
Vinnytska	1.21	1.52	0.85	1.06	0.81	1.61	−0.60	0.05	−0.70	0.04	1.23	1.88
Volynska	0.91	1.64	0.87	1.14	0.95	1.51	−0.57	−0.44	−0.77	−0.52	1.65	2.16
Dnipropetrovska	0.39	0.57	0.19	0.58	0.08	0.55	−0.42	−0.03	−0.53	−0.19	0.77	1.14
Donetska	0.30	0.48	0.20	0.40	−0.19	−0.04	−0.78	−0.14	−0.64	−0.47	0.67	1.14
Zhytomyrska	0.82	1.41	0.70	1.04	0.75	1.22	−0.78	−0.59	−1.02	−0.69	1.14	1.98
Zakarpatska	0.64	1.53	0.25	0.47	0.11	0.97	−0.54	−0.54	−0.99	−0.68	0.82	2.18
Zaporizka	0.41	0.60	0.25	0.53	0.09	0.17	0.14	0.20	−0.43	−0.26	0.79	1.17
Ivano-Frankivska	0.91	1.47	0.54	0.82	0.19	0.30	−1.11	−1.12	−1.36	−1.25	1.41	1.93
Kyivska	1.01	1.12	0.64	0.85	0.41	1.47	−0.99	−0.04	−1.08	−0.03	1.04	1.95
Kirovohradska	0.47	0.63	0.16	0.47	−0.57	0.83	−1.36	−0.15	−1.46	−0.20	0.28	1.30
Luhanska	0.37	0.56	0.22	0.56	0.00	0.10	−0.54	0.00	−0.55	−0.33	0.70	1.03
Lvivska	0.82	1.19	0.68	0.92	0.41	0.75	−1.08	−0.90	−1.25	−1.10	1.15	2.15
Mykolaivska	0.48	0.67	0.33	0.57	0.34	0.94	−0.21	0.19	−0.32	0.19	0.78	1.40
Odeska	0.62	0.79	0.51	0.70	0.39	1.01	0.12	0.31	−0.21	0.26	0.83	1.16
Poltavska	0.73	1.21	0.49	0.66	0.11	1.23	−1.09	−0.22	−1.13	−0.21	0.80	1.60
Rivnenska	1.04	1.55	0.83	1.11	0.65	1.14	−0.67	−0.58	−0.94	−0.65	1.46	2.08
Sumska	0.59	1.06	0.24	0.47	0.17	0.74	−1.10	−0.86	−1.21	−0.92	0.61	1.24
Ternopilska	0.81	1.09	0.48	0.73	0.39	0.78	−0.93	−0.87	−1.12	−0.89	1.13	1.64
Kharkivska	0.63	0.85	0.22	0.61	−0.99	0.66	−1.85	−0.43	−1.72	−0.41	0.07	1.38
Khersonska	0.70	0.84	0.43	1.00	1.34	1.45	0.62	0.59	−0.15	0.11	0.66	1.27
Khmelnytska	1.26	1.38	0.94	1.08	0.41	1.53	−0.94	−0.08	−1.03	−0.11	1.07	1.70
Cherkaska	1.05	1.31	0.64	0.80	0.35	1.55	−1.12	−0.09	−1.05	−0.11	1.00	1.82
Chernivetska	1.25	1.39	0.49	0.65	0.55	1.28	−0.23	0.14	−0.39	0.17	1.13	1.91
Chernihivska	0.56	1.29	0.44	0.64	0.04	0.67	−1.34	−1.11	−1.53	−1.21	0.81	1.57

Note: B—BAU scenario, I—INNO scenario. Source: own calculations based on [40].

The positive balance of humus indicates the possibility of removing part of the crop residues in compliance with agro-environmental requirements. In the cases of soybean and sunflower, even within the INNO scenario, assuming the almost maximum allowable approximation of the yield to the potential level (and therefore an increase in the yield of the crop and its residues), there is a deficit of organic matter recovery in the soil. This is due to the relatively lower amount of crop root residues of soybean, and—for sunflower—a relatively lower coefficient of humification of crop stalks (0.2–0.25 for cereals and legumes, and 0.14 for sunflower). Thus, the average amount of straw that can be removed for bioenergy generation purposes varies depending on the crop, the region, and assumptions according to the developed scenarios. For example, according to the BAU scenario, the amount of crop residues available for bioenergy generation purposes varies from 0 to 3 t/ha, with a national average of 0.8 t/ha. Under the INNO scenario, the average and maximum values increase to 2 and 3.8 t/ha, respectively (Figure 6).

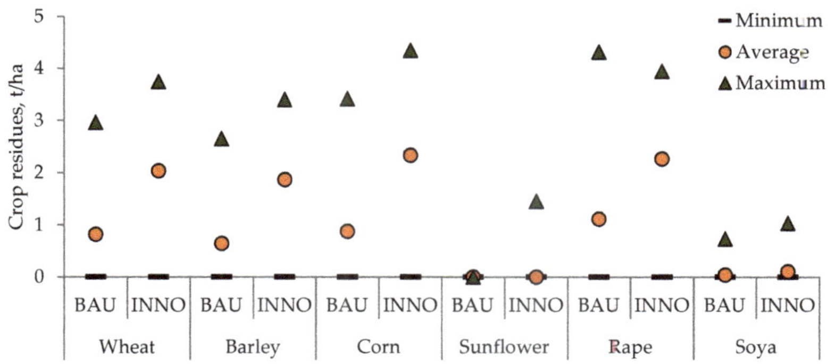

Figure 6. Amount of crop residues available for bioenergy generation purposes while satisfying agro-environmental requirements. Source: own calculations.

In addition to satisfying the agro-environmental requirements of land use, a proportion of cereals' residues (primarily wheat and barley straw) is used for livestock production purposes: the estimated demand for these purposes equals 3.6 and 2.7 million tons for the BAU and INNO scenarios, respectively. The lower demand for crop residues from livestock production corresponds with the current level in EU countries specializing in livestock. In particular, the maximum amount of straw used for livestock production purposes does not exceed 3.9 million tons in Poland, 3.8 million tons in France, and 2.9 million tons in Denmark [25].

3.4. *Bioenergy Generation Potential (National and Regional)*

Both developed scenarios (BAU and INNO) confirm the output growth of the investigated crops, meaning a higher potential loss of humus and nutrients (in particular due to crop harvest). Thus, taking into account the requirement to satisfy a positive balance of humus and nutrients in the soil, the total bioenergy generation potential of the analyzed crop residues in 2030 has been assessed. It is expected to equal 3643 ktoe within the BAU scenario, while, in the case of the INNO scenario, it is forecast to reach 10,723 ktoe (Table 7).

Table 7. Economic potential of bioenergy generation from the main crop residues in Ukraine in 2030.

Crops	BAU Scenario		INNO Scenario	
	ktoe	%	ktoe	%
Wheat	1294	35.5	3892	36.3
Barley	442	12.1	1655	15.4
Corn	1257	34.5	3172	29.6
Sunflower	21	0.6	87	0.8
Rape	0	0.0	342	3.2
Soya	629	17.3	1575	14.7
Total	**3634**	**100.0**	**10,723**	**100.0**

Source: own calculations.

The estimated total economic potential of crop residues available for bioenergy generation purposes is unevenly distributed throughout Ukraine. Within the BAU scenario, the largest amounts of crop residues for bioenergy generation purposes would be available in the regions of Kyivska, Cherkaska, Poltavska, Vinnytska, and Ternopilska (200–350 ktoe each). Following the assumptions of the INNO scenario, the regional distribution of future potential does not vary significantly, but the increase in the amounts of biomass available for renewable energy purposes makes it more promising in most regions of Ukraine (Figure 7).

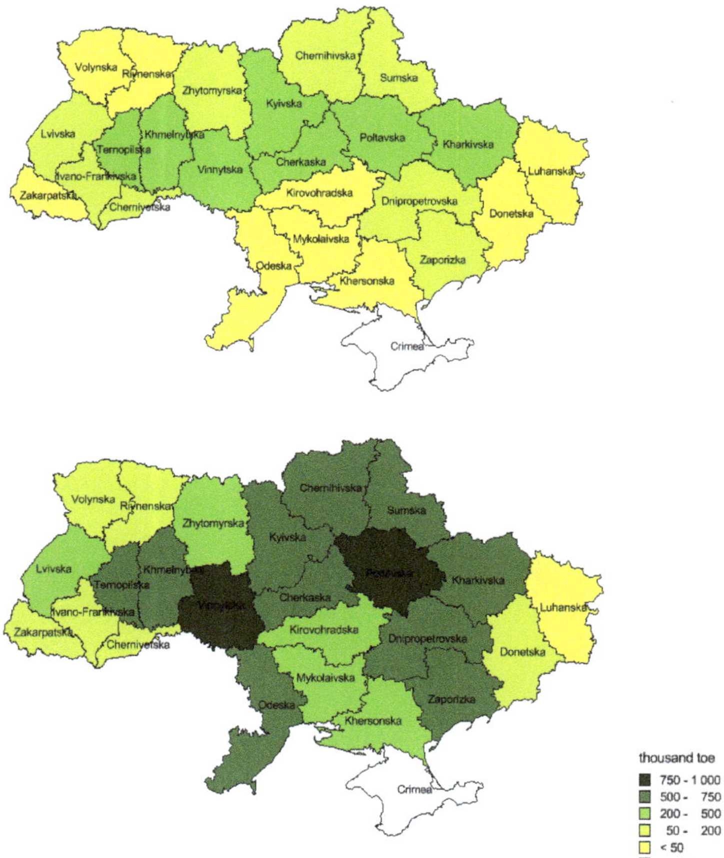

Figure 7. Assessment of economic potential for bioenergy generation purposes from the main crop residues within the BAU (upper map) and INNO (bottom map) scenarios. Note: no data available for the Autonomous Republic of Crimea. Source: own calculations.

There is an understanding that these results reveal a slightly lower (compared to real availability) level of economic potential due to a number of aforementioned constraints applied within the estimations, as the calculations are based on average regional values of parameters and coefficients, as well as assumptions about the utilized technologies. If the same calculations were to be conducted using local data, the estimates of economic bioenergy generation potential would be higher, as they would involve less generalized constraints and reveal specific local situations.

Additionally, the availability of local data concerning actual crop rotations would facilitate greater precision with the estimation of the accumulation of nutrients and humus in the soil, including residues from predecessor crops, in turn increasing the possibility of extracting crop residues for the next crop rotation (e.g., sunflower, which is less demanding in regard to the Law of Return). Such an absence of nutrient flows and humus changes within the crop rotation, despite being a limitation, would have minimal impact on the results obtained. Another limitation of the research concerns the absence of actual data and assumptions regarding the regional land management systems (e.g., conventional, mini- or no-till), which affect the level of nutrients and humus changes, albeit not significantly.

4. Discussion

The Ukrainian expert community has formed an opinion about the possibility of extracting a significant proportion of agricultural residues for energy generation purposes [4,31,32]. Typically, such conclusions for Ukraine are based on the application of a scientifically established 0.25 coefficient for crop residues' utilization for energy purposes [30]. In addition, along with increasing yield levels, this ratio can grow to up to 0.3–0.4 depending on the specific crop. It has been noted [29] that the value of the coefficient is based on compliance with the principles of meeting the crop residue demand of livestock production (for feed, bedding) and the return to the soil of some of the organic matter withdrawn with the harvest. In our understanding, the use of a more flexible coefficient would be more precise. Thus, in this study, the coefficient varies by region and crop type, and primarily depends on requirements to satisfy the nutrient and humus balance in the soil. However, in order to achieve greater precision, data on future trends in land erosion, soil quality, the spatial supply of manure and other agronomical aspects would be beneficial.

This study's results correspond to a large extent with other similar studies for Ukraine, yet differ in terms of several key elements pertaining to the methods used for estimation. For example, the value of assessed economic potential under the developed INNO scenario corresponds with the future trajectory of energy potential from crop residues developed by the Bioenergy Association of Ukraine. In particular, the latter's assessment of residue energy generation potential for 2050 equals 10.8 Mtoe [29]. This amount includes the residue potential of all grain and oilseed crops, with the share of the six selected crops investigated in our study slightly exceeding 95%. More modest forecasts of the potential of crop residues are included in the Energy Strategy of Ukraine [1], totaling 5.3 Mtoe in 2030.

The results obtained regarding the economic potential of crop residues are also close to those of the International Renewable Energy Agency (IRENA) [58] analysis, which confirmed the possible availability of 11.0–18.3 Mtoe of agricultural residues, including 6.0–9.6 Mtoe of field residues in 2030 depending on the scenario: reference or REmap (Renewable Energy Roadmap). Nevertheless, the implementation of the REmap scenario assumes the several measures: the development of collection systems for agricultural residues, establishing the practice of long-term biomass supply contracts between producers and consumers, and the compulsory inclusion of a biogas plant in major new projects by agro-food companies. Implementation of these measures would currently be highly limited due to the difficult economic situation in Ukraine.

Another study [36] evaluating straw and stubble availability alone has shown that, within Europe, based on 2012 data, Ukraine has the second largest potential for energy generation from agricultural residues, and the largest potential based on a projection for 2030. It is important to note that Ukraine is seeing growth in the volumes of crop residues, while a general declining trend for straw and stubble supply for energy production was also outlined in this study.

In contrast to most of the abovementioned studies, the present investigation suggests going beyond typical projections based on current fixed yields (and therefore fixed RPRs) and, instead of assuming a general national coefficient for crop residues that need to be left on the fields, to take into account regional agro-environmental limitations based on the available soil nutrients. Thus, it becomes possible to understand which regions have the most bioenergy generation potential from crop residues, and to use it for future substantiations of regional economic and energy development strategies. Such estimations could be taken further, achieving greater precision at the regional level if specific soil fertility data could be obtained and used for calculations.

Discussions have recently taken place as to the future of bioenergy, as the limitations of its development are becoming more visible based on the experiences of economically developed countries. One study [8] (p. 274) argues that "land intensive bioenergy makes the most sense as a transitional element of the global energy mix, playing an important role over the next few decades and then fading, probably after mid-century". We can agree with this statement, although, in our opinion, differences should also be expected depending on the particular region and the level of economic development of the country being analyzed. While the aforementioned study has presented a general global perspective, locally each country would still tend toward the most feasible option to ensure energy security and utilize the available potential. High differentiation would be present depending on economic development, technological advancement, and technical efficiency.

In this light, Ukraine can still be considered in an early development stage in regard to bioenergy generation, with a total 4.6% share (or 4.3 Mtoe) of renewable sources in the total energy supply (including 3.2 Mtoe from biofuels and waste energy combined) as of 2018 [2]. The country's dependency on foreign energy supply also remains critical (considering its unstable relations with its main long-time energy supplier, the Russian Federation [59]) as, in the same year, 36.5% of Ukraine's total energy supply was imported [60]. Decarbonization, being another key goal of the global renewable energy development goes in line with the growth of biomass utilization, yet presents itself as a complex issue [61] that needs to be considered and resolved wisely. Therefore, for Ukraine, the utilization of bioenergy generation potential is among the key perspective development directions available, and only under conditions of the efficient transformation of energy structure and the appearance of more beneficial (both economically and environmentally) energy generation technologies compared to biomass-based ones will further changes be possible.

Furthermore, the high probability of the projected intensification of crop production (within the six main types outlined in the research) can be identified for Ukraine based on the growing influence of large agriholdings in the past decade and their specialization. Focusing, in most cases, on crop production, these agriholdings are gradually accumulating agricultural land, and working to increase yields and therefore output volumes. Aiming primarily at exports, they managed to increase the exports of the crops investigated in this study over three-fold between 2005 and 2017: grains by 3.4 times (from 12.7 million tons in 2005 to 42.5 million tons in 2017), and oil crops by 6.7 times (from 0.9 to 6.0 million tons) [62]. Therefore, the existing evidence supports the projected growing capacities for the production of crop residues.

Another issue that needs to be highlighted here concerns the differences between the results of the BAU scenario developed within the GLOBIOM and the results derived from the AGMEMOD model, which was used for reference (Figures 3 and 5). The differences can be explained by the contrasting approaches taken concerning the projections of crops' yields and spatial allocation. In particular, AGMEMOD is based on the econometric function of yield, depending on the logarithm trend of yield (2008 = 0) and the expected income from crops in the regions (more details on AGMEMOD in [39]. The GLOBIOM uses the simulated crop yields from the EPIC model, which depend on biophysical yield potential and technological progress (calculated on the elasticities between yield and GDP growth).

5. Conclusions

Growth in the production of the main agricultural crops in Ukraine has been intensifying since the beginning of the 2000s, and is expected to continue in the next decade. This will be accompanied by increasing amounts of available crop residues, which could potentially result in the intense development of energy generation from agricultural biomass. These assumptions demanded in-depth analysis, utilizing several assessment approaches modified by the authors to increase the precision of the analysis and to show which elements in the existing methods could be improved.

The study has projected cultivation trends for six selected crops (wheat, barley, corn, sunflower, rape, and soya), these being the key specializations of the Ukrainian agricultural sector in the past two decades. Estimations based on the GLOBIOM (verified against the openly available AGMEMOD results) for the year 2030 included two agricultural development scenarios (traditional BAU and innovative INNO), allowing us to project the future crop production volumes and yields for the selected crops. The target year results have revealed a growth in crop output volumes, with higher rates in the INNO scenario (assuming the implementation of intensive production technologies). The data obtained within the developed scenarios have enabled us to carry out projections regarding the expected volumes of the applicable crop residues.

The previous literature regarding crop residues has shown that despite the generally accepted guidelines for assessing the energy potential of biomass, there are many uncertainties that significantly affect estimations of the feasible level of field biomass removal, referred to in this article as the economic potential of crop residues for bioenergy generation. The main uncertainties are caused by the nonlinearity of RPR, the uncertainty of the harvesting technologies used, weather, and climatic risks. Therefore, to estimate projected crop residue volumes, yield-dependent RPRs drawn from the literature have been used, allowing for improved precision of assessment as well as taking into account national crop cultivation specificities.

The key agro-environmental limitation of the utilization of crop residues as biomass for energy generation in Ukraine is the need to preserve soil productivity through the restoration of its fertility. Under the limited availability of traditional organic fertilizers, crop residues currently represent the main source for humus recovery and the return of the nutrients withdrawn from the soil during the cultivation process. The traditional approach to assessing the bioenergy potential of crop residues is based on the assumption of an up to 25% technical availability of biomass for energy generation purposes; however, taking into account the agro-environmental requirements (primarily the Law of Return), this level varies greatly depending on the region and the crop type.

Therefore, this study has gone further to improve the assessment method for the potential withdrawal of crop residues, and has aimed to calculate the projected availability of soil nutrients (nitrogen, phosphorus, potassium and humus) for all regions of Ukraine. This then served as the basis for the estimation of the sustainable removal limits of crop residues for bioenergy generation. The conclusions based on this analysis demonstrate the need to improve the applied methodological approach to biomass potential assessment that is being used as the basis for national strategic documents in the energy policy of Ukraine.

Regarding the results, the assessment of the level of economically feasible bioenergy generation potential from crop residues for the year 2030 has indicated the possibility of extracting biomass in the equivalent of 3.6–10.7 Mtoe, depending on the scenario of agricultural development (traditional vs. innovative). This does not exceed 24% (varying between 8–24%) of the theoretical potential (total crop residue amount). The projections have also shown that, within the INNO scenario, wheat, corn and barley combined are expected to provide up to 81.3% of the bioenergy generation potential from crop residues. Although these results are comparable to several other studies, the approach utilized here offers less generalized assumptions and the possibility to take into account national crop cultivation peculiarities, as well as regional soil quality conditions.

As for the practical use of the results obtained, it needs to be stated that an expansion in bioenergy generation potential from crop residues in Ukraine while complying with agro-environmental limitations requires the intense implementation of organizational and technical innovations. In particular, it is crucial to ensure the circulation of biomass between agriculture and bioenergy generation, which is still poorly developed, thereby influencing the low output of the bioenergy generation sector and its slow development, as is apparent in terms of its share in total energy output. The specificity of such exchange determines the necessity and feasibility for development of local and regional bioenergy systems. These results demonstrate the need for the implementation of policy measures for the use of local renewable sources of energy, as planned by the Energy Strategy of Ukraine until 2035 [1]. At the same time, this study warns against the excessive removal of crop residues in Ukraine, necessitating the monitoring and maintenance of soil fertility, a function that could be overseen by the State Service of Ukraine for Geodesy, Cartography and Cadastre.

Author Contributions: Conceptualization, S.K., O.B. and V.K.; methodology, S.K. and O.B.; formal analysis, S.K., V.K.; investigation, S.K., V.K., O.B. and A.W.; data curation, S.K.; writing—original draft preparation, S.K., V.K.,

O.B.; writing—review and editing, S.K., V.K., O.B., A.W.; visualization, S.K. and V.K. All authors have read and agreed to the published version of the manuscript.

Funding: This research received no external funding.

Conflicts of Interest: The authors declare no conflict of interest.

References

1. *The Energy Strategy of Ukraine until 2035 "Safety, Energy Efficiency, Competitiveness"*; Cabinet of Ministers of Ukraine: Kyiv, Ukraine, 2017. Available online: https://zakon.rada.gov.ua/laws/show/605-2017-%D1%80 (accessed on 20 August 2020).
2. *Energy Supply Based on Renewable Sources*; State Statistics Service of Ukraine: Kyiv, Ukraine, 2019. Available online: http://www.ukrstat.gov.ua/operativ/operativ2016/sg/ekolog/ukr/esp_vg_u.htm (accessed on 2 August 2020).
3. Bioenergy Association of Ukraine. Available online: https://uabio.org/en (accessed on 7 September 2020).
4. Geletuha, G.G.; Zheliezna, T.A.; Dragniev, S.V. *Analysis of Barriers in Production of Energy from Agrobiomass in Ukraine: Analytical Report*; BAU: Kyiv, Ukraine, 2019; Volume 21. Available online: https://saf.org.ua/library/198 (accessed on 22 July 2020).
5. Iqbal, Y.; Lewandowski, I.; Weinreich, A.; Wippel, B.; Pforte, B.; Hadai, O.; Tryboi, O.; Spöttle, M.; Peters, D. *Maximising the Yield of Biomass from Residues of Agricultural Crops and Biomass from Forestry*; Ecofys: Berlin, Germany, 2016. Available online: https://ec.europa.eu/energy/sites/ener/files/documents/Ecofys%20-%20Final_%20report_%20EC_max%20yield%20biomass%20residues%2020151214.pdf (accessed on 22 July 2020).
6. Puri, M.; Rincon, L. *Egypt, Turkey and Ukraine: Sustainable Bioenergy Options from Crop and Livestock Residues*; FAO: Rome, Italy; EBRD Crop.: London, UK, 2017. Available online: http://www.fao.org/3/a-i8150e.pdf (accessed on 11 August 2020).
7. *Bioenergy in the Global Energy Context. GBEP: Global Bioenergy Partnership*; FAO: Rome, Italy, 2007. Available online: http://www.fao.org/3/a1348e/a1348e02.pdf (accessed on 22 July 2020).
8. Reid, W.V.; Ali, M.K.; Field, C.B. The future of bioenergy. *Glob. Chang. Biol.* **2019**, *26*, 274–286. [CrossRef] [PubMed]
9. Dhanya, B.S.; Mishra, A.; Chandel, A.K.; Verma, M.L. Development of sustainable approaches for converting the organic waste to bioenergy. *Sci. Total Environ.* **2020**, *723*. [CrossRef] [PubMed]
10. Kaltschmitt, M. Biomass as renewable source of energy, possible conversion routes. In *Encyclopedia of Sustainability Science and Technology*; Meyers, R.A., Ed.; Springer: New York, NY, USA, 2012. [CrossRef]
11. Maltsoglou, I.; Koizumi, T.; Felix, E. The status of bioenergy development in developing countries. *Glob. Food Secur.* **2013**, *2*, 104–109. [CrossRef]
12. DeCicco, J.M.; Schlesinger, W.H. Opinion: Reconsidering bioenergy given the urgency of climate protection. *Proc. Natl. Acad. Sci. USA* **2018**, *115*, 9642–9645. [CrossRef] [PubMed]
13. *How2Guide for Bioenergy*; IEA: Paris, France; FAO: Rome, Italy, 2017; Available online: http://www.fao.org/3/a-i6683e.pdf (accessed on 6 August 2020).
14. Fischer, G.; Schrattenholzer, L. Global bioenergy potentials through 2050. *Biomass Bioenergy* **2001**, *20*, 151–159. [CrossRef]
15. Van Vuuren, D.P.; van Vliet, J.; Stehfest, E. Future bio-energy potential under various natural constraints. *Energy Policy* **2009**, *37*, 4220–4230. [CrossRef]
16. Ladanai, S.; Vinterbäck, J. *Global Potential of Sustainable Biomass for Energy*; Sveriges Lantbruksuniversitet (SLU): Uppsala, Sweden, 2009. Available online: https://pub.epsilon.slu.se/4523/1/ladanai_et_al_100211.pdf (accessed on 22 July 2020).
17. Kappas, M. Estimation of global bioenergy potentials and their contribution to the world's future energy demand—A short review. In *Sustainable Bioenergy Production—An Integrated Approach*; Ruppert, H., Kappas, M., Ibendorf, J., Eds.; Springer: Dordrecht, The Netherlands, 2013. [CrossRef]
18. Deng, Y.Y.; Koper, M.; Haigh, M.; Dornburg, V. Country-level assessment of long-term global bioenergy potential. *Biomass Bioenergy* **2015**, *74*, 253–267. [CrossRef]
19. Usmani, R.A. Potential for energy and biofuel from biomass in India. *Renew. Energy* **2020**, *155*, 921–930. [CrossRef]

20. Streimikiene, D.; Mikalauskiene, A.; Mikalauskas, I. Comparative assessment of sustainable energy development in the Czech Republic, Lithuania and Slovakia. *J. Compet.* **2016**, *8*, 31–41. [CrossRef]
21. Gradziuk, P.; Gradziuk, B.; Trocewicz, A.; Jendrzejewski, B. Potential of straw for energy purposes in Poland—Forecasts based on trend and causal models. *Energies* **2020**, *13*, 5054. [CrossRef]
22. Poppa, J.; Lakner, Z.; Harangi-Rákos, M.; Fári, M. The effect of bioenergy expansion: Food, energy, and environment. *Renew. Sustain. Energy Rev.* **2014**, *32*, 559–578. [CrossRef]
23. Erb, K.H.; Haberl, H.; Plutzar, C. Dependency of global primary bioenergy crop potentials in 2050 on food systems, yields, biodiversity conservation and political stability. *Energy Policy* **2012**, *47*, 260–269. [CrossRef] [PubMed]
24. Sarkodie, S.A.; Strezov, V.; Weldekidan, H.; Asamoah, E.F.; Owusu, P.A.; Doyi, I.N.Y. Environmental sustainability assessment using dynamic autoregressive-distributed lag simulations—Nexus between greenhouse gas emissions, biomass energy, food and economic growth. *Sci. Total Environ.* **2019**, *668*, 318–332. [CrossRef] [PubMed]
25. Scarlat, N.; Martinov, M.; Dallemandet, J.-F. Assessment of the availability of agricultural crop residues in the European Union: Potential and limitations for bioenergy use. *Waste Manag.* **2010**, *10*, 1889–1897. [CrossRef]
26. Rocha-Meneses, L.; Bergamo, T.F.; Kikas, T. Potential of cereal-based agricultural residues available for bioenergy production. *Data Brief* **2019**, *23*, 103829. [CrossRef]
27. Muth, D.J., Jr.; Bryden, K.M.; Nelson, F.G. Sustainable agricultural residue removal for bioenergy: A spatially comprehensive US national assessment. *Appl. Energy* **2013**, *102*, 403–417. [CrossRef]
28. *Statistical Yearbook of Ukraine 2018*; State Statistics Service of Ukraine: Kyiv, Ukraine, 2019. Available online: http://www.ukrstat.gov.ua/druk/publicat/kat_u/2019/zb/11/zb_yearbook_2018.pdf (accessed on 3 Septembe 2020).
29. Liashchenko, V.; Geletuha, G.; Dragniev, S.; Kucheruk, P.; Matvieyev, Y. *Use of Biomass as A Fuel in the Municipal Sector of Ukraine: A Practical Guidebook*; UNDP: Kyiv, Ukraine. Available online: http://bioenergy.in.ua/media/filer_public/f5/9c/f59c3f7f-8eca-4b6d-94cd-ffda1150f3ae/biofin.pdf (accessed on 15 August 2020).
30. Dubrovin, V.O.; Golub, G.A.; Dragniev, S.V.; Geletuha, I.T.; Zhelieznaya, T.A.; Kucheruk, P.P.; Matveyev, Y.B.; Kudria, S.O.; Zabarnyi, G.M.; Masliukova, Z.V. Method of general evaluation of technically achievable biomass energy potential. *Viol Print* **2013**, 25.
31. Jiang, Y.; Havrysh, V.; Klymchuk, O.; Nitsenko, V.; Balezentis, T.; Streimikiene, D. Utilization of crop residue for power generation: The case of Ukraine. *Sustainability* **2019**, *11*, 7004. [CrossRef]
32. Halchynska, Y.M. Development of biomass potential of by-products of agricultural crops in the agrarian sector of economy. *Ekon. APK* **2019**, *5*, 15–26. [CrossRef]
33. Report on Energy Potential of Myrhorod Rayon and Assessment of Currently Available Calculation Methodologies. Institute of Local Development in Cooperation with USAID (Agreement AID-121-A-13-00002). 2013. Available online: http://myrgorod.pl.ua/files/images/Madem/2.pdf (accessed on 2 August 2020).
34. Morozov, R.V.; Fedorchuk, Y.M. Assessment of bioenergy potential of crop residues and energy crops in agriculture of Khersonska oblast. *Sci. J. Kherson State Univ.* **2015**, *10*, 111–117.
35. Bezuglyi, M.D.; Bulgakov, V.M.; Grynyk, I.V. Scientific-practical approaches to use of straw and crop residues. *Her. Agrar. Sci.* **2010**, *3*, 5–8. Available online: http://nbuv.gov.ua/j-pdf/vaan_2010_3_3.pdf (accessed on 5 August 2020).
36. Dees, M.; Elbersen, B.; Fitzgerald, J.; Vis, M.; Anttila, P.; Forsell, N.; Ramirez-Almeyda, J.; Garcia, D.; Monti, A.; Glavonjic, B.; et al. *Atlas with Regional Cost Supply Biomass Potentials for EU 28, Western Balkan Countries, Moldavia, Turkey and Ukraine*; Project Report. S2BIOM—A Project Funded under the European Union 7th Framework Programme for Research. Grant Agreement No608622; Chair of Remote Sensing and Landscape Information Systems, Institute of Forest Sciences, University of Freiburg: Breisgau, Germany, 2017. Available online: http://s2biom.alterra.wur.nl/doc/S2Biom_D1_8_v1_1_FINAL_19_04_2017_CP.pdf (accessed on 4 September 2020).
37. Walsh, M.E. Biomass Resource Assessment. In *Reference Module in Earth Systems and Environmental Sciences*; University of Tennessee: Knoxville, TN, USA, 2014. [CrossRef]
38. Havlík, P.; Schneider, U.; Schmid, E.; Bőttcher, H.; Fritz, S.; Skalsky, R.; Aoki, K.; De Cara, S.; Kindermann, G.; Kraxner, F.; et al. Global land-use implications of first and second generation biofuel targets. *Energy Policy* **2011**, *39*, 5690–5702. [CrossRef]

39. Bogonos, M.; Stepaniuk, O. *Agrarian Perspectives of Ukraine 2017–2030*; German-Ukrainian Agropolicy Dialogue, 2017. Available online: https://apd-ukraine.de/images/APD_APR_06-2017_AGMEMOD_Baseline_ukr.pdf (accessed on 4 August 2020).
40. Grekov, V.O.; Datsko, L.V.; Zhylkin, V.A.; Maystrenko, M.I.; Datsko, M.O.; Melnyk, A.I.; Zinchuk, M.I.; Makarova, G.A.; Fandaliuk, A.V.; Kolodiy, N.M.; et al. Methodic recommendations on soil protection. *State Sci. Technol. Cent. Soil Fertil. Prot.* **2011**, 107.
41. Camia, A.; Robert, N.; Jonsson, R.; Pilli, R.; Garcia-Condado, S.; Lopez-Lozano, R.; vander Velde, M.; Ronzon, T.; Gurria, P.; M'Barek, R.; et al. *Biomass Production, Supply, Uses and Flows in the European Union. First Results from an Integrated Assessment*; Publications Office of the European Union: Luxembourg, 2018. Available online: https://publications.jrc.ec.europa.eu/repository/handle/JRC109869 (accessed on 5 August 2020).
42. USAID. Evaluation of Current Energy Potential of Clean Energy Sources of Sumska Region. 2016. Available online: https://merp.org.ua/images/Docs/USAID_MERP_Sumy_Oblast_ASEpotential.pdf (accessed on 11 August 2020).
43. Bezhenar, I.M. The state of specialization in agricultural business structures. *Glob. Natl. Probl. Econ.* **2016**, *10*, 128–133. Available online: http://global-national.in.ua/archive/10-2016/28.pdf (accessed on 2 September 2020).
44. Tsyliuryk, O. Minimal Soil Cultivation and No-Till System. Agro Bus. Today. 2019. Available online: http://agro-business.com.ua/agro/ahronomiia-sohodni/item/12869-minimalnyi-obrobitok-hruntu-ta-systema-notill.html (accessed on 5 August 2020).
45. Kravtsov, D. System of Zero Soil Tillage or "No-Till" Technology. Available online: https://uhbdp.org/ua/eco-technologies/articles/1865-systema-nulovoi-obrobky-gruntu-abo-no-till-tekhnolohiia (accessed on 29 July 2020).
46. Peiretti, R.A.; Dumanski, J. The transformation of agriculture in Argentina through soil conservation. *Int. Soil Water Conserv. Res.* **2014**, *2*, 14–20. [CrossRef]
47. Shao, Y.; Wang, J.; Preto, F.; Zhu, J.; Xu, C. Ash deposition in biomass combustion or co-firing for power/heat generation. *Energies* **2012**, *5*, 5171–5189. [CrossRef]
48. State Service of Ukraine for Geodesy, Cartography and Cadastre. Available online: https://land.gov.ua (accessed on 1 September 2020).
49. Use of Inorganic and Organic Fertilizers in 1990–2019. State Statistics Service of Ukraine. 2020. Available online: http://www.ukrstat.gov.ua/operativ/operativ2018/sg/vmod/vmod1990-2019_e.xls (accessed on 29 August 2020).
50. National Report on the State of Soil Fertility in Ukraine. Ministry of Agrarian Policy of Ukraine. 2010. Available online: http://www.iogu.gov.ua/wp-content/uploads/2013/07/stan_gruntiv.pdf (accessed on 2 September 2020).
51. Crop Production in Ukraine. State Statistics Service of Ukraine. 2019. Available online: http://www.ukrstat.gov.ua/druk/publicat/kat_u/2018/zb/04/zb_rosl2017pdf.zip (accessed on 7 September 2020).
52. Ryabchenko, O.; Nonhebel, S. Assessing wheat production futures in the Ukraine. *Outlook Agric.* **2016**, *45*, 165–172. [CrossRef]
53. Alexandratos, N.; Bruinsma, J. *World Agriculture towards 2030/2050: The 2012 Revision*; ESA Working Paper No 12-03; Food and Agriculture Organization (FAO): Rome, Italy, 2012.
54. Global Yield Gap Atlas. Available online: http://www.yieldgap.org/web/guest/yieldgaps (accessed on 7 September 2020).
55. Record Harvest of Cereals in 2018. BusinessViews. 2018. Available online: https://businessviews.com.ua/ru/economy/id/urozhaj-2018-1915 (accessed on 17 August 2020).
56. The Harvest of Cereals Reached the Record 66 Million Tons in 2016. UKRINFORM. 2017. Available online: https://www.ukrinform.ua/rubric-economy/2157010-vrozaj-zernovih-u-2016-roci-sagnuv-rekordnih-66-miljoniv-tonn-minagro.html (accessed on 22 July 2020).
57. Sobkevych, O.V.; Rusan, V.M.; Yurchenko, A.D.; Kovaliova, O.V.; Zhurakovska, L.A. Resource Potential of Agrarian Sector: Problems and Objectives for Efficient Utilization. National Institute for Strategic Studies, 2013. Available online: http://old2.niss.gov.ua/content/articles/files/Agro_sfera-cf5ec.pdf (accessed on 3 August 2020).
58. Gielen, D.; Saygin, D.; Wagner, N. REmap 2030. Analysis for Ukraine. IRENA. 2015. Available online: http://banisaenergy.com/sites/default/files/remap-2030-analysis-for-ukraine-120315.pdf (accessed on 5 August 2020).
59. Van de Graaf, T.; Colgan, J.D. Russian gas games or well-oiled conflict? Energy security and the 2014 Ukraine crisis. *Energy Res. Soc. Sci.* **2017**, *24*, 59–64. [CrossRef]

60. Energy balance of Ukraine 2007–2018. State Statistics Service of Ukraine, 2019. Available online: http://www.ukrstat.gov.ua/operativ/operativ2012/energ/en_bal/arh_2012.htm (accessed on 29 July 2020).
61. Sadik-Zada, E.R.; Loewenstein, W. Drivers of the CO2-Emissions in the fossil-fuel abundant settings: (pooled) mean group and nonparametric panel analyses. *Energies* **2020**, *13*, 3956. [CrossRef]
62. Agriculture of Ukraine in 2018: Statistical Yearbook. State Statistics Service of Ukraine. 2019. Available online: http://www.ukrstat.gov.ua/druk/publicat/kat_u/2018/zb/09/zb_sg2017_pdf.pdf (accessed on 7 September 2020).

Publisher's Note: MDPI stays neutral with regard to jurisdictional claims in published maps and institutional affiliations.

energies

MDPI

Article

Renewable Energy Utilization in Rural Residential Housing: Economic and Environmental Facets

Aleksandra Siudek [1], Anna M. Klepacka [1,*], Wojciech J. Florkowski [2] and Piotr Gradziuk [3]

1 Department of Economics and Organisation of Enterprises, Institute of Economics and Finance, Warsaw University of Life Sciences, 02-787 Warsaw, Poland; aleksandra_siudek@sggw.edu.pl
2 Department of Agricultural and Applied Economics, University of Georgia, Griffin, GA 30223-1797, USA; wojciech@uga.edu
3 Polish Academy of Sciences Institute of Rural and Agricultural Development, 00-330 Warsaw, Poland; pgradziuk@irwirpan.waw.pl
* Correspondence: anna_klepacka@sggw.edu.pl

Received: 24 November 2020; Accepted: 15 December 2020; Published: 16 December 2020

Abstract: Energy and climate policies benefit from modernized construction technology and energy supply source choices. Energy-efficiency improvement and CO_2 emission reduction will result from renewable energy (RE) utilization in new and retrofit single-family houses in rural Poland. Several house construction scenarios and heating energy sources comparing building costs and potential emission reduction are based on already existing structures calculated for a 100 m^2 dwelling corresponding to the average rural home. With the addition of thermal insulation and RE-generating equipment, construction costs increase, but the energy costs of operating the home dramatically shrink between a conventional and energy-neutral house. The latter scenario includes thermal solar panels and a heat pump as heating energy sources as well as electricity-generating PV panels. Replacing coal with environmentally-friendly RE reduces CO_2 emissions by about 90% annually. Additionally, lower dependence on coal lessens other GHG emissions leading to immediate air quality improvement. New house building regulations guide homeowner construction and heating energy choice, but even larger gains could result from retrofitting existing rural houses, expanding environmental benefits and generating energy bill savings to households. However, the varying climate throughout Poland will require the purchase of energy in winter to assure residents' comfort.

Keywords: renewable energy; rural residential housing; emission reduction; construction regulations; Poland

1. Introduction

The decarbonization of the economy requires multipronged efforts encouraging wide adoption of energy-efficient technologies and the increased utilization of renewable energy (RE) sources. Poland's policies fall in line with European Union (EU) goals and pursue the decarbonization of the economy [1]. The reduction of fossil fuels is part of the energy and climate change policy in Poland, the country most dependent on coal for the production of energy in the European Union [2]. Lowering the use of coal is also the goal of the country's air quality improvement policy because of the substantial negative health outcomes caused by toxic emissions [3]. The dramatic re-structuring of industry in Poland in the last decades led to large reductions in using coal and lowered GHG emissions, especially CO_2 which decreased from 11.164 ton per capita in 1989 to 7.876 ton per capita in 2016 [4]. However, climatic conditions necessitate heating living spaces in homes and apartments from fall until early spring, and even during cool periods at other times of the year. The needs for heating energy vary across the country [5], but are particularly important to occupants of single-family homes, which prevail in rural areas of Poland and are heavy users of fossil fuels. Rural residents lack opportunities to access piped heating used by 40% of urban households in 2018 [6]. Improvement of life quality and utilization of locally available resources such as RE are the goals of the Strategy of Sustainable Development on Rural Areas, Agriculture and Fisheries for the period 2012–2020 see for example [7]. This strategy reflects the commitment to sustainable development stated in the Polish Constitution (1997, Chapter 1, art. 5).

Heating energy accounts for the largest portion of all household energy expenditures. In the past decade, programs supporting RE utilization by households provided generous subsidies for the purchase and installation of thermal solar panels and rural households participated in the program motivated by energy cost savings [8]. However, the utilization of solar energy installations for single-family homes remains relatively limited. In 2018, one in 52 households used solar energy in Poland [6]. More recently, homeowners could take advantage of subsidies and low-cost loans through participation in a program involving the replacement of coal-using boilers with energy-efficient heating systems including wood pellet furnaces [9]. Such programs reduce heating energy needs in single-family homes, but those needs strongly depend on the construction technology used to build the house. In this context, the adoption of new construction technology mandated by new building regulations takes the concept of sustainable development to a new level. The regulations change construction methods which result in reduction of energy needs of households.

The new regulations are one element of policies aimed at increasing energy security, GHG emission reduction, utilization of RE, and encouraging modern thermal insulation methods in the construction of single-family homes in Poland. The regulations resulted from the EPBD Directive on the energy efficiency of buildings and become compulsory in 2021. The regulations require all newly built houses to achieve zero-energy status, i.e., show zero energy consumption. The EPBD also applies to old buildings. According to the Directive, every potential homeowner receiving a building permit in 2021 has to meet the technical conditions (WT 2021). The new standards apply to all construction projects, including modernization or expansion of an existing building, although some details have not been finalized [10,11].

New regulations are a part of a multipronged approach at increasing energy efficiency that sustain efforts to modernize older rural housing to permanently improve living conditions, local air and environment quality, and assure energy cost savings to households. The requirements in new construction regulations improve the coherence of government programs and allow current or future homeowner to take advantage of several separate programs aiming at energy-efficiency and RE utilization. Additionally, the mandatory building regulations eliminate the resistance to RE, while the subsidy and low-cost loans encourage homeowners who expressed opinions that subsides for RE-utilizing equipment were important in their decision-making [8]. These approaches lessen rural resident opposition to the use of RE and stronger building regulations while implementing constitutionally-mandated sustainable development.

To provide insight about gains from energy-efficiency enhancement of single-family houses, this study compares investment costs, heating energy costs, and the amount of emitted CO_2 and other GHGs under alternative construction technology scenarios and several renewable and non-renewable energy use setups to heat the living space and domestic hot water. Space and water heating are the main uses of heating energy by Polish households [12]. Heating energy needs result from climatic conditions which are quite complex because of the country's geographical location [13]. The current study assumes the perspective of a predominantly rural homeowner in a country where GDP per capita in PPS is below the EU average [14] complementing research on the European decarbonization pathways analyzing emission reduction and the use of RE at the aggregate level, for example, see [15].

The scenarios include a traditional house heated with coal, traditional house using natural gas, traditional house heated with wood pellets, a house that utilizes RE and thermal insulation, and an energy-neutral house built with the latest construction technology. An average rural house had approximately 108 m^2 of living space in 2018 [16] traditionally heated with coal. Starting in 2021, households and the homeowners will face construction regulations requiring all newly constructed buildings to be passive in terms of energy use and equipped with RE micro-installations [17]. Owners of existing houses constructed with outdated technology and using heating systems emitting large amounts of air pollutants can enjoy major energy savings by retrofitting their dwellings by taking advantage of currently implemented government programs. Results from this study provide information for public education campaigns illustrating the cost differences of operating houses using different types of energy as well as the associated reduction in air pollution. Lower toxic emissions instantaneously improve air quality in the immediate neighborhood enhancing life quality and health outcomes and serving as another argument for convincing homeowners to use the best available technical solutions.

The study expands the existing literature that includes research on energy performance of multifamily buildings [18] and the use of gas-powered boilers to heat residential houses in northern Poland [19] or the use of hybrid central heating systems using RE in southern Poland [20]. Moreover, the study considers multiple construction technologies and alternative heating systems. New construction scenarios are supplemented by a discussion of retrofitting an existing rural home with a central heating system utilizing biomass, the most common form of RE in the EU [21]. The selected biomass heating system utilizes wood pellets, a relatively new form of sustainable fuel gaining popularity in Poland.

The study is justified by constant new construction and the existence of a large number of detached houses in rural areas of Poland (more than three million) and emphasizes opportunities for higher energy efficiency, a key element in transforming the energy system in the EU [22]. The focus on single-family houses is also motivated by the portion of households residing in single-family homes increasing by more than 3% between 2005 and 2016 reaching 38.2% [23]. The trend to live in a detached house in Poland, defies the trend towards apartment living in multi-family housing observed in western EU country-members. The trend reflects the generally smaller average living space in Poland than in many other EU countries, but the larger detached house space involves critical decisions regarding energy-efficiency and the choice of heating energy. The consideration of alternative scenarios of building technology and heating energy sources offers insights about newly constructed homes, but also provides information for educating the owners of existing detached rural houses about economic gains from investing in energy efficiency and RE utilization. The study supplements previous research showing the crucial role of engaging owners of single-family houses to lower heating energy needs by using some RE [24]. The case study also develops alternative scenarios estimating the amount of CO_2 emission reduction resulting from new building regulations.

An earlier study found that the greater the benefits to rural communities in Spain, the greater the social acceptance of projects involving the use of RE [25], while a sizable portion of the public did not see obvious benefits from the long-term economic feasibility of RE use in Finland [26]. The absence of public consultation before imposing new construction regulations in Poland forcing the use of RE coincides with the limited information on the benefits and may be interpreted as lacking impartiality. Although the public has generally favorable views of RE in Poland, once the costs of RE use affect the household, attitudes change. The distinction between the general support and local perceptions should be considered suggested a study of German public [27].

2. Rural Housing and Sustainable Growth

Rural areas cover 93% of Poland's territory [28] and rural residents account for 39.9% of the total population and that share has increased 0.7% since 2018. Rural areas are associated with farming and farming dominates the rural economy, but housing construction has been rapidly growing and contributing to local economies. The drivers of the housing construction sector are the desire of many Poles to enjoy their own individual family house as well as the replacement or renovation of existing homes. The first phenomenon results from strict regulations limiting apartment and house size under the former centrally-planned economy and a chronic shortage of accommodation for the expanding population. The generation of "baby boomers" was forced to live in cramped apartments in urban areas, or share rural houses with parents and grandparents. The never-ending shortage of construction materials severely constrained the ability to enlarge or build new houses even if the size of a rural property could allow such an expansion. The transition to a market-driven economy since 1989 removed restraining regulations, while eliminating scarcities of construction materials and lack of access to updated building technology. The new limitation is the availability of real estate in urban areas, which led to migration to nearby rural areas, where land was less expensive. The majority of the approximately 80,000 new homes built every year in Poland is located in rural areas.

Simultaneously, the abundance of construction material permitted rural residents to either replace their old house or retrofit and enlarge the existing structure. Single-family homes represent 86.3% of all housing in rural areas [29], while multi-family housing accounts for 76.5% of all urban housing [29]. The booming construction in rural areas creates new demands on the energy supply. The construction sector uses about 40% of the world's energy [30]. An average household uses about 65% of purchased energy for space heating and 16.6% for heating water for daily use in Poland. The share of energy used for lighting and cooking is relatively small, 9.8% and 8.5%, respectively. The typical rural household uses more energy than an urban household because of the difference in size. For example, the average size of a rural house living area was 108.1 m^2 versus 67 m^2 in urban areas in 2018 [29]. Although many rural residents enjoy large living areas, their incomes are often below those of urban residents which drives their search for the lowest possible energy costs [31].

With scattered settlements of low-density housing, rural areas pose a challenge in the supply of heating energy in Poland. Although 45.8% of multi-family housing in urban areas receive heat from centralized heat-generating plants, the share among rural households was 2.9% in 2018. For example, natural gas was used for space and domestic water heating only in 7.3% of rural households in 2018. The typical rural home is heated with solid fuels, primarily coal used, by 86.2% of households. Rural households used 9.9 million tons of coal and thousands of tons of wood to start the coal fire. Some households using coal-fired furnaces burn plastic and other burnable waste increasing air pollution [32]. A recently enacted regulation allows local government representatives to enter homes in Poland to verify what is being burned in boilers [31]. Inadequate insulation and inefficient furnaces contribute to heat loss and house construction technology is a major factor determining the heating energy requirements.

A sizable share of rural homes was constructed before 1961 when regulations allowed the thermal efficiency of $k < 0.87$. Rural homes built between 1961 and 1995 represent 51.8% of housing and had to meet higher requirements of $k < 0.3$. Since 1996, another 14.5% of houses were built in rural areas, still under the requirement of $k < 0.3$. New regulations placed in 2008 increased thermal efficiency requirements to $k < 0.25$ [33]. Since then, new regulations follow the guidelines adopted by the European Commission [34].

The new construction requirements provide strong incentives to use RE, e.g., solar thermal panels, wood pellet boilers, and heat pumps for home heating systems and PV panels for generating electricity. The recently introduced programs offer subsidies and low-interest loans for replacement of home heating systems and are specifically addressed to single-family homeowners and those building new homes [9]. However, the response to the program operating since January 2018 has been minor [6]. The results of this study provide evidence of the substantial long-term economic gains through cost reduction of operating a house and can be applied to popularize the program. An increased participation in the program directly achieves the goal of local air quality improvement and extends contributions to national and EU energy and climate policy implementation. As a result, the consideration of alternative building technologies with a focus on the type of energy used involves the economic, social, and environmental aspects of sustainability.

The social acceptance of new building regulations determines the future compliance and the use of RE in new homes. Moreover, once the homes constructed using the modern technology guided be recent regulations [35] appear in rural landscape, the owners of the existing houses are more likely to undertake thermal modernization of their residencies. The living comfort in an energy neutral house, the convenience of purchasing heat-generating energy, and largely eliminated disposal of ash have an unquestionable appeal. With the social aspect of sustainability in the background, the focus shifts to the economic and environmental aspects.

3. Methodology

3.1. House Construction Scenarios

The case study presents three scenarios involving the use of different construction methods. All of the scenarios involve fully completed and closed structures and include the installation of all windows, external doors and the door to the garage. Additionally, the scenarios also include insulation of the roof. However, the energy-efficiency of the insulating materials varies depending on the scenario. Each of the single-family house scenarios is equipped with heating systems utilizing fossil fuels and different RE sources. The comparisons also include the use of electricity supplied from the grid.

3.2. Building Model

The average living space of a single-family rural home according to the National Census summary was almost 97 m^2 in 2002 and increased to 101.8 m^2 in 2011 [36]. The average living space has been gradually increasing over time and reached 108.1 m^2 in 2018. However, many newly built detached houses have a floor plan much larger than the existing homes as indicated by the national average of 143.5 m^2 in 2018 [37]. The ever-changing living space alters heating needs, although regional climate variation may shorten or extend those needs in Poland. The current study assumes a single-family home with 100 m^2 of living space and the calculations provide a benchmark that allows for adjustments for specific homes. Another simplification is the application of a house plan that is a rectangle and includes a garage as a part of the building. The house is a duplex with a functional second story (attic) and a gabled roof. Such design is common in rural areas among newly built and existing homes. The typical house occupies a flat and open terrain; most newly constructed homes are not on farms and contrast with older rural houses of farming families that are typically surrounded by buildings on the same property. The model building is heated using radiators mounted on the walls. The space heating equipment also provides the hot domestic water. The building plan does not include a basement or a cellar.

3.3. Construction Costs

Information about construction costs, costs of RE micro-installations, costs of a coal-fired furnace, and exterior wall and roof insulation were obtained from publicly available sources (see Appendix A). The scenarios include separate estimates of the annual costs of supplying the family with hot domestic water and heating the living space, the two main energy-related expenses for households in Poland. Calculations use energy prices reported by the Central Statistical Office (GUS). The alternative construction scenarios include the same major elements: building the shell of the house, installing windows, doors, roof insulation, and the heating and energy supply fitting. Although prices of construction and insulation materials vary between regions, those variations are usually negligible. Any reductions in costs that suppliers can achieve through market segmentation generally do not exceed transportation costs. Some potential homeowners in rural areas, for example, may save on labor costs if they choose to perform some tasks such as the installation of roof insulation or doors, but the savings are relatively small. The summarized costs are for the same building plan, but differ in the amount, type and energy-efficiency of selected materials and, wherever applicable, the associated labor costs.

3.4. Evaluation of CO_2 Reduction Emission

The indicators of CO_2 emissions for various sources of energy used by a single-family dwelling were obtained from the National Center for Emission Accounting and Management [3]. In the case of coal, the emission indicator was 94.7 kg/GJ and for natural gas it was 56.1 kg/GJ. The indicator associated with electricity that a house will have to purchase, especially during the long heating season, was 93.87 kg/GJ. The study follows the Ministry of Infrastructure and Development methodology for establishing energy features of a house or a part of a house issued on 27 February 2015 [38].

3.5. Calculation of Reduction of Other GHG and Particulate Emissions

Currently, the majority of rural homes use coal in inefficient residential stoves to heat the living space and domestic water. Burning coal emits NO_2 and SO_2 [39] as well as particulate matter, a well-recognized problem in Poland [40]. The emissions negatively affect health [41,42], including that of children [43]. The reduction estimates of selected toxic emissions supplement the measures of environmental benefits associated with the new building regulations.

4. Results

4.1. Construction and Thermal Insulation Costs

The construction of the shell of the building includes the ground preparation, construction of foundations, external walls, chimneys, ceilings, roof construction and cover, and gutters (Table 1). Sustainable growth in single-family home construction begins with the initial investment. The use of conventional construction technology is less expensive than the energy-neutral technology, but the differences in material used to construct external walls reveal energy-efficiency gains.

The conventionally-built home uses a brick type characterized by the heat transfer coefficient $U = 0.35$ W/m^{2}K and drops to $U = 0.31$ W/m^{2}K for the other two scenarios (where the lower value implies better insulating properties) (Table 1). The least expensive scenario includes thin external wall thermal insulation, and the costs increase for other scenarios. The energy-efficient home has an insulation layer of material 12 cm thick, while the energy-neutral home's wall insulation is 18 cm thick with higher insulating value. The material costs increase by 50% in the energy-efficient house variant and nearly triple (an increase of 283.3%) in the energy-neutral scenario. Labor costs increase by 89.5% for the two scenarios as compared to the conventional construction, but the energy efficiency improves substantially as the heat transfer coefficient decreases from 0.28 W/m^{2}K to 0.15 W/m^{2}K.

Once the shell of the house is completed, the major factors contributing to the total construction costs are related to the energy efficiencies of the three types of homes (Table 1). Windows, doors and the garage door installation completes the house and allows for the interior work on the house (not considered here). A conventional house is equipped with PCV double-pane windows with the heat transfer coefficient $k = 1.4$. More energy efficient widows with $k = 1.1$ are installed under the next scenario, but triple pane windows with $k = 0.85$ are installed in the energy neutral house. The cost differences are substantial and compared to the traditional home, the energy-efficient scenario lists window costs, respectively, 206.6% and the energy neutral 412.8% higher. Labor costs are only 25% higher in the energy neutral case because of the more involved installation. The door selected for the conventional house costs about a third of the entrance door installed in the energy neutral home and almost 60% of the door in an energy efficient house.

The labor cost difference in mounting the garage door is large in relative terms, 33.3%, but small in absolute terms between the first and the other two scenarios (Table 1). The garage doors installed in the three scenarios differ in their insulating capacity ($U = 1.6$ W/m^{2}K vs. $U = 1.1$ W/m^{2}K vs. $U = 0.9$ W/m^{2}K) and the model, a single vs. segmented door. The interior thermal insulation of the attic differs only in the cost of the material since the labor cost is the same under all scenarios.

The last item of closing the structure is the cost of insulating the roof. The heat transfer coefficient of the insulation, $U = 0.036$ W/m^{2}K, is identical for the two scenarios and there is a difference in the thickness of the mineral wool layer (Table 1). The energy-neutral house uses a different insulation, polyurethane (PUR) foam, characterized by $U = 0.023$ W/m^{2}K. With the labor costs identical for the three scenarios, the cost difference is in the amount and type of materials, and the cost more than triples (327.7% higher) in the case of the energy neutral house as compared to the conventional scenario.

Table 1. Building stage and costs of materials and labor for three construction technologies applying different thermal insulation and heating and supplementary energy supply systems for a single-family home with 100 m^2 of living space, Poland, 2020.

Construction Stage	Traditional Construction	Energy-Efficient Construction	Energy Neutral Construction
Ground preparation, foundations, insulated floor on the ground	Concrete sleeper on compacted soil, moisture barrier, a 5 cm Styrofoam layer, floor screed M = 16,000 PLN; L = 22,000 PLN; M + L = 38,000 PLN (8877 EUR)	Concrete base on compacted soil, moisture barrier, a 15 cm extruded polystyrene EXP layer, floor screed M = 27,500 PLN; L = 22,000 PLN; M + L = 49,500 PLN (11,563 EUR)	Foundation plate with thermal insulation performing the floor function set on the ground surface M = 31,000 PLN; L = 20,000 PLN; M + L = 51,000 PLN (11,915 EUR)
External walls	Single layer wall made of blocks of autoclave aerated concrete (AAC) 36.5 cm in length with U = 0.035 W/m^2K and covered with cement-gypsum M = 12,000 PLN; L = 9500 PLN; M + L = 21,500 PLN (5023 EUR); External wall U = 0.28 W/m^2K	Double layer wall of ceramic blocks 25 cm P + W with U = 0.31 m^2K + 12 cm Styrofoam with U = 0.035 W/m^2K + thin-layer plaster M = 25,000 PLN; L = 18,000 PLN; M + L = 43,000 PLN (10,045 EUR); External wall U = 0.23 W/m^2K	Double layer wall of ceramic blocks 25 cm P + W, U = 0.31 W/m^2K + layer of graphite polysterene 18 cm thick and U = 0.031 W/m^2K + thin-layer plaster M = 34,000 PLN; L = 18,000 PLN; M + L = 52,000 PLN (12,147 EUR); External wall U = 0.15 W/m^2K
Chimneys	M = 6000 PLN; L = 2000 PLN; L + M = 8000 PLN (1869 EUR)	M = 3500 PLN; L = 500 PLN; L + M = 4000 PLN (934 EUR)	M = 3500 PLN; L = 500 PLN; L + M = 4000 PLN (934 EUR)
Monolithic ceiling, reinforced concrete	M = 26,000 PLN; L = 9000 PLN; L + M = 35,000 PLN (8176 EUR)	M = 26,000 PLN; L = 9000 PLN; L + M = 35,000 PLN (8176 EUR)	M = 26,000 PLN; L = 9000 PLN; L + M = 35,000 PLN (8176 EUR)
Timber roof truss	M = 7500 PLN; L = 8000 PLN; L + M = 15,500 PLN (3621 EUR)	M = 7500 PLN; L = 8000 PLN; L + M = 15,500 PLN (3621 EUR)	M = 7500 PLN; L = 8000 PLN; L + M = 15,500 PLN (3621 EUR)
Roof initial covering, soffit boards, gutters, etc.	M = 4500 PLN; L = 5500 PLN; M + L = 10,000 PLN (2336 EUR)	M = 4500 PLN; L = 5500 PLN; M + L = 10,000 PLN (2336 EUR)	M = 4500 PLN; L = 5500 PLN; M + L = 10,000 PLN (2336 EUR)
Roofing	Steel tile M = 6150 PLN; L = 4100 PLN; L + M = 10,250 PLN (2394 EUR)	Ceramic tile M = 9500 PLN; L = 9500 PLN; L + M = 19,000 PLN (4439 EUR)	Ceramic tile M = 9550 PLN; L = 9500 PLN; L + M = 19,000 PLN (4439 EUR)

Table 1. *Cont.*

Construction Stage	Traditional Construction	Energy-Efficient Construction	Energy Neutral Construction
	Closed house shell		
Windows	Double-pane PCV windows, k = 1.4 M = 7480 PLN; L = 2500 PLN; M + L = 9980 PLN (2331 EUR)	Double-pane PCV windows, k = 1.1 M-12,950 PLN; L = 2500 PLN L + M = 15,450 PLN (3610 EUR)	Triple-pane PCV windows, k = 0.85 M-24,050 PLN; L = 3000 PLN; (warm assembly), L + M = 31,550 PLN (7370 EUR)
Entrance door	Door with k = 1.7 M = 1850 PLN; L = 450 PLN L + M = 2300 PLN (537 EUR)	Door with k = 1.5 M-3100 PLN; L = 600 PLN L + M = 3700 PLN (864 EUR)	Door with k = 1.3 M = 5500 PLN; L = 600 PLN; L + M = 6100 PLN (1425 EUR)
Garage door	Up-and-over garage door with a drive, U = 1.6 W/m^{2}K M = 2250 PLN; L = 1000 PLN, M + L = 3250 PLN (759 EUR)	Sectional garage door with a drive, U = 1.1 m^{2}K M = 4650 PLN; L = 1000 PLN, M + L = 5650 PLN (1320 EUR)	Sectional garage door with a drive, U = 0.9 m^{2}K M = 9600 PLN; L = 1000 PLN M + L = 10,600 PLN (2475 EUR)
Attic insulation	Mineral wool, 10 cm layer, U = 0.036 m^{2}K M = 2500 PLN; L = 4350 PLN L + M = 6850 PLN (1600 EUR)	Mineral wool, 15 cm layer, U = 0.033 m^{2}K M = 3200 PLN; L = 4350 PLN L + M = 7550 PLN (1765 EUR)	Rigid PIR foam board 10 cm thick, U = 0.023 m^{2}K M = 10,693 PLN; L = 4350 PLN; L + M = 15,043 PLN (3515 EUR)
		Heating and supplementary energy equipment installation	
Central heating	Central heating installation with a gas boiler for natural gas M = 7200 PLN; L = 8000 PLN L + M = 15,200 PLN (3551 EUR)	Central heating system with a gas condensing natural gas boiler M = 13,000 PLN; L = 8000 PLN L + M = 21,000 PLN (4905 EUR)	Central heating installation with a ground heat pump M = 28,000 PLN; L = 8000 PLN L + M = 36,000 PLN (8411 EUR)
Supplementary energy supply sources	None	Thermal solar panels M = 22,000 PLN; L = 5000 PLN M + L = 27,000 PLN (6310 EUR)	PV panels, 6 kW capacity M = 28,000 PLN; L = 7000 M + L = 35,000 PLN (8175 EUR)
Radiators	Radiators M = 4500 PLN; L = 3000 PLN L + M = 7500 PLN (1750 EUR)	Radiators M = 4500 PLN; L = 3000 PLN L + M =7500 PLN (1750 EUR)	Low-temperature radiators, preferably flat M = 8000 PLN; L = 3000 PLN L + M = 11,000 PLN (2570 EUR)
Ventilation	Gravity ventilation L + M = 1000 PLN (233 EUR)	Mechanical ventilation M = 11,000 PLN; L = 3000 PLN; M + L = 14,000 PLN (3271 EUR)	Mechanical ventilation + heat exchangerM = 11,000 PLN + 9420 PLN; L = 3000 PLN; M + L = 23,420 PLN (5472 EUR)

Note: L—Labor, M—Material. Source: Calculations based on [44–57].

4.2. *Heating System and Electricity Supply*

The primary source of heating energy and hot water is a natural gas-operated boiler in the conventional home. The energy-efficient home is equipped with a gas-fired double function condensing boiler. Both scenarios imply that a rural resident has access to piped natural gas. Access to piped gas in rural areas is increasing but still infrequent in Poland. In some regions, especially those with local natural gas deposits, the use of gas-fired boilers is realistic. Rural residents could use LPG tanks that must be periodically refilled, but the weather pattern in winter months determines the refilling frequency and two gas explosions in November 2020 in homes heated by LPG indicate the possible problems in operating such systems. The differences in costs of these systems are not considered in the current discussion. The energy neutral home obtains heat energy using the geothermal heat pump, which is the primary reason for the nearly three-fold increase in equipment costs. Specifically, the boiler cost of 7200 PLN (1692 EUR) in the conventional house scenario increases to 28,000 PLN (6542 EUR), or 288.9% more, when choosing a geothermal heat pump. Interestingly, labor costs are basically the same regardless of the homeowner's choice of the heating system. Space heating involves radiators in the case of the conventional and energy-efficient house, and flat, low-temperature radiators in the case of the energy-neutral house.

The energy-efficient and energy neutral homes utilize RE in the form of solar radiation. The energy-efficient home uses thermal solar panels, while the energy neutral house uses PV panels with the capacity of 6 kW. The panels are intended as the supplementary source of energy to power the heat supply system. Under Poland's climatic conditions and depending on the region, an energy neutral house will likely require a purchase of electricity during the months when the demand for heat is particularly high because of the scarcity of solar radiation. In Poland the available solar radiation is most scarce during the periods of highest demand for space heating [58]. The energy neutral house will generate surplus electricity in other periods because the solar radiation is typically higher, while the heating needs are limited to the use of hot water. On balance, the home will offset electricity purchase with the supply of electricity to the grid.

Finally, the costs of house ventilation are lowest in the case of using a gravitational system in the conventional house, but 11 times higher when the energy-efficient house uses mechanical ventilation (Table 1). The mechanical ventilation system in an energy neutral house is even more pricey and includes the heat exchanger for the total cost 24 times higher than in the conventional home.

4.3. *Total Cost Differences*

Table 2 summarizes the total costs of building a single-family 100 m^2 house using the three construction technologies and three choices of the central heating system. The costs for various construction stages are listed in Polish zloty and euro. The cost of construction of the unfinished energy-efficient house is 44.5% higher than a house using traditional technology. The costs are 182.8% higher in the case of an energy neutral house. The cost of the heating system and the supplementary electricity source for an energy neutral single-family house is a staggering 353.2% greater than that of a rural house having access to piped natural gas, which reaches a fraction of the rural population.

Table 2. Construction costs of a single-family house using alternative construction technologies and energy, in Polish zloty (PLN)/euro.

Construction Costs	Traditional House	Energy-Saving House	Energy-Neutral House
Closed unfinished house	22,380/5229	32,350/7558	63,293/14,788
Heating and supplementary energy equipment installation	23,700/5537	69,500/16,238	107,420/25,100
Total cost [a]	183,830/42,951	272,850/63,750	356,713/81,650

[a] Assumes a single-family house has 100 m^2. Note: Exchange rate as of November 15, 2020: 1 euro = 4.2807 Polish zloty [59].

4.4. Heating Energy Needs under Alternative Construction Scenarios and Retrofit Options

In the conventional house common in rural areas, the heating system utilizes coal and often serves a dual purpose of heating the space and domestic water. Maintaining the room temperature requires constant monitoring and adding coal. Coal not only generates a sizable volume of ash, but ash disposal involves additional fees. However, the annual cost of heating space and domestic water is lowest, slightly outperforming the use of natural gas (Table 3). The use of natural gas does not require constant monitoring and eliminates the removal and disposal cost of ash. Given the lack of access to natural gas in rural areas, a wood pellet boiler offers an alternative. Wood pellet is a rather novel energy source available for household use and the specialized boilers require less frequent monitoring than the coal boiler does, while the amount of ash is a fraction of that resulting from coal burning. Wood pellets generate substantial environmental benefits because they are a locally available RE supplied by manufacturers located mostly in rural areas. Moreover, the wood pellet ash can be readily applied as fertilizer [60] in landscape surrounding a single-family home. The convenience of wood pellet use is countered by the higher total costs of supplying the house with heat energy as compared to coal (30.5% more) or natural gas (22% more) (Table 3).

Table 3. Energy generation costs for domestic water and central heating systems for homes built using alternative technologies and using different energy sources scenarios in PLN and euro.

Heating Purpose	Traditional House (Coal)	Traditional House (Natural Gas)	Traditional House (Wood Pellet)	Energy Saving House (Natural Gas + Solar Panels)	Energy-Neutral House (Heat Pump + PV)
Domestic water system	4404/1029	3779/883 [a]	5118/1196	3779/883	148/34
Central heating system	4007/936	5217 [b]/1219	5860/1369	864/198	766/179

Note: The exchange rate on 15 November 2020 was 1 euro = 4.2807 PLN [59]. [a] Price of 1 kWh generated from natural gas is 0.25 PLN as listed by Viessman. [b] Price of 1 kWh from electricity is 0.65 based on [16].

The annual costs of heating energy are substantially less in the case of the energy-efficient house. Those costs are 44.8% less than the coal-using traditional house, the least expensive scenario (Table 3). In the case of an energy-neutral house the energy costs of heating space and domestic water amount to only 879 PLN (205 EUR), or 10.5% of the cost of heating the single-family traditional house with coal. The calculations in the current study disregard the possible costs of routine maintenance.

The heating energy needs vary dramatically for various types of houses (Appendix B, Tables A1 and A2). A traditionally built house that uses a coal-fired boiler is estimated to require 31,909 kWh heating energy per year. By switching to the use of natural gas as the energy source, the requirements drop by 37.6%. An energy efficient house that is equipped with a RE installation requires 83.4% less heating energy then the coal-heated house. Those needs drop by 86% in the case of an energy neutral house (Table A1). The cost of the annual needs of heating energy depend on prices suggesting that heating with coal is (15.6%) less expensive than using the environmentally friendly natural gas coal has been traditionally a secure and affordable energy in Poland [61]. The calculations do not account for the convenience associated with the use of natural gas and assume that the pipped gas is available at rural locations. The energy savings are slightly larger when the traditional house is heated with natural gas, but since that option is available only to a fraction of rural homes such comparison is less realistic.

The boiler heating water requires an electric pump to force water circulation. The traditional house with a coal-fired boiler necessitates 334 kWh annually of auxiliary electricity supply, 8% less than when using natural gas (Table 1). Under the considered construction scenarios, the corresponding costs of energy production drop from 4404 PLN (1029 EUR) in a traditional house that depends on coal to 766 PLN (179 EUR) for the energy efficient and energy neutral houses, or 86.4% less.

4.5. Changes in CO_2 Emission

The sustainability principle is well served by the reduction in emissions stemming from the use of modern construction technology and heating energy source. The traditional coal heated house emits about 165% times more CO_2 than a similar house heated with natural gas (Table 4). The traditional coal-using house considered in this study already includes energy-efficiency supporting upgrades such as insulated windows, doors, external wall insulation and an insulated roof. However, among more than three million single-family houses in rural areas, many still have not completed such upgrades, while using the inefficient coal-burning boilers.

Table 4. Annual emissions for single-family rural house construction scenarios.

Construction Technology	CO_2 kg/m²/year	SO_2 kg/year	NO_2 kg/year	$PM_{2.5}$ kg/year	PM_{10} kg/year
Traditional house (coal)	199	23.8	20.4	6.7	7.6
Traditional house (natural gas)	75	15.1	6.4	6.5	8.3
Traditional house (wood pellet)	0.8	0	0	1.9 [a]	2.1 [a]
Energy-saving house (natural gas and passive solar panels)	25	10.5	5.25	4.7	6.3
Energy-neutral house (geothermal heat pump and PV panels)	5	0	0	0	0

[a] g/kg of burned wood pellet.

Switching to a wood pellet boiler nearly eliminates CO_2 emissions because the burning recycles the gas already absorbed by wood from the atmosphere [62]. The only emissions associated with the use of wood pellet boiler is the electricity needed to operate it causing that heating RE energy option to emit 0.8 kg/m^2/year.

A house built with energy-efficiency in mind and enabled to use RE in the form of thermal solar panels generates 87.5% less CO_2 than a traditional house heated with coal (Table 4). The energy neutral house releases 3.5% of CO_2 volume emitted by a traditional house heating with coal but more than a house equipped with a wood pellet boiler (Table 4).

4.6. Changes in Toxic GHG and Particulate Emissions

Rural houses in Poland that use the inefficient coal furnaces or boilers are a source of emissions including SO_2, NO_2, $PM_{2.5}$, and PM_{10}. The health effects of those emissions are wide in scope and well established. For example, rural children exposed to GHG and particulate matter (PM) have higher risk of developing Type 1 diabetes [63]. The energy neutral house does not emit any toxic gases to its immediate neighborhood. The amount and composition of emissions associated with that type of a house depend on the energy source used to supply the house with electricity. However, the energy efficient house or a traditional house heated with natural gas generates gases other than CO_2.

Of particular interest to rural homeowners is the use of biomass in the form of wood pellet. Compared to coal, a kilogram of wood pellet emits 22.5% less NO_2 [64]. Finally, wood pellet emits 96% less particulate matter than coal. The actual differences in GHG emission reduction depend on the energy-efficiency of the specific model of the boiler. The use of natural gas, infrequent in rural areas, also substantially reduces GHG emissions. Heating with the gas condensing natural gas boiler emits 0.001% of SO_2 as compared to a coal-fired boiler and virtually no particulate matter [65].

5. Discussion

The desire to own a family home will drive Poland's potential homeowners to build their new house, primarily in rural areas due to space availability. The calculated building costs (Table 1) must be considered in the context of the household ability to finance the construction. The cost difference between the single-family home built using the conventional technological solutions and the energy-neutral house is much larger than similar differences reported in studies in other countries. For example, the cost difference between the standard house built in accordance with Belgian regulations and a low-energy house (roughly comparable to the energy-efficient house considered here) was 4% and a passive (energy neutral) house 16%, respectively [66], while the difference between a standard and a passive house amounted to 10% in Germany [67]. Both studies suggest a considerably smaller relative differences between a conventional house and its energy-efficient alternatives established in the scenarios considered in the current study.

Mortgage financing has a bad reputation in Poland since the financial crisis of 2009–2010 because although the continuing GDP growth contrasted with the global malaise, many families suffered [68]. Prior to the financial crisis, banks offered mortgages priced in Swiss francs. The Swiss franc rapidly appreciated during the crisis and dramatically increased the mortgage debt servicing. The repercussions of that phenomenon are still felt today. Such recent memories combined with the shock in the ongoing COVID-19 pandemic and the induced economic slowdown affect households' attitudes discouraging the long-term credit-financed investment.

A number of future homeowners may not qualify for mortgage financing given the average income in rural households in general. Regional income disparities also persist. The lack of access to mortgage financing suggests the already observed prolonged construction because a rural households and their owners is likely to accumulate savings and then use them to finance the construction in stages. For example, the first stage will be limited to the unfinished house without the external wall insulation, windows and door. The last stage could involve the central heating system installation. An extended construction potentially delays the occurrence of gains to the household and the environment.

The large initial investment in the energy neutral house is expected to eliminate the cost of energy purchase to heat space and water. Those costs account of the about 60–70% of the cost of total energy purchase by an average household in Poland. The share is much larger than in many other EU countries or in the United States. An earlier study showed that the motive to save on the monthly energy bill was a major motive in the rural household investment in the thermal solar panels [8]. However, the rural homeowners mentioned the cost of RE utilizing equipment as a constrained and viewed the subsidy as important.

The currently operating furnace replacement program offers subsidies for qualifying households if they choose to replace their old furnace or boiler. The subsidy is matched to per capita income in the household and is proportionally larger for those with least income. The program aims at the improvement of energy-efficiency by households and includes, besides furnace replacement, subsidies to window and door replacement as well as external wall and roof insulation. The program also offers low interest loans for those who do not qualify for a grant. The program is more attractive to retrofitting an existing single-family home and less to future homeowners.

The comparison of the three construction technologies and the use of various insulating materials provides important knowledge to consider by future homeowners, but also by the owners of the existing rural family homes. Splitting the various construction stages (Table 1) demonstrates the costs associated with the use of alternative insulation. A retrofit of an existing house could involve specific projects. Among the largest is the resignation from a fossil-fuel based space and water heating system and its replacement by the RE-based installation such as wood pellet furnace. A retrofit generates potentially substantial savings to the household and the largest environmental benefits. The reduction of CO_2 emission contributes to the national and EU climate policy and represents a direct contribution from households. The energy neutral house eliminates the emissions of NO_2, SO_2 and particulate matter comparison to the coal-heated single-family dwelling. The obtained energy savings, environmental benefits, and economic gains support the conclusion that the thermo-modernization of a single-family rural residence and the modernization of the heating source bring direct economic gains to a household, indirect benefits in the local air quality improvement, and long term benefits in health benefits and environmental quality.

Limitations of the Study

The building model considered in this study assumes a simple floor plan and is limited in size. Many newly constructed houses are larger and there is a great variety of floor plans. Variation in the age, construction technology, and floor plans pose a challenge for thermal modernization of existing rural family homes. Although some parts of the house may be thermally insulated-for example the roof - other parts, like the internal partitions, would have to be rebuilt creating a domino effect forcing the renovation of several areas of the house. Improving the thermal insulation between various floors excludes portions of the house from use for the duration of the project. As a result, the thermal insulation project for an older house may cost more than a similar project for a new house.

The analyzed construction costs use the price lists reflect the asking price of local suppliers. Ultimately, prices of all building materials and heating systems may be both negotiable and region-specific. Consequently, the actual construction costs are likely to deviate somewhat from those considered here. Another source of price variation are changes in price level, which may have been affected by the ongoing pandemic and altered market conditions.

Another source of limitations is the use of energy prices from the publicly available sources and for a particular time period. Those prices will change over time and some, for example wood pellet or coal prices, will vary with respect to fuel quality which is ultimately decided by the homeowner. Homeowner choices of the heating energy type will influence the volume of GHG emissions, but will not reverse the general trend in emission reduction due to the shift away from using coal.

The current study did not account for the cost of the parcel, which is likely to be highly variable across rural areas in the country. A closer examination of household incomes and construction costs will enable the forecasting of new house construction, and potentially, regional economic growth and real estate market development. Earlier studies linked the energy performance of dwellings with market valuation [69,70] once the suitable data became available. Also, the scenarios presented in the current paper can be supplemented by financial analyses, which can account for regional household income variation and different types of mortgage loans. Finally, as new construction technology and heating systems become available, a future study will be necessary to update the data on energy savings and environmental benefits.

6. Conclusions

Improved energy-efficiency of single-family homes in Poland is required by the new building regulations, while the retrofitting of the existing detached houses in supported by government programs aiming at enhancing air quality as well as energy and climate policies. The energy demand reduction and environmental benefits occur primarily in rural areas because the detached houses dominate the rural landscape. Results from this study show that switching away from coal, still the primary energy sources in rural areas, required by new construction regulations leads to a substantial increase in construction costs of the shell of a house and even larger costs in installing the RE utilizing equipment to heat space and domestic water. The costs increase for a model single-family home considered here far exceed the previous estimates for other EU countries. The increased costs may deny the opportunity of residing in own home to many rural households due to debt servicing. However, retrofitting an existing house improving its energy-efficiency also generates substantial savings to households on their energy bill. Retrofitting split into smaller projects, as permitted by household savings, is realistic and likely to continue in rural areas.

The use of RE is required in all new detached housing, but encouraged in the existing single-family homes by the subsidy programs. The scenarios considered in this study used wood pellet, solar radiation and geothermal energy. For rural households, the RE in the form of wood pellet may be more appealing because it can reliably heat the house, while being locally accessible. The intensity of solar radiation is inversely related to the heating energy needs of households in Poland determined by climate. The regional variations in climate will have to be considered by households investing in a new home since the winter temperatures can substantially vary. Regional considerations are often ignored in the "one-size-fits-all" regulations, and are verified by the actual site-specific conditions.

The environmental sustainability is well served by the new construction regulations and the scenarios considered in this study. Gains in reduction of CO_2 are impressive once a household uses any amount of RE in comparison to a traditional rural house the uses coal. Retrofitting the existing house with the wood pellet burning boiler nearly eliminates CO_2 and SO_2 emissions, while substantially lowers other noxious gases and particulate matter. The effects are instantaneous in improving local air quality, while the broader effects benefit the implementation of the decarbonization efforts and help achieve goals of the climate policy.

Author Contributions: Conceptualization: A.S. and A.M.K.; Data curation: A.S.; Formal analysis: A.S.; Investigation: A.S., A.M.K. and W.J.F.; Methodology: A.S., A.M.K. and W.J.F.; Supervision: A.M.K.; Writing—original draft: A.S., A.M.K. and W.J.F.; Writing—review & editing: A.S., A.M.K., W.J.F., P.G. All authors have read and agreed to the published version of the manuscript.

Funding: This research received no external funding.

Conflicts of Interest: The authors declare no conflict of interest.

Appendix A

Construction materials used in the various scenarios were selected by the architectural design firm Archiko, Kornelia Lisowska-Siudek, Manager, and the labor costs were estimated by the construction company collaborating with the design firm. Construction cost estimates were also based on the estimations developed by the company, "Dobredomy".

Appendix B

Table A1. Annual heating energy needs, energy production costs, auxiliary electricity needs and total costs for a rural detached residence central heating system.

Cost Category	Units	Traditional House (Coal)	Traditional House (Natural Gas)	Traditional House (Wood Pellet)	Energy-Saving House (Natural Gas + Solar Panels)	Energy-Neutral House (Heat Pump + PV)
Heating energy	kWh/year	31,909	19,923	27,350	5300	4481
Price per kWh from a given heat source	PLN	0.13	0.25	0.21	0.12	0.12
Energy production costs	PLN	4148	4981	5744	636	538
Auxiliary electrical power supply	kWh/year	334	363	179	351	351
Price per kWh [a]	PLN	0.65	0.65	0.65	0.65	0.65
Energy production costs	PLN	255	234	116	228	228
Total costs	PLN/EUR	4404/1039	5217/1219	5860/1369	864/202	766/179

Source: Based prices listed by [71] and data from [16]. [a] Price from [16].

Table A2. Energy generation costs for the domestic water heating system.

Cost Category	Units	Traditional House (Coal)	Traditional House (Natural Gas)	Traditional House (Wood Pellet)	Energy-Saving House	Energy-Neutral House
Heat demand	kWh/year	29,396	14,409	23,706	14,409	4105.5
Price 1kWh from a given heat source	PLN	0.13	0.25	0.21	0.25	0
Cost of Energy production	PLN	3821	3602	4978	3602	0
Auxiliary electrical power supply	kWh/year	169	272	215	272	247
Price for 1 kWh of (GUS, 2016)	PLN	0.65	0.65	0.65	0.65	0.65
Energy production costs	PLN	110	177	140	177	161
Total costs	PLN/EUR	4007/936	3779/883	5118/1196	3779/883	148/34

Source: Calculations based on company price list [71], and data from [16].

References

1. Gawlikowska-Fyk, A. Coping with the challenges of decarbonization and diversification. In *New Political Economy of Energy in Europe*; International Political Economy Series; Palgrave Macmillan: London, UK, 2018; pp. 195–2014.
2. European Commission National Energy and Climate Plans. Executive Summary of Poland's National Energy and Climate Plan for the Years 2021–2030. 2020. Available online: https://ec.europa.eu/energy/sites/ener/files/documents/pl_final_necp_summary_en.pdf (accessed on 28 September 2020).
3. KOBiZE. *Wartości Opałowe (WO) i Wskaźniki Emisji CO2 (WE) w Roku 2015 do Raportowania w Ramach Systemu Handlu Uprawnieniami do Emisji za Rok 2018*; KOBiZE: Warszawa, Poland, 2017.
4. Emissions (Metric Tons per Capita)–Poland. 2020. Available online: https://data.worldbank.org/indicator/EN.ATM.CO2E.PC?locations=PL (accessed on 27 October 2020).
5. Canales, F.A.; Jadwiszczak, P.; Jurasz, J.; Wdowikowski, M.; Ciapała, B.; Kaźmierczak, B. The impact of long-term changes in air temperature on renewable energy in Poland. *Sci. Total Environ.* **2020**, 729. [CrossRef] [PubMed]
6. Czym Polacy Ogrzewają Domy? Najnowsze Dane GUS. 2019. Available online: https://www.gramwzielone.pl/trendy/102011/czym-polacy-ogrzewaja-domy-najnowsze-dane-gus (accessed on 22 October 2020).
7. Kamińska, W.; Heffner, K. *Polityka Spójności UE a Rozwój Obszarów Wiejskich. Stare Problemy i Nowe Wyzwania*; KPZK PAN: Warszawa, Poland, 2014; pp. 267–282.
8. Klepacka, A.M.; Florkowski, W.J.; Meng, T. Clean, Accessible, and Cost-Saving: Reasons for Rural Household Investment in Solar Panels in Poland. *Resour. Conserv. Recy.* **2018**, *139*, 338–350. [CrossRef]
9. Część Druga Programu dla Beneficjentów Uprawnionych do Podwyższonego Poziomu Dofinansowania. Available online: http://czystepowietrze.gov.pl/wez-dofinansowanie/ (accessed on 14 August 2020).

10. Directive (EU) 2018/844 of the European Parliament and of the Council of 30 May 2018 amending Directive 2010/31/EU on the Energy Performance of Buildings and Directive 2012/27/EU on Energy Efficiency. Available online: https://eur-lex.europa.eu/legal-content/pl/TXT/?uri=CELEX%3A32018L0844 (accessed on 27 October 2020).
11. Act of 13 February 2020 Amending the Act-Building Law and Certain Other Acts. Available online: https://isap.sejm.gov.pl/isap.nsf/DocDetails.xsp?id=WDU20200000471 (accessed on 27 October 2020).
12. Gałązka, T. *Krajowy Plan Mający Na Celu Zwiększenie Liczby Budynków O Niskim Zużyciu Energii*; Wydawnictwa Uczelniane UTP w Bydgoszczy: Bydgoszcz, Poland, 2015.
13. Kundzewicz, Z.W.; Matczak, P. Climate change regional review: Poland. *Wires Clim. Chang.* **2012**, *3*, 297–311. [CrossRef]
14. Eurostat. GDP per Capita in PPS. 2020. Available online: https://ec.europa.eu/eurostat/databrowser/view/tec00114/default/table?lang=en (accessed on 23 October 2020).
15. Capros, P.; Paroussos, L.; Fragkos, P.; Tsani, S.; Boitier, B.; Wagner, F.; Busch, S.; Resch, G.; Blesl, M.; Bollen, J. European decarbonisation pathways under alternative technological and policy choices: A multi-model analysis. *Energy Strateg. Rev.* **2014**, *2*, 231–245. [CrossRef]
16. GUS. *Zużycie Energii w Gospodarstwach Domowych w 2018 Roku*; GUS: Warszawa, Poland, 2019.
17. Porozumienie Branżowe na Rzecz Efektywności Energetycznej POBE. Jak Spełnić Wymagania, jakim Powinny Odpowiadać Budynki od 2021 Roku? *Poradnik dla Architektów, Projektantów i Inwestorów.* Available online: https://www.nibe.pl/Documents/nibe_pl/12_Poradnik_POBE_WT_2021_jak_spelnic_wymagania.pdf (accessed on 14 August 2020).
18. Grudzińska, M.; Jakusik, E. Energy performance of buildings in Poland on the basis of different climatic data. *Indoor Built Environ.* **2017**, *26*, 561–566. [CrossRef]
19. Krawczyk, D.A. International Scientific Conference "Environmental and Climate Technologies", CONECT 2015, 14–16 October 2015, Riga, Latvia Analysis of energy consumption for heating in a residential house in Poland. *Energy Procedia* **2015**, *95*, 216–222. [CrossRef]
20. Pater, S. Field measurements and energy performance analysis of renewable energy source devises in a heating and cooling system in a residential building in southern Poland. *Energy Build.* **2019**, *199*, 115–125. [CrossRef]
21. Pantoutsou, C.; Bauen, A.; Böttcher, H.; Alexopoulou, E.; Fritsche, U.; Uslu, A.; van Stralen, J.N.P.; Elbersen, B.; Kretschmer, B.; Capros, P.; et al. Biomass futures: An integrated approach for estimating the future contribution of biomass value chains to the European energy system and inform future policy formation. *Biofuel Bioporducts Biorefin.* **2013**, *7*, 106–114. [CrossRef]
22. European Commission. *Energy Road Map 2050*; Publications Office of the European Union: Luxembourg, 2012.
23. Mocny Trend. Polacy Przeprowadzają się do Domów i Szeregowców. 2017. Available online: https://tvn24.pl/biznes/z-kraju/gus-spada-ilosc-gospodarstw-domowych-w-domach-wielorodzinnych-ra775548-4486182 (accessed on 26 October 2020).
24. Bauermann, K. German Energiewende and the heating market–Impact and limits of policy. *Energy Policy* **2016**, *94*, 235–246. [CrossRef]
25. Del Río, P.; Burguillo, M. An empirical analysis of the impact of renewable energy deployment on the local sustainability. *Renew. Sust. Energ. Rev.* **2009**, *13*, 1314–1325. [CrossRef]
26. Moula, M.M.E.; Maula, J.; Hamdy, M.; Fang, T.; Jung, N.; Lahdelma, R. Researching social acceptability of renewable energy technologies in Finland. *Int. J. Sustain. Built Environ.* **2013**, *2*, 89–98. [CrossRef]
27. Bertsch, V.; Hall, M.; Weinhardt, C.; Fichtner, W. Public acceptance and preferences related to renewable energy and grid expansion policy: Empirical insights from Germany. *Energy* **2016**, *114*, 465–477. [CrossRef]
28. GUS. *Obszary Wiejskie w Polsce w 2018 Roku*; GUS: Warszawa, Poland, 2020.
29. GUS. *Zużycie Energii w Gospodarstwach Domowych w 2015 r*; GUS: Warszawa, Poland, 2017.
30. Bochnia, K. Budynki Niemal Zeroenergetyczne. *Builder* **2017**, *65*, 64–65.
31. Tomsej, T.; Horak, J.; Tomsejova, S.; Krpec, K.; Klanova, J.; Dej, M.; Hopan, F. The impact of co-combustion of polyethylene plastics and wood in a small residential boiler on emissions of gaseous pollutants, particulate matter, PAHs and 1,3,5-triphenylbenzene. *Chemosphere* **2018**, *196*, 18–24. [CrossRef] [PubMed]
32. Stecuła, K.; Tutak, M. Causes and effects of low-stack emission in selected regions of Poland. 18th International Multidisciplinary Scientific Geo-Conference SGEM 2018. *Sect. Air Pollut. Clim. Chang.* **2018**, *18*, 357–364. [CrossRef]

33. Krause, P.; Steidl, T.; Wojewódka, D. Kierunki Termomodernizacji. *Builder* **2015**, *46*, 46–48.
34. Dziennik Ustaw. Minister of Transportation, Construction and Maritime Regulation of 5 July 2013, Regarding the Technical Conditions Pertaining to Buildings and Their Location. Item 926, August 2013. Available online: http://prawo.sejm.gov.pl/isap.nsf/DocDetails.xsp?id=WDU20130000926 (accessed on 30 April 2020).
35. Dziennik Ustaw. Obwieszczenie Marszałka Sejmu Rzeczypospolitej Polskiej z Dnia 7 Czerwca 2018 r. w Sprawie Ogłoszenia Jednolitego Tekstu Ustawy-Prawo Budowlane. Item 1202. 2018. Available online: http://isap.sejm.gov.pl/isap.nsf/DocDetails.xsp?id=WDU20180001202 (accessed on 8 September 2020).
36. GUS. *Zamieszkane Budynki. Narodowy Spis Powszechny Ludności i Mieszkań w 2011*; GUS: Warszawa, Poland, 2013.
37. GUS. *Budownictwo w 2020 Roku*; GUS: Warszawa, Poland, 2020.
38. Dziennik Ustaw. Minister of Infrastructure and Development Regulation of 27 February 2015, Regards the Methodology of Determining Energy Features of Buildings and Buildings Parts and Building Features certification. Item 376, 18 March 2015. Available online: http://isap.sejm.gov.pl/isap.nsf/download.xsp/WDU20150000376/O/D20150376.pdf (accessed on 10 November 2020).
39. Krochmal, D.; Kalina, A. Measurements of nitrogen dioxide and sulphur dioxide concentrations in urban and rural areas of Poland using passive sampling method. *Environ. Pollut.* **1997**, *96*, 401–407. [CrossRef]
40. Hławiczka, S.; Kubica, K.; Zielonka, U. Partitioning factor of mercury during coal combustion in low capacity domestic heating units. *Sci. Total Environ.* **2003**, *302*, 261–265. [CrossRef]
41. Tainio, M.; Juda-Rezler, K.; Reizer, M.; Warchałowski, A.; Trapp, W.; Skotak, K. Future climate and adverse health effects caused by fine particulate matter air pollution: Case study for Poland. *Reg. Environ. Chang.* **2013**, *13*, 705–715. [CrossRef]
42. Jędrak, J.; Konduracka, E.; Badyda, A.J.; Dąbrowiecki, P. *Influence of air pollution on health. Polskiego Towarzystwa Medycyny Środowiskowej, Wojskowego Instytutu Medycznego, Polskiej Federacji Stowarzyszeń Chorych na Astmę, Alergię i POChP*; Krakowski Alarm Smogowy: Krakow, Poland, 2017; ISBN 978-83-943065-0-2.
43. Wang, Q.; Xu, X.; Zeng, Z.; Zheng, X.; Ye, K.; Huo, X. Antioxidant alterations link polycyclic aromatic hydrocarbons to blood pressure in children. *Sci. Total Environ.* **2020**, 732. [CrossRef]
44. Kalkulator Budowlany-Cennik Płyt Fundamentowych 2020 w Ponad 150 Miastach w Polsce. Available online: https://kb.pl/cenniki/plyta-fundamentowa/ (accessed on 15 November 2020).
45. Kalkulator Budowlany-Styropian EPS czy Styropian XPS? A Może Grafitowy? Available online: https://kb.pl/porady/styropian-eps-czy-styropian-xps-moze-grafitowy-sprawdzamy/ (accessed on 15 November 2020).
46. Retail Price List of Blachotrapez. Available online: https://www.blachotrapez.eu/pl/pliki/cennik-detaliczny-blachotrapez?gclid=Cj0KCQiAnb79BRDgARIsAOVbhRpeXyhqgmnfebqFBiWhrlcKc4hLyg4yJvmL3V2gzxgNxoQrv7UpzvMaAuS7EALw_wcB (accessed on 15 November 2020).
47. Budujemy Dom-Dach Kryty Dachówką Ceramiczna Przykładowa Wycena. Available online: https://budujemydom.pl/stan-surowy/dachy-i-rynny/a/17293-dach-kryty-dachowka-ceramiczna-przykladowa-wycena (accessed on 12 November 2020).
48. Kalkulator Budowlany-Okna Trzyszybowe czy Dwuszybowe-Porady i Ceny. Available online: https://kb.pl/porady/okna-trzyszybowe-czy-dwuszybowe-porady-i-ceny-dla-kazdego/ (accessed on 10 November 2020).
49. Kalkulator Budowlany-Montaż Drzwi- Cennik po Województwach. Available online: https://kb.pl/porady/montaz-drzwi-cennik-po-wojewodztwach/ (accessed on 9 November 2020).
50. Kalkulator Budowlany-Cennik Ocieplenia Poddasza Użytkowego 2020 w Ponad 150 Miastach w Polsce. Available online: https://kb.pl/cenniki/ocieplenie-poddasza-uzytkowego/ (accessed on 10 November 2020).
51. Available online: https://www.hurtowniastyropianu.pl/sklep/styropian/plyty-pir/ (accessed on 10 November 2020).
52. Strop Monolityczny–Ile Kosztuje? *Aktualne Ceny i Charakterystyka.* Available online: https://jakbudowac.pl/Strop-monolityczny-ile-kosztuje-Aktualne-ceny-icharakterystyka (accessed on 10 November 2020).
53. Koszty Różnych Rodzajów Kominów. Available online: https://www.instalacjebudowlane.pl/6533-23-12396-koszty-roznych-rodzajow-kominow.html (accessed on 10 November 2020).
54. Strona Internetowa Firmy Regulus Wraz z Cennikami. Available online: https://www.regulus.com.pl/news/grzejniki-niskotemperaturowe (accessed on 13 November 2020).
55. Wentylacja Grawitacyjna- co to Jest i Ile Kosztuje? Available online: https://wentylacja.com.pl/news/wentylacja-grawitacyjna-co-to-jest-i-ile-kosztuje-67245.html (accessed on 10 November 2020).

56. Wentylacja Grawitacyjna czy Mechaniczna? Available online: https://www.rekuperatory.pl/wentylacja-grawitacyjna-czy-mechaniczna (accessed on 13 November 2020).
57. Kalkulator budowlany-Koszt Budowy Domu 2020- Wyliczenia Krok po Kroku. Available online: https://kb.pl/porady/koszt-budowy-domu-dokladne-wyliczenia/ (accessed on 12 November 2020).
58. Lorenc, H. *Atlas Klimatyczny Polski*; Instytut Meteorologii i Gospodarki Wodnej: Warszawa, Poland, 2005.
59. NBP. Tabela A kursów Średnich Walut Obcych. Tabela nr 221/A/NBP/2019 z Dnia 2019-11-15. 2020. Available online: http://rss.nbp.pl/kursy/TabRss.aspx?n=2019/a/19a221 (accessed on 15 November 2020).
60. Park, N.D.; Rutherford, P.M.; Thring, R.W.; Helle, S.S. Wood pellet fly ash and bottom ash as an effective liming agent and nutrient source for rye grass (*Lolium perenne* L.) and oats (*Avena sativa*). *Chemosphere* **2012**, *86*, 427–432. [CrossRef] [PubMed]
61. Kerimray, A.; Rojas-Solorzano, L.; Torkmahalleh, M.A.; Hopke, P.H.; O Gallachoir, B.P. Coal use for residential heating: Patterns, health implications and lessons learned. *Energy Sustain. Dev.* **2017**, *40*, 19–30. [CrossRef]
62. Petersen Raymer, A.K. A comparison of avoided greenhouse gas emissions when using different kinds of woo energy. *Biomass Bioenerg.* **2006**, *30*, 605–617. [CrossRef]
63. Michalska, M.; Zorena, K.; Waz, P.; Bartoszewicz, M.; Brandt-Varma, A.; Slezak, D.; Robakowska, M. Gasous pollutant and particulate matter in ambient air and the number of new cases of Type 1 diabetes in children and adolescents in the Pomeranian Voivodeship, Poland. *Biomed Res. Int.* **2020**. [CrossRef]
64. Właściwości Paliwa. 2020. Available online: http://www.pelet.info.pl/x.php/1,439/Wlasciwosci-paliwa.html%202020 (accessed on 19 November 2020).
65. Available online: https://vaillant-partner.pl/kalkulatory-on-line/kalkulator-emisji-zanieczyszczen/ (accessed on 19 November 2020).
66. Audenaert, A.; De Cleyn, S.H.; Vanckerkhove, B. Economic analysis of passive houses and low-energy houses compared with standard houses. *Energy Policy* **2008**, *36*, 47–55. [CrossRef]
67. Schnieders, J.; Hemerlink, A. CEPHEUS results: Measurements and occupants' satisfaction provide evidence for Passive Houses being an option for sustainable building. *Energy Policy* **2006**, *34*, 151–171. [CrossRef]
68. Szopiński, T. Kontrowersje wokół "walutowych" kredytów hipotecznych. *Konsumpcja Rozw.* **2018**, *22*, 79–84.
69. Manganelli, B.; Morano, P.; Tajani, F.; Salvo, F. Affordability assessment of energy-efficient building construction in Italy. *Sustainability* **2019**, *11*, 249. [CrossRef]
70. De Ruggiero, M.; Forestiero, G.; Manganelli, B.; Salvo, F. Buildings energy performance in a market comparison approach. *Buildings* **2017**, *7*, 16. [CrossRef]
71. Cennik Firmy Viessmann. 2020. Available online: https://www.broetje.pl/wp-content/uploads/2017/04/Cennik-Brotje-2019-wydrukowany.pdf (accessed on 24 October 2020).

Publisher's Note: MDPI stays neutral with regard to jurisdictional claims in published maps and institutional affiliations.

energies

MDPI

Article

Energy Efficiency of Maize Production Technology: Evidence from Polish Farms

Anita Konieczna [1], Kamil Roman [2,*], Monika Roman [3], Damian Śliwiński [4] and Michał Roman [3]

1 Department of Economic and Energy Analysis, Institute of Technology and Life Sciences in Falenty, Warsaw Branch, 32 Rakowiecka St., 02-532 Warsaw, Poland; a.konieczna@itp.edu.pl
2 Institute of Wood Sciences and Furniture, Warsaw University of Life Sciences, 166 Nowoursynowska St., 02-787 Warsaw, Poland
3 Institute of Economics and Finance, Warsaw University of Life Sciences, 166 Nowoursynowska St., 02-787 Warsaw, Poland; monika_roman@sggw.edu.pl (M.R.); michal_roman@sggw.edu.pl (M.R.)
4 Institute of Technology and Life Sciences, 05-090 Falenty, Poland; d.sliwinski@itp.edu.pl
* Correspondence: kamil_roman@sggw.edu.pl

Abstract: The purpose of this work is to determine the impact of selected silage maize cultivation technologies, including energy inputs in the production chain (cultivation, harvesting, heap placing), on energy efficiency. The analysis of energy inputs, energy efficiency for the silage maize production technology were estimated. The research was performed for 13 farms producing silage maize. The data from the farms covered all the activities and the agrotechnical measures performed. The calculations of energy inputs made for the silage maize production for selected technologies were performed using the method developed by the Institute of Construction, Mechanization and Electrification for Agriculture (IBMER), once the method was verified and adapted to the needs and conditions of own research. Based on the accumulated energy production and the energy accumulated in the yield, energy efficiency index values for 13 silage maize cultivation technologies were calculated. The greatest impact on the results of energy efficiency calculations was shared by fertilizer and fuel inputs. In conclusion, it can be stated that, in terms of energy efficiency, maize cultivation is justified and it can generate energy benefits.

Keywords: energy efficiency; energy accumulated; crop production; silage maize; biomass; farms

Citation: Konieczna, A.; Roman, K.; Roman, M.; Śliwiński, D.; Roman, M. Energy Efficiency of Maize Production Technology: Evidence from Polish Farms. *Energies* **2021**, *14*, 170. https://doi.org/10.3390/en14010170

Received: 25 November 2020
Accepted: 25 December 2020
Published: 31 December 2020

1. Introduction

Crop cultivation is of special importance for covering the demand for consumption and animal feed and, to a growing extent, also for energy [1]. The socioeconomic progress, scientific and technical advancements, and hence the economic development result in an ongoing increase in electricity and transport fuel consumption globally, which triggers an increase in the concentration of pollution and environmental degradation (water, soil, air) [2–4]. The danger that is associated with this matter is the continued increase of unemployment and famine factors, unless intensive and preventive tasks are introduced, especially in saving agro-systems transformation. Past actions were based on previous generations' experience. The increased level of development in various fields requires us to use the results of interdisciplinary research. At the same time, the rapid changes in the conditions of the agro-systems environment and the growing demand for new, more effective technologies require the simultaneous contribution of knowledge not only in the field of food production, but also in the field of the quality of newly created products, their marketing and maintaining the ethical principles of their acquisition, processing and distribution. In particular, it is about the links between the humanities, technical, and agronomic sciences, creating an interdisciplinary consilium of experts on the transformation of agro-systems.

Considerations on the effectiveness of agro-systems transformations, both in the past and in the long-term perspective, lead to the knowledge and explanation of the mechanisms of influence of various environmental factors on the anticipated changes in the parameters describing the energy-technological condition of the considered objects. The development of agriculture in line with the paradigm of sustainable development has become particularly important for industrialized countries, which previously based the development of the agricultural sector on the industrial model. Especially in Europe, the agricultural model based on too intensive fertilization, mechanization, and concentration led to the deterioration of the quality of the natural environment (Stoate et al., 2009). One of the elements of this research from the microeconomic perspective (at the farm level) is the comprehensive energy consumption estimated with the so-called pull method was started at the end of the 1970s [5]. In Europe, the precursor of such research was Pellizzi [6,7]. The energy performance indicators of respective means of production sometimes differ quite considerably depending on the authors, the indicator value changes with time, which is due to the changes in industrial production methods, more complete and more accurate energy consumption research in various areas of human life and production activity [5,8]. The energy technology method to assess the food economy transformation effectiveness was proposed by Nowacki [9], presenting the reasons for the macroeconomic approach to the food system. He covers the problem of the relationships of the economic measures, energy and sociological management effectiveness. As the technological level of an agricultural facility grows, human labor inputs decrease, which is the cause of a significant outflow of people employed in agriculture to other professions.

In the microeconomic approach, the research of the accumulated production energy consumption and efficiency is performed to evaluate the management quality in the enterprise, including an agricultural farm. The evaluation of the outcomes and economic methods management effectiveness often fails as the prices imposed often do not correspond to the cash value of the goods or energy offered. Hence, an increase in the importance of the evaluation of the energy consumption and energy efficiency method based on the values expressed in reference energy units is seen, allowing for their comparison irrespective of the place, time, and price relations. Bearing in mind the preferences, payments and grants for agricultural production, consumables and raw materials, services and credits applied on the EU market, the study of the energy efficiency becomes of special importance even though it will not replace the economic analyses which, under more complete market economy conditions, are definitely the best and the simplest business activity evaluation methods [10].

Because of a growing energy consumption, new methods of energy generation are searched for, the existing ones are improved, and the participation of the renewable resources of energy is increased in the energy balance. The rational use of the resources is nowadays associated with a permanent use of renewable resources, which means using them in such amounts in which their increase occurs [11]. The topic of renewable energy resources is one of the many aspects referring to the limited resources of fossil fuels, a considerable share of the energy sector in the greenhouse gas (GHG) emissions, which contributes to climate changes and an increase in energy security. In 2007 the Member States accepted the so-called climate and energy package, the assumptions of which are, e.g., limiting the GHG emissions and enhancing the energy security. One of the ways to accomplish those objectives is to increase the share of energy from the renewable resources in its total consumption. A special attention must be given to, e.g., agricultural biogas, the gas produced in the process of methane fermentation of agricultural raw materials, agricultural by-products, liquid or solid animal feces, by-products, waste or the remains from the processing of the products of agricultural origin or forest biomass of plant biomass collected from the areas other than recorded as agricultural or forest, except for the biogas produced from materials derived from sewage treatment plants and landfill sites [12]. That renewable energy is considered to show a potential as it is a stable and predictable source (important in terms of energy security) meeting a number of positive functions not only for the electrical power system, next to the energy and economic benefits as well as the environmental ones; it decreases the GHG emissions and provides global and local social benefits. It helps activate the rural areas, it creates new jobs, it enhances the investment attractiveness of the region [13–15].

In Poland, to produce the agricultural biogas, most frequently a mixture of animal feces with energy crops or with by-products of agricultural origin is used. Applying co-substrate with a higher content of dry weight, as compared with its content in animal feces, enhances the production of biogas and the process economic effectiveness [16]. The right combination is conditioned by the biogas potential of each component as well as component interaction [17]. An excellent supplement to the fermentation mass in terms of technology is, e.g., maize silage. Literature provides many reports on the use of maize for energy purposes or [18–20] the relations between the maize prices and the prices of energy materials [21,22].

One must note, however, that growing maize requires high energy inputs, hence a need to perform research to increase the production energy efficiency. Energy efficiency must be defined as a ratio of the energy value of the biomass yield (the accumulated energy contained in biomass) to the total energy inputs (the accumulated energy required to produce the biomass) [10,23]. The cultivation technologies applied affect the environment to a varied extent and so, in terms of maize cultivation for energy purposes, calculating the energy efficiency becomes essential [24]. Estimating the energy consumption and energy efficiency of agricultural materials is indispensable for energy crops in a form of renewable energy. According to the requirements of Directive 2009/28/WE (RED) on the promotion of the use of energy from renewable sources, processors of biofuels are required to prove that the production from agricultural raw materials and the whole process of liquid biofuels production meet the sustainability criteria. It can be demonstrated by the LCA (life cycle assessment) method, that proves the reduction of GHG emissions along the entire chain of production. [25,26]. Those considerations are based on analyses at the microeconomic level concerning only the energy efficiency of various technologies of the maize for silage production in use of raw material for processing into biogas as an energy carrier. The analysis of the drag method by Pellizzi, adapted to Polish conditions by Wójcicki, allows the comparison of results whether the place, time, and price relationship was used. The drag method is successfully used by many authors and institutions, for example IBMER-ITP. The biomass production efficiency is very important in terms of increasing its share in the energy production. Because of the fears of a competition between plant production for energy purposes and crops for human consumption, actions are taken and research is performed to decrease the energy consumption of plant production owing to the optimal planning and possibly the most effective use of the land allocated for cultivation, compliant with the principles of sustainable development. An example of such research can be found, e.g., in the reports by Houshzyara et al. [27,28].

The primary objective of the paper is to determine the energy efficiency of maize with the use of various technologies. With that in mind, the analysis of energy inputs was made and energy efficiency was calculated for the silage maize production technology. To achieve the objective, the results of own research performed on 13 agricultural farms were applied.

The respective sections of the article present the theoretical grounds for the use of maize for energy purposes followed by the experimental part of the energy efficiency analysis. After the introduction, chapter 2 discusses in detail the source material and research methods. The third chapter covers the results of the analyses. With the basic information on the maize market, calculations were made on accumulated energy, the structure of energy inputs and energy efficiency. The last part of the article provides discussion and results.

2. Maize as an Energy Crop

Maize, similarly as most energy crops, is mostly used as a starch material derived from seed and as a material mostly for producing bioethanol [29] and as biomass including leaves, stems, and blade apex. Biomass can be used to produce bioethanol of the second generation, for incineration [30–32] or as silage for biogas production [33,34]. Interestingly, the maize acreage in Poland since 2007 has increased by 146% and in 2018 it was 0.65 m ha (Figure 1), which accounted for 8% of the total acreage of crops in the EU.

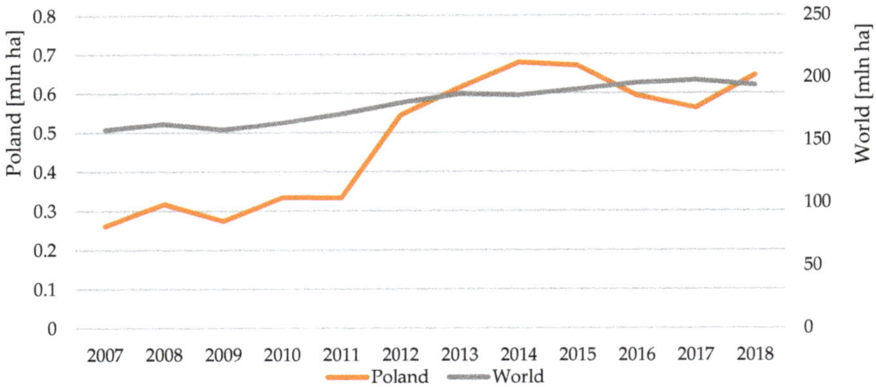

Figure 1. Maize production in 2007–2018, Source: own elaboration based on [35].

Silage maize is a roughage used for cattle feeding. In fact, for the entire year of the fodder crop field harvest structure, maize shows the highest acreage, in 2016 the green maize yields accounted for 73.5% of the total fodder crop yields (Figure 2). Next to the silage allocation to animal feed for livestock or milk production, there also appeared a possibility of using it for energy purposes, as a valuable substrate for methane fermentation bacteria for biogas production. One of the most frequently applied substrates of agricultural origin is slurry, with varied properties depending on the feeding method or the animal species. A relatively low content of dry weight requires supplementation with substrates, e.g., plant substrates. An excellent supplement for the fermentation mass in terms of technology is maize silage which, according to the National Centre for Agriculture Support (KOWR), in 2018, accounted for 12% of the total substrates used in agricultural biogas plants (Figure 3).

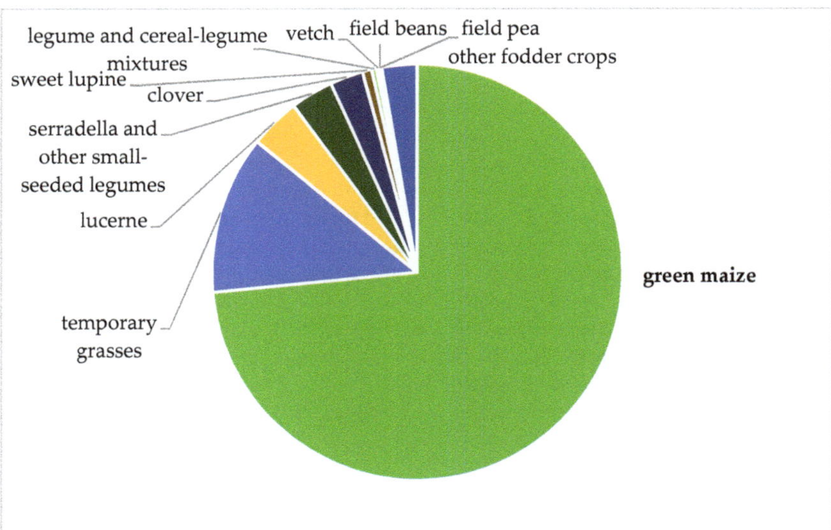

Figure 2. Structure of field green fodder crops production (Total production of field green fodder crops = 100. Source: own elaboration based on [36].

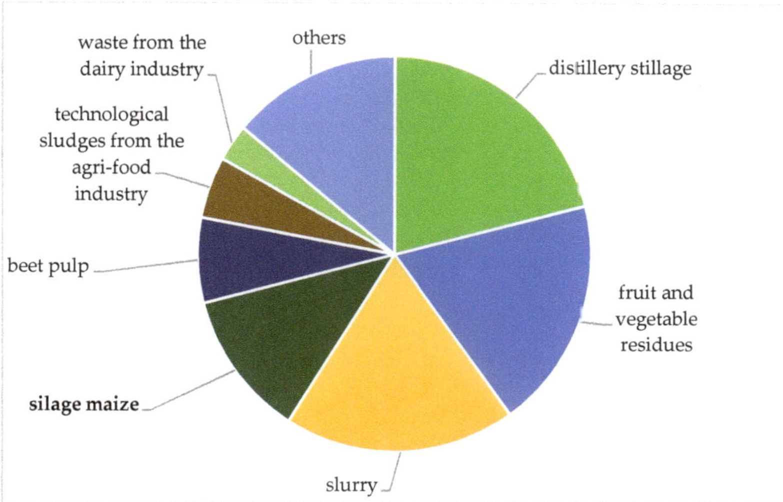

Figure 3. Share of substrates used for agricultural biogas production in 2018. Source: own elaboration based on [37].

The key criterion of the applicability of maize silage for biogas production is the share of dry weight from 28 to 35%, to much extent dependent on the right harvest date [38]. One of the biggest assets of maize is its high yields (photosynthesis C4); most frequently from 30 to 50 t/ha. For comparison, the average rye yield under Poland's conditions is 2.8 t per hectare, wheat—4.7 t per hectare for winter wheat and 3.6 t per hectare for spring wheat; the 2011–2015 means. One hectare of silage maize can produce from 4050 to 6750 m^3 of biogas, which can generate from 87 to 145 GJ of energy (Table 1). Biogas production from 1 ton of silage can reach 200 m^3, and from 1 ton of dry weight of silage the average of 550–650 m^3 of biogas is produced. The amount of methane production ranges from 300 to 400 m^3 per ton of dry silage weight [39–41].

Table 1. Average yields, biogas production, and energy from silage maize.

Average Yield of Fresh Weight, ($t \cdot ha^{-1}$)	Average Biogas Production, ($m^3 \cdot ha^{-1}$)	Average Energy Production, ($GJ \cdot ha^{-1}$)
30–50	4050–6750	87–145

Source: own elaboration based on [42,43].

Maize shows, e.g., high yields of green weight per area unit, a high biogas yield, a good ensilaging capacity [44]. The productivity of photosynthesis in C4 plants is about 1.5–2 times higher than in C3 plants; hence a high interest in those plants to be used for energy purposes [45–48]. The foreseen climate changes have and will have a high impact on crop cultivation conditions. Maize must be definitely considered a species which because of its physiology, gets fast adapted to unfavorable climate changes [49]. Table 2 presents the maize adaptation to climate changes.

The maize cultivation technologies, irrespective of the direction of use, should consider the economic effectiveness, energy efficiency as well as, while facing climate changes, the environmental effectiveness, to alleviate and to decrease the rate of environmental changes, and to limit the GHG emissions [50]. Accomplishing those goals is considered feasible owing to lowering the energy consumption of production technologies and increasing the efficiency. The plant production, including the production of silage maize, requires performing many agrotechnical treatments. The factors affecting the silage maize cultivation success, acquiring a high-quality material in terms of its ensilaging applicability are the adequate agrotechnical practices, the cultivar selection adequate for the climate zone and the stand [39,51]. The basic principles for making maize silage are accurate crushing, adding a preservative, fast placing of the heap or filling the silo, hermetic coverage, and the adequate pick-up, which affects the silage quality and limiting losses [52]. As for an inaccurate packing, the remaining oxygen makes ensilaging longer, it can lead to the development of undesired aerobic microorganisms. The hermetic coverage of the heap with foil prevents from the rainwater penetrating into the silage, and the load—the right ensilage straw deposition [53–55]. The tillage system and the dependent material and energy inputs, the frequency of practices, the dates of the agrotechnical practices performed, the harvest at the optimal date with a minimum level of losses during the practices are the key factors of the production energy efficiency.

Table 2. Maize adaptation and response to an anticipated climate change.

Climate Change	Mazie Adoption
Warming	→Thermophilic plant
Extension of the growing season	→Relatively low water needs
Less rainfall in the summer	→Less risk of crop failure →Longer growing season; later forms have a greater yielding potential

Source: own elaboration based on [49].

3. Materials and Methods

3.1. Source Material and Object Characteristics

The quality of agricultural enterprise management can be estimated with a balance sheet for the production period, breaking down the revenues and inputs and the incomes which can be expressed in a cash unit or the balance sheet for the business activity can be developed by breaking down the inputs and incomes in reference energy units (MJ) and reference grain units (JZ) [10]. The efficiency is a quotient of the outcome to the input [56].

The research was performed on 13 farms in the Podlaskie voivodeship (in south-eastern Poland). The climate of this region is moderate with huge continental influence. This voivodeship is dominated by agriculture, which is the main branch of the region's economy. The fodder area is approx. 55% of the agricultural area. Over 31% stands for sown area on the arable land are fodder plants. In the studied farms the silage maize was grown in real farming conditions. The crop acreage ranged from 2.0 to 13.0 ha. The fields were 0.05 to 2.5 km away from the habitation. The yields varied and ranged from 45 to 80 t·ha^{-1} (Table 3).

Table 3. Selected elements characteristic for silage maize cultivation.

Technology Number	Crop Acreage (ha)	Distance from the Habitat (km)	Yield (t·ha^{-1})
1	3.4	0.8	72
2	2.0	0.1	60
3	13.0	1.5	50
4	5.1	0.5	55
5	3.3	0.6	65
6	4.5	1.0	80
7	2.0	2.5	50
8	3.5	2.5	70
9	10.0	2.2	67
10	5.0	1.6	75
11	8.0	2.2	75
12	5.0	1.0	50
13	3.21	1.5	45

Source: own study.

The data from the agricultural farms on selected silage maize cultivation technologies were provided in the elaborations and process sheets, breaking down all the factors and agrotechnical practices (record of treatments and practices as well as production inputs), especially:

- Types and technical parameters of the machinery, tools and tractors used;
- Machinery aggregate performance;
- Labour inputs;
- Consumables and raw materials and fuel consumption.

With the method of direct interview with farmers, made twice over the vegetation period, there were determined the levels of the agrotechnical factors applied, which provided the data on the means-of-production inputs for the technologies investigated, following the silage maize cultivation technologies applied on a given farm and the consumption of the real sowing material, natural and artificial fertilizers, plant protection agents, and the yields per hectare.

The selected agricultural farms varied in terms of the type and amount of the fertilization applied. As for 12 out of 13 cultivation variants, natural fertilization was involved in 6 variants—manure only (no 3, 4, 9, 10, 11, 13) at the doses from 12.5 $t \cdot ha^{-1}$ (no 13) to 41.4 $t \cdot ha^{-1}$ (no 4), in 1—only slurry at the dose of 20 $t \cdot ha^{-1}$ (no 6), in 5—manure and slurry (no 1, 2, 5, 8, 12), manure—from 30 (no 8, 12) to 47.1 $t \cdot ha^{-1}$ (no 1) and slurry—from 14 $t \cdot ha^{-1}$ (no 5, 8) to 20 $t \cdot ha^{-1}$ (no 12), respectively. The cultivation technology marked with number 7 did not involve natural fertilization, whereas technology 7—used mineral fertilization only.

In the objects under study there were also considerable differences in the tractors and machinery used. The tractors engaged in agrotechnical practices and actions, harvest, technology transport, or placing a heap varied in terms of power and weight. Depending on the type of the work performed, carrying out the actions with own tractors, machinery, and tools or outsourced as services, the power and weight of tractors ranged from 22.4 kW for Ursus C330 of 1675 kg to 114 kW U1634 with 5190 kg. The harvest was made using the aggregate of a tractor with a tractor-operated chaff cutter (8 plantations) and with forage harvesters (5 plantations) with the power of up to 300 kW. For forage kneading and placing a heap of silage, tractors with weight added reaching the weight of up to 6 tons (ZT 232A) were used.

3.2. Silage Maize Production Energy Consumption

The energy inputs for silage maize production for selected technologies were calculated with the calculation method developed by IBMER [57,58] following a verification and adapting it to the needs and conditions of own research. The accumulated energy consumption stands for the total consumables and raw material and energy inputs in silage maize production technologies. To calculate it, the following dependence was used:

$$E_{pro} = \sum E_{mat} + \sum E_{tm} + \sum E_{ON} + \sum E_{l}, \quad (1)$$

where, E_{pro}—the sum of energy inputs incurred on the silage maize production, $[MJ \cdot ha^{-1}]$, E_{mat}—energy consumption of consumables and raw materials engaged in production, $[MJ \cdot ha^{-1}]$, E_{tm}—energy consumption of tractors, machinery, and tools $[MJ \cdot ha^{-1}]$, E_{ON}—energy consumption of fuel, $[MJ \cdot ha^{-1}]$, E_{l}—energy consumption generated by human labor, $[MJ \cdot ha^{-1}]$.

The total value of accumulated energy consumption includes: energy consumption of consumables and raw materials engaged in production, energy consumption of the use of tractors, machinery and tools, the energy consumption of the fuel, and the energy consumption of the labor. The respective components of accumulated production energy consumption were calculated following the formulae: (E_{mat}, E_{tm}, E_{ON}, E_{l}).

Energy consumption of materials involved in production:

$$E_{mat} = E_{s} + E_{f} + E_{pch}, \quad (2)$$

where, $E_{s} = M_{s} * I_{s}$—energy contained in the seeds of maize, $[MJ \cdot ha^{-1}]$, M_{s}—seed weight, $[kg \cdot ha^{-1}]$, I_{s}—unit energy consumption index of maize seeds, $[MJ \cdot kg^{-1}]$, E_{f}—energy contained in fertilizers, $[MJ \cdot ha^{-1}]$,

$$E_f = E_{nf} + E_{mf}, \tag{3}$$

where, E_{nf}—energy contained in natural fertilizers, [MJ·ha^{-1}],

$$E_{nf} = E_{nfm} + E_{nfs}, \tag{4}$$

where:

$E_{nfm} = M_{nfm} * I_{nfm}$—energy contained in manure, [MJ·ha^{-1}],
M_{nfm}—manure mass, [kg·ha^{-1}],
I_{nfm}—unit manure energy consumption index, [MJ·kg^{-1}],
$E_{nfs} = M_{nfs} * I_{nfs}$—the energy contained in the slurry, [MJ·ha^{-1}],
M_{nfs}—slurry mass, [kg·ha^{-1}],
I_{nfs}—unit energy consumption index of slurry, [MJ·kg^{-1}],
E_{mf}—energy contained in mineral fertilizers, [MJ·ha^{-1}],

$$E_{mf} = E_{mfN} + E_{mfP} + E_{mfK} + E_{mfCa}, \tag{5}$$

where:

EmfN = MmfN * ImfN—energy contained in nitrogen fertilizers, [MJ·ha^{-1}],
MmfN—mass of nitrogen fertilizer, [kg·ha^{-1}],
ImfN—unit energy consumption index of nitrogen fertilizers, [MJ·kg^{-1}],
EmfP = MmfP * ImfP—energy contained in phosphorus fertilizers, [MJ·ha^{-1}],
MmfP—mass of phosphorus fertilizer, [kg·ha^{-1}],
I_{mfP}—unit energy consumption index of phosphorus fertilizers, [MJ·kg^{-1}],
$E_{mfK} = M_{mfK} * I_{mfK}$—energy contained in potash fertilizers, [MJ·ha^{-1}],
M_{mfK}—mass of potash fertilizer, [kg·ha^{-1}],
I_{mfK}—unit energy consumption index of potash fertilizers, [MJ·kg^{-1}],
$E_{mfCa} = M_{mfCa} * I_{mfCa}$—energy contained in calcium fertilizers, [MJ·ha^{-1}],
M_{mfCa}—mass of calcium fertilizer, [kg·ha^{-1}],
I_{mfCa}—unit energy consumption index of calcium fertilizers, [MJ·kg^{-1}],
$E_{pch} = M_{pch} * I_{pch}$—energy contained in plant protection chemicals, [MJ·ha^{-1}],
M_{pch}—mass of plant protection chemicals, [kg·ha−1],
I_{pch}—unit energy consumption index of plant protection chemicals, [MJ·kg^{-1}],

Energy consumption of using tractors and machines was calculated according to the formula:

$$E_{tm} = E_t + E_m, \tag{6}$$

where, E_t—energy consumption of tractors, [MJ·ha^{-1}], E_m—energy consumption of the machine/s, [MJ·ha^{-1}].

$$E_t = \frac{M_t * I_t + M_{sp} * I_{sp}}{I_{pet} * I_{oe}}, \tag{7}$$

where:

M_t—mass of the tractor, [kg],
I_t—unit tractor energy consumption index, [MJ·kg^{-1}],
M_{sp}—mass of worn spare parts on the tractor, [kg],
I_{sp}—unit index of energy consumption of spare parts, [MJ·kg^{-1}],
I_{pet}—exploitation potential (standard number of hours of operation of the tractor during its use, [h]
I_{oe}—operational efficiency of the machine when performing a given procedure, [ha·h^{-1}].

$$E_m = \frac{M_m * I_m + M_{sp} * I_{sp}}{I_{pem} * I_{oe}}, \tag{8}$$

where:

M_m—machine weight, [kg],
I_m—unit energy consumption index of the machine, [MJ·kg^{-1}],
M_{sp}—mass of used spare parts in the machine, [kg],
I_{pem}—exploitation potential (standard number of hours of operation of the machine during its use, [h]

Energy intensity brought in the form of human labor:

$$E_l = \frac{N_t * I_{to}}{I_{oe}}, \quad (9)$$

where, N_t—number of employed tractor drivers, machine operators, I_{to}—unit index of energy consumption of work by tractor driver, machine operator, [MJ·rbh^{-1}],

Energy consumption of used fuel:

$$E_{ON} = Z_{ON} \cdot I_{ON}, \quad (10)$$

where, C_{ON}—fuel (diesel) consumption of tractors and self-propelled machines, [dm3·ha^{-1}], I_{ON}—unit fuel energy consumption index, [MJ·kg^{-1}].

Formulas (1)–(10) include unitary indicators of energy consumption, assuming different values reported by respective authors, which is related to the changes in the production methods, living standards, etc. To calculate the energy inputs related to direct energy carriers, mineral fertilizers, agrochemicals, the application of tractors, machinery, and human labor, the accumulated energy consumption indicators for respective energy resources were used (Table 4) [5,10,59–61].

Table 4. Unitary energy consumption indicators.

Indicator	Symbol	Unit of Measure	Value
	Investment measures		
Tractors	I_t	MJ·kg^{-1}	125
Machines	I_m	MJ·kg^{-1}	110
Tractor tools	I_n	MJ·kg^{-1}	100
Spare parts and repair materials	I_{sp}	MJ·kg^{-1}	85
	Human labour		
Tractor drivers, machine operators	I_{to}	MJ·rbh^{-1}	80
	Materials		
Manure	I_{nfm}	MJ·kg^{-1}	0.3
Slurry	I_{nfs}	MJ·kg^{-1}	0.2
Nitrogen fertilizers	I_{mfN}	MJ·kg^{-1} N	77
Phosphorus fertilizers	I_{mfP}	MJ·kg^{-1} P_2O_5	15
Potash fertilizers	I_{mfK}	MJ·kg^{-1} K_2O	10
Calcium fertilizers	I_{mfCa}	MJ·kg^{-1} CaO	6
Plant protection chemicals	I_{pch}	MJ·kg^{-1} SA	300
Maize seeds	I_s	MJ·kg^{-1}	9
	Direct energy carrier		
Diesel	I_{ON}	MJ·kg^{-1}	48
	Agricultural products		
Silage maize	I_{maize}	MJ·kg^{-1}	0.8

Source: own elaboration based on [5,30,60–62].

The unitary energy consumption indicators used for the calculations express the energy equivalent of the unit of given means of production engaged in the production for a given silage maize cultivation technology, e.g., 1 kg of tractor or machinery, 1 kg of the raw materials used, 1 dm3 of fuel and 1 man-hour of human labor.

3.3. Silage Maize Production Energy Efficiency

Energy efficiency of the production of specific yield must be considered a ratio of the energy value of the product to the amount of energy consumed for the production. The energy efficiency index value was calculated following the dependence provided by Harasim [63], Kuś [64] and expressed as a dimensionless coefficient:

$$Ep = Ev/Epro, \tag{11}$$

where, E_p—energy efficiency index of silage maize, E_v—energy value of the maize yield per 1 ha, [MJ·ha^{-1}], E_{pro}—the sum of energy inputs incurred on the silage maize production, [MJ·ha^{-1}].

Using the unitary energy index for silage maize, 0.8 MJ·kg^{-1} [29], the value of the energy of the maize yield was calculated from the dependence:

$$Ev = I_{maize} \cdot Y_m, \tag{12}$$

where, I_{maize}—unit energy index maize silage, [MJ·kg^{-1}], Y_m—maize yield for silage, [kg·ha^{-1}].

The estimates of the energy consumption for the production of respective crops, including silage maize, are applied to determine the energy consumption of respective kinds of agricultural biomass for energy use. The methodology for investigating the energy consumption of plant production, silage maize, is applied to study the outcomes and energy efficiency of agricultural material for human consumption or for energy purposes.

4. Results

4.1. Accumulated Energy

One of the elements affecting the energy consumption of the production process is the energy inputs resulting from the consumption of traditional energy carriers (ON) in the technologies applied. For that group of energy carriers and for the cultivation technologies analyzed and investigated, the diesel oil was considered.

The accumulated energy consumption of the energy carriers in a form of diesel oil for the technologies studied varied. In extreme cases the differences were more than double. The data provided in Figure 4 show that the accumulated energy consumption of energy carriers ranged from 4684.11 (technology 7) to 17,162.61 MJ·ha^{-1} (technology 10). The average value of accumulated energy consumption of the energy carriers for the silage maize cultivation technologies was 10,561.90 MJ·ha^{-1}.

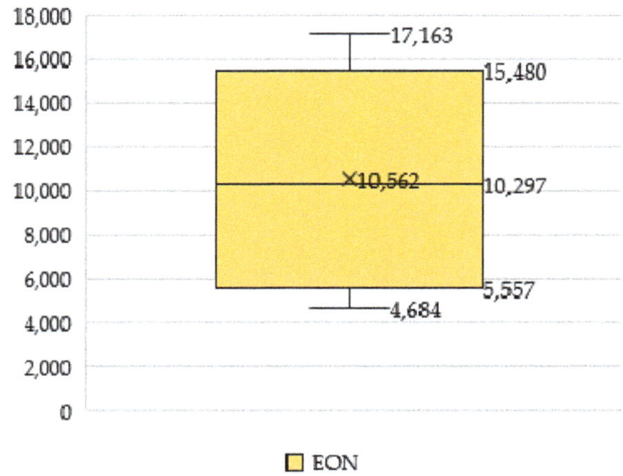

Figure 4. Total energy for the diesel oil consumed [MJ·ha^{-1}]. Source: own study.

The energy consumption for the cultivation technology is also affected by the accumulated energy related to the use of tractors, machinery, and tools for silage maize production (Figure 5). The calculations show that the share of the input of energy accumulated in tractors, machinery, and tools in the total accumulated energy ranged from 524.26 MJ·ha^{-1}, which accounted for 1% (technology 1), to 12,196.21 MJ·ha^{-1}, which accounted for 23.4% (technology 5). The mean value for the calculated accumulated energy consumption in tractors, machinery, and tools was 3886.28 MJ·ha^{-1}. The share of energy generated by human labor is shown in Figure 6.

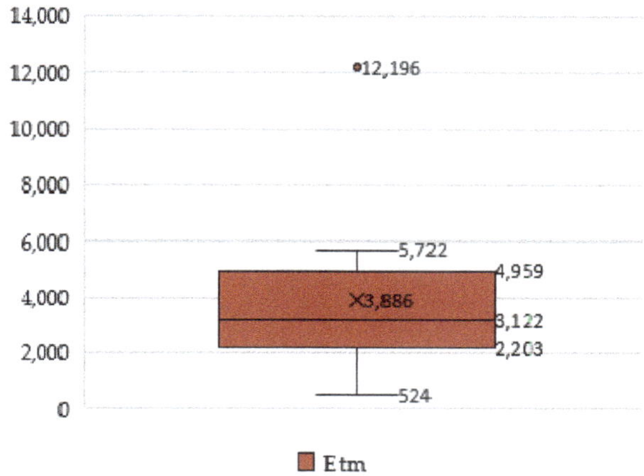

Figure 5. Total energy accumulated in the tractors, machines and tools used [MJ·ha^{-1}]. Source: own study.

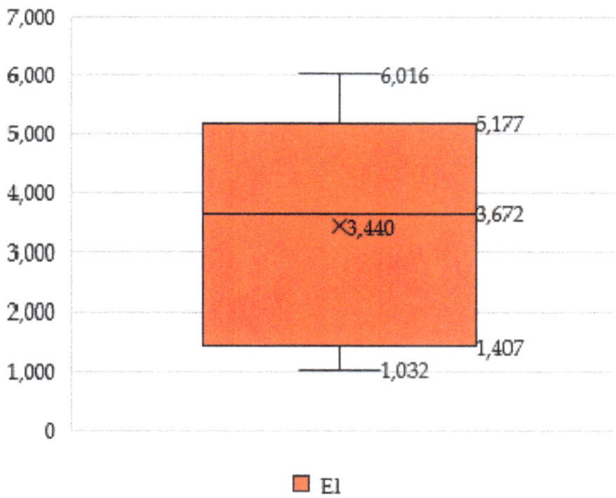

Figure 6. Amount of energy generated by human labor [MJ·ha^{-1}]. Source: own study.

As for the silage maize cultivation technologies investigated, the energy input considered was the consumption of consumables and raw materials. The average value of the energy consumption accumulated in the consumables for the technologies analyzed was 21,051.22 MJ·ha^{-1}. The share of the input of energy accumulated in the consumables in the total accumulated energy ranged from 3924.32 MJ·ha^{-1}, which accounted for 32.1% (technology 13), to 31,845.65 MJ·ha^{-1}, which accounted for 63.0% (technology 1) (Figure 7).

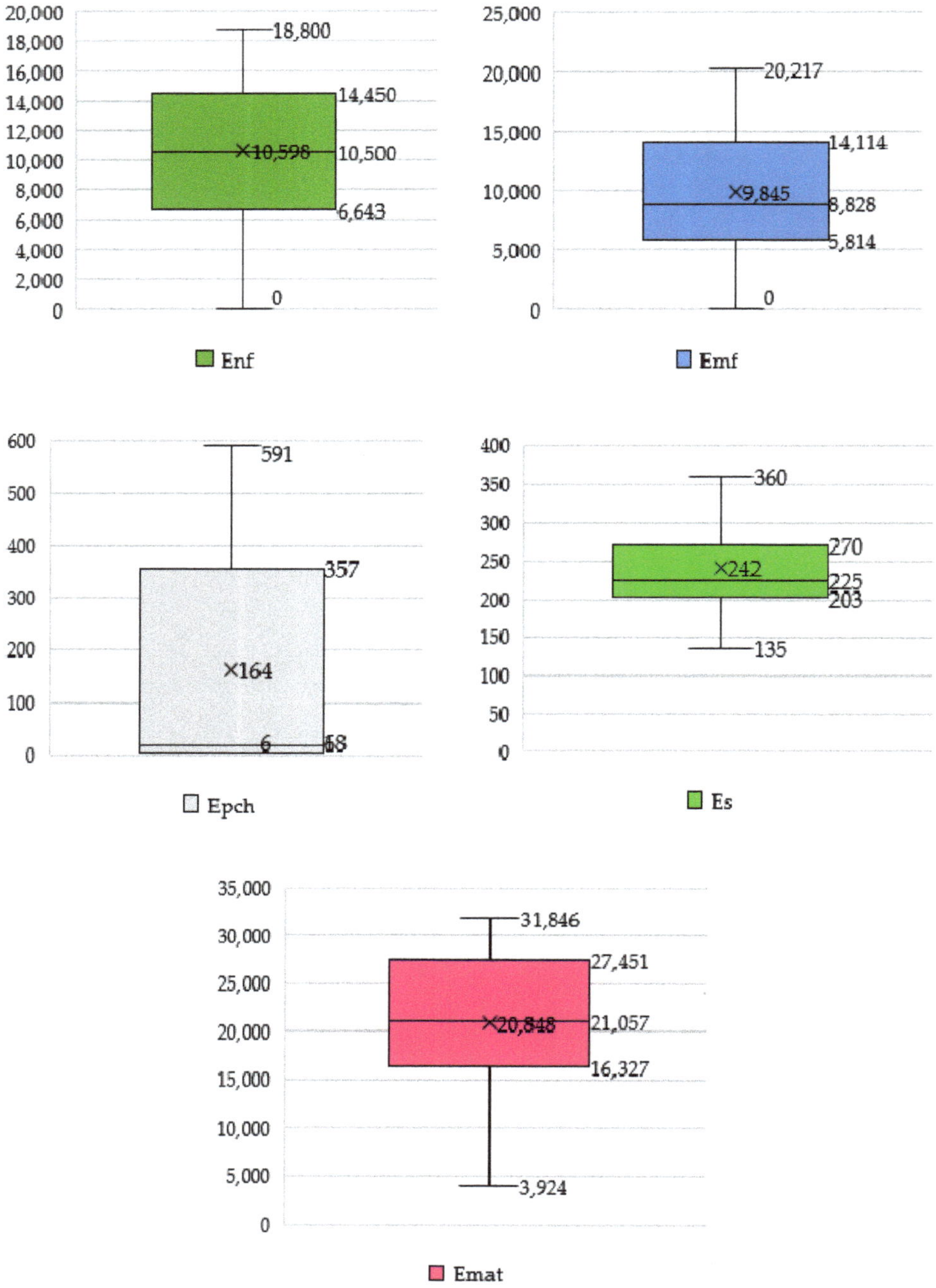

Figure 7. Total energy accumulated in the natural and mineral fertilizers, plant protection products, and seeds used [MJ·ha^{-1}]. Source: own study.

4.2. Energy Inputs Structure

The structure of the energy consumption accumulated for technologies was calculated for respective energy inputs, namely the energy carriers, inputs of labor, the consumables, and raw materials as well as tractors, machinery, and tools. Figure 8 presents the share of accumulated material—energy inputs for the silage maize cultivation technologies from four energy inputs: in tractors, machinery, and tools, means of transport, as well as in the spare parts and materials used for the repairs of that equipment, the direct energy carrier, namely the diesel oil, the consumables, and raw materials used for production and the human labor inputs.

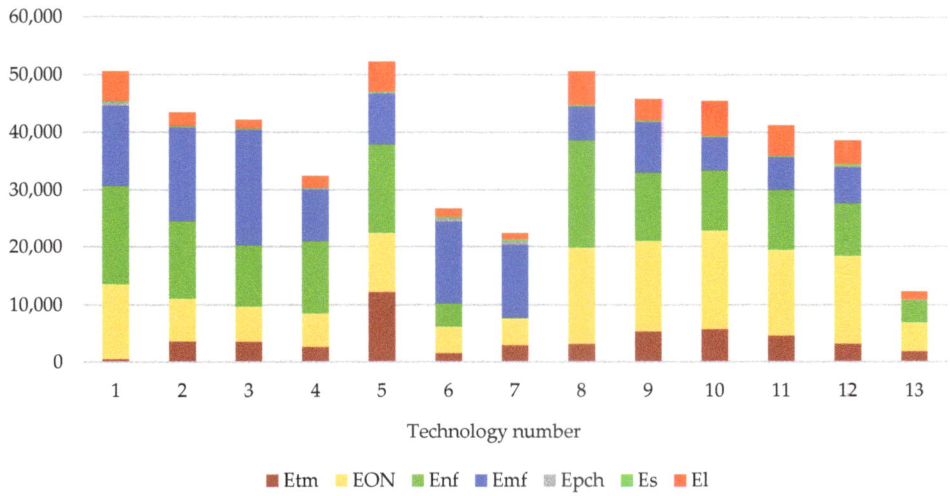

Figure 8. Energy intensity accumulated in the technologies of maize silage from individual energy inputs. [$MJ \cdot ha^{-1}$]. Source: own study.

Of all the energy inputs analyzed, the greatest share in the total accumulated energy consumption was recorded for the consumables and raw materials; from 32.1% for technology 13 to 73.5% for technology 3. On average the share of the energy accumulated in the consumables and raw materials accounted for 53.7%. A lower share in the total energy consumption was recorded for the energy inputs related to the use of diesel oil for agrotechnical practices and jobs in the production chain. The values ranged from 14.6% (technology 3) to 42.2% (technology 13). The mean share of energy accumulated in the energy carriers accounted for 27.5%. Another element affecting the total production energy consumption was the energy accumulated in tractors, machinery, and tools. The average share of inputs from that input accounted for 10.2%. As for the technologies analyzed, the energy consumption values for tractors, machinery, and tools ranged from 1.0% (technology 1) to 23.4% (technology 5). The lowest share in the total accumulated energy consumption for production was the energy accumulated in human labor. The share of inputs from that energy input ranged from 3.4% (technology 3) to 13.2% (technology 10), and the mean value accounted for 8.6%. Table A1 in the Appendix A provides the results of the calculations of the energy accumulated for all the inputs of energy for each silage maize technology studied. The percentage share of accumulated energy for each of the inputs for the technologies researched has also been given.

4.3. Energy Efficiency

Table 5 demonstrates the calculated index of energy efficiency calculated for 13 silage maize cultivation technologies. The values range from 0.95 (technology 3) to 2.94 (technology 13). As for 11 out of 13 silage maize-growing technologies, the index value was higher than 1. It means that for the biomass produced the value of the accumulated energy was higher than the energy in the energy inputs made.

Table 5. Energy expenditure in means of production, energy accumulated in yield, and energy efficiency index.

Technology Number	Accumulated Energy			E_p
	E_{pro}		E_v	
	[MJ·ha^{-1}]	%	[MJ·ha^{-1}]	
1	50,587.01	100	57,408.00	1.13
2	43,436.57	100	48,000.00	1.11
3	42,116.45	100	40,000.00	0.95
4	32,400.24	100	44,000.00	1,36
5	52,229.49	100	51,696.00	0.99
6	26,724.37	100	64,000.00	2.39
7	22,338.06	100	40,000.00	1.79
8	50,567.99	100	56,000.00	1.11
9	45,736.91	100	53,600.00	1.17
10	45,440.50	100	60,000.00	1.32
11	41,161.42	100	60,000.00	1.46
12	38,593.77	100	40,000.00	1.04
13	12,233.01	100	36,000.00	2.94

Source: own study.

The most favorable technological variant in terms of energy efficiency was variant 13 for which, assuming as the reference 100% to be the mean value of the inputs of energy accumulated in the consumables and raw materials (20,847.75 MJ·ha^{-1}); the energy accumulated in that input was 81.2% lower than the average value, which had the greatest effect on the result despite a considerably low yield (45 t·ha^{-1}). Assuming the mean value of the inputs of energy accumulated in tractors, machinery, and tools for the technologies to account for 100% (3886.28 MJ·ha^{-1}), for technology 13 those inputs were 53.6% lower than the average. The analysis of the human labor inputs made for production demonstrated that, as for technology 13, they are 61% lower than the mean for the technologies investigated. As compared with the mean value of the inputs of energy accumulated in direct sources of energy; diesel oil (10,561.90 MJ·ha^{-1}, assumed as 100%) for the technologies under study the inputs of the accumulated energy from that input in technology 13 were 51.1% lower than the mean value.

For two plantations of all those investigated, the energy efficiency index value was below 1, which means that the technological solutions applied for the crops marked 3 and 5 showed a lack of energy efficiency and the inputs of accumulated energy used to produce 1 biomass yield unit were higher than the energy accumulated in the yield, which was an unjustifiable solution in terms of energy. The greatest impact on the energy efficiency calculation results, 0.99 for technology 5, was recorded for the energy accumulated in tractors, machinery and tools, and for technology 3 (0.95), the inputs of energy accumulated in the consumables and raw materials.

5. Discussion and Conclusions

According to the union market preferences like grants and subsidies for agricultural production (materials, services, and loans) the studies in the field of energy efficiency have huge importance. Plant production, including silage maize, requires performing many agrotechnical practices. Currently there are undergoing analyses about the possibility of sustainability development use and the reduction of energy inputs to reduce the negative impact on the environment, water, soil, air. It can be afforded by the reduction of the number of treatments performed in order to, among others, replacing the conventional tillage system with simplified or direct sowing [25,65]. The tillage system and the resulting material and energy inputs, the frequency of jobs, the dates of agrotechnical practices, the harvest at the optimum date with a minimum level of losses while performing the jobs are the key factors of production energy efficiency [66].

The plant production energy efficiency has been researched by many authors who referred the index to single agrotechnical practices [67,68], tillage systems [69,70], the entire technologies and agricultural products [71–74], and to elements of crop rotation [67,70,75].

As for the research of silage maize production technologies in the structure of energy inputs made for production, it was found that the highest share is recorded for the inputs of energy accumulated in the consumables and raw materials; 53.7% on average, and for the energy input the highest share is noted for fertilizers (98% on average, assuming 100% mean energy accumulated in consumables and raw materials). Gorzelany et al. [76] and Budzyński et al. [74] also claim that the highest share in the structure of the inputs of energy made for producing silage maize is accounted for the consumables and raw materials, 56%, respectively, and, depending on the technology traditional 76.5%, including fertilizers 71.5%, integrated, including fertilizers—63.8%. Similarly, as provided in the results of this research, the lowest share was found for the inputs of energy in a form of human labor, depending on the technology, 0.6 and 0.9%. The analysis of the results of the research of the technologies covered by these considerations also demonstrates the lowest share of energy of the human labor in the structure of inputs, on average 8.6%. It can be slightly higher for the results reported by the above authors as they did not consider the operation time of the tractors and machinery, and the time of driving to the field. In those analyses the time was factored in and considered essential in terms of performance of the sets of machinery and tools as well as fuel consumption. The inputs of accumulated energy from all the inputs for the technologies studied ranged from 12,233.01 to 52,229.49 MJ·ha^{-1}, respectively for technologies 3 and 5; 38,735.83 MJ·ha^{-1} on average. Szempliński and Dubis [77] generated 22,000–24,000 MJ·ha^{-1} of the energy inputs made for silage maize production technologies, Budzyński et al. [74]—23,900—18,700 MJ·ha^{-1} depending on the input intensity, Gorzelany et al. [76] 24,305 MJ·ha^{-1}, and considering the energy produced in the yield—37,800 MJ·ha^{-1}; the energy efficiency index was 1.5. In the present study the mean result of energy efficiency for the silage maize production technologies was 1.44, the minimum value—1.04 (disregarding the technologies with no such efficiency), the maximum value—2.94. Wielogórska et al. [78] reported on the research performed to evaluate the silage maize-growing technologies on the farms with the agricultural land acreage of at least 5 ha, show that the mean energy inputs made for cultivation were 22,200 MJ·ha^{-1}, and the technologies investigated recorded a high energy consumption index which was the crops mean of 3.2.

The research performed for 13 technologies for the plantation acreage from 2 to 13 has shown that the most favorable technology solution in terms of energy efficiency was variant 13. With the results in mind, it can be claimed that a relatively low natural fertilization and a lack of mineral fertilization decreased the maize silage production energy consumption (12,233.01 MJ·ha^{-1}) enough for, despite the lowest yield for all the technologies, the highest value of the energy efficiency. Assuming that 100% is the mean value of the energy accumulated in the consumables and raw materials for those technologies (20,847.75 MJ·ha^{-1}) in variant 13 the energy accumulated from that input was 81.2% lower than the mean value, which significantly affected the silage maize production energy efficiency index value. The energy accumulated in the yield for that technology was 28.1% lower than the mean value (50,054.15 MJ·ha^{-1}). As for that technology, the value of the energy efficiency index was 182% higher than the value for the technology least favorable in terms of energy, however, with the index value above 0 (1.04 technology 12).

For two plantations of all those investigated, the energy efficiency index value was lower than 1 (0.95 and 0.99). It means that the technological solutions applied in the variants marked 3 and 5, respectively, recorded a lack of energy efficiency, the accumulated energy inputs made to produce a biomass yield unit were higher than the energy accumulated in the yield. A higher yielding and the accumulated energy of the yield, 40,000.00 MJ·ha^{-1}

and 51,696.00 MJ·ha^{-1} (for technologies 3 and 5, respectively) did not compensate for the high energy inputs made for their accomplishment; 42,116.45 MJ·ha^{-1} for technology 3 and 52,229.49 MJ·ha^{-1} for technology 5. Those were the solutions which were unjustified in terms of energy.

When deciding on a given technological solution, it should be remembered that the increase in yields is not linear to the increase in energy inputs. Depending on this, there is a point to which increasing the level of expenditure is justified. Above the optimal value adjusted to e.g., type of crop, farm size, cultivation system, increasing energy inputs are not compensated in the yield.

In agricultural practice, due to a decrease in the inputs and a negative impact on the environment, various production technology modification methods, also for plant production, for simplified tillage systems and limiting the inputs are being searched for [79–81]. The energy calculation should be an essential element for the assessment of plant production, which is frequently limited to economic and production criteria [70,74].

To recapitulate, one can state that, in terms of energy efficiency, growing maize is justifiable and it can trigger energy benefits. The main factors dividing the cultivation technologies and harvesting maize for silage in typical Polish farm are size of plantation, diversity of crops (according to soil and climatic conditions), the level of farmer education, technical procedure and knowledge, machine park and machine services availability, possibility of products managing (using and demanding), types of activities in the adjacent areas, preferences and involvement of production units, and possibility of usage in energy production e.g., in biogas plants. The development of agricultural biogas plants contributes to the new jobs opportunities in rural areas, which enables the diversification of farmers income sources. According to KOWR (31 August 2017), there are nine agricultural biogas plants registered in the Podlaskie voivodeship. The adequately selected silage maize production technologies make it a very attractive crop in terms of energy, which becomes of special importance while facing the need of using the energy from renewable resources. The analyses covered the maize production technology variants: a field production stage, through harvest to the preparation of raw material for ensilaging, and placing a heap. The energy efficiency of a further use will depend on the processing technology applied in the biogas plant, which will be a continuation of the previous considerations and support the research results obtained at this stage. Other limitations resulting from the production of bioenergy from biomass should also be considered. The intensive cultivation of certain crops requires large areas of cultivation because of their future energy potential (so-called energy crops like maize). This is related to the acquisition of a significant amount of the plant and may be associated with the excessive use of fertilizers and other substances polluting the soil and water, and the reduction of areas for food production [82].

According to the EIA (Energy Information Administration) report from 2015, by the year of 2040 world energy consumption will increase by 56%, world energy consumption will increase by 56%, which will lead to an increase in the world CO_2 emissions up to 46% [82]. According to above, the estimation of GHG emissions released into the atmosphere is an important issue, which is planned as a continuation of the present research.

Moreover, in the literature there are analyzes of the relationship between the prices of energy raw materials and the prices of grains and oils [83]. In connection with this, an interesting research issue will also be the analysis of the relationship between the prices of maize and the prices of energy resources.

Author Contributions: Conceptualization, A.K., K.R., M.R. (Monika Roman), D.Ś., M.R. (Michał Roman); methodology, A.K., K.R., M.R. (Monika Roman), software, A.K., K.R., M.R. (Monika Roman), D.Ś., M.R. (Michał Roman); validation, A.K., K.R., M.R. (Monika Roman), D.Ś., M.R. (Michał Roman); formal analysis, A.K., K.R., M.R. (Monika Roman), D.Ś., M.R. (Michał Roman); resources, A.K., M.R. (Monika Roman), K.R.; data curation, A.K.; writing—original draft preparation, A.K., K.R., M.R. (Monika Roman), D.Ś., M.R. (Michał Roman); writing—review and editing, A.K., M.R. (Monika Roman); visualization, A.K., M.R. (Monika Roman); supervision, A.K., K.R., M.R. (Monika Roman), D.Ś., M.R. (Michał Roman); project administration, M.R. (Michał Roman); funding acquisition, M.R.

(Michał Roman), M.R. (Monika Roman) All authors have read and agreed to the published version of the manuscript.

Funding: This research received no external funding.

Institutional Review Board Statement: Not applicable.

Informed Consent Statement: Not applicable.

Data Availability Statement: Not applicable.

Conflicts of Interest: The authors declare no conflict of interest.

Appendix A

Table A1. Value of accumulated energy consumption and the percentage share of accumulated energy in individual energy inputs for the technologies studied.

Technology Number	Accumulated Energy									
	E_{tm}		E_{ON}		E_{mat}		E_l		E_{pro}	
	$MJ·ha^{-1}$	%	$MJ·ha^{-1}$	%	$MJ·ha^{-1}$	%	$MJ·ha^{-1}$	%	$MJ·ha^{-1}$	%
1	524.26	1.0	13,042.99	25.8	31,845.65	63.0	5174.12	10.2	50,587.01	100
2	3606.39	8.3	7443.88	17.1	30,006.30	69.1	2380.00	5.5	43,436.57	100
3	3563.23	8.5	6166.14	14.6	30,960.00	73.5	1427.08	3.4	42,116.45	100
4	2602.04	8.0	5948.59	18.4	21,814.11	67.3	2035.50	6.3	32,400.24	100
5	12,196.21	23.4	10,297.07	19,7	24,581,75	47.1	5154.46	9.9	52,229.49	100
6	1509.79	5.6	4684.11	17.5	19,143.80	71.6	1386.67	5.2	26,724.37	100
7	2862.77	12.8	4844.29	21.7	13,599.00	60.9	1032.00	4.6	22,338.06	100
8	3121.92	6.2	16,722.50	33.1	24,895.00	49.2	5828.57	11.5	50,567.99	100
9	5344.52	11.7	15,662.99	34.2	21,057.40	46.0	3672.00	8.0	45,736.91	100
10	5722.01	12.6	17,162.61	37.8	16,539.88	36.4	6016,00	13.2	45,440.50	100
11	4572.82	11.1	14,868.73	36.1	16,539.88	40.2	5180.00	12.6	41,161.42	100
12	3092.38	8.0	15,296.26	39.6	16,113.71	41.8	4091.43	10.6	38,593.77	100
13	1803.28	14.7	5164.60	42.2	3924.32	32.1	1340.81	11.0	12,233.01	100

Source: own study.

References

1. Jeziorska-Thöle, A.; Rudnicki, R.; Kluba, M. Development of energy crops cultivation for biomass production in Poland. *Renew. Sustain. Energy Rev.* **2016**, *62*, 534–545. [CrossRef]
2. Wang, T.; Lin, B. Fuel consumption in road transport: A comparative study of China and OECD countries. *J. Clean. Prod.* **2019**, *206*, 156–170. [CrossRef]
3. Balsalobre-Lorente, D.; Shahbaz, M.; Roubaud, D.; Farhani, S. How economic growth, renewable electricity and natural resources contribute to CO_2 emissions? *Energy Policy* **2018**, *113*, 356–367. [CrossRef]
4. Katircioglu, S.T. International tourism, energy consumption, and environmental pollution: The case of Turkey. *Renew. Sustain. Energy Rev.* **2014**, *36*, 180–187. [CrossRef]
5. Szeptycki, A. *Efektywność Postępu Technicznego w Technologiach Towarowej Produkcji Ziemniaków*; Polskie Towarzystwo Inżynierii Rolniczej: Warszawa, Poland, 2002.
6. Pellizzi, G. Use of energy and labour in Italian agriculture. *J. Agric. Engng. Res.* **1992**, *52*, 111–119. [CrossRef]
7. Pellizzi, G. Sul risparmio energetico nella meccanizzazione agricola. *Mondo Macch.* **2000**, *2*, 14–22.
8. Fathollahi, H.; Mousavi-Avval, S.H.; Akram, A.; Rafiee, S. Comparative energy, economic, and environmental analyses of forage production systems for dairy farming. *J. Clean. Prod.* **2018**, *182*, 852–862. [CrossRef]
9. Nowacki, T. Paradygmaty Energotechnologicznej Efektywności Transformacji Gospodarki Żywnościowej. In *Problemy Rozwoju Wsi i Rolnictwa*; IRWiR PAN: Warszawa, Poland, 1999; ISBN 83-85369-43-0.
10. Wójcicki, Z. Metodyczne problemy badania energochłonności produkcji rolniczej. *Probl. Inżynierii Rol.* **2005**, *1*, 5–12.
11. Górka, K. Zasoby Naturalne Jako Czynnik Rozwoju Społeczno-Gospodarczego. In *Gospodarka w Praktyce i Teorii*; Nr 3(36)2014; Instytut Ekonomii, Uniwersytet Łódzki: Łódź, Poland, 2014.
12. Dz.U. 2015 poz. 478. Ustawa o Odnawialnych Źródłach Energii z Dnia 20 Stycznia 2015r. Available online: http://isap.sejm.gov.pl/isap.nsf/DocDetails.xsp?id=wdu20150000478 (accessed on 30 September 2020).
13. Żygadło, M.; Madejski, R. The Conversion of Biomass into Energy in Farm Biogas Plant. *Archiwum Gospodarki Odpadami i Ochrony Środowiska* **2016**, *18*, 55–66.

14. Ogryzek, M.P.; Rząsa, K.; Ciski, M. The role of the National Support Centre for Agriculture in the process of revitalization and renewal of the rural areas. *J. Water Land Dev.* **2019**, *42*, 143–149. [CrossRef]
15. KOBiZE 2018. Poland's National Inventory Report 2018. In *Greenhouse Gas Inventory for 1988–2016*; Instytut Ochrony Środowiska—Państwowy Instytut Badawczy, Krajowy Ośrodek Bilansowania i Zarządzania Emisjami: Warszawa, Poland, 2018.
16. Curkowski, A.; Oniszk-Popławska, A.; Mroczkowski, P.; Wiśniewski, G.; Zowsik, M. *Poradnik dla Inwestorów Zainteresowanych Budową Biogazowni Rolniczych*; Instytut Energetyki Odnawialnej, Ministerstwo Gospodarki: Warszawa, Poland, 2011. Available online: http://www.mg.gov.pl/files/upload/13229/poranik%20biogazowy.pdf (accessed on 12 December 2018).
17. Voytovych, I.; Malovanyy, M.; Zhuk, V.; Mukha, O. Facilities and problems of processing organic wastes by family-type biogas plants in Ukraine. *J. Water Land Dev.* **2020**, *45*, 185–189. [CrossRef]
18. Menardo, S.; Airoldi, G.; Cacciatore, V.; Balsari, P. Potential biogas and methane yield of maize stover fractions and evaluation of some possible stover harvest chains. *Biosyst. Eng.* **2015**, *129*, 352–359. [CrossRef]
19. Križan, M.; Krištof, K.; Angelovič, M.; Jobbágy, J. The use of maize stalks for energy purposes and emissions measurement during their combustion. *Agron. Res.* **2017**, *15*, 456–467.
20. Skoufogianni, E.; Solomou, A.; Charvalas, G.; Danalatos, N. Maize as Energy Crop. In *Maize—Production and Use, Akbar Hossain*; IntechOpen: London, UK, 2019; pp. 1–16. [CrossRef]
21. Jadidzadeh, A.; Serletis, A. The global crude oil market and biofuel agricultural commodity prices. *J. Econ. Asymmetries* **2018**, *18*, e00094. [CrossRef]
22. Coronado, S.; Rojas, O.; Romero-Meza, R.; Serletis, A.; Chiu, L.V. Crude Oil and Biofuel Agricultural Commodity Prices. In *Uncertainty, Expectations and Asset Price Dynamics*; Jawadi, F., Ed.; Springer: Cham, Switzerland, 2018; Volume 24. [CrossRef]
23. Dobek, T. Efektywność ekonomiczna produkcji ziemniaków jadalnych w wybranych gospodarstwach. *Inżynieria Rol.* **2006**, *2*, 247–254.
24. Roman, K.K.; Konieczna, A. Evaluation of a different fertilisation in technology of corn for silage, sugar beet and meadow grasses production and their impact on the environment in Poland. *Afr. J. Agric. Res.* **2015**, *10*, 1351–1358.
25. Krasuska, E.; Faber, A.; Pudełko, R.; Jarosz, Z.; Borzecka-Walker, M.; Kozyra, J.; Syp, A. Emission saving opportunities for corn cultivation for ethanol in Poland. *J. Food Agric. Env.* **2013**, *11*, 3–4.
26. Jarosz, Z.; Księżak, J.; Faber, A. Ocena wielkości emisji gazów cieplarnianych systemów uprawy stosowanych w kukurydzy wykorzystywanej do produkcji bioetanolu. *Roczniki* **2017**, *XIX*, 1. [CrossRef]
27. Houshyar, E.; Azadi, H.; Almassi, M.; Davoodi, M.J.S.; Witlox, F. Sustainable and efficient energy consumption of corn production in Southwest Iran: Combination of multi-fuzzy and DEA modeling. *Energy* **2012**, *44*, 672–681. [CrossRef]
28. Houshyar, E.; Zareitard, H.R.; Grundmann, P.; Smith, P. Determining efficiency of energy input for silage corn production: An econometric approach. *Energy* **2015**, *93*, 2166–2174. [CrossRef]
29. Persson, T.; Garcia, A.; Paz, J.; Jones, J. Maize ethanol feedstock production and net energy value as affected by climate variability and crop management practices. *Agric. Syst.* **2009**, *100*, 11–21. [CrossRef]
30. Karp, A.; Shield, I. Bioenergy from plants and the sustainable yield challenge. *New Phytol.* **2008**, *179*, 15–33. [CrossRef] [PubMed]
31. Balat, M. Production of bioethanol from lignocellulosic materials via the biochemical pathway: A review. *Energy Convers. Manag.* **2011**, *52*, 858–875. [CrossRef]
32. Golub, M.; Martinov, M.; Bojic, S.; Viskovic, M.; Djatkov, D.; Dragutinovic, G.; Dallemand, J.F. Investigation on Possibilities for Sustainable Provision of Corn Stover as an Energy Sourse: Case Study for Vojvodina. In *Agricultural in Asia, Africa and Latin America*; Shin-Norinsha Co: Japan, Osaka, 2016; Volume 47, pp. 7–15. ISSN 0084-5841.
33. Bernacki, Z. Biomass production of maize (*Zea mays* L.) cropping in exceptionally advantageous conditions in central Wielkopolska (Poland). *Biomass Bioenergy* **2018**, *110*, 25–32. [CrossRef]
34. Pokój, T.; Gusiatin, Z.M.; Bułakowska, K.; Dubis, B. Production of biogas using maize silage supplemented with residual glycerine from biodiesel manufacturing. *Arch. Environ. Prot.* **2014**, *40*, 17–29. [CrossRef]
35. FAO. Production Crops. Available online: http://www.fao.org/faostat/en/#data/QC (accessed on 20 October 2020).
36. GUS 2017. Wyniki produkcji roślinnej w 2016 r. *Inf. Opracowania Stat. Warszawa* **2017**, *82*, 46–49.
37. KOWR. 2018. Available online: http://bip.kowr.gov.pl/uploads/pliki/oze/biogaz/Surowce_w_2018_r..pdf (accessed on 29 March 2020).
38. Neumann, M.; Venancio, B.J.; Horst, E.H.; Braga Cristo, F.; Petkowicz, K.; Bobato Pontarolo, G.; Mendez, M.C.; Antonietti Martins, M.B. Corn hybrid silage quality according to harvesting time. *Semin. Ciências Agrárias* **2020**, *41*, 369–382. [CrossRef]
39. Podkówka, W. Biopaliwa Dziś i Jutro. *Kukurydza* **2007**, *2007*, 4–35.
40. Podkówka, Z.; Podkówka, W. *Substraty dla Biogazowni Rolniczych*; Agro Serwis: Warszawa, Poland, 2010.
41. Michalski, T. Biogazownia w każdej gminie—Czy wystarczy surowca? *Wieś Jutra* **2009**, *3*, 12–14.
42. Braun, R. Biogas and bioenergy system developments towards bio-refineries Trends in a central European context. In Proceedings of the Baltic Biorefinery Conference, Esbjerg, Denmark, 26–28 May 2005.
43. Gostomczyk, W. Efektywność Substratów Wykorzystywanych do Produkcji Biogazu. In *Energetyka Biogaz, Red*; Cenian, J., Ed.; 2012; Available online: https://www.imp.gda.pl/bf2020/BF2012/prezentacje/p141.pdf (accessed on 29 September 2020).
44. Garcia, A. Chapter 18: Corn Silage Production and Utilization. In *iGROW Corn: Best Management Practices*; Clay, D.E., Clay, S.A., Byamukama, E., Eds.; South Dakota State University: Brookings, SD, USA, 2016.

45. Sulewska, H. Wymagania Środowiskowe Kukurydzy i Możliwości Jej Uprawy w Polsce. In *Technologia Produkcji Kukurydzy*; Dubas, A., Ed.; Wieś Jutra: Warszawa, Poland, 2004; pp. 16–23. ISBN 83-89503-17-4.
46. Kaniuczak, Z.; Pruszyński, S. *Integrowana Produkcja Kukurydzy*; Instytut Ochrony Roślin: Poznań, Poland, 2007; ISBN 978-83-89867-16-2.
47. Drożak, A.; Wasilewska, W.; Buczyńska, A.; Romanowska, E. Fotosynteza typu C4. *Pol. Tow. Biochem.* **2012**, *58*, 44–53.
48. Walerowska, M. *Kukurydza w Mistrzowskiej Uprawie*; Polskie Wydawnictwo Rolnicze Sp. Z O.O.: Poznań, Poland, 2012; ISBN 978-83-61078-48-7.
49. Michalski, T. Kukurydza—Roślina z Przyszłością. In *Materiałach z Konferencji: Kukurydza i Sorg—Produkcja, Wykorzystanie, Rynek*; 2012; Available online: https://docplayer.pl/15780082-Materialy-z-konferencji-kukurydza-i-sorgo-produkcja-wykorzystanie-rynek.html (accessed on 30 September 2020).
50. Konieczna, A.; Borek, K.; Mazur, K.; Wardal, W.J. Emisje tlenku azotu(I) i ditlenku węgla z aplikacji nawozów nieorganicznych i naturalnych w wybranych technologiach upraw kukurydzy na kiszonkę. *Przemysł Chem.* **2019**, *98*, 1000–1005. [CrossRef]
51. Komainda, M.; Taube, F.; Kluß, C.; Antje, H. Effects of catch crops on silage maize (*Zea mays* L.): Yield, nitrogen uptake efficiency and losses. *Nutr. Cycl. Agroecosyst.* **2018**, *110*, 51–69. [CrossRef]
52. Borreani, G.; Tabacco, E.; Schmidt, R.J.; Holmes, B.J.; Muck, R.E. Silage review: Factors affecting dry matter and quality losses in silages. *J. Dairy Sci.* **2018**, *101*, 3952–3979. [CrossRef] [PubMed]
53. Podkówka, W.; Podkówka, Z. Technologie Produkcji Kiszonki z Całych Roślin Kukurydzy i Jej Wykorzystanie w Żywieniu Zwierząt. In *Technologia Produkcji Kukurydzy*; Wieś Jutra: Warszawa, Poland, 2004.
54. Podkówka, W.; Podkówka, Z. *Wpływ Poziomu Suchej Masy na Zawartość Białka Ogólnego, NDF i ADF w Kiszonce z Kukurydzy*; Annals of Warsaw Agricultural University: Warszawa, Poland, 2014; pp. 111–116.
55. Bereś, P.K.; Mrówczyński, M.; Kierzek, R.; Węgorek, P.; Sosnowska, D.; Korbas, M.; Sulewska, H.; Adamczyk, J.; Szulc, P.; Warzecha, R.; et al. *Metodyka Integrowanej Ochrony i Produkcji Kukurydzy dla Doradców*; IOR—PIB: Poznań, Poland, 2016; ISBN 978-83-64655-23-4.
56. Fried, H.O.; Lovell, K.; Schmidt, S.S. *The Measurement of Productive Efficiency and Productivity Growth*; OXFORD University Press: New York, NY, USA, 2008; Available online: https://books.google.pl/books?hl=pl&lr=&id=djMh3iuB1EIC&oi=fnd&pg=PA92&dq=efficiency+analysis&ots=XHZodICuO1&sig=bPz683s-yq_NI5CGVRVtKAVlKPg&redir_esc=y#v=onepage&q=efficiency%20analysis&f=false (accessed on 29 September 2020).
57. Anuszewski, R.; Pawlak, J.; Wójcicki, Z. Energochłonność Produkcji Rolniczej. In *Metodyka Badań Energochłonności Produkcji Surowców Żywnościowych*; IBMER: Armonk, NY, USA, 1979.
58. Wójcicki, Z. *Metody Badania i Ocena Przemian w Rozwojowych Gospodarstwach Rodzinnych*; Wydawnictwo PTIR: Kraków, Poland, 2001; p. 136. ISBN 83-862264-74-8.
59. Szeptycki, A.; Wójcicki, Z. *Postęp Technologiczny i Nakłady Energetyczne w Rolnictwie Do 2020 r*; IBMER: Warszawa, Poland, 2003; p. 242.
60. Wójcicki, Z. *Metodyka Badań Postępu Technologicznego w Gospodarstwach Rodzinnych*; Monografia IBMER: Warszawa, Poland, 2008.
61. Stolarski, M.; Szczukowski, S.; Tworkowski, J. *Efektywność Energetyczna Produkcji Biomasy Wierzby w Systemie Eko-Salix*; Fragmenta Agronomica; Polskie Towarzystwo Agronomiczne, Uniwersytet Przyrodniczy w Poznaniu: Poznań, Poland, 2011; Volume 28, pp. 62–69. ISSN 0860-4088.
62. Wójcicki, Z. Metodyka badania energochłonności produkcji rolniczej. *Probl. Inżynierii Rol.* **2015**, *4*, 17–29.
63. Harasim, A. Możliwości kompensacji ujemnego wpływu stanowiska na plonowanie i efektywność produkcji pszenicy ozimej. II Efektywność ekonomiczna i energetyczna. *Pam. Puł.* **1997**, *111*, 73–87.
64. Kuś, J. *Efektywność Energetyczna Produkcji Biopa liw Płynnych*; Wieś Jutra: Warszawa, Poland, 2002; Volume 9, pp. 8–10.
65. Król, A.; Księzak, J.; Kubińska, E.; Rozakis, S. Evaluation of sustainability of maize cultivation in Poland. A prospect theory—PROMETHEE approach. *Sustainability* **2018**, *10*, 4263.
66. Jaskulski, D.; Jaskulska, I. *Współczesne Sposoby i Systemy Uprawy Roli w Teorii i Praktyce Rolniczej*; Centrum Doradztwa Rolniczego w Brwinowie, Oddział w Poznaniu: Poznań, Poland, 2016; ISBN 978-83-60232-75-0.
67. Harasim, A. *Kompleksowa Ocena Płodozmianów z Różnym Udziałem Roślin Zbożowych i Okopowych: Rozprawa Habilitacyjna*; Dział Upowszechniania i Wydawnictw IUNiG: Puławy, Poland, 2002.
68. Kamionka, J. Wpływ techniki na efektywność pogłównego nawożenia zbóż. *Inżynieria Rol.* **2005**, *9*, 1–106.
69. Czyż, E.; Tomaszewska, J.; Sawa, J. Efektywność produkcyjna i energetyczna różnych systemów uprawy roli. *Fragm. Agron.* **1995**, *12*, 20–27.
70. Harasim, A.; Bujak, K.; Frant, M. Wpływ uproszczeń w uprawie roli i poziomu nawożenia mineralnego na efektywność energetyczną produkcji roślinnej w płodozmianie 4-polowym. *Fragm. Agron.* **2007**, *24*, 64–71.
71. Dopka, D. Efektywność energetyczna zróżnicowanej uprawy przedsiewnej na przykładzie pszenżyta ozimego. *Sectio E. Agric.* **2004**, *59*, 2071–2077.
72. Nasalski, Z.; Sadowski, T.; Stępień, A. Produkcyjna, ekonomiczna i energetyczna efektywność produkcji jęczmienia ozimego przy różnych poziomach nawożenia azotem. *Acta Sci. Pol. Agric.* **2004**, *3*, 83–90.
73. Czarnocki, S.; Starczewski, J.; Turska, E. Ocena wybranych technologii uprawy pszenżyta jarego. *Fragm. Agron.* **2006**, *2*, 287–297.

74. Budzyński, W.; Szempliński, W.; Parzonka, A.; Sałek, T. Rolnicza, Energetyczna i Ekonomiczna Efektywność Produkcji biomasy Wybranych Gatunków Roślin z Przeznaczeniem na Biogaz. In *Modelowe kompleksy agroenergetyczne: Technologie pozyskania i kondycjonowania biomasy rolniczej i wodnej dla biogazowni i zgazowarki*; Gołaszewski, J., Ed.; Wydawnictwo UWM: Olsztynie, Poland, 2014; pp. 11–282. ISBN 978-83-7299-873-6.
75. Kordas, L. Energy and economic effects of reduced tillage in crop rotation. *Acta Sc. Pol. Agric.* **2005**, *4*, 51–59.
76. Gorzelany, J.; Puchalski, C.; Malach, M. Ocena Kosztów i Nakładów Energetycznych w Produkcji kukurydzy na Ziarno i Kiszonkę. *Inż. Roln.* **2011**, *8*, 135–141.
77. Szempliński, W.; Dubis, B. Wstępne badania nad plonowaniem i wydajnością energetyczną wybranych roślin uprawianych na cele biogazowe. *Fragm. Agron.* **2011**, *28*, 77–86.
78. Wielogórska, G.; Turska, E.; Czarnocki, S. Ocena technologii stosowanych w gospodarstwach uprawiających kukurydzę na kiszonkę w rejonie środkowowschodniej Polski. *Pamiętnik Puławski* **2008**, *147*, 203–213.
79. Morris, N.L.; Miller, P.C.H.; Orson, J.H.; Froud-Williams, R.J. The adaptation of non-inversion tillage systems in the United Kingdom and the agronomic impact on soil, crops and the environment—A review. *Soil Tillage Res.* **2010**, *108*, 1–15. [CrossRef]
80. Smagacz, J. Produkcyjno-Ekonomiczne i Środowiskowe Skutki Różnych Systemów Uprawy Roli. In *Problemy Zrównoważonego Gospodarowania w Produkcji Rolniczej*; Studia i Raporty IUNG-PIB: Puławy, Poland, 2012; Volume 2, pp. 121–133. ISBN 978-83-7562-111-2.
81. Huseyin Ozturk, H. An Assessment of Conventional and Conservation Tillage Systems in Terms of Carbon Dioxide Emissions in Corn Production. *Agric. Mech. Asia Afr. Lat. Am.* **2019**, *50*, 7–18.
82. IPCC 2020. Climate Change and Land. Summary for Policymakers. Available online: https://www.ipcc.ch/site/assets/uploads/sites/4/2020/02/SPM_Updated-Jan20.pdf (accessed on 13 December 2020).
83. Roman, M.; Górecka, A.; Domagała, J. The linkages between crude oil and food prices. *Energies* **2020**, *13*, 6545. [CrossRef]

energies

MDPI

Article

Determinants of Energy Cooperatives' Development in Rural Areas—Evidence from Poland

Jakub Jasiński [1,*], Mariusz Kozakiewicz [2] and Maciej Sołtysik [3]

1 Institute of Rural and Agricultural Development, Polish Academy of Sciences, 72 Nowy Świat St., 00-330 Warsaw, Poland
2 Collegium of Economic Analysis, Warsaw School of Economics, Madalińskiego 6/8 St., 02-513 Warsaw, Poland; mariusz.kozakiewicz@sgh.waw.pl
3 Faculty of Electrical Engineering, Częstochowa University of Technology, Armii Krajowej St. 17, 42-200 Częstochowa, Poland; maciej.soltysik@pcz.pl
* Correspondence: jjasinski@irwirpan.waw.pl

Citation: Jasiński, J.; Kozakiewicz, M.; Sołtysik, M. Determinants of Energy Cooperatives' Development in Rural Areas—Evidence from Poland. *Energies* **2021**, *14*, 319. https://doi.org/10.3390/en14020319

Received: 13 December 2020
Accepted: 6 January 2021
Published: 8 January 2021

Abstract: The strategies, plans and legislation on energy market development and decarbonization in the European Union (EU) developed in recent years, such as the directives implementing the package "Clean energy for all Europeans", aim at promoting not only renewable energy sources, but also new institutions that involve the development of local energy markets and a greater role for citizens in managing their own energy generation. At the same time, Poland remains the economy most dependent on coal and one of the largest air polluters in the EU. In order to minimize this problem and to meet the direction of energy development in the EU, Poland decided to establish, among other things, an energy cooperative. It is intended to fill the gap in the development of the civil dimension of energy on a local scale and at the same time improve efficiency in the use of the potential of renewable energy sources in rural areas. The authors of the paper seek to verify the extent to which this new institution, which is part of the idea of a local energy community, one of the driving forces for the implementation of the objectives and directions of development of "clean energy" set by the EU, has a chance to develop. The research took into account the characteristics of energy producers and consumers in rural areas, economic preferences provided for by law, relating to the functioning of an energy cooperative and the existing alternative solutions dedicated to prosumers. A dedicated mathematical model in the mixed integer programming technology was used to optimize the functioning of an energy cooperative, and more than 5000 simulations were carried out, with a typical optimization task performed as part of the research with about 50,000 variables. The conclusions and simulations make it possible to confirm the thesis that profitable energy cooperatives can be established in rural areas, with the objective of minimizing the sum of energy purchases from the distribution network and losses on the energy deposit (virtual network storage) (the energy deposit (or network deposit) should be understood as energy introduced to the grid during generation surpluses for its subsequent consumption, taking into account the discount factor).

Keywords: energy cooperatives; renewable energy sources; rural areas; renewable energy community; mixed integer programming

1. Introduction

Decentralization of large-scale energy, replacing it with pro-ecological, distributed generation sources and building the civil dimension of energy [1,2], are the objectives of the energy transformation in the European Union (EU) [3,4]. EU legislation does not impose a precise formula for achieving these objectives, providing the freedom to individual member states [5,6]. Building energy self-sufficiency at local level is possible on the basis of institutions called energy communities (EC) [7]. The first of these is the energy community defined in the REDII Directive (Renewable Energy Directive II) [8], focused on a renewable energy community (REC) [9]. The second form of activity is the citizens energy community (CEC) [10], introduced by the market directive [11]. Both these concepts serve the development of locally distributed energy, have a legal personality and are characterized by voluntary and open participation [12]. Their primary purpose is to provide economic and environmental benefits at local level. The main differences between the REC and the CEC [13] are that large and medium-sized enterprises are excluded from the CEC, whereas energy generation capacity must be available in an REC (it must be from a renewable energy source—RES) and there are requirements for investment [14].

The EU direction of energy market transformation has also been reflected in Polish law, where, similarly to the community provisions, two institutions were created that introduce the civil dimension of energy. These include energy clusters [15] and energy cooperatives—the latter, being the newest [16] form of support for distributed civil energy, are the subject of consideration by the authors of this paper. The activity of an energy cooperative may be conducted in the territory of a rural or urban-rural municipality or in no more than three such municipalities directly adjacent to each other. The energy cooperative relies on:

- the generation of electricity or biogas, or heat in RES systems;
- the balancing of the demand for the auxiliaries of the energy cooperative and its members.

The provision of opportunities to build local energy communities [14,17] on the basis of cooperatives [18,19] can be very important especially in rural areas [20]—including by developing the "smart villages" concept [21,22]. This is where the greatest potential exists for the use of renewable energy sources (including biomass and biogas), including those of a waste nature that are part of the characteristics of a circular economy [23,24]. Unfortunately, although solutions concerning energy cooperatives have been in place in Poland since mid-2019, by the end of 2020 no energy cooperative had yet been established [25]. The considerations of different options for the cost-effectiveness of creating an energy cooperative as a Polish response to the development of local energy communities in the EU [26] are the core of the research presented in the paper. However, it should be noted that it is not the purpose of this paper to assess the extent to which the institution of an energy cooperative meets Poland's obligation with respect to REC and CEC.

The main objective of our paper is to check where generation (number, type and capacity of installed sources) and consumption (energy demand) configurations the energy cooperative will be a cost-effective solution for the potential members of cooperatives. Thus an attempt will be made to answer the question of the production and consumer structure and the number of members that will be necessary for a form of self-organization such as an energy cooperative to develop. All restrictions and conditions resulting directly from the Polish law will be taken into account, and an analysis of the cost-effectiveness of establishing cooperatives in relation to the preferences that apply to individual prosumers will be analyzed. According to the authors, due, among other things, to different preferences for individual prosumers [27] and energy cooperatives (the so-called 1:0.8 or 1:0.7 discount rate for individual prosumers (this is a type of support scheme for prosumers in Poland: owners of micro-installations (capacity up to 50 kW) are allowed to exchange the surplus of energy produced under favourable conditions for gaps in energy production. The ratio is 1 to 0.8 for capacity up to 10 kW and 1 to 0.7 in the case of micro-installations between 10 and 50 kW) [28]) [29] and 1:0.6 for energy cooperatives [16]), the establishment of a cooperative will not always be economically justified. All the models of cooperatives proposed and researched in this paper, likewise data concerning energy production and consumption in households which are potential members of cooperatives, are anonymized real data from rural areas of Poland.

Our paper fills in a literature gap in the following ways. Firstly, our research, based on real data, allowed us to create several model energy cooperatives taking into account the conditions for the establishment and development of these institutions in Poland. Owing to this, the authors do not limit themselves to examining one specific example, but seek to show the most optimal model of operation of the citizens' energy community, which includes the energy cooperative. Secondly, our paper fills a methodological gap [30], as the methodological approach presented can easily be used in relation to other distributed energy institutions (especially the various energy communities), which have been and will be established in Europe and on other continents. Thirdly, it is novel at the national level as it investigates a problem not yet explored for Poland.

The contribution of the paper to scientific literature consists in proving that the energy communities' efficiency may be considered using optimization models in the mixed-integer programming technology, which are a basic tool for operational research applicable to decision-support processes in all areas of the economy. The paper also contributes to policy-related literature, because the economic effects of the functioning of energy communities are underestimated in decisions regarding the creation of such new institutions [30–32], where legal [33], administrative [34,35] and technological aspects prevail [36–38].

The paper is structured as follows. The second section provides the rules for the functioning of energy cooperatives and the main model assumptions arising from Polish law and those made by the authors. The next section describes the output data and their selection, as well as the mathematical optimization method used in the research process. Then the results are presented and discussed step by step. The final section presents the conclusions that have made it possible to conclude the entire research work and to present its universal character in relation to the methodology of examining the cost-effectiveness of creating institutions that are part of the idea of energy communities, which is gaining increasing recognition all over the world.

2. The Background of Energy Cooperatives

2.1. Energy Cooperatives as a Response to Renewable Energy Community (REC) and Citizens Energy Community (CEC)

As a country whose energy sector is still mainly dependent on coal [39], Poland is unlikely to meet the minimum 15% national target for renewable energy sources (RES) [40] in the final balance of energy consumption in 2020 (the European Union is committed to achieving 20% of energy from renewable sources in gross final energy consumption in 2020. All member states have individual targets to ensure that the EU plan is met. Poland has committed itself to achieving 15%. For comparison, Sweden has the highest level of commitment (49%) and Malta the lowest (10%) [41]) [42] (official EUROSTAT data will be available in 2022). One of the solutions enabling the acceleration of Poland's green transformation is the popularization of local energy communities and the resulting decentralization of energy. In 2019, the Polish government introduced the institution of an energy cooperative in order to meet the objectives of the EU directives on the development of RES and citizens' energy communities. Energy cooperatives are also intended to:

- increase the energy independence of rural areas;
- improve living and business conditions in rural areas, including increasing the competitiveness of the agri-food sector;
- increase the use of local renewable resources.

Rural areas make up more than 93% of Poland, with almost 40% of the country's population. The increase in energy demand in these areas, coupled with increased consumption by agriculture, is forcing rural people to use it more efficiently and politicians to develop energy security strategies for rural areas [43]. This is possible by creating a sustainable energy policy using renewable energy sources. Rural areas are largely associated with food production and processing, where agricultural holdings are important. They should now be seen on the one hand as energy users and on the other as producers of energy or final energy components based on renewable energy sources [44], including those concentrated and developed in energy cooperatives.

The definition of an energy cooperative appeared in Polish law during the amendment of the Act on Renewable Energy Sources. Pursuant to this, an energy cooperative is one within the meaning of the provisions of the Cooperative Law [45] and the Act on Farmers' Cooperatives [46] the object of which is to generate electricity, heat or biogas (exclusively for the auxiliaries of the energy cooperative and its members). Energy cooperatives may be established (conditions to be met cumulatively):

- in a rural or urban-rural municipality or in no more than three neighboring municipalities of this kind;
- in the area of operation of one distribution-system operator. The area of operation of an energy cooperative is to be determined on the basis of the places where the generators and consumers who are members of the cooperative are connected to the electricity distribution network or gas distribution network or district heating network;
- as part of low- and medium-voltage networks.

2.2. The Functioning and Billing Rules in an Energy Cooperative

Energy cooperatives operate on the basis of a prosumer system consisting of energy billing on the basis of "discounts". With the energy cooperative the energy seller (a licensed seller of a given type of energy, designated by the Energy Regulatory Office (https://www.ure.gov.pl/en) in a given area) accounts only for the difference between the amount of electricity fed into the power distribution network and the amount of electricity drawn from it for the auxiliaries of the cooperative (its members) in the ratio corrected by the quantitative coefficient 1 to 0.6 (in the case of prosumers, the coefficients 1 to 0.8 or 1 to 0.7 apply in Poland, depending on the system capacity). In other words, for one MWh of energy generated by the cooperative and not used at a given moment by its members, i.e., fed into the distribution network (the network in this situation operates as a deposit (storage) for energy unused by the cooperative), 0.6 MWh (600 kWh) of energy can be drawn from the distribution network. This may happen at any time during the billing period when the cooperative's generation sources do not cover the electricity demand generated by its members. This billing concerns electricity fed into and drawn from the distribution network by all electricity generators and consumers who are members of the energy cooperative. The same applies if heat or gas is the subject of the cooperative's operation.

Therefore, it is reasonable to assume that the more the members of the cooperative "synchronize" the amount of energy generated and consumed at any given time so as not to discharge energy surpluses into the network, the greater the economic effects of the energy cooperative will be. It can be said that in such a situation the distribution network will, in a way, only "protect" the internal energy economy of the cooperative. For this reason, the authors considered that the production mix should be optimized for the pre-set demand, with the minimization of energy not drawn from the network deposit and that additionally purchased from the network, resulting from possible shortages of energy generated inside the cooperative.

As a prosumer the energy cooperative operates in the power system under a comprehensive agreement with an external energy seller. This regulates both the distribution and sale of possible energy shortages to cooperatives. For an energy seller, an energy cooperative is a single, collective final consumer subject to single billing. For internal billing of an energy cooperative between its individual members, the seller indicates the amount of energy fed into and drawn from the network by its individual members. The cooperative accounts for them in accordance with internally accepted rules. The amount of unused energy remains to be taken (compensated) within a given billing period, which is usually 12 months from the last day of the month in which the surplus occurred. The energy cooperative does not pay the following fees for the amount of electricity thus billed:

- a fee for energy billing;
- a variable distribution fee;
- costs of commercial balancing.

These costs are covered by the energy seller as part of the value of energy at its disposal, i.e., 40% of the energy fed into the distribution network by the energy cooperative. (For the amount of electricity drawn from the network storage (with a coefficient of 0.6), the energy cooperative does not pay variable fees for the distribution service and does not pay the seller the billing fees. In addition, the RES, capacity and co-generation fee is not charged and collected for the amount of electricity generated in all RES systems of the energy cooperative, and subsequently consumed by all consumers of that cooperative, including the amount of energy billed. Energy cooperatives also do not have to obtain property rights arising from certificates of origin in order to redeem them, and fulfill the obligations relating to energy efficiency and capacity fees).

2.3. Legal Conditions and Assumptions for Energy Cooperatives

The institution of an energy cooperative is subject to a number of requirements and assumptions arising directly from the legislation [16], which must be met in order for an entity to be considered an energy cooperative. According to the authors, in order to better understand the subject of the analysis and the clarity of the arguments presented in the following sections, it is necessary to specify all requirements for the establishment and functioning of energy cooperatives:

1. Energy cooperatives may be established only in rural or urban-rural municipalities.
2. The total capacity of the cooperative's RES system must cover no less than 70% of its auxiliaries.
3. The maximum capacity generated by the energy cooperative is not to exceed 10 MW (30 MW for heat).
4. The maximum number of members is 999.
5. The energy cooperative generates electricity (as well as biogas or heat) exclusively for its auxiliaries and the auxiliaries of its members.
6. The cooperative discharges the surplus to the common distribution network. The billing of the provision and consumption of energy to and from the network is carried out in the system of discounts at the ratio of 1:0.6—i.e., with the possibility of recovery by the cooperative of 60% of previously produced (and unused) energy.
7. Individual prosumers may benefit from discounts at the ratio of 1 0.8 or 1:0.7, depending on the capacity of their sources.
8. The "external" balancing of cooperatives with the seller and the distribution system operator takes place during the annual billing period.
9. The "internal" balancing of energy between the members of the cooperative is carried out within one hour. From the sum of energy taken within an hour, the sum of energy fed in at the same time is subtracted. Thus for billing purposes only the result of this calculation is regarded as energy fed into or drawn from the network (depending on the result), while the rest is treated as self-consumption, which is not subject to the system of discounts or charges.
10. The internal billing model can be run for any period—e.g., from an hour to a year.
11. The difference in the amount of energy fed in or drawn out in the different phases is irrelevant, as the amount of energy is added to the net amount in one hour and is thus balanced. Single-phase and three-phase systems are treated the same.
12. The surplus of energy fed into the network in relation to the energy drawn out at a given moment is accumulated in the network deposit during the annual billing period. After 12 months, the stock is reduced to zero.

3. Materials and Methods (Optimization Model)

3.1. Assumptions for Creating a Sample of the Energy Cooperative for Simulation Purposes

In order to carry out the simulation and study the hypotheses, it was necessary to prepare several simulation scenarios that reflect the possible reality of energy cooperatives. Based on actual data on energy producers and consumers in rural areas in Poland, five types of energy cooperative were created for simulation. They were developed so as to reproduce different: (i) locational nature, (ii) level (scale) of electricity demand, (iii) nature of economic activity of the participants in the cooperative, (v) profile of electricity consumption of each member of the cooperative, (vi) generation potential among the members of the cooperative, (vii) level of supply voltage of the members of the cooperative and (viii) size.

The structure of energy cooperatives also takes account of formal and legal aspects resulting from the regulations in force. In particular, the location criterion for the allocation of members in up to three neighboring rural or rural-urban municipalities and low or medium voltage supply was maintained. The criterion for the selection of the generation structure by the optimizer took account at least 70% of the energy demand within the annual billing period and different types of generation.

Due to unfavorable hydrological conditions in Poland, which translate into stagnation in the construction of new hydro power plants [47], it was assumed that a maximum of one hydro power plant [48] may operate within the energy cooperative. The members of the cooperative were selected so that there was a watercourse in their municipalities that could be adapted for the construction of a small hydro power plant. A practical assumption was adopted, stating that a small hydro power plant is characterized by low capacity of between several dozen and several hundred kW. The simulation, therefore, took account of the capacity limits of a single source from 0 to 500 kW with increments of 50 kW. Discreet increments make the simulation realistic because a source with continuous capacity cannot currently be installed.

In Poland there are moderately favorable solar conditions, but the prosumer energy is practically based 100% on photovoltaic sources. The structure of photo-voltaic (PVPP) sources is currently the most popular and fastest-growing method to achieve energy self-sufficiency in Poland [49]. According to the data from the Ministry of Development, Labor and Technology [50], at the end of September 2020, there were about 357,000 micro-systems in Poland (increase of 35.5% compared to the end of the 2nd quarter of 2020 and as much as 131% compared to the end of 2019) [51]. The total PVPP capacity in the Polish power grid is about 3420 MW [52]. The dynamics of micro-systems growth is influenced by numerous aid programs (The Importance of Renewable Energy Sources in Poland's Energy Mix) [53]. The development of photovoltaic sources is also influenced by the economic aspect and the constantly decreasing unit cost of energy generation (LCOE) [54] for this type of source. In view of the above, for the simulations it was assumed that at least 25% of energy production of the members of the cooperative is from solar energy. In addition, capacity limits for a single PVPP farm from 0 to 1000 kW with increments of 50 kW were adopted.

Rural and rural-urban areas are very often undeveloped or have low-rise buildings. These factors support the construction of low-mast wind sources with low and medium capacity. The efficiency of wind generation is about twice as high for Polish wind conditions as for photovoltaic sources, which makes this type of generation attractive in terms of efficiency and cost [55]. For analysis and simulation, the possibility of cooperative participants establishing sources with a capacity from 0 to 1000 kW with increments of 250 kW was assumed.

The development of energy cooperatives must ensure that they are at least 70% self-sufficient in energy per annum and that they have a stable daily and hourly generation profile. The achievement of these indicators is determined not only by the level of installed capacity and the efficiency of generation, but also by its stability and the resultant consumer and generation profile. Taking into account the location criterion when establishing the cooperative was also intended to take advantage of the agricultural character and potential of the regions. In this context, it was assumed that members of the cooperative can also build generation sources with a stable generation profile based on biomass and biogas. For the simulation, the capacity both of biomass and biogas sources was limited to 0 to 600 kW with increments of 200 kW. The presence of generation sources of both stochastic (PVPP, wind) and stable (biomass, biogas) generation in energy cooperatives will result in flattening of the profile and reduction in generation differences between seasons of the year or times of day.

Due to the fact that the installation of new sources involves significant costs, which are not analyzed in the model presented in the paper, the assumption was made that for the optimal balance of demand in the cooperative one member has at most two energy generation sources, which does not exclude a situation where not all members have them and are thus energy producers. The discount nature of the operation of the energy cooperative and its members means that the loss of some energy on its introduction into the distributor's network and its subsequent consumption should be balanced by a slight increase in the installed capacity of the source. For the simulation, it was assumed that the total annual energy production of each member of the cooperative could not exceed

120% of the annual energy demand. This level ensures that each member of the cooperative is fully balanced at an individual level and allows for developing self-sufficiency at an aggregated cooperative level. The discount model is also characterized by the fact that the temporary production surplus fed into the network is continuously accumulated in the network in a follow-up manner, making it possible to use this energy and consume it during periods when the demand is not covered by the current generation.

3.2. Characteristics of Energy Cooperatives Adopted for Simulation Purposes

Energy cooperatives were established on the basis of current measurement data and profiles of consumers and generation for each type of renewable energy source. The purpose of selecting the participants of the cooperative was to reflect:

- the location character—the simulation was made for participants in two southern voivodeships (administrative divisions), Małopolskie and Śląskie, and the selection took account of different locations of municipalities within the voivodeships. The selection of two different voivodeships was also aimed at reflecting potentially different solar levels and thus the efficiency of generation.
- a different level of electricity demand—this resulted in cooperatives with demand ranging from 762 MWh/year to 9759 MWh/year. Within this criterion, participants were also selected taking account of the diversity of their individual energy demands. The cooperative included participants with negligible consumption, oscillating around one MWh/year, up to 3.5 GWh.
- the nature of participants' business activity—the selection of participants reflected the division in Polish law according to PKD codes (Polish Classification of Activities) relevant for typical agricultural activities, i.e., crop, vegetable, cereal production, raising of poultry, pigs and cattle as well as services for the agricultural sector. The complete classification is shown in Table 1.
- the electricity consumption profile of each member of the cooperative—the full range and variety of tariffs applicable in Poland—was taken into account, which may exist among the members of energy cooperatives. All analytical scenarios included entities belonging to one-, two- or three-zone tariffs, thus mapping the diverse nature of energy consumption. The shape of the profiles of the participants in the cooperatives is shown in Figure 1, and the belonging of differently profiled members to specific cooperatives in Table 1.
- the generation potential of members of the cooperatives—the selection of municipalities took account of the possibility of building renewable energy sources in each technology: wind, photovoltaic, biogas, biomass, water.
- the voltage supply of the members of the cooperatives—within each of the five cooperatives, the participants were consumers connected to the network at both medium and low voltages.
- the size—the aim was also to map cooperatives of different sizes, from 11 to 19 members.

Table 1. Characteristics of each analytical scenario.

	Cooperative 1	Cooperative 2	Cooperative 3	Cooperative 4	Cooperative 5
Voivodeship	Śląskie	Małopolskie	Małopolskie	Śląskie	Śląskie
Number of members of the cooperative	11	15	11	15	16
Profile of agricultural activity [1] and number of members (pcs)	01.46.Z; (3) 01.13.Z; (3) 01.47.Z; (5)	01.11.Z; (4) 01.13.Z; (1) 01.19.Z; (4) 01.47.Z; (3) 01.50.Z; (3)	01.11.Z; (2) 01.13.Z; (2) 01.19.Z; (1) 01.47.Z; (1) 01.50.Z; (5)	01.13.Z; (4) 01.19.Z; (1) 01.43.Z; (1) 01.47.Z; (3) 01.49.Z; (1) 01.50.Z; (5)	01.11.Z; (2) 01.13.Z; (2) 01.19.Z; (3) 01.47.Z; (6) 01.50.Z; (2) 01.62.Z; (1)
Voltage (LV/MV) and number of members (pcs)	LV (4) MV (7)	LV (10) MV (5)	LV (10) MV (1)	LV (8) MV (7)	LV (9) MV (7)

Table 1. *Cont.*

	Cooperative 1	Cooperative 2	Cooperative 3	Cooperative 4	Cooperative 5
Tariff group [2] and number of members (pcs)	C11 (2) C12a (1) C22b (1) B21 (1) B23 (6)	C11 (4) C12b (1) C21 (3) C22a (2) B21 (3) B23 (2)	C11 (4) C12a (1) C12b (2) C22a (2) C22b (1) B11 (1)	C11 (6) C21 (2) B21 (3) B23 (4)	C11 (6) C21 (2) C22a (1) B21 (1) B22 (2) B23 (4)
Energy demand [MWh/year]	9757	3559	762	3383	5922
Minimum, mean and maximum energy consumption by a member of the cooperative [MWh/year]	min.: 52 mean: 887 max.: 3574	min.: 0 mean: 237 max.: 1045	min.: 3 mean: 69 max.: 312	min.: 6 mean: 214 max.: 1258	min.: 1 mean: 328 max.: 1542

[1] Agricultural activity profile: 01.11.Z—Growing of cereals, leguminous crops and oil plants, for seeds, except rice; 01.13.Z—Growing of vegetables and melons, roots and tubers; 01.19.Z—Growing of other non-perennial crops; 01.43.Z—Raising of horses and other equines; 01.46.Z—Raising of pigs; 01.47.Z—Raising of poultry; 01.50.Z—Mixed farming; 01.62.Z—Support activities for farm-animal production. [2] Tariff group: The first character (C, B) refers to the tariff type, C—low voltage, B—medium voltage; The second character (1 or 2) refers to the installed capacity level, 1—up to 40 kW, 2—above 40 kW; The third character (1, 2 or 3) indicates the number of time zones; The fourth character, if any, indicates how to account for the time zones, a—division into peak and off-peak, b—division into day and night.

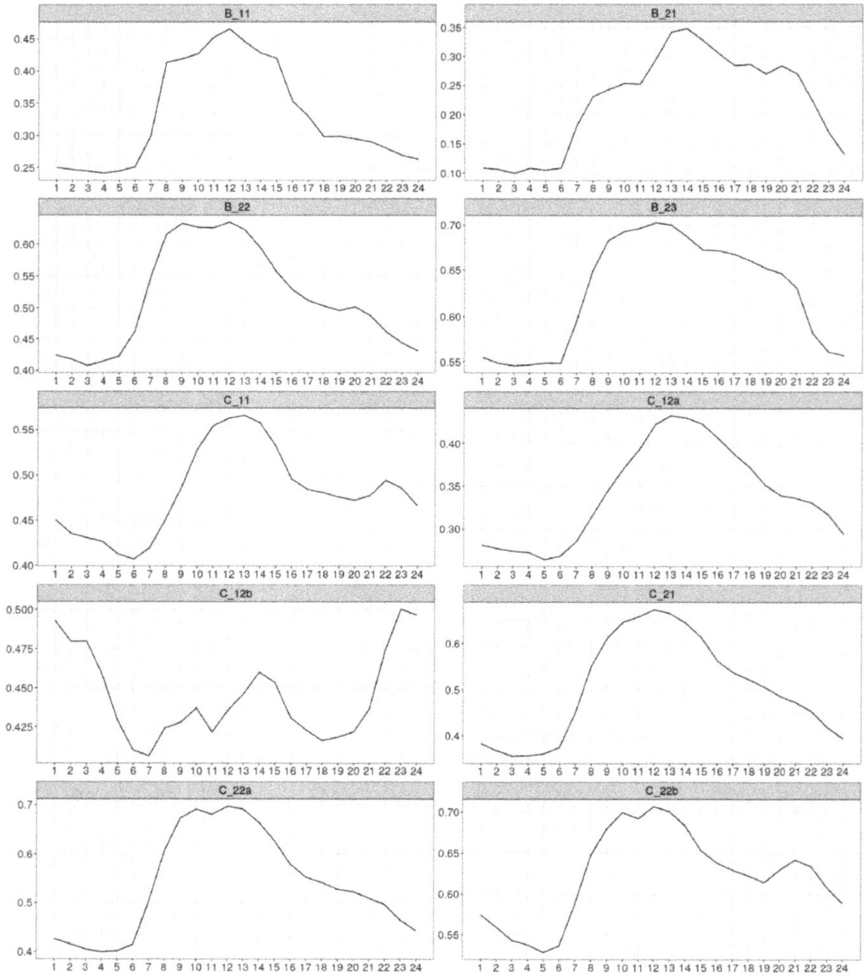

Figure 1. Average daily-hour profiles for each tariff.

The imposition of the above criteria made it possible to map the structures that combine all the above features in order to reproduce the real conditions for the establishment and operation of cooperatives as faithfully as possible and to examine the hypotheses formulated. Table 1 shows the characteristic and most important features of each of the cooperatives developed for the simulation.

3.3. Optimization Model

The paper's results were obtained on the basis of data from a simulation of a dedicated mathematical model (the mathematical model has been developed and included as Supplementary Materials (see "mip_model.pdf" and "mip_model.tex" files). The mixed integer programming technique [56] was used for modeling. GLPK software was used in particular for modeling the high-level GMPL language [57] (this is open-source software). The COIN-OR/CBC software [58] (also open-source) was used to solve the individual optimization tasks. The model's basic assumptions will be discussed below01 and fragments of the GMPL model will be illustrated.

The input data for the model comprised a two-year horizon data in hourly granulation. The calculation sessions used current data from several dozen consumers from different billing tariffs. The current two-year generation profiles of the following electricity sources were used: a small hydro power plant, a wind power plant, a photovoltaic power plant, wastewater- and biomass-based biogas plants.

Generally, the task of the optimization model was to select the optimum production mix for the pre-set demand, minimizing the energy not taken from the network deposit and purchased from the network. The energy demand, depending on the calculation scenario, was created by individual consumers or aggregated consumers within predefined cooperatives. The energy mix is to be understood as the vector of discrete factors scaling the generation profiles of the energy producers considered. The coordinates of this vector are fixed during the optimization period. The business process modeled connected the consumers with the sources on a proprietary basis (the consumer was the source owner/prosumer). Properly produced energy could be a discrete multiple of the profile adopted. Below is a relevant fragment of a mathematical model in GMPL language.

```
subject to def_ProductionMultiplier{e in EnergySources}:
  ProductionMultiplier[e] = ProductionDiscretization[e]*ProductionDiscretizationLevel[p];
subject to def_Production{e in EnergySources, h in Hours}:
  Production[h,e] = ProductionMultiplier[e]*ProductionProfile[h,e];
```

The ProductionMultiplier[e] is an integer variable defining a multiple of the standardized production of a specific source at hour h, i.e., ProductionProfile[h,e]. Total production of the source e is expressed by the variable Production[h,e].

The values of the total integer vector ProductionMultiplier[e] were concretized as part of the optimization task, for each energy source e. The Production[h,e] vector expressed the energy production by the source e at hour h. The energy produced could be consumed as part of own demand, or could be sent to the network to be recovered from it when needed, with a specific discount. In addition to using energy from self-consumption production and energy previously accumulated in a network deposit, the consumer could buy additional energy in the case of an absence of energy in the network deposit. The energy balance equation in the GMPL modeling language is presented below.

```
subject to def_EnergyBalance{h in Hours}:
  EnergyDemand[h]
  =
  BuyFromNetwork[h]
  +
  sum{e in EnergySources} Production[e,h]
  -
  SendToNetwork[h]
  +
  PickUpFromNetwork[h];
```

where EnergyDemand[h] is energy demand at hour h, BuyFromNetwork[h] is energy purchase at hour h, Production[h] is energy production at hour h, SendToNetwork[h] is energy sending at hour h, and PickUpFromNetwork[h] is energy collected at hour h.

The model assumes that it is not possible to send energy to the operator and take energy from it or collect energy previously sent at the same hour h. The relevant model equations are as follows:

```
subject to constr_SingleComponentFlow{h in Hours}:
SendToNetworkIndicator[h] + BuyFromNetworkIndicator[h] + PickUpFromNetworkIndicatorl[h] <= 1;
```

The variables SendToNetworkIndicator[h], BuyFromNetworkIndicator[h], PickUpFromNetworkIndicatorl[h] are binary variables indexed by the hours of the optimization horizon, which took the value 1 for non-zero values of the corresponding current variables and the value 0 for zero flows.

The equations modeling the operation of a network deposit of energy produced by prosumers and sent to the network for later recovery are as follows:

```
subject to def_EnergyStorage{h in Hours}:
  Storage[h] =
  if(h in StartsOfBillingPeriods) then
    0
  else
  (
    Storage[h-1]
    -
    PickUpFromNetwork[h]
    +
    Discount*SendToNetwork[h]
  );
```

where the set StartsOfBillingPeriods was a set of indexes containing the beginnings of billing periods—especially h = 1, i.e., the beginning of optimization, and Storage[h] is the state of the network energy deposit (storage) at hour h.

The optimization objective function was the sum of two components—energy taken from the network and energy produced but not consumed. Optimization was to minimize the following objective function.

```
minimize objective:sum{h in EndsOfBillingPeriods}
  Storage[h]
+
sum{h in Hours}
  BuyFromNetwork[h]
;
```

The set EndsOfBillingPeriods covered the last hours of billing periods.

The optimization covered a two-year horizon and the results are from the first year of optimization. This operation was aimed at avoiding the "end of the world problem"—this is manifested by non-intuitive results in the final optimization period (e.g., zero energy deposit (storage) states in final optimization intervals).

A typical optimization task carried out as part of the research covered approximately 50,000 variables, of which approximately 10,000 were total variables. The limitations were 150% of the number of variables, and the optimization task matrix comprised about 300,000 non-zero factors. Due to the volume of output data (many variables indexed by hours within a year), the analysis of the results of a single optimization session is a complex task. Without additional analytical tools, drawing conclusions from the results of thousands of optimization sessions is an impossible task. For this reason, an algorithm of regression trees [59] was used to analyze the profitability of a cooperative depending on the input parameters. Aggregates based on the correlation of weekly profiles of the cooperative members' demand, aggregates based on the correlation of their annual demand profiles, and types and capacity of installed sources were used to describe the cooperative's profit.

4. Results and Discussion

The following section presents the results of two series of experiments. The first examined the profitability of five specific cooperatives described in detail in Section 3. As part of the second series of experiments, which attempted to generalize the results, a total of 5000 cooperatives made up of a set of randomly selected participants were analyzed.

4.1. Hypothetical Cooperatives (Energy Cooperatives Adopted for Simulation)

Table 2 presents a summary of the most important information concerning farms that are prosumers or energy consumers and that have become members of the energy cooperatives established for the purpose of the research. The generation structure in prosumer farms was based on various renewable technologies and fuels. The average daily production level in prosumer farms varied between 13 and 100 kWh. It is worth noting that photovoltaic sources dominated the generation structure, whose production characteristics resulted in generation shortages at night and capacity surplus at noon The variability of the daily generation level of individual members of the cooperative ranged between 0 and 1075 kWh and the average daily energy demand was between 0 and 651 kWh.

Table 2. Information concerning farms (prosumers and energy consumers) before the establishment of the cooperative.

		Cooperative CP1-Members	Cooperative CP2-Members	Cooperative CP3-Members	Cooperative CP4-Members	Cooperative CP5-Members
Members		11	15	11	15	16
Capacity [kW]	PVPP	3810	1815	550	1635	2605
	SHPP	200	200	200	200	200
	WPP	3750	500	0	1000	1250
	BMPP	400	800	0	400	1600
	BGPP	600	200	0	400	600
Production	Total [kWh/year]	8,760,000	3,515,000	750,000	3,635,000	6,255,000
	Average [kWh/day]	100	33	13	35	47
	Max [kWh/day]	1075	394	120	685	891
	Min [kWh/day]	0	0	0	0	0
	Std. dev [kWh/year]	78	25	10	25	34
Consumption	Total [kWh/year]	9,757,041	3,558,868	761,883	3,383,286	5,921,660
	Average [kWh/day]	101	27	8	26	42
	Max [kWh/day]	651	234	66	285	366
	Min [kWh/day]	2	0	0	0	0
	Std. dev [kWh/year]	26	13	2	11	14

PVPP—Photovoltaic power plant; SHPP—Small hydro power plant; WPP—Wind power plant; BMPP—Biomass power plant; BGPP—Biogas power plant.

For each member of the five cooperatives analyzed, the optimization task was solved and the production mix, total energy consumption and energy loss within the network deposit—unused in the annual billing period—was determined. The simulation results for energy cooperatives and the aggregated results for the individual members are presented in Table 3.

Table 3. Simulation results for cooperatives compared to the results obtained by aggregating partial results of the individual farms.

		Consumption from the Network (Outside the Network Deposit) [kWh/Year]	Sending to the Network Deposit [kWh/Year]	Collection from the Network Deposit [kWh/Year]	Loss on the Network Deposit [kWh/Year]
Cooperative CP1	(1)	2,073,952	2,522,978	1,446,067	67,737
Cooperative CP1-Members	(2)	2,143,778	2,915,817	1,769,080	272,049
	(3) = (1) − (2)	−69,826	−392,839	−323,014	−204,313
	(4) = (3)/(2)	−3%	−13%	−18%	−75%
Cooperative CP2	(5)	487,893	1,091,142	647,116	7571
Cooperative CP2-Members	(6)	483,684	1,289,791	849,975	52,898
	(7) = (5) − (6)	4209	−198,650	−202,858	−45,327
	(8) = (7)/(6)	1%	−15%	−24%	−86%
Cooperative CP3	(9)	149,797	344,764	206,850	9
Cooperative CP3-Members	(10)	122,274	352,138	241,746	4750
	(11) = (10) − (9)	27,523	−7374	−34,896	−4741
	(12) = (11)/(10)	23%	−2%	−14%	−100%
Cooperative CP4	(13)	290,354	1,153,711	611,642	80,591
Cooperative CP4-Members	(14)	332,716	1,476,989	892,559	141,363
	(15) = (13) − (14)	−42,362	−323,278	−280,916	−60,773
	(16) = (15)/(14)	−13%	−22%	−31%	−43%
Cooperative CP5	(17)	430,477	1,787,588	1,023,770	48,835
Cooperative CP5-Members	(18)	469,204	2,034,755	1,232,211	192,167
	(19) = (17) − (18)	−38,727	−247,167	−208,440	−143,332
	(20) = (19)/(18)	−8%	−12%	−17%	−75%

The analysis of the calculation results leads to several key conclusions:

- In each of the cases of energy cooperatives analyzed, there is a significant reduction in the loss on the network deposit, i.e., the energy accumulated in it is used almost entirely within the set billing period. This is particularly evident in the case of Cooperative CP3, where at the end of the year there is only 9 kWh left in the deposit. If the cooperative were not established and its members were not accounted for on an individual basis, the total stock of their deposits would be 4750 kWh. Thus a nearly 100% reduction in lost volume was achieved. The worst results allow for a reduction in energy loss, as much as by 43% from 141,363 kWh to 80,591 kWh.
- In each of the cases of energy cooperatives analyzed, there is an increase in self-consumption, i.e., the consumption of electricity by the cooperative during which the electricity production occurs. This effect is shown in Table 4. The increase in self-consumption translated into a reduction in sending of energy to the network deposit and a reduction in the volume of energy drawn from the deposit. The reduction in the volume of energy fed into the deposit by the cooperative compared to the sum of individual use of the deposit by its members ranged from 2% to 22%. It should be emphasized that the energy fed into the network deposit is drawn from it, taking account of the discount, so minimizing the volume of energy that is sent there is a desirable phenomenon.
- In three out of five cases it was possible to parameterize the optimization model in order to achieve a reduction in the energy consumption from the network. In the case of cooperative CP2, no significant change in the amount collected was observed between the scenario of the aggregate of individual functioning of farms and the cooperative thereby established. The establishment of cooperative CP3 proved to be ineffective from this perspective, since the volume drawn from the network increased by 23%. The optimization objective set in the task, consisting in a minimization of the deposit loss and electricity consumption from the distribution network, is also very important from the perspective of aspects of economic rationality not analyzed within the framework of the paper, but extremely important. Energy consumption from the network outside the network deposit is billed each time at full purchase cost

including both the electricity component as a commodity and full distribution fees relating to its delivery. The loss of electricity in the deposit after the billing period is closed is of a similar nature.

- In order to present the economic benefits of the operation of energy cooperatives, Table 5 presents the results of analyses for the example cooperative CP4 (Source data, calculation formulas and results are available as Supplementary Materials—at public source file "Analysis_CP4.xls") taking account of the costs of energy, costs of distribution and power (capacity market) fee, broken down into: (i) costs incurred individually by farms, (ii) costs incurred by farms as prosumers, (iii) costs incurred by farms—members of cooperatives. The calculations were made based on the actual tariff rates [60]. Due to the complexity of the economic analyses, their complete picture is an area of separate analyses and publications conducted by the authors' team.

Table 4. Energy self-consumption at the level of the energy cooperative and the individual members.

	Cooperative CP1	Cooperative CP2	Cooperative CP3	Cooperative CP4	Cooperative CP5
Self-consumption in a cooperative [%]	71	69	54	68	71
Average level of self-consumption among members of an energy cooperative [%]	57	43	37	48	52

Table 5. Economic account results for energy cooperative (CP4).

Stage	Users	Energy Cost	Distribution Cost	Capacity Market	Total
		PLN	PLN	PLN	PLN
1	Farm1	193,958	64,035	14,923	272,916
	Farm2	592,559	195,632	45,591	833,782
	Farm3	118,430	18,588	10,730	147,748
	Farm4	148,143	23,252	13,422	184,817
	Farm5	97,014	15,227	8790	121,031
	Farm6	59,344	6591	4706	70,641
	Farm7	66,762	7414	5295	79,471
	Farm8	61,525	6833	4879	73,238
	Farm9	86,834	9644	6886	103,364
	Farm10	15,708	4928	1073	21,708
	Farm11	2826	887	193	3906
	Farm12	13,937	4372	952	19,260
	Farm13	3315	1040	226	4582
	Farm14	15,262	4788	1042	21,091
	Farm15	72,372	22,703	4942	100,017
	Total (1)	1,547,988	385,933	123,651	2,057,572
2	Prosumer1	15,920	5256	1162	22,338
	Prosumer2	21,844	7212	1595	30,650
	Prosumer3	16,426	2578	1658	20,662
	Prosumer4	24,614	3863	2367	30,843
	Prosumer5	8502	1334	757	10,594
	Prosumer6	11,567	1264	731	13,561
	Prosumer7	13,003	1412	851	15,266
	Prosumer8	12,237	1330	807	14,374
	Prosumer9	417	40	29	486
	Prosumer10	3174	996	186	4355
	Prosumer11	836	262	47	1145
	Prosumer12	2869	900	164	3932
	Prosumer13	1294	406	73	1773
	Prosumer14	3572	1121	210	4903
	Prosumer15	15,475	4854	861	21,190
	Total (2)	151,747	32,829	11,498	196,074
3	Member1	15,059	4546	1186	20,792
	Member2	46,006	13,890	3624	63,520
	Member3	12,637	1983	1185	15,805
	Member4	15,807	2481	1482	19,770
	Member5	10,352	1625	970	12,947
	Member6	5047	1061	383	6491
	Member7	5678	1187	431	7295
	Member8	5232	1114	397	6744
	Member9	7385	241	560	8186
	Member10	1349	423	96	1869
	Member11	243	76	17	336
	Member12	1197	375	86	1658
	Member13	285	89	20	394
	Member14	1311	411	94	1815
	Member15	6215	1950	444	8609
	Cooperative (3)	133,801	31,454	10,976	176,231
	(2) − (1)	−1,396,241	−353,104	−112,153	−1,861,497
	(3) − (1)	−1,414,187	−354,479	−112,675	−1,881,340
	(3) − (2)	−17,946	−1375	−522	−19,843

The economic analysis of the selected energy cooperative (presented in Table 5) proves the profitability of its establishment and operation. The analysis does not take account of the fixed distribution and settlement fees, as these are the same at each stage. Additionally, similar to the assumptions of the model, the analysis of investment outlays was not included. Investments in generation sources and the related cost depend not only on the technological solutions used, but also on many ways of financing. It is possible, for example, to obtain subsidies or loans under national and regional support schemes, to participate in RES auctions, to participate in the feed-in tariff or feed-in premium mechanisms. The parameterization of grants and loans is conditioned by many (frequently changing) factors—the above makes it impossible to carry out a synthetic analysis and include it in the article [61].

- Self-consumption can be considered as a parameter for optimal adaptation of the generation profile to the consumption profile. In the analytical scenario before the establishment of energy cooperatives, the average self-consumption in prosumer farms ranged from 37% to 57%. The establishment of the cooperative made it possible to achieve simultaneous generation and consumption at 54–71%.

The solution to the optimal task within the production resources held by its individual members consisted in determining for each cooperative the equivalent of total energy consumption from the network and total uncollected energy within the network deposit, as shown in Table 6.

Table 6. Evaluation of optimization results.

	Consumption from the Network (Outside the Network Deposit) [kWh/Year]	Loss on the Network Deposit [kWh/Year]	Subject of Optimization
	(1)	(2)	(3) = (1) + (2)
Cooperative CP1	2,073,952	67,737	2,141,689
Cooperative CP1-Members	2,143,778	272,049	2,415,827
Profitability of the Cooperative CP1 vs. Members scenario			11.3%
Cooperative CP2	487,893	7571	495,464
Cooperative CP2-Members	483,684	52,898	536,582
Profitability of the Cooperative CP2 vs. Members scenario			7.7%
Cooperative CP3	149,797	9	149,806
Cooperative CP3-Members	122,274	4750	127,024
Profitability of the Cooperative CP3 vs. Members scenario			−17.9%
Cooperative CP4	290,354	80,591	370,945
Cooperative CP4-Members	332,716	141,363	474,079
Profitability of the Cooperative CP4 vs. Members scenario			21.8%
Cooperative CP5	430,477	48,835	479,312
Cooperative CP5-Members	469,204	192,167	661,371
Profitability of the Cooperative CP5 vs. Members scenario			27.5%

It is worth noting that in the case of cooperatives CP1, CP2, CP4 and CP5, the profitability was of 7.7% to 27.5%. At −17.93% cooperative CP3 turned out to be unprofitable. This result was different to the authors' expectations and was the starting point for designing and conducting the second series of experiments.

4.2. *Random Cooperatives*

As part of this experiment, the profitability of cooperatives selected at random from the previously prepared data of about 100 prosumers was analyzed. As in the first experiment, the individual prosumers' data were prepared on the basis of current data. The composition of the cooperative was drawn from this set and then the optimal tasks for the individual members and the cooperative were solved. Cooperatives with 10, 20, 30, 40, 50 members were considered. The main result of the experiment is shown in Figure 2. A single point describes the profitability of the cooperative. On the horizontal axis, the total installed capacity of the cooperative and on the vertical axis, the profitability of the cooperative as a percentage value are provided. The experimental points corresponding to cooperatives with a certain number of members are marked in one color. Additionally, regression straight lines were applied to each such group of points.

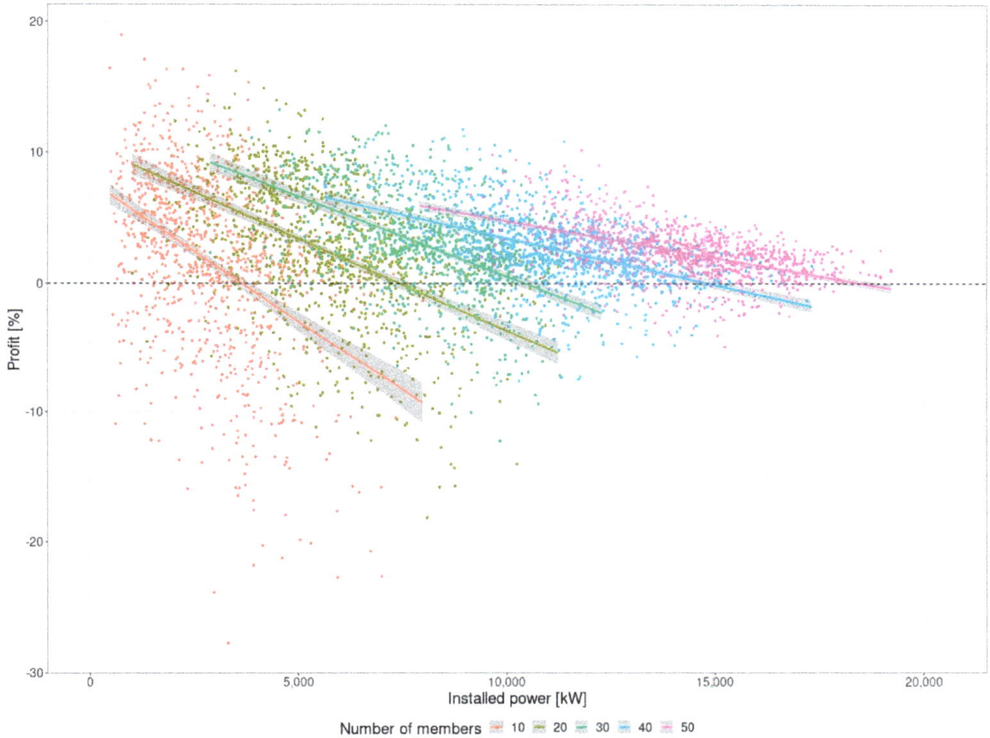

Figure 2. Visualization of the profitability of the energy cooperative by number of members.

The statistical results determining the level of profitability as percentages are presented in Table 7. The first column indicates the number of members of the cooperative. The following lines contain statistics corresponding to a cooperative with a certain number of members. The second column contains information on the minimum value of the profitability of the cooperative. This is followed by the 5%, 25%, 50%, 75%, 95% quantiles and the average profitability. The final columns comprise the maximum value, standard deviation and number of observations greater than zero.

Table 7. Statistics of the profitability of the cooperative depending on the number of members.

Number of Members	Min.	Quantile 5%	Quantile 25%	Median	Mean	Quantile 75%	Quantile 95%	Max	Standard Deviation	Greater than Zero
10	−27.70	−10.93	−1.63	2.42	1.73	6.59	11.18	18.98	6.86	683
20	−18.16	−7.51	−0.40	2.51	2.38	6.36	9.84	16.20	5.21	718
30	−12.21	−4.03	0.62	2.66	2.77	4.97	9.05	13.24	3.90	695
40	−7.63	−2.05	0.83	2.30	2.51	4.00	7.79	11.76	2.86	847
50	−4.95	−1.10	1.16	2.32	2.39	3.50	6.06	10.16	2.08	899

Table 7 shows that the average profitability of cooperatives with different numbers of members is similar and amounts to around 2%, but the risk of losses varies greatly. As the number of members increases, the respective profitability distributions become more concentrated—see Figure 3.

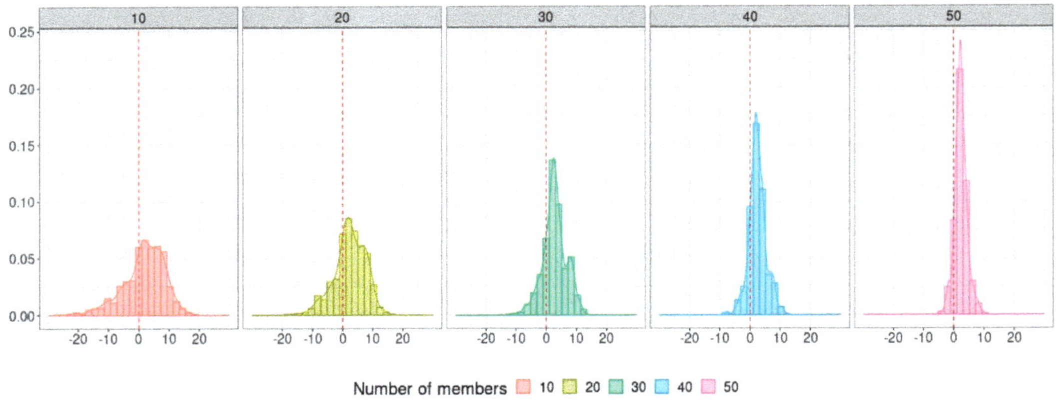

Figure 3. Distribution of the profitability of the cooperative by number of members.

Experimental data were analyzed using a regression-tree algorithm in order to detect the rules governing cooperatives' profitability. Separate trees were created for cooperatives with the same number of members. As variables describing profitability, the aggregates of correlation matrices of average weekly profiles in hourly granulation and average annual profiles in weekly granulation were assumed; information on the structure and capacity of sources installed were also used. The set of decision-making rules makes it possible to conclude that in order to maximize the profit of the cooperative:

- for small cooperatives, participants with a similar aggregated energy demand whose daily/weekly profiles are negatively correlated as much as possible should be selected;
- for larger cooperatives, aggregated energy demand may vary greatly, but the correlation of annual demand trends should be as close as possible to 1;
- diversification of generation sources should be pursued; optimal results were obtained for cooperatives which had a small hydro-power plant and a wind-power plant, but with a share in production not exceeding 30%.

It is worth noting that as the number of members increases the risk of losses (negative profitability) decreases. As the number of members increases, the minimum profitability increases and the standard deviation decreases.

5. Conclusions

The results of the research and simulations confirm that the profitability of energy cooperatives is highly dependent on the nature and supply and demand profile of its members. The analyses unequivocally confirm that the more numerous an energy cooperative in which the daily-hour profiles of its participants are maximally negatively correlated, the higher is the probability of positive profitability, understood as the minimization of the sum of energy consumption from the network and energy loss in deposit. However, positive profitability in such a scenario means lower profitability than would be the case for cooperatives with fewer members. The level of profitability is, therefore, limited, but the probability of achieving it increases.

The establishment of energy communities on the basis of energy cooperatives has not yet been practically analyzed in Poland. In the authors' opinion, this phenomenon is caused by the necessity to carry out a dedicated and non-trivial optimization analysis each time confirming the profitability of establishing a cooperative, which is proved by this research. However, the analyses suggest that in most cases there is a solution that guarantees the creation of a profitable cooperative. This is confirmed by the results of the simulation, in which the objective for the first group of simulations was achieved in four out of five cases. The second group of calculations each time concerned 1000 simulations, for each of the five variants with 10, 20, 30, 40 and 50 members of the cooperative. The results obtained—respectively 683, 718, 695, 847 and 899—also confirm this thesis.

It is worth noting that the legislative framework determining the functioning of energy cooperatives in Poland imposes relatively unfavorable boundary conditions for their development. In particular, the high discount rate of 1/0.6 is a limitation for the establishment of cooperatives. It limits their profitability in the context of the possibility of independent functioning of farms in the prosumer area and the achievement of discount rates of 1/0.7 or 1/0.8. It is also worth emphasizing that the negative impact of a high discount factor (ratio) can be limited by the physical storage of surplus energy and minimizing its flow through the operator's network. Work is currently underway in Poland to develop and launch support schemes for the construction of energy storage facilities, which will certainly have a positive impact on the increase in the profitability of energy cooperatives. In the authors' opinion, the legal solution in Poland and—as shown in the paper—giving measurable effects in most cases, may also be applicable and scalable in other EU countries obliged to build distributed and civil energy (corresponding to the Community framework imposed by institutions such as REC and CEC) [8,11].

In order to fully assess the profitability of cooperatives, financial aspects not covered by this paper (excepting the analysis for CP4—see Table 5) should also be taken into account, including both the amount of capital expenditure and operating costs. Market conditions, including current electricity prices and transmission and distribution fees, are also becoming crucial in this context—as well as energy storage (batteries) and e-mobility chargers. Due to their complexity, these elements will constitute an area of further exploration, research and publication by the authors' team.

Supplementary Materials: The data presented in this study can be found here, https://www.mdpi.com/1996-1073/14/2/319/s1.

Author Contributions: Conceptualisation, J.J., M.K., M.S.; methodology, M.K.; formal analysis, J.J., M.K., M.S.; investigation, J.J., M.K., M.S.; resources, M.S.; writing—original draft preparation, J.J., M.K., M.S.; writing—review and editing, J.J.; visualisation, M.K.; supervision, J.J., M.S. All authors have read and agreed to the published version of the manuscript.

Funding: This research received no external funding.

Conflicts of Interest: The authors declare no conflict of interest.

References

1. Horstink, L.; Wittmayer, J.M.; Ng, K.; Luz, G.P.; Marín-González, E.; Gährs, S.; Campos, I.; Holstenkamp, L.; Oxenaar, S.; Brown, D. Collective Renewable Energy Prosumers and the Promises of the Energy Union: Taking Stock. *Energies* **2020**, *13*, 421. [CrossRef]
2. Miller, C.A.; Iles, A.; Jones, C.F. The Social Dimensions of Energy Transitions. *Sci. Cult.* **2013**, *22*, 135–148. [CrossRef]
3. Clean Energy for All Europeans Package. Available online: https://ec.europa.eu/energy/topics/energy-strategy/clean-energy-all-europeans_en (accessed on 18 November 2020).
4. International Renewable Energy Agency; European Commission. *Renewable Energy Prospects for the European Union*; International Renewable Energy Agency (IRENA): Abu Dhabi, UAE, 2018; ISBN 978-92-9260-007-5.
5. Lüth, A.; Weibezahn, J.; Zepter, J.M. On Distributional Effects in Local Electricity Market Designs—Evidence from a German Case Study. *Energies* **2020**, *13*, 1993. [CrossRef]
6. Candelise, C.; Ruggieri, G. Status and Evolution of the Community Energy Sector in Italy. *Energies* **2020**, *13*, 1888. [CrossRef]
7. Gregg, J.S.; Nyborg, S.; Hansen, M.; Schwanitz, V.J.; Wierling, A.; Zeiss, J.P.; Delvaux, S.; Saenz, V.; Polo-Alvarez, L.; Candelise, C.; et al. Collective Action and Social Innovation in the Energy Sector: A Mobilization Model Perspective. *Energies* **2020**, *13*, 651. [CrossRef]
8. *Directive (EU) 2018/2001 of the European Parliament and of the Council of 11 December 2018 on the Promotion of the Use of Energy from Renewable Sources*; (OJ L 328, 21.12.2018); European Union: Brussels, Belgium, 2018.
9. Soeiro, S.; Dias, M.F. Renewable Energy Community and the European Energy Market: Main Motivations. *Heliyon* **2020**, *6*, e04511. [CrossRef]
10. Position Paper: Citizens Energy Communities: Recommendations for a Successful Contribution to Decarbonisation; Eurelectric—Dépôt Légal: D/2019/12.105/13. 2019. Available online: https://www.apren.pt/contents/publicationsothers/eurelectric--citizens-energy-communities.pdf (accessed on 28 December 2020).

11. *Directive (EU) 2019/944 of the European Parliament and of the Council of 5 June 2019 on Common Rules for the Internal Market for Electricity and Amending Directive 2012/27/EU*; (OJ L 158, 14.6.2019); European Union: Brussels, Belgium, 2019.
12. Sokołowski, M.M. Renewable and Citizen Energy Communities in the European Union: How (Not) to Regulate Community Energy in National Laws and Policies. *J. Energy Nat. Resour. Law* **2020**, *38*, 289–304. [CrossRef]
13. Report: Energy Community Definitions; H2020-824424 COMPILE: Integrating Community Power in Energy Islands. 2019. Available online: https://www.compile-project.eu/wp-content/uploads/Explanatory-note-on-energy-community-definitions.pdf (accessed on 28 December 2020).
14. Ghiani, E.; Giordano, A.; Nieddu, A.; Rosetti, L.; Pilo, F. Planning of a Smart Local Energy Community: The Case of Berchidda Municipality (Italy). *Energies* **2019**, *12*, 4629. [CrossRef]
15. Tomaszewski, K. Klastry energii w Polsce—W kierunku poprawy bezpieczeństwa energetycznego. *Koncepcje i uwarunkowania* **2020**, *4/2019*, 68–78. (In Polish)
16. Ustawa z Dnia 19 Lipca 2019 r. o Zmianie Ustawy o Odnawialnych Źródłach Energii Oraz Niektórych Innych Ustaw (Act of 19 July 2019 Amending the Act on Renewable Energy Sources and Certain Other Acts) (DZ.U. z 2019 r. Poz. 1524). Available online: http://isap.sejm.gov.pl/isap.nsf/DocDetails.xsp?id=WDU20190001524 (accessed on 28 December 2020).
17. Dóci, G.K. Renewable Energy Communities: A Comprehensive Study of Local Energy Initiatives in the Netherlands and Germany. Ph.D. Thesis, Vrije Universiteit Amsterdam, Amsterdam, The Netherlands, September 2017.
18. Debor, S. The Socio-Economic Power of Renewable Energy Production Cooperatives in Germany: Results of an Empirical Assessment. *Wupp. Pap.* **2014**, *187*, 25.
19. Neagu, B.-C.; Ivanov, O.; Grigoras, G.; Gavrilas, M.; Istrate, D.-M. New Market Model with Social and Commercial Tiers for Improved Prosumer Trading in Microgrids. *Sustainability* **2020**, *12*, 7265. [CrossRef]
20. Future of Rural Energy in Europe (FREE) Initiative. Rural Energy Matters. Report and Recommendations for Policymakers. 2016. Available online: https://www.rural-energy.eu/wp-content/uploads/2018/10/Summary-Report-Rural-Energy-Matters.pdf (accessed on 28 December 2020).
21. European Commission Smart Villages and Renewable Energy Communities. Available online: https://enrd.ec.europa.eu/sites/enrd/files/enrd_publications/smart_villages-capacity_tools-renewable_energy_communities-v08.pdf (accessed on 21 December 2020).
22. Watson, J.K.R. Energy Diversification and Self-Sustainable Smart Villages. In *Smart Villages in the EU and beyond*; Visvizi, A., Lytras, M.D., Mudri, G., Eds.; Emerald Publishing Limited: Bingley, UK, 2019; pp. 99–109. ISBN 978-1-78769-846-8.
23. Safari, M.A.M.; Masseran, N.; Jedi, A.; Mat, S.; Sopian, K.; Bin Abdul Rahim, A.; Zaharim, A. Rural Public Acceptance of Wind and Solar Energy: A Case Study from Mersing, Malaysia. *Energies* **2020**, *13*, 3855. [CrossRef]
24. Agnoletti, M.; Baptista, R.; Henriques, R.; Nogueira, P.; Costa Pinto, L.M.; Ramísio, P.J.; Santoro, A.; Teixeira, J.C.; Vaz, E. A Green City: Impossible Dream or Necessity. In *Environmental History in the Making*; Joanaz de Melo, C., Vaz, E., Costa Pinto, L.M., Eds.; Environmental History; Springer International Publishing: Cham, Switzerland, 2017; Volume 7, pp. 355–376. ISBN 978-3-319-41137-8.
25. Wykaz Spółdzielni Energetycznych (List of Energy Cooperatives). Available online: http://www.kowr.gov.pl/odnawialne-zrodla-energii/spoldzielnie-energetyczne/wykaz-spoldzielni-energetycznych (accessed on 18 November 2020).
26. Proposal for a Directive of the European Parliament and of The Council on Common Rules for the Internal Market in Electricity (Recast) COM(2016) 864 Final/2—2016/0380(COD). Available online: https://eur-lex.europa.eu/legal-content/EN/TXT/?uri=CELEX%3A52016PC0864R%2801%29 (accessed on 11 November 2020).
27. Description of Development of Prosumer Energy Sector in Poland. *Polityka Energetyczna-Energy Policy J.* **2018**. [CrossRef]
28. Krzysztof Ignaciuk—RES LEGAL Europe Legal Sources on Renewable Energy (Promotion in Poland—Summary of Support Schemes). Available online: http://www.res-legal.eu/search-by-country/poland/tools-list/c/poland/s/res-e/t/promotion/sum/176/lpid/175/ (accessed on 18 November 2020).
29. Ustawa z Dnia 20 Lutego 2015 r. o Odnawialnych Źródłach Energii (Act of February 20, 2015 on Renewable Energy Sources) (Dz.U. 2020 Poz. 261). Available online: http://isap.sejm.gov.pl/isap.nsf/DocDetails.xsp?id=wdu20150000478 (accessed on 28 December 2020).
30. Hoppe, T.; Coenen, F.H.J.M.; Bekendam, M.T. Renewable Energy Cooperatives as a Stimulating Factor in Household Energy Savings. *Energies* **2019**, *12*, 1188. [CrossRef]
31. Sifakis, N.; Savvakis, N.; Daras, T.; Tsoutsos, T. Analysis of the Energy Consumption Behavior of European RES Cooperative Members. *Energies* **2019**, *12*, 970. [CrossRef]
32. Grève, Z.D.; Bottieau, J.; Vangulick, D.; Wautier, A.; Dapoz, P.-D.; Arrigo, A.; Toubeau, J.-F.; Vallée, F. Machine Learning Techniques for Improving Self-Consumption in Renewable Energy Communities. *Energies* **2020**, *13*, 4892. [CrossRef]
33. Campos, I.; Pontes, L.G.; Marín-González, E.; Gährs, S.; Hall, S.; Holstenkamp, L. Regulatory challenges and opportunities for collective renewable energy prosumers in the EU. *Energy Policy* **2020**, *138*. [CrossRef]
34. Lowitzsch, J.; Hoicka, C.E.; van Tulder, F.J. Renewable Energy Communities under the 2019 European Clean Energy Package—Governance Model for the Energy Clusters of the Future? *Renew. Sustain. Energy Rev.* **2020**, *122*, 109489. [CrossRef]
35. Wagemans, D.; Scholl, C.; Vasseur, V. Facilitating the Energy Transition—The Governance Role of Local Renewable Energy Cooperatives. *Energies* **2019**, *12*, 4171. [CrossRef]

36. Sikorski, T.; Jasiński, M.; Ropuszyńska-Surma, E.; Węglarz, M.; Kaczorowska, D.; Kostyla, P.; Leonowicz, Z.; Lis, R.; Rezmer, J.; Rojewski, W.; et al. A Case Study on Distributed Energy Resources and Energy-Storage Systems in a Virtual Power Plant Concept: Technical Aspects. *Energies* **2020**, *13*, 3086. [CrossRef]
37. Giordano, A.; Mastroianni, C.; Menniti, D.; Pinnarelli, A.; Sorrentino, N. An Energy Community Implementation: The Unical Energy Cloud. *Electronics* **2019**, *8*, 1517. [CrossRef]
38. Giordano, A.; Mastroianni, C.; Scarcello, L. Optimization Model for IoT-Aware Energy Exchange in Energy Communities for Residential Users. *Electronics* **2020**, *9*, 1003. [CrossRef]
39. Eurostat: Renewable Energy Statistics in the EU in 2019. Available online: https://ec.europa.eu/eurostat/statistics-explained/index.php?title=Renewable_energy_statistics (accessed on 22 December 2020).
40. UE Blisko Celu OZE, Polsce Sporo Brakuje (The EU Is Close to the RES Target, Poland Is Far from Achieving It). Available online: https://www.cire.pl/item,209253,1,0,0,0,0,0,ue-blisko-celu-oze-polsce-sporo-brakuje.html (accessed on 22 December 2020).
41. Directive 2009/28/EC of the European Parliament and of the Council of 23 April 2009 on the Promotion of the Use of Energy from Renewable Sources and Amending and Subsequently Repealing Directives 2001/77/EC and 2003/30/EC (Text with EEA Relevance) (OJ L 140, 5.6.2009). Available online: https://eur-lex.europa.eu/legal-content/EN/TXT/PDF/?uri=CELEX:32009L0028&from=EN (accessed on 28 December 2020).
42. Raport "Rozwój Sektora Odnawialnych Źródeł Energii" (Report: "Development of the Renewable Energy Sector") Najwyzsza Izba Kontroli 171/2017/P/17/020/KGP. Available online: https://www.nik.gov.pl/kontrole/wyniki-kontroli-nik/pobierz,kgp~{}p_17_020_201709181219251505737165~{}01,typ,kk.pdf (accessed on 17 November 2020).
43. Piwowar, A.; Dzikuć, M. Development of Renewable Energy Sources in the Context of Threats Resulting from Low-Altitude Emissions in Rural Areas in Poland: A Review. *Energies* **2019**, *12*, 3558. [CrossRef]
44. Woźniak, M. Sustainable energy economy in rural areas in Poland (Zrównoważona gospodarka energetyczna na obszarach wiejskich w Polsce). *Polityka Energ.* **2018**, *T. 21*, 69–84.
45. Ustawa z Dnia 16 Września 1982 r.—Prawo Spółdzielcze (Act of September 16, 1982—Cooperative Law) (Dz. U. z 2018 r. Poz. 1285 Oraz z 2019 r. Poz. 730, 1080, 1100). Available online: https://isap.sejm.gov.pl/isap.nsf/DocDetails.xsp?id=WDU19820300210 (accessed on 28 December 2020).
46. Ustawa z Dnia 4 Października 2018 r. o Spółdzielniach Rolników (Act of October 4, 2018 on Farmers' Cooperatives) (Dz. U. Poz. 2073). Available online: https://isap.sejm.gov.pl/isap.nsf/DocDetails.xsp?id=WDU20180002073 (accessed on 28 December 2020).
47. Energy Regulatory Office Narodowy Potencjał OZE w Liczbach—Stan na 31 Grudnia 2019 r. (The Domestic Renewable Energy Potential in Numbers as of 31 December 2019). Available online: https://www.ure.gov.pl/pl/oze/potencjal-krajowy-oze/5753,Moc-zainstalowana-MW.html (accessed on 18 November 2020).
48. Paska, J. Polish Hydropower Resources and Example of Their Utilization. *Przegląd Elektrotechniczny* **2020**, *1*, 3–7. [CrossRef]
49. Księżopolski, K.; Drygas, M.; Pronińska, K.; Nurzyńska, I. The Economic Effects of New Patterns of Energy Efficiency and Heat Sources in Rural Single-Family Houses in Poland. *Energies* **2020**, *13*, 6358. [CrossRef]
50. Ministerstwo Rozwoju, Pracy i Technologii Energetyka Prosumencka i Rozproszona (Prosumer and Distributed Energy). Available online: https://www.gov.pl/web/rozwoj-praca-technologia/energetyka-prosumencka-i-rozproszona (accessed on 12 November 2020).
51. Gramwzielone.pl Sp. z o. o. The Number of Prosumers in Poland Exceeds 357,000. It's Time for Energy Storage. Available online: https://www.gramwzielone.pl/energia-sloneczna/104268/liczba-prosumentow-w-polsce-przekracza-357-tys-pora-na-magazyny-energii (accessed on 28 December 2020).
52. Polish Press Agency. Polish Solar Farm Has Exceeded the Level of 3400 MW of Capacity. Available online: https://www.wnp.pl/budownictwo/polska-fotowoltaika-przekroczyla-poziom-3400-mw-mocy,440020.html (accessed on 28 December 2020).
53. Marks-Bielska, R.; Bielski, S.; Pik, K.; Kurowska, K. The Importance of Renewable Energy Sources in Poland's Energy Mix. *Energies* **2020**, *13*, 4624. [CrossRef]
54. Ministry of Energy Polityka Energetyczna Polski do 2040 r.: Wnioski z Analiz Prognostycznych dla Sektora Paliwowo—Energetycznego (Poland's Energy Policy Until 2040: Conclusions from Prognostic Analyzes for the Fuel and Energy Sector). Available online: https://www.gov.pl/attachment/cff9e33d-426a-4673-a92b-eb4fb0bf4a04 (accessed on 17 November 2020).
55. Dolega, W. Problems, Barriers and Perspectives of RES Development in Poland. In *Engineering and Industry*; Mavromatakis, F., Siderakis, K., Eds.; Trivent Publishing: Budapest, Hungary, 2016; Volume 3, ISBN 978-615-80340-7-4.
56. Williams, H.P. *Model Building in Mathematical Programming*, 5th ed.; Wiley: Hoboken, NJ, USA, 2013; ISBN 978-1-118-44333-0.
57. GLPK (GNU Linear Programming Kit). Available online: https://www.gnu.org/software/glpk/ (accessed on 18 November 2020).
58. COIN-OR/CBC. Available online: https://projects.coin-or.org/Cbc (accessed on 18 November 2020).
59. Luger, G.F. *Artificial Intelligence: Structures and Strategies for Complex Problem Solving*, 6th ed.; Pearson Addison-Wesley: Boston, MA, USA, 2009; ISBN 978-0-321-54589-3.
60. TAURON Dystrybucja S.A. Taryfa dla Energii Elektrycznej na Rok 2020 (Tariff for Electricity for 2020). Available online: https://www.tauron.pl/-/media/offer-documents/firma/taryfa-msp/gze-podstawowe/taryfa_dystrybucyjna.ashx (accessed on 28 December 2020).
61. Wróbel, J.; Sołtysik, M.; Rogus, R. Selected Elements of the Neighborly Exchange of Energy—Profitability Evaluation of the Functional Model. *Polityka Energ. Energy Policy J.* **2019**, *22*, 53–64. [CrossRef]

energies

Article

Complex Valuation of Energy from Agricultural Crops including Local Conditions

Václav Voltr [1,*], Martin Hruška [1] and Luboš Nobilis [2]

[1] Institute of Agricultural Economics and Information, 120 00 Praha 2, Czech Republic; hruska.martin@uzei.cz
[2] ECO Trend s.r.o., 147 00 Praha 4, Czech Republic; nobilis@ecotrend.cz
* Correspondence: voltr.vaclav@uzei.cz; Tel.: +420-60-773-7414

Abstract: This paper provides values of economic, energy and environmental assessments of 20 crops and assesses the relationships of soil-climatic conditions in the example of the Czech Republic. The comparison of main soil quality indicators according to the configuration of land and climate regions is performed on the basis of energy and economic efficiency as well as a comparison of the level of environmental impacts. The environmental impacts are identified based on the assessment of emissions from production and also in the form of soil compaction as an indicator of the relationship to soil quality. As concerns soil properties, of major importance is soil skeleton, slope of land and the depth of soil, which cause an increase in emissions from the energy produced. Substantially better emission parameters per 1 MJ through energy crops, the cultivation of perennial crops and silage maize has been supported. Among energy crops, a positive relationship with the quality of soil is seen in alfalfa, with a significant reduction in soil penetrometric resistance; energy crops are also politically justifiable in competition with other crops intended for nutrition of population. The main advantage of energy crops for the low-carbon economy is their CO_2 production to MJ, which is almost half, especially in marginal areas with lower soil depths, slopes and stoniness, which can be included in the new agricultural policy.

Keywords: energy crops; gross margin; local conditions; climate; soil; modeling; LCA

Citation: Voltr, V.; Hruška, M.; Nobilis, L. Complex Valuation of Energy from Agricultural Crops including Local Conditions. *Energies* **2021**, *14*, 1415. https://doi.org/10.3390/en14051415

Academic Editor: Vitaliy Krupin

Received: 31 December 2020
Accepted: 27 February 2021
Published: 4 March 2021

1. Introduction

The relationship between food production, energy and the environment is currently an essential issue faced by agriculture [1]. Soil as a means of crop production is subject to many relationships associated with nutrition of population, environmental cleanliness as well as the need to ensure a sustainable source of energy [2]. The problem is escalating due to the necessity to secure food for the growing population [3–5], while responding to changes in farming conditions as a result of climate change [6,7]. A need arises to more accurately specify the production that will be politically justifiable. It turns out that the priority for political decision-making is the food and nutrition security of the population, but a wide-ranging discussion has emphasized the additional potential of energy generation from agricultural products. The entire process has not been adequately specified as yet due to insufficient knowledge of the context of agricultural production with regard to the referred to aspects and diverse conditions [8]. Discussions have been opened up on the use of straw for energy purposes [9], but there are also other matters to be addressed apart from the sufficient volume of production, which also concern the organic matter in soil [10,11].

Agricultural production has diverse impacts on the environment, economy and energy production depending on the relevant conditions [12]. Determination of soil and climatic conditions for crop production and their impact on energy and the environment are crucial for drafting the supporting documents for the purpose of analyzing the relationship between the energy production possibilities and the environment as well as for agricultural policy-making. The existing data provide good quality information regarding the individual crops in the form of a case study helping to identify mutual relationships. However, a systematic overview of crop production as a whole, including energy production and links to the environment, is missing. The submitted paper uses the comprehensive information on land resources in the Czech Republic and draws up mutual interactions of economic and energy nature that are dependent on the main characteristics of climatic and soil conditions in the Czech Republic. The paper determines the energy and economic margin based on the cost–benefit analysis by crop production technology, and also a comprehensive relationship to the environment based on the life cycle assessment (LCA) analysis for a total of 18 impacts in line with the midpoint analysis. Matters of evaluation of production with respect to decision-making on support, location and potential of agricultural crop production also for energy generation are addressed by numerous recent publications [13–15]. Multiple issues arise that are difficult to resolve by the agricultural policy unless the main economic, energy and environmental context is known well in advance [8].

These matters are extensively covered by the literature. Nonetheless, for the evaluation of all relationships, no comprehensive data are available on the entire territory and production structure. These issues are therefore mostly reflected only with respect to the production of one or more crops based on a model solution or a case study [14].

All the explored parameters, namely the economic, energy as well as environmental parameters, should be subjected to a comprehensive analysis in order to find the optimal future use of crops. Razm [16] in his production assessment used the LCA model in order to achieve the Pareto optimality in assessing the environmental and social impacts of crop production. This procedure can also be applied to seeking the optimal production when making decisions on the use of biofuels.

Some countries have searched for the missing framework for crop production assessment—e.g., development of supporting documents for agricultural crop production in Denmark should be based on a review and an assessment of publicly available databases, inventory reports and scientific literature on measures in the field of governance and their effectiveness with respect to legislation, agreements, conventions and standardizations (Bentsen [17]). The main reason behind this is the necessity to promote the environmental sustainability represented by greenhouse gas emissions from the agricultural sector, soil carbon sequestration, water quality, and biological diversity.

The bioeconomy plays an important role in replacing fossil fuels and is the key factor for sustainability. Wohlfahrt [8] stresses the socio-ecological concept of its exploration, the importance of knowledge of individual territories, flexibility of business activities of subsystems and local regulatory instruments. This justifies the necessity to develop an integrated model approach with various subsystems and heterogeneity. This builds on the assessment of agricultural land composition and its configuration.

2. Materials and Methods

The energy plan of the Czech Republic provisionally estimates that with a decrease in the production of biofuels from 11,093 TJ in 2020 to 9276 TJ in 2030—i.e., a decrease by approximately 15%—that biogas production should fall from 22,856 to 20,166 TJ—i.e., by 12% [18]. This plan states that the value of agricultural production is very uncertain in the future and depends mainly on the setting of the rules of the Common Agricultural Policy (CAP). Further development of the trend is based on a careful evaluation of all aspects of energy production, including in terms of the function of energy production in the landscape.

Overall energy effectiveness of production is conditional on the choice of crops in the given location, while respecting local conditions which depend on the particular type of farming of agricultural holdings and may vary according to the needs of animal production. Fundamental studies necessary to derive energy indicators have been addressed by a number of authors [19–21], and data for the Czech Republic from the IAEI survey and Preininger [22] were used in this analysis.

In the Czech Republic, a permanent monitoring system of agricultural production was developed based on the evaluation of evaluated soil ecological units (ESEUs, in the Czech Republic called BPEJ, introduced in the Czech Republic in 1970). The system is based on the classification of climatic factors within the climatic region (Appendix C, Table A2, the main soil unit (MSU (HPJ)), describing the main pedological characteristics of soil, and on the description of terrain configuration: slope (°), deep of the soil (cm), stoniness (%), and by the area in the Czech Republic [23]. MSUs are laid down in a decree [24], but more information is provided by the tracking of the Research Institute for Soil and Water Conservation [25]. In the Czech Republic, a total of 2199 ESEUs have been classified that are mutually compatible throughout the territory of the Czech Republic. Based on the definition of these units, a paper was elaborated in order to cast more light on the links between the production and soil-climatic conditions—e.g., the modification of economic indicators based on production functions [26]. Even though the evidence of ESEU is domestic, the obtained generalized information can also be used for assessments in other countries.

Environmental indicators are essential to assess any production. The LCA assessment of crop growing impacts is described, e.g., by [27–30], and preparations are carried out for individual evaluation of technologies for the size of emissions, especially CO_2 [31].

This study is based on the values included in the Agri-footprint 4.0 database [32] and impact categories of the ReCiPe method were used. Model processes, based on Agri-footprint database processes, were modified on the basis of specific data for the Czech Republic. The adequacy of the modified processes was verified by comparison with the results of the original Agri-footprint processes and the Ecoinvent 3 database.

The data included in the national database of soil economic information were used to set the main yields and inputs and are subsequently updated in line with the soil and climatic conditions in the Czech Republic. The statistical survey is based on a sample cost survey of approximately 250 agricultural enterprises and the results are processed according to the IAEA methodology [33,34].

Information on crop yields and costs on individual soil-climatic conditions was used to calculate total emissions in individual categories according to the Agri-footprint database [32] and to calculate crop production in MJ. The resulting ratio was assessed against the description of soils in the Czech Republic [35].

The current papers add more information on ESEU and thus offer a comprehensive picture of mutual inter-relationship of economic, energy and environmental aspects [36–38]. Figure 1 shows a comprehensive monitoring system based on available data, which are validated and specified against individual ESEUs. The diagram shows the sources for processing the economic and energy data of crop yields, technologies and their costs as well as the composition of crops, including links to the calculation of emissions from Agri-footprint data.

The main scheme of calculation is given in Appendix A. It is used to calculate the economic, energy and environmental data.

The basic approach consists in the cost–benefit analysis of production of individual crops.

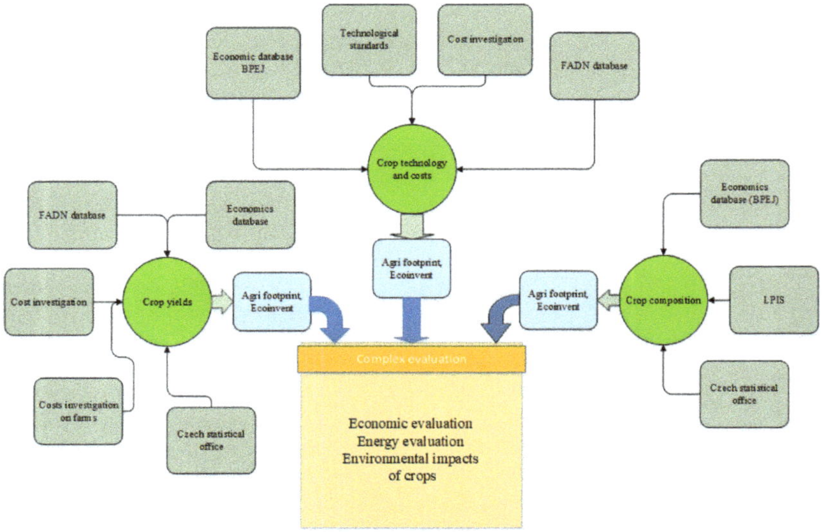

Figure 1. Processing of economic, energy and environmental data for calculation.

2.1. Method to Determine the Economic and Energy Values

When assessing the production, the indicators of economic and energy gain (Gross margin including overheads) were determined based on the production value once the necessary operating and overhead costs were deducted [39].

$$GMo_{i,p} = SO_{i,p} - COGSo_{i,p} \quad (1)$$

where: $GMo_{,i,p}$ is a Gross Margin with overheads for crops p and soil-climatic conditions i, $COGSo$ = cost of goods sold including overheads (EUR·ha^{-1}) and SO is a standardized output of crop products (EUR·ha^{-1}). More details are provided in Appendix B.

This procedure was opted for due to the need to calculate the total costs of production for the purpose of assessing the economics of farms according to individual crops by the Institute of Agricultural Economics and Information (IAEI) and it is reflected in all the supporting documents [40].

The overheads are derived from the economic data on agricultural holdings ascertained by the IAEI survey and, with respect to energy, the same value was used as the costs of working operations in in the overhead costs to direct variable crop production cost ratios. The reason thereof is primarily the burdening of production of some crops (e.g., potatoes) with high overhead costs of postharvest treatment and storage.

The market price of agricultural production resulting from the IAEI statistical survey and the resulting price depends on the yields corresponding to the given soil and climatic units according to ESEU.

Energy gross margin including overheads 2 is similar:

$$EGMo_{i,p} = ENS_{i,p} - ECOGSo_{i,p} \quad (2)$$

where: *EGMo,i,p* is the energy gross margin with overheads for crops p and soil-climatic conditions i, *ECOGSo* = energy of cost of goods sold (GS) including overheads (Tables A4–A7 MJ) and *ENSi,p* is the standardized output in MJ (Table A3).

The energy values of *EGMo, i, p* production were evaluated on the same inputs and outputs as *GMo* (1). The primary energy values of the costs are derived from weight of the machines in kilograms listed in the database according to the example in Table A1, where the weights (in kilograms) of machines needed for the production inputs are described.

The value of primary energy per kilogram of weight (Table A5) is divided by the number of years of depreciation and by the number of hectares processed per year. Energy of fuel (Table A6) is given by fuel consumption for work operations and for maintenance on the basis of fuel consumption equivalent, fuel consumption for the transport of materials and technological equipment, according to the energy of organic and inorganic fertilizers (Tables A6 and A7) [34], and protective equipment [41]. The costs were calculated by the Institute of Agricultural Economics and Information (IAEI) [36]. The costs of transport of material were calculated for the standard distance of 5 km between the farm and the land. The calculation of costs also included the labor costs based on the average labor rates in agriculture in the last 5 years.

Soil conditions were determined by soil classification in the ESEU system. Data on slope, soil depth and percentage of stones over 2 mm in the soil were evaluated in the physical units.

The economic values of yields and inputs into the soil for soil-climatic conditions in the Czech Republic were compiled according to the database of ESEUs rated [33,35,36]. Earlier data on revenues for ESEU based on the data of 1970 have been updated by a detailed survey of 529 plots conducted over a period of 9 years (2002–2010). The yield (Y) design was based on the production functions of the dependence of yields on natural and technological conditions [38] according to Equation (3).

$$Y = f(Wt; S;\ Z;\ P:\ L; T) \tag{3}$$

where Y: yield of crops, W: variables of temperature, precipitations and soil moisture, S: soil type, sort and conditions, Z: nutrition of nitrogen, phosphorus and kalium; P: number of chemical protection operations, L: cultivation of soil and T: progress of technology.

The underlying values for yields and similarly for nitrogen dosing and the chemical treatment application were compiled according to the statistical valuation of the given environmental conditions [39].

According to the identified functions, the yields were standardized to the remaining soil-climatic conditions. Subsequently, the proposed value of standard yields was validated with the current value of yields under the given conditions, and a new standardized value of yields was proposed for ESEU. A similar function such as the derivation of the yield (Relationship 3) was compiled by the dependence of nitrogen doses on soil-climatic influences.

An example of a comparison of actual and standardized results of production functions for yield of winter wheat is shown in Figure 2.

The compiled standardized yield values correspond to the categories of soil-climatic conditions for which the values are determined. The climatic factors that are most important for the achieved yields are therefore always calculated for the relevant classification scale and its values—i.e., for climatic regions 0–9 (Table A2). When evaluating specific yield conditions, there is always a deviation from the standardized values due to the achieved weather values, which similarly applies to the monitored soil values. Nevertheless, the database is based on the balanced properties given by long-term observation and statistical analysis of individual effects. For the purpose of this article, the data are sufficiently informative even if they do not meet the requirements of directly measured values, and thus the statistical results are affected by a certain similarity of climatic and soil influences within specific groups of conditions.

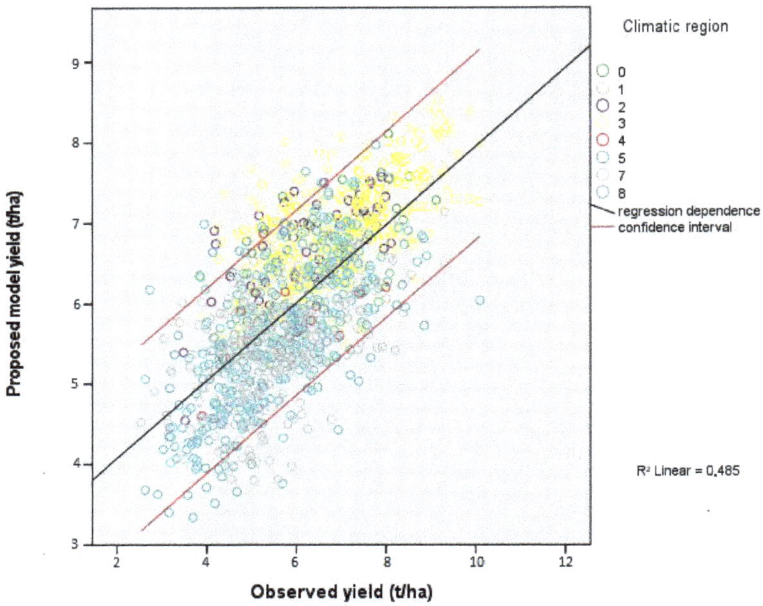

Figure 2. Comparison of winter wheat yields and yield prediction using the production function [38].

For the economic evaluation of the cultivated crop, the economic efficiency ηEp was determined according to the Equation (4).

$$\eta Ep_{i,p} = SO_{i,p} / COGSo_{i,p} \quad (4)$$

as the ratio of the value of output (SO) to the value of input ($COGS$).

Energy efficiency ηEnp is computed similarly in Equation (5).

$$\eta Enp_{i,p} = ENS_{i,p} / ECOGSo_{i,p}. \quad (5)$$

2.2. Assessing the Environmental Impacts of Crop Production

The assessment comprises the methods determining the formation of emissions from crop production on soil, water and air as well as physical impacts of production on the quality of soil in the form of soil compaction.

Determining the Formation of Emissions on the Environment

The content of the evaluation in this article is mainly the ratio of individual types of emissions and the achieved energy of outputs, which evaluates the relationship of individual ESEUs and crops to emissions. The main indicator assessing the environmental impact of production used in this paper was the ratio of total emissions of individual types of indicators to the total crop production including the by-product [42].

Environmental impacts were added on the basis of a description of specific emissions (midpoint) and the system allows global life cycle impact (endpoint) for selected crops [39]. Values are based on the results of the ReCiPe method assessment of primary data for the Czech Republic and the secondary results are based on the Agri-footprint LCI database.

In this context, the stages of the product life cycle are divided into: upstream—processes preceding the actual manufacturing of the product, core—actual manufacturing of the product and downstream—processes following the manufacturing of the product.

2.3. Model Processes to Determine Emissions

The model processes are based on the Agri-footprint database; they were modified using the specific data of IAEI and are based on the data for energy evaluation—the weight of machinery, repairs (equivalent in l/ha of consumed diesel), transport costs (energy requirements of transport in MJ/ha), consumption of fuels and chemical protection (necessary technology in MJ/ha and weighted dose of pesticides in kg/ha) were expressed as diesel consumed by diesel engine of an agricultural machine (energy, from diesel burned in machinery/RER economic) [41,43–45]. The impact of fertilizers was calculated for crop inputs of N, P2O5, K2O, MgO, CaO, and S, and the emissions factors were derived from Agri-footprint database. The emission size relationship is based on the source data of the Agri-footprint database [32,46]. Organic fertilizers were calculated as manure in accordance with the database documents at the level of ESEU. Emissions to air mainly include nitrous oxide, ammonia and pesticide residues, carbon dioxide emissions, which as a reaction of soil with limestone and urea were not included due to the lack of specific data on consumption. Based on the specific data, emissions from minerals, livestock manure and pesticides were recalculated and adjusted. Emissions from crop residues remained the same as in the original process. Emissions to water from mineral fertilizers and livestock manure and pesticides were recalculated and adjusted on the basis of specific data, and emissions from crop residues and heavy metals were assessed according to the original process. Emissions to from soils were based on the specific data; emissions from pesticide residues were recalculated and adjusted and heavy metal emissions were used from the original process.

The ratio of total emissions of individual types of indicators and the energy contained in the total production of the crop, including the by-product, was used as the main indicator for assessing the ecological impact of production. An overview of the average energy efficiency of crops in the Czech Republic is given in Figure 1. To assess the impact of emission to MJ (*EmMJ*) of produced energy, the specific value of emissions per unit of output energy in MJ 6 was used:

$$EmMJ_{i,p} = EMmidp_{i,p}/ENS_{i,p} \quad (6)$$

where *EMmidp* is the emission of midpoint classifications as a sum of all included partial emissions of operations, fertilization, chemical inputs and transport, for crops p and soil-climatic conditions i.

Evaluation of the significance and influence of individual ecological indicators is a separate issue beyond the scope of this work. For their complex evaluation, it is possible to use more methods based on the evaluation of the meaning and weights of individual indicators. Due to the large number of indicators and their various possible interpretations and due to the simplification of the issue, the methodology of multicriteria decision-making was chosen to determine the total emission value per MJ *EEm* according to Equation (7)

$$EEm_i = \sum_{E=1}^{k} \left(\sum_{i=1}^{l} Em_o/l\right)/k \quad (7)$$

where Em_o is an order of the value of emission for *ith* crop on ESEU, E is a sort of emission, k is a number of calculated emissions and l is a number of ESEU.

3. Method of Processing

Databases are maintained in MS Access and MS SQL databases. For each ESEU, selected technology and crop, standardized values of economic and energy efficiency ηEp ηEnp as well as Em_o and $EmMJ$ for each environmental indicator were processed.

The supporting documents were elaborated in line with the technological procedures and verified yields of individual crops under the given ESEU. In the system, it is possible to compile procedures for different antierosion methods of soil treatment and for different nitrogen inputs. To compare soil-climatic conditions, the plowing method of tillage was chosen. All individual ESEU categories were always evaluated during the processing. To evaluate the average conditions of crops, weighted averages of indicators of crops were calculated according to the area of ESEU representation in the Czech Republic.

4. Method of Assessment of Crop Impacts on Soil Environment

The impact of crops on soil compaction was evaluated from a survey conducted in the years 2002–2011; the assessment of the impacts on soil was based on the penetrometric resistance of soil, which is an appropriate indicator of the overall conditions of agricultural land [47,48], bearing in mind the need to obtain information on deeper layers of soil on large areas. Soil penetrometric resistance is closely related to soil-organic carbon (SOC) formation, where soil resistance decreases with higher soil content [49].

The underlying principles of penetrometer measurements are described in the paper by Lhotský [50]. The methodology has been modified to have one sample point for approximately 5 ha. There were three sample points on a plot with the area of up to 10 ha, with another sample point that always added an additional area of up to 5 ha; the sufficient number of sample points, however, was 10. The location of these points was chosen so that they were equally distributed across the entire land plot and were not located in the headland. During each measurement, the probe was pushed into the soil at a constant speed and the penetrometer was reset in cases where the probe hit a stone. Soil samples were collected in each plot in order to determine the soil moisture—namely, from no deeper than the soil tillage depth and from the subsoil layer.

The obtained values of penetrometric resistance are expressed in the form of the mean resistance of three layers—namely, 0–18 cm, 19–38 cm and 39–72 cm.

The results were assessed based on the correlation analysis and show, as well as the direct impact of the crop on the resistance, the general relationships, which determine the effects of penetrometric reistance in the respective soil layers.

5. Results

The results of the comprehensive assessment of economic, energy and environmental impacts of crop growing are based on the determination of individual soil and climatic parameters of the locations where the crops are cultivated. Altogether, the assessment covered a selection of 20 crops and different options for their use. The assessment of economic indicators builds on the calibrated economic results of agricultural holdings—namely, on the average of the last 5 years. Therefore, the results are stable and independent of the respective year. The economic and energy indicators are based on the cost–benefit analysis, which facilitates the evaluation of the absolute profit per hectare of the agricultural land in monetary or energy terms. These indicators are shown in the figures and tables as the attained efficiencies according to Relationships (4) and (5). The environmental indicators are related to the produced energy in production including straw.

5.1. Relation of Economic, Energy and Environmental Characteristics to Soil-Climatic Conditions in the Czech Republic

To determine more detailed effects of weather and soil conditions in the Czech Republic on the achieved economic, energy and environmental parameters, the available database data of individual crops and environmental indicators in the database were processed.

The overview of average energy and economic efficiency of crops in the Czech Republic in comparison to CO_2 emissions is provided in Figure 3 and Table A8 and the average terrestrial ecotoxicity values are described in Table A9, with individual data provided in a separated file for all emissions [35].

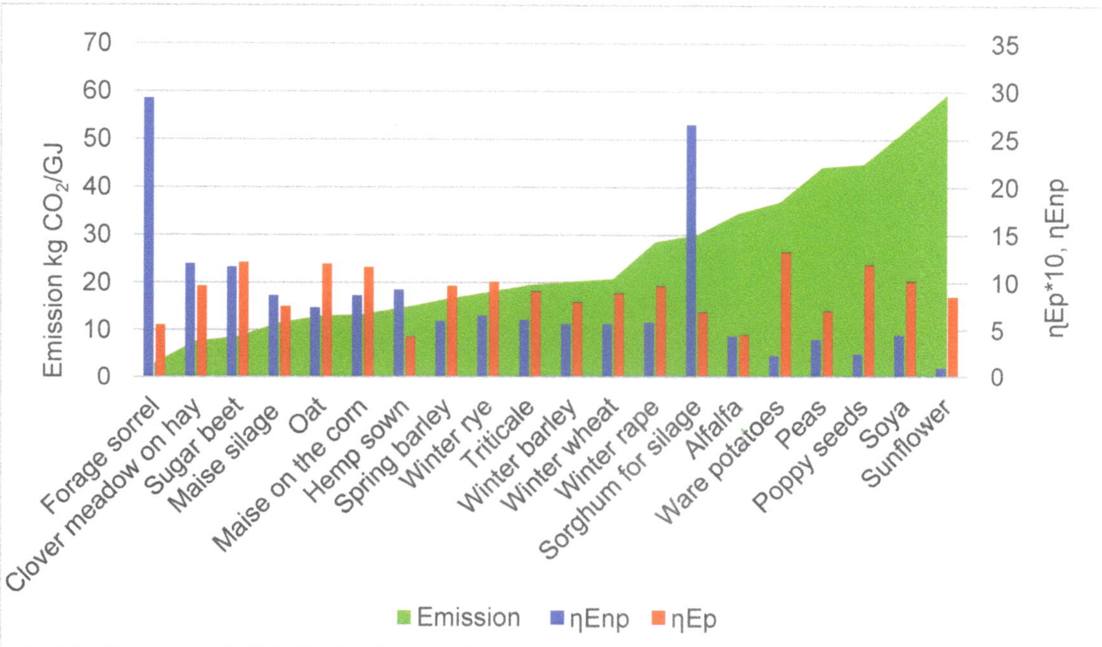

Figure 3. Average values of ηEnp, ηEp and CO_2 emission per GJ of energy in the product.

The system enables a comparison of results of ηEnp, ηEp and $EmMJ$ in the same soil-climatic conditions as well as all the other monitored inputs and outputs. The results of individual crops show the lowest emission load for CO_2 per MJ produced for forage bulk crops, the largest load is achieved for crops with food use, where economic efficiency also prevails over energy efficiency. The achieved environmental results depend very much on the technologies used for growing crops and harvesting. For example, alfalfa has almost the same cultivation technology as clover, but its environmental impact reflects a high consumption of diesel fuel, when the silage mass is harvested by high-performance and high-consumption cutters instead of using solar energy for drying. The different value of the energy balance between energy and food crops also provides a new perspective on emissions from animal production, which consumes bulk feeds with better energy efficiency than food production.

In the following section, the main soil-climatic indicators according to the ESEU system were used individually.

The obtained values in line with the ESEU code are included in Figure 4.

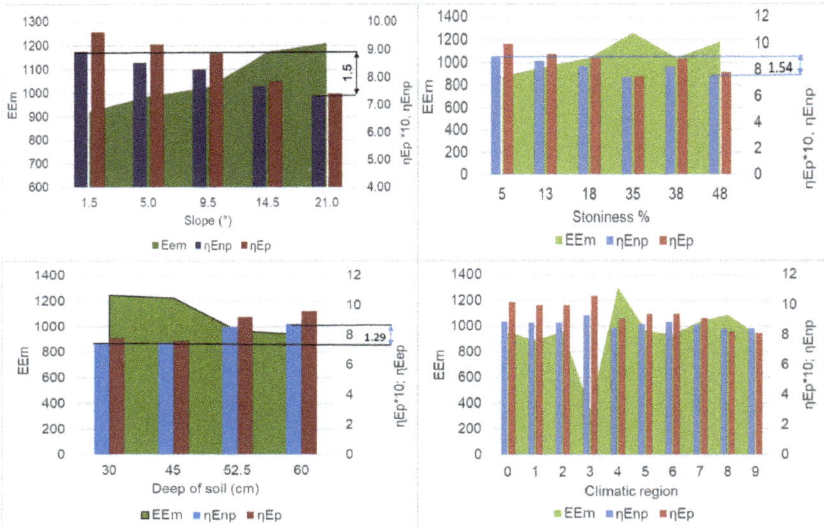

Figure 4. Main configuration and climatic properties and EGMo, GMo and EEm in average of crops.

From the above dependencies, the importance of land and climate configuration indicators is obvious. Due to the slope of soil, the energy efficiency decreases most significantly, namely, by 1.58, while due to the stoniness of soil there is a decrease of 1.54, and due to the depth of soil, of 1.29. Due to the difference in climate regions of 0.94, the difference in economic efficiency decreases in similar relations, and in absolute values less significantly (values in the graphs are multiplied by 10), but the percentage of the decrease is more pronounced. The percentage changes of all indicators are given in Tables A10–A13.

The results for the main types of soil are shown in Figure 5.

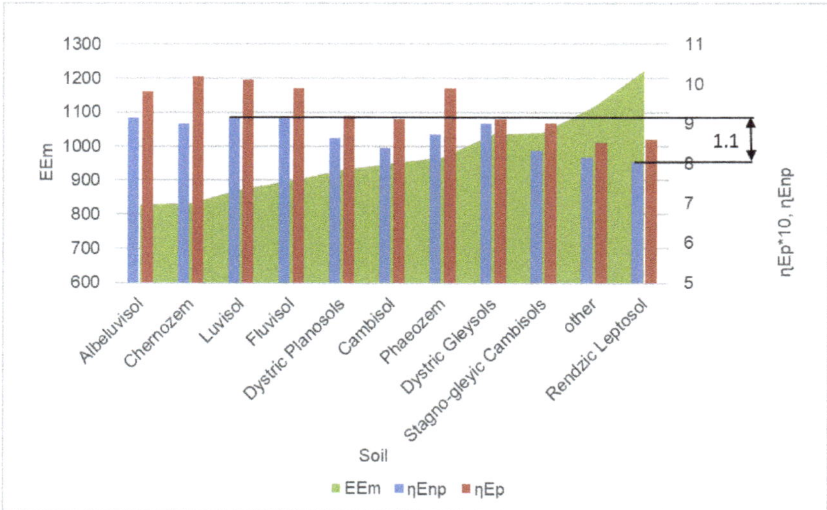

Figure 5. *EEm*, *ηEnp*, and *ηEp* among the main types of soil in average of crops.

The results show the main difference value of *ηEnp* is 1.1; there is also an interesting difference between the energy and economic efficiency in chernozem, which is mainly caused by growing economically favorable crops on fertile areas. Higher emissions per energy of outputs correspond to the lower economic and energy efficiency.

5.2. Assessment of Impacts of Chosen Crops on Penetrometric Resistance

In order to assess the relationship to soil compaction, a survey was carried out measuring the penetrometer resistance by frequency of crops grown on the plots. The assessment also included cases when more than three values of penetrometer resistance for the respective crop were obtained. The correlation analysis (Table A13) indicates the main dependence of the value of resistance in the monitored crop, determining the effects of penetrometer resistance in the respective soil layers.

The results of the survey of penetrometric resistances from the years 2002–2011 [38] are shown in Figure 6.

Crop	Crop repetition (Nr.)	Penetrometric resistance 0-18 cm (MPa)	Penetrometric resistance 19- 38 cm (MPa)	Penetrometric resistance >39 cm (MPa)
Alfalfa	0	1.68	4.06	4.97
	1	1.63	4.04	5.63
	2	1.29	3.06	4.58
	3	1.36	3.05	4.68
Spring barley	0	1.63	4.11	5.18
	1	1.75	4.09	4.95
	2	1.63	3.36	4.22
	3	1.40	3.38	3.90
Poppy	0	1.72	4.14	5.07
	1	1.08	2.58	4.06
	2	1.05	2.26	3.42
Winter wheat	0	1.78	4.04	4.83
	1	1.53	3.97	5.15
	2	1.62	3.83	4.84
Cereals (all)	0	1.31	3.25	3.92
	1	1.53	3.67	5.25
	2	1.39	3.66	4.69
	3	1.69	4.02	4.69
	4	1.90	4.39	4.74
	5	1.81	4.67	5.85
Triticale	0	1.63	3.94	4.61
	1	1.99	4.61	5.20
	2	2.44	5.35	5.67
Winter raps	0	1.46	3.52	4.35
	1	1.76	4.29	4.73
	2	1.93	4.63	5.43
	3	4.08	6.58	5.76
Silage maise	0	1.65	3.78	4.74
	1	1.72	4.44	4.64
	2	1.56	4.15	4.24
	3	1.85	4.37	4.92

Figure 6. The penetrometric resistance according to the number of crop repetitions depending on the depth of the soil.

The results of penetrometer measurements and the identified trends in soil compaction are included in Table A14, with plotted significant dependences of the penetrometric resistance on the crop, the positive effects of alfalfa in subsoil, and the negative impact of winter rapeseed growing across the soil profile as well as of maize in the bottom layer at the depth of more than 39 cm. The values in subsoil are crucial for the assessment of effects of penetrometer resistance. The values of penetrometric resistance at the depth of more than 39 cm can be influenced by penetrometer measurements that ended prematurely due to the solid bedrock. The largest effect on subsoil compaction is seen in poppy seed (difference of 1.88 MPa) and alfalfa (1.1 MPa). The highest compaction, on the contrary, is reported for winter rapeseed (3.06 MPa) and triticale (1.4 MPa).

6. Discussion

The main contribution of this article is a comprehensive view of the economic, energy and emission context of the production of individual crops according to soil and climatic conditions. The evaluation of individual factors is based on the standardized values of inputs and outputs of individual crop processing technologies up to the level of work operations. The work thus enables a systematic view of the production structure of farms in their soil-climatic conditions, and thus enables better planning and management of land use in local conditions. The existing information in the literature is fragmented into partial cases under specific conditions, which are difficult to combine into one framework to find complex contexts. The literature presents analyses of individual energy and economic balance of crops, especially according to higher territorial units and countries, or on the basis of a partial calculation of technology data and simulation of operating conditions [51,52]. This issue is addressed on the basis of data of technological processes individually also according to the yield of straw [9,53] or biomass of selected crops [54]; however, the overall crop balance depending on local conditions for energy, economic and environmental concepts is not addressed. Specific conditions by territorial units are determined on the basis of statistical surveys without functional interdependence [9]. This work does not address the individual technological context of the use of new technological procedures, but the basic standardized framework, by which the newly obtained data can be evaluated. The way in which emission data are processed by ordering ESEUs within individual crops allows the impact of emissions on specific businesses and for specific input choices to be adapted. The way in which emission data are processed by ordering ESEUs within individual crops allows the impact of emissions on specific businesses and for specific input choices to be adapted. Environmental impacts are based on the results of ReCiPe method assessment of specific data for Czech Republic combined with model processes of the LCI database Agri-footprint. A system approach to derive emissions based on this database makes it possible to evaluate individual soil-climatic conditions based on the full impact of technologies. The basis is a complete evaluation of emissions according to primary energy in the manufacture of machinery, according to fuel consumption, fertilizers and protective equipment depending on the doses of material and the performance of kits in individual operations in specific soil and climatic conditions. Emission sources are therefore assessed comprehensively and compared to some other sources, which only evaluate some emission components [55]. The standard LCA database evaluation approach allows for crop-specific evaluation but without the choice of individual emission items according to machine aggregations [4,56]. This division makes it possible to adapt the emission factors for the individual difficulty conditions.

A comprehensive evaluation of individual crops shows significant differences between energy and food crops. Higher economic efficiency of food crops is accompanied by increased costs per unit of energy and higher emissions (e.g., soybeans, poppy, sunflower). Higher energy efficiency of feed crops and lower emissions of energy produced can contribute to a further discussion on the focus of food in relation to animal production as well as to discussions on energy production. There are conflicting views on this topic and a detailed LCA analysis of the whole process is needed [57–59]. An important context of the relationship between emissions *EmMJ* from the production of feed crops and grains for human consumption is given in Table 1.

Table 1. Emissions per GJ of produced energy between crops for food and energy production.

Crops for	Emissions	Unit	Unit/GJ	%	Dif %
Food	Freshwater ecotoxicity	kg 1.4-DCB	0.8142	100	
Fodder	Freshwater ecotoxicity	kg 1.4-DCB	0.4080	50.11	−49.89
Food	Global warming	kg CO2 eq	30.1074	100	
Fodder	Global warming	kg CO2 eq	15.4570	51.34	−48.66
Food	Human carcinogenic toxicity	kg 1.4-DCB	1.2271	100	
Fodder	Human carcinogenic toxicity	kg 1.4-DCB	0.7158	58.34	−41.66
Food	Terrestrial ecotoxicity	kg 1.4-DCB	99.5932	100	
Fodder	Terrestrial ecotoxicity	kg 1.4-DCB	50.2889	50.49	−49.51

The table shows that emissions produced from energy crops (fodder: clover grass, clover hey, maize silage) are 42–50% lower per GJ of energy produced than those from food crops (winter wheat, spring barley, peas).

The identified connections between energy, economic and environmental impacts of agricultural crop production show a very significant dependence on soil-climatic conditions. The article separately evaluated the individual properties of land on the operational indicators of crops. The soil depth affects the energy efficiency of crops in the Czech Republic by 15%, the economic efficiency by 21% and the overall order of emissions by 33%. The land slope affects the energy efficiency of crops in the Czech Republic by 18%, the economic efficiency by 23% and the overall order of emissions by 31%. The stoniness affects the energy efficiency of crops in the Czech Republic by 14%, the economic efficiency by 18% and the overall order of emissions by 31%. The results depend on long-term observations of the IAEA and the identification of crop production functions.

Climatic indicators are a factor acting together with soil indicators and according to their specific compositions, overall results can be derived. The interaction is mainly due to the achieved crop yields in specific conditions. From the point of view of the suitability of crops for production, the dependences found show that marginal soils with a shallow soil depth, high stoniness and slope, even on less fertile soils, have higher relative emissions from crop production to 1 MJ. Consequently, there is a need to grow crops in these conditions without large emission effects, especially perennial energy crops, which can be used for both animal production and energy production.

The local conditions also cover the effects on the environment in soil based on the mechanical effects of crop growing on soil. The obtained results suggest major impacts of individual crops on soil conditions. The penetrometric resistance of the soil depends mainly on the content of organic matter in the soil and on the method of farming. The content of organic matter in the soil is ensured both by organic fertilization and in deeper layers, above all by the decomposition of the root system of crops. According to the performed penetrometric survey, less compaction of subsoil and subsoil is found in alfalfa and some springs, spring barley and poppy. In terms of lasting effect on improving the condition of the soil in the deeper layers of the soil, alfalfa is very important crop [60–62]. Global biogas (methane) production needs new opportunities for production using legumes on arable land, as they do not significantly degrade soil quality compared to other crops [63], unlike the cultivation of sown maize [64]. Under the new climatic conditions, there is a significant relationship to precipitation, where alfalfa is highly profitable in dry conditions, while clover in humid conditions [65]. A very important advantage is the high production of roots in depth with a positive effect on the soil structure, the content of soil organic matter (SOM) and consequently also on the productivity of the stand [64]. This makes it possible to improve the sustainability and resilience of the natural environment, in particular with regard to reduced external inputs, improved humus balance (carbon, energy and nutrient cycle), reduced greenhouse gas emissions and the general positive impact of fodder and catch crops in crop production practices [63].

For the purpose of aligning the growing of crops for food and energy purposes, according to the effects of selected crops on soil ascertained based on the obtained values of penetrometric resistance of individual crops, alfalfa is a highly suitable crop since it improves subsoil compaction and at the same time provides good energy gain. The area under alfalfa, however, substantially decreased in recent years due to the reduction in cattle breeding and has reached its minimum in the Czech Republic. The current need to improve the subsoil conditions together with the need to increase the energy crop capacities, with the concurrent pressure to reduce the cultivation of maize for silage, speaks in favor of its production. Alfalfa can easily be used in all the existing biogas plants, up to a share of 20%, for pellet production and cattle fattening.

The system can analyze 22 environmental indicators in the endpoint category and 18 environmental indicators in the midpoint category [35] and can be combined with the physical effects of crop growing on soil. The physical effects of crop growing on soil constitute an equal impact on the environment as the emissions and assume the form of numerous impacts, especially on soil erosion, soil fertility, resilience to drought as well as water contamination in the case of topsoil wash off. The subsoil compaction keeps increasing as a result of a change in crop composition and climate change, with a decrease in the number of frost days causing soil swelling (frost heaving), as well as an increase in crop yields that have to be harvested and transported from the land by heavy machinery.

7. Conclusions

The paper describes the process of developing the system of assessment of soil and climatic impacts on individual crops with respect to economic, energy and environmental indicators for the classified unit of soil and climatic properties—i.e., ESEU in the Czech Republic. The main indicator that was selected to compare the individual conditions was the ratio of the value of individual types of emissions per energy output in MJ. Apart from this indicator, other usual indicators were also set such as the energy of production and economic efficiency of production. The statistical results can also be defined for all the other indicators. Aside from direct classification of soil and climatic conditions, other soil properties, available from the monitoring of the Research Institute for Soil and Water Conservation, were subjected to regression analysis.

With respect to emission impacts, perennial energy crops (silage sorghum, sorrel, hemp), should be encouraged. The current status among major energy crops is of the corn silage with good emission characteristics, but it is necessary to ensure the proper growing conditions with regard to soil quality. The exploration of energy outputs diminishes the nutritional properties of food crops. In spite of this, the analysis shows that in terms of emissions the energy crops bring more benefits when grown under marginal conditions, if the cultivation of these crops under the respective conditions is possible. From the point of view of impact on the soil and sustainable development, justified cultivation of alfalfa with a proven influence on the amelioration of compacted soils is crucial. Alfalfa has increased emission effects compared to clover due to harvesting with a high-power cutter. In the case of alfalfa harvesting on hay, its emissions are comparable to clover. Due to the increasingly difficult search for suitable biomass for energy production while respecting the requirements for food production, alfalfa production is a suitable solution for ensuring the quality of soil and replacement biomass for current energy crops. From the point of view of sustainable development, this solution is very essential for obtaining biomass from agricultural sources. The overall use of results should be based on the evaluation of Pareto optimality [16] in line with the current production options and requirements determined by policies and thorough knowledge of territorial aspects of production. For the sake of further development, the use of maps with the impacts of production on emissions under specific conditions is expected. In the future, it is possible to consider a comprehensive assessment of emission effects in agriculture [66].

The article provides a comprehensive view of the joint impact of natural factors on energy, economic and environmental indicators, and thus provides a better picture of their impact on measures for further development of energy in regions and for agricultural policy. As Wohlfart [8] writes, a comprehensive assessment of all contexts is always important for further assessment of a bioeconomy, and therefore also for energy policy. For further development, it is important to compare modeled and measured results in connection with local land conditions for a real evaluation of the conditions of the whole region.

One of the best examples of aligned energy generation and food production is the use of alfalfa as a sanitary crop to address subsoil compaction and as a crop that can help reduce maize silage on the soils at risk of erosion and emissions impact [11,67]. Deep-rooting crops are a desirable source of carbon in the deeper layers of the soil, where they also ensure the stability of soil aggregates and sufficient soil permeability. Knowledge of local conditions and their appropriate agricultural use should also become part of the Green for Europe strategy [68], which assumes keeping global warming below 1.5 °C while still reducing greenhouse gas emissions. The main advantage of energy crops for the low-carbon economy is their potentially lower CO_2 production, especially in marginal areas with less soil depth, slope and stoniness. Higher variability of biomass production in the field, taking into account the requirements of sustainable energy, can also lead to higher deregulation and liberalization of the energy market. See [69] for case of ensuring sufficient biomass capacities.

Author Contributions: Conceptualization, V.V. and L.N.; methodology, M.H. and V.V.; software, validation, formal analysis, investigation, data curation, project administration, funding acquisition V.V. All authors have read and agreed to the published version of the manuscript.

Funding: This research was funded by Národní Agentura pro Zemědělský Výzkum, grant no QK1710307.Institutional Review Board: Not applicable.

Informed Consent Statement: Informed consent was obtained from all subjects involved in the study.

Data Availability Statement: Not applicable.

Acknowledgments: The paper was elaborated in the framework of the project of MoA QK1710307 "Economic support for strategic and decision-making processes at national and regional level leading to the optimal use of renewable energy sources, especially biomass, while respecting food self-sufficiency and soil protection". The authors acknowledge the support of the project. The authors also thank Maria Macháčková from the IAEI for preparing the materials for the search, Václav Hofman for help in preparing the data and also all other persons who participated in the validation of the input data used for the preparation of materials for the article including Ivo Pokorný for the MS SQL services.

Conflicts of Interest: The authors declare no conflict of interest.

Appendix A

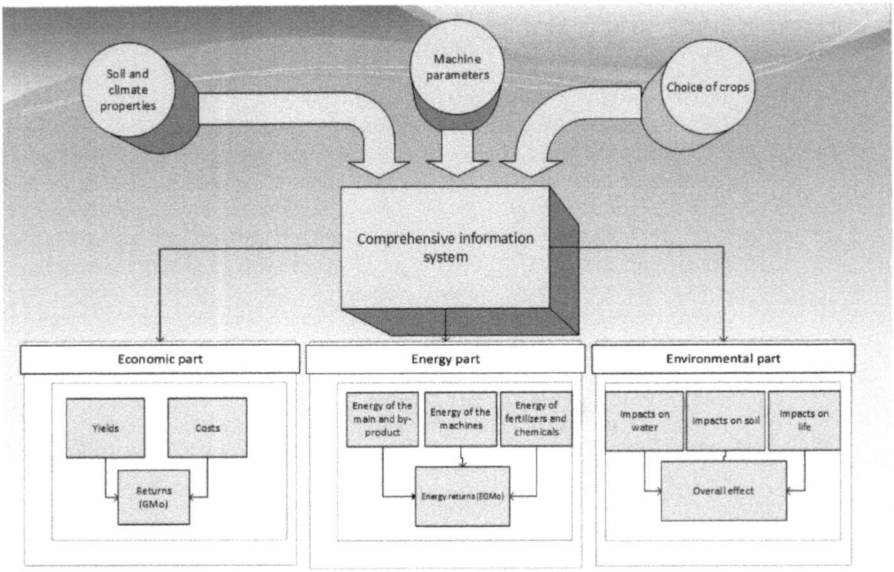

Figure A1. Scheme of the Comprehensive Information System.

Appendix B. Calculation of Costs and Outputs

Appendix B.1. Costs

The evaluation of costs of parameterized production is based on the sum of costs that have to be spent to achieve the production of crop under ESEU, for plowing technology with operations factored in. The variable costs, VCosts 8, were calculated based on the standardized technological procedures for all main crops according to the Institute of Agricultural Economics and Information.

$$VCosts_{i,p,r} = \left(WO_{i,p,r} + TC_{i,p,r} + CM_{i,p,r} + LC_{i,p,r}\right) \times CVC_{i,p} \tag{A1}$$

where: *i* evaluated soil-climatic unit ESEU; *p*—evaluated crop; *r*—number of operations; *WOi,p,r* = unit costs of work operation in line with technologies proposed by Research Institute of Agricultural Engineering, p.r.i., (EUR/ha); *TCi,p,r* = transport costs (EUR/ha); *CMi,p,r* = costs of material, fertilizers, plant protection products and auxiliary products (EUR/ha); *LCi,p,r* = unit labor costs of per cultivation technology and crop under the given soil and climatic conditions based on the five-year average costs (EUR/hour); *CVCi,p,r* = coefficient of variable costs derived from the IAEI cost survey for ESEU, crop and operation.

The indirect costs of producing of crops are determined with the coefficient *ICfc* 9, which is determined according to the IAEI cost survey as a share of indirect *ICi,p,r* and direct variable crop production costs.

$$ICfc = IC_{i,p,r}/VCosts_{i,p,r} \tag{A2}$$

Appendix B.2. Outputs

The price of the parameterized production 10 was determined for standardized yields on ESEU and is composed of the production of the main product and by-product:

$$PO_{i,p} = Y_{i,p} \times P_pCR + Yb_{i,p} \times Pb_p \tag{A3}$$

where: $Y_{i,p}$ = yield of parameterized production of the main product for the p-th crop, which is the corrected normative natural yield of individual main agricultural crops (p) for individual ESEU (i) [33] (t/ha); the yield is updated annually according to the five-year average of crop yields in the IAEA cost survey and the FADN survey; $PpCR$ = normative prices of the main product of individual p-th crops differentiated according to climatic regionalization (EUR/t); it is updated annually according to the five-year average of agricultural crop prices from the IAEI and the Czech statistical office (CZSO) survey; Y_b = Yield of by-product (straw) on ESEU and crop; P_b = a normative price of by-product of crop.

Appendix C

Table A1. The case of technological operations by soya.

Operation			5-Undermining (Chiseling) 500 mm		10-Transport and Spreading of Limestone (1.5 t/ha to 2 t/ha))		15-Transport and Spreading of Manure and Compost (30 t/ha)	
	Number of Operations/Year		0.1		0.1		0.2	
weight	Tractor	Kg	10,800		6400		7000	
	Machine	Kg	2800		6200		9150	
Material inputs	Name				Limestone, finely ground		Manure	
	Quantity	MJ/ha	0		2 t		30 t	
	Price	EUR/MJ	0		23.9		11.4	
	Costs	EUR/ha	0		4.77		68.2	
Technical description of the operation	Set		TK 200 kW	Chisel bar 3 m	TK 120 kW	Spread it out. semitrailer 12,000 L	TK 130 kW	Manure spreader 16 t
	Normatives	h/ha	0.56		0.5		1.33	
		l/ha	21		5.5		25	
		EUR/ha	49.3		22.3		106.2	
	Costs	EUR/ha	4.93		2.22		21.2	
Variable costs total EUR/ha			4.93		7		89.4	

Source: Research Institute of Agricultural Engineering, p.r.i., 2018.

Table A2. Description of the climatic regions.

Region Numeric Code	Temperature Sum °C/Year	Temperature Average °C/Year	Rainfall Average mm/Year	Risk of Dry Years Coefficient	Moisture Security 1 Minimum, 10 Maximum
0	3680	10.08	550	0.40	1.5
1	3430	9.40	450	0.50	1
2	3430	9.40	550	0.25	3
3	3380	9.26	600	0.15	5.5
4	3230	8.85	500	0.35	2
5	3080	8.44	600	0.23	7
6	3330	9.12	800	0.05	10
7	3030	8.30	700	0.10	10
8	2830	7.75	750	0.03	10
9	2680	7.34	850	0.00	10

Source: IAEI new calculation of temperature and rainfall [24].

Table A3. Energy of crop production.

Crop	Product	Dry Matter Main Product %	Energy of Main Production (MJ/kg of Product)	By-Product Energy (MJ/kg of Product)
Potatoes	potatoes	22	3.45	3
Early potatoes	potatoes	22	3.45	3
Seedling potatoes	potatoes	22	3.45	3
Sugar beet	bulbs	23	3.89	1.76
Barley spring	grain	85	15.93	13.73
Barley spring malt	grain	85	15.93	13.73
Winter barley	grain	85	15.48	13.73
Clover on green	green matter	21	3.07	
Clover on hay	hay	85	13.06	
Clover grass	hay	85	13.13	
Corn for silage	corn silage 32% of moisture	32	5.984	
Corn on the cob	grain	85	16.21	13.5
Grass	hay	85	13.19	
Poppy	grain	85	15.48	13.69
Oat	grain	85	17.45	13.38
Nonfood wheat	grain	85	15.82	13.46
Food wheat	grain	85	15.82	13.46
Winter rape	grain	85	25.22	13.64
Triticale	grain	85	16.22	13.46
Alfalfa	silage 40%	40	6.25	
Rye	grain	85	15.48	13.46
Mustard	green matter	21	2.67	
Pea	grain	85	14.15	13.63
Hemp	dry matter	85	13.69	
Buckwheat mixture bundles	green matter	21	3.07	
Soya	grain	85	17.66	13.63
Sunflower	grain	85	12.41	
Bundle	green matter	21	3.07	
Sorghum	green matter	21	3.07	
Sorrel	dry matter	85	19.17	

Source: Preininger [22].

Table A4. Used unit costs of materials.

Inputs	Unit Price (EUR/Unit)	Unit
Oil	0.98	EUR/L
Work	9.09	EUR/hod
N	1.55	EUR/Kg
P2O5	2.05	EUR/Kg
K2O	1.09	EUR/Kg
MgO	1.68	EUR/Kg
CaO	0.36	EUR/Kg
Sulfur	0.48	EUR/Kg
Chemicals	89.73	EUR/Kg
Manure	40.36	EUR/t

Source: IAEI.

Table A5. Conversion factors for calculating the energy contained in machines

Machine Category	Value	Unit
Tractors	95.7	MJ/kg
Tillage machines	99.2	MJ/kg
Seeders	95.4	MJ/kg
Spreaders and sprayers	95.4	MJ/kg
Combine harvester	83.5	MJ/kg
Straw harvest	95.4	MJ/kg
Traffic machine	83.5	MJ/kg

Source: Preininger [22].

Table A6. Energy of used materials.

Inputs	Value (MJ/Unit)	Unit
oil	40.7	dm^3dm^3
benzine	41.5	dm^3dm^3
propane butane	50.8	kg
natural gas	33.8	m^3
lubricants	45.2	dm^3dm^3
electrical energy	9.6	kWh
coal	27.8	kg
cereal seed	8	kg of seeds
oilseeds, rape, flax	5.7	kg of seeds
seed potatoes	2	kg of seed
beet seed (batch of 100,000 seeds)	172	dose
corn seed (50,000 seeds)	16.2	dose
pea seed bean	7	kg of seeds
N	82.5	kg
P_2O_5	17.7	kg
K_2O	9.6	kg

Source: Preininger [22].

Table A7. Price and energy of nutrients in cow manure.

Nutrients	Nutrients kg/t [2]	Price EUR/kg [1]	Price Total EUR/t [1]	Energy MJ/t [3]
N	5	1.55	7.73	
P_2O_2	3.1	2.05	6.34	
K_2O	7.1	1.09	7.75	
Mg	1.5	1.68	2.52	
Ca	4.5	0.36	1.62	
S	1	0.48	0.48	
	Total		26.43	463

Source: [1] IAEI; [2,3] Preininger [22].

Table A8. Average emissions global warming and efficiency of crops.

Crop	Emission	ηEn	$\eta E \times 10$	kg C02/GJ
Forage sorrel	global warming	29.24	5.53	2.96
Clover meadow on hay	global warming	11.97	9.59	7.58
Sugar beet	global warming	11.62	12.12	8.51
Maize silage	global warming	8.62	7.46	11.42
Oat	global warming	7.36	11.95	12.84
Maize on the corn	global warming	8.63	11.60	13.19
Hemp sown	global warming	9.25	4.27	14.70
Spring barley	global warming	5.94	9.64	16.41
Winter rye	global warming	6.49	10.10	17.93
Triticale	global warming	6.05	9.05	19.45
Winter barley	global warming	5.64	7.90	20.17
Winter wheat	global warming	5.64	8.86	20.76
Winter rape	global warming	5.78	9.65	28.53
Sorghum for silage	global warming	26.53	6.88	30.01
Alfalfa	global warming	4.37	4.53	34.45
Ware potatoes	global warming	2.29	13.24	36.86
Peas	global warming	4.01	7.01	44.19
Poppy seeds	global warming	2.50	11.89	44.80
Soya	global warming	4.45	10.15	51.88
Sunflower	global warming	0.93	8.49	59.34

Table A9. Average emissions of terrestrial ecotoxicity and efficiency of crops.

Crop	Emission	ηEn	$\eta E \times 10$	kg 1,4-DCB/GJ
Forage sorrel	Terrestrial ecotoxicity	29.241	5.533	9.510
Clover meadow on hay	Terrestrial ecotoxicity	11.971	9.593	24.442
Sugar beet	Terrestrial ecotoxicity	11.624	12.122	27.214
Maize silage	Terrestrial ecotoxicity	8.620	7.460	36.668
Oat	Terrestrial ecotoxicity	7.363	11.952	43.309
Maize on the corn	Terrestrial ecotoxicity	8.628	11.596	43.632
Hemp sown	Terrestrial ecotoxicity	9.249	4.271	47.801
Spring barley	Terrestrial ecotoxicity	5.945	9.639	54.442
winter rye	Terrestrial ecotoxicity	6.489	10.096	60.899
Triticale	Terrestrial ecotoxicity	6.045	9.050	65.772
Winter barley	Terrestrial ecotoxicity	5.639	7.899	67.811
Winter wheat	Terrestrial ecotoxicity	5.636	8.864	70.070
Winter rape	Terrestrial ecotoxicity	5.775	9.655	97.490
Sorghum for silage	Terrestrial ecotoxicity	26.525	6.879	98.681
Alfalfa	Terrestrial ecotoxicity	4.366	4.532	112.083
Ware potatoes	Terrestrial ecotoxicity	2.289	13.236	117.159
Peas	Terrestrial ecotoxicity	4.010	7.011	143.674
Poppy seeds	Terrestrial ecotoxicity	2.503	11.895	153.232
Soya	Terrestrial ecotoxicity	4.454	10.146	166.673
Sunflower	Terrestrial ecotoxicity	0.932	8.490	197.321

Table A10. Average emissions and efficiency of crops according to the climatic regions.

CR	ηEnp		ηEp		EEm	
	Value	%	Value	%	Value	%
0	8.88	95.18	10.2	96.23	954	272.36
1	8.81	94.43	10	94.34	891	254.53
2	8.81	94.43	10	94.34	959	273.78
3	9.33	100.00	10.6	100.00	350	100.00
4	8.49	91.00	9.1	85.85	1301	371.52
5	8.72	93.46	9.4	88.68	966	275.75
6	8.86	94.96	9.4	88.68	933	266.46
7	8.7	93.25	9.1	85.85	1036	295.91
8	8.39	89.92	8.2	77.36	1087	310.49
9	8.45	90.57	8.1	76.42	941	268.69
min	8.39	89.92	8.1	76.42	350	100.00
max	9.33	100.00	10.6	100.00	1301	371.52

Table A11. Average emissions and efficiency of crops according to the slope.

Slope (°)	ηEnp		ηEp		EEm	
	Value	%	Value	%	Value	%
1.5	8.93	100.00	9.63	100.00	922.04	100.00
5.0	8.54	95.68	9.20	95.53	987.03	107.05
9.5	8.29	92.86	8.87	92.12	1025.44	111.21
14.5	7.68	86.04	7.85	81.52	1174.31	127.36
21.0	7.35	82.31	7.42	77.03	1211.89	131.44
min	7.35	82.31	7.42	77.03	922.04	100.00
max	8.93	100.00	9.63	100.00	1211.89	131.44

Table A12. Average emissions and efficiency of crops according to the depth of the soil.

Depth of Soil (cm)	ηEnp		ηEp		EEm	
	Value	%	Value	%	Value	%
30	7.5	85.32	7.8	81.25	1246.73	132.765
45	7.53	85.67	7.6	79.17	1223.34	130.2742
52.5	8.57	97.50	9.2	95.83	968.6	103.1468
60	8.79	100.00	9.6	100.00	939.05	100
min	7.50	85.32	7.60	79.17	939.05	100
max	8.79	100.00	9.60	100.00	1246.73	132.765

Table A13. Average emissions and efficiency of crops according to the stoniness.

Stoniness	ηEnp		ηEp		EEm	
	Value	%	Value	%	Value	%
5	9.04	104.27	10	108.70	874.42	90.68
12.5	8.67	100.00	9.2	100.00	964.27	100.00
17.5	8.29	95.62	8.9	96.74	1040.84	107.94
35	7.46	86.04	7.5	81.52	1260.94	130.77
37.5	8.24	95.04	8.8	95.65	1036.69	107.51
47.5	7.5	86.51	7.8	84.78	1181.26	122.50
min	7.46	86.04	7.50	81.52	964.27	100.00
max	8.67	100.00	9.20	100.00	1260.94	130.77

Table A14. Correlation analysis of penetrometric resistance according to the frequency of crops on the plot.

		Alfalfa	Spring Barley	Poppy	Cereals	Triticale	Raps Winter	Maize Silage	Winter Wheat
Penetrometric resistance 0–18 cm	Pearson Correlation	−0.076	0.006	−0.159 **	0.172 **	0.078	0.202 **	0.006	0.023
	Sig. (2-tailed)	0.185	0.917	0.005	0.003	0.175	0.000	0.919	0.688
	N	306	306	306	306	306	306	306	306
Penetrometric resistance 19–38 cm	Pearson Correlation	−0.118 *	−0.106	−0.215 **	0.164 **	0.077	0.236 **	0.095	0.023
	Sig. (2-tailed)	0.038	0.065	0.000	0.004	0.178	0.000	0.097	0.683
	N	306	306	306	306	306	306	306	306
Penetrometric resistance >39 cm	Pearson Correlation	−0.044	−0.225 **	−0.219 **	0.083	0.134 *	0.176 **	0.247 **	−0.003
	Sig. (2-tailed)	0.476	0.000	0.000	0.179	0.029	0.004	0.000	0.958
	N	264	264	264	264	264	264	264	264

** Correlation is significant at the 0.01 level (2-tailed). * Correlation is significant at the 0.05 level (2-tailed).

References

1. Monforti-Ferrario, F.; Dallemand, J.-F.; Pascua, I.P.; Motola, V.; Banja, M.; Scarlat, N.; Medarac, H.; Castellazzi, L.; Labanca, N.; Bertoldi, P.; et al. *Energy Use in the EU Food Sector: State of Play and Opportunities for Improvement*; Publications Office of the European Union: Luxembourg, 2015. [CrossRef]
2. Mcmichael, A.J.; Powles, J.W.; Butler, C.D.; Uauy, R. Series Energy and Health 5 Food, livestock production, energy, climate change, and health. *Lancet* **2007**, *370*, 1253–1263. [CrossRef]
3. Máté, D.; Rabbi, M.F.; Novotny, A.; Kovács, S. Grand challenges in Central Europe: The relationship of food security, climate change, and energy use. *Energies* **2020**, *13*, 5422. [CrossRef]
4. Colantoni, A.; Picchio, R.; Marucci, A.; Di Mattia, E.; Cristofori, V.; Recanatesi, F.; Villarini, M.; Monarca, D.; Cecchini, M. WP3—Innovation in Agriculture and Forestry Sector for Energetic Sustainability. *Energies* **2020**, *13*, 5985. [CrossRef]
5. Maggio, A.; Van, T.; Jean, C.; Malingreau, P. *JRC Science and Policy Reports-Global Food Security 2030—Assessing trends with a View to Guiding Future EU Policies*; Publications Office of the European Union: Luxembourg, 2015. [CrossRef]
6. Nicholson, C.F.; Stephens, E.C.; Kopainsky, B.; Thornton, P.K.; Jones, A.D.; Parsons, D.; Garrett, J. Food security outcomes in agricultural systems models: Case examples and priority information needs. *Agric. Syst.* **2021**, *188*, 103030. [CrossRef]
7. Al Blooshi, L.S.; Ksiksi, T.S.; Aboelenein, M.; Gargoum, A.S. The impact of climate change on agricultural and livestock production and groundwater characteristics in Abu Dhabi, UAE. *Nat. Environ. Pollut. Technol.* **2020**, *19*, 1945–1956. [CrossRef]
8. Wohlfahrt, J.; Ferchaud, F.; Gabrielle, B.; Godard, C.; Kurek, B.; Loyce, C.; Therond, O. Characteristics of bioeconomy systems and sustainability issues at the territorial scale. A review. *J. Clean. Prod.* **2019**, *232*, 898–909. [CrossRef]
9. Gradziuk, P.; Gradziuk, B.; Trocewicz, A.; Jendrzejewski, B. Potential of straw for energy purposes in Poland—Forecasts based on trend and causal models. *Energies* **2020**, *13*, 5054. [CrossRef]
10. Arthurson, V. Closing the global energy and nutrient cycles through application of biogas residue to agricultural land—Potential benefits and drawbacks. *Energies* **2009**, *2*, 226–242. [CrossRef]
11. Mann, L.; Tolbert, V.; Cushman, J. Potential environmental effects of corn (*Zea mays* L.) stover removal with emphasis on soil organic matter and erosion. *Agric. Ecosyst. Environ.* **2002**, *89*, 149–166. [CrossRef]

12. Hettenhaus, J. Achieving sustainable production of agricultural biomass for biorefinery feedstock. *Ind. Biotechnol.* **2006**, *2*, 257–274. [CrossRef]
13. Panoutsou, C.; Alexopoulou, E. Costs and profitability of crops for bioeconomy in the EU. *Energies* **2020**, *13*, 1222. [CrossRef]
14. Gaydon, D.S.; Balwinder-Singh; Wang, E.; Poulton, P.L.; Ahmad, B.; Ahmed, F.; Akhter, S.; Ali, I.; Amarasingha, R.; Chaki, A.K.; et al. Evaluation of the APSIM model in cropping systems of Asia. *Field Crops Res.* **2017**, *204*, 52–75. [CrossRef]
15. Hercher-Pasteur, J.; Loiseau, E.; Sinfort, C.; Hélias, A. Energetic assessment of the agricultural production system. A review. *Agron. Sustain. Dev.* **2020**, *40*, 1–23. [CrossRef]
16. Razm, S.; Nickel, S.; Sahebi, H. A multi-objective mathematical model to redesign of global sustainable bioenergy supply network. *Comput. Chem. Eng.* **2019**, *128*, 1–20. [CrossRef]
17. Bentsen, N.S.; Larsen, S.; Stupak, I. Sustainability governance of the Danish bioeconomy—The case of bioenergy and biomaterials from agriculture. *Energy. Sustain. Soc.* **2019**, *9*, 40. [CrossRef]
18. Ministry of Industry and Trade of the Czech Republic. *The Czech Republic 's National Energy and Climate Plan (Vnitrostátní Plán České Republiky v Oblasti Energetiky a Klimatu)*; MPO: Prague, Czech Republic, 2020.
19. Fluck, R.C. *Energy in Farm Production*; Elsevier: Amsterdam, The Netherlands, 2001.
20. Slesser, M.; Wallace, F. *Energy Consumption per Tonne of Competing Agricultural Products Available to the EC*; Office for Official Publications of the European Communities, Wisconsin Unioversity: Madison, WI, USA, 1982; p. 82.
21. Patterson, M.G. What is energy efficiency?: Concepts, indicators and methodological issues. *Energy Policy* **1996**, *24*, 377–390. [CrossRef]
22. Preininger, M. Energetic Evaluation of Production Process in Crop Production (Energetické Hodnocení Výrobních Procesů v Rostlinné Výrobě). Praha: Ústav vědeckotechnických Informací pro Zemědělství. 1987 Available online: https://www.cistbrno.cz/Record/mzk.MZK01-000535029/UserComments#tabnav (accessed on 3 March 2021).
23. Voltr, V. Description on ESEU. Available online: https://www.uzei.cz/data/usr_001_cz_soubory/210210_Description_of_ESEU.pdf (accessed on 18 February 2021).
24. 227/2018 Sb. Decree on the Characteristics of Rated Soil Ecological Units and the Procedure for Their Management and Updating (Vyhláška o Charakteristice Bonitovaných Půdně Ekologických Jednotek a Postupu Pro Jejich Vedení a Aktualizaci). Available online: https://www.zakonyprolidi.cz/cs/2018-227 (accessed on 18 February 2021).
25. Voltr, V. Characteristic of the Main Soil Units. Available online: https://www.uzei.cz/data/usr_001_cz_soubory/210218_characteristic_of_the_Main_Soil_Units.pdf (accessed on 18 February 2021).
26. Voltr, V. Concept of soil fertility and soil productivity: Evaluation of agricultural sites in the Czech Republic. *Arch. Agron. Soil Sci.* **2012**, *58*. [CrossRef]
27. Brandão, M.; Clift, R.; Milà, L.C.; Basson, L. A life-cycle approach to characterising environmental and economic impacts of multifunctional land-use systems: An integrated assessment in the UK. *Sustainability* **2010**, *2*, 3747–3776. [CrossRef]
28. Baum, R.; Bieńkowski, J. Eco-efficiency in measuring the sustainable production of agricultural crops. *Sustainability* **2020**, *12*, 1418. [CrossRef]
29. Vatsanidou, A.; Kavalaris, C.; Fountas, S.; Katsoulas, N.; Gemtos, T. A life cycle assessment of biomass production from energy crops in crop rotation using different tillage system. *Sustainability* **2020**, *12*, 6978. [CrossRef]
30. Kiesel, A.; Wagner, M.; Lewandowski, I. Environmental performance of miscanthus, switchgrass and maize: Can C4 perennials increase the sustainability of biogas production? *Sustainability* **2017**, *9*, 5. [CrossRef]
31. Al-Mansour, F.; Jejcic, V. A model calculation of the carbon footprint of agricultural products: The case of Slovenia. *Energy* **2017**, *136*, 7–15. [CrossRef]
32. Durlinger, B.; Consultants, B.; Koukouna, E.; Broekema, R.; Van Paassen, M.; Scholten, J. *Agri-Footprint 4.0*; Agri Footprint: Gouda, The Netherlands, 2017.
33. Voltr, V. Economy of Crops on ESEU. Available online: https://www.uzei.cz/data/usr_001_cz_soubory/210210_Economy_of_crops_on_ESEU.pdf (accessed on 18 February 2021).
34. Voltr, V. Fertilizers on ESEU and Crop. Available online: https://www.uzei.cz/data/usr_001_cz_soubory/210218_Fertilizers_on_ESEU.pdf (accessed on 18 February 2021).
35. Voltr, V. Ecologic Indicators on ESEU. Available online: https://www.uzei.cz/data/usr_001_cz_soubory/midp_ind_En.pdf (accessed on 18 February 2021).
36. Voltr, V.; Hruška, M.; Nobilis, L.; Fuksa, P. *Methodology for Economic, Energy and Environmental Assessment of Crop Production. (Metodika Ekonomického, Energetického a Environmentálního Hodnocení Výroby Plodin)*; ÚZEI: Prague, Czech Republic, 2019; ISBN 978-80-7271-242-7.
37. Voltr, V.; Hruška, M.; Šařec, P.; Leština, J.; Froněk, P. *The Methodology for the Valuated Soil-é Soil-Ecological units (BPEJ). (Metodika Ocenění Půdy Pro Bonitované Půdně-Ekologické Jednotky (BPEJ))*; ÚZEI, Ed.; ÚZEI: Prague, Czech Republic, 2012.
38. Voltr, V.; Bartlová, J.; Brtnický, M.; Denešová, O.; Froněk, P.; Honz, J.; Hlavsa, T.; Hruška, M.; Khel, T.; Kohút, M.; et al. *Evaluation of Land in Environmental Conditions (Hodnocení Půdy v Podmínkách Ochrany Životního Prostředí)*; ÚZEI: Prague, Czech Republic, 2012; ISBN 978-80-86671-86-4.
39. Voltr, V.; Hruška, M.; Nobilis, L.; Fuksa, P.; Pokorný, J. Procedure of Economic, Energy and Environmental Evaluation of Crop Production in the Czech Republic. *Eur. Biomass Conf. Exhib. Proc.* **2020**, *251409*, 55–61. [CrossRef]

40. Jansky, J.; Zivelova, I.; Polackova, J.; Boudny, J.; Redlichova, R. Trend Analysis of Revenues and Costs within the Chosen Commodities under the Conditions of Organic Agriculture. *ZEMED. EKON. PRAHA* **2006**, *52*, 437. Available online: https://doi.org/10.17221/5047-AGRICECON (accessed on 3 March 2021).
41. Voltr, V. Emission of Pesticides. Available online: https://www.uzei.cz/data/usr_001_cz_soubory/210218_pesticides_emission.pdf (accessed on 18 February 2021).
42. Camargo, G.G.T.; Ryan, M.R.; Richard, T.L. Energy Use and Greenhouse Gas Emissions from Crop Production Using The Farm Energy Analysis Tool. *Bioscience* **2013**, *63*, 263–273. [CrossRef]
43. Voltr, V. Emissions of Annorganic Fertilizers. Available online: https://www.uzei.cz/data/usr_001_cz_soubory/210218_Anorganic_fertilizers_emission.pdf (accessed on 18 February 2021).
44. Voltr, V. Emission of Organic Fertilizers. Available online: https://www.uzei.cz/data/usr_001_cz_soubory/210218_organic_fertilizers.pdf (accessed on 18 February 2021).
45. Voltr, V. Diesel Emission. Available online: https://www.uzei.cz/data/usr_001_cz_soubory/210218_Diesel_emission.pdf (accessed on 18 February 2021).
46. Huijbregts, M.A.J.; Steinmann, Z.J.N.; Elshout, P.M.F.; Stam, G.; Verones, F.; Vieira, M.; Zijp, M.; Hollander, A.; Van Zelm, R. ReCiPe2016: A harmonised life cycle impact assessment method at midpoint and endpoint level. *Comment. Discuss. Artic.* **2017**, *22*, 138–147. [CrossRef]
47. Gallardo-Carrera, A.; Léonard, J.; Duval, Y.; Dürr, C. Effects of seedbed structure and water content at sowing on the development of soil surface crusting under rainfall. *Soil Tillage Res.* **2007**, *95*, 207–217. [CrossRef]
48. Nunes, M.R.; Karlen, D.L.; Denardin, J.E.; Cambardella, C.A. Corn root and soil health indicator response to no-till production practices. *Agric. Ecosyst. Environ.* **2019**, *285*, 106607. [CrossRef]
49. Baker, J.M.; Ochsner, T.E.; Venterea, R.T.; Griffis, T.J. Tillage and soil carbon sequestration-What do we really know? *Agric. Ecosyst. Environ.* **2007**, *118*, 1–5. [CrossRef]
50. Lhotský, J. *Soil Compaction and Measures Against it. System of Measures for the Reclamation of Agricultural Land (Zhutňování Půd a Opatření Proti Němu. Soustava Opatření ke Zúrodňování Zemědělských Půd)*; ÚVTIZ: Prague, Czech Republic, 1984.
51. Yu, S.; Tao, J. Economic, energy and environmental evaluations of biomass-based fuel ethanol projects based on life cycle assessment and simulation. *Appl. Energy* **2009**, *86*, S178–S188. [CrossRef]
52. Hanegraaf, M.; Bioenergy, E.B.-B. Undefined Assessing the ecological and economic sustainability of energy crops. *Biomass Bioenergy* **1998**, *15*, 345–355. [CrossRef]
53. Ren, J.; Yu, P.; Xu, X. Straw utilization in China-status and recommendations. *Sustainability* **2019**, *11*, 1762. [CrossRef]
54. Dornburg, V.; Termeer, G.; Faaij, A.P.C. Economic and greenhouse gas emission analysis of bioenergy production using multi-product crops—Case studies for the Netherlands and Poland. *Biomass Bioenergy* **2005**, *28*, 454–474. [CrossRef]
55. Börjesson, P.; Prade, T.; Lantz, M.; Björnsson, L. Energy crop-based biogas as vehicle fuel-the impact of crop selection on energy efficiency and greenhouse gas performance. *Energies* **2015**, *8*, 6033–6058. [CrossRef]
56. Arodudu, O.; Helming, K.; Wiggering, H.; Voinov, A. Bioenergy from low-intensity agricultural systems: An energy efficiency analysis. *Energies* **2017**, *10*, 29. [CrossRef]
57. Place, S.E.; Mitloehner, F.M. Beef production in balance: Considerations for life cycle analyses. *Meat Sci.* **2012**, *92*, 179–181. [CrossRef] [PubMed]
58. Cooprider, K.L.; Mitloehner, F.M.; Famula, T.R.; Kebreab, E.; Zhao, Y.; van Eenennaam, A.L. Feedlot efficiency implications on greenhouse gas sustainability. *J. Anim. Sci.* **2011**, *89*, 2643–2656. [CrossRef]
59. Place, S.E.; Stackhouse, K.R.; Wang, Q.; Mitloehner, F.M. Mitigation of greenhouse gas emissions from U.S. beef and dairy production systems. In *Understanding Greenhouse Gas Emissions from Agricultural Management*; ACS Symposium Series; American Chemical Society Publications: Washington, DC, USA, 2011; Volume 1072, pp. 443–457. [CrossRef]
60. Voltr, V.; Hruška, M.; Havelka, F. Comparison of the energy and economic balance of crop production. *Agric. Eng. Int. CIGR J.* **2020**, *22*, 138–150.
61. Fathollahi, H.; Mousavi-Avval, S.H.; Akram, A.; Rafiee, S. Comparative energy, economic and environmental analyses of forage production systems for dairy farming. *J. Clean. Prod.* **2018**, *182*, 852–862. [CrossRef]
62. Wurster, P.; Maneta, M.; Beguería, S.; Cobourn, K.; Maxwell, B.; Silverman, N.; Ewing, S.; Jensco, K.; Gardner, P.; Kimball, J.; et al. Characterizing the impact of climatic and price anomalies on agrosystems in the northwest United States. *Agric. For. Meteorol.* **2020**, *280*, 107778. [CrossRef]
63. Stinner, W.; Schmalfuß, T. *Perennial and Intercrop Legumes as Energy Crops for Biogas Production*; Springer: Singapore, 2018; pp. 139–171. [CrossRef]
64. Hakl, J.; Fuksa, P.; Konečná, J.; Plant, L.P. Effect of applied cultivation technology and environmental conditions on lucerne farm yield in the Central Europe. *Plant Soil Environ.* **2014**, *60*, 475–480. [CrossRef]
65. Peterson, G.A.; Schlegel, A.J.; Tanaka, D.L.; Jones, O.R. Precipitation Use Efficiency as Affected by Cropping and Tillage Systems. *J. Prod. Agric.* **1996**, *9*, 180–186. [CrossRef]
66. Chameides, W.L.; Yu, H.; Liu, S.C.; Bergin, M.; Zhou, X.; Mearns, L.; Wang, G.; Kiang, C.S.; Saylor, R.D.; Luo, C.; et al. Case study of the effects of atmospheric aerosols and regional haze on agriculture: An opportunity to enhance crop yields in China through emission controls? *Proc. Natl. Acad. Sci. USA* **1999**, *96*, 13626–13633. [CrossRef] [PubMed]

67. Voltr, V.; Klir, J.; Hruška, M. Soil productivity and its relation to the environment in the Czech Republic. In Proceedings of the IOP Conference Series: Earth and Environmental Science, Hangzhou, China, 26–29 June 2018; Volume 185.
68. A European Green Deal | European Commission. Available online: https://ec.europa.eu/info/strategy/priorities-2019-2024/european-green-deal_en (accessed on 18 February 2021).
69. Liberalisation of the Energy Market (Electricity and Gas) | Energy. Available online: https://ec.europa.eu/energy/content/liberalisation-energy-market-electricity-and-gas_en (accessed on 18 February 2021).

energies

MDPI

Article

Renewable Energy Generation Gaps in Poland: The Role of Regional Innovation Systems and Knowledge Transfer

Patrycjusz Zarębski [1], Vitaliy Krupin [2] and Dominika Zwęglińska-Gałecka [2,*]

[1] Department of Economics, Koszalin University of Technology, Kwiatkowskiego 6E, 75-343 Koszalin, Poland; patrycjusz.zarebski@tu.koszalin.pl

[2] Institute of Rural and Agricultural Development, Polish Academy of Sciences, Nowy Świat 72, 00-330 Warsaw, Poland; vkrupin@irwirpan.waw.pl

* Correspondence: dzweglinska@irwirpan.waw.pl

Abstract: Aim of the research is to analyze regional gaps in terms of renewable energy generation across Poland. For this purpose, four types of regions were outlined based on two indicators: the existing renewable energy generation capacity and the current regional energy demand revealed through the number of residents. This classification allowed to reveal regions in Poland that have distinct features of energy gaps and peripherality, while also more successful regions with renewable energy surpluses and distinct sustainable energy potential. For each of the region type key potential systemic problems in terms of renewable energy generation development were given. To understand how peripheral regions and regions with energy gaps could be supported in their development of renewable energy generation the regional innovation systems, social networks, knowledge and technology transfer and diffusion were substantiated. Results of the research can serve as an aid in development of national and regional energy policies, helping to understand peculiarities of local renewable energy generation and the influence of enabling environment peculiar to the specific region, including the regional innovation systems and intensity of knowledge transfer and diffusion.

Keywords: renewable energy; energy policy; regional innovation system; social network; knowledge transfer

Citation: Zarębski, P.; Krupin, V.; Zwęglińska-Gałecka, D. Renewable Energy Generation Gaps in Poland: The Role of Regional Innovation Systems and Knowledge Transfer. *Energies* **2021**, *14*, 2935. https://doi.org/10.3390/en14102935

Academic Editor: Gianfranco Chicco

Received: 30 April 2021
Accepted: 16 May 2021
Published: 19 May 2021

1. Introduction

Renewable sources of energy are among key sustainable development elements that are perceived as the future of our planet and can help in its environmental protection while satisfying the energy needs of global society. Due to the progressive economic development and the growth of the world's population the supplies of various fossil fuels necessary for energy generation are rapidly declining. Yet what is as important—the impact of energy generation from fossil fuels is polluting the environment and causing an intensifying climate change, thus leading to unpredictable consequences and shifts in various areas: weather events, climate zoning, plant growth patterns and biodiversity [1–3], and last but not least—human health [4]. These issues and the fact that their cause is mainly anthropogenic [5–7] is the reason numerous state governments and international organizations are taking actions to intensify shifting the societies and their economies toward sustainable development patterns and thus reducing or even stopping the unfavorable environmental and economic trends.

The European Union is especially active in this sense implementing policies aiming at large-scale transformations toward renewable energy generation, reduction of greenhouse gas emissions, and limiting the pollution of the environment. EU's Climate and Energy Framework [8] is one of such policies setting 2030 as their target year. Its basis is the recast Renewable Energy Directive (RED II) [9] with its key joint EU goals being: (a) at least 32% share for renewable energy, (b) at least 32.5% improvement in energy efficiency, and (c) 40% cuts in greenhouse gas emissions (compared to 1990 levels).

To reach the joint renewable energy goal each EU country has set their 2030 targets within their National Energy and Climate Plans (NECP) [10], ranging from 10% in Malta to 49% in Sweden, with an average 22% share in the final consumption among all 27 states. Yet, due to the recent introduction of the European Green Deal these targets are to be updated, as in order to achieve the planned climate-neutrality by 2050 a reduction of the core indicator—the greenhouse gas emissions—needs to reach 55% by 2030, thus requiring changes in the corresponding elements of the energy sector. In terms of the share of renewable energy in the final consumption the European Commission expects the target to increase from 32% to 38–40% at the EU-27 level [11].

Such ambitions will require intensified efforts from each EU member state, yet some of them are already struggling to reach the first stage levels—the ones set for the 2020 within the initial Renewable Energy Directive (RED I) and corresponding National Renewable Energy Action Plans. Our analysis of the current progress toward achieving the 2020 national renewable energy goals shows (Figure 1) that by the 2019 (more recent data is still not available [12]) nine countries were behind the schedule. The largest gaps were manifested by France (5.8 p.p.), the Netherlands (5.2 p.p.), Ireland (4 p.p.), and Luxembourg (4 p.p.). Among these countries is also Poland being 2.8 p.p. short of the 2020 goal, while the 2030 target of 21% of renewables in the final energy consumption implies the need of additional 6 p.p. to reach it. Future updates due to the European Green Deal will increase this obligation further.

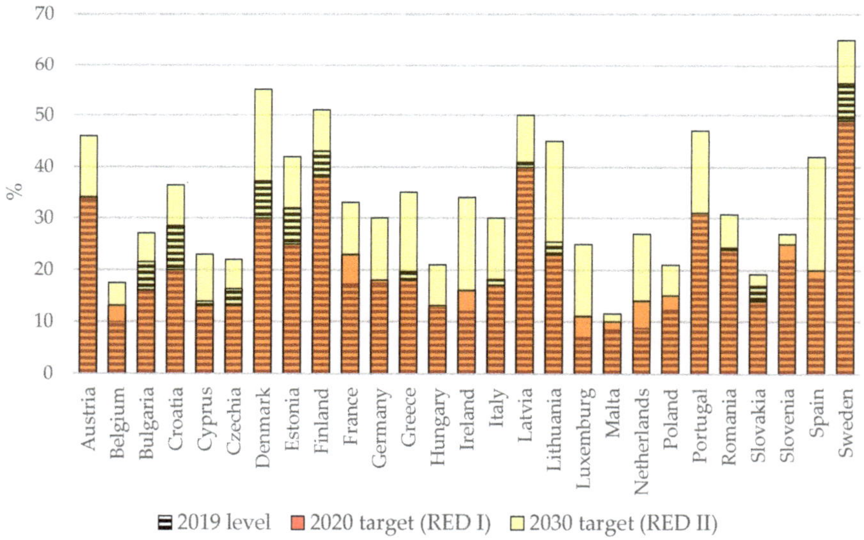

Figure 1. Actual (2019) and targeted (RED I by 2020 and RED II by 2030) shares of renewable energy in gross final energy consumption in the EU-27 countries. Source: own compilation based on [10,12,13].

These expectations will require intensification of efforts from both Poland and other EU countries. Overall, the pace and efficiency of changes in the energy generation structure are determined by various economic factors related to the possibility of financing and conducting the modernization of energy infrastructure, both at the level of energy suppliers and consumers. However, apart from economic factors, non-economic factors of a social and organizational nature also deserve attention. This differentiation is further deepened by regional peculiarities and varying levels of socio-economic development, thus leading to forming of permanent leaders and outsiders among regions, thus deepening the differentiation further due to occurrence of a system archetype [14] called "success to the successful", having an adverse impact on peripheral regions.

Therefore, the article aims to define the existing gaps in renewable energy generation across districts of Poland and to broaden the knowledge about the combined role of regional innovation systems, networks of enterprises and institutions in the process of knowledge and technology transfer and diffusion. It is put in the context of renewable energy generation development taking into account specific conditions of four outlined types of regions in regard to their energy generation capacities and further generation potential. The general theoretical framework is based on the theory of regional innovation systems [15,16], social networks [17] and the concept of "gatekeepers of knowledge" [18]. While this study is based on the case study of Poland, it helps to deliver new knowledge about the processes of renewable energy development typical for other countries as well, especially in the Central and Eastern Europe.

The need to intensify the further spread of technologies and their implementation in various regions arises not only from the necessity to increase the share of renewable energy generation to achieve the set targets. Decentralization of energy generation has also been emphasized upon by the RED II [9] stating that "the move towards decentralized energy production has many benefits, including the utilization of local energy sources, increased local security of energy supply, shorter transport distances and reduced energy transmission losses. Such decentralization also fosters community development and cohesion by providing income sources and creating jobs locally". It is also supporting the transformation toward ensuring of local and sustainable energy access [19], and minimizing possible energy access risks.

The conducted research revealed that in the diffusion of technologies and innovations in the field of renewable energy generation, the flow of knowledge and information is important, as it helps to raise awareness about the necessity of transformation toward renewable energy, as well as simplifies the access to innovational solutions and technologies. Such knowledge relates, inter alia, to the general characteristics of renewable energy sources, which are the resources obtained within natural processes that are not depleting limited natural supplies and are an alternative to conventional non-renewable energy sources. This knowledge is relatively well communicated and understandable to the majority. However, it can be difficult to understand the process of conversion between energy sources and how to engage the transformation in practice. As technologies and solutions constantly develop based on ongoing scientific research, there's a necessity to make this information more accessible and understandable. Currently, nine different sources of renewable energy are used in Poland, the main ones being the solid and liquid biofuels, as well as the wind energy (Figure 2). It is also worth mentioning that the structure of renewable energy generation in Poland is not yet highly diversified and differs significantly from the corresponding structure of the European Union.

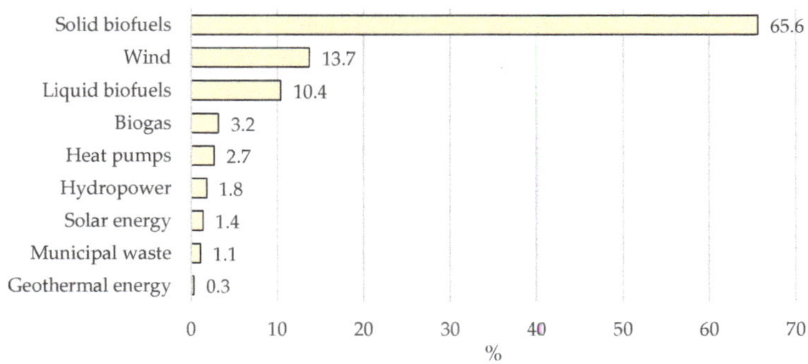

Figure 2. Structure of generated renewable energy by types of installations in Poland in 2019. Source: [20].

There are various factors causing the existing disproportions in the presented structure of renewable sources of energy. These include, among other, the state support and permit systems, investment costs, local conditions (e.g., territorial, infrastructural). Yet among other crucial factors are also knowledge level and rate of diffusion of innovations and technologies. When looking at regional level, the energy generation there takes place under specific conditions depending on particular region, which could be unbeneficial in case of unfavorable economic and social conditions. This is especially visible in case of peripheral areas, which are characterized by a number of negative phenomena caused by weak economic and social structure, low population density, low level of human capital, low infrastructural development, lack of available and adequate financing sources for implementation of new advanced technologies. Therefore, such areas require a separate approach in the context of renewable energy generation development. In the process of developing the energy innovations, apart from traditional factors related to the development of technologies and their financing, non-economic factors related to social relations and knowledge-flow networks are equally important. In the literature on the subject [21,22], the concept that has already gained a large group of supporters and constitutes the theoretical framework for the implementation of innovation and the flow of knowledge is the existence of regional innovation systems. So, what purpose can they serve in regard to renewable energy generation?

The knowledge and practical skills in the field of renewable energy generation are often developed and shared within limited scientific environments or enterprises involved in production of dedicated equipment. The knowledge gap arising from this fact is not the fault of these professional communities, yet is appearing due to technical limitations, complexity of issues and solutions, and lack of direct networking platforms involving stakeholders of various types. This knowledge gap is among the reasons why the rate of implementation of new technologies in the economy is too slow and falls behind the expected levels. Lack of knowledge about technologies and benefits, not only economic, but also environmental and social, is a significant barrier to sustainable development. Proper knowledge flow and its diffusion in various socio-economic environments can significantly accelerate development processes. One of the concepts that explains the mechanisms of creating and diffusing new technologies is one of the regional innovation system [23]. Its main underlying mechanism are the networks of relations within which their stakeholders exchange knowledge and cooperate in various projects. This concept embraces a multi-disciplinary approach, enabling to understand not only the role of technology in energy generation, but also social and policy conditions that are necessary to boost the processes of its implementation, especially the knowledge diffusion based on social networks and relations.

The emphasis on the importance of intensification of renewable energy generation results from a paradigm shift from the classic perception of energy policy [24] to its perception in terms of sustainable development. While concentrating the following analysis on the energy policy and its implications at local level, it is emphasized by the authors that this approach is not derived from the classic energy policy, but the so-called sustainable energy policy [25], being an overall long-term improvement of social welfare by striving to maintain a balance between the following: energy security, satisfaction of social needs, competitiveness of economy, and environmental protection. Thus, it is more than just ensuring the energy supply. There is a need to balance other crucial socio-economic elements, the condition and quality of which may significantly affect the conditions and the existing potential for development of renewable energy generation.

To conclude the introduction, the article consists of eight sections. Following the Introduction, Section 2 delivers information and literature review regarding peculiarities of renewable energy generation development in Central and Eastern EU countries. Section 3 explains the method used to carry out the analysis of regional renewable energy generation gaps in Poland, substantiates the use of own typology of regions according to their renewable energy generation potential, provides information about the data sources and peculiarities of approach in terms of regional dimension. Section 4 presents the research results, including: (a) an analysis of existing renewable energy generation capacities across districts of Poland aimed to understand the occurrence and scale of gaps at regional level, (b) an analysis of four outlined types of regions in regard to their renewable energy generation capacities, (c) recommendations for renewable energy development policy across the four outlined types of regions based on the influence of regional innovation systems and potential knowledge transfer. Section 5 covers the substantiation of regional innovation systems, social networks and knowledge transfer from the standpoint of stimulation of regional renewable energy generation in regions demonstrating energy peripherality and gaps, which is based on the literature review providing its synthesis and own arguments. Section 6 covers discussion between the obtained results and other studies, while the Section 7 is devoted to conclusions, implications of presented research and other possible research directions.

2. Peculiarities of Renewable Energy Generation Development in Central and Eastern EU Countries

Renewable energy sources (wind, solar, hydroelectric, ocean, geothermal, biomass and biofuels) are alternatives to fossil fuels and contribute to reducing greenhouse gas emissions, diversifying energy supply and reducing dependence on uncertain and unstable fossil fuel markets, especially oil and gas. Global development dynamics of energy generation from renewable sources in recent decades indicate that combined they are the fastest growing exploited source of energy. This is clearly visible in economies of such countries as USA, India or China. For example, the key slogan for current India's economic development is "Go Green", which is actively implemented through development of renewable energy generation, zero-emission public transport, other "green" technologies [26].

Yet while the development of renewable energy generation is becoming a global trend, there are differences between countries and macroregions due to specificities of socio-economic past and present development. The report "Global trends in renewable energy investment 2020" indicates that on a global scale, investments in the renewable energy sector in developed economies increased by 2% in 2019 only (compared to the previous year)—to USD 130 billion. At the same time, it is stressed that there were sharp increases in outlays in the USA, Spain, the Netherlands and Poland, and big falls in the UK, Germany, Australia and Belgium [27].

Renewable energy resources and the level of their utilization should be assessed primarily through the prism of a country's energy supply and demand. Consequently, the concept of a renewable energy resource is a purely economic concept and is derived from the function it delivers. And it should be remembered that the amount of renewable energy supply may increase according to the changes in the energy needs and along the growing knowledge and technological possibilities of its conversion into exploitable energy. Although on the EU-27 scale the share of renewable energy in gross final energy consumption has increased from 9.6% (2004) to 19.7% (2019) [12], it is precisely the above-mentioned aspects that can be indicated as the reasons why the "renewable energy revolution" [28] takes place at a different pace in different countries. The use of natural resources for energy generation depends on many factors that could be categorized collectively as: economic, social, legislative, and—to a lesser extent—technological [29]. Although renewable energy generation will definitely be one of key development elements of future long-term EU strategies, their implementation to an equal extent in all EU countries will always be a difficult task due to differences in national priorities and conditions. In Western Europe, the focus on decarbonisation, slowing the climate change and building a single energy market is well underway, while in countries of Central and Eastern EU, where large share of energy is still derived from conventional fuels, a rapid transformation replacing them with renewable energy sources will be more difficult.

The EU member states, in the name of the common good and to limit the climate change can support one energy source at the expense of abandoning other. Yet to implement such choices key factor is the social support. Therefore, an awareness of climate change challenges facing all economic sectors is an increasingly discussed issue among the academic circles and general public, which is the case in the European Union countries, and especially those in Central and Eastern Europe. Within these debates researchers emphasize the importance of natural, spatial [30], technological [31], social, and political [32,33] conditions for the development of renewable energy generation. Researchers indicate potentials for such generation and determine the possible development paths—both renewable energy generation in general and broken down into specific types of renewable energy sources, also taking into account the interrelationships between sources that can be used for generation of electricity, heat and transport fuels [34].

In the literature it is indicated that countries of Central and Eastern EU have substantial problems in steering the process of setting and achieving long-term energy policy goals due to the short-term way of thinking, characteristic for countries undergoing economic transformation [35]. Not without significance are the past experiences of communism common for the countries of Central and Eastern Europe, which still manifest specific impact on local development conditions, including relatively higher corruption, low trust in political elites and innovative solutions [36]. For example, in Poland, on one side, a strong political force is aiming to keep the traditional framework in the energy sector (e.g., large-scale generation, use of coal), and on the other side, there is an understanding of alternativeless need to intensify the shift to renewable energy generation, which is not only an important element of the global climate protection movement and part of EU policies, but is also cheaper, more efficient and increases local energy security. Yet it is indicated that level of participation of the public in creation of the energy policy is still not sufficient, which may cause the deficiency in development of the prosumer model of renewable energy generation [37]. These issues overall impact the support for renewable energy generation and implementation of policy measures aimed at such development, as well as cause implementation of support mechanisms to be highly volatile, thus making this sector relatively less attractive for investors and entrepreneurs. Existing barriers increase investment risks, which translate directly onto the costs of energy obtained from these sources, which could be higher compared to energy obtained from conventional energy sources. It is also indicated that renewable energy generation technologies are characterized by an uneven pace of energy supply over time or even intermittent operation process, meaning their generation level could be at times inconsistent with the level of energy demand. At the same time, researchers emphasize that the need to take action is forced both by international public obligations in the field of climate change mitigation adopted by all the EU countries (including the Central and Eastern ones), as well as economic, social and environmental considerations, which results in the necessity to undertake costly investments not only in the energy generation sector, but also in energy transmission infrastructure.

3. Materials and Methods

According to the legislation in Poland [38] a renewable energy generation installation is an installation that constitutes a separate set of devices used to generate energy, described by technical and commercial data, in which energy is produced from renewable energy sources, or construction facilities and equipment constituting a technical and utility unit used to generate agricultural biogas—as well as an energy storage connected to this unit, including an agricultural biogas storage facility.

While the sources of renewable energy vary, generated output can be converted into either electrical or thermal energy. This article, due to data limitations, deals strictly with generation of electrical energy, thus it is meant across the article whenever generation of energy is mentioned.

Processes of innovation and technology diffusion are conditioned by spatial factors and depend on peculiarities of local socio-economic systems. Due to this, an analysis of spatial allocation's differentiation regarding existing technological capacities for generation of energy from renewable sources was carried out in relation to the regional energy demand expressed by a number of people living in a given administrative area. Presented herewith study aims to understand the differences between the regions in development of renewable energy generation capacities and define what mechanisms determined this process. Description of the research stages adopted for this study can be found in Figure 3. First, a literature review was carried out to explore the renewable energy generation specificities across Poland and define what factors could influence decisions to undertake the development of such generation capacities. In the next stage (the empirical part), a spatial analysis of allocation of renewable energy generation capacities was carried out. It was conducted based on indicators assigned to individual districts in accordance with the "Jenks natural breaks optimization" method [39] and presented on maps. This method is used to present heterogeneous data sets as it aggregates analytical units into groups with similar values. Grouping of values in different classes is carried out by the function aiming to minimize the mean deviation of each class from the mean class, while maximizing the deviation of each class from the mean of other groups. The next step was to develop a typology of regions based on the relationship of the two aforementioned datasets. The first one represented the aggregate index of technological capacity, while the second one represented the population quantity in a given area. Understanding the occurrence of districts manifesting energy gaps and energy peripherality was the basis to go deeper into research of regional innovation systems (RIS) and their possible influence upon intensification of renewable energy generation development. The regional innovation systems therefore, constitute the theoretical framework for considerations and explain such phenomena as social embedding of innovation processes and institutional relations, which play an important role in knowledge and technology transfer processes.

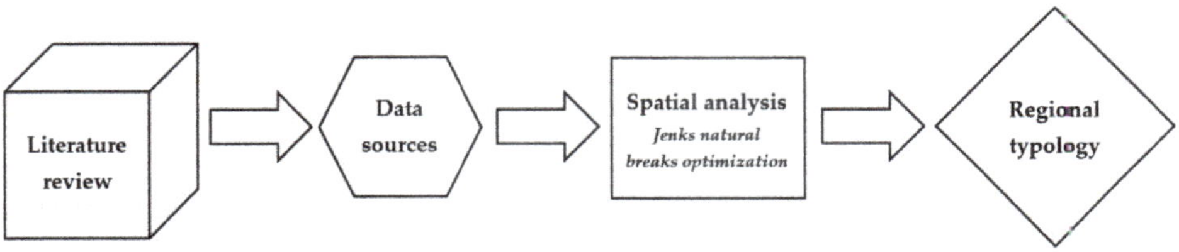

Figure 3. Algorithm of the performed research. Source: own substantiation with the use of [39].

In terms of the covered types of renewable energy sources (RES), a region's relative renewable energy generation potential depends mainly on the capacity of generation installations based on renewable energy sources in relation to the energy demand expressed by number of residents in a given region. Large disproportions in this relation cause regional energy surpluses or deficits (energy gaps), meaning in some regions an imbalance between supply and demand for energy is present. In order to better recognize this phenomenon, statistical data was collected about the regional capacities of installations aimed for renewable energy generation, utilizing the following:

- Biogas (BG),
- Biomass (BM),
- Solar energy (PVA),
- Wind energy (WIL),
- Hydropower (WO),
- Technology of co-combustion of biomass, biogas or bioliquids combined with other fuels (fossil fuels and biomass/biogas/bioliquids) (ITPO).

Therefore, the aggregate energy generation capacity from renewable sources (EGC) is calculated as the total of generation capacities of all available technologies:

$$BG_i + BM_i + PVA_i + WIL_i + WO_i + ITPO_i = EGC_i$$

where:

i—spatial unit number,

EGC—energy generation capacities from all renewable sources.

Due to existence of diversified technologies for electricity generation the data on different types of renewable energy installations is derived from various sources and covers those entities and individuals which have: a license to generate electricity, an entry in the regulated activity register kept by the President of the Energy Regulatory Office [40] (register of small installation energy producers [41]), an entry in the regulated activity register kept by the General Director of the National Agricultural Support Center [42] (register of agricultural biogas producers [43]), as well as micro-installations generating electricity under the "certificate of origin" system, the guaranteed feed-in tariff system, or the auction support system [44]. The main source of data used within the study were those published by the Central Statistical Office in Poland (GUS), additionally some of the data was obtained from available statistical and scientific publications.

The regional renewable energy generation capacities are presented in the study at a district level, which in Poland are referred to as powiat, being units of the second-degree state administrative division, each consisting of several municipalities (gmina) and having its own administrative body. The total number of such districts in Poland equals 314.

In the scientific literature it is said that deficits and problems in the functioning of regional innovation systems can be related to the conditions that prevail in specific types of regions, such as peripheral regions (facing organizational thinness), old industrial areas (facing cognitive blockades) and some metropolitan regions (facing fragmentation of interactions and networks) [45]. It should be noted that in many cases the problems of regional innovation development are in fact similar, but there may be some dominant innovation problems specific to a given type of region. Therefore, the authors propose four types of regions that differ in terms of population numbers and renewable energy generation capacities.

In the Figure 4, the X-axis holds two types of population quantity expressed through the number of inhabitants and the Y-axis represents the renewable energy generation capacity. The four quarters are constructed on the basis of the average results of assessments of these two parameters. The average number of inhabitants for the collected data across the regions equaled 82,185 people, while the average renewable energy generation capacity was at the level of 28.5 MW. These values are the cut-off points that determine the energy region types.

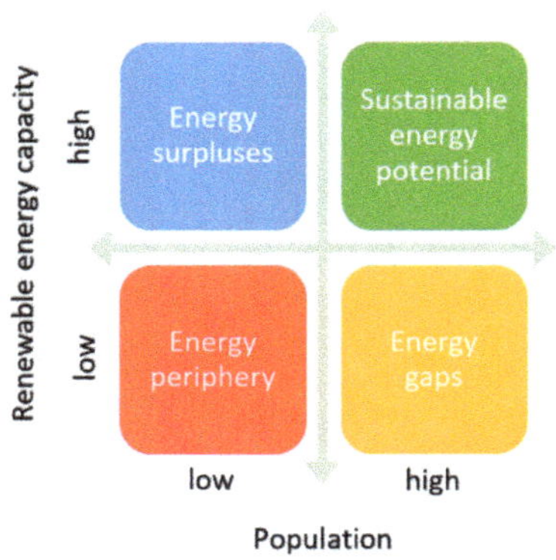

Figure 4. Outlined region types according to the energy/population parameters. Source: own substantiation.

The research questions to be explored within the article include:

- What disproportions occur between regions concerning renewable energy generation, which regions can be considered peripheral in this regard and which manifest existence of technological gaps?

- What systemic problems block the development of renewable energy generation technologies region-wise?
- Can regional innovation systems support and intensify the innovation and technology deployment process in regions with energy deficits and regions with peripheral features?
- What is the role of social and organizational networks in the policy of innovation and technology knowledge diffusion in regard to renewable energy generation?

4. Defining the Renewable Energy Generation Gaps at Regional Level

Determining the capacity of renewable energy generation installations and the population quantity in districts allowed to develop spatial distribution of energy generation from available renewable sources across Poland. Results (Table 1) revealed that overall energy generation from renewable energy installations within the 2005–2020 increased over ninefold. The quantity and capacity of solar energy installations have shown the most dynamic growth rates. State support programs and subsidies encouraging to invest in such installations are a key driver [46] and a clear long-term state policy toward development of renewable energy generation sources is of high importance [29].

Table 1. Dynamics of installed renewable energy generation capacities within 2005–2020 in Poland [in MW].

Types of Installations	Years			
	2005	2010	2015	2020
Biogas	31.972	82.884	212.497	255.699
Biomass	189.790	356.190	1122.670	1512.885
Solar energy	-	0.033	71.031	887.434
Wind energy	83.280	1180.272	4582.036	6347.111
Hydropower	852.495	937.044	981.799	976.047
Total	1157.537	2556.423	6970.033	9979.176

Source: Energy Regulatory Office in Poland.

In-depth analysis of the collected data split by the types of renewable energy generation installations in Poland is presented in Table 2, while Figure 5 presents their density distributions.

Table 2. Descriptive statistics for the analysis of renewable energy generation installations in Poland.

Parameters	Types of Installations						
	Biogas	Solar Energy	Wind Energy	Hydropower	Biomass	Co-Combustion	All Sources
Valid	156	276	208	191	29	4	314
Missing	158	38	106	123	285	310	0
Mean	1.267	3.152	30.495	4.144	23.584	12.447	28.457
Std. Error of Mean	0.091	0.226	4.037	1.332	10.228	2.170	3.033
Median	1.000	2.015	10.915	0.220	2.200	11.400	9.450
Mode *	1.000	1.000	0.600	0.030	0.050	8.990	1.000
Std. Deviation	1.131	3.749	58.230	18.415	55.077	4.340	53.740
Variance	1.278	14.055	3390.683	339.114	3033.454	18.833	2887.993
Skewness	1.856	2.804	6.185	8.504	2.853	0.745	5.887
Std. Error of Skewness	0.194	0.147	0.169	0.176	0.434	1.014	0.138
Kurtosis	6.097	11.191	56.055	83.765	7.891	−1.865	54.389

Table 2. *Cont.*

Parameters	Types of Installations						
	Biogas	Solar Energy	Wind Energy	Hydropower	Biomass	Co-Combustion	All Sources
Std. Error of Kurtosis	0.386	0.292	0.336	0.350	0.845	2.619	0.274
Range	7.620	25.710	631.130	207.170	229.950	9.010	634.940
Minimum	0.040	0.020	0.010	0.010	0.050	8.990	0.000
Maximum	7.660	25.730	631.140	207.180	230.000	18.000	634.940
Sum	197.660	869.960	6342.970	791.410	683.950	49.790	8935.400
25th percentile	0.378	0.935	2.000	0.080	0.950	8.998	3.020
50th percentile	1.000	2.015	10.915	0.220	2.200	11.400	9.450
75th percentile	1.903	4.190	42.045	1.125	5.800	14.850	30.230

Notes: * More than one mode exists, only the first is reported. Source: own calculations.

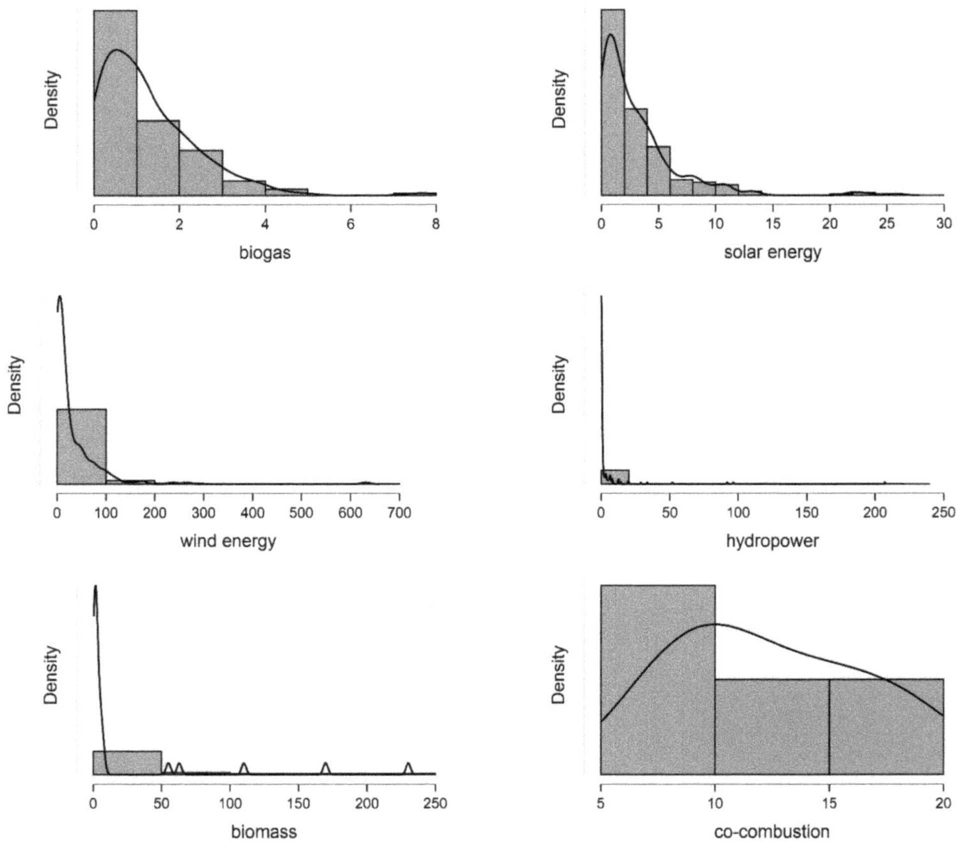

Figure 5. Density distribution of renewable energy generation installations in districts of Poland in 2020. Source: own calculations.

On the national scale in Poland, among the analyzed types of renewable energy generation capacities the highest density is present in case of installations for solar energy generation, which are located mainly in the north and east of the country (Figure 6). Districts located in the Pomorskie region (north-western part of the country, with access to the Baltic Sea coast) are national leaders in their construction. In turn, the areas of central Poland are showing the lowest densities or even lacking such installations whatsoever.

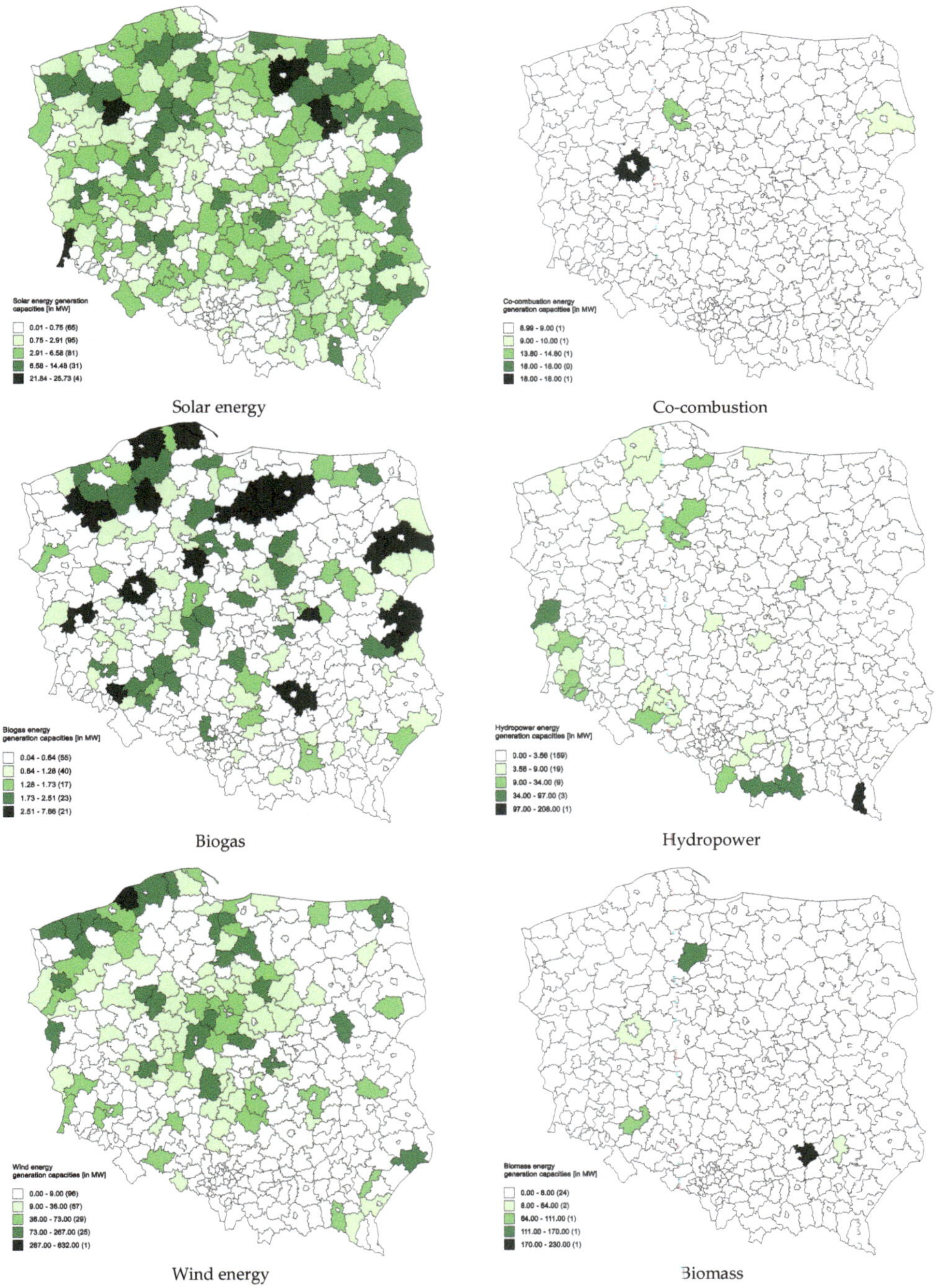

Figure 6. Renewable energy generation capacity by types of installations in districts of Poland in 2020. Source: own calculations and presentation.

It should be noted that the renewable energy installations are overall scattered across the nation, and such approach is one of the pillars of state energy policy. The purpose of having decentralized energy sources is to provide energy supplies to less urbanized areas and rural areas, as well as to guarantee sustainable local development of those areas. The main reason for the development of decentralized energy sources is the technological progress, which contributes to the reduction of costs of energy generation from renewable sources, as well as the possibility to utilize energy resources available locally.

Figure 7 presents the combined data about renewable energy generation capacities for installations of all analyzed types, while also revealing the distribution of population among the districts, which served to proceed with typologization of regions.

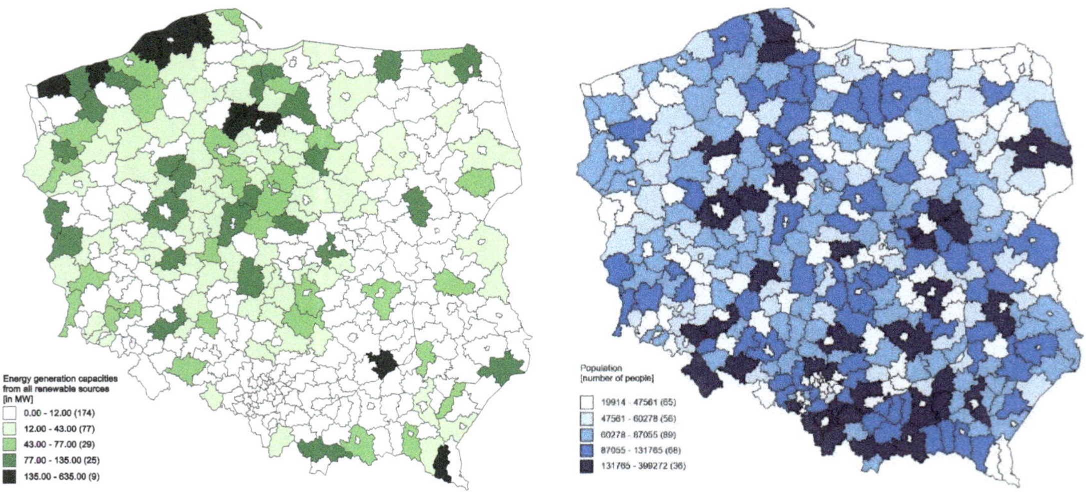

Figure 7. Renewable energy generation capacity of all installation types combined (**left**) and population (**right**) in districts of Poland in 2020. Source: own calculations and presentation.

The conducted analysis made it possible to identify four types of regions. Their spatial arrangement across the country has a mosaic pattern (Figure 8). The most numerous are the energy periphery regions (Table 3), which can be described as problematic regions with low energy generation capacity, while also manifesting low population numbers. And as the research conducted in Poland shows [47], the low population is generally accompanied by overall low socio-economic development and deficiencies in infrastructure. This also concerns the energy network infrastructure, which hinders and sometimes even blocks possibility for the development of new energy sources. Low level and quality of internal factors (both traditionally understood as soft factors) are noted here. Districts of this type are characterized by a low level of key conditions essential for the development of innovation. This is referred to in the literature as organizational thinness [45]. The problem of the periphery regions is also the underdeveloped network and connections of specialized knowledge providers, such as universities and research organizations [48]. This type of region occurs numerously throughout the country, but is especially highly concentrated in the north-east and mid-west of Poland. In terms of demographic determinants, the situation in these areas can be assessed as a range from bad to average. They are characterized by a negative migration balance and outflow of local residents, as well as by struggle with the problems of ageing population. Residents in these regions are primarily employed in small-scale agriculture, and have lower average education degree compared to residents of other regions [47]. Compared to Poland in general, these areas are the least economically developed, and the phenomenon of energy poverty occurs to a high extent in these regions.

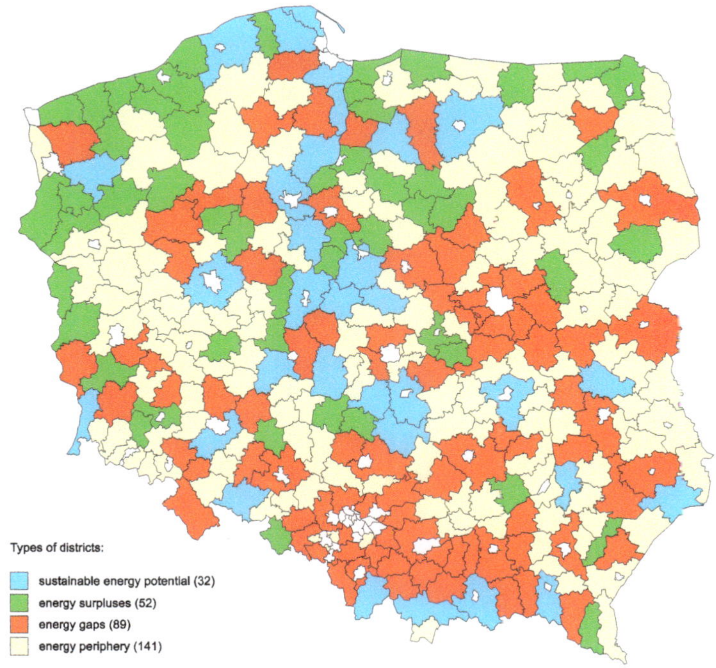

Figure 8. Spatial distribution of districts according to outlined types in 2020. Source: own calculations and presentation.

Table 3. Descriptive statistics for the analysis of districts according to their renewable energy generation type.

Parameters	District/Region Types			
	Sustainable Energy Potential	Energy Surpluses	Energy Gaps	Energy Periphery
Valid	32	52	89	141
Missing	0	0	0	0
Mean	75.844	91.047	7.799	7.658
Std. Error of Mean	8.968	12.856	0.735	0.600
Median	59.065	67.945	5.770	5.120
Mode *	29.090	28.870	0.210	1.000
Std. Deviation	50.728	92.704	6.929	7.122
Variance	2573.356	8594.022	48.018	50.718
Skewness	2.567	4.282	0.973	0.972
Std. Error of Skewness	0.414	0.330	0.255	0.204
Kurtosis	8.305	23.379	0.019	−0.075
Std. Error of Kurtosis	0.809	0.650	0.506	0.406
Range	251.500	606.070	27.660	27.450
Minimum	29.090	28.870	0.000	0.000
Maximum	280.590	634.940	27.660	27.450
Sum	2427.020	4734.460	694.090	1079.830

Notes: * More than one mode exists, only the first is reported. Source: own calculations.

The second most abundant type are the regions with energy gaps, which boasting a relatively high population are simultaneously characterized by low renewable energy generation capacities. These types of districts are located mainly in central and southern Poland. It should be noted that this type of region could transform rather quickly into the type with sustainable energy potential. This may be due to the existing population potential, since rural areas of this type are showing a positive demographic trend and are a migration destination [47], thus additionally stimulating their development. Areas of this type are often located near large cities that determine population development trend, but at the same time the low renewable energy generation capacity may indicate insufficient development of infrastructure networks (including energy networks). Unlike the periphery regions, they face the reverse problem of over-clustering because they are over-specialized in mature industries hit by decline [49]. This type of regions often has a highly developed and specialized system of knowledge generation and diffusion; however, their problem is too much focus on traditional industries and fields of technology (e.g., the region of Silesia) [50]. With regard to functioning networks of relations, a key feature of the old industrial regions is that they suffer from various forms of "lock-in" which significantly block the development potential and the possibilities of diffusion of innovation and knowledge. Such blockades are a consequence of overly rigid networks established between enterprises and the policy, and links between public and private entities, which hinder the process of industrial restructuring. However, these districts have a substantial positive potential for the development of prosumer energy, as the population with relatively higher income levels is more eager to install renewable energy installations [51]. Residents of areas described as regions with energy gaps are relatively well educated. These are people who maintain permanent contacts with the city—mainly by being employed there. It is possible that the environment these people dwell in, contacts with other educated and highly aware citizens positively affect their pro-environmental attitudes, which in turn leads to growing popularization of the prosumer model of energy generation in the areas of their primary residence.

Another type are the regions with energy surpluses. This type is characterized by low population potential and high energy generation potential—which is not used due to low population density. Spatially, such districts are located primarily in the north of the country—mainly in places where wind energy generation is being developed. Nevertheless, this type is also highly dispersed nationally. It can be seen that the surplus of energy results from the specificity of the sources located there. The energy generation capacities are not always located at the densely populated areas. The key task of such regions is to create energy transmission grids with regions of high energy needs. Financial resources and systemic solutions are needed so that these regions could flawlessly transmit energy. There is a lack of organizations that have commercialized this energy generation potential and there are no institutions connecting the regions.

The last and least numerous type are the regions with sustainable energy potential, which are characterized by high population numbers and high renewable energy generation capacities. Districts of this type are highly dispersed throughout the country. However, the sequence of communes running from Pomorskie to the central part of the country is quite characteristic. These regions are the least problematic because there is a balance between their needs and generation capacities. Yet political decisions, unfavorable for further development and the lack of continuation of the energy policy may pose a threat. Energy generation leaders in these regions block small producers, which makes it difficult to implement the idea of decentralized energy generation based on individual prosumers.

In the light of the differences in outlined types of regions an important issue is to understand possible implications that may have a systemic character and hinder the potential development of renewable energy generation. For this purpose a typology of systemic problems [52] has been combined with types of regions outlined in this study (Table 4). While the types of regions differ according to their renewable energy generation potentials, all four types may encounter mild or severe potential systemic problems, which might either slow down their positive development, or block possibilities of such. Again, an intensive transfer and wide diffusion of knowledge in this matter could serve as an important factor ensuring minimization of potential negative implications.

Table 4. Potential systemic problems in regions depending on their renewable energy generation.

Region Types	Potential Systemic Problems According to Selected Dimensions (Based on Typology of Systemic Problems of [52])
Energy periphery	Capabilities/capacities: – Lack of technological knowledge of policy makers and engineers, – Lack of ability of entrepreneurs to pack together, to formulate clear message, to lobby to the government, – Lack of users to formulate demand, – Lack of skilled staff.
Energy gaps	Knowledge infrastructure: – Wrong focus or no specific courses at universities and knowledge institutes, – Gap/misalignment between knowledge produced at universities and what needed in practice. Lack of legitimacy: – Different actors opposing change, – Individualistic entrepreneurs. Too weak interactions: – Individualistic entrepreneurs, – No networks, no platforms, – Lack of knowledge diffusion between actors, – Lack of attention for learning by doing. Hard institutions: – "Misalignment" between policies on sector level such as agriculture, waste, and energy, and on governmental levels, i.e., EU, national, regional level and other.
Energy surpluses	Hard institutions: – "Stop and go policy": lack of continuity and long-term regulations; inconsistent policy and existing laws and regulations, – "Attention shift": policy makers only support technologies if they contribute to the solving of a current problem, – "Valley of Death": lack of subsidies, feed-in tariffs, tax exemption, laws, emission regulations, venture capital to move technology from experimental phase towards commercialization.
Sustainable energy potential	Too strong interactions: – Strong dependence on government action or dominant partners (incumbents), – Network allows no access to new entrants. Market structures: – Large-scale criteria, – Incremental/near-to-market innovation, – Incumbents' dominance.

Source: own substantiation utilizing the typology of systemic problems from [52].

5. Regional Innovation Systems, Social Networks and Knowledge Transfer

The concept of regional innovation system (RIS) is popular among scientists of various disciplines as its theoretical framework responds to the needs of researching the phenomena and mechanisms related to the emergence of innovation and knowledge transfer [45]. The main idea behind the RIS is that the innovation efficiency in the economy depends on the innovative capabilities of enterprises and research institutions, and on how they interact with each other and public institutions [53]. The specificity of the concept lies in the fact that it shows overlapping dimensions, i.e., broadly understood institutional infrastructure and production system, and then explains the mechanisms of relations that arise between them based on established rules and regional policy. In this part of the study, authors search to answer what theoretical perspective the concept of RIS originates from, what are its main components and mechanisms, and whether there are different forms of RIS in the context of regional diversification of the periphery center.

The concept of RIS was created as a result of evolution of views on the functioning of the national innovation system (NIS), which was described in the works of Edquist [54], Lundvall [55], and Nelson [56]. It should be noted that it is quite difficult to identify the differences between NIS and RIS. The rationale for the emergence of the RIS from the general concept of NIS was that researchers wondered to what extent regions differ from one another in terms of their potential and processes that take place in the creation and absorption of innovation. It became the reason to propose a new concept, which to a greater extent takes into account the regional disparities of potentials for the creation of innovations. One of the first attempts to define and describe the concept of regional innovation systems can be found in Cooke et al. [15] who define the RIS as a system in which companies and other organizations are systematically engaged in interactive learning through an institutional environment characterized by embeddedness. In addition, Asheim and Isaksen [16] noted that a (regional) innovation system consists of a production structure (techno-economic structures) and institutional infrastructure (political-institutional structures).

In the definition of the concept of a regional innovation system, there are three important aspects that are key to its understanding. First is the expression "interactive learning", which means an interactive process by which knowledge is transferred, then combined and, as a result, constitutes a knowledge base as a shared resource of various entities cooperating in the system. Knowledge is the basic factor used not only in the process of creating innovation, but also in the process of absorption of innovation/technology and building cognitive, organizational and social closeness. Second, the term "environment" being an open territorial complex that includes principles, norms, values, and human and material resources. It is a set of territorial conditions which combined create a potential for the functioning of a system specific for a given area. The third aspect that definitely distinguishes the discussed concept is paying attention to social closeness. Economic relations are to some extent always embedded in a social context, while social ties or relations influence economic performance [57]. Social closeness is related to the term "embeddedness" which covers all economic and knowledge creation processes, and then its duplication in business environment and beyond. The process of learning and absorption of innovation is often based on trust, therefore social relations facilitate the exchange of tacit knowledge, which is by nature more difficult to communicate, exchange and trade through markets [58].

The concept of RIS assumes that innovation is a process in which enterprises use both internal and external resources of a material and institutional nature. It should be emphasized that the functioning of regional innovation systems depends not only on knowledge resources created by enterprises and institutions, but also on the strength and structure of relations, created networks, which are the platform of cooperation with the environment. Innovations do not arise in isolation and cannot thrive solely based on given enterprise's internal resources, but are rather a result of synergy of numerous factors and processes. The environment mentioned earlier can therefore be perceived as a network of entities and institutions that form the framework for innovative activity and interactive learning. Thus, the interaction between educational organizations, which can be defined in terms of knowledge and information flows, investment flows, networking and other partnerships, is the most important process driving the evolution and strengthening of RIS [53]. In conclusion, RIS is primarily a social system that involves systematic interactions between different groups of private and public sector institutions and individuals in order to increase localized learning opportunities in the region.

The innovation system requires defining its main components, i.e., institutions that play an important role in the innovation process. Lundvall [55] lists the basic elements of such system, which are: internal organization of enterprises, relations between enterprises, role of the public sector, institutional structure of the financial sector, research and development intensity, and research and development organizations. In general, the main elements encompassing RIS are enterprises, institutions, and the knowledge infrastructure and innovation policy.

An innovation-oriented policy is an important regulator of the processes that take place in the regional innovation systems. Its direction and scale of activities increase, among other things, the possibilities of learning and diffusion of knowledge. As practice shows, the optimal level to implement an innovation policy is one of a region, which was confirmed in the policy carried out by the European Commission [59,60]. The genesis of these policies dates back to the 1980s, when a group of OECD experts developed the concept of a dynamic approach to international competitiveness as an alternative to a static, cost-based view of the theory of international trade and competitiveness. The concept developed at that time assumes that international competitiveness can be achieved by promoting learning and innovation development in societies. Also in other dimensions, the approach to innovation systems represents an important theoretical and political progress. Identifying innovation as a key factor of economic growth emphasizes the role of interactive learning processes between multiple entities and organizations [61]. Such RIS policies aim to improve the interaction and collaboration between the knowledge infrastructure, companies and institutions. Moreover, these policies respond to individual and collective innovation needs. In other words, strategies are developed to support the endogenous institutional capacity of regions by encouraging the diffusion of technology on a regional scale [62]. Innovative policy tools typically include: managing the scientific knowledge base; providing financial incentives for innovation efforts, technology dissemination policies and initiatives; promoting programs and companies leading the implementation of new technologies; and the creation and maintenance of intangible assets and legal regulations that favor innovation and technology transfer.

RIS concept distinguishes four main internal mechanisms that explain the efficiency and success of the system, being the: interactive learning, knowledge generation, proximity, and social embeddedness. At a heart of the concepts of RIS and knowledge transfer is the concept of embeddedness, which is based on relationships and social networks and requires an understanding of the institutional and cultural context [55]. The concept of embedding appeared in social theory and works of James Coleman [63], which is the main propagator of the concept of social capital theory. According to this theory, the concept of rooting refers to resources embedded in the structure of social relations (networks). In social capital theory, the concept of "social embedding" describes a situation in which economic activities and behavior are related to or depend on non-economic institutions and activities such as culture, social networks, politics and religion [17]. The social structure is made up of connections with social networks, the key element of which is shaping of this rooting, as well as cohesion, integration and social support [64]. These are the features that are not subject to market rules, cannot be duplicated or sold, but are crucial for interactive learning [65]. From the perspective of diversifying development potentials, embedding occurs in regions where there is a significant concentration of enterprises and institutions, a high degree of shared social and cultural values, and various resources that can be used to generate new production and processes.

Within RIS, embedding concerns mainly the relationship between interactive and collective learning and the nature of knowledge exchange between enterprises and their institutional environment that supports innovation processes and knowledge transfer. It follows that networks of social and organizational relations constitute a key dimension of embeddedness. For shaping the policy of innovation and diffusion of knowledge, it is important to indicate which network structures are created under the RIS in regions with specific conditions influencing the development of innovations, as well as understanding who are the gatekeepers of knowledge, what are the mechanisms of information transfer and what are the abilities of the potential recipients to the knowledge.

In knowledge-based economies, innovation is considered to be the key driving force behind economic development [66]. Nowadays, they can rarely be developed by single entities or individuals. Their creation and success require the activation of the broadly understood innovative potential, located in the private sector of the economy, but also the one accumulated in the public and civic sector, so that as a result it is possible to engage the potential of creativity and innovation on a mass scale [67]. For this reason, intangible assets, in particular relational capital, play a special role.

This capital, apart from human capital (competences, education) and structural capital (structure/organization), is one of the components of intellectual capital [68]. It is defined as resources related to interpersonal relations, the ability to establish and maintain close and lasting relationships, building one's own social network [69]. Relational capital enables the creation of a network of contacts and long-term cooperation relations. The high level of this capital influences the connection of entities operating in the networks. It also affects the quality of information flow between network links and joint activities undertaken by all or only some of the entities [70].

These "social networks" contain two important components. The first is the network, which is essentially considered a structure formed by entities (primarily actors/entities) and their connections. The social nature of these connections, taking the form of interactions, relationships and ties, is the second component. Functioning in networks allows you to reach various resources through the exchange of knowledge. It enables the acquisition of external knowledge and combining it with individual/organizational and tacit knowledge. Nowadays, in a complex environment requiring a variety of reactions and stimuli, even large companies and organizations find it difficult to gather all competences and skills in one place [71]. An important element is also the interpenetration of different areas of knowledge. In addition, collaborative networking enables greater freedom and security through sharing experiences and sharing risks. Some researchers also argue that network structures can accelerate trust building in R&D cooperation, which typically requires mutual disclosure of knowledge related to competition [72,73].

On the one hand, networks enable the flow of knowledge based on direct relations, and on the other hand, they can contribute to the exchange of knowledge through indirect connections. In the case of indirect knowledge exchange, innovation brokers and gatekeepers of knowledge play an important role. Brokers are network actors that transfer knowledge between organizations that are not directly related [72]. They play the role of an information intermediary between information resources and people/organizations that need information. On the other hand, gatekeepers of knowledge absorb knowledge scattered on a global scale and introduce it to innovation processes—both at the regional and local level [66,71,74]. Their tasks, according to Wesley Cohen and Daniel Levinthal [75], are to monitor the external environment and translate technical information into a form that is understandable to local stakeholders. As a result, gatekeepers contribute to the popularization of new ideas and the transfer of new knowledge to the regional and local level [66]. Both brokers and gatekeepers act as knowledge repositories and contribute to the use of knowledge they derive from different contexts [76]. As a result, they do not so much control the flow of information, as influence it, among other by interpreting the message or giving it a specific meaning. In a sense, they decide which information will circulate and which will not.

6. Discussion

The study attempts to use the concept of regional innovation system to understand the dynamics of renewable energy generation development and the role and peculiarities of energy policy aimed at its development. Scarce attempts have been made so far to explore how the concept of RIS and social networks can support the diffusion of knowledge related to renewable energy sources. Moreover, the research conducted so far focuses mainly on regions with strong centrality features and well-developed economic and research infrastructure. Recent studies, however, suggest paying more attention to deficit and peripheral regions and their determinants for the creation and absorption of innovation and new technologies [77–79]. A particularly important area of research in which there is a knowledge gap are the problems of innovation systems occurring in regions with an energy deficit and peripheral regions with specific development characteristics, which can provide an appropriate framework for shaping energy policy focused on the development of renewable energy sources.

In order to understand the mechanisms underlying the appearance of regions characterized by energy gaps and energy peripherality, one should refer to the scientific paradigm related to the systemic nature of innovation [80]. This paradigm explains that the speed, direction and success of innovation processes are highly influenced by the environment, i.e., the regional innovation system in which innovations arise. Such a system as a complex structure of various institutions and their relations and principles of functioning, may encounter many emergency situations that hinder the processes taking place within it. Understanding these problems will help to better understand the occurrence of regional disproportions in the rate and level of investment for installations related to the production of renewable energy.

The main feature of large technological systems, including energy systems, is a strong interconnection with the economic system [81]. This dependence means that the transformation of the energy system will affect all elements of a sustainable economic system and requires an excessively high modernization effort, which many economies have not dealt with so far [82]. Literature review shows various theoretical and practical examples of problems related to the functioning of systems, such as: problems with the market structure, infrastructure problems, institutional problems, problems with interaction and problems with opportunities and local potential [52]. The systemic context allows, first of all, to identify directions of policy and public support, and to indicate areas with deficiencies in energy innovation and gaps in knowledge about them.

The new technology may face problems resulting from the market structure and competitive substitutes, which may be cheaper than the introduced innovations or have low utility when, for example, there are no externalities in the network. Moreover, when some actors dominate and control the market, the customer selection processes are limited [83]. Practical examples in Denmark show that in case of renewable energy generation installations, the small-scale wind energy generation technologies have been implemented successfully [84], as opposed to large-scale energy projects such as biomass [85], gasification and heat pumps [86], which in practice hampered the dissemination of the technology.

Infrastructural problems may concern equipping the region with physical facilities necessary for the functioning of society or enterprises in economic structures. These are, for example, electrical energy, natural gas transmission/distribution networks and communication networks such as high-speed ICT infrastructure and highways. Another dimension of infrastructural problems is equipping the region with a physical knowledge infrastructure, which includes highly specialized buildings (laboratories and research facilities) and equipment, as well as intangible infrastructure related to scientific and applied knowledge. The implementation of investments related to RES generation installations requires the transformation of large technical systems, such as the energy system. This is associated with high investment costs for the expansion of new infrastructure and coordination problems, and the entire process often requires government intervention [87].

Institutional problems relate to institutional mechanisms that may hinder innovation processes in the region. Institutions are the main constituent of innovation systems, and the institutional context defines this system and provides a structural framework. Formal institutions are consciously created and are characterized by clearly articulated and written rules of conduct, while informal institutions function as established rules of the game rooted in local social and cultural structures. Together, these two dimensions of institutions create the environment in which companies, knowledge institutions and the government itself are embedded [87]. In practice, institutional problems caused by instability in regulations and subsidy systems are often encountered. Once adopted, activities and support programs are withdrawn to be restarted after a few years. Such situations for micro-CHP, wind, PV, biomass and marine energy have been observed in the UK [88] or with solar collectors in Sweden [89]. Another observed phenomenon is the shift in political priorities with regard to the technology or its application context. An example of such activities is the implementation of solar cells in the Netherlands. A policy was adopted in the 1970s–1980s that focused on countries with the highest theoretical solar energy potential and developing countries as a priority, followed by a sudden increase in interest in PV technology in the 1990s, which due to climate change was also seen as an opportunity for the countries of North-Western Europe. The following years saw a change in policy and PV technology is no longer considered a viable option due to its high cost [90].

Another frequently mentioned problem in the functioning of institutions is the lack of coherent actions between different administrative levels in a given country. An example are biofuels and PVs in the Netherlands, both supported by provincial (regional) governments, while at the national level the government hinders the development and diffusion of these technologies [91,92]. Informal institutions, on the other hand, are responsible for the legitimacy of the implementation of new technologies and their social acceptance and observance through the prism of a given institution [93], in case of new technologies, obtaining legitimacy is often a slow and tedious process.

The functioning of formal and informal institutions often results in problems with network interaction. Actors of the innovation system, such as: enterprises, knowledge institutions, government—all interact with each other, including regarding product development and design, knowledge exchange and diffusion of new technologies. Interference and inefficiencies in the functioning of the network can be caused by either too strong or too weak interactions.

Network failure occurs when dominant gatekeepers and knowledge brokers fail to fulfill their role and consequently fail to provide the required knowledge. The network can also be too closed to external interactions, which means that actors are reluctant to leave the group or let new participants into it. Another situation concerns the imprisonment of relationships and the inability to develop the network, it results from the same costs of such changes as well as the possibility of establishing relationships with new partners. Network failure may also be caused by poor connectivity of network actors with new technologies, which prevents the process of learning, adaptation to new technological developments and innovations. Moreover, little involvement in cooperation in the system may lead to the lack of a common vision of technology development in the future, which in turn may hinder the coordination of research efforts and investments [87].

The company's abilities in the form of the lack of competences and resources to modernize and implement a new technology may also prove to be a significant problem [94,95]. The possibilities of searching for new solutions are significantly limited by the knowledge gaps of enterprises and the long cognitive and geographical distance, which is why they are often not aware of the existing opportunities and do not include innovation in their development vision [83].

Another issue are the interaction problems that affect the dissemination of knowledge in a multi-stakeholder regional system. Networking of different actors facilitates knowledge flows, accelerates technology development, reduces uncertainty and creates demand. Research in this area shows that weak or excessively strong connections and disturbances in the network are a mechanism blocking renewable energy generation technologies.

The case of Swedish producers of small biofuel boilers shows the lack of cooperation within the network and the weakness of the relationship [96]. There are only two or three producers of large biofuel boilers in Sweden, therefore the lack of cooperation may be due to the lack of potential partners. On the other hand, the weakness of relations and connectivity within the system may partially result from an information gap about other entities being potential partners in the region. Another issue is the considerable individualism of small companies, which means that these companies do not want to cooperate and share their knowledge with other companies. In addition, some companies, rooted for some time in local economic and social structures, are reluctant to new entities and create distance instead of building relationships. Summarizing, the case of Sweden shows that cooperation networks within the framework of energy generation technologies are characterized by poor connectivity and a lack of willingness to cooperate and share knowledge with other companies.

In terms of ways to assess the efficiency of regional innovation systems and their influence upon creation of innovations one of them is to take into account the number of patents. A research [97] was conducted within 194 countries to assess how different renewable energy support policies affect innovation in solar and wind energy generation technologies. This substantial work shows that a more comprehensive portfolio of renewable energy support policies increases the number of patents in the particular field, as well as there is a definite positive impact on patent activity, which is increasing significantly over time along with the growing duration of research programs and achievement of R&D objectives.

A different approach to understanding of opportunities and constraints that are created by the development of renewable energy generation capacities is mentioned by van Zalk and Behrens [98] in the U.S.A. context. Namely, in their opinion the issue of land use in this regard is not in favor of renewable energy development, as "the surface area required for renewable energy systems is greater than that for non-renewable systems, exacerbating existing environmental policy challenges, from increasing land competition, to visual impacts". While this is certainly true and needs to be taken into account, the overall environmental benefits from decreased pollution are still prevailing.

7. Conclusions

The Polish energy sector is currently facing serious challenges, the currently defined directions of the state energy policy are to a large extent interdependent. With the clear goals to increase the share of renewable energy generation in the next decade intense transformations are needed to reach them, as in the past years its growth rates have been falling behind "the schedule". These can be achieved by simultaneous increase of renewable energy generation shares and by improving the energy efficiency. The later reduces the increase in demand for fuels and energy, contributing to increased energy security, as a result of reducing dependence on imports, and also works to improve the environmental impact of energy by reducing emissions. Similar effects are brought by the development of renewable energy generation, including the use of biofuels and non-pollutant technologies.

However, the sector also faces numerous challenges resulting from, inter alia, permanently high energy demand, inadequate level of development of fuel and energy production and transport infrastructure, high dependence on external supplies of natural gas, nearly total dependence on imported supply of crude oil, and environmental protection obligations, including climate change mitigation. These intensify the necessity to implement decisive measures to prevent deterioration of the economic situation of fuel and energy consumers. Therefore, in order to fully use the energy potential in Poland, it is necessary to use innovations and knowledge, especially expert knowledge. This is of exceptional importance for the energy sector, which operates in a complex legal, economic and technological environment.

In the energy sector nowadays, as in other sectors of economies, one of the conditions for functioning and development is the systemic use of knowledge to solve emerging problems, including creating and facilitating innovations. The fulfillment of this condition requires an incorporation of knowledge into management. The use of knowledge acquired at the local and regional level, within the company, as well as in contacts between various organizations obtained through networking activities can serve the process of gaining knowledge through the exchange of experiences, mutual evaluation of one's models and practices, exchange of views and ideas, and conducting joint experiments [18,70]. Network participants who are in contact with each other ignite discussions and generate new ideas, define stimulants and share experiences. A special role here is played by entities such as knowledge gatekeepers and brokers, which are responsible for providing practically useful knowledge to solve key problems of an organization (this type of knowledge can be defined as informal knowledge or unclassified knowledge), as well as information that contributes to the creation and implementation of innovative solutions. Gatekeepers and brokers often act as liaisons who influence the creation and connection of various sets of knowledge, both within the company and from external sources. These are people and entities with networking skills, with a high level of social and communication skills.

It is worth noting that while in highly developed, innovative areas both the networks themselves and the entities intermediating in acquiring are quite "dense", their shortage in peripheral areas may be another element hindering the development of innovation. In regional systems, the strive to increase innovation and related entrepreneurship contributes to changes in the functional structure of regions towards strongly developing regional centers and less growing peripheral areas [99]. However, it is commonly assumed that firms in peripheral regions benefit less from local knowledge transfer than firms located in agglomerations or industrial clusters [100]. This is due to the fact that the peripheral regions are characterized by a weaker supply of local knowledge transfer than the key regions. The literature indicates that companies and organizations from peripheral regions can be innovative to the extent that they are able to compensate for the missing possibilities of spreading and absorbing knowledge [101]. It is the collaboration in networks and with intermediaries for acquiring knowledge and skills that is a potential compensation mechanism, as they establish the organizational framework (organizational proximity) that enables interactive learning processes.

To conclude, the conducted analysis on the level of districts has revealed that the energy generation capacities in Poland are spatially unevenly distributed. Most of the surveyed districts are problematic areas which can be defined as energy periphery or energy gaps. These are areas that, regardless of the population and demographic situation, or the level of socio-economic development, are characterized by a low energy generation potential. Numerous densely populated areas are characterized by a scarcity of energy generation installations, thus also represent low energy generation potential. This may indicate the fact that the development of the network infrastructure—in this case, above all, the energy infrastructure—is not keeping up with the influx of people as well as social and population changes. Nevertheless, the investment potential in most of Poland is high. Thus, investment opportunities could be looked for in virtually every area of the market.

The authors focused on the characteristics of Polish regions regarding generation of renewable energy and its dependence on regional innovation systems and knowledge transfer. However, detailed definition of renewable energy generation development in light of local development conditions in various regions requires further study—including, among other, field research. It may be especially valuable to learn about non-economic factors, including relational capital and its role in the processes of energy generation knowledge/technology transfer and collective learning. Another important aspect that relates to regional innovation systems and renewable energy is the issue of existing policies for innovation and energy generation development. This would help to understand if their directions and implemented measures become catalysts for each other and if this accelerates the processes of transformation of the energy generation system achieving synergistic effect.

An interesting aspect for further analyzes could also be a research of the structure of energy generated from renewable sources, yet broken down into its types and specific sources. Investment chances could be analyzed separately, which are needed on every stage of product creation, in this case the generation and distribution of energy. Consequently, this may translate into energy security studies, whether on national or regional levels, either of which is a strategic issue for each country. Production and transmission of energy is an economic bloodstream, which, in addition to the transport system, determines the efficient functioning of the economy. Economic development of all countries depends on access to energy, which is why its sustainable development is so important.

Author Contributions: Conceptualization, P.Z., V.K. and D.Z.-G.; methodology, P.Z.; writing—original draft preparation, P.Z., V.K. and D.Z.-G.; writing—review and editing, P.Z., V.K. and D.Z.-G.; visualization, P.Z., V.K. and D.Z.-G.; supervision, P.Z. All authors have read and agreed to the published version of the manuscript.

Funding: This research received no external funding.

Institutional Review Board Statement: Not applicable.

Informed Consent Statement: Not applicable.

Data Availability Statement: Not applicable.

Conflicts of Interest: The authors declare no conflict of interest.

References

1. International Monetary Fund. The Economics of Climate. Finance and Development 2019. Available online: https://www.imf.org/external/pubs/ft/fandd/2019/12/pdf/fd1219.pdf (accessed on 5 March 2021).
2. Lelieveld, J.; Klingmüller, K.; Pozzer, A.; Burnett, R.T.; Haines, A.; Ramanathan, V. Effects of fossil fuel and total anthropogenic emission removal on public health and climate. *Proc. Natl. Acad. Sci. USA* **2019**, *116*, 7192–7197. [CrossRef] [PubMed]
3. Shen, N.; Wang, Y.; Peng, H.; Hou, Z. Renewable energy green innovation, fossil energy consumption, and air pollution—spatial empirical analysis based on China. *Sustainability* **2020**, *12*, 6397. [CrossRef]
4. Vohra, K.; Vodonos, A.; Schwartz, J.; Marais, E.A.; Sulprizio, M.P.; Mickley, L.J. Global mortality from outdoor fine particle pollution generated by fossil fuel combustion: Results from GEOS-Chem. *Environ. Res.* **2021**, *195*, 110754. [CrossRef] [PubMed]
5. Meneganzin, A.; Pievani, T.; Caserini, S. Anthropogenic climate change as a monumental niche construction process: Background and philosophical aspects. *Biol. Philos.* **2020**, 35. [CrossRef]
6. Burrell, A.L.; Evans, J.P.; De Kauwe, M.G. Anthropogenic climate change has driven over 5 million km2 of drylands towards desertification. *Nat. Commun.* **2020**, *11*, 3853. [CrossRef] [PubMed]
7. Taube, O.; Ranney, M.A.; Henn, L.; Kaiser, F.G. Increasing people's acceptance of anthropogenic climate change with scientific facts: Is mechanistic information more effective for environmentalists? *J. Environ. Psychol.* **2021**, *73*, 101549. [CrossRef]
8. European Commission. 2030 Climate & Energy Framework. Available online: https://ec.europa.eu/clima/policies/strategies/2030_en (accessed on 12 April 2021).
9. Directive (EU) 2018/2001 of the European Parliament and of the Council of 11 December 2018 on the Promotion of the Use of Energy from Renewable Sources, 11 December 2018. Available online: http://data.europa.eu/eli/dir/2018/2001/oj (accessed on 28 March 2021).
10. European Commission. National Energy and Climate Plans (NECPs). Available online: https://ec.europa.eu/energy/topics/energy-strategy/national-energy-climate-plans_en (accessed on 15 March 2021).
11. European Commission. Driving Forward the Green Transition and Promoting Economic Recovery through Integrated Energy and Climate Planning. Brussels, 17 September 2020. Available online: https://eur-lex.europa.eu/legal-content/EN/TXT/?uri=COM:2020:564:FIN (accessed on 2 April 2021).
12. Eurostat. Share of Energy from Renewable Sources (nrg_ind_ren). 2021. Available online: https://appsso.eurostat.ec.europa.eu/nui/show.do?dataset=nrg_ind_ren&lang=en (accessed on 3 April 2021).
13. European Commission. National Renewable Energy Action Plans 2020. Available online: https://ec.europa.eu/energy/topics/renewable-energy/national-renewable-energy-action-plans-2020_en (accessed on 17 March 2021).
14. Kim, D. System Archetypes I: Diagnosing Systemic Issues and Designing High-Leverage Interventions. 2000. Available online: https://thesystemsthinker.com/systems-archetypes-i-diagnosing-systemic-issues-and-designing-interventions (accessed on 28 April 2021).
15. Cooke, P.; Uranga, M.G.; Etxebarria, G. Regional systems of innovation: An evolutionary perspective. *Environ. Plan. A Econ. Space* **1998**, *30*, 1563–1584. [CrossRef]
16. Asheim, B.; Isaksen, A. Location, agglomeration and innovation: Towards regional innovation systems in Norway? *Eur. Plan. Stud.* **1997**, *5*, 299–330. [CrossRef]
17. Granovetter, M. Economic action and social structure: The problem of embeddedness. *Am. J. Sociol.* **1985**, *91*, 481–510. [CrossRef]
18. Allen, T.J. *Managing the Flows of Technology: Technology Transfer and the Dissemination of Technological Information within the R&D Organization*; Mit Press: Cambridge, MA, USA, 1977.
19. O'Sullivan, M.; Overland, I.; Sandalow, D. The Geopolitics of Renewable Energy. Working Paper, June 2017. Available online: https://energypolicy.columbia.edu/sites/default/files/CGEPTheGeopoliticsOfRenewables.pdf (accessed on 2 March 2021).
20. Statistics Poland. *Energy from Renewable Sources in 2019*; GUS: Warsaw, Poland, 2020.
21. Braczyk, H.J.; Cooke, P.N.; Heidenreich, M.; Krauss, G. *Regional Innovation Systems: The Role of Governances in a Globalized World*; Psycholoy Press: London, UK, 1988.
22. Cooke, P. Regional innovation systems, clusters, and the knowledge economy. *Ind. Corp. Chang.* **2001**, *10*, 945–974. [CrossRef]
23. Cooke, P.; Uranga, M.G.; Etxebarria, G. Regional innovation systems: Institutional and organisational dimensions. *Res. Policy* **1997**, *26*, 475–491. [CrossRef]
24. Zito, A.R.; Burns, C.; Lenschow, A. *The Future of European Union Environmental Politics and Policy*; Routledge: London, UK, 2020.
25. Lu, Y.; Khan, Z.A.; Alvarez-Alvarado, M.S.; Zhang, Y.; Huang, Z.; Imran, M. A Critical review of sustainable energy policies for the promotion of renewable energy sources. *Sustainability* **2020**, *12*, 5078. [CrossRef]

26. Sukhatme, S.P.; Nayak, J.K. *Solar Energy: Principles of Thermal Collection and Storage*; McGraw-Hill Education: Gautam Buddha Nagar, India, 2009.
27. Global Trends in Renewable Energy Investment. 2020. Available online: https://www.fs-unep-centre.org/wp-content/uploads/2020/06/GTR_2020.pdf (accessed on 11 May 2021).
28. Mathews, J.A. Developing Countries and the Renewable Energy Revolution. OECD Development Centre. Available online: https://www.oecd.org/dev/developing-countries-and-the-renewable-energy-revolution.htm (accessed on 12 May 2021).
29. Tomaszewski, K.; Sekściński, A. Odnawialne źródła energii w Polsce—Perspektywa lokalna i regionalna. *Rynek Energii* **2020**, *4*, 10–19.
30. Olszewski, M. O energetyce przyjaznej środowisku prawie wszystko. *Warszawa Fund. Inst. Na Rzecz Ekorozw.* 2012. Available online: https://www.cire.pl/pliki/tematyczne/6/Leksykon_INE_DLA%20DZIENNIKARZY.pdf (accessed on 15 March 2021).
31. Lehner, B.; Czisch, G.; Vassolo, S. The impact of global change on the hydropower potential of Europe: A model-based analysis. *Energy Policy* **2005**, *33*, 839–855. [CrossRef]
32. Varan, C.; Cretan, R. Place and the spatial politics of intergenerational remembrance of the Iron Gates displacements in Romania, 1966–1972. *Area* **2018**, *50*, 509–519. [CrossRef]
33. Cretan, R.; Vesalon, L. The political economy of hydropower in the communist space: Iron Gates revisited. *Tijdschr. Voor Econ. En. Soc. Geogr.* **2017**, *108*, 688–701. [CrossRef]
34. Sovacool, B.; Hancock, K. International political economy and renewable energy: Hydroelectric power and the resource curse. *Int. Stud. Rev.* **2018**, *20*, 615–632. [CrossRef]
35. Andonova, L.B. The Europeanization of environmental policy in central and eastern Europe. In *The Europeanization of Central and Eastern Europe*; Schimmelfennig, F., Sedelmeier, U., Eds.; Cornell University Press: New York NY, USA, 2005.
36. Batory, A. Why do anti-corruption laws fail in central eastern Europe? A target compliance perspective. *Regul. Gov.* **2012**, *6*, 66–82. [CrossRef]
37. Księżopolski, K.; Kotlewski, D.; Maśloch, G. *Energetyka Odnawialna—Wyzwanie dla Krajów Europy Środkowo-Wschodniej*; Raport; SGH Karpacz: Warsaw, Poland, 2020; pp. 129–165.
38. Min, R.M.; Gospodarki, W.D. *Law of Poland on Renewable Sources of Energy (Ustawa z dnia 20 Lutego 2015 r. o Odnawialnych źródłach Energii)*; ISAP: Warsaw, Poland, 2015; p. 478.
39. Jenks, G.F. The data model concept in statistical mapping. *Int. Yearb. Cartogr.* **1967**, *7*, 186–190.
40. Energy Regulatory Office. Available online: https://www.ure.gov.pl/pl/oze/potencjal-krajowy-oze/8108,Instalacje-odnawialnych-zrodel-energii-wg-stanu-na-dzien-31-grudnia-2019-r.html (accessed on 4 February 2021).
41. Energy Regulatory Office. Rejestr wytwórców ENERgii w Małej Instalacji (Register of Small Installation Energy Producers). Available online: https://bip.ure.gov.pl/bip/rejestry-i-bazy/wytworcy-energii-w-male/2138,Rejestr-wytworcow-energii-w-malej-instalacji.html (accessed on 11 February 2021).
42. National Agricultural Support Center. Register of Small Installation Energy Producers. Available online: https://www.kowr.gov.pl (accessed on 2 February 2021).
43. National Agricultural Support Center. Rejestr Wytwórców Biogazu Rolniczego (Register of Agricultural Biogas Producers). Available online: https://www.kowr.gov.pl/odnawialne-zrodla-energii/biogaz-rolniczy/wytworcy-biogazu-rolniczego/rejestr-wytworcow-biogazu-rolniczego (accessed on 2 February 2021).
44. DWF. Quick Guide to the Polish Auction System for Renewables: Q&A. 2020. Available online: http://psew.pl/en/wp-content/uploads/sites/2/2020/06/QUICK-GUIDE-TO-THE-2020-POLISH-AUCTION-SYSTEM-FOR-RENEWABLES.pdf (accessed on 1 March 2021).
45. Tödtling, F.; Trippl, M. One size fits all?: Towards a differentiated regional innovation policy approach. *Res. Policy* **2005**, *34*, 1203–1219. [CrossRef]
46. Ministerstwo Aktywów Państwowych. Krajowy Plan na Energii i Klimatu na Lata 2021–2030: Założenia i Cele Oraz Polityki i Działania. 2019. Available online: https://www.gov.pl/attachment/df8c4c37-808c-44ff-9278-676fb94add88 (accessed on 15 March 2021).
47. Stanny, M.; Rosner, A.; Komorowski, Ł. *Monitoring Rozwoju Obszarów Wiejskich. Etap III. Struktury Społeczno-Gospodarcze, ich Przestrzenne Zróżnicowanie i Dynamika*; Fundacja Europejski Fundusz Rozwoju Wsi PolskiejInstytut Rozwoju Wsi i Rolnictwa PAN: Warsaw, Poland, 2018.
48. Landabaso, M.; Mouton, B. Towards a Different Regional Innovation Policy: Eight Years of European Experience through the European Regional Development Fund Innovative Actions. Draft for Publication in Greenwood Publishing 2003, Brussels. Available online: http://www.fundacaofia.com.br/pgtusp/eventos/pag_simp2002/arquivos_simp2002/conferencia_final_lessons.doc (accessed on 18 April 2021).
49. Tichy, G. Regionale Kompetenzzyklen–Zur Bedeutung von Produktlebenszyklus-und Clusteransätzen im regionalen Kontext. *Z. Für Wirtsch.* **2001**, *45*, 181–201. [CrossRef]
50. Kaufmann, A.; Todtling, F. Systems of innovation in traditional industrial regions: The case of Styria in a comparative perspective. *Reg. Stud.* **2000**, *34*, 29–40. [CrossRef]
51. Grabher, G. The weakness of strong ties; the lock-in of regional development in Ruhr area. In *The Embedded Firm; on the Socioceonomics of Industrial Networks*; Grabher, G., Ed.; Routledge: Oxfordshire, UK, 1993; pp. 255–277.

52. Negro, S.O.; Alkemade, F.; Hekkert, M.P. Why does renewable energy diffuse so slowly? A review of innovation system problems. *Renew. Sustain. Energy Rev.* **2012**, *16*, 3836–3846. [CrossRef]
53. Doloreux, D. What we should know about regional systems of innovation. *Technol. Soc.* **2002**, *24*, 243–263. [CrossRef]
54. Edquist, C. *Systems of Innovation: Technologies, Institutions and Organizations*; Pinter: London, UK, 1997.
55. Lundvall, B.A. *National Systems of Innovation: Towards a Theory of Innovation and Interactive Learning*; Pinter: London, UK, 1992.
56. Nelson, R. *National Systems of Innovation: A Comparative Study*; Oxford University Press: New York, NY, USA, 1993.
57. Boschma, R. Proximity and innovation: A critical assessment. *Reg. Stud.* **2005**, *39*, 61–74. [CrossRef]
58. Maskell, P.; Malmberg, A. The competitiveness of firms and regions. 'Ubiquitification' and the importance of localized learning. *Eur. Urban Reg. Stud.* **1999**, *6*, 9–25. [CrossRef]
59. European Commission. National/Regional Innovation Strategies for Smart Specialisation. 2014. Available online: http://ec.europa.eu/regional_policy/sources/docgener/informat/2014/smart_specialisation_en.pdf (accessed on 29 April 2021).
60. European Commission. Regional Innovation Scoreboard. European Union: Brussels. 2017. Available online: https://ec.europa.eu/growth/industry/policy/innovation/regional_en (accessed on 1 April 2021).
61. Asheim, B.T. Smart specialisation, innovation policy and regional innovation systems: What about new path development in less innovative regions? *Innov. Eur. J. Soc. Sci. Res.* **2019**, *32*, 8–25. [CrossRef]
62. Hassink, R. Regional innovation policies compared. *Urban Stud.* **1993**, *30*, 1009–1024. [CrossRef]
63. Coleman, J.S. Social capital in the creation of human capital. *Am. J. Sociol.* **1988**, *94*, 95–120. [CrossRef]
64. Kwon, H.W. A social embeddedness perspective on turnover intention: The role of informal networks and social identity evidence from South Korea. *Public Pers. Manag.* **2017**, *46*, 263–287. [CrossRef]
65. Storper, M. *The Regionalized World: Territorial Development in a Global Economy*; Guilford Press: New York, NY, USA, 1997.
66. Kauffeld-Monz, M.; Fritsch, M. Who are the brokers of knowledge in regional systems of innovation? A multi-actor network analysis. *Reg. Stud.* **2013**, *47*, 669–685. [CrossRef]
67. Pike, S.; Roos, G.; Marr, B. Strategic management of intangible assets and value drivers in R&D organizations. *RD Manag.* **2005**, *35*, 111–124. [CrossRef]
68. Lane, M.B.; Ross, H.; Dale, A.P. Social impact research: Integrating the technical, political and planning paradigms. *Hum. Organ.* **1997**, *56*, 302–310. [CrossRef]
69. Schrader, S. Informal technological transfer between firms: Cooperation through information trading. *Res. Policy* **1991**, *20*, 153–170. [CrossRef]
70. Morrison, A. Gatekeepers of knowledge within industrial districts: Who they are, how they interact. *Reg. Stud.* **2008**, *42*, 817–835. [CrossRef]
71. Fritsch, M.; Kauffeld-Monz, M. The impact of network structure on knowledge transfer: An application of social network analysis in the context of regional innovation networks. *Ann. Reg. Sci.* **2010**, *44*, 21. [CrossRef]
72. Nooteboom, B. Problems and solutions in knowledge transfer. In *Cooperation, Networks and Institutions in Regional Innovation Systems*; Fornahl, D., Brenner, T., Eds.; Edward Elgar Publishing: Cheltenham, UK, 2003; p. 384.
73. Das, T.K.; Teng, B.S. Trust, control and risk in strategic alliances: An integrated framework. *Organ. Stud.* **2001**, *22*, 251–283. [CrossRef]
74. Bathelt, H.; Malmberg, A.; Maskell, P. Clusters and knowledge: Local buzz, global pipelines and the process of knowledge creation. *Prog. Hum. Geogr.* **2004**, *28*, 31–56. [CrossRef]
75. Cohen, W.M.; Levinthal, D.A. Absorptive capacity: A new perspective on learning and innovation. *Adm. Sci. Q.* **1990**, 128–152. [CrossRef]
76. Hargadon, A.; Sutton, R.I. Technology brokering and innovation in a product development firm. *Adm. Sci. Q.* **1997**, *42*, 716–749. [CrossRef]
77. Virkkala, S. Innovation and networking in peripheral areas—A case study of emergence and change in rural manufacturing. *Eur. Plan. Stud.* **2007**, *15*, 511–529. [CrossRef]
78. Fitjar, R.D.; Rodríguez-Pose, A. Innovating in the periphery: Firms, values and innovation in Southwest Norway. *Eur. Plan. Stud.* **2011**, *19*, 555–574. [CrossRef]
79. Petrov, A. Beyond Spillovers. Interrogating innovation and creativity in the peripheries. In *Beyond Territory. Dynamic Geographies of Knowledge Creation, Diffusion, and Innovation*; Bathelt, H., Feldman, M.P., Kogler, D., Eds.; Routledge: London, UK, 2011; pp. 149–167.
80. Smith, K. Innovation as a systemic phenomenon: Rethinking the role of policy. *Enterp. Innov. Manag. Stud.* **2000**, *1*, 73–102. [CrossRef]
81. Hughes, T.P. *Networks of Power: Electrification in Western Society, 1880–1930*; Johns Hopkins University Press: Baltimore, MD, USA, 1983.
82. Geels, F.; Hekkert, M.; Jacobsson, S. The dynamics of sustainable innovation journeys. *Technol. Anal. Strateg. Manag.* **2008**, *20*, 521–536. [CrossRef]
83. Jacobsson, S.; Johnson, A. The diffusion of renewable energy technology: An analytical framework and key issues for research. *Energy Policy* **2000**, *28*, 625–640. [CrossRef]
84. Raven, R.; Gregersen, K.H. Biogas plants in Denmark: Successes and setbacks. *Renew. Sustain. Energy Rev.* **2007**, *11*, 116–132. [CrossRef]

85. Negro, S.O.; Suurs, R.A.A.; Hekkert, M.P. The bumpy road of biomass gasification in the Netherlands: Explaining the rise and fall of an emerging innovation system. *Technol. Forecast. Soc. Chang.* **2008**, *75*, 57–77. [CrossRef]
86. Raven, R.; Verbong, G.P.J. Ruling out innovations—Technological regimes, rules and failures: The cases of heat pump power generation and bio-gas production in The Netherlands. *Innov. Manag. Policy Pract.* **2004**, *6*, 178–198. [CrossRef]
87. Woolthuis, K.W.; Lankhuizen, M.; Gilsing, V. A system failure framework for innovation policy design. *Technovation* **2005**, *25*, 609–619. [CrossRef]
88. Foxon, T.J.; Gross, R.; Chase, A.; Howes, J.; Arnall, A.; Anderson, D. UK innovation systems for new and renewable energy technologies: Drivers, barriers and systems failures. *Energy Policy* **2005**, *33*, 2123–2137. [CrossRef]
89. Jacobsson, S.; Bergek, A. Transforming the energy sector: The evolution of technological systems in renewable energy technology. *Ind. Corp. Chang.* **2004**, *13*, 815–849. [CrossRef]
90. Verbong, G.; Geels, F. The ongoing energy transition: Lessons from a sociotechnical, multi-level analysis of the Dutch electricity system (1960–2004). *Energy Policy* **2007**, *35*, 1025–1037. [CrossRef]
91. Suurs, R.A.A.; Hekkert, M.P. Cumulative causation in the formation of a technological innovation system: The case of biofuels in the Netherlands. *Technol. Forecast. Soc. Chang.* **2009**, *76*, 1003–1020. [CrossRef]
92. Negro, S.O.; Vasseur, V.; van Sark, W.G.J.H.M.; Hekkert, M.P. Solar eclipse—The rise and 'dusk' of the Dutch PV innovation system. *Int. J. Technol. Policy Manag.* **2012**, *12*, 135–157. [CrossRef]
93. Bergek, A.; Jacobsson, S.; Sanden, B. Key processes and policy challenges in the formation and early growth of a technology-specific innovation system: Lessons from the field of renewable energy. In Proceedings of the Workshop Understanding Processes in Sustainable Innovation Journeys, Utrecht, The Netherlands, 2–3 October 2006.
94. Afuah, A.N.; Utterback, J.M. Responding to structural industry changes: A technological evolution perspective. *Ind. Corp. Chang.* **1997**, *6*, 183–202. [CrossRef]
95. Anderson, P.; Tushman, M.L. Technological discontinuities and dominant designs: A cyclical model of technological change. *Adm. Sci. Q.* **1990**, 35. [CrossRef]
96. Johnson, A.; Jacobsson, S. Inducement and blocking mechanisms in the development of a new industry: The case of renewable energy technology in Sweden. In *Technology and the Market. Demand, Users and Innovation*; Coombs, R., Green, K., Richards, A., Walsh, V., Eds.; Edwar Elgar Publishing Ltd.: Cheltenham, UK, 2000; pp. 89–111.
97. Hille, E.; Althammer, W.; Diederich, H. Environmental regulation and innovation in renewable energy technologies: Does the policy instrument matter? *Technol. Forecast. Soc. Chang.* **2020**, *153*, 119921. [CrossRef]
98. Van Zalk, J.; Behrens, P. The spatial extent of renewable and non-renewable power generation: A review and meta-analysis of power densities and their application in the US. *Energy Policy* **2018**, *123*, 83–91. [CrossRef]
99. Heffner, K.; Gibas, P. Obszary funkcjonalne ośrodków regionalnych w Polsce. In *Zrozumieć Terytorium. Idea i Praktyka. Łódź*; Nowakowska, A., Ed.; Uniwersytet Ekonomiczny w Katowicach: Katowice, Poland, 2013; pp. 217–238.
100. Grillitsh, M.; Nilsson, M. Innovation in peripheral regions: Do collaborations compensate for a lack of local knowledge spillovers? *Ann. Reg. Sci.* **2015**, *54*, 299–321. [CrossRef]
101. Lagendijk, A.; Lorentzen, A. Proximity, knowledge and innovation in peripheral regions. On the intersection between geographical and organizational proximity. *Eur. Plan. Stud.* **2007**, *15*, 457–466. [CrossRef]

energies

MDPI

Article

Energy Self-Subsistence of Agriculture in EU Countries

Tomasz Rokicki [1,*], Marcin Ratajczak [2], Piotr Bórawski [3], Aneta Bełdycka-Bórawska [3], Barbara Gradziuk [4], Piotr Gradziuk [5] and Agnieszka Siedlecka [6]

1 Institute of Economics and Finance, Warsaw University of Life Sciences, ul. Nowoursynowska 166, 02-787 Warsaw, Poland
2 Management Institute, Warsaw University of Life Science, ul. Nowoursynowska 166, 02-787 Warszawa, Poland; marcin_ratajczak@sggw.edu.pl
3 Department of Agrotechnology and Agribusiness, Faculty of Agriculture and Forestry, University of Warmia and Mazury in Olsztyn, ul. Oczapowskiego 2, 10-719 Olsztyn, Poland; pboraw@uwm.edu.pl (P.B.); aneta.beldycka-borawska@uwm.edu.pl (A.B.-B.)
4 Department of Management and Marketing, Faculty of Agrobioengineering, University of Life Sciences in Lublin, ul. Akademicka 13, 20-950 Lublin, Poland; barbara.gradziuk@up.lublin.pl
5 Institute of Rural and Agricultural Development, Polish Academy of Sciences, ul. Nowy Świat 72, 00-330 Warsaw, Poland; pgradziuk@irwirpan.waw.pl
6 Department of Economy, Faculty of Economic Sciences, Pope John Paul II State School of Higher Education, Sidorska 95/97, 21-500 Biała Podlaska, Poland; a.siedlecka@dydaktyka.pswbp.pl
* Correspondence: tomasz_rokicki@sggw.edu.pl; Tel.: +48-22-59-342-59

Abstract: The paper's main purpose was to identify the level and factors influencing the consumption of bioenergy of agricultural origin in agriculture in EU countries. All EU countries were deliberately selected for research, as of 31 December 2018. The research period covered the years 2004 to 2018. The sources of materials were the subject literature, Eurostat data, and IEA (International Energy Agency) data. The following methods were used for the analysis and presentation of materials: descriptive, tabular, graphical, Gini concentration coefficient, Lorenz concentration curve, descriptive statistics, Kendall's tau correlation coefficient and Spearman's rank correlation coefficient. In the EU, there was a high level of concentration of renewable energy consumption in several countries. There was also no change in the use of bioenergy of agricultural origin in agriculture, but the concentration level was low. The degree of concentration has not changed for both parameters of renewable energy over a dozen or so years, which proves a similar pace of development of the use of renewable energy sources in individual EU countries. Higher consumption of bioenergy of agricultural origin in agriculture was shown to occur in economically developed countries, but with high agricultural production. There was a strong correlation between the consumption of bioenergy of agricultural origin in agriculture for the entire EU and individual economic parameters in the field of energy and agriculture. The relations were positive for all economic parameters, for total renewables and biofuels consumption and for agricultural production parameters. Negative relations concerned the total energy consumption and parameters related to the area of agricultural crops.

Keywords: renewable energy sources; agriculture; energy policy; energy in agriculture; bioenergy of agricultural origin

Citation: Rokicki, T.; Ratajczak, M.; Bórawski, P.; Bełdycka-Bórawska, A.; Gradziuk, B.; Gradziuk, P.; Siedlecka, A. Energy Self-Subsistence of Agriculture in EU Countries. *Energies* **2021**, *14*, 3014. https://doi.org/10.3390/en14113014

Academic Editors: Talal Yusaf and Francesco Calise

Received: 5 April 2021
Accepted: 21 May 2021
Published: 23 May 2021

1. Introduction

Preserving the natural environment for future generations is one of the most important goals facing the world [1–3]. This issue was presented at many conferences and discussions at the global and regional level [4–6]. This problem has also been dealt with in the European Union. In December 2008, the Council of the European Union adopted assumptions on counteracting climate change. The EU plan is commonly known as "3 × 20", but there were four proposals [7,8]. According to them, by 2020, the European Union should reduce greenhouse gas emissions by 20% (compared to 1990), increase the share of energy from renewable sources (RES) in its total consumption to 20%, and increase efficiency by 20% energy. It was also assumed that the share of biofuels in the total consumption of transport fuels would increase by at least 10%. For individual countries, different target shares of energy from renewable sources in gross final energy consumption have been defined for 2020 [9,10]. Its highest share was expected in Sweden (increase from 39.8% in 2005 to 49.0% in 2020), Latvia (from 32.6% to 40%) and Finland (from 28.5% to 38%). In other countries, the target share ranged from 10% in Malta, to 15% in Poland, to 31% in Portugal. In total, this share was to reach 20% or more in 12 countries, and up to 15% in 10 countries. In the remaining ones, it was lower than 15%. The next challenges posed by the European Union are even more ambitious, as they involve achieving a reduction by at least 40% of greenhouse gas emissions by 2030 (compared to the level from 1990), increasing the share of renewable energy in its total consumption to a minimum of 32% and an increase of at least 32.5% in energy efficiency [11–13]. The European Green Deal published by the Commission has set out a clear vision of how to achieve climate neutrality by 2050. It proposes to increase the EU's climate ambition for 2030 and 2050, a zero-pollution ambition for a toxic-free environment, supplying clean, affordable and secure energy, mobilizing industry for a clean and circular economy, building and renovating in an energy and resource efficient way, preserving and restoring ecosystems and biodiversity, a fair, healthy and environmentally friendly food system, accelerating the shift to sustainable and smart mobility [14].

For millennia, mankind has mostly used natural, reproducible energy sources. These were plants that provide food and fuel, animal products (including oils), and to some extent, water, wind, and especially the sun. Along with the socio-economic development, raw materials obtained from the depths of the earth, such as coal, oil and natural gas, were increasingly important. These are non-renewable resources and their combustion products pollute the environment. In this situation, mankind is forced to look for such energy sources that are constantly recreated. Such sources include solar energy [15], wind energy, water [16], including river currents, sea and ocean waves, nuclear energy [17], energy from biomass [18], biogas or bioliquids [19]. Renewable energy also includes the heat obtained from the ground (heat pumps, geothermal energy), air (aerothermal) and water (hydrothermal energy) [20–24].

The most important feature of renewable energy is its inexhaustibility. In addition, renewable energy sources have a much lower negative environmental impact than conventional fossil energy technologies. Most of the expenditure related to renewable energy is related to the materials and labor needed to build and maintain facilities. However, there are no costs of importing energy [25–27]. The main advantages of renewable energy include ensuring energy security for the future, which means a continuous and uninterrupted supply of energy necessary to run the economy [28]. There is a strong relationship between the level of energy intensity and socio-economic development. Renewable energy also creates additional jobs [20,29,30]. Another advantage of renewable energy sources is their availability. These sources are scattered around the world, making them easily accessible to everyone. Renewable energy reduces the electrification gap between rural and urban areas. As a result, it can affect the development of rural areas [31]. Where connection to the power grid is almost impossible, renewable energy is the most effective solution [32–34]. RES also reduces the negative effects on the environment and health. The emission of harmful substances to the atmosphere is reduced and the carbon footprint is reduced. This reduces the risk to human health as most diseases are related to air pollution [35–37].

Agriculture provides many types of renewable energy. It is easiest to identify energy produced only in agriculture, such as solid and liquid biofuels and biogas. Solid biofuels are defined as any plant material that is used directly as fuel or transformed into other forms prior to combustion. This includes many wood materials produced by an industrial process or supplied directly by forestry and agriculture (firewood, wood chips, bark, sawdust, shavings; sulphite lyes, also known as black liquor; animal materials/waste, industrial waste (renewable) and others solid biofuels). Charcoal is not included in this category. Biogas are gases consisting mainly of methane and carbon dioxide, produced either by anaerobic digestion of biomass or by thermal processes. Liquid biofuel is primarily biodiesel, mixed or replaced with fossil gas or diesel fuel [38–41].

Agriculture is a significant producer of renewable energy and is considered by policy makers in this respect. Meanwhile, the consumption of renewable energy in this sector must also be encouraged. Ideally, this energy should be produced on farms themselves. The undertaken research topic is important as it shows the other and still underestimated side of renewable energy in agriculture, i.e., its consumption. Thus, the article fills the research gap.

The paper's main purpose was to identify the level and factors influencing the consumption of bioenergy of agricultural origin in agriculture in EU countries. Additional objectives were to define the conditions for the development of renewable energy sources in the EU, determine the directions of changes and the importance of renewable energy in individual EU countries, and present the consumption of bioenergy of agricultural origin in agriculture in EU countries.

Two research hypotheses were formulated in the paper:

Hypothesis 1. *The processes of the concentration of bioenergy of agricultural origin in agriculture entails a greater concentration of this energy consumption in the countries that are the largest agricultural producers in the EU.*

Hypothesis 2. *The use of bioenergy of agricultural origin in agriculture was closely correlated with the parameters of agricultural production.*

The organization of this paper is as follows: in Section 2, the literature review is elaborated. Section 3 proposes methods to identify the level and factors influencing bioenergy consumption of agricultural origin in agriculture. In Section 4, the results of the research were presented. In Section 5, Discussion, reference is made to other research results that dealt with the relationships tested. Finally, Section 6 concludes this paper.

2. Literature Review

Agriculture has many functions. In this sector, natural resources are somewhat limited, as is the land stock. Thus, there is a competition between the use of land for food production and energy purposes. The production and use of renewable energy represent a secondary transformation in the agricultural sector [42–47]. The production and consumption of renewable energy may result from the reluctance of farmers to use environmentally harmful fuels and as a way to diversify agricultural production. The biophysical features of the farm are also important, as they determine the investment in the production of renewable energy [48–51]. In addition, for example, the production of agricultural biogas or electricity from agricultural biogas is a regulated activity requiring the registration of energy companies operating in the production of agricultural biogas. The external factor for developing small-scale agricultural biogas plants is the system of subsidies on this account. Financial aid in the EU is granted to farmers for projects to diversify into non-agricultural activities, including agricultural biogas production and energy production from agricultural biogas. Internal factors concern human and financial resources and the level of marketization of agriculture and the agrarian structure in a given area. The size of farms mainly determines the development of agricultural biogas plants. It is easier to obtain the raw material for agricultural biogas production in large farms [52,53]. One way to produce and consume renewable energy in the agricultural sector is to use crop residues from existing crops [54]. Biomass production can, in principle, apply to all types of agricultural products. The potential for energy use is therefore very large [55].

Biofuels can be one of the sources that meet the global energy demand. Their advantage is environmental neutrality [56–58]. Agriculture is one of the most important sectors that supply various forms of biofuels. The production of first-generation biofuels relies heavily on energy crops such as maize and sugarcane. In Europe, biodiesel production is dominant, such as in the USA—of ethanol [59,60]. Second-generation biofuels are made of cellulose, hemicellulose or lignin. The lignocellulosic raw materials are mainly maize straw, rice husk, wheat straw and sugar cane bagasse [61]. Second generation biofuels can be blended with gasoline, which can be burned in internal combustion engines and distributed via existing infrastructure or engines slightly modified for internal combustion. An example of second-generation biofuel is cellulose ethanol, which is produced biochemically [62]. Third-generation biofuels come from algae biomass. The production of biofuel from algae is usually dependent on the lipid content. Algae are used, among others, for the production of biodiesel [63–67]. Fourth-generation biofuels use inexhaustible, cheap and widely available raw materials to convert solar energy into solar biofuels. The production of photobiological solar biofuel or electrofuel uses the synthetic biology of algae and cyanobacteria [68–70].

Biofuel production systems competing with farmland may to some extent threaten food production, but also increase environmental pressure and affect biodiversity and ecosystem services [71,72]. The production of biofuels by agriculture is part of the concept of sustainable agriculture [73]. On the other hand, agriculture should also use the energy produced in this sector, e.g., biofuels. Then, in a sense, the agricultural sector would supply itself with energy [74,75]. Such activities would also be beneficial for sustainable agriculture and the environment [76,77]. In agriculture, renewable energy can be used for heating and cooling. The problem here is the existence of low-capacity installations that convert this energy and supply the farm directly. An example is the use of biogas, the production of which is unlimited by climatic conditions. Biogas is used in agriculture for cooking and heating. Another possibility is to produce biofuels in small factories on the farm. Then, such fuel can be directly used in agricultural machinery operating on a farm. Optionally, it can be used as a percentage additive to conventional fuel [78–83].

Increasing the use of renewable energy can be achieved by initiating effective policies by governments. Policymakers should encourage domestic and foreign investors to invest in renewable energy projects, including providing tax breaks to produce renewable energy [84,85]. Investments are the main factor driving the increase in renewable energy consumption in all sectors, including agriculture [86,87].

3. Materials and Methods

EU countries were deliberately selected for research, as of 31 December 2018. The research period covered the years 2004 to 2018. The sources of materials were the literature on the subject, Eurostat data, and IEA (International Energy Agency) data.

The first stage presents issues related to renewable energy in the EU. The aim was to show the similarities and differences between EU countries. The differentiation between individual countries regarding the declared share of renewable energy in total energy until 2020 was presented. Subsequently, the degree of concentration of renewable energy consumption in the EU and changes in this regard were determined. Gini's associate was used for this purpose. The degree of concentration is measured by the amount of renewable energy consumed in the EU. If these values concern only one country, the coefficient would be 1. If they are spread over more countries, the coefficient becomes lower; the closer to 0, the more even the distribution of the volume of renewable energy consumption among EU countries. The Lorenz curve is a graphical representation of the degree of concentration of the volume of renewable energy consumption in EU countries.

The Gini coefficient is a measure of unevenness (concentration) of distribution of a random variable. When the observations are sorted in ascending order, the coefficient can be represented by the formula [88]:

$$G(y) = \frac{\sum_{i=1}^{n}(2i - n - 1) \times y_i}{n^2 \times \overline{y}} \tag{1}$$

where:

n—number of observations,
y_i—value of the "i-th" observation,

$\overline{y}$—the average value of all observations, i.e., $\overline{y} = \frac{1}{n}\sum_{i=1}^{n} y_i$.

The Lorenz curve determines the degree of concentration of a one-dimensional random variable distribution [89]. With sorted observations y_i, which are non-negative values $0 \le y_1 \le y_2 \le \cdots \le y_n$, $\sum_{i=1}^{n} y_i > 0$, the Lorenz curve is a polyline whose apexes (x_h, z_h), for $h = 0, 1, \ldots, n$, have the following coordinates:

$$x_0 = z_0 = 0, \quad x_h = \frac{h}{n}, \quad z_h = \frac{\sum_{i=1}^{h} y_i}{\sum_{i=1}^{n} y_i} \tag{2}$$

The Gini coefficient determines the area between the Lorenz curve and the diagonal of a unit square multiplied by 2.

EU countries were deliberately selected for research, as of 31 December 2018. The research period covered the years 2004 to 2018. The sources of materials were the literature on the subject, Eurostat data, and IEA (International Energy Agency) data.

In the second stage of the research, descriptive statistics relating to the share of renewable energy in individual EU countries were presented. This part of the research aimed to obtain information on regularities occurring in individual EU countries and in the entire EU. Statistics analyzed include the average, median, minimal, maximal, standard deviation, coefficient of variation, skewedness, curtosis.

The third stage focused on the use of renewable energy in agriculture. As the Eurostat data do not contain precise data on renewable energy consumption in individual sectors (including agriculture), it was decided to use IEA (International Energy Agency) data. The consumption of renewable energy in the agricultural sector is presented, but it refers only to renewable energy produced in agriculture (primary solid biofuels, biogases and liquid biofuels). Originally, data was collected for all EU countries. After verification, it turned out that in six countries the data was incomplete. In Croatia, Ireland, Malta and Slovenia, the consumption of bioenergy of agricultural origin in agriculture was not recorded but was generated in this sector. In Cyprus and Portugal, the consumption of bioenergy of agricultural origin in agriculture was not recorded in the first years of the period considered. Therefore, it was decided not to include these countries in the analysis. As a result, 22 EU countries were subjected to the study. At this stage of the research, the degree of concentration of bioenergy of agricultural origin consumption in agriculture in individual EU countries was shown. The Gini coefficient was used for this purpose. Graphically, the concentration level is represented by the Lorenz curve. These methods have been described earlier.

In the fourth stage, the dynamics of changes in bioenergy consumption of agricultural origin in agriculture in individual EU countries were determined. As a result, the following trends were observed. Additionally, the research period was divided into three- to four-year periods. As a result, changes in particular periods were more visible. The dynamics indices with a constant base were used for the research. The constant-based dynamics index has the following formula [90]:

$$i = \frac{y_n}{y_0} \text{ or } i = \frac{y_n}{y_0} \times 100\% \tag{3}$$

where:

y_n—the level of the phenomenon in a certain period,
y_0—level of the phenomenon during the reference period.

In the fifth stage, descriptive statistics concerning the share of renewable energy consumption in agriculture (coming only from primary solid biofuels, biogases and liquid biofuels) in the total consumption of renewable energy from this sector were presented. Thanks to this, it is possible to identify regularities occurring in individual countries, as in the entire EU. Agriculture contributes to renewable energy production but is generally responsible for the low consumption of this energy type, especially produced in this sector.

In the sixth stage of the research, non-parametric tests were used to establish the correlation between the variables. The first is Kendall's tau correlation coefficient. It is based on the difference between the probability that two variables fall in the same order (for the observed data) and the probability that they are different. This coefficient takes values in the range <−1, 1>. Value 1 means full match, value 0 means no match of ordering, and value -1 means complete opposite. The Kendall coefficient indicates not only the strength but also the direction of the relationship. It is a good tool for describing the similarity of the data set orderings. Kendall's tau correlation coefficient is calculated by the formula [91]:

$$\tau = P[(x_1 - x_2)(y_1 - y_2) > 0] - P[(x_1 - x_2)(y_1 - y_2) < 0] \quad (4)$$

The given formula estimates Kendall's tau based on a statistical sample. All possible pairs of the sample observations are combined, and then the pairs are divided into three possible categories:

P—compatible pairs, when the compared variables within two observations fluctuate in the same direction, i.e., either in the first observation both are greater than in the second, or both are smaller,

Q—incompatible pairs, when the variables change in the opposite direction, i.e., one of them is greater for this observation in the pair, for which the other is smaller,

T—related pairs when one of the variables has equal values in both observations.

The Kendall tau estimator is then calculated from the formula:

$$\tau = \frac{P - Q}{P + Q - T} \quad (5)$$

Additionally,

$$P + Q + T = \binom{N}{2} = \frac{N(N-1)}{2} \quad (6)$$

where:

N—sample size.

The pattern can be represented as:

$$\tau = 2\frac{P - Q}{N(N-1)} \quad (7)$$

The second non-parametric test is Spearman's rank correlation coefficient. It is used to describe the strength of the correlation of two features. It is used to study the relationship between quantitative traits for a small number of observations. Spearman's rank correlation coefficient is calculated according to the formula [92]:

$$r_S = 1 - \frac{6\sum_{i=1}^{n} d_i^2}{n(n^2 - 1)} \quad (8)$$

where:

d_i—differences between the ranks of the corresponding features x_i and feature y_i ($i = 1, 2, \ldots, n$).

The correlation coefficient takes values in the range $-1 \leq r_s \leq +1$. A positive sign of the correlation coefficient indicates a positive correlation, while a negative sign indicates a negative correlation. The closer the modulus (absolute value) of the correlation coefficient is to one, the stronger the correlation between the examined variables.

The following methods were used to present the materials: descriptive, tabular and graphic.

4. Results

In 2019, most of the energy in the world came from crude oil—33.1% (in 2010 it was 34.7%), then coal—27% (29.8%) and natural gas—24% (22.4%), while nuclear energy provided 4.3% (5.2%). Renewable energy accounted for 11.4% (7.8% in 2010) of sources, of which 6.4% (unchanged) was hydropower. Overall, it can be said that changes in the structure of global energy consumption are small, but are in the generally desirable direction, i.e., a decrease in the share of fossil fuels (by 5.5 percentage points) and an increase in renewable sources (by 3.6%) [93,94].

4.1. Renewable Energy in EU Countries

There was a large variation in the share of energy from renewable sources in the total energy consumption in the EU countries (Figure 1). Countries were using renewable energy to a very large extent (Sweden, Latvia, Finland), but also to a small extent (Malta, Luxembourg, The Netherlands). Each country submitted declarations of achieving a certain share of renewable energy in total energy consumption. Based on the 2018 data, it can be concluded that slightly more than half of the countries will achieve their targets. Natural and economic factors may cause the existing differentiation in goals and the possibility of achieving them.

Figure 1. Share of energy from renewable sources in gross final energy consumption in EU countries in 2018.

Then, the degree of concentration of renewable energy consumption in the EU countries was determined. For this purpose, the Gini coefficient was used. This coefficient is a correct and commonly used measure of inequality because it meets all the postulated axioms in this respect. It assumes values in the range from 0 to 1. A result close to 1 means that there is a very high concentration of energy consumption in one country, and close to 0 means that consumption is dispersed across many countries. The number of observations was 28 (all EU countries). The results are presented for the consumption of renewable energy. The Gini coefficient for total renewable energy consumption in 2004, calculated from the sample, was 0.57, and the estimated coefficient for the population was 0.59. This meant quite a high concentration of renewable energy consumption in several EU countries. In the case of repeating the research for 2018, the results were virtually identical (sample coefficient 0.56 and estimated for the population 0.58). Therefore, there have been no significant changes in the distribution of renewable energy consumption in EU countries. The existing differentiation was also presented by means of the Lorenz concentration curve (Figure 2). In 2018, most renewable energy was consumed in Germany, France, Italy, Sweden and Spain. In these five countries, the combined use of renewable energy accounted for 54% of total renewable energy consumption in the EU. In total, the top 10 countries used 79% of the total EU renewable energy consumption. As a rule, most renewable energy was consumed in economically developed countries and the largest countries in terms of socio-economic potential. Concentration ratios were also calculated for the earlier periods, with a frequency of every three or four years. As a result, the results concern the years 2005 to 2018. Such a combination allows to determine the direction and pace of changes in the concentration of renewable energy consumption. Generally, it can be noticed that the concentration of renewable energy consumption is maintained in several countries (Table 1). One of the reasons may be a fairly stable rate of increase in the consumption of renewable energy in individual countries and the use of technologies that ensure similar energy efficiency.

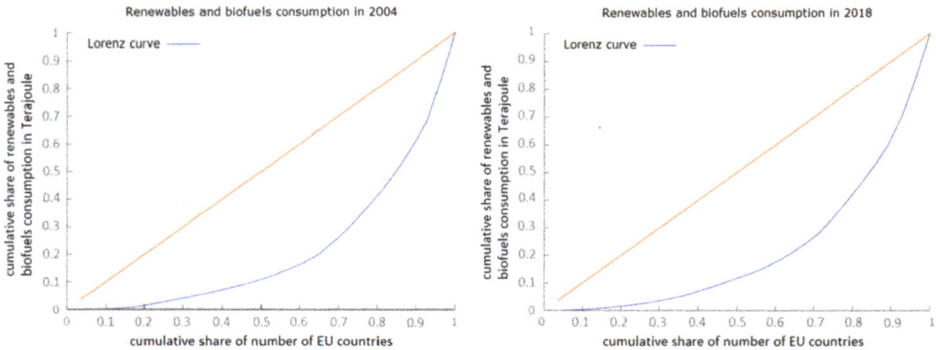

Figure 2. Lorenz concentration curves for renewables and biofuels consumption in the EU countries in 2004 and 2018.

Table 1. Gini coefficients for renewables and biofuels consumption in the EU countries in 2004 to 2018.

Type of Coefficient	Gini Coefficients in Years				
	2004	2008	2011	2014	2018
from the sample	0.57	0.58	0.56	0.56	0.56
estimated	0.59	0.60	0.58	0.58	0.58

The share of energy from renewable energy sources changed from 2004 to 2019 (Table 2). The highest average share of energy from renewable energy sources were in Sweden (48.45%), Latvia (34.99%) and Finland (34.90%). It is worth mentioning that the average in Poland was 9.74% and it increased 21.22% (Table 1). The lowest share of energy from renewable energy sources was found in the analyzed period in Malta (3.14%), Luxemburg (3.85%) and The Netherlands (4.73%). The average share of energy from renewable sources for the EU 28 was 13.93%. It is worth mentioning that Iceland (70.55%) and Norway (64.66%) achieved a much higher share than the EU.

We have also analyzed the minimal share of energy from renewable sources. As we can see from Table 1, the lowest minimal share of energy from renewable sources was in 2004 in Malta (0.10%), Luxemburg (0.90%), and United Kingdom (1.10%). The highest minimal share of energy from renewable sources in 2004 was in Latvia (29.62%), Finland (28.81%), and Sweden (38.68%). These countries also had the highest maximal share of energy from renewable sources in 2019, respectively (40.98%, 43.08%, and 56.39%).

The coefficient of variation informs about the changes that were in the analyzed variable. The biggest changes were observed in Malta (99.00%), United Kingdom (67.32%), Luxemburg (56.98%), and Ireland (45.68%). The smallest changes were found in Slovenia (6.75%), Croatia (9.71%), and Latvia (10.78%).

Skewedness was positive in the following countries: Denmark, Ireland, Greece, France, Cyprus, Latvia, Lithuania, Luxemburg, Malta, The Netherlands, Slovakia, Finland, and

United Kingdom. It means that the tail on the right side of the distribution is longer than the left side. Other countries reached negative skewedness.

Kurtosis is also an asymmetry measure. The data proved that the kurtosis reached a positive value for 2004 to 2019 only in Luxemburg and Slovakia. The vast majority of the EU countries achieved a negative value, indicating that the measure was different in 2004 to 2019 in relation to the mean.

Table 2. Descriptive statistics of share of energy from renewable sources (%) in the EU in 2004 to 2019. Red font indicates the lowest results. Bold font indicates the highest scores.

Countries	Average	Median	Minimal	Maximal	Range	Standard Deviation	Coefficient of Variation	Skewedness	Curtosis
Austria	30.65	32.11	22.55	33.81	11.25	3.58	11.70	−1.08	−0.11
Belgium	6.16	6.68	1.89	9.92	8.03	2.76	44.76	−0.25	−1.39
Bulgaria	14.88	14.99	9.10	21.56	12.47	4.50	30.24	−0.08	−1.51
Croatia	25.73	26.07	21.99	28.97	6.98	2.50	9.71	−0.21	−1.53
Cyprus	7.48	6.70	3.07	13.90	10.83	3.51	46.96	**0.43**	−0.81
Czechia	11.70	11.88	6.77	16.24	9.47	3.41	29.11	−0.18	−1.55
Denmark	25.05	24.43	14.84	37.20	22.36	7.53	30.07	**0.18**	−1.35
Estonia	24.11	25.33	15.97	31.89	15.92	5.13	21.29	−0.27	−1.24
Finland	**34.90**	33.50	**28.81**	**43.08**	14.27	4.88	13.99	**0.25**	−1.44
France	12.95	12.97	9.34	17.22	7.88	2.63	20.34	**0.05**	−1.32
Germany	12.37	13.00	6.21	17.35	11.15	3.34	26.97	−0.34	−0.92
Greece	12.45	12.45	7.16	19.68	12.52	4.35	34.92	**0.13**	−1.48
Hungary	11.76	12.68	4.36	16.21	11.84	3.50	29.78	−0.71	−0.69
Ireland	6.76	6.79	2.38	11.98	9.61	3.09	**45.68**	**0.10**	−1.22
Italy	13.79	14.23	6.32	18.27	11.95	4.11	29.78	−0.49	−1.15
Latvia	**34.99**	35.01	**29.62**	**40.98**	11.36	3.77	10.78	**0.01**	−1.33
Lithuania	21.24	20.69	16.48	26.04	9.56	3.61	17.00	**0.04**	−1.54
Luxembourg	3.85	3.02	0.90	8.97	8.07	2.19	**56.98**	**0.82**	**0.01**
Malta	3.14	2.36	0.10	8.49	8.39	3.10	**99.00**	**0.47**	−1.28
The Netherlands	4.73	4.59	2.03	8.77	6.74	1.81	38.22	**0.55**	−0.18
Poland	9.74	10.66	6.89	12.16	5.28	2.07	21.22	−0.39	−1.54
Portugal	25.64	24.59	19.21	30.87	11.66	4.21	16.42	−0.02	−1.39
Romania	21.88	22.83	16.81	25.03	8.22	2.98	13.63	−0.63	−1.14
Slovenia	20.68	20.93	18.37	22.86	4.49	1.40	6.75	−0.40	−0.82
Slovakia	10.07	10.24	6.36	16.89	10.53	2.81	27.88	**0.60**	**0.26**
Spain	13.71	14.07	8.34	18.35	10.01	3.49	25.48	−0.31	−1.31
Sweden	**48.45**	49.08	**38.68**	**56.39**	17.71	5.46	11.27	−0.32	−1.09
United Kingdom	5.47	4.43	1.10	12.34	11.24	3.69	**67.32**	**0.49**	−1.05
European Union	13.93	14.03	8.56	18.88	10.32	3.35	24.06	−0.18	−1.27

4.2. *Consumption of Bioenergy of Agricultural Origin in Agriculture*

Agriculture is one of the many sectors that can benefit from renewable energy. The article presents energy consumption in agriculture, but coming from primary solid biofuels, biogases, and liquid biofuels, i.e., agricultural energy. The Gini coefficient was used to determine the degree of concentration of such energy consumption in the agricultural sector. In 2004, the Gini coefficient calculated from the sample was 0.32, and the estimated coefficient for the population was 0.33. This meant a relatively low concentration of bioenergy of agricultural origin consumption in agriculture in several EU countries. In the case of repeating the research for 2018, the results were identical. Therefore, there have been no significant changes in the distribution of bioenergy of agricultural origin consumption in agriculture in the EU countries. The existing differentiation was also presented by means of the Lorenz concentration curve (Figure 3). In 2018, most bioenergy of agricultural origin was used in agriculture in Germany, Poland, France, Finland and The Netherlands. In these five countries, combined use of bioenergy of agricultural origin in agriculture accounted for 65% of total renewable energy consumption in the EU agricultural sector. The top 10 countries used 90% of total bioenergy of agricultural origin consumption in agriculture in the EU. As a rule, most renewable energy in agriculture was used in economically developed countries and countries with large agricultural production and those developing energy from non-renewable sources. The latter factor may even be decisive. Concentration coefficients were also calculated for the earlier periods, with a frequency of every three or four years. As a result, the results relate to the years 2004 to 2018. Such a statement allows determining the direction and pace of changes in the concentration of bioenergy of agricultural origin consumption in agriculture. Generally, it can be noticed that the concentration of bioenergy of agricultural origin consumption in agriculture is maintained in a dozen or so countries, of which usually two to three countries consumed this energy the most (Table 3). Individual countries changed at the positions of leaders. In 2004, Sweden consumed the most bioenergy of agricultural origin in agriculture; in 2008 and 2011, it was Poland; in 2014 and 2018, it was Germany. Despite these changes between countries, the level of concentration has remained unchanged. One of the reasons may be a certain stabilization in agricultural production and a fairly stable pace of growth in renewable energy consumption in individual countries. This is because countries use technology that provides similar energy efficiency.

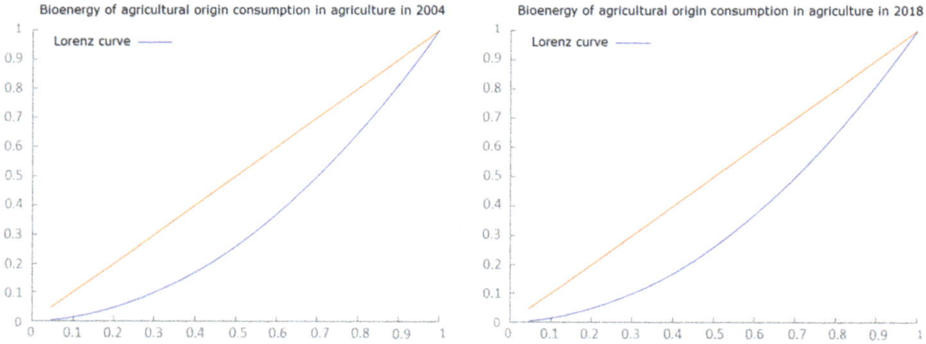

Figure 3. Lorenz concentration curves for bioenergy of agricultural origin consumption in agriculture in the EU countries in 2018.

Table 3. Gini coefficients for bioenergy of agricultural origin consumption in agriculture in the EU countries in 2004 to 2018.

Type of Coefficient	Gini Coefficients in Years				
	2004	2008	2011	2014	2018
from the sample	0.32	0.32	0.32	0.32	0.32
estimated	0.33	0.33	0.33	0.33	0.33

The use of bioenergy of agricultural origin in agriculture varied across countries. The dynamics of changes was also different (Table 4). The use of bioenergy of agricultural origin in agriculture was the fastest in The Netherlands, Italy and Belgium. In the years 2004 to 2018, in these countries, there was an increase of several dozen times. Of these, only The Netherlands has achieved a high volume of bioenergy of agricultural origin consumption in agriculture. In Italy and Belgium, the starting level in 2004 was low, so despite the high dynamics, the level in 2018 was still relatively low. Only in Sweden has the consumption of bioenergy of agricultural origin in agriculture decreased. Despite this, the country was characterized by a high volume of bioenergy of agricultural origin consumption in agriculture. High growth dynamics were achieved among the countries with the highest volume of bioenergy of agricultural origin consumption in agriculture, i.e., in Germany (increase by 427%) and Poland (by 187%). In most countries, the highest increases in the consumption of bioenergy of agricultural origin in agriculture were recorded in 2004 to 2008, and the lowest in 2014 to 2018. It can therefore be concluded that bioenergy of agricultural origin is introduced more and more slowly in agriculture. In some countries, there was stagnation or a decrease in the consumption of this energy.

Table 4. Dynamics indicators for the consumption of bioenergy of agricultural origin in agriculture in the EU countries in 2004 to 2018.

Countries	Dynamics of Changes in the Years				
	2004–2008	2011–2008	2014–2011	2018–2014	2004–2018
Austria	92.24	110.50	109.34	100.19	103.54
Belgium	1357.14	188.27	155.99	90.83	170.31
Bulgaria	63.38	166.67	192.00	71.18	139.96
Czechia	86.98	280.73	235.99	110.42	428.73
Denmark	110.72	94.34	97.35	101.83	526.89
Estonia	118.75	107.24	151.53	88.26	348.70
Finland	103.35	108.93	126.80	98.04	132.86
France	271.62	122.21	108.55	118.98	6230.43
Germany	212.08	164.10	136.15	111.19	182.64
Greece	176.81	189.51	89.45	116.34	190.09
Hungary	91.84	124.22	68.69	169.53	232.91
Italy	113.04	273.08	1259.15	160.29	7594.44
Latvia	56.04	123.23	138.52	190.93	286.50
Lithuania	124.02	127.12	95.05	126.85	335.38
Luxembourg	245.57	87.11	63.91	170.37	1674.32
The Netherlands	2238.89	157.77	138.88	154.81	540.69
Poland	260.00	126.34	81.30	107.27	39.85
Romania	863.08	35.47	129.65	84.50	212.90
Slovakia	154.05	730.70	28.57	520.59	228.95
Spain	237.96	208.21	112.82	96.74	103.54
Sweden	104.30	46.87	91.97	88.64	170.31
United Kingdom	189.99	112.14	291.59	34.27	139.96
EU 22	158.40	118.28	123.25	99.15	428.73

Individual EU countries differed in terms of the level of bioenergy of agricultural origin consumption in agriculture. Indicators can also be used to determine the importance of bioenergy of agricultural origin in agriculture. One of them is the share of bioenergy of agricultural origin consumption in agriculture (coming only from primary solid biofuels, biogases and liquid biofuels) in the total consumption of renewable energy from this sector. Particular descriptive statistics allowed for the identification of regularities occurring in individual countries and the entire EU (Table 5). In 2004 to 2018, the highest average share of bioenergy of agricultural origin in agriculture was recorded in The Netherlands and Poland (over 10%). In turn, it was the lowest in Italy, Romania and Bulgaria. In most EU countries, the median was most often close to the arithmetic mean value. The lowest minimum share of bioenergy of agricultural origin consumption in agriculture in the total consumption of renewable energy from agriculture was in Italy (0.01%) and Romania (0.05%), and the highest in Luxembourg (4.75%) and Poland (4.52%). In the case of the maximum values, the lowest share was in Italy (0.52%) and Romania (1.16%), and the highest in The Netherlands (20.56%) and the United Kingdom (16.72%). The smallest difference between the maximum and minimum value was in the case of Italy (0.52 percentage points) and Romania (1.11), and the largest in The Netherlands (20.12) and the United Kingdom (13.35).

Table 5. Descriptive statistics of share of energy consumption from bioenergy of agricultural origin in agriculture (%) in the EU in 2004 to 2019. Red font indicates the lowest results. Bold font indicates the highest scores.

Countries	Average	Median	Minimal	Maximal	Range	Standard Deviation	Coefficient of Variation	Skewedness	Curtosis
Austria	5.05	5.13	4.37	6.24	1.88	0.48	9.47	**0.85**	**1.43**
Belgium	2.37	2.81	0.21	3.77	3.56	1.26	53.30	−0.80	−0.77
Bulgaria	0.62	0.48	0.29	1.73	1.44	0.38	60.89	**2.14**	**5.06**
Czechia	3.24	2.53	1.01	5.99	4.98	1.99	61.31	**0.18**	−1.96
Denmark	5.01	4.94	4.03	6.87	2.84	0.73	14.58	**1.10**	**1.84**
Estonia	1.00	0.90	0.25	2.50	2.25	0.51	51.10	**1.81**	**5.35**
Finland	3.13	3.13	2.63	3.91	1.29	0.37	11.85	**0.38**	−0.36
France	1.51	1.51	0.47	2.03	1.57	0.40	26.29	−1.21	**2.25**
Germany	4.18	4.06	1.79	6.71	4.92	1.75	41.83	**0.17**	−1.38
Greece	2.14	2.15	0.89	3.37	2.48	0.81	37.85	−0.07	−1.39
Hungary	0.86	0.86	0.43	1.89	1.45	0.34	39.64	**1.88**	**5.62**
Italy	0.17	0.02	0.01	0.52	0.52	0.23	**133.36**	**0.87**	−1.35
Latvia	1.50	1.44	0.76	3.01	2.25	0.59	39.11	**1.28**	**1.92**
Lithuania	1.63	1.70	0.92	2.43	1.51	0.44	27.04	**0.08**	−0.06
Luxembourg	7.78	7.92	**4.75**	11.71	6.96	1.85	23.74	**0.02**	**0.34**
The Netherlands	**11.36**	**12.01**	0.44	**20.56**	**20.12**	6.26	55.13	−0.44	−0.83
Poland	**10.75**	**11.13**	**4.52**	12.67	8.15	1.92	17.82	−2.65	**8.72**
Romania	0.30	0.18	0.05	1.16	1.11	0.34	**112.68**	**2.21**	**3.77**
Slovakia	3.11	1.52	0.54	7.87	7.33	2.94	94.54	**0.70**	−1.36
Spain	1.32	1.67	0.38	1.95	1.57	0.59	44.98	−0.77	−1.34
Sweden	4.41	3.86	1.90	7.31	5.41	2.07	47.02	**0.12**	−2.03
United Kingdom	9.67	9.74	3.38	**16.72**	**13.35**	4.56	47.15	−0.05	−1.25
EU 22	3.25	3.34	1.63	4.06	2.43	0.63	19.55	−1.28	**1.89**

The variability of the examined index of the share of bioenergy of agricultural origin in agriculture was also determined. The greatest stabilization was in Austria (the coefficient of variation was 9.47%) and Finland (11.85%), and the greatest in Italy (133.36%) and Romania (112.68%). In most EU countries, volatility was very high. For the entire EU, the coefficient of variation was around 20%.

Skewness was positive in most EU countries, which means that the results were higher than the average for most of the years studied. Particularly high results were achieved in Romania and Bulgaria. On the other hand, most of the below-average results were achieved in Poland and France. Kurtosis is a measure of how results are concentrated around the mean. Results were positive in about half of the countries and negative in another half. A lot of results concentrated around the average were recorded in Poland and Hungary, and the lowest in Sweden and the Czech Republic. In general, it can be stated that there was a large variation between EU countries in the share of bioenergy of agricultural origin consumption in agriculture (coming only from primary solid biofuels, biogases and liquid biofuels) in the total consumption of renewable energy from this sector.

To establish the relationship between the amount of energy consumption from bioenergy of agricultural origin in agriculture in the EU countries and the basic parameters of the economy, energy and agriculture, Kendall's tau correlation coefficient and Spearman's rank correlation coefficient were calculated (Table 6). $p = 0.05$ was adopted as the border value of the significance level. Significant results are marked in bold in the table. Correlation coefficients were calculated for all EU countries (28 countries) for the entire 2004 to 2018 period. The study tried to check the correlation, which does not indicate that a given factor affects another, but a strong or weak relationship between them.

Table 6. Kendall's tau correlation coefficients and Spearman's rank correlation coefficients between the volume of bioenergy of agricultural origin consumption in agriculture in the EU countries and the parameters of the economy, energy and agriculture. Bold font indicates the highest scores.

Tested Parameters	Kendall's Tau Correlation Coefficient		Spearman's Rank Correlation Coefficient	
	τ	***p*-Value**	**r_s**	***p*-Value**
Correlation coefficients between bioenergy of agricultural origin consumption in agriculture and				
Value of GDP	**0.810**	**0.001**	**0.925**	**0.001**
Final consumption expenditure of households	**0.810**	**0.001**	**0.925**	**0.001**
Export of goods and services	**0.810**	**0.001**	**0.929**	**0.001**
Import of good and services	**0.771**	**0.001**	**0.911**	**0.001**
GDP per capita	**0.790**	**0.001**	**0.914**	**0.001**
Final consumption expenditure of households per capita	**0.790**	**0.001**	**0.914**	**0.001**
Total energy consumption	**−0.771**	**0.001**	**−0.925**	**0.001**
Total renewables and biofuels consumption	**0.867**	**0.001**	**0.932**	**0.001**
Total energy consumption in agriculture	−0.105	0.553	−0.136	0.010
Gross value added of agriculture, forestry and fishing	**0.657**	**0.001**	**0.821**	**0.001**
Area of agricultural crops	**−0.733**	**0.001**	**−0.857**	**0.001**
Area of grain sowing	**−0.505**	**0.008**	**−0.686**	**0.001**
Raw cows' milk delivered to dairies	**0.329**	**0.001**	**0.946**	**0.001**

In the case of Kendall's tau correlation, significant positive relations were found for almost all parameters with the amount of bioenergy of agricultural origin consumption in agriculture in the EU. The strength of the relationship was very great for the economic parameters. These relationships were very strong for both the global performance and per-capita performance parameters. Energy-related parameters were also strongly correlated with the consumption of bioenergy of agricultural origin in agriculture. The only exception was the parameter on total energy consumption in agriculture. The dependencies were varied, as a positive correlation was found in relation with the total consumption of renewable energy, and a negative correlation in the total energy consumption. This means that changes in the consumption of bioenergy of agricultural origin in agriculture follow the same direction as changes in the renewable energy consumption in the economy. There is an overall reduction in energy consumption in the EU; therefore, there was a negative correlation for this parameter. The parameters related to agriculture had less correlation with bioenergy consumption of agricultural origin in the agricultural sector. A strong positive relationship was observed for gross value added of agriculture, forestry and fishing, and a very strong positive one for raw cows' milk delivered to dairies. Both parameters showed an upward trend. Negative strong and average relations were found for the total agricultural area and agricultural area of grain, respectively. Overall, these areas slightly decreased, while the consumption of bioenergy of agricultural origin in agriculture increased. The presented correlation results indicate that there were very strong relationships between the volume of bioenergy of agricultural origin consumption in agriculture and the economic potential and the level of economic development. The general situation in the economy was more decisive. When favorable, it also fueled agriculture and favored more work. In turn, the economic crisis also affected agriculture and led to a reduction in production. In land-related parameters, these relationships were negative, because land resources do not increase but even decrease. In turn, the consumption of bioenergy of agricultural origin in agriculture grew, including the more and more common use of this type of energy, cheaper technologies, and the promotion of renewable energy. In animal production, milk production increased, which was positively correlated with bioenergy consumption of agricultural origin in agriculture. It should also be noted that there were also various correlations with different parameters of energy consumption in the economy. It all depended on the existing trend. Total energy consumption was falling, so it was negatively correlated with bioenergy of agricultural origin in agriculture. In turn, the consumption of renewable energy in the economy increased, which meant a positive correlation.

The analysis carried out with the use of Spearman's rank correlation coefficients gave very similar results. However, the strength of the relation was much greater. Both tests confirm the close relation between bioenergy of agricultural origin consumption in agriculture and economic and energy-related parameters, and a smaller one with agricultural parameters.

5. Discussion

Kazar and Kazar [95] stated that, in the long term, economic development will lead to the production of renewable energy. In a short time, there is a two-way causal link between renewable energy production and economic development. The study covered 154 countries between 1980 and 2010. In turn, Apergis and Payne [96] found that this relationship is bidirectional in both the short and long term. The study was conducted for a panel of 20 OECD countries over the period of 1985 to 2005. Sadorsky [97] also found such relationships based on a study of 18 economies of developing countries. Over the long term, a 1% increase in real per capita income has increased the per capita renewable energy consumption by around 3.5% in these economies. Similar two-way relationships were found in the studies by Pao and Fu [98] examining Brazil, Lin and Moubarak [99] studying China, Shahbaz et al. [100] examining Pakistan, and Khoshnevis Yazdi [101] studying Iran. The one-way causality between renewable energy and economic growth was stated by Leitão [102] in the Portuguese economy. This study was conducted for the period of 1970 to 2010, using time series (OLS, GMM, unit root test, VEC model, and Granger causality). The Granger causality reports a unidirectional causality between renewable energy and economic growth. Bhattacharya et al. [103] carried out studies on 38 countries with the highest renewable energy consumption. Renewable energy consumption has had a significant positive impact on economic performance in most of the countries surveyed. Similar results were obtained in the study by Saidi and Omri [104] on the example of 15 countries with the highest consumption of renewable energy. The fully modified ordinary least square (FMOLS) and the vector error correction model (VECM) techniques were used. A bidirectional causality between economic growth and renewable energy in the short- and long-run for both models was found. Menegaki [105] performed research on a sample of 27 European countries. The empirical results do not confirm a causal relationship between renewable energy consumption and GDP. However, this study covered the period of 1997 to 2007, before the targets for 20% of renewable energy in total energy were set. A study from 2004 to 2017 by Busu [106] confirmed the relationship between renewable energy and economic growth in 28 EU countries. Biomass production has had the most significant impact on economic growth of all renewable energy sources. According to the author, in the analyzed period, an increase in the basic production of biomass by 1% would affect the economic growth by 0.15%. Similar results were obtained by Armeanu et al. [107] for the years 2003 to 2014. The 1% increase in primary production of solid biofuels increased GDP per capita by 0.16%. The vast majority of studies have found a two-way relationship between renewable energy production and economic growth. In particular, such results were achieved when the share of renewable energy in the studied countries was already significant. In addition, it must also be remembered that almost all of its production is spent on domestic consumption in the case of renewable energy. There was little trade in this type of energy.

Similarly, as in other sectors, agriculture is also closely related to energy. The increase in agricultural production is positively correlated with energy consumption. In most EU countries, the technical and technological modernization of agriculture directly affects the lower energy consumption of production [108–112]. In the case of agriculture, the studies by Alola and Alola [113] found a one-sided relationship between the use of agricultural land and the consumption of renewable energy. There was no feedback. The research results concerned 16 countries of the Mediterranean coast in the years 1995 to 2014. Ben Jebli and Ben Youssef [114] found in their research a long-term two-way causal relationship between renewable energy consumption and agricultural value added (AVA). The study concerned Tunisia in the years 1980 to 2011. The research was repeated in Morocco [115] and five North

African countries [116]. In this case, too, a two-way relationship in consumption between renewable energy consumption and agricultural value added (AVA) was found. Khan et al. [117] obtained similar results on in their study of Pakistan. They showed a multilateral relationship between AVA, renewable energy consumption, and carbon dioxide emissions. The research covered the years 1981 to 2015. According to these authors, Rehman et al. [118] and Ali et al. [119], the government in Pakistan should support the growth of the AVA, because it will contribute to a greater use of renewable energy and, consequently, lower emissions of pollutants into the environment. Additionally, it is necessary to introduce modern technologies [120]. The authors suggest that increasing international economic exchange will allow the agricultural sector to develop and benefit from the transfer of renewable energy technologies. Aydoğan and Vardar [121] argue that increasing the share of renewable energy may increase production in the agricultural sector. The balanced panel data set of E7 countries (Emerging Seven—Brazil, China, India, Indonesia, Mexico, Russia and Turkey) over the period 1990 to 2014 was used. Liu et al. [122] suggest that the development of sustainable agriculture can promote renewable energy. The research in four selected countries of the Association of Southeast Asian Nations (ASEAN-4: Indonesia, Malaysia, the Philippines, and Thailand) in 1970 to 2013 was made.

Jebli and Youssef [123], using data from Argentina in 1980 to 2013, found that agriculture and renewable energy production are substituting and competing for land use. Such substitution and competition should be limited by encouraging R&D in the production of second- or third-generation biofuels and new technologies for renewable energy, or the increase in agricultural productivity per unit area. Quite a controversial statement was made by Al-Mulali et al. [124]; according to these authors, the production of renewable energy increases the inefficiency of land and water use. The research was performed in 58 developed and developing countries in the years 1980 to 2009. Destek and Sinha [125] thought the opposite. Consuming renewable energy reduces environmental impacts. The authors studied OECD countries in 1980 to 2014. Similar results were obtained in the study by Destek et al. [126], which concerned the EU countries in 1980 to 2013. In the studies of Solarin et al. [127], it was noted that there are differences between types of renewable energy. To reduce carbon dioxide emissions, there is a need to replace fossil fuels with other renewable energy sources (e.g., hydropower) rather than energy from biomass. The research covered 80 developed and developing countries in the period 1980 to 2010. In a study by Wang [128] on Brazil, Russia, India, China and South Africa (BRICS countries), a high impact of biomass in reducing environmental pollution was found. Energy from biomass was treated as a clean energy source. The research covered the years 1992 to 2013. Based on the presented review, it can be concluded that the perception of renewable energy also depended on the period covered by the research. More recent data show that renewable energy has a significant environmental impact. Therefore, it is purposeful to promote this energy on farms instead of conventional energy. An example is the largest biogas market in the world. The development of the German biogas sector has mainly been triggered and driven by consecutive versions of the Renewable Energy Act (REA) and accompanying regulations [129]. It was similar in the USA, where new tax incentives were introduced [130]. Piwowar [131] stated that in Poland, institutional support is necessary and increases the awareness of farmers. The low propensity of farmers to use renewable energy technologies was also found in other EU countries. An example was, among others, Ireland [132]. Energy self-subsistence of agriculture is particularly desirable in European agriculture, where the individual farm sizes are average [133].

6. Conclusions

Maintaining the natural environment in a proper condition requires an energy transformation. There is a need for an increased use of renewable energy sources. The EU has adopted targets for the share of these energy sources in total energy consumption. Each country has different limits that it has pledged to meet by 2020. The declarations resulted from the current state of development of renewable energy in a given country and technical possibilities and the level of economic growth. In the years 2004 to 2019, the production of renewable energy in the EU was relatively high and increased in each country at a similar pace. However, the starting potentials were different, and there are still significant differences between countries.

There has been a relatively high concentration of renewable energy consumption in several countries. The level of concentration has not changed, but there have been some changes in the order of countries. The reason is the development of renewable energy production in all countries and the use of similar technology. There was some stabilization in countries with a high share of renewable energy in total energy consumption. The dynamics of changes were slight, but these countries already had an established position. Countries with less renewable energy increased their consumption volumes very quickly but started far below the leaders. Therefore, very high dynamics of changes can be misleading. As a result, there were still disproportions.

The use of bioenergy of agricultural origin in agriculture, but produced in this sector, was quite dispersed across many countries. The level of concentration has not changed. There have been some changes in the positions of individual countries. Thus, the first research hypothesis was not confirmed. As a rule, most renewable energies was used in agriculture in economically developed countries, but with large agricultural production, such as Germany, Poland, France, Finland, and Spain. The factor related to the development of obtaining energy from non-renewable sources can be a decisive factor. One of the reasons was changes in the level of concentration of non-renewable energy consumption in agriculture. It may stabilize in agricultural production and a relatively stable rate of increase in renewable energy consumption in individual countries. The agricultural sector changes relatively slowly compared to other sectors of the economy. For faster changes, large investment outlays are necessary, but also greater project support.

There were very large differences between EU countries in terms of basic statistics on the share of bioenergy of agricultural origin in agriculture but coming from energy produced in this sector. The average percentage in a few countries was above 10%, and in a few countries, it was below 1%. In addition, in the years 2004 to 2018, there were large differences between the maximum and minimum share of bioenergy of agricultural origin in agriculture in countries such as The Netherlands and the United Kingdom, and small ones in Italy and Romania. Interestingly, Italy and Romania were the countries with the most significant variability in the share of bioenergy of agricultural origin in agriculture. The reason was the very low shares in these countries, not exceeding 1.2%. There were also differences between countries in terms of measures of asymmetry and concentration. Overall, there was a very wide variation between EU countries in the share of bioenergy of agricultural origin in agriculture. It was more significant than for the entire sector of the economy. This may indicate that the agricultural sector is less innovative and less willing to introduce changes than the whole economy.

A correlation was found between the consumption of bioenergy of agricultural origin in agriculture for the entire EU and individual economic parameters in the field of energy and agriculture. The strength of the relationship varied. Only in the case of total energy consumption in agriculture was there no relationship. There was a high positive correlation in the relationship between bioenergy of agricultural origin consumption in agriculture and economic parameters. This means that economic development contributes to the greater use of bioenergy of agricultural origin in agriculture. In the case of energy-related parameters, the relationships were not clear. Total energy consumption, i.e., a parameter that tended to decrease, was negatively correlated with bioenergy consumption of agricultural origin in agriculture. The parameter of total renewables and biofuels consumption, which showed an upward trend, was positively correlated. There was also a differentiation in the parameters related to agriculture. The dependencies here were similar to those for energy. A positive correlation was found for gross value added of agriculture, forestry, and fishing, as well as for raw cows' milk delivered to dairies. The production parameters of agriculture increased. On the other hand, the negative correlation was with parameters with a downward trend, such as the area of agricultural crops and area of grain sowing. The second hypothesis was confirmed, according to which the consumption of bioenergy of agricultural origin in agriculture was closely related to the parameters of agricultural production. However, it should be added that the relations were positive or negative depending on the trend in the case of a given parameter related to agriculture.

Overall, it must be stated that the consumption of bioenergy of agricultural origin in agriculture from this sector was at a low level and was growing very slowly. This is due to the low propensity to innovate in the agricultural sector. It can also be stated that the potential of agriculture related to the production of renewable energy for energy consumption on a farm is not used. There is much to be carried out in this regard. Apart from the appropriate information campaign, it is necessary to financially support initiatives contributing to the self-supply of farms with energy.

The limitation of the research is the poor availability of data. Information is aggregated. It is difficult to obtain data on the type of renewable energy used in agriculture that is produced in agriculture. For example, solar energy can be used in a farmer's household and on a farm. There is a need to perform micro-level research concerning farms in one locality or individual farms. The results may differ from one EU country to another.

Author Contributions: Conceptualization, T.R.; methodology, T.R.; software, T.R.; validation, T.R.; formal analysis, T.R., P.G.; investigation, T.R.; resources, T.R.; data curation, T.R.; writing—original draft, T.R., M.R., P.B., A.B.-B., B.G., P.G., A.S.; writing— review and editing, T.R., M.R., P.B., A.B.-B., B.G., P.G., A.S.; supervision, T.R.; funding acquisition, T.R., M.R. All authors have read and agreed to the published version of the manuscript.

Funding: This research received no external funding.

Institutional Review Board Statement: Not applicable.

Informed Consent Statement: Not applicable.

Data Availability Statement: Not applicable.

Conflicts of Interest: The authors declare no conflict of interest.

References

1. Le, T.H.; Nguyen, C.P. Is energy security a driver for economic growth? Evidence from a global sample. *Energy Policy* **2019**, *129*, 436–451. [CrossRef]
2. Le, T.H.; Nguyen, C.P.; Su, T.D.; Tran-Nam, B. The Kuznets curve for export diversification and income inequality: Evidence from a global sample. *Econ. Anal. Policy* **2020**, *65*, 21–39. [CrossRef]
3. Fang, D.; Shi, S.; Yu, Q. Evaluation of sustainable energy security and an empirical analysis of China. *Sustainability* **2018**, *10*, 1685. [CrossRef]
4. International cooperation to accelerate sustainable development in developing countries and related domestic policies. In Proceedings of the Report of the United Nations Conference on Environment and Development, Rio de Janeiro, Brazil, 3–14 June 1992; United Nations: New York, NY, USA, 1993; Volume 1.
5. Tuxworth, B. From environment to sustainability: Surveys and analysis of Local Agenda 21 process development in UK local authorities. *Local Environ.* **1996**, *1*, 277–297. [CrossRef]
6. Biermann, F. Earth System Governance-Beyond Traditional Environmental Policy. *GAIA Ecol. Perspect. Sci. Soc.* **2018**, *27*, 337. [CrossRef]
7. Fouquet, D.; Johansson, T.B. European renewable energy policy at crossroads—Focus on electricity support mechanisms. *Energy Policy* **2008**, *36*, 4079–4092. [CrossRef]
8. Jacobsson, S.; Bergek, A.; Finon, D.; Lauber, V.; Mitchell, C.; Toke, D.; Verbruggen, A. EU renewable energy support policy: Faith or facts? *Energy Policy* **2009**, *37*, 2143–2146. [CrossRef]
9. Directive 2009/28/EC of the European Parliament and of the Council of 23 April 2009 on the Promotion of the Use of Energy from Renewable Sources and Amending and Subsequently Repealing Directives 2001/77/EC and 2003/30/EC. *Off. J. Eur. Union* **2009**, *140*, 16–62.
10. Klessmann, C.; Held, A.; Rathmann, M.; Ragwitz, M. Status and perspectives of renewable energy policy and deployment in the European Union—What is needed to reach the 2020 targets? *Energy Policy* **2011**, *39*, 7637–7657. [CrossRef]
11. European Commission. *Green Paper A 2030 Framework for Climate and Energy Policies*; EC European Commission: Brussels, Belgium, 2013.
12. Klessmann, C. The evolution of flexibility mechanisms for achieving European renewable energy targets 2020—ex-ante evaluation of the principle mechanisms. *Energy Policy* **2009**, *37*, 4966–4979. [CrossRef]

13. Paska, J.; Surma, T. Electricity generation from renewable energy sources in Poland. *Renew. Energy* **2014**, *71*, 286–294. [CrossRef]
14. The European Green Deal. Communication from The Commission to The European Parliament, The European Council, The Council, The European Economic and Social Committee and The Committee of The Regions. COM/2019/640 Final. Available online: https://ec.europa.eu/info/sites/info/files/european-green-deal-communication_en.pdf (accessed on 31 March 2021).
15. Sanz-Casado, E.; Lascurain-Sánchez, M.L.; Serrano-Lopez, A.E.; Larsen, B.; Ingwersen, P. Production, consumption and research on solar energy: The Spanish and German case. *Renew. Energy* **2014**, *68*, 733–744. [CrossRef]
16. Meyer, N.I. Learning from wind energy policy in the EU: Lessons from Denmark, Sweden and Spain. *Eur. Environ.* **2007**, *17*, 347–362. [CrossRef]
17. Brook, B.W.; Alonso, A.; Meneley, D.A.; Misak, J.; Blees, T.; van Erp, J.B. Why nuclear energy is sustainable and has to be part of the energy mix. *Sustain. Mater. Technol.* **2014**, *1*, 8–16. [CrossRef]
18. Flamos, A.; Georgallis, P.G.; Doukas, H.; Psarras, J. Using biomass to achieve European Union Energy Targets—A review of biomass status, potential, and supporting policies. *Int. J. Green Energy* **2011**, *8*, 411–428. [CrossRef]
19. Hashemi, H.; Safari, M.; Ebrahimi, A.; Samaei, M.R.; Khodabakhshi, A. Feasibility of large amounts biogas production from garbage bioliquid. *Int. J. Health Syst. Disaster Manag.* **2015**, *3*, 147–150.
20. Owusu, P.A.; Asumadu-Sarkodie, S. A review of renewable energy sources, sustainability issues and climate change mitigation. *Cogent Eng.* **2016**, *3*, 1167990. [CrossRef]
21. Liang, X. Emerging power quality challenges due to integration of renewable energy sources. *IEEE Trans. Ind. Appl.* **2016**, *53*, 855–866. [CrossRef]
22. Tareen, W.U.K.; Anjum, Z.; Yasin, N.; Siddiqui, L.; Farhat, I.; Malik, S.A.; Aamir, M. The prospective non-conventional alternate and renewable energy sources in Pakistan—A focus on biomass energy for power generation, transportation, and industrial fuel. *Energies* **2018**, *11*, 2431. [CrossRef]
23. Rodríguez-Monroy, C.; Mármol-Acitores, G.; Nilsson-Cifuentes, G. Electricity generation in Chile using non-conventional renewable energy sources—A focus on biomass. Renew. Sustain. *Energy Rev.* **2018**, *81*, 937–945.
24. Khan, K.A.; Hasan, M.; Islam, M.A.; Alim, M.A.; Asma, U.; Hassan, L.; Ali, M.H. A study on conventional energy sources for power production. *Int. J. Adv. Res. Innov. Ideas Educ.* **2018**, *4*, 214–228.
25. Shahzad, U. The need for renewable energy sources. *Energy Int. J. Inf. Technol. Electr. Eng.* **2012**, *2*, 16–18.
26. Inglesi-Lotz, R. The impact of renewable energy consumption to economic growth: A panel data application. *Energy Econ.* **2016**, *53*, 58–63. [CrossRef]
27. Gradziuk, P.; Gradziuk, B. Renewable energy sources as a development opportunity for peripheral areas. *Econ. Reg. Stud.* **2020**, *13*, 184–198. [CrossRef]
28. Timmons, D.; Harris, J.M.; Roach, B. *The Economics of Renewable Energy*; Global Development and Environment Institute; Tufts University: Medford, MA, USA, 2014; Volume 52, pp. 1–52.
29. Ray, P. Renewable energy and sustainability. *Clean Technol. Environ. Policy* **2019**, *21*, 1517–1533. [CrossRef]
30. Gradziuk, P.; Gradziuk, B. Employment impacts of renewable energy in UE. In Proceedings of the 2018 International Scientific Conference Economic Sciences for Agribusiness and Rural Economy, Warsaw, Poland, 7–8 June 2018; Warsaw University of Life Sciences–SGGW Faculty of Economic Sciences: Warsaw, Poland, 2018; Volume 1, pp. 259–267.
31. Gradziuk, P.; Bańkowska, K. Renewable Energy–Implications for Agriculture and Rural Development in Poland. *Wieś Rol.* **2017**, *3*, 121–146.
32. Malek, A.A.; Hasanuzzaman, M.; Abd Rahim, N. Prospects, progress, challenges and policies for clean power generation from biomass resources. *Clean Technol. Environ. Policy* **2020**, *22*, 1–25.
33. Saha, S.; Ruslan, A.R.M.; Morshed, A.M.; Hasanuzzaman, M. Global prospects and challenges of latent heat thermal energy storage: A review. *Clean Technol. Environ. Policy* **2020**, *23*, 1–29.
34. Roy, N.K.; Das, A. Prospects of renewable energy sources. In *Renewable Energy and the Environment*; Springer: Singapore, 2018; pp. 1–39.
35. Eckstein, D.; Künzel, V.; Schäfer, L.; Winges, M. Global Climate Risk Index. Germanwatch. 2020. Available online: https://germanwatch.org/sites/germanwatch.org/files/20-2-01e%20Global,20 (accessed on 17 March 2021).
36. Uğurlu, E. Renewable Energy Strategies for Sustainable Development in the European Union. In *Renewable Energy*; Palgrave Macmillan: Cham, Switzerland, 2019; pp. 63–87.
37. Uğurlu, E. Greenhouse gases emissions and alternative energy in the Middle East. In *Climate Change and Energy Dynamics in the Middle East*; Springer: Cham, Switzerland, 2019; pp. 259–291.
38. Renewables Information. 2020 Edition, Database Documentation. IEA. Available online: https://iea.blob.core.windows.net/assets/a1bd577e-14d0-4d4c-8b01-4e20ae1ded07/REN_Documentation1.pdf (accessed on 17 March 2021).
39. Correa, D.F.; Beyer, H.L.; Fargione, J.E.; Hill, J.D.; Possingham, H.P.; Thomas-Hall, S.R.; Schenk, P.M. Towards the implementation of sustainable biofuel production systems. *Renew. Sustain. Energy Rev.* **2019**, *107*, 250–263. [CrossRef]
40. Rodionova, M.V.; Poudyal, R.S.; Tiwari, I.; Voloshin, R.A.; Zharmukhamedov, S.K.; Nam, H.G.; Zayadan, B.K.; Bruce, B.D.; Hou, H.J.M.; Allakhverdiev, S.I. Biofuel production: Challenges and opportunities. *Int. J. Hydrog. Energy* **2017**, *42*, 8450–8461. [CrossRef]
41. Raud, M.; Kikas, T.; Sippula, O.; Shurpali, N.J. Potentials and challenges in lignocellulosic biofuel production technology. *Renew. Sustain. Energy Rev.* **2019**, *111*, 44–56. [CrossRef]

42. Sutherland, L.A.; Peter, S.; Zagata, L. Conceptualising multi-regime interactions: The role of the agriculture sector in renewable energy transitions. *Res. Policy* **2015**, *44*, 1543–1554. [CrossRef]
43. Pigford, A.A.E.; Hickey, G.M.; Klerkx, L. Beyond agricultural innovation systems? Exploring an agricultural innovation ecosystems approach for niche design and development in sustainability transitions. *Agric. Syst.* **2018**, *164*, 116–121. [CrossRef]
44. El Bilali, H. The Multi-Level Perspective in Research on Sustainability Transitions in Agriculture and Food Systems: A Systematic Review. *Agriculture* **2019**, *9*, 74. [CrossRef]
45. Melchior, I.C.; Newig, J. Governing Transitions towards Sustainable Agriculture—Taking Stock of an Emerging Field of Research. *Sustainability* **2021**, *13*, 528. [CrossRef]
46. Zambon, I.; Cecchini, M.; Mosconi, E.M.; Colantoni, A. Revolutionizing Towards Sustainable Agricultural Systems: The Role of Energy. *Energies* **2019**, *12*, 3659. [CrossRef]
47. Fortunati, S.; Morea, D.; Mosconi, E.M. Circular economy and corporate social responsibility in the agricultural system: Cases study of the Italian agri-food industry. *Agric. Econ.* **2020**, *66*, 489–498.
48. Ruiz-Fuensanta, M.J.; Gutiérrez-Pedrero, M.-J.; Tarancón, M.-Á. The Role of Regional Determinants in the Deployment of Renewable Energy in Farms. The Case of Spain. *Sustainability* **2019**, *11*, 5937. [CrossRef]
49. Pérez-Neira, D.; Soler-Montiel, M.; Gutiérrez-Peña, R.; Mena-Guerrero, Y. Energy Assessment of Pastoral Dairy Goat Husbandry from an Agroecological Economics Perspective. A Case Study in Andalusia (Spain). *Sustainability* **2018**, *10*, 2838. [CrossRef]
50. Saganeiti, L.; Pilogallo, A.; Faruolo, G.; Scorza, F.; Murgante, B. Territorial Fragmentation and Renewable Energy Source Plants: Which Relationship? *Sustainability* **2020**, *12*, 1828. [CrossRef]
51. Morea, D.; Poggi, L.A. An innovative model for the sustainability of investments in the wind energy sector: The use of green sukuk in an Italian case study. *Int. J. Energy Econ. Policy* **2017**, *7*, 53–60.
52. Chodkowska-Miszczuk, J.; Szymańska, D. Update of the review: Cultivation of energy crops in Poland against socio-demographic factors. *Renew. Sustain. Energy Rev.* **2011**, *15*, 4242–4247. [CrossRef]
53. Chodkowska-Miszczuk, J.; Szymańska, D. Agricultural biogas plants—A chance for diversification of agriculture in Poland. *Renew. Sustain. Energy Rev.* **2013**, *20*, 514–518. [CrossRef]
54. Monforti, F.; Bódis, K.; Scarlat, N.; Dallemand, J.F. The possible contribution of agricultural crop residues to renewable energy targets in Europe: A spatially explicit study. *Renew. Sustain. Energy Rev.* **2013**, *19*, 666–677. [CrossRef]
55. Avcıoğlu, A.O.; Dayıoğlu, M.A.; Türker, U. Assessment of the energy potential of agricultural biomass residues in Turkey. *Renew. Energy* **2019**, *138*, 610–619. [CrossRef]
56. Kour, D.; Rana, K.L.; Yadav, N.; Yadav, A.N.; Rastegari, A.A.; Singh, C.; Negi, P.; Singh, K.; Saxena, A.K. Technologies for biofuel production: Current development, challenges, and future prospects. In *Prospects of Renewable Bioprocessing in Future Energy Systems*; Springer: Cham, Switzerland, 2019; pp. 1–50.
57. Ahorsu, R.; Medina, F.; Constantí, M. Significance and challenges of biomass as a suitable feedstock for bioenergy and biochemical production: A review. *Energies* **2018**, *11*, 3366. [CrossRef]
58. Voloshin, R.A.; Rodionova, M.V.; Zharmukhamedov, S.K.; Veziroglu, T.N.; Allakhverdiev, S.I. Biofuel production from plant and algal biomass. *Int. J. Hydrog. Energy* **2016**, *41*, 17257–17273. [CrossRef]
59. Eckert, C.T.; Frigo, E.P.; Albrecht, L.P.; Albrecht, A.J.P.; Christ, D.; Santos, W.G.; Berkembrock, E.; Egewarth, V.A. Maize ethanol production in Brazil: Characteristics and perspectives. *Renew. Sustain. Energy Rev.* **2018**, *82*, 3907–3912. [CrossRef]
60. Manochio, C.; Andrade, B.R.; Rodriguez, R.P.; Moraes, B.S. Ethanol from biomass: A comparative overview. *Renew. Sustain. Energy Rev.* **2017**, *80*, 743–755. [CrossRef]
61. Bajwa, D.S.; Peterson, T.; Sharma, N.; Shojaeiarani, J.; Bajwa, S.G. A review of densified solid biomass for energy production. *Renew. Sustain. Energy Rev.* **2018**, *96*, 296–305. [CrossRef]
62. Alvira, P.; Tomás-Pejó, E.; Ballesteros, M.; Negro, M.J. Pretreatment technologies for an efficient bioethanol production process based on enzymatic hydrolysis: A review. *Bioresour. Technol.* **2010**, *101*, 4851–4861. [CrossRef] [PubMed]
63. Chisti, Y. Biodiesel from microalgae beats bioethanol. *Trends Biotechnol.* **2008**, *26*, 126–131. [CrossRef] [PubMed]
64. Hu, Q.; Sommerfeld, M.; Jarvis, E.; Ghirardi, M.; Posewitz, M.; Seibert, M.; Darzins, A. Microalgal triacylglycerols as feedstocks for biofuel production: Perspectives and advances. *Plant J.* **2008**, *54*, 621–639. [CrossRef] [PubMed]
65. Gouveia, L.; Oliveira, A.C. Microalgae as a raw material for biofuels production. *J. Ind. Microbiol. Biotechnol.* **2009**, *36*, 269–274. [CrossRef]
66. Lardon, L.; Hélias, A.; Sialve, B.; Steyer, J.-P.; Bernard, O. Life-Cycle Assessment of Biodiesel Production from Microalgae. *Environ. Sci. Technol.* **2009**, *43*, 6475–6481. [CrossRef]
67. Dragone, G.; Fernandes, B.; Vicente, A.; Teixeira, J. Third generation biofuels from microalgae. *Curr. Res. Technol. Educ. Top. Appl. Microbiol. Microb. Biotechnol.* **2010**, 1355–1366.
68. Hays, S.G.; Ducat, D.C. Engineering cyanobacteria as photosynthetic feedstock factories. *Photosynth. Res.* **2015**, *123*, 285–295. [CrossRef]
69. Berla, B.M.; Saha, R.; Immethun, C.M.; Maranas, C.D.; Moon, T.S.; Pakrasi, H.B. Synthetic biology of cyanobacteria: Unique challenges and opportunities. *Front. Microbiol.* **2013**, *4*, 1–14. [CrossRef]
70. Scaife, M.A.; Nguyen, G.T.D.T.; Rico, J.; Lambert, D.; Helliwell, K.E.; Smith, A.G. Establishing Chlamydomonas reinhardtii as an industrial biotechnology host. *Plant J.* **2015**, *82*, 532–546. [CrossRef] [PubMed]

71. Liu, J.; Mooney, H.; Hull, V.; Davis, S.J.; Gaskell, J.; Hertel, T.; Li, S. Systems integration for global sustainability. *Science* **2015**, *347*, 1258832. [CrossRef]
72. Xu, Z.; Chen, X.; Liu, J.; Zhang, Y.; Chau, S.; Bhattarai, N.; Li, Y. Impacts of irrigated agriculture on food–energy–water–CO_2 nexus across metacoupled systems. *Nat. Commun.* **2020**, *11*, 1–12. [CrossRef] [PubMed]
73. Pandey, V.C.; Bajpai, O.; Singh, N. Energy crops in sustainable phytoremediation. *Renew. Sustain. Energy Rev.* **2016**, *54*, 58–73. [CrossRef]
74. Liguras, D.K.; Verykios, X.E. A novel, highly efficient and environmentally friendly process for combined heat and power production from biomass. In Proceedings of the 8th International Conference on Environmental Science and Technology, Lemnos Island, Greece, 8–10 September 2003.
75. Martinho, V.J.P.D. Interrelationships between renewable energy and agricultural economics: An overview. *Energy Strategy Rev.* **2018**, *22*, 396–409. [CrossRef]
76. Chel, A.; Kaushik, G. Renewable energy for sustainable agriculture. *Agron. Sust. Dev.* **2011**, *31*, 91–118. [CrossRef]
77. Martinho, V.J.P.D. Agricultural entrepreneurship in the European Union: Contributions for a sustainable development. *Appl. Sci.* **2020**, *10*, 2080. [CrossRef]
78. Bayrakcı, A.G.; Koçar, G. Utilization of renewable energies in Turkey's agriculture. *Renew. Sustain. Energy Rev.* **2012**, *16*, 618–633. [CrossRef]
79. Koçar, G.; Civaş, N. An overview of biofuels from energy crops: Current status and future prospects. *Renew. Sustain. Energy Rev.* **2013**, *28*, 900–916. [CrossRef]
80. Igliński, B.; Buczkowski, R.; Iglińska, A.; Cichosz, M.; Piechota, G.; Kujawski, W. Agricultural biogas plants in Poland: Investment process, economical and environmental aspects, biogas potential. *Renew. Sustain. Energy Rev.* **2012**, *16*, 4890–4900. [CrossRef]
81. Mekhilef, S.; Faramarzi, S.Z.; Saidur, R.; Salam, Z. The application of solar technologies for sustainable development of agricultural sector. *Renew. Sustain. Energy Rev.* **2013**, *18*, 583–594. [CrossRef]
82. Nacer, T.; Hamidat, A.; Nadjemi, O. A comprehensive method to assess the feasibility of renewable energy on Algerian dairy farms. *J. Clean. Prod.* **2016**, *112*, 3631–3642. [CrossRef]
83. Goncalves, F.A.; dos Santos, E.S.; de Macedo, G.R. Use of cultivars of low cost, agroindustrial and urban waste in the production of cellulosic ethanol in Brazil: A proposal to utilization of microdistillery. *Renew. Sustain. Energy Rev.* **2015**, *50*, 1287–1303. [CrossRef]
84. Paramati, S.R.; Ummalla, M.; Apergis, N. The effect of foreign direct investment and stock market growth on clean energy use across a panel of emerging market economies. *Energy Econ.* **2016**, *56*, 29–41. [CrossRef]
85. Paramati, S.R.; Apergis, N.; Ummalla, M. Financing clean energy projects through domestic and foreign capital: The role of political cooperation among the EU, the G20 and OECD countries. *Energy Econ.* **2017**, *61*, 62–71. [CrossRef]
86. Lee, J.W. The contribution of foreign direct investment to clean energy use, carbon emissions and economic growth. *Energy Policy* **2013**, *55*, 483–489. [CrossRef]
87. Sbia, R.; Shahbaz, M.; Hamdi, H. A contribution of foreign direct investment, clean energy, trade openness, carbon emissions and economic growth to energy demand in UAE. *Econ. Model.* **2014**, *36*, 191–197. [CrossRef]
88. Dixon, P.M.; Weiner, J.; Mitchell-Olds, T.; Woodley, R. Erratum to 'Bootstrapping the Gini Coefficient of Inequality. *Ecology* **1988**, *69*, 1307. [CrossRef]
89. Dagum, C. The Generation and Distribution of Income, the Lorenz Curve and the Gini Ratio. *Econ. Appliquée* **1980**, *33*, 327–367.
90. Starzyńska, W. *Statystyka Praktyczna*; Wydawnictwo Naukowe PWN: Warszawa, Poland, 2002.
91. Kendall, M.G. *Rank Correlation Methods*; Griffin: London, UK, 1955; Volume 19.
92. Spearman, C. The proof and measurement of association between two things. *Am. J. Psychol.* **1904**, *15*, 72–101. [CrossRef]
93. Statistical Review of World Energy 2020; 69th ed. Available online: https://www.bp.com/content/dam/bp/business-sites/en/global/corporate/pdfs/energy-economics/statistical-review/bp-stats-review-2020-full-report.pdf (accessed on 19 December 2020).
94. Statistical Review of World Energy. Available online: https://www.bp.com/en/global/corporate/energy-economics/statistical-review-of-world-energy.html (accessed on 19 December 2020).
95. Kazar, G.; Kazar, A. The renewable energy production-economic development nexus. *Int. J. Energy Econ. Policy* **2014**, *4*, 312–319.
96. Apergis, N.; Payne, J.E. Renewable energy consumption and economic growth: Evidence from a panel of OECD countries. *Energy Policy* **2010**, *38*, 656–660. [CrossRef]
97. Sadorsky, P. Renewable energy consumption and income in emerging economies. *Energy Policy* **2009**, *37*, 4021–4028. [CrossRef]
98. Pao, H.T.; Fu, H.C. Renewable energy, non-renewable energy and economic growth in Brazil. *Renew. Sustain. Energy Rev.* **2013**, *25*, 381–392. [CrossRef]
99. Lin, B.; Moubarak, M. Renewable energy consumption–economic growth nexus for China. *Renew. Sustain. Energy Rev.* **2014**, *40*, 111–117. [CrossRef]
100. Shahbaz, M.; Loganathan, N.; Zeshan, M.; Zaman, K. Does renewable energy consumption add in economic growth? An application of auto-regressive distributed lag model in Pakistan. *Renew. Sustain. Energy Rev.* **2015**, *44*, 576–585. [CrossRef]
101. Khoshnevis Yazdi, S.; Shakouri, B. The globalization, financial development, renewable energy, and economic growth. *Energy Sourcespart. B Econ. Plan. Policy* **2017**, *12*, 707–714. [CrossRef]

102. Leitão, N.C. Economic growth, carbon dioxide emissions, renewable energy and globalization. *Int. J. Energy Econ. Policy* **2014**, *3*, 391–399.
103. Bhattacharya, M.; Paramati, S.R.; Ozturk, I.; Bhattacharya, S. The effect of renewable energy consumption on economic growth: Evidence from top 38 countries. *Appl. Energy* **2016**, *162*, 733–741. [CrossRef]
104. Saidi, K.; Omri, A. The impact of renewable energy on carbon emissions and economic growth in 15 major renewable energy-consuming countries. *Environ. Res.* **2020**, *186*, 109567. [CrossRef]
105. Menegaki, A.N. Growth and renewable energy in Europe: A random effect model with evidence for neutrality hypothesis. *Energy Econ.* **2011**, *33*, 257–263. [CrossRef]
106. Busu, M. Analyzing the impact of the renewable energy sources on economic growth at the EU level using an ARDL model. *Mathematics* **2020**, *8*, 1367. [CrossRef]
107. Armeanu, D.Ş.; Vintilă, G.; Gherghina, Ş.C. Czy energia odnawialna napędza trwały wzrost gospodarczy? Wielowymiarowe dowody danych panelowych dla krajów UE-28. *Energies* **2017**, *10*, 381. [CrossRef]
108. Rokicki, T.; Perkowska, A.; Klepacki, B.; Bórawski, P.; Bełdycka-Bórawska, A.; Michalski, K. Changes in Energy Consumption in Agriculture in the EU Countries. *Energies* **2021**, *14*, 1570. [CrossRef]
109. Rokicki, T.; Perkowska, A. Diversity and Changes in the Energy Balance in EU Countries. *Energies* **2021**, *14*, 1098. [CrossRef]
110. Rokicki, T.; Perkowska, A. Changes in Energy Supplies in the Countries of the Visegrad Group. *Sustainability* **2020**, *12*, 7916. [CrossRef]
111. Rokicki, T.; Perkowska, A.; Klepacki, B.; Szczepaniuk, H.; Szczepaniuk, E.K.; Bereziński, S.; Ziółkowska, P. The Importance of Higher Education in the EU Countries in Achieving the Objectives of the Circular Economy in the Energy Sector. *Energies* **2020**, *13*, 4407. [CrossRef]
112. Inumula, K.M.; Singh, S.; Solanki, S. Energy Consumption and Agricultural Economic Growth Nexus: Evidence from India. *Int. J. Energy Econ. Policy* **2020**, *10*, 545–552. [CrossRef]
113. Alola, A.A.; Alola, U.V. Agricultural land usage and tourism impact on renewable energy consumption among Coastline Mediterranean Countries. *Energy Environ.* **2018**, *29*, 1438–1454. [CrossRef]
114. Ben Jebli, M.; Ben Youssef, S. Renewable energy consumption and agriculture: Evidence for cointegration and Granger causality for Tunisian economy. *Int. J. Sustain. Dev. World Ecol.* **2017**, *24*, 149–158. [CrossRef]
115. Jebli, M.B.; Youssef, S.B. Renewable Energy, Arable Land, Agriculture, CO_2 Emissions, and Economic Growth in Morocco. MPRA Paper No. 767980. 2017. Available online: https://mpra.ub.uni-muenchen.de/76798/1/MPRA_paper_76798.pdf (accessed on 31 March 2021).
116. Jebli, M.B.; Youssef, S.B. The role of renewable energy and agriculture in reducing CO_2 emissions: Evidence for North Africa countries. *Ecol. Indic.* **2017**, *74*, 295–301. [CrossRef]
117. Khan, M.T.I.; Ali, Q.; Ashfaq, M. The nexus between greenhouse gas emission, electricity production, renewable energy and agriculture in Pakistan. *Renew. Energy* **2018**, *118*, 437–451. [CrossRef]
118. Rehman, A.; Ozturk, I.; Zhang, D. Związek przyczynowy między emisjami CO_2 a wydajnością rolnictwa w Pakistanie: Dowody empiryczne z autoregresywnego podejścia do testowania granic opóźnienia. *Appl. Sci.* **2019**, *9*, 1692. [CrossRef]
119. Ali, S.; Ying, L.; Shah, T.; Tariq, A.; Ali Chandio, A.; Ali, I. Analysis of the Nexus of CO2 Emissions, Economic Growth, Land under Cereal Crops and Agriculture Value-Added in Pakistan Using an ARDL Approach. *Energies* **2019**, *12*, 4590. [CrossRef]
120. Ali, S.; Gucheng, L.; Ying, L.; Ishaq, M.; Shah, T. The Relationship between Carbon Dioxide Emissions, Economic Growth and Agricultural Production in Pakistan: An Autoregressive Distributed Lag Analysis. *Energies* **2019**, *12*, 4644. [CrossRef]
121. Aydoğan, B.; Vardar, G. Evaluating the role of renewable energy, economic growth and agriculture on CO2 emission in E7 countries. *Int. J. Sustain. Energy* **2020**, *39*, 335–348. [CrossRef]
122. Liu, X.; Zhang, S.; Bae, J. The impact of renewable energy and agriculture on carbon dioxide emissions: Investigating the environmental Kuznets curve in four selected ASEAN countries. *J. Clean. Prod.* **2017**, *164*, 1239–1247. [CrossRef]
123. Jebli, M.B.; Youssef, S.B. Investigating the interdependence between non-hydroelectric renewable energy, agricultural value added, and arable land use in Argentina. *Environ. Modeling Assess.* **2019**, *24*, 533–546. [CrossRef]
124. Al-Mulali, U.; Solarin, S.A.; Sheau-Ting, L.; Ozturk, I. Does moving towards renewable energy cause water and land inefficiency? An empirical investigation. *Energy Policy* **2016**, *93*, 303–314. [CrossRef]
125. Destek, M.A.; Sinha, A. Renewable, non-renewable energy consumption, economic growth, trade openness and ecological footprint: Evidence from organisation for economic Co-operation and development countries. *J. Clean. Prod.* **2020**, *242*, 118537. [CrossRef]
126. Destek, M.A.; Ulucak, R.; Dogan, E. Analyzing the environmental Kuznets curve for the EU countries: The role of ecological footprint. *Environ. Sci. Pollut. Res.* **2018**, *25*, 29387–29396. [CrossRef]
127. Solarin, S.A.; Al-Mulali, U.; Gan, G.G.G.; Shahbaz, M. The impact of biomass energy consumption on pollution: Evidence from 80 developed and developing countries. *Environ. Sci. Pollut. Res.* **2018**, *25*, 22641–22657. [CrossRef]
128. Wang, Z. Does biomass energy consumption help to control environmental pollution? Evidence from BRICS countries. *Sci. Total Environ.* **2019**, *670*, 1075–1083.
129. Thrän, D.; Schaubach, K.; Majer, S.; Horschig, T. Governance of sustainability in the German biogas sector—adaptive management of the Renewable Energy Act between agriculture and the energy sector. *Energy Sustain. Soc.* **2020**, *10*, 1–18. [CrossRef]

130. Gan, J.; Stupak, I.; Smith, C.T. Integrating policy, market, and technology for sustainability governance of agriculture-based biofuel and bioeconomic development in the US. Energy. *Sustain. Soc.* **2019**, *9*, 1–25.
131. Piwowar, A. Biogaz rolniczy—ważny element okrężnego i niskoemisyjnego rozwoju w Polsce. *Energies* **2020**, *13*, 1733. [CrossRef]
132. O'Connor, S.; Ehimen, E.; Pillai, S.C.; Power, N.; Lyons, G.A.; Bartlett, J. An Investigation of the Potential Adopment of Anaerobic Digestion for Energy Production in Irish Farms. *Environments* **2021**, *8*, 8. [CrossRef]
133. O'Connor, S.; Ehimen, E.; Pillai, S.C.; Black, A.; Tormey, D.; Bartlett, J. Biogas production from small-scale anaerobic digestion plants on European farms. *Renew. Sustain. Energy Rev.* **2021**, 110580. [CrossRef]

energies

MDPI

Article

The Effectiveness of Energy Cooperatives Operating on the Capacity Market

Jakub Jasiński [1,*], Mariusz Kozakiewicz [2] and Maciej Sołtysik [3]

[1] Institute of Rural and Agricultural Development, Polish Academy of Sciences, 72 Nowy Świat St., 00-330 Warsaw, Poland
[2] Collegium of Economic Analysis, Warsaw School of Economics, Madalińskiego 6/8 St., 02-513 Warsaw, Poland; mariusz.kozakiewicz@sgh.waw.pl
[3] Faculty of Electrical Engineering, Częstochowa University of Technology, Armii Krajowej St. 17, 42-200 Częstochowa, Poland; maciej.soltysik@pcz.pl
* Correspondence: jjasinski@irwirpan.waw.pl

Abstract: The European Green Deal aims to make Europe the world's first climate-neutral continent by 2050 by shifting to a clean circular economy, combating biodiversity loss and reducing pollution levels. In Poland, whose economy invariably remains one of the most dependent on coal consumption in Europe, institutional responses to the above EU objectives have taken the shape of energy cooperatives aimed at filling the gaps in the development of the civic dimension of energy on a local scale and the use of potential renewable energy sources in rural areas, including in relation to the agricultural sector. This article is a continuation of the authors' previous research work, which has so far focused on the analysis of the development of profitability of Polish institutions that fit into the European idea of a "local energy community", which includes energy cooperatives. In this research paper, they present the results of subsequent research work and analyses performed on the basis of it which, on the one hand, complement the previously developed optimization model with variables concerning actual energy storage and, on the other hand, analyze the profitability of the operation of energy cooperatives in the conditions of the "capacity market". The latter was actually introduced in Poland at the beginning of 2021. The research took account of the characteristics of energy producers and consumers in rural areas of Poland, the legally defined rules for the operation of the capacity market and the institutional conditions for the operation of energy cooperatives that can use the potential of energy storage. A dedicated mathematical model in mixed integer programming technology was used, enriched with respect to previous research, making it possible to optimize the operation of energy cooperative with the use of actual energy storage (batteries). Conclusions from the research and simulation show that the installation of energy storage only partially minimizes the volume of energy drawn from the grid in periods when fees related to the capacity market are in force (which should be avoided due to higher costs for consumers). The analysis also indicates that a key challenge is the proper parameterization of energy storage.

Keywords: energy cooperatives; capacity market; energy storage; rural areas; mixed integer programming

Citation: Jasiński, J.; Kozakiewicz, M.; Sołtysik, M. The Effectiveness of Energy Cooperatives Operating on the Capacity Market. *Energies* **2021**, *14*, 3226. https://doi.org/10.3390/en14113226

Academic Editor: Talal Yusaf

Received: 26 April 2021
Accepted: 27 May 2021
Published: 31 May 2021

1. Introduction

Decreasing amounts of raw material and constantly increasing pro-environmental pressure make it necessary to look for solutions to increase the efficiency of the use of resources and the optimization of their use [1]. Based on social relationships, the global sharing economy trend is changing fundamental organizational and distribution models and is built on a network of integrated individuals and communities [2]. This phenomenon, which is based on the human tendency to cooperate [3], to share and exchange resources, is beginning to encompass more and more spheres of social life, including the electricity market sector [4,5]. The EU policy imposes a direction for the reorganization of the generating sector by supporting the creation of self-balancing energy areas (regions and communities) [6,7], where energy generation is based on renewable sources (Clean Energy Package (CEP)) [8].

Building local energy independence manifests itself in the creation of cooperatives, enabling benefits to be drawn by cooperating demand and supply side entities. In the EU, creating energy self-sufficiency at the local level is possible on the basis of institutions called energy communities (EC) [9,10]. Their structure and operating model correspond with the guidelines set out in (i) the REDII directive [11], where the focus has been on the *Renewable Energy Community* [12], and it is also a result of (ii) the "electricity market directive" [13], where the *Citizens Energy Community (CEC)* is promoted. Creating opportunities for building local energy communities [14] can be of great importance, especially in rural areas. This is where the greatest potential exists in terms of the use of renewable energy sources (including biomass and biogas), as well as potential waste.

The Polish responses [15] to the development of local energy communities promoted in the EU are characterized by energy clusters [16] and the energy cooperatives analyzed in this article, which the creators of this form of energy cooperation intend to be established in rural areas. The object and the scope of activity of energy cooperatives, as well as the conditions affecting their creation and operating profitability, have already been assessed and described by the authors of this research paper [17]. The considerations and analyses carried out earlier focused on illustrating the benefits seen from the perspective of integrating the supply side (producers) and the demand side (consumers) into the structure of a cooperative. However, they did not correspond with the change in energy market activity in Poland from an energy-only market to a capacity market [18]. In Poland, capacity-market mechanisms have been implemented since January 2021, which significantly affect the operating models and market strategies of both power suppliers and consumers, who are the payers of capacity fees [19]. From this perspective, it is reasonable to extend the analyses carried out so far to energy cooperatives, with elements related to their operation in the energy market and the capacity market [20].

The creation of the capacity market in 2018 was one of the largest changes in the Polish power sector in recent years [21]. From 2021 onwards, those in Poland will pay not only for the electricity generated but also for the available capacity of the power system [22]. This means that power plants will be paid both for electricity production and standby capacity, i.e., full availability [23]. On 30 November 2020, Poland's Energy Regulatory Office (URE) announced the electricity rates for consumers [24] that Poles would have seen on their bills from January 2021 [25]. A new item on the bill for electricity supplies is the capacity fee which, for recipients other than households, will be PLN 76.20/MWh (EUR 1 is around PLN 4.5) [26]. The capacity fee provides financing for the capacity market, i.e., mainly for maintaining capacity in readiness and for modernization and construction of new conventional power plants. This fee will depend on energy consumption between 7 am and 10 pm on weekdays but, given that these are standard working hours, the vast majority of energy consumption will be covered by it [27].

The aim of this paper is to present the results of the research concerning the assessment of the actual impact of energy storage (not only based on the virtual network deposit) on the operating efficiency of energy cooperatives, the increase in the degree of energy independence in the conditions of the capacity market and the minimization of energy consumption during the capacity-fee hours. Thus, an attempt will be made to answer the question: with what production, storage and consumer structure and with what number and configuration of cooperative members does a form of self-organization such as an energy cooperative have a chance to develop and improve its operating efficiency? The models developed by the authors so far [17] will be supplemented by the key element for these considerations—batteries (energy stores) [28]—which may determine not only the character and structure of emerging cooperatives but also, in some cases, become a factor determining their profitability [29,30]. The research objectives set by the authors are the analysis and assessment of: (i) whether, through an appropriate choice of generation sources and an energy storage facility, it is possible to avoid consuming energy from the grid (outside the virtual network deposit) [17,31], which makes it possible to avoid capacity fees; (ii) whether it is worth installing high-capacity energy storage, or whether a measurable effect of improving the optimization target can already be achieved by storage with a lower capacity and power; (iii) to what extent it is possible to estimate the volumetric savings in consuming energy from the virtual network deposit through the use of an actual energy store; and (iv) whether it is reasonable to build an actual store for each cooperative and, if not, which cooperatives it would be reasonable for.

All models of cooperative proposed and studied in this paper, as well as the data on production and energy consumption on farms that are potential members of a cooperative, are anonymized real data obtained from rural areas of Poland. In addition, real data and parameters of energy stores were used in the analyses, so that the conclusions and recommendations from the analyses take on a real and practical dimension.

The paper is structured as follows. The second section provides the main characteristics of the capacity-market model in Poland. The institutional description and legally defined rules for the operation of energy cooperatives can be found in the research team's previous paper [17]; they have therefore not been reproduced in detail in this article. The next section describes the assumptions for the selection of the research sample, taking into account the formal conditions for the establishment of energy cooperatives and the specific character of farms in rural areas in Poland, the input data and their selection, as well as the optimization method, developed in relation to previous analyses, which was used in the research process. The results of the research are then presented and discussed in the subsequent section of the article. In conclusion, it was possible to summarize all the completed work and research, and potential areas of further research interest to the authors have been indicated.

2. Background—The Capacity Market in Poland

The diagnosis of a permanent and continuously growing power shortage, seen both in the short- and long-term scenario [32], was clearly stated by the Polish Transmission System Operator (TSO) in 2014–2015 [33]. In the context of the problems identified at that time, it became a strategic objective to plan changes in the functional model of the energy market in Poland in such a way as to prioritize the security of electricity supply in the Polish Power System (PPS) [34]. The need for changes was an effect of the situation in the regulatory and market environment, which resulted in the permanent exclusion of some centrally dispatched generating units that are critical from a PPS point of view [35]. Maintaining the energy market as an energy-only market would clearly lead to energy supply disruptions and, consequently, to high costs for the economy for not supplying energy [36]. Two main phenomena that influenced the implementation of capacity mechanisms were diagnosed:

- Missing money;
- Missing capacity.

The problem of missing money was due to the fact that the revenues from the units critical to the safety of system operations did not cover their operating and capital costs. An analysis of the short-run marginal cost (SRMC) amounting to about PLN 155–160/MWh (for 2015) for a typical generation system in Poland, i.e., a 200 MW coal-fired power unit, and the average level of wholesale prices on the energy exchange (TGE S.A.) [37], showed that these values were at similar levels. This resulted in the inability to cover the fixed operating costs of the generating unit and a lack of investment impulses for modernization of the existing energy sources and construction of new ones. In 2015, the average operating time of a 200 MW coal-fired power unit was only 3817 h/year (43%).

The phenomena of the low wholesale price and lower volume of energy sold resulted from the conditions of the energy-only market. This model means that the operators of generating units are remunerated based on the production of electricity, which is evaluated on the wholesale market. The valuation of energy is not affected by the type of technology by which it was produced. Low wholesale prices were a consequence of:

- Supporting renewable energy sources (RES) outside the wholesale market area;
- RES operating on low variable costs;
- RES performance characteristics not always ensuring energy security.

The above factors caused the displacement of coal-fired units, the limitation of their operating duration and, consequently, the inability to fully cover the costs of operation and energy production. The forecast of a long-term money-shortage problem clearly indicated a lack of investment incentives over the long term, resulting in a shortfall in capacity.

With the introduction of the capacity market, the "capacity obligation" service was implemented [38]. The generating units covered by this are required to be ready to supply power and to deliver it to the system during a period of threatened shortage, with adequate remuneration. The capacity market is closely related to the development of demand-side response (DSR) services, which consist in the temporary reduction in electricity consumption by consumers or the postponement of its consumption (demand-side management) at the request of the TSO in exchange for remuneration [39].

The need to generate units to remain on standby and supply power to the system during an emergency is related to a cost allocated to all electricity consumers. For households, the cost is of a lump-sum nature and is dependent on the average annual level of energy demand. The rates for this group of consumers range from PLN 1.87/month for consumers with a consumption up to 0.5 MWh/year to PLN 10.46/month for consumers with a consumption above 2.8 MWh/year. Other types of consumers are charged a single rate of PLN 76.20/MWh, calculated for energy consumed on working days between 7 am. and 10 pm. It is worth noting that the capacity fee accounts for approximately 10–15% of the total cost of energy, calculated together with the distribution service [24].

The capacity fee is a component of the distribution fee, and it is therefore related to the energy consumed directly from the grid. In order to reduce the capacity fee, energy consumption from the grid should therefore be reduced. The answer is to find ways to reduce energy consumption or to generate, self-consume and store energy from renewable sources.

3. Materials and Methods (Optimization Model)

3.1. *Assumptions for the Creation of a Sample of Energy Cooperatives for Simulation Purposes*

An assessment of the impact of capacity-market implementation on the level of consumers' costs, on consumers' behavior and on the rationality of building local energy communities required simulations and hypothesis-testing related to simulation scenarios by mapping actual energy cooperatives. For this purpose, real data on electricity production and energy demand in rural areas in Poland was used and five types of energy cooperative were created for simulation purposes. An additional requirement was to represent the diversity of: (i) the locational nature, (ii) the level (scale) of electricity demand, (iii) the nature of economic activity of the cooperative participants, (iv) the electricity consumption profile of each member of the cooperative, (v) the generation potential among the members of the cooperative, (vi) the level of voltage supply from the members of the cooperative, (vii) population size, and (viii) the energy storage capacity.

The selection of members of energy cooperatives took account of locational constraints, i.e., the allocation of members in up to three neighboring rural or rural–urban municipalities. The criteria for the selection of the generation structure by the optimizer took account of at least 70% of the energy demand within the annual billing period, which depended on different types of generation sources and different storage capacities.

The unfavorable hydrological conditions in Poland significantly affect the possibility of using hydro-power for electricity production. For the analyses, it was assumed that a maximum of one hydro-power plant may operate within an energy cooperative, and that there is at least one watercourse that could be adapted for energy generation purposes in the areas of the municipalities where the energy cooperatives were simulated. A practical assumption was adopted, stating that a small hydro-power plant is characterized by low capacity at the level of between several dozen and several hundred kW. The simulation therefore took account of the capacity limits of a single source from 0 to 500 kW, with increments of 50 kW. Discreet increments make the simulation realistic because a source with any continuous capacity cannot be installed.

In Poland, the development of prosumer sources based practically 100% on photovoltaic sources is ongoing and still accelerating. The construction of PV sources is currently the most popular and fastest growing method to achieve energy self-sufficiency in Poland [40]. According to the data from the Ministry of Development, Labor and Technology, at the end of December 2020 [41], there were more than 457,000 micro-systems in Poland (an increase of 28.1% compared to the end of Q3 2020, and as much as 196% more compared to the end of 2019) [42] with a total capacity of about 3006 MW [43]. The dynamics and trends of micro-system growth are influenced by numerous aid programs [44]. In view of the above, for the simulations, it was assumed that at least 25% of energy production of the members of the cooperatives is from solar energy. In addition, capacity limits for an individual PVPP farm from 0 to 1000 kW with increments of 50 kW were adopted, which has a practical justification, since the capacity limit for a micro-system according to the Polish law is 50 kW.

Energy cooperatives can be established in rural and rural–urban areas, i.e., in sparsely urbanized areas [45]. These factors support the construction of low-mast wind sources with low and medium capacity. The efficiency of wind generation is about twice as high for Polish wind conditions as for photovoltaic sources, which makes this type of generation attractive in terms of efficiency and cost [46]. For the analysis and simulation, the ability of cooperative participants to establish sources with a capacity from 0 to 1000 kW with increments of 250 kW was assumed.

Taking account of the location criterion when establishing the cooperative was also intended to take advantage of the agricultural character and potential of the regions, particularly in the context of the stability of the generation profile based on biomass and biogas sources. For the simulation, the capacity limitations of these sources were assumed to be from 0 to 600 kW, with increments of 200 kW. The presence of generation sources of both stochastic (PVPP, wind) and stable (biomass, biogas) generation in energy cooperatives will result in a flattening of the profile and a reduction in generation differences between seasons of the year or times of day.

Additionally, an assumption was adopted indicating that, in the selection of generation sources for the optimal balance of demand in the cooperative, one member has at most two energy generation sources, which does not exclude a situation where not all members have them and are thus energy producers.

The discount nature of the operation of the energy cooperative and its members means that the loss of some energy on its introduction into the distributor's network and its subsequent consumption should be balanced by a slight increase in the installed capacity of the source. For the simulation, it was assumed that the total annual energy production of each member of the cooperative could not exceed 120% of the annual energy demand. This assumption ensures that each member of the cooperative is fully balanced at an individual level and the surplus that occurs further allows the development of self-sufficiency at an aggregated cooperative level. As energy prosumers, cooperatives benefit from a discount model that allows them to manage temporary energy surpluses and shortages. Improving cooperation with the DSO [30] and the efficiency of this mechanism, as well as and increasing real-time energy self-consumption, are further enabled by real energy storage. The addition of real energy storage (batteries) to the model is one of the key elements of the study described in this paper.

3.2. *Characteristics of Energy Cooperatives and Energy Storage Used in the Study*

The sample of energy cooperatives used in the study was constructed using actual measurement data and customer and generation profiles for each type of renewable energy source. The purpose of selecting the participants of the cooperative was to reflect:

- The location character—the simulation was made for participants in two southern voivodeships (administrative divisions) of Poland, Małopolskie and Śląskie, and the

selection took account of different locations of municipalities within the voivodeships. The selection of two different voivodeships was also intended to reflect potentially different insulation levels and thus the efficiency of generation.
- A different level (scale) of electricity demand—this resulted in cooperatives with a demand ranging from 762 MWh/year to 9759 MWh/year. Within this criterion, participants were also selected taking account of the diversity of their individual energy demands. The cooperative consisted of participants with negligible consumption, oscillating around one MWh/year, up to 3.5 GWh.
- The nature of participants' business activities—the selection of participants reflected the division in Polish law according to PKD codes (Polish Classification of Activities) relevant to typical agricultural activities, i.e., crop, vegetable, cereal production, raising of poultry, pigs and cattle, as well as services for the agricultural sector. The complete classification is shown in Table 1.
- The electricity consumption profile of each member of the cooperative—the full range and variety of possible tariffs applicable in Poland for the members of energy cooperatives was taken into account. All analytical scenarios included entities belonging to one-, two- or three-zone tariffs, thus mapping the diverse nature of energy consumption. The affiliation of differently profiled members to specific cooperatives is shown in Table 1.
- Moreover, for the simulations and research, and in order to generalize the results and reflect the energy effects seen from the perspective of minimizing the sum of energy taken from the network and unused energy within the network storage (virtual network deposit), reference models of one hundred energy cooperatives were constructed, each with a variable population size from 10 to 50 members with increments of 10.

Table 1. Characteristics of each analytical scenario.

	Cooperative 1	Cooperative 2	Cooperative 3	Cooperative 4	Cooperative 5
Voivodship	Śląskie	Małopolskie	Małopolskie	Śląskie	Śląskie
Number of members of the cooperative	11	15	11	15	16
Profile of agricultural activity [1] and number of members (pcs)	01.46.Z; (3) 01.13.Z; (3) 01.47.Z; (5)	01.11.Z; (4) 01.13.Z; (1) 01.19.Z; (4) 01.47.Z; (3) 01.50.Z; (3)	01.11.Z; (2) 01.13.Z; (2) 01.19.Z; (1) 01.47.Z; (1) 01.50.Z; (5)	01.13.Z; (4) 01.19.Z; (1) 01.43.Z; (1) 01.47.Z; (3) 01.49.Z; (1) 01.50.Z; (5)	01.11.Z; (2) 01.13.Z; (2) 01.19.Z; (3) 01.47.Z; (6) 01.50.Z; (2) 01.62.Z; (1)
Voltage (LV/MV) and number of members (pcs)	LV (4) MV (7)	LV (10) MV (5)	LV (10) MV (1)	LV (8) MV (7)	LV (9) MV (7)
Tariff group [2] and number of members (pcs)	C11 (2) C12 a (1) C22 b (1) B21 (1) B23 (6)	C11 (4) C12 b (1) C21 (3) C22 a (2) B21 (3) B23 (2)	C11 (4) C12 a (1) C12 b (2) C22 a (2) C22 b (1) B11 (1)	C11 (6) C21 (2) B21 (3) B23 (4)	C11 (6) C21 (2) C22 a (1) B21 (1) B22 (2) B23 (4)
Consumption [MWh/year]: Total min average max	9757 52 887 3574	3559 0 237 1045	762 3 69 312	3383 6 214 1258	5922 1 328 1542

Table 1. *Cont.*

	Cooperative 1	Cooperative 2	Cooperative 3	Cooperative 4	Cooperative 5
Capacity [3] [kW]:	3810	1815	550	1635	2605
PVPP:	200	200	200	200	200
SWPP	3750	500	0	1000	1250
WPP	400	800	0	400	1600
BMPP	600	200	0	400	600
BGPP					
Energy storage [4] [MWh]					
min_s	250	50	50	250	250
expert level	4660	1824	608	1216	2432
optimal level	6000	900	300	6000	6500
max_s	25,000	25,000	3500	15,500	15,000
Production [MWh/year]	8760	3515	750	3635	6255

[1] Agricultural activity profile: 01.11.Z—growing of cereals, leguminous crops and oil plants for seeds, except rice; 01.13.Z—growing of vegetables and melons, roots and tubers; 01.19.Z—growing of other non-perennial crops; 01.43.Z—raising of horses and other equines; 01.46.Z—raising of pigs; 01.47.Z—raising of poultry; 01.50.Z—mixed farming; 01.62.Z—support activities for farm-animal production. [2] Tariff group: The first character (C, B) refers to the tariff type, C—low voltage, B—medium voltage; the second character (1 or 2) refers to the installed capacity level, 1—up to 40 kW, 2—above 40 kW; the third character (1, 2 or 3) indicates the number of time zones; the fourth character, if any, indicates how to account for the time zones, a—division into peak and off-peak, b—division into day and night. [3] PVPP—photovoltaic power plant; SHPP—small hydro power plant; WPP—wind power plant; BMPP—biomass power plant; BGPP—biogas power plant. [4] min_s—minimum energy storage capacity; expert level—energy storage capacity —expert recommendation; optimal level—energy storage capacity — optimal value; max_s—maximum energy storage capacity.

For the simulation and research, sample energy storage facilities (batteries) [47] with real parameters and operation profiles were mapped in the structures of cooperatives. The parameters of a minimum and maximum battery capacity and computational step were selected for each cooperative. For modeling, it was assumed that the change in the nature of the storage operation (charging/discharging) could occur at hourly intervals. Furthermore, it was assumed that unlimited charging and discharging is possible throughout the 24 h period. Recommendations from an energy storage expert were also used in the analyses. The expert selected capacity parameters and charging and discharging powers based on the demand-supply profile of each of the five simulated cooperatives and at the request of the research team. A summary of the parameters is shown in Table 2. This data served as a comparative element for the energy storage (battery) capacities determined during the optimization process, used further in the analysis, and presented in Table 1.

Table 2. Energy storage parameters.

Designation of the Cooperative	Storage Type	Capacity [kWh]	Charging Power [kW]	Discharging Power [kW]
CP1	TPS-E	4660	540	675
CP2	TS HV 70	1824	360	450
CP3	TS HV 70	608	120	150
CP4	TS HV 70	1216	240	300
CP5	TS HV 70	2432	480	600

3.3. *The Optimization Model*

The results presented in this paper were obtained on the basis of data from a simulation of a dedicated mathematical model. The mixed-integer programming technique [48] was used for modeling [49]. GLPK software was used for modeling, particularly the shared high-level GMPL language (this is an open-source software) [50]. COIN-OR/CBC software (also an open-source software) was used to solve the individual optimization tasks [51]. The basic assumptions of the model are discussed below, and parts of the model in the GMPL language are illustrated.

The input data for the model comprised a two-year horizon data in an hourly granulation. The calculation sessions used real data from several dozen consumers from different billing tariffs. The real two-year generation profiles of the following electricity sources were used: a small hydro-power plant, a wind-power plant, a photovoltaic power plant, wastewater- and biomass-based biogas plants.

As the research progressed, the model described in the authors' earlier joint paper was developed [17] by adding an actual electricity store (battery) to it which, through appropriate parameterization, offered the possibility of being used in two scenarios. The first scenario assumed that the cooperative had no electricity storage (i.e., it had a storage with a maximum capacity of 0 kWh). The task was to select the optimum production mix for the pre-set demand, minimizing the energy not taken from the virtual network deposit and purchased from the network. The energy demand, depending on the calculation scenario, was created by individual consumers or aggregated consumers within predefined cooperatives. The energy mix is to be understood as the vector of discrete factors scaling the generation profiles of the energy producers considered. The coordinates of this vector are fixed during the optimization period. The business process modeled connected the consumers with the sources on a proprietary basis (the consumer was the source owner/prosumer). The first scenario is consistent with that already analyzed in the research team's previous article [17]. After solving the task without the real storage (with 0 kWh capacity storage), tasks were solved where the use of batteries was possible for the generation structure obtained in the first scenario. Properly produced energy could be a discrete multiple of the profile adopted.

The energy balance equation in the GMPL modeling language is presented in Algorithm 1.

Algorithm 1.

```
subject to def_EnergyBalance{h in Hours}:
EnergyDemand[h]
=
BuyFromNetwork[h]
+
sum{e in EnergySources} Production[e,h]
+
PickUpFromNetwork[h] - SendToNetwork[h]
+
PickUpFromBattery[h] - SendToBattery[h]
;
```

EnergyDemand[h] is the energy demand at hour *h*; *BuyFromNetwork[h]* is the energy purchase at hour *h*; *Production[h]* is the energy production at hour *h*; *SendToNetwork[h]* is the energy sending at hour *h* and *PickUpFromNetwork[h]* is the energy collected at hour *h*; *SendToBattery[h]* is the energy sending at hour *h to the real battery*; and *PickUpFromBattery[h]* is the energy collected from the real battery at hour *h*.

The model assumes that it is not possible to simultaneously send energy to the real battery and collect energy from it in the same hour *h*. The relevant model equations are in Algorithm 2:

Algorithm 2.

```
subject to constr_SingleComponentBatteryFlow{h in Hours}:
SendToBatteryIndicator[h] + PickUpFromBatteryIndicator[h] < = 1;
```

The *SendToBatteryIndicator[h]* and *PickUpFromBatteryIndicator[h]* variables are binary variables indexed by the hours of the optimization horizon, which takes the value 1 for non-zero values of the corresponding real variables and the value 0 for zero flows.

The algorithms modeling the operation of a real storage of energy produced by prosumers are in Algorithm 3:

Algorithm 3.

```
subject to def_EnergyBattery{h in Hours}:
Battery[h] =
if(h=1) then
0
else
(
Battery[h-1]
-
PickUpFromBattery[h]
+
SendToBattery[h]
);
```

In the analyses, it was assumed that, for $h = 1$, i.e., at the beginning of the optimization, the battery was not charged, i.e., *Battery[0] = 0*.

The optimization objective function was the sum of two components—energy taken from the network and energy produced but not consumed. In Algorithm 4, the optimization was to minimize the following objective function.

Algorithm 4.

```
minimize   objective:
sum{h in EndsOfBillingPeriods}
Storage[h]
+
sum{h in Hours}
BuyFromNetwork[h]
;
```

The *EndsOfBillingPeriods* set covered the last hours of billing periods.

The optimization covered a two-year horizon and the results presented refer to the first year of optimization.

4. Results and Discussion

The effect of energy storage can be considered on multiple levels. Given the availability of the actual metering data, the authors simulated the effect of energy storage separately for the specific five energy cooperatives and five energy cooperatives of different sizes, for which a random selection of members was made to obtain reference scenarios and conclusions. The results of the storage effect were shown in the volumetric dimension. The omission of financial aspects introduces universality into the approach, as it avoids the lack of offer transparency and avoids comparisons with electricity market prices specific to a particular country.

4.1. Simulation Results for Dedicated and Reference Agricultural Energy Cooperatives

A simulation was carried out in which the capacity of the energy storage operating within the cooperative was increased from 0 kWh to 50 MWh, with increments of 100 kWh to 1000 kWh, 500 to 10,000 kWh and 1000 kWh to 50,000 kWh. Each time, the charging power as well as the discharging power was assumed to be equal to 10% of the total capacity expressed in kW.

As a result of the simulations of the behavior of five agricultural energy cooperatives, it is possible to evaluate and analyze the results for specific configurations, taking into account the construction and operation of a real energy storage with the capacity and making it possible to achieve half of the maximum optimization effect. The rationality of choosing the storage capacity, obtaining the desired measurability of the storage effect and the impact on the Demand Side Management (DSM) is also dictated by the results of available studies [52,53]. An illustration of the results is shown in Figure 1.

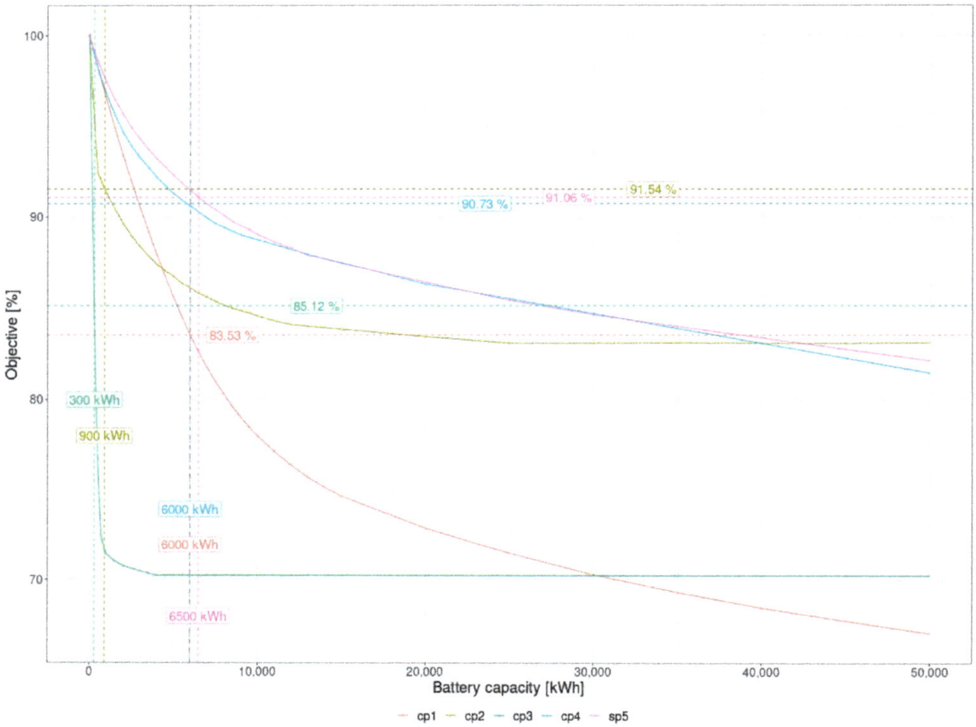

Figure 1. Simulation results of the optimization effect of the sum of energy taken from the network and energy remaining in the virtual network deposit after the billing period, depending on the battery capacity for individual energy cooperatives.

The results allow the following conclusions to be drawn:

- As energy storage capacity increases, the optimization product, which is the sum of energy taken from the network and unused energy within the network deposit at the end of the billing period, decreases nonlinearly.
- The increase in the number of members of the cooperative does not directly relate to the dynamics and profile of dependency of the optimization effect as a function of storage (battery) capacity. This is exemplified by the results for CP1 and CP4, for which the sequence characteristics are very similar, despite differences in cooperative sizes, energy consumption levels and generation-source capacity levels.
- Noteworthy is the fact of different dynamics of the optimization effect in the context of different structures of energy generation within cooperatives. The highest dynamics of the optimization effect is observed for cooperative CP3, where the generation is based on PV only. As capacity increases in profile-stable generation sources, the dynamics decrease, e.g., CP2.
- Regardless of the number of cooperative members, it can be observed that, for small storage capacities, the optimization effect increment is the largest. On this basis, it can be concluded that there is no justification for increasing the storage capacity beyond a specified inflection point of the curve, which is particularly evident in the case of CP3 or CP2.

- The average improvement in the optimization effect relative to the scenario without the energy storage varies and, depending on the size of the cooperative, ranges from about 15% for CP2 to about 35% for CP1.
- Half of the average optimization effect can be obtained for the following energy storage capacities:
  - 6000 kWh; CP1; half of the average effect: 16.5%;
  - 900 kWh; CP2; half of the average effect: 8.5%;
  - 300 kWh; CP3; half of the average effect: 14.9%;
  - 6000 kWh; CP4; half of the average effect: 9.3%;
  - 6500 kWh; CP5; half of the average effect: 8.9%.

The next stage of the analysis assumed a simulation of the effect of a real energy store for reference scenarios reproducing a random drawing of cooperative structures while maintaining the criterion of a specified number of members. For each such reference cooperative, a simulation was carried out in which the capacity of the energy storage within the cooperative was increased from 0 kWh to 50 MWh, with increments of 100 kWh to 1000 kWh, 500 to 10,000 kWh and 1000 kWh to 50,000 kWh. Each time, the charging power, as well as the discharging power, was assumed to be equal to 10% of the total capacity, expressed in kW. The results are shown in Figure 2.

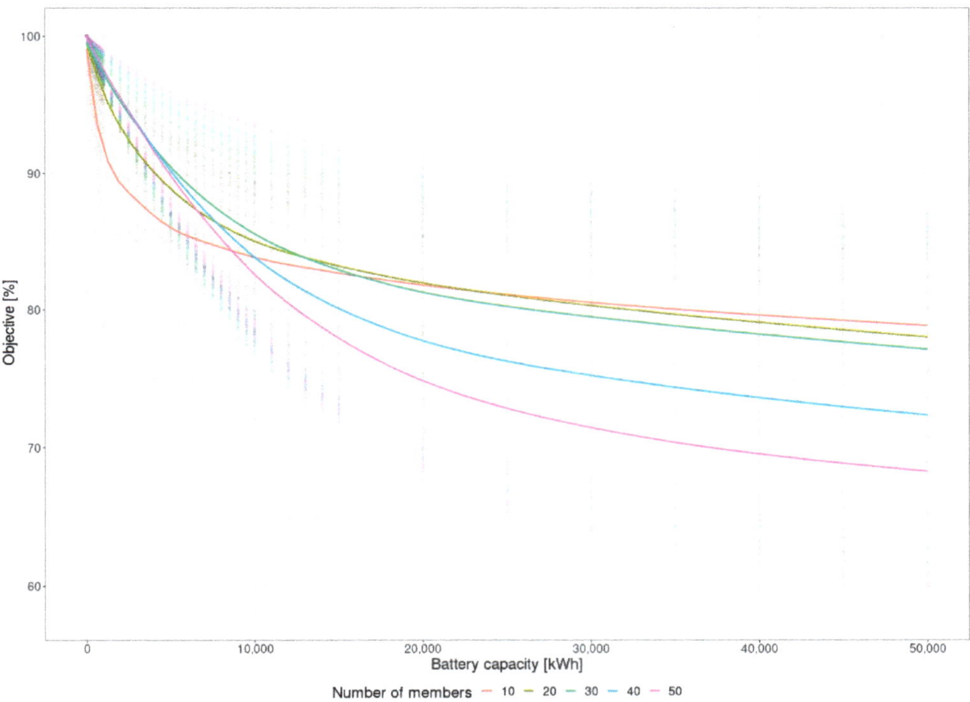

Figure 2. Simulation results of a battery addition effect depending on the battery capacity for reference energy cooperatives. Also seen are measurement points and an average value for cooperatives with a specified number of members.

The simulation results are illustrated in Figure 2, which shows the percentage of the cooperative objective function depending on battery capacity. Solid lines represent the average effect of the measurement points obtained. The condition obtained reflects the effect of providing the cooperative with an energy storage relative to the condition before its installation (capacity: 0 kWh, objective function 0%). The results for cooperatives with different numbers of members are marked with points in different colors. The solid line corresponds to the average measurement points for cooperatives with a specified number of members.

The results allow the following conclusions to be drawn:

- As the energy storage capacity increases, the optimization product, which is the sum of energy taken from the network and unused energy within the network deposit at the end of the billing period, decreases nonlinearly.
- The increase in the number of members of the cooperative affects the dynamics and profile of dependency of the optimization effect as a function of the storage capacity. The higher the number of cooperative members, the milder and shallower is the effect for small capacities and the deeper it is for large storage capacities. The differences between the average level of the optimization effect for the maximum storage capacity analyzed reach 10%.
- Regardless of the number of cooperative members, it can be observed that, for small storage capacities, the optimization effect increment is the largest. On this basis, it can be concluded that there is no justification for increasing the storage capacity beyond the established inflection point of the curve.
- The average improvement in the optimization effect relative to the scenario without the energy storage varies and, depending on the size of the cooperative, ranges from about 30% for 50 cooperative members to about 20% for 10 cooperative members.
- The application of a real energy store makes it possible to limit the sum of energy introduced into the network deposit and the energy lost at the end of the billing period by a value within the range of 10.93% to 41.83%. However, achieving this effect would require the use of storage systems with large power and capacity, which is currently not economically justified. It was therefore assumed that the optimum storage capacity would reflect the achievable half of the maximum optimization effect, and the average benefit values for this scenario are shown in Figure 3.
- The analyses and simulation results show that the real energy store should only have an on-demand role. The operation of an energy cooperative based on the discount model and the temporary deposition of energy in the operator's network enables effective volumetric balancing in the long term. Both storage environments, i.e., the real one (energy storage) and the virtual one (virtual deposit) within the operator's network, are complementary, which makes it possible to significantly improve the volumetric balance that simultaneously burdens the distribution network and maximizes self-consumption. The proposed approach is one of the possible ways to benefit from the storage system. The results of alternative studies show the importance of the predictions made and the optimized target function [54].

In addition, a detailed analysis of the cooperative size scenarios presented in Table 1 and Figure 2 allows the following conclusions:

- Half of the average optimization effect can be obtained for the following energy storage capacities and set size scenarios:
    - 2000 kWh; size: 10 members; half of the average effect: 10.6%;
    - 5000 kWh; size: 20 members; half of the average effect: 11.0%;
    - 6500 kWh; size: 30 members; half of the average effect: 11.4%;
    - 8000 kWh; size: 40 members; half of the average effect: 13.8%;
    - 8500 kWh; size: 50 members; half of the average effect: 15.8%.
- In the context of the potential benefits of energy storage, the use of high-capacity storage systems is not justified. Half of the optimization effect is obtained for storage capacities between 4% (for cooperatives with 10 members) and 17% (for cooperatives with 50 members) of the maximum capacity analyzed.
- As the number of members of energy cooperatives increases, the average value of the benefit identical to minimizing the sum of the purchase of energy from the network, as well as the condition of the network deposit at the end of the billing period, approaches the maximum possible effects of optimization.

- The increase in size also results in a smoothing of the dependence profile of the optimization effect as a function of storage capacity. For a cooperative with 50 members, the dynamics of changes in characteristics in a range between 0 and 8500 kWh are significantly lower than for a cooperative with 10 members and a range between 0 and 2000 kWh.
- The benefit of applying real storage systems is greater in the case of the dominance of wind and PV sources; this conclusion will be justified using data-mining techniques in Section 3 of this article.

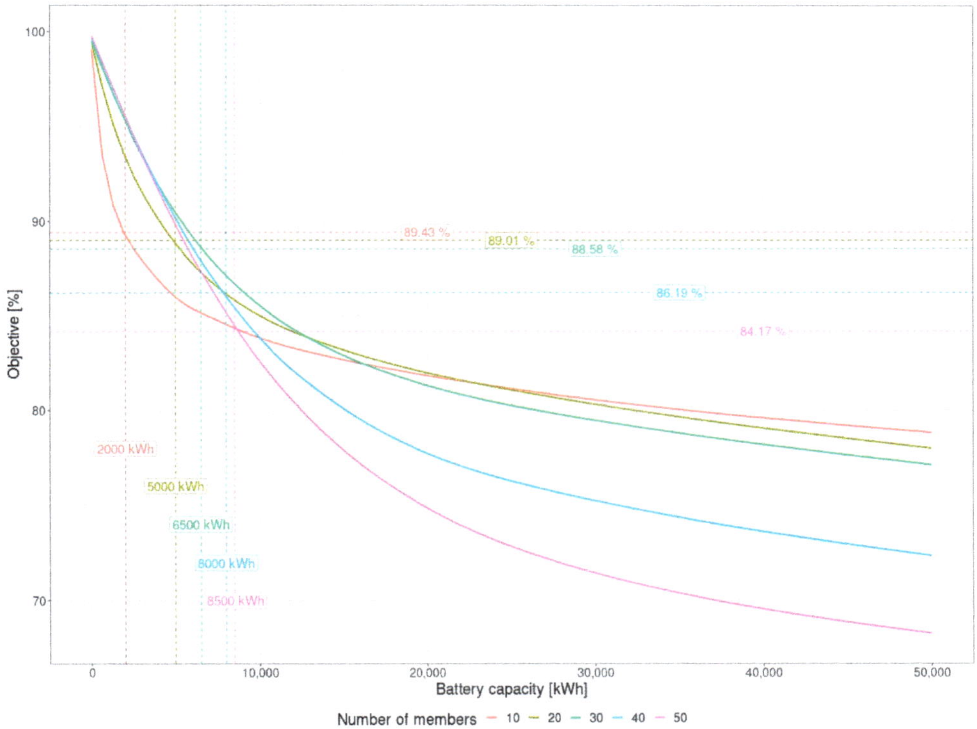

Figure 3. Simulation results of the optimization effect of the sum of energy taken from the network and energy remaining in the network deposit after the billing period, depending on the battery capacity and the number of cooperative members. Also seen is the average effect and battery capacity per average effect.

4.2. Results and Evaluation of the Volumetric Effect of Energy Storage

The simulation of the energy storage process was carried out according to two scenarios. Respectively, they assumed the evaluation of the impact of real energy storage in CP1 – CP5 agricultural energy cooperatives and reference models of one hundred energy cooperatives with population sizes from 10 to 50 members. Due to the dynamic development of the storage sector and the potential difficulty in rationally estimating the current and appropriate level of capital outlays on storage construction, the analyses focused exclusively on the volumetric effect of storage. The results for both scenarios are presented in Tables 3 and 4.

Table 3. Volumetric effects of real energy storage in cooperatives (CP).

	CP1	CP2	CP3	CP4	CP5
Consumption [MWh/year]:	9757	3559	762	3383	5922
Capacity [kW]:	3810	1815	550	1635	2605
Production [MWh/year]:	8760	3515	750	3635	6255
Energy storage—optimal level of capacity [kWh]	6000	900	300	6000	6500
Volumetric storage effect [MWh/year]					
Reduction in energy consumption from the network	348.7	41.3	23.3	31	38.6
Reduction in energy consumption subject to the capacity fee	151.1	23.7	13.8	20.6	19.7
Increase in self-consumption	522.4	61.6	35	46.1	57
in relation to consumption [%]					
Reduction in energy consumption from the network	3.6	1.2	3.1	0.9	0.7
Reduction in energy consumption subject to the capacity fee	1.5	0.7	1.8	0.6	0.3
Increase in self-consumption	5.4	1.7	4.6	1.4	1

Table 4. Volumetric effects of real energy storage in reference cooperatives with varying numbers of members (Mxx).

	CP_M10	CP_M20	CP_M30	CP_M40	CP_M50
Consumption [MWh/year]:					
Min	512	2567	4561	8440	11,810
Average	2792	6248	9743	13,440	16,569
Max	6217	11,224	16,920	18,920	20,122
Reduction in energy consumption from the network [MWh/year]:	23.9	24	52.6	65.7	61.7
Min	36.3	122	188.7	314.8	411.5
Average	114.2	301.3	425.1	519.4	556.8
Max					
In relation to consumption [%]:	4.7	0.9	1.2	0.8	0.5
Min	1.3	2	1.9	2.3	2.5
Average	1.8	2.7	2.5	2.7	2.8
Max					
Reduction in energy consumption subject to the capacity fee [MWh/year]:	12.5	13.9	22.9	43.9	35
Min	16.7	45.7	70.4	123.7	172.1
Average	30.2	128	147.2	176.5	277.3
Max					
In relation to consumption [%]:	2.4	0.5	0.5	0.5	0.3
Min	0.6	0.7	0.7	0.9	1
Average	0.5	1.1	0.9	0.9	1.4
Max					
Increase in self-consumption [MWh/year]:	38.9	112.2	444.5	545.8	694.9
Min	54.4	182.4	282.4	470.3	615.2
Average	85.6	260.9	288.6	250.3	337.6
Max					
In relation to consumption [%]:	7.6	4.4	9.7	6.5	5.9
Min	1.9	2.9	2.9	3.5	3.7
Average	1.4	2.3	1.7	1.3	1.7
Max					

The analysis of the results allows the following conclusions to be drawn:

- The reduction in the average percentage level of energy consumption subject to the capacity fee increases with the size of the reference cooperative from 0.6% for 10 members to 1.0% for 50 members.
- The increment in the average level of self-consumption increases with the size of the reference cooperative from 1.9% for 10 members to 3.7% for 50 members.

- The average level of reduction in energy consumption from the network increases with the size of the reference cooperative by 1.3% for 10 members and by 2.5% for 50 members.
- The maximum difference in reduction levels of energy consumption from the network in relation to demand ranges from 0.5% to 4.7%.
- The analysis of the results for cooperatives CP1-CP5 does not provide unambiguous conclusions and regularities because it refers to specific, individual cases characterized by a different production and consumption structure and the level of the storage capacities analyzed.

4.3. Application of Decisions Trees to Assess the Effect of Energy Storage

Based on the analysis of the profitability of specific cooperatives CP1-CP5, it was not possible to draw conclusions on regularity. For this reason, the focus was on serial randomized experiments. The results of serial experiments, in which the composition of cooperatives was randomized, were analyzed using decision trees [55] available in the R-Project software [56]. The relevant information was obtained in two modeling scenarios.

The first scenario assumed that the variable modeled was the average effect of the savings achieved after using a battery with a capacity equivalent to the half of the effect, discretized to three values: WeakImpact, MediumImpact and StrongImpact. In the experiment, random cooperatives with varying numbers of members achieved different types of percentage savings. In terms of savings achieved, cooperatives were flagged with the WeakImpact flag if the benefit of the battery use ranked in the lower $^1/_3$ of all possible percentage savings achieved during the series of calculations. The StrongImpact flag was used to designate cooperatives for which $^1/_3$ of the highest savings were achieved, and all others were marked with MediumImpact.

The following variables were used as explanatory variables: the number of members of the cooperative (NumberOfMembers, which could assume one of the following values 10, 20, 30, 40, 50), the percentage total share of wind and photovoltaic sources (marked as WindAndPV) and other sources (marked as OtherEnergySourceThanWindAndPV).

Figure 4 shows a decision tree describing the class of savings size depending on the type of sources used and the number of cooperative members. The set of observations was divided into eight segments by the decision-tree algorithm. The leftmost branch of the tree was considered to explain the reading method. It ends with a green leaf marked MediumImpact. In the classification process, observations for which the number of cooperative members was less than 35 and, at the same time, fewer than 25 were included in this segment. In addition, the share of wind and photovoltaic sources in the production was at least 70%. This leaf accounted for 22% of all observations. In all, 75% of the population of this leaf constitute MediumImpact class elements, and 7% and 17% are StrongImpact and WeakImpact class elements, respectively.

The rightmost leaf was marked WeakImpact. In the classification process, observations for which the number of cooperative members was greater than or equal to 35 and in which the share of sources other than wind and photovoltaic in the production was at least 28% were included in this segment. This leaf accounted for 13% of all observations. The elements of the MediumImpact class constitute 6% of the population of this leaf, and the elements of the StrongImpact and WeakImpact classes constitute 38% and 56%, respectively. The set of decision rules of the tree visualized in Figure 4 is recorded in Table 5.

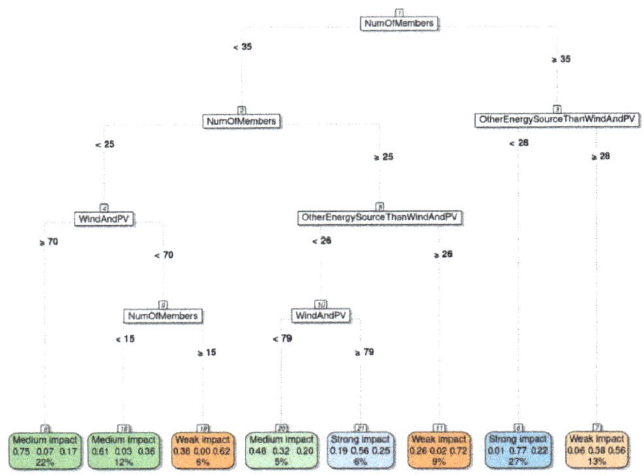

Figure 4. Decision tree quantifying the magnitude of potential benefits after the implementation of a battery corresponding to the "half effect" relative to the distribution of sources and number of members.

Table 5. Decision-tree rules quantifying the magnitude of potential benefits after the implementation of a battery corresponding to the "half effect" relative to the distribution of sources and number of members. The color scheme used in column two corresponds to that of the leaves in Figure 4.

NumOfMembers < 35 and NumOfMembers ≥ 25 and OtherEnergySourceThanWindAndPV ≥ 25.5	Weak Impact
NumOfMembers < 35 and NumOfMembers < 25 and WindAndPV < 69.5 and NumOfMembers ≥15	Weak Impact
NumOfMembers ≥ 35 and OtherEnergySourceThanWindAndPV > = 27.5	Weak Impact
NumOfMembers < 35 and NumOfMembers ≥ 25 and OtherEnergySourceThanWindAndPV < 25.5 and WindAndPV ≥ 78.5	Medium Impact
NumOfMembers ≥35 and OtherEnergySourceThanWindAndPV < 27.5	Medium Impact
NumOfMembers < 35 and NumOfMembers < 25 and WindAndPV < 69.5 and NumOfMembers < 15	Strong Impact
NumOfMembers < 35 and NumOfMembers ≥ 25 and OtherEnergySourceThanWindAndPV < 25.5 and WindAndPV < 78.5	Strong Impact
NumOfMembers < 35 and NumOfMembers < 25 and WindAndPV ≥ 69.5	Strong Impact

The second modeling scenario did not include the size of cooperatives in the explanatory variables. Relevant rules are more aggregated than those shown in Figure 4 but are more easily interpretable in business terms. They show that the greatest effect from using a real battery can be obtained if the percentage production from wind and PV sources is in the 72% to 83% range (StrongImpact). This is illustrated in Figure 5. The MediumImpact level is reached when the percentage of production share from wind and PV sources is greater than 83% (other sources generate no more than 17%), and the effect is the lowest when the share of production from wind and PV sources is less than 72% (other sources generate at least 28%).

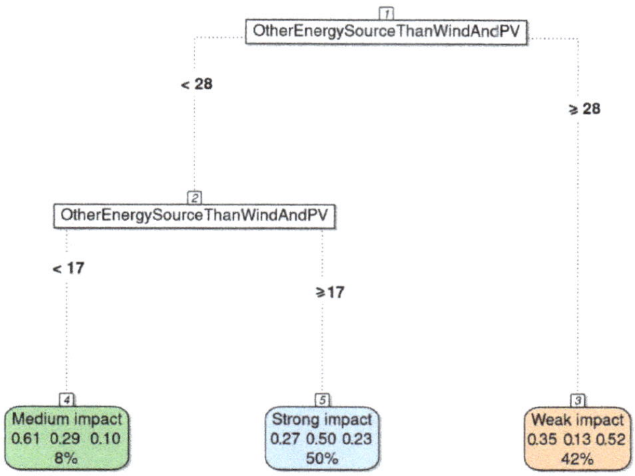

Figure 5. Simulation results in the regression-tree modeling without a variable specifying the number of cooperative members.

It seems that a good approximation of the dependencies obtained is the statement that the maximization of the effect from the application of batteries takes place when the production mix is 75% for wind and PV sources and 25% for other sources. This conclusion seems reasonable, given that sources considered other than PV and wind have a more stable generation profile. Thus, the sources with unstable profiles interact more effectively with real energy storage systems.

5. Conclusions

1. Both the analyses of five specific energy cooperatives and the simulations designed to represent the reference nature of cooperatives confirm that, in each case, the installation of real energy storage systems allows only for a partial minimization of the volume of energy taken from the network during the hours and days when the capacity charges are in force. Depending on the scenario analyzed, an improvement in the average level of volume reduction from 0.6% to 1.0% was achieved, and the maximum value was 2.4%.
2. The results confirm that oversizing the storage capacity has no or a negligible effect. Obtaining half of the effect as the aim of optimization was already possible at the capacity of a few or several percentage points of the maximum capacity analyzed.
3. The use of real energy storage systems makes it possible to reduce the level of energy taken from the network and not from one's own production. The maximum differences of reduction levels of energy consumption from the network in relation to demand range from 0.5% to 4.7%. In addition, the regularity indicating that the increase in this effect corresponds to the increase in the size of the reference cooperative seems to be important.
4. The ability to change the nature of the use of the storage operation in the hourly interval (charging/discharging) further makes it possible to improve the self-consumption rate, which varied between 1.9% and 3.7% for 10 and 50 cooperative members, respectively.
5. Based on the volumetric results, it can be concluded that the construction of a real energy store cannot be treated as a "universal" and appropriate tool for improving the efficiency of cooperatives.
6. The study also makes it possible to evaluate the optimization effect, taking into account both the real energy storage and the nature of generation. The storage effect is

maximized when up to 75% of the generation is from wind and photovoltaic sources, and only 25% from hydro, biogas and biomass sources.

7. The real storage of energy within energy cooperatives, integrating the already optimally selected participants, does not result in a significant improvement in the objective function. The reduction in energy intake from the network and the increase in self-consumption always occur but, in the authors' opinion, the scale of the phenomenon is not satisfactory. The storage analysis should be carried out individually for each configuration of the cooperative. This conclusion indicates that the main purpose of the article, which was to present the results of the assessment of the actual impact of energy storage on operational efficiency, was fairly presented. Energy storage is therefore an interesting area for further in-depth exploration and research, and sensitivity analyses should take into account (i) different charging and discharging time regimes; (ii) mapping investment outlays and operating costs as a function of time; (iii) leveled cost of electricity (LCOE); (iv) predicted improvement in storage efficiency due to advancing technology, quantum batteries [57,58] and propensity to change social behavior [59]; and (v) the fact that effect and prices in the dual market for energy and capacity seem to be particularly valuable. These topics will be the subject of further research by the authors.

Author Contributions: Conceptualization, J.J., M.K. and M.S.; methodology, M.K.; formal analysis, J.J., M.K. and M.S.; investigation, J.J., M.K. and M.S.; resources, M.S.; writing—original draft preparation, J.J., M.K. and M.S.; writing—review and editing, J.J.; visualization, M.K.; supervision, J.J. and M.S. All authors have read and agreed to the published version of the manuscript.

Funding: This research received no external funding.

Institutional Review Board Statement: Not applicable.

Informed Consent Statement: Not applicable.

Data Availability Statement: Data available in a publicly accessible repository that does not issue DOIs. The data presented in this study can be found here: https://drive.google.com/drive/folders/1JKvGCnHBZabmodsTrMUuxp6eXhagI0K0?usp=sharing.

Conflicts of Interest: The authors declare no conflict of interest.

References

1. European Commission. *Communication from the Commission to the European Parliament, the Council, the European Economic and Social Committee and the Committee of the Regions*; European Climate Pact—COM(2020) 788 Final; European Commission: Brussels, Belgium, 2020.
2. Wojcik-Jurkiewicz, M.; Lubicz-Posochowska, A.; Czarnecka, M.; Kinelski, G.; Sadowska, B. Legal aspects of sharing economy: The case of games' platforms. *Eur. Res. Stud. J.* **2021**, *24*, 1196–1210. [CrossRef]
3. European Commission. New Circular Economy Action Plan. Available online: https://ec.europa.eu/commission/presscorner/detail/en/ip_20_420 (accessed on 12 April 2021).
4. Cuenca, J.; Jamil, E.; Hayes, B. Energy communities and sharing economy concepts in the electricity sector: A survey. In Proceedings of the 2020 IEEE International Conference on Environment and Electrical Engineering and 2020 IEEE Industrial and Commercial Power Systems Europe (EEEIC/I&CPS Europe), Madrid, Spain, 9–12 June 2020; pp. 1–6.
5. European Commission. Sustainable Batteries for a Circular and Climate Neutral Economy. Available online: https://ec.europa.eu/commission/presscorner/detail/en/ip_20_2312 (accessed on 12 April 2021).
6. European Commission. European Green Deal. Available online: https://ec.europa.eu/info/strategy/priorities-2019-2024/european-green-deal_en (accessed on 12 April 2021).
7. European Commission. State of the Union: Commission Raises Climate Ambition and Proposes 55% Cut in Emissions by 2030. Available online: https://ec.europa.eu/commission/presscorner/detail/en/ip_20_1599 (accessed on 12 April 2021).
8. Clean Energy for All Europeans Package. Available online: https://ec.europa.eu/energy/topics/energy-strategy/clean-energy-all-europeans_en (accessed on 18 November 2020).
9. Energy Community. Available online: https://www.emissions-euets.com/internal-electricity-market-glossary/2095-energy-community (accessed on 12 April 2021).
10. Mucha-Kuś, K.; Sołtysik, M.; Zamasz, K.; Szczepańska-Woszczyna, K. Coopetitive nature of energy communities—The energy transition context. *Energies* **2021**, *14*, 931. [CrossRef]

11. The European Parliament and The Council of the European Union. Directive (EU) 2018/2001 of the European Parliament and of the council of 11 December 2018 on the promotion of the use of energy from renewable sources. *Off. J. Eur. Union* **2018**, *328*, 82–209.
12. Lowitzsch, J.; Hoicka, C.E.; van Tulder, F.J. Renewable energy communities under the 2019 European clean energy packae—Governance model for the energy clusters of the future? *Renew. Sustain. Energy Rev* **2020**, *122*, 109489. [CrossRef]
13. The European Parliament and The Council of the European Union. Directive (EU) 2019/944 of the European Parliament and of the Council of 5 June 2019 on common rules for the internal market for electricity and amending directive 2012/27/EU. *Off. J. Eur. Union* **2018**, *158*, 125–199.
14. Ghiani, E.; Giordano, A.; Nieddu, A.; Rosetti, L.; Pilo, F. Planning of a smart local energy community: The case of Berchidda municipality (Italy). *Energies* **2019**, *12*, 4629. [CrossRef]
15. Kucharska, A.P. "Czysta energia dla wszystkich Europejczyków"—Konsekwencje wdrażania polityk europejskich dla polskich inicjatyw energetycznych (EN: "Clean Energy for All Europeans" package—Consequences of implementing European policies for Polish energy initiatives). *Energ. Rozproszona* **2020**, *2*, 37–44.
16. Czarnecka, M. Development of energy clusters in Poland—General remarks. *Stud. Prawno Ekon.* **2018**, *109*. [CrossRef]
17. Jasiński, J.; Kozakiewicz, M.; Sołtysik, M. Determinants of energy cooperatives' development in rural areas—Evidence from Poland. *Energies* **2021**, *14*, 319. [CrossRef]
18. Tucki, K.; Orynycz, O.; Wasiak, A.; Świć, A.; Dybaś, W. Capacity market implementation in Poland: Analysis of a survey on consequences for the electricity market and for energy management. *Energies* **2019**, *12*, 839. [CrossRef]
19. Ministry of Climate. Poland Polish Implementation Plan (Amended Version). Available online: https://ec.europa.eu/energy/sites/default/files/polish_implementation_plan_final_courtesy_translation.pdf (accessed on 14 February 2021).
20. Mucha-Kuś, K.; Sołtysik, M.; Zamasz, K. Capacity market in contemporary economic policy. In *Capacity Market in Contemporary Economic Policy*; Difin: Warsaw, Poland, 2015; ISBN 978-83-7930-967-2.
21. Janosz, A.; Sołtysik, M. Customer in the capacity market in the context of qualitative regulation model. In *Capacity Market in Contemporary Economic Policy*; Difin: Warsaw, Poland, 2015; ISBN 978-83-7930-967-2.
22. Zamasz, K. *Efektywność Ekonomiczna Przedsiębiorstwa Energetycznego w Warunkach Wprowadzenia Rynku Mocy*; (Economic Efficiency of an Energy Company in the Conditions of Operation on the Capacity Market); PWN: Warsaw, Poland, 2015; ISBN 978-83-01-18378-3.
23. Gawlikowska-Fyk, A. Capacity Market for Review. Analysis of the Results of Three Auctions. Available online: https://www.forum-energii.eu/en/analizy/rynek-mocy-do-przegladu (accessed on 14 February 2021).
24. Poland's Energy Regulatory Office (URE). Electricity Rates in Poland in 2021. Available online: https://www.ure.gov.pl/pl/urzad/informacje-ogolne/aktualnosci/9163,Jak-zmienia-sie-nasze-rachunki-za-energie-elektryczna-w-przyszlym-roku.html (accessed on 14 February 2021).
25. Dolatowski, M. *Capacity Fee: A Significant Increase in Electricity Prices from 2021*; Wardyński & Partners: Warsaw, Poland, 2020.
26. Bloomberg: EUR to PLN Exchange Rate. Available online: https://www.bloomberg.com/quote/EURPLN:CUR (accessed on 12 April 2021).
27. Ministerstwo Klimatu i Środowiska. *Rozporządzenie Ministra Klimatu i Środowiska z Dnia 9 Listopada 2020 r. w Sprawie Pobierania Opłaty Mocowej i Wyznaczania Godzin Doby Przypadających Na Szczytowe Zapotrzebowanie Na Moc w Systemie*; (Regulation of the Minister of Climate and Environment of 9 November 2020 on Charging the Capacity Fee and Determining the Hours of the Day Corresponding to the Peak Demand for Power in the System); (Dz. U. z 2020 r. Poz. 2009); Ministerstwo Klimatu: Warsaw, Poland, 2020.
28. Andrychowicz, M. RES and ES integration in combination with distribution grid development using MILP. *Energies* **2021**, *14*, 383. [CrossRef]
29. Matthiss, B.; Momenifarahani, A.; Binder, J. Storage placement and sizing in a distribution grid with high PV generation. *Energies* **2021**, *14*, 303. [CrossRef]
30. Andrychowicz, M. Comparison of the use of energy storages and energy curtailment as an addition to the allocation of renewable energy in the distribution system in order to minimize development costs. *Energies* **2020**, *13*, 3746. [CrossRef]
31. Sim, J.; Kim, M.; Kim, D.; Kim, H. Cloud energy storage system operation with capacity P2P transaction. *Energies* **2021**, *14*, 339. [CrossRef]
32. PSE S.A. *Plan Rozwoju w Zakresie Zaspokojenia Obecnego i Przyszłego Zapotrzebowania Na Energię Elektryczną Na Lata 2021–2030*; (Development Plan to Meet the Present and Future Electricity Demand for 2021–2030); Polish Power Grids: Konstancin-Jeziorna, Poland, 2020.
33. PSE S.A. *Prognoza Pokrycia Zapotrzebowania Szczytowego Na Moc w Latach 2016–2035*; (Forecast of Peak Demand Coverage for Power in 2016–2035); Polish Power Grids: Konstancin-Jeziorna, Poland, 2016.
34. PSE S.A. Polish Power System—General Information. Available online: https://www.pse.pl/obszary-dzialalnosci/krajowy-system-elektroenergetyczny/informacje-o-systemie (accessed on 21 April 2021).
35. Majchrzak, H.; Kozioł, M. Analysis of various options for balancing power systems' peak load. *Energies* **2021**, *14*, 513. [CrossRef]
36. Stamtsis, G.; Doukas, H. Cooperation or localization in European capacity markets? A coalitional game over graph approach. *Energies* **2018**, *11*, 1473. [CrossRef]

37. Towarowa Giełda Energii—Notowania; (Polish Power Exchange—Quotations). Available online: https://tge.pl/en-home (accessed on 21 April 2021).
38. Ministerstwo Właściwy DS Energii Rada Ministrów. *Ustawa z Dnia 8 Grudnia 2017 r. o Rynku Mocy*; (The Act of December 8, 2017 on the Capacity Market); (Dz. U. z 2020 r. Poz. 247); Ministerstwo Właściwy DS Energii Rada Ministrów: Warsaw, Poland, 2018.
39. Jeżyna, B.; Lis, M.; Mesjasz-Lech, A. The DSR scheme in the capacity market of industrial enterprises in Poland—Participation determinants. *Energies* **2020**, *13*, 5462. [CrossRef]
40. Księżopolski, K.; Drygas, M.; Pronińska, K.; Nurzyńska, I. The economic effects of new patterns of energy efficiency and heat sources in rural single-family houses in Poland. *Energies* **2020**, *13*, 6358. [CrossRef]
41. Ministerstwo Rozwoju, Pracy i Technologii. Energetyka Prosumencka i Rozproszona. (Prosumer and Distributed Energy). Available online: https://www.gov.pl/web/rozwoj-praca-technologia/energetyka-prosumencka-i-rozproszona (accessed on 12 November 2020).
42. Gramwzielone.pl. The Number of Prosumers in Poland Exceeds 357,000. It's Time for Energy Storage. Available online: https://www.gramwzielone.pl/energia-sloneczna/104268/liczba-prosumentow-w-polsce-przekracza-357-tys-pora-na-magazyny-energii (accessed on 28 December 2020).
43. Polish Press Agency. Polish Solar Farm Has Exceeded the Level of 3400 MW of Capacity. Available online: https://www.wnp.pl/budownictwo/polska-fotowoltaika-przekroczyla-poziom-3400-mw-mocy,440020.html (accessed on 28 December 2020).
44. Olczak, P.; Kryzia, D.; Matuszewska, D.; Kuta, M. "My Electricity" program effectiveness supporting the development of PV installation in Poland. *Energies* **2021**, *14*, 231. [CrossRef]
45. SEJM. *Ustawa z Dnia 4 Października 2018 r. o Spółdzielniach Rolników*; (Act of October 4, 2018 on Farmers' Cooperatives); (Dz. U. Poz. 2073); SEJM: Warsaw, Poland, 2018.
46. Marks-Bielska, R.; Bielski, S.; Pik, K.; Kurowska, K. The importance of renewable energy sources in Poland's energy mix. *Energies* **2020**, *13*, 4624. [CrossRef]
47. Tesvolt—Battery Storage Systems. Available online: https://www.tesvolt.com/en/ (accessed on 21 April 2021).
48. Williams, H.P. *Model Building in Mathematical Programming*, 5th ed.; Wiley: Hoboken, NJ, USA, 2013; ISBN 978-1-118-44333-0.
49. Tomlin, J.A. Special ordered sets and an application to gas supply operations planning. *Math. Program.* **1988**, *42*, 69–84. [CrossRef]
50. GLPK (GNU Linear Programming Kit). Available online: https://www.gnu.org/software/glpk/ (accessed on 18 November 2020).
51. COIN-OR/CBC. Available online: https://projects.coin-or.org/Cbc (accessed on 18 November 2020).
52. Gitizadeh, M.; Fakharzadegan, H. Battery capacity determination with respect to optimized energy dispatch schedule in grid-connected photovoltaic (PV) systems. *Energy* **2014**, *65*, 665–674. [CrossRef]
53. Ke, B.-R.; Ku, T.-T.; Ke, Y.-L.; Chuang, C.-Y.; Chen, H.-Z. Sizing the battery energy storage system on a university campus with prediction of load and photovoltaic generation. *IEEE Trans. Ind. Appl.* **2015**, *52*, 1136–1147. [CrossRef]
54. Chen, P.; Liu, M.; Chen, C.; Shang, X. A battery management strategy in microgrid for personalized customer requirements. *Energy* **2019**, *189*, 116245. [CrossRef]
55. Luger, G.F. *Artificial Intelligence: Structures and Strategies for Complex Problem Solving*, 6th ed.; Pearson Addison-Wesley: Boston, MA, USA, 2009; ISBN 978-0-321-54589-3.
56. The R Project for Statistical Computing. Available online: https://www.r-project.org/ (accessed on 20 April 2021).
57. Maksymych, V.; Całus, D.; Ivashchyshyn, F.; Pidluzhna, A.; Chabecki, P.; Shvets, R. Quantum energy accumulation in semiconductor <ionic liquid > layered clathrates. *Appl. Nanosci.* **2021**. [CrossRef]
58. Grygorchak, I.; Całus, D.; Pidluzhna, A.; Ivashchyshyn, F.; Hryhorchak, O.; Chabecki, P.; Shvets, R. Thermogalvanic and local field effects in SiO_2 structure. *Appl. Nanosci.* **2020**, *10*, 4725–4731. [CrossRef]
59. Colasant, A.; D'Adamo, I.; Marone, P. Nudging for the increased adoption of solar energy? Evidence from a survey in Italy. *Energy Res. Soc. Sci.* **2021**, *74*, 101978. [CrossRef]

energies

MDPI

Article

Production Profile of Farms and Methane and Nitrous Oxide Emissions

Zofia Koloszko-Chomentowska [1,*], Leszek Sieczko [2] and Roman Trochimczuk [3]

1 Department of Management Economy and Finance, Bialystok University of Technology, 15-351 Bialystok, Poland

2 Department of Biometry, Warsaw University of Life Sciences—SGGW, 02-787 Warsaw, Poland; leszek_sieczko@sggw.edu.pl

3 Department of Automatic Control and Robotics, Bialystok University of Technology, 15-351 Białystok, Poland; r.trochimczuk@pb.edu.pl

* Correspondence: z.koloszko@pb.edu.pl

Abstract: The negative impact of agricultural production on the environment is manifested, above all, in the emission of greenhouse gases (GHG). The goals of this study were to estimate methane and nitrous oxide emissions at the level of individual farms and indicate differences in emissions depending on the type of production, and to investigate dependencies between greenhouse gas emissions and economic indicators. Methane and nitrous oxide emissions were estimated at three types of farms in Poland, based on FADN data: field crops, milk, and mixed. Data were from 2004–2018. Statistical analysis confirmed the relationship between greenhouse gas emissions and economic performance. On milk farms, the value of methane and nitrous oxide emissions increased with increased net value added and farm income. Milk farms reached the highest land productivity and the highest level of income per 1 ha of farmland. On field crops farms, the relationship between net value added and farm income and methane and nitrous oxide emissions was negative. Animals remain a strong determinant of methane and nitrous oxide emissions, and the emissions at milk farms were the highest. On mixed farms, emissions result from intensive livestock and crop production. In farms of the field crops type, emissions were the lowest and mainly concerned crops.

Keywords: agricultural production; emission; methane; nitrous oxide; dairy cows; field crops; agricultural production; family farm income; land productivity

Citation: Koloszko-Chomentowska, Z.; Sieczko, L.; Trochimczuk, R. Production Profile of Farms and Methane and Nitrous Oxide Emissions. *Energies* **2021**, *14*, 4904. https://doi.org/10.3390/en14164904

Academic Editors: Vitaliy Krupin and Roman Podolets

Received: 15 April 2021
Accepted: 5 August 2021
Published: 11 August 2021

1. Introduction

The 2030 Agenda for Sustainable Development, adopted in 2015, is a comprehensive plan of development for the world established by the United Nations (UN). All UN member states committed to taking action toward creating adequate living conditions and conditions for economic progress while simultaneously protecting the environment and counteracting climate change. Climate change is progressing due to increased greenhouse gas (GHG) emissions, including carbon dioxide (CO_2). In 2020, annual CO_2 emissions increased by 20% globally compared to 2005 [1]. East Asian and Pacific countries emitted more CO_2 than in 2005 (by 50%), whereas emissions decreased in North America (by 13%) and in European and Central Asian countries (by 9%) [1]. The amount of greenhouse gases emitted annually by the EU decreased by 12% compared to 2010, while Poland emits over 400 million tons of greenhouse gases annually, which makes up 9.8% of the EU's emissions [1]. It is necessary to take action over the next several years to reduce the risk of irreversible effects of climate change, particularly since the Earth will continue to react to increases in greenhouse gas emissions for a long time after they are reduced [2]. Increasing the use of renewable energy sources is one measure that can contribute to the reduction of greenhouse gas emissions. Poland is involved in actions aimed at limiting climate change that are being undertaken by the international community. It is one of the signatories of the UN Framework Convention on Climate Change (UNFCCC) since 1992 and the Kyoto Protocol since 2002 [3].

Agriculture is one of the sectors of the economy that has a strong relationship with the natural environment. The technological, biological, and organizational progress that is being made affords access to increasingly modern production technologies. This leads to improvements in the technical and economic efficiency of agricultural production. However, these changes are generating a series of threats to the natural environment. In relation to the growth of the global population and a growing demand for food, there is pressure to increase the magnitude of agricultural production. Today, technical capabilities with regard to increasing the scale of production are not a limitation, as this process is accompanied by economic benefits; however, an environmental barrier does arise. The negative impact of agricultural production on the environment is manifested, above all, in the emission of greenhouse gases (GHGs), mainly nitrous oxide (N_2O) and methane (CH_4) [4–6].

The Kyoto Protocol lists CO_2 as one of the gases that has an influence on the greenhouse effect. This is the most important factor in climate change, and is covered in most studies. However, some researchers are voicing the opinion that basing estimates solely on CO_2 emissions and omitting other gases in the balance associated with agriculture, particularly in rural and urban–rural municipalities, leads to underestimation of GHG emissions from Polish agriculture [7–9]. Research by Wiśniewski [7] shows that over half of the total emissions from agriculture in Poland is associated with animal raising and breeding. This is confirmed by the results of many studies. GHG emissions from agriculture in Africa are showing some of the highest rates of growth in the world, the greatest source of which is animal production on farms [6]. The case is similar in EU member states, where the largest amount of emissions also comes from animal production. During 2004–2017, GHG emissions were highly concentrated in several EU member states; these were the countries with the most developed agriculture: France, Germany, Spain, and the United Kingdom [10].

There is a strong emphasis on the need to reduce greenhouse gas emissions in agriculture and to incorporate agriculture in actions against unfavorable climate change [2,11]. According to the Ministry of Climate, in Poland in 2018, agriculture was responsible for 8% of greenhouse gas emissions (in CO_2 equivalent) with respect to the base in 1988 [12]. Although it is necessary, reducing GHG emissions in this sector remains an enormous challenge. This is because there is a specific conflict of interest in this area. Agricultural holdings are subject to competition in the food market, and reconciling economic and environmental interests is a problem. Unfortunately, the magnitude of GHG emissions from agriculture is disturbing. In Poland in 2018, a 7.2% increase in GHG emissions from agriculture was recorded with respect to 2015 [13]. The search for effective tools for production technology management in order to consume fewer resources and reduce the environmental impact is ongoing [14]. It seems that only deep changes in the structure of the entire agri-food system can reduce greenhouse gas emissions in the agricultural sector [15]. This is not only about the practices employed in agricultural production, but also changes in consumers' nutritional habits; for example, research conducted in Mediterranean regions indicates that reducing meat and dairy consumption by 40% could reduce GHG emissions by 20–30% [16].

Transforming the economy into a low-emissions economy is currently one of the most important challenges facing the modern world. A circular, low-emissions economy plays a critical role in the development of agriculture, as it is an opportunity to improve both the quality of the environment and economic well-being. The social aspect of the low-emissions economy is highlighted. Limiting greenhouse gas emissions brings about benefits in terms of human health regardless of the level of prosperity, as the benefits apply to both rich societies and less affluent ones [17]. The economic dimension of the relationship between agriculture and climate change is also important. A slight reduction in GHG emissions resulting from the growth of value added in agriculture and renewable energy was observed in studies conducted in Pakistan [18]. The authors of that research suggest that increasing the value added of agriculture and consumption of renewable energy could counterbalance the increased GHG emissions resulting from the consumption of coal-generated electricity. Zafeiriou et al. [19] obtained divergent results in their research on the relationship between greenhouse gas emissions from agriculture and per capita income in the agricultural sectors of different EU countries. The results indicated that if CO_2 emissions rise, so would income from agriculture, which was confirmed in the case of Spain. However, the authors expressed a reservation regarding the nonlinear relationship between agricultural income and CO_2 emissions. Other studies indicate a positive influence of direct foreign investment in agriculture on the CO_2 emission equivalent in developing countries [20]. The economic aspects of greenhouse gas reduction are rarely raised in studies. A report by the Centre for Climate and Energy Analyses unequivocally shows that reducing methane and nitrous oxide emissions from agriculture in Poland causes changes in farmers' level of production and income, and should be considered through the lens of economic effects [21]. At the same time, the report's authors are aware of how difficult it is to reach a compromise between these two objectives.

The assessment of agriculture's environmental impact is part of the concept of sustainable development. Studies concerning the impact of farms with different production profiles on the environment are an important part of this. It seems that such assessment is important because the impact of an agricultural holding on the environment depends on its specialization. Specialized farms are the ones that determine the basic trend of transformation in Polish agriculture. Specialization is a factor that fosters improvement of farming efficiency; however, there are environmental limitations linked to the growth of such farms. In such cases, activity is associated with the concentration of resources and intensity of production, so the environmental impact assessment is multi-dimensional. Various environmental and economic sustainability indicators are taken into account in such assessments [22–27]. The choice of indicator generally depends on the availability of data. In most assessments, greenhouse gas emissions are either omitted or treated as a side note. The relationships between environmental practices and economic results have also been insufficiently investigated. The present paper broadens the knowledge in this scope. Farmers, even those with the highest environmental awareness, will always be motivated by an economic objective in their activity. Thus, it is necessary to account for economic aspects in analyses of greenhouse gas emissions. Doing so can provide a broader picture of the dependencies existing between the farmer's choice of agricultural practices and the realization of the environmental objective, i.e., reducing methane and nitrous oxide emissions from agriculture.

The goals of the study were to estimate methane and nitrous oxide emissions at the level of an individual farm and indicate differences in emissions depending on the type of production, and to investigate dependencies between greenhouse gas emissions and economic indicators. The authors' intent is to present estimates of CH_4 and N_2O for three types of specialized agricultural holdings and to indicate the relationships between the economic objectives that motivate farms and the environmental objectives that arise from concern for the natural environment.

2. Methodology

Data concerning farms were obtained from the Farm Accountancy Data Network (FADN), published by the Institute of Agricultural and Food Economics, Polish Research Institute [28]. The data used in this research are not available in other databases. They concern agricultural accountancy, and hence are focused mainly on economic categories and the financial situation of individual farms; however, they can also be used for environmental analyses [29–31]. For the purposes of this study, we adopted the methodology described by Wiśniewski [7], who proposed assessing the magnitude of greenhouse gas emissions based on data from public statistics. The proposed solution complies with the methodology and standard indicators of the Intergovernmental Panel on Climate Change [32] and accounts for emission indicators developed by the National Centre for Emissions Management [33]. Other authors have also used data from public statistics to estimate emissions, including methane and nitrous oxide emissions [34–36].

Although the applied methodology is a simplified solution, it makes it possible to utilize generally available data on agricultural holdings and assess the impact of farming on the environment. Such a solution makes it possible to assess the variability of emissions and compare farms with respect to criteria such as farm size, production system, and type of production. Dick et al. [37] point to the advantages of such a solution, mainly from a practical perspective. Above all, it enables farmers to apply the best practices, select a method of production, and choose the means of its implementation.

The data come from 2004 to 2018. Three types of agricultural holdings were considered in the analyses: those that specialize in field crops, specialize in milk production (dairy cattle), or have a mixed production profile. These are the main types of farms in Poland. Data on the number and basic characteristics of farms are presented in Tables A1 and A2 (Appendix A) It should be noted that the number of farms changes every year, which is due to the selection of the sample included in the FADN system. Every year, some farms remain outside of FADN's area of observation, and other farms enter the sample.

Research was focused on the three main sources of greenhouse gas emissions, emitted directly over the course of agricultural production: gastrointestinal fermentation in farm animals (main source of methane emissions), animal feces (source of methane and nitrous oxide emissions), and nitrous oxide emissions from the use of mineral fertilizers.

Estimates of the magnitude of methane and nitrous oxide emissions from animal production were made based on the number of livestock and emission coefficients. In the case of cattle, available national gut fermentation CH_4 emission coefficients applied by KOBiZE are used to prepare annual inventory reports. They are prepared based on daily energy demand for selected categories of cattle and coefficients of conversion to methane (share of energy in fodder converted to methane). Methane emission indicators from the livestock's gut fermentation is estimated based on the more general, default indicators recommended by the IPCC [32]. The level of nitrous oxide is estimated based on default indices of nitrogen content in animal feces and default N_2O-N emission coefficients for different methods of animal feces management [32]. The following emissions coefficients were applied (kg per animal per year): CH_4 from gastrointestinal fermentation: dairy cows, 122.0; other cattle, 49.65; swine, 1.5; CH_4 from feces: dairy cows, 11.87; other cattle, 2.15; swine, 3.07; nitrogen excreted in feces: dairy cows, 70.26; other cattle, 49.95; swine, 30.22 [7]. In FADN data, animals are counted as livestock units, which was why there was a need to convert these units into physical headcounts. This was done according to coefficients for conversion of cattle and swine, with the following coefficients adopted: dairy cows, 1.0; swine, 0.25; other cattle, 0.40 (mean value determined for heifers and calves) [38]. Poultry was omitted in the calculations due to the lack of IPCC guidelines.

The amounts of methane and nitrogen emissions were then calculated per 1 ha of farmland. The reference to farmland area was made for two reasons. First, when conducting a comparative analysis of three types of specialized farms, one needs to accept a single point of reference, and that is farmland area. Second, this made it possible to investigate dependencies between methane and nitrous oxide emissions and economic results, which was the intended goal of this work.

The average consumption of fertilizers per ha of farmland was adopted as the basis for estimating nitrous oxide emissions from mineral fertilizers. There is no information in the generally available FADN data about the consumption of mineral fertilizers, which is why this quantity was estimated indirectly based on average NPK consumption at individual farms in the country according to the Central Statistical Office [39]. The quantity of NPK consumption in the studied agricultural holdings was corrected by the indicator representing the general production and economic advantage of the studied farms over individual farms in Poland, collectively, as applied by the Institute of Agricultural and Food Economics [40]. This indicator was determined for every year based on a comparison of the production value per 1 ha of farmland of the most important products (basic cereals, potatoes, milk, and pig livestock) of the studied farms, with the production value of these products for the collective of farms, according to the Central Statistical Office, accepted as 1.

It is accepted that farms that conduct agricultural accountancy achieve higher production and economic results than average farms in the region and in the country.

When estimating amounts of emissions from the use of mineral fertilizers, the default nitrous oxide emission coefficient of 0.01 kg N_2O-N per 1 kg N was accepted. The mass of nitrogen originating from the application of mineral fertilizers was corrected by the amounts of ammonia and nitrous oxides emitted [7].

The following indicators were also applied to evaluate the economic situation of agricultural holdings: net value added (PLN $\cdot AWU^{-1}$), family farm income (PLN), family farm income per 1 ha of farmland (PLN), and land productivity per 1 ha of farmland (PLN).

The arithmetic mean, minimum, maximum, and standard deviation were used to present results from 15 years of observation. Based on economic indicators originating from the described farms and emission values calculated for the selected greenhouse gases, an attempt was made to present the variation of these indicators over the course of those 15 years. Thirteen features describing the economic and agrarian characteristics of farms, along with the number of farms with a given agricultural production profile taking part in FADN studies, were taken as variables and subjected to reduction during analysis. The number of farms is not a feature associated with emissions, and in normal studies with repetitions, it should not be taken for analysis. Here, however, FADN studies were based on a variable number of farms, therefore, this feature could justify variation within the very short time period of one year. Ten indicators of GHG emissions were also taken as variables for analysis. The set of input data consisted of 23 features, representing the dimensions of the three described types of farms over 15 years, which were treated as objects in the analysis (Table 1).

Table 1. Values of variables used in the analysis (for three types of farms).

Specification	A–Field Crops		B–Milk		C–Mixed	
	Average	Standard Deviation	Average	Standard Deviation	Average	Standard Deviation
Economic Indicators						
X_1—Number of farms	3058.20	680.01	1881.67	870.35	4011.20	500.67
X_2—Utilized agricultural area (ha)	30.51	10.38	19.13	2.96	16.83	1.32
X_3—Total livestock unit (LU)	2.42	1.22	20.17	4.50	12.47	1.09
X_4—Total output (PLN)	128,122.07	54,478.53	115,330.67	38,910.40	82,903.47	17,700.99
X_5—Total inputs (PLN)	110,920.67	43,201.69	88,252.33	30,940.41	77,282.40	20,966.41
X_6—Fertilizers (PLN)	21,068.53	10,469.06	7914.53	3043.14	7583.80	2310.94
X_7—Energy (PLN)	13,366.07	6218.51	9342.87	3264.85	7747.00	2085.39
X_8—Total intermediate consumption (PLN)	76,011.67	30,406.09	65,534.33	22,908.43	57,296.93	14,988.95
X_9—Total inputs (PLN ha^{-1})	3593.80	503.97	4468.67	1006.58	4598.40	944.21
X_{10}—Land productivity (PLN ha^{-1})	4117.87	514.65	5873.73	1236.60	4890.93	734.48
X_{11}—Farm net value added (PLN AWU^{-1})	34,984.33	17,023.93	29,603.73	11,186.76	17,554.33	4079.47
X_{12}—Family farm income (PLN)	46,797.33	28,032.23	48,952.87	19,791.13	22,283.53	5290.34
X_{13}—Family farm income (PLN$\cdot ha^{-1}$)	1442.53	372.45	2480.40	714.25	1325.73	313.05
Indicators of GHG sources						
Z_1—Dairy cattle CH_4 (kg y^{-1})	55.16	42.16	1825.33	287.15	337.35	65.07
Z_2—Dairy cattle N_2O (kg y^{-1})	28.95	22.13	958.11	150.87	177.05	34.15
Z_3—Other cattle CH_4 (kg$\cdot y^{-1}$)	86.67	28.41	797.43	326.89	446.06	133.44
Z_4—Other cattle N_2O (kg y^{-1})	83.58	27.39	769.60	315.28	430.13	128.67
Z_5—Pigs CH_4 (kg y^{-1})	19.49	12.33	-	-	107.41	10.86
Z_6—Pigs N_2O (kg y^{-1})	123.37	76.15	-	-	709.30	71.78
Z_7—Total emissions CH_4(kg y^{-1})	154.07	86.20	2622.75	611.07	890.82	102.46
Z_8—emissions CH_4 (kg ha^{-1})	5.96	4.07	135.94	14.25	76.37	6.80
Z_9—total emissions N_2O (kg y^{-1})	239.89	116.65	1727.71	464.10	1317.15	122.23
Z_{10}—emissions N_2O (kg ha^{-1})	9.27	5.65	89.04	12.15	78.30	4.26

Source: Own calculation based on FADN data [28].

Three independent analyses were carried out for each type of farm. Factor analysis was conducted using principal component analysis (PCA) [41,42]. To facilitate the interpretation of results, varimax rotation was applied. This involves rotation of the X- and Y-axes (linear combination) so as to maximize the variance of loadings between factors and minimize their variance within the new factor called a component here.

3. Results

The level of greenhouse gas emissions was dependent on the production profile and was characterized by high variation during the studied period (Figure 1). Farms specializing in milk production emitted the most CH_4 and N_2O. This is the result of high livestock density and intensive production technology. Farms specializing in milk production surpassed other types of farms in terms of the amount of income from the farm and land productivity (Table A2, Appendix A). At field crop farms, the levels of methane and nitrous oxide emissions were the lowest among the studied farms. This is due to the specialization adopted and consistent reduction of animals on the farm. During the period of study, changes in the levels of CH_4 and N_2O occurred at all farms and were associated with organizational changes at the farms. There is a positive correlation between methane and nitrous oxide emissions and economic results measured at the level of family farm income. Milk farms reached the highest land productivity and the highest income level per 1 ha of farmland.

On farms that specialize in field crops, the average methane emissions amounted to 5.96 $kg \cdot ha^{-1}$, varying within a range of 0.25 to 11.3. Within the studied time interval, several periods of lesser and greater CH_4 emissions can be distinguished (Figure 1A). After a period of slight decrease in the level of emissions during 2005–2006, there was an increase during the next two years (2008–2009) to a level of 11.31 $kg \cdot ha^{-1}$. In the following years, a declining tendency can be seen, and in 2018, total CH_4 emissions per 1 ha of farmland was more than three times lower than in 2004. During the studied time interval, the period of 2010–2012 is noteworthy as greenhouse gas emissions were very low during that time. The factor responsible for this was the selection of farms that were within FADN's area of observation during those years. These were much larger farms and the average area of farmland was twice as large as in other years.

The level of emissions should also be considered against the backdrop of organizational changes in agricultural holdings, particularly with regard to animal production. In 2004, the average number of animals on a farm in livestock units amounted to 3.51 LU (1.6 of cattle and 1.91 of swine). This number decreased every year after that (R^2 = 0.7706). In 2018, the number of animals was reduced to 0.09 LU of dairy cattle (which can be considered as total elimination), 0.55 LU of other cattle and 0.43 LU of swine. Organizational changes in the studied group of farms indicate progressing specialization. These farms specialize in field crops. The first few years were a period of adaptation to the selected production profile and many holdings continued to raise animals. In every year that followed, farms reduced animal production in favor of field crops according to the specialization they adopted.

In the case of dairy cattle farms, CH_4 emissions were substantially higher ranging from 119.90 to 157.73 $kg \cdot ha^{-1}$ depending on the year. Throughout the entire period of study, the level of emissions stayed at a relatively constant level and systematically increased starting from 2013 (Figure 1B). The causes of this situation are understandable. These holdings specialize in milk production, and over the course of successive years, farmers increased their herds of dairy cows (R^2 = 0.9663). The average number of cows on a farm in 2004 amounted to 10.69 LU, and in 2018, the number was 17.12 LU. This is the basic production herd, and in the case of farms specializing in milk production, the scale of production is fundamentally important in terms of farming economics. Besides cattle in the basic herd, other functional groups of cattle were also present, most likely constituting a replacement herd. This population of cattle also increased from year to year. In the case of this group of holdings, the total magnitude of emissions originated from cattle raising (there were no other animal species).

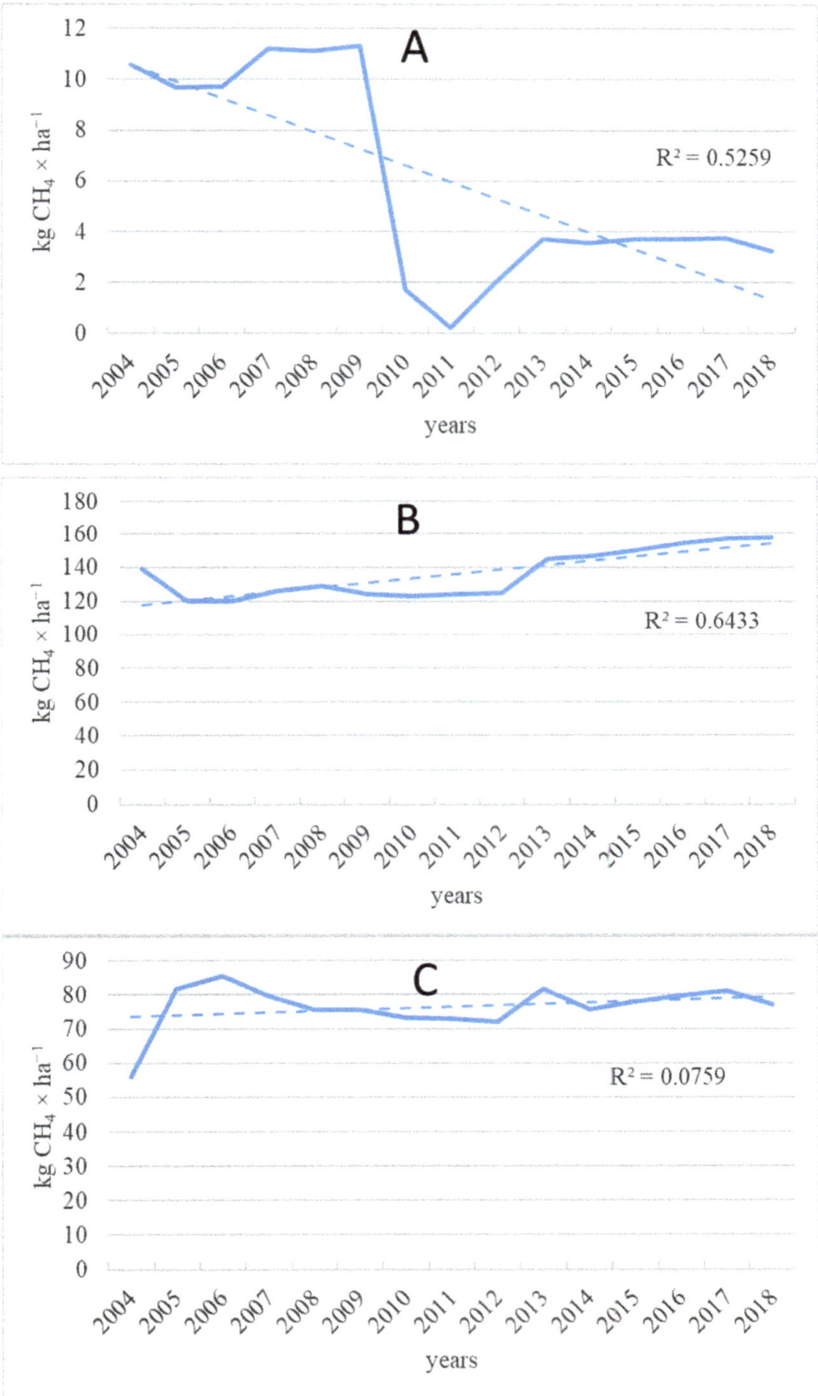

Figure 1. Emission levels of CH_4 (kg·ha^{-1}) at three types of farms: (**A**) Field crops; (**B**) Milk; (**C**) Mixed.

For holdings with a mixed profile, CH_4 emissions were at a moderate level. If we accept the level of CH_4 emissions at dairy cow holdings to be 1, then farms specializing in field crops were at a level of 0.045 on average during the studied period and mixed holdings were at a level of 0.57. During the studied period, the amount of CH_4 emissions changed from 56.09 to 85.67 kg ha^{-1}. The level of this gas exceeded 80 kg·ha^{-1} in only four years (Figure 1C).

The amount and changes of CH_4 emissions should be considered against the backdrop of the way animal production was organized. During 2004–2005, over half of CH_4 emissions originated from dairy cattle, and the number of cattle was the greatest during those years. However, starting from 2006, the number of cows was successively reduced and herds of beef cattle were enlarged. These changes are reflected in the structure of CH_4 emission sources. In 2018, over 65% of emissions originated from the raising of beef cattle. In total, cattle were responsible for 85–90% of CH_4 emissions depending on the year. In terms of swine herds, changes in herd populations were small, and they were responsible for approximately 12% of methane emissions on average during the studied period; this level declined in the years that followed.

The analysis performed indicates that holdings with a mixed profile raised both cattle and swine, however, in recent years, they became more oriented toward raising beef cattle. These changes were reflected in greenhouse gas emissions. Regardless of the production profile, gastrointestinal fermentation in cattle is mainly responsible for methane emissions. Cattle were responsible for 82.5% of methane emissions on mixed farms and up to 92% on dairy farms. Accordingly, 8–17.5% of methane emissions originated from animal feces. Animal production is also a source of nitrous oxide emissions via feces. N_2O emissions were higher for farms with larger animal herds (Figure 2).

Farms that specialize in milk production produce the most nitrous oxide emissions, taking into account nitrous oxide from animal feces and mineral fertilization at a level of 90.06 kg N_2O ha^{-1} (Figure 2B). Meanwhile, on mixed farms, the value of nitrous oxide emissions was 78.96 kg N_2O·ha^{-1}, and on farms specializing in field crops it was 9.65 kg N_2O ha^{-1}. These data indicate that cattle are the main emitters of not only methane but also nitrous oxide. On dairy farms, 93–95% of N_2O emissions originated from animal feces. On field crop farms, N_2O emissions mainly originated from mineral fertilization. Animals were kept solely for the family's own needs (0.09 LU dairy cattle and 0.55 LU other cattle in 2018).

Data on nitrous oxide emissions from the application of mineral fertilizers indicate that the greatest emissions came from farms that specialize in milk production. This is the result of intensive fertilization of cultivated plants. Corn, which constitutes the main feed base for cattle, is dominant in the crop structure. During the studied period, the area of corn cultivation increased from 0.3 ha in 2004 to 14.24 ha in 2018, while the area of cereals changed from 5.39 ha to 7.49 ha. During this period, consumption of mineral fertilizers increased from 163 to 271 kg NPK·ha^{-1}.

The amount of nitrous oxide emissions from mineral fertilizers on mixed farms was 65% of the amount on dairy cattle farms. During the studied period, the level of N_2O emissions was variable, and it is difficult to unequivocally identify a trend (Figure 2C). The highest level of emissions was recorded in 2006 at 86.31 kg N_2O ha^{-1} and the lowest in 2012 at 72.64 kg N_2O ha^{-1} per farm. Consumption of mineral fertilizers increased during the studied period from 132 to 152 kg NPK·ha^{-1} per farm.

The lowest level of N_2O emissions from mineral fertilizers was noted on farms that specialize in field crops (Figure 2A). Emissions from field crop farms amounted to 35% of dairy farm emissions and 55% of mixed farm emissions. Field crop farms applied less mineral fertilization than the other studied groups. Fertilizer consumption ranged from 76 to 98 kg NPK ha^{-1} depending on the year.

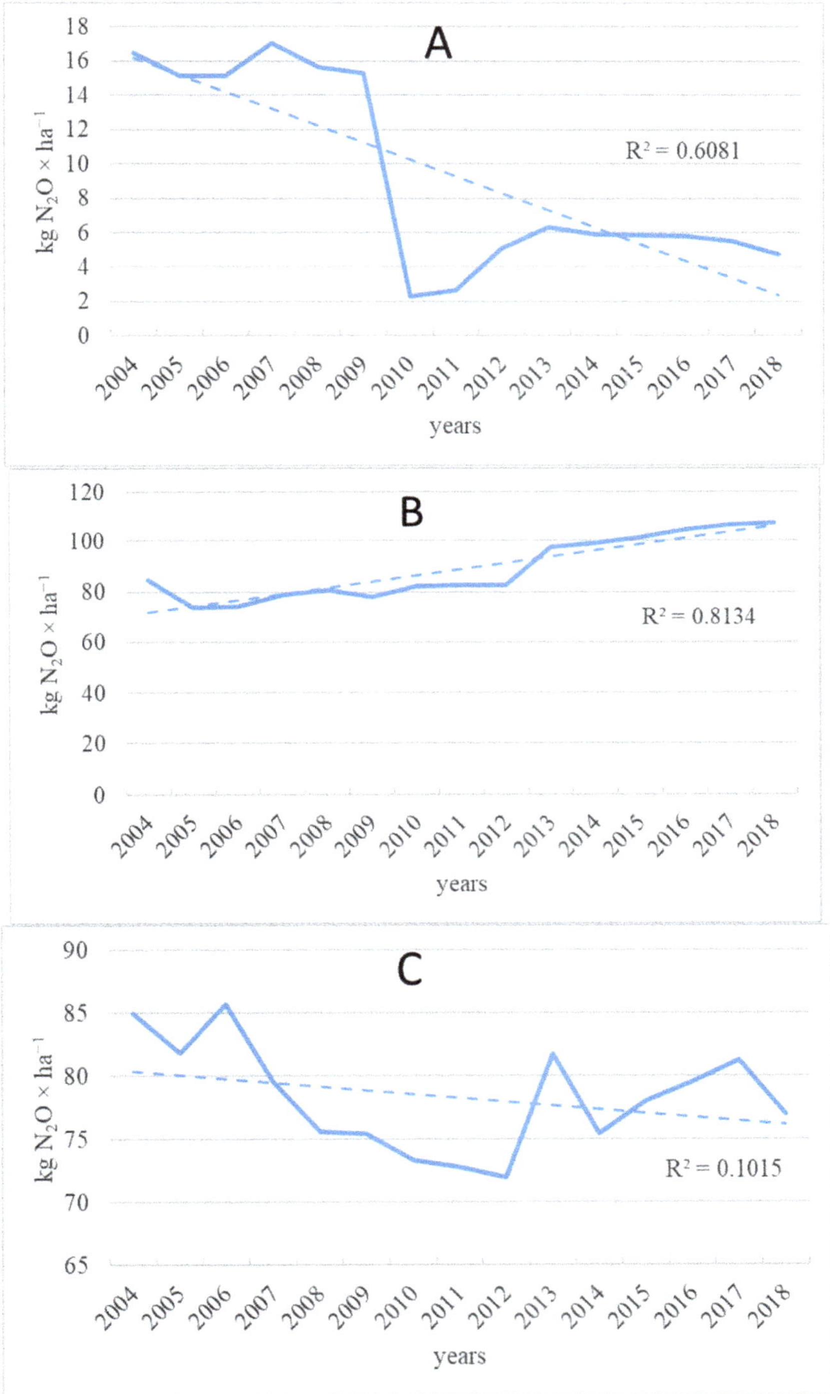

Figure 2. Emission levels of N_2O ($kg{\cdot}ha^{-1}$) at different types of farms: (**A**) Field crops; (**B**) Milk; (**C**) Mixed.

Reducing the 23 dimensions representing the primary features introduced for analysis in the case of dairy cattle farms, 21 dimensions (absence of swine herd) distinguished just one principal component responsible for 91.35% of total variation. This demonstrates that these farms are highly specialized in this production. All undertakings associated with activities described by the studied features had an equally strong impact on emissions of selected greenhouse gases over the course of the 15-year period. After applying varimax rotation, two principal components were distinguished: PC1 (57.39%) and PC2 (38.26%) (Figure 3, Table A3, Appendix A). After varimax rotation, the assignment of the majority of features to PC1 remained unchanged with the exception of CH_4 and N_2O emissions per 1 ha.

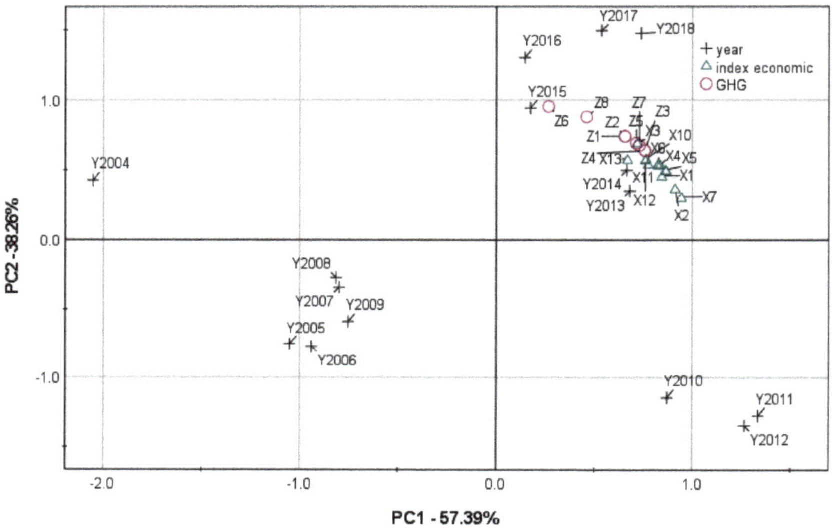

Figure 3. Relationships of locations of studied sources of GHG emissions (Z_1 .... Z_4 Z_7 .... Z_{10}) and economic indicators (X_1 .... X_{13}) for farms with milk production in the studied years (Y2004 .... Y2018) in the space of the first two components, PC1 and PC2.

Similar to farms with a mixed production profile, field crop farms are characterized by diverse factors that influence total GHG emissions. Principal component analysis clearly differentiated the studied features into three principal components describing total variation. The first component was responsible for 41.19% of total variation and the features most strongly correlated with it included CH_4 and N_2O emissions from each group of animals, total CH_4 and N_2O emissions, and number of animals.

The second component explained 39.64% of total variation (Figure 4, Table A4, Appendix A) and was most strongly correlated with the following economic indicators: total family farm income (PLN) and income per 1 ha of farmland, as well as net value added (PLN·AWU^{-1}), total output (PLN), farmland area (ha), energy (PLN), intermediate consumption (PLN), total inputs (PLN), and mineral fertilizers (PLN). The third component explained 15.64% of total variation and was most strongly correlated with land productivity (PLN ha^{-1}) and total inputs (PLN ha^{-1}).

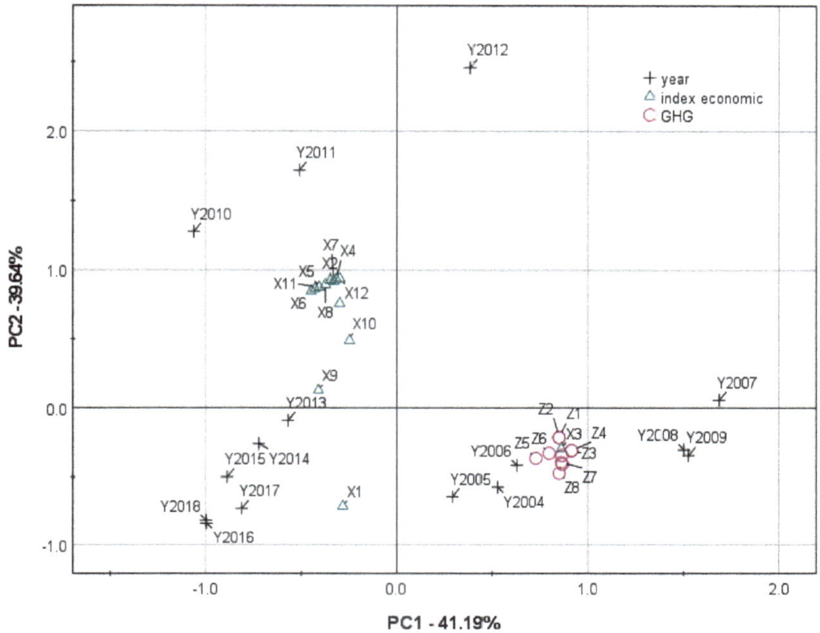

Figure 4. Relationships of locations of examined indices (Z_1 .... Z_{10}) of GHG emissions and economic indices (X_1 .... X_{13}) for field crop farms in the studied years (Y2004 .... Y2018) in the space of the first two components, PC1 and PC2.

Mixed production farms are characterized by diverse factors influencing total CH_4 and N_2O emissions. Principal component analysis clearly differentiated the studied features into four principal components describing total variation. The first component explained 48.83% of total variation and was most strongly correlated with variables including farmland area (ha), livestock units (LU), total production (PLN), total inputs (PLN), mineral fertilizers (PLN), energy (PLN), intermediate consumption (PLN) and total inputs (PLN·ha^{-1}) as well as land productivity (PLN·ha^{-1}), net value added (PLN·AWU^{-1}), CH_4 and N_2O emissions originating from cattle groups other than dairy cows, and total CH_4 and N_2O emissions. The second component explained 22.55% of total variation and was most strongly correlated with variables including livestock units (LU) and CH_4 and N_2O emissions from dairy cattle and swine (Figure 5, Table A5, Appendix A). The variation explained by the first two components amounts to 71.38% of the total variation. The next components distinguished in the analysis were PC3, explaining 14.95%. and PC4, explaining 8.33% of the total variation.

Statistical analysis confirmed the dependency between CH_4 (Z_8) and N_2O (Z_{10}) emissions and economic results: net value added (X_{11}) and family farm income per 1 ha (X_{13}). On dairy cattle farms, the value of CH_4 and N_2O emissions grew as the values of economic indicators increased. Net value added and family farm income (PLN·ha^{-1}) were positively correlated with CH_4 (r = 0.700 and 0.700) and N_2O (r = 0.802 and 0.774) emissions (Table A7, Appendix A). On field crop farms, the dependency between net value added and CH_4 and N_2O emissions was negatively correlated (r = −0.814 and −0.785). Similarly, there was a negative dependency between family farm income (PLN·ha^{-1}) and emissions of the studied greenhouse gases (r = −0.695 and −0.676) (Table A6, Appendix A). On mixed farms, the dependency between economic indicators and CH_4 and N_2O emissions was negative, except the dependency between net value added and CH_4 emissions (r = 0.272) (Table A8, Appendix A).

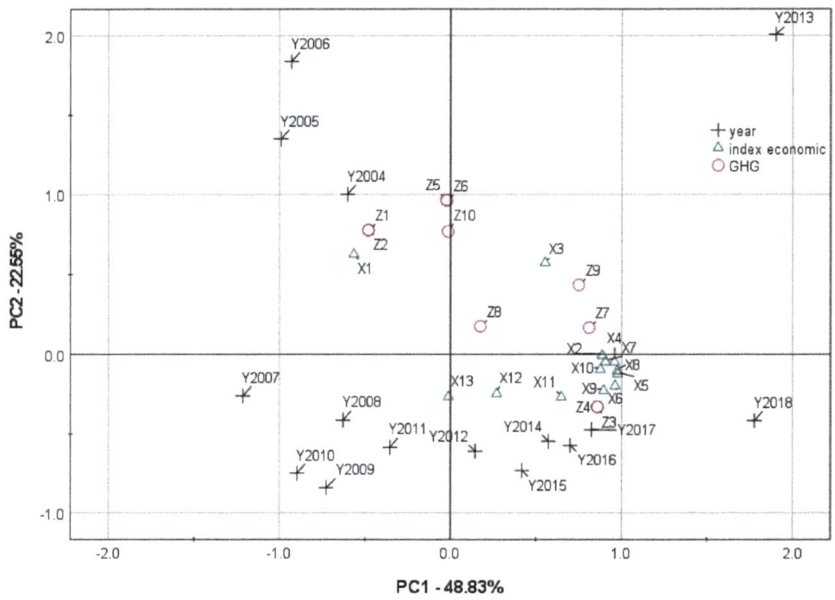

Figure 5. Relationships of locations of studied GHG emission sources (Z_1 .... Z_{10}) and economic indicators (X_1. .... X_{13}) for mixed production farms in the studied years (Y2004 .... Y2018) in the space of the first two components, PC1 and PC2.

4. Discussion and Conclusions

The growing demand for food requires intensification of agricultural production, which has a negative impact on the environment. This impact contributes to depletion of energy carriers, global warming, and reduction of air quality [43,44]. In order to ensure sustainable development, we need to search for solutions that can conserve environmental values while enabling the achievement of economic goals. The agricultural ecosystem both emits and absorbs greenhouse gases, and because of this, we use the concept of net greenhouse gases [45].

The analysis results indicate that animals remain a strong determinant of GHG emissions. A particularly high level of emissions is associated with cattle raising, which was the case for dairy cattle holdings. The level of nitrous oxide emissions was also high, as a result of the application of intensive feed production technologies. This was also confirmed by principal component analysis (PCA). This indicates that on dairy cattle farms, the organization of both animal and plant production is completely subordinated to milk production. Emissions on mixed farms are the result of intensive animal and plant production. Meanwhile, on field crop farms, where animal production was successively reduced, emissions were the lowest and mainly pertained to crops. This was also confirmed by analyses conducted at the regional level [46]. In Poland, regions with a larger share of large agricultural holdings and animal production (the northeastern part of the country and the Wielkopolska region) are characterized by higher emissions levels [7,36].

The values obtained in this research are higher than those obtained by other authors. According to Wiśniewski [7], 42% of emissions originate from gastrointestinal fermentation in rural and urban–rural municipalities. In their investigations of emissions from agriculture in Africa, Tongwane et al. [6] determined that gastrointestinal fermentation was responsible for over half of all emissions originating from agriculture. Studies conducted in Ireland showed that 49% of emissions originated from gastrointestinal fermentation [47]. However, it should be noted that, in general, greenhouse gas emissions are estimated based on data for an average agricultural holding in the region or country and include all greenhouse gases. Meanwhile, our studies examined commodity farms that apply intensive technologies linked to specialization of production. In this case, we are dealing with concentrated means of production and high livestock density. These farms are distinguished against a background of so-called average farms by significantly higher production and economic results, but at the same time, they exert greater pressure on the environment. This hypothesis was confirmed by the research of Wysocka-Czubaszek et al. [36] concerning CH_4 and N_2O emissions in Poland. According to those authors, 51% of CH_4 and 37% of N_2O is emitted by three voivodeships where there is intensive agriculture: the Masovian and Podlaskie voivodeships, leading producers of milk and beef, and another voivodeship characterized by intensive production of animals and plants. The release of large amounts of methane and nitrous oxide is therefore the result of specialization which is associated with the concentration of agricultural holdings' resources.

The economic results obtained for field crop farms are concurrent with the results obtained by Khan et al. [18]. Growth of net value added and farm income per 1 ha of farmland caused a reduction of CH_4 and N_2O emissions. Meanwhile, on dairy cow farms, dependencies between economic results and gas emissions are different, confirming the results of Zafeirou et al. [19]. In this case, as value added and farm income increase, so do CH_4 and N_2O emissions. Syp and Osuch [31] obtained similar results in their investigations of organic and conventional farms. In their research, higher productivity was found on milk farms and was associated with higher GHG emissions. The view that farms which have more animals (conventional farms had more animals than organic farms) emphasize economic objectives, and that productivity is prioritized over environmental objectives, was also confirmed.

The results obtained indicate that the direction of dependencies between greenhouse gas emissions and economic results is determined by the presence of animal production. particularly cattle. Cattle are responsible for the highest emissions of CH_4 [48].

The example of the three types of agricultural holdings described in this study confirms the hypothesis of the relationship between specialization of agricultural production and CH_4 and N_2O emissions. Dairy farms are the most harmful to the environment. Compared to farms of other types (field crop and mixed), they emit the highest amounts of CH_4 and N_2O. This is the result of a high concentration of animals on the farm and intensive plant production for use as fodder. Farms of this type successfully implement their economic goals, with the highest net value added and farm income per area unit. Field crop farms are less harmful to the environment. Farms of this type have successively reduced their livestock production, resulting in lower CH_4 and N_2O emissions. In this case, the quality of the soil may deteriorate due to the lack of organic fertilization.

Intensive agriculture does not have to be a threat to the environment. Countries that have achieved sustainable agriculture have done so by developing large farms and a high level of mechanization [49]. It is expected that agriculture will satisfy the needs of the growing global population while contributing to the reduction of GHG emissions. Achieving this goal will require intensification of production with higher emissions per unit of land area but lower emissions per unit of agricultural production [50,51].

Reducing greenhouse gas emissions from agriculture requires the introduction of innovative technologies and tools to increase the efficiency of agricultural production. One effective method of limiting methane emissions is to use a cattle nutrition strategy. Studies confirm that methane emissions have been reduced as a result of the application of high-starch diets or exogenic enzymes. Supplementation with fats also yields good results. This indicates that appropriate diets can be implemented for dairy and beef cattle in order to reduce methane emissions without reducing productivity [52].

According to Hoglund-Isaksson et al. [53], the possibilities of reducing methane and nitrous oxide emissions from agriculture are limited and technological solutions are insufficient. Hence, they propose the introduction, by 2050, of institutional reforms and changes to human nutritional habits on a broad scale, in addition to the implementation of technological solutions. Meanwhile, Ockoet al. [54] believe that achieving a reduction of methane and nitrous oxide emissions by changing the human diet is less realistic than implementing technological strategies.

Specialization fosters the development of farms and builds competitive advantage. As research indicates, specialization is deepening, and economic goals are the decisive factor in the adoption of areas of specialization by agricultural holdings [55]. Farms with intensive animal production have the strongest impact with respect to the environment. Solutions that make it possible to reduce the pressure of agriculture on the environment while maintaining food security are already known. These are, above all, good agricultural practices, including no-till farming, breeding progress, and effective fertilizer management. Good management practices may reduce the burden on the environment and the costs of agricultural production. Economic instruments are also indicated for strategies to limit emissions, e.g., in the form of compensation for income lost due to reduced production intensity [56].

Wąs et al. [21] presented several scenarios of reduced methane and nitrous oxide emissions based on data describing the Polish agricultural sector in the base year 2015. Changes in income levels are an important indicator from the perspective of analyzing the potential economic consequences of various scenarios of reduced greenhouse gases from agriculture. The process of reducing emissions in agriculture using currently known and available technologies is highly complex and inevitably leads to drops in production and income. According to these scenarios, the greatest drops in production and income are observed in the case of beef and dairy cattle. Plant production is the least sensitive to restrictions with respect to emissions.

The present research contributes to agricultural science and environmental economics by broadening the knowledge on the subject of relationships between intensive agricultural production and the environment, including the economic aspect, a subject raised infrequently in the literature. The present paper broadens the knowledge concerning the relationships between methane and nitrous oxide emissions and the economic results of agricultural holdings with different specializations. This research can serve as a basis for creating models for the development of agriculture. Research conducted until now has been on a regional scale [7,36]. Based on previous research, we can only make approximate inferences about the applied technologies in terms of their relationship with methane and nitrous oxide emissions. The advantages of the research in this paper, at the level of individual farms, are that it increases the accuracy of estimating greenhouse gas emissions and makes it possible to determine dependencies between emissions and economic results. Few analyses have raised these issues. Methods of reducing GHG emissions by changing production technology are necessary. These methods should contribute to improving environmental protection and reducing production costs. Specialized farms are and will continue to be the foundation of the country's food economy. The economic objective is the motivating factor for adopting a specialization. Therefore, a deeper understanding of environmental and economic relationships in agricultural production will make it possible to promote technological innovations leading to low emissions. Consequently, integrating all aspects of the low-emissions economy will contribute to raising the competitiveness of agricultural holdings.

This research also has practical value. It makes it possible to evaluate greenhouse gas emissions from agricultural production in a relatively simple way and to verify practices applied at the level of individual agricultural holdings and environmental protection.

5. Limitations

The authors are aware of the limitations of this analysis. One limitation is due to the choice of research subjects. The research subjects are commodity farms that are achieving higher results than average farms in Poland, which hinders the ability to generalize and make inferences. Another limitation is that the estimation of greenhouse gas emissions was done indirectly due to the lack of detailed data. There is a clear need to supplement databases with data allowing for environmental assessment of agricultural holdings. This has also been noted by other authors [29,30].

Yet another limitation is the limited number of production profiles. Further research should account for all production profiles adopted by farms and for their economic results. This is important with regard to the efficacy of methods of reducing methane and nitrous oxide emissions in agriculture.

Author Contributions: Conceptualization: Z.K.-C.; Methodology: L.S. and Z.K.-C.; Data curation: Z.K.-C.; Formal analysis: Z.K.-C. and L.S.; Investigation: L.S.; Z.K.-C. and R.T.; Supervision: Z.K.-C.; Writing—original draft: Z.K.-C. and L.S.; Writing—review and editing: Z.K.-C., L.S. and R.T. All authors have read and agreed to the published version of the manuscript.

Funding: This research received no external funding.

Data Availability Statement: Publicly available datasets were analyzed in this study. The data can be found here: https://fadn.pl/publikacje/wyniki-standardowe-2/wyniki-standardowe-srednie-wazone/ (accessed on 13 March 2021).

Conflicts of Interest: The authors declare no conflict of interest.

Appendix A

Table A1. Number of farms in 2004–2018.

Specification	Year														
	2004	2005	2006	2007	2008	2009	2010	2011	2012	2013	2014	2015	2016	2017	2018
Field crops	2573	2650	2622	2800	3241	3287	2047	2177	2303	3215	3342	3411	3893	4049	4263
Milk	785	895	877	817	891	952	2319	2271	2302	2652	2735	2782	2749	2659	2539
Mixed	4937	4614	4430	4470	4288	3967	3862	3785	3517	4282	4083	3942	3446	3352	3193

Source: FADN data [28].

Table A2. Descriptive statistics of farms.

Specification	Average	Min.	Max.	SD
	Field Crops			
Utilized agricultural area (ha)	30.51	22.90	40.70	10.38
Total livestock unit (LU)	2.42	1.20	4.00	1.22
Total output (PLN)	128,122.07	77,603.00	261,535.00	54,478.53
Total inputs (PLN)	110,920.67	64,499.00	204,878.00	43,201.69
Fertilizers (PLN)	21,068.53	10.157.00	43.319.00	10,469.06
Energy (PLN)	13,366.07	7222.00	27,537.00	6218.51
Total intermediate consumption (PLN)	76,011.67	44,856.00	143,873.00	30,406.09
Total inputs (PLN ha^{-1})	3593.80	2647.00	4092.00	514.65
Land productivity (PLN ha^{-1})	4117.87	3117.00	5158.00	514.65
Farm net value added (PLN AWU^{-1})	34,984.33	15,352.00	73,098.00	17,023.93
Family farm income (PLN)	46,797.33	21,135.00	113,721.00	28,032.23
Family farm income (PLN ha^{-1})	1442.53	849.00	2243.00	372.45
	Milk			
Utilized agricultural area (ha)	19.13	12.90	22.50	2.96
Total livestock unit (LU)	20.17	14.10	27.00,	4.50
Total output (PLN)	115,330.67	60,928.00	175,076.00	30,940.41
Total inputs (PLN)	88,252.33	42,916.00	129,854.00	30,940.41

Table A2. *Cont.*

Specification	Average	Min.	Max.	SD
Fertilizers (PLN)	7914.53	3748.00	11,778.00	3043.14
Energy (PLN)	9342.87	4315.00	13,013.00	3264.85
Total intermediate consumption (PLN)	65,534.33	31,653.00	97,070.00	22,908.43
Total inputs (PLN ha^{-1})	4468.67	2869.00	5771.00	1006.58
Land productivity (PLN ha^{-1})	5873.73	4185.00	7832.00	1236.60
Farm net value added (PLN·AWU^{-1})	29,603.73	12,874.00	49,893.00	11,186.76
Family farm income (PLN)	48,952.87	20,069.00	85,401.00	19,971.13
Family farm income (PLN ha^{-1})	2480.40	1342.00	3917.00	714.25
	Mixed			
Utilized agricultural area (ha)	16.83	14.80	19.90	1.32
Total livestock unit (LU)	12.47	10.90	14.30	1.09
Total output (PLN)	82,903.47	63,110.00	117,397.00	17,700.99
Total inputs (PLN)	77,282.40	50,150.00	114,631.00	20,966.41
Fertilizers (PLN)	7583.80	4612.00	11,363.00	2310.94
Energy (PLN)	7747.00	4724.00	12,142.00	2085.39
Total intermediate consumption (PLN)	57,296.93	37,961.00	85,167.00	14,988.95
Total inputs (PLN ha^{-1})	4598.40	3254.00	6163.00	944.21
Land productivity (PLN ha^{-1})	4890.93	3802.00	6312.00	734.48
Farm net value added (PLN·AWU^{-1})	17,554.33	10,344.00	22,836.00	4079.47
Family farm income (PLN)	22,283.53	14,696.00	31,387.00	5290.34
Family farm income (PLN ha^{-1})	1325.73	969.00	1949.00	313.05

Source: Own calculation based on FADN data [28].

Table A3. Eigenvalues and proportions of total variance in 15 years as explained by the first two principal components for original traits and correlation coefficients between these traits and the first three PCs on milk production farms.

Rotated Component Matrix.		
Indicators	Component	
	1	2
X_1—Number of farms	0.845	0.450
X_2—Utilized agricultural area (ha)	0.912	0.360
X_3—Total livestock (LU)	0.717	0.688
X_4—Total output (PLN)	0.837	0.531
X_5—Total inputs (PLN)	0.864	0.493
X_6—Fertilizers (PLN)	0.828	0.538
X_7—Energy (PLN)	0.944	0.299
X_8—Total intermediate consumption (PLN)	0.866	0.491
X_9—Total inputs (PLN ha^{-1})	0.824	0.530
X_{10}—Land productivity (PLN ha^{-1})	0.764	0.574

Table A3. *Cont.*

Rotated Component Matrix.		
Indicators	**Component**	
	1	**2**
X_{11}—Farm net value added (PLN AWU^{-1})	0.774	0.537
X_{12}—Family farm income (PLN)	0.759	0.566
X_{13}—Family farm income (PLN ha^{-1})	0.669	0.570
Z_1—Dairy cattle CH_4	0.656	0.742
Z_2—Dairy cattle N_2O	0.656	0.742
Z_3—Other cattle CH_4	0.756	0.642
Z_4—Other cattle N_2O	0.759	0.639
Z_7—Total emissions CH_4 (kg y^{-1})	0.713	0.692
Z_8—Emissions CH_4 (kg ha^{-1})	0.268	0.954
Z_9—Total emissions N_2O (kg y^{-1})	0.729	0.676
Z_{10}—Emissions N_2O (kg ha^{-1})	0.461	0.880
Total variance explained—rotation sums of squared loadings		
Total	12.052	8.035
% of variance	57.388	38.264
Cumulative %	57.388	95.652

Extraction Method: Principal Component Analysis.
Rotation Method: Varimax with Kaiser Normalization.

Source: Own calculation based on FADN data [28].

Table A4. Eigenvalues and proportions of total variance in 15 years as explained by the first three principal components for original traits and correlation coefficients between these traits and the first three PCs on field crop farms.

Rotated Component Matrix			
Indicators	**Component**		
	1	**2**	**3**
X_1—Number of farms	−0.281	−0.711	0.625
X_2—Utilized agricultural area (ha)	−0.346	0.927	−0.066
X_3—Total livestock (LU)	0.861	−0.294	−0.412
X_4—Total output (PLN)	−0.315	0.933	0.155
X_5—Total inputs (PLN)	−0.405	0.878	0.224
X_6—Mineral fertilizers (PLN)	−0.447	0.850	0.224
X_7—Energy (PLN)	−0.329	0.921	0.148
X_8—Total intermediate consumption (PLN)	−0.372	0.895	0.209
X_9—Total inputs (PLN ha^{-1})	−0.409	0.129	0.885
X_{10}—Land productivity (PLN·ha^{-1})	−0.246	0.490	0.766
X_{11}—Farm net value added (PLN AWU^{-1})	−0.427	0.868	0.226
X_{12}—Family farm income (PLN)	−0.297	0.939	0.112
X_{13}—Family farm income (PLN·ha^{-1})	−0.297	0.760	0.445

Table A4. *Cont.*

Rotated Component Matrix			
Indicators	Component		
	1	2	3
Z_1—Dairy cattle CH_4	0.849	−0.214	−0.469
Z_1—Dairy cattle N_2O	0.849	−0.214	−0.469
Z_3—Other cattle CH_4	0.915	−0.308	0.029
Z_4—Other cattle N_2O	0.915	−0.308	0.029
Z_5—Pigs CH_4	0.728	−0.364	−0.518
Z_6—Pigs N_2O	0.797	−0.327	−0.436
Z_7—Total emissions CH_4 ($kg{\cdot}y^{-1}$)	0.861	−0.395	−0.251
Z_8—Emissions CH_4 ($kg{\cdot}ha^{-1}$)	0.850	−0.472	−0.214
Z_9—Total emissions N_2O ($kg{\cdot}y^{-1}$)	0.864	−0.345	−0.359
Z_{10}—Emissions N_2O (kg ha^{-1})	0.866	−0.408	−0.270
Total variance explained—rotation sums of squared loadings			
Total	9.473	9.117	3.597
% of variance	41.188	39.641	15.638
Cumulative %	41.188	80.829	96.466
Extraction Method: Principal Component Analysis. Rotation Method: Varimax with Kaiser Normalization.			

Source: Own calculation based on FADN data [28].

Table A5. Eigenvalues and proportions of total variance in 15 years as explained by the first four principal components for original traits and correlation coefficients between these traits and the first four PCs on mixed production farms.

Rotated Component Matrix				
Indicators	Component			
	1	2	3	4
X_1—Number of farms	−0.566	0.626	−0.235	−0.385
X_2—Utilized agricultural area (ha)	0.887	−0.007	−0.031	0.367
X_3—Total livestock (LU)	0.554	0.573	−0.403	0.413
X_4—Total output (PLN)	0.959	−0.048	0.265	0.047
X_5—Total inputs (PLN)	0.976	−0.123	0.102	0.108
X_6—Fertilizers (PLN)	0.962	−0.197	0.078	0.076
X_7—Energy (PLN)	0.906	−0.047	0.348	0.000
X_8—Total intermediate consumption (PLN)	0.975	−0.103	0.130	0.089
X_9—Total inputs (PLN ha^{-1})	0.895	−0.227	0.290	−0.024
X_{10}—Land productivity ($PLN{\cdot}ha^{-1}$)	0.876	−0.095	0.408	−0.117
X_{11}—Farm net value added (PLN AWU^{-1})	0.648	−0.268	0.662	0.212
X_{12}—Family farm income (PLN)	0.270	−0.246	0.908	0.081
X_{13}—Family farm income ($PLN{\cdot}ha^{-1}$)	−0.011	−0.265	0.946	−0.018

Table A5. *Cont.*

Rotated Component Matrix				
Indicators	**Component**			
	1	**2**	**3**	**4**
Z_1—Dairy cattle CH_4	−0.481	0.778	−0.293	−0.133
Z_2—Dairy cattle N_2O	−0.481	0.778	−0.293	−0.133
Z_3—Other cattle CH_4	0.859	−0.328	−0.130	0.347
Z_4—Other cattle N_2O	0.859	−0.328	−0.130	0.347
Z_5—Pigs CH_4	−0.025	0.969	−0.049	0.141
Z_6—Pigs N_2O	−0.021	0.963	−0.053	0.142
Z_7—Total emissions CH_4 (kg y^{-1})	0.810	0.169	−0.360	0.382
Z_8—Emissions CH_4 (kg ha^{-1})	0.174	0.178	0.121	0.845
Z_9—Total emissions N_2O (kg·y^{-1})	0.750	0.436	−0.249	0.412
Z_{10}—Emissions N_2O (kg·ha^{-1})	−0.014	0.769	−0.413	0.153
Total variance explained—rotation sums of squared loadings				
Total	11.231	5.187	3.439	1.917
% of variance	48.831	22.551	14.954	8.333
Cumulative %	48.831	71.382	86.336	94.669
Extraction Method: Principal Component Analysis. Rotation Method: Varimax with Kaiser Normalization.				

Source: Own calculation based on FADN data [28].

Table A6. Correlation matrix of variables describing field crop farms.

Variables	X_1	X_2	X_3	X_4	X_5	X_6	X_7	X_8	X_9	X_{10}	X_{11}	X_{12}	X_{13}	Z_1	Z_2	Z_3	Z_4	Z_5	Z_6	Z_7	Z_8	Z_9	Z_{10}
X_1.	1.000	−0.598	−0.293	−0.470	−0.364	−0.338	−0.462	−0.397	0.570	0.186	−0.351	−0.508	−0.191	−0.375	−0.375	−0.015	−0.015	−0.269	−0.261	−0.115	−0.038	−0.226	−0.122
X_2	−0.598	1.000	−0.544	0.964	0.949	0.931	0.960	0.951	0.203	0.456	0.935	0.965	0.752	−0.459	−0.458	−0.574	−0.574	−0.579	−0.577	−0.652	−0.717	−0.603	−0.670
X_3	−0.293	−0.544	1.000	−0.611	−0.699	−0.729	−0.613	−0.669	−0.753	−0.667	−0.720	−0.582	−0.661	0.990	0.990	0.866	0.866	0.945	0.962	0.951	0.955	0.994	0.976
X_4	−0.470	0.964	−0.611	1.000	0.984	0.972	0.987	0.987	0.385	0.659	0.978	0.988	0.861	−0536	−0.536	−0.575	−0.575	−0.645	−0.624	−0.676	−0.739	−0.649	−0.691
X_5	−0.364	0.949	−0.699	0.984	1.000	0.993	0.989	0.998	0.492	0.674	0.975	0.959	0.843	−0.632	−0.632	−0.611	−0.611	−0.739	−0.724	−0.763	−0.812	−0.734	−0.775
X_6	−0.338	0.931	−0.729	0.972	0.993	1.000	0.980	0.993	0.515	0.676	0.967	0.942	0.829	−0.667	−0.667	−0.650	−0.650	−0.743	−0.743	−0.781	−0.835	−0.755	−0.796
X_7	−0.462	0.960	−0.613	0.987	0.989	0.980	1.000	0.994	0.404	0.625	0.962	0.966	0.826	−0.541	−0.541	−0.567	−0.567	−0.647	−0.626	−0.700	−0.756	−0.651	−0.703
X_8	−0.397	0.951	−0.669	0.987	0.998	0.993	0.994	1.000	0.473	0.668	0.970	0.960	0.837	−0.600	−0.600	−0.592	−0.592	−0.708	−0.692	−0.739	−0.790	−0.704	−0.747
X_9	0.570	0.203	−0.753	0.385	0.492	0.515	0.404	0.473	1.000	0.828	0.469	0.321	0.565	−0.786	−0.786	−0.379	−0.379	−0.795	−0.755	−0.642	−0.608	−0.710	−0.648
X_{10}	0.186	0.456	−0.667	0.659	0.674	0.676	0.625	0.668	0.828	1.000	0.706	0.625	0.852	−0.664	−0.664	−0.437	−0.437	−0.704	−0.638	−0.593	−0.597	−0.641	−0.589
X_{11}	−0.351	0.935	−0.720	0.978	0.975	0.967	0.962	0.970	0.469	0.706	1.000	0.981	0.919	−0.662	−0.662	−0.657	−0.657	−0.744	−0.719	−0.751	−0.814	−0.753	−0.785
X_{12}	−0.508	0.965	−0.582	0.988	0.959	0.942	0.966	0.960	0.321	0.625	0.981	1.000	0.886	−0.512	−0.512	−0.565	−0.565	−0.619	−0.587	−0.637	−0.711	−0.625	−0.669
X_{13}	−0.191	0.752	−0.661	0.861	0.843	0.829	0.826	0.837	0.565	0.852	0.919	0.886	1.000	−0.639	−0.639	−0.547	−0.547	−0.701	−0.629	−0.643	−0.695	−0.676	−0.676
Z_1	−0.375	−0.459	0.990	−0.536	−0.632	−0.667	−0.541	−0.600	−0.786	−0.664	−0.662	−0.512	−0.639	1.000	1.000	0.826	0.826	0.929	0.951	0.917	0.920	0.975	0.951
Z_2	−0.375	−0.458	0.990	−0.536	−0.632	−0.667	−0.541	−0.600	−0.786	−0.664	−0.662	−0.512	−0.639	1.000	1.000	0.826	0.826	0.929	0.951	0.917	0.920	0.975	0.951
Z_3	−0.015	−0.574	0.866	−0.575	−0.611	−0.650	−0.567	−0.592	−0.379	−0.437	−0.657	−0.565	−0.547	0.826	0.826	1.000	1.000	0.719	0.766	0.890	0.910	0.873	0.885
Z_4	−0.015	−0.574	0.866	−0.575	−0.611	−0.650	−0.567	−0.592	−0.379	−0.437	−0.657	−0.565	−0.547	0.826	0.826	1.000	1.000	0.719	0.766	0.890	0.910	0.873	0.885
Z_5	−0.269	−0.579	0.945	−0.645	−0.739	−0.743	−0.647	−0.708	−0.795	−0.704	−0.744	−0.619	−0.701	0.929	0.929	0.719	0.719	1.000	0.961	0.910	0.898	0.954	0.935
Z_6	−0.261	−0.577	0.962	−0.624	−0.724	−0.743	−0.626	−0.692	−0.755	−0.638	−0.719	−0.587	−0.629	0.951	0.951	0.766	0.766	0.961	1.000	0.925	0.924	0.968	0.957
Z_7	−0.115	−0.652	0.951	−0.676	−0.763	−0.781	−0.700	−0.739	−0.642	−0.593	−0.751	−0.637	−0.643	0.917	0.917	0.890	0.890	0.910	0.925	1.000	0.987	0.967	0.979
Z_8	−0.038	−0.717	0.955	−0.739	−0.812	−0.835	−0.756	−0.790	−0.608	−0.597	−0.814	−0.711	−0.695	0.920	0.920	0.910	0.910	0.898	0.924	0.987	1.000	0.971	0.989
Z_9	−0.226	−0.603	0.994	−0.649	−0.734	−0.755	−0.651	−0.704	−0.710	−0.641	−0.753	−0.625	−0.676	0.975	0.975	0.873	0.873	0.954	0.968	0.967	0.971	1.000	0.991
Z_{10}	−0.122	−0.670	0.976	−0.691	−0.775	−0.796	−0.703	−0.747	−0.648	−0.589	−0.785	−0.669	−0.676	0.951	0.951	0.885	0.885	0.935	0.957	0.979	0.989	0.991	1.000

Source: Own calculation based on FADN data [28].

Table A7. Correlation matrix of variables describing milk production farms.

Variables	X_1	X_2	X_3	X_4	X_5	X_6	X_7	X_8	X_9	X_{10}	X_{11}	X_{12}	X_{13}	Z_1	Z_2	Z_3	Z_4	Z_7	Z_8	Z_9	Z_{10}
X_1	1.000	0.942	0.929	0.921	0.964	0.964	0.952	0.958	0.947	0.866	0.830	0.830	0.731	0.880	0.880	0.953	0.954	0.923	0.675	0.934	0.811
X_2	0.942	1.000	0.922	0.934	0.971	0.953	0.957	0.968	0.931	0.850	0.872	0.867	0.766	0.889	0.889	0.940	0.941	0.921	0.573	0.928	0.733
X_3	0.929	0.922	1.000	0.951	0.963	0.969	0.880	0.961	0.951	0.909	0.901	0.910	0.833	0.993	0.993	0.997	0.997	1.000	0.844	1.000	0.937
X_4	0.921	0.934	0.951	1.000	0.975	0.963	0.946	0.979	0.966	0.981	0.970	0.972	0.918	0.934	0.934	0.958	0.959	0.951	0.731	0.955	0.847
X_5	0.964	0.971	0.963	0.975	1.000	0.994	0.971	0.999	0.990	0.927	0.903	0.905	0.814	0.934	0.934	0.977	0.978	0.962	0.709	0.968	0.841
X_6	0.964	0.953	0.969	0.963	0.994	1.000	0.953	0.993	0.991	0.918	0.885	0.890	0.800	0.944	0.944	0.981	0.981	0.968	0.746	0.973	0.867
X_7	0.952	0.957	0.880	0.946	0.971	0.953	1.000	0.973	0.957	0.898	0.865	0.862	0.771	0.833	0.833	0.907	0.909	0.877	0.555	0.888	0.712
X_8	0.958	0.968	0.961	0.979	0.999	0.993	0.973	1.000	0.991	0.935	0.908	0.911	0.824	0.934	0.933	0.973	0.974	0.959	0.708	0.965	0.838
X_9	0.947	0.931	0.951	0.966	0.990	0.991	0.957	0.991	1.000	0.938	0.882	0.888	0.804	0.926	0.926	0.963	0.964	0.950	0.748	0.956	0.863
X_{10}	0.866	0.850	0.909	0.981	0.927	0.918	0.898	0.935	0.938	1.000	0.963	0.969	0.943	0.898	0.898	0.912	0.913	0.910	0.764	0.912	0.853
X_{11}	0.830	0.872	0.901	0.970	0.903	0.885	0.865	0.908	0.882	0.963	1.000	0.998	0.979	0.896	0.896	0.900	0.901	0.902	0.700	0.903	0.802
X_{12}	0.830	0.867	0.910	0.972	0.905	0.890	0.862	0.911	0.888	0.969	0.998	1.000	0.981	0.908	0.908	0.906	0.907	0.911	0.725	0.911	0.821
X_{13}	0.731	0.766	0.833	0.918	0.814	0.800	0.771	0.824	0.804	0.943	0.979	0.981	1.000	0.841	0.842	0.823	0.824	0.836	0.700	0.834	0.774
Z_1	0.880	0.889	0.993	0.934	0.934	0.944	0.833	0.934	0.926	0.898	0.896	0.908	0.841	1.000	1.000	0.981	0.980	0.995	0.872	0.991	0.949
Z_2	0.880	0.889	0.993	0.934	0.934	0.944	0.833	0.933	0.926	0.898	0.896	0.908	0.842	1.000	1.000	0.981	0.980	0.994	0.872	0.991	0.949
Z_3	0.953	0.940	0.997	0.958	0.977	0.981	0.907	0.973	0.963	0.912	0.900	0.906	0.823	0.981	0.981	1.000	1.000	0.996	0.815	0.998	0.920
Z_4	0.954	0.941	0.997	0.959	0.978	0.981	0.909	0.974	0.964	0.913	0.901	0.907	0.824	0.980	0.980	1.000	1.000	0.995	0.813	0.998	0.918
Z_7	0.923	0.921	1.000	0.951	0.962	0.968	0.877	0.959	0.950	0.910	0.902	0.911	0.836	0.995	0.994	0.996	0.995	1.000	0.846	1.000	0.938
Z_8	0.675	0.573	0.844	0.731	0.709	0.746	0.555	0.708	0.748	0.764	0.700	0.725	0.700	0.872	0.872	0.815	0.813	0.846	1.000	0.836	0.977
Z_9	0.934	0.928	1.000	0.955	0.968	0.973	0.888	0.965	0.956	0.912	0.903	0.911	0.834	0.991	0.991	0.998	0.998	1.000	0.836	1.000	0.932
Z_{10}	0.811	0.733	0.937	0.847	0.841	0.867	0.712	0.838	0.863	0.853	0.802	0.821	0.774	0.949	0.949	0.920	0.918	0.938	0.977	0.932	1.000

Source: Own calculation based on FADN data [28].

Table A8. Correlation matrix of variables describing mixed production farms.

Variables	X_1	X_2	X_3	X_4	X_5	X_6	X_7	X_8	X_9	X_{10}	X_{11}	X_{12}	X_{13}	Z_1	Z_2	Z_3	Z_4	Z_5	Z_6	Z_7	Z_8	Z_9	Z_{10}
X_1	1.000	−0.666	−0.054	−0.645	−0.675	−0.702	−0.596	−0.661	−0.675	−0.581	−0.784	−0.584	−0.398	0.882	0.882	−0.804	−0.804	0.557	0.546	−0.428	−0.235	−0.274	0.521
X_2	−0.666	1.000	0.667	0.853	0.904	0.886	0.827	0.897	0.751	0.671	0.630	0.247	−0.061	−0.454	−0.454	0.879	0.879	0.046	0.055	0.862	0.431	0.822	−0.059
X_3	−0.054	0.667	1.000	0.414	0.462	0.400	0.316	0.456	0.218	0.203	0.046	−0.297	−0.524	0.257	0.257	0.494	0.494	0.614	0.616	0.872	0.436	0.949	0.666
X_4	−0.645	0.853	0.414	1.000	0.975	0.956	0.963	0.982	0.942	0.958	0.820	0.514	0.253	−0.576	−0.576	0.822	0.822	−0.087	−0.086	0.695	0.243	0.648	−0.147
X_5	−0.675	0.904	0.462	0.975	1.000	0.991	0.945	0.999	0.939	0.897	0.744	0.379	0.100	−0.605	−0.605	0.891	0.891	−0.135	−0.133	0.762	0.293	0.685	−0.160
X_6	−0.702	0.886	0.400	0.956	0.991	1.000	0.935	0.987	0.938	0.884	0.722	0.360	0.088	−0.661	−0.661	0.890	0.890	−0.195	−0.189	0.719	0.254	0.634	−0.218
X_7	−0.596	0.827	0.316	0.963	0.945	0.935	1.000	0.955	0.926	0.923	0.804	0.542	0.292	−0.562	−0.562	0.719	0.719	−0.078	−0.075	0.572	0.261	0.549	−0.285
X_8	−0.661	0.897	0.456	0.982	0.999	0.987	0.955	1.000	0.941	0.910	0.752	0.397	0.121	−0.588	−0.588	0.872	0.872	−0.124	−0.122	0.750	0.290	0.677	−0.163
X_9	−0.675	0.751	0.218	0.942	0.939	0.938	0.926	0.941	1.000	0.940	0.807	0.532	0.309	−0.694	−0.694	0.786	0.786	−0.265	−0.262	0.555	0.198	0.474	−0.305
X_{10}	−0.581	0.671	0.203	0.958	0.897	0.884	0.923	0.910	0.940	1.000	0.837	0.614	0.411	−0.602	−0.602	0.698	0.698	−0.172	−0.175	0.508	0.123	0.460	−0.206
X_{11}	−0.784	0.630	0.046	0.820	0.744	0.722	0.804	0.752	0.807	0.837	1.000	0.886	0.713	−0.735	−0.735	0.649	0.649	−0.289	−0.289	0.348	0.272	0.306	−0.423
X_{12}	−0.584	0.247	−0.297	0.514	0.379	0.360	0.542	0.397	0.532	0.614	0.886	1.000	0.951	−0.587	−0.587	0.242	0.242	−0.287	−0.287	−0.088	0.106	−0.079	−0.525
X_{13}	−0.398	−0.061	−0.524	0.253	0.100	0.088	0.292	0.121	0.309	0.411	0713	0.951	1.000	−0.471	−0.471	−0.028	−0.028	−0.319	−0.322	−0.370	−0.014	−0.349	−0.537
Z_1	0.882	−0.454	0.257	−0.576	−0.605	−0.661	−0.562	−0.588	−0.694	−0.602	−0.735	−0.587	−0.471	1.000	1.000	−0.682	−0.682	0.727	0.718	−0.176	−0.074	−0.013	0.659
Z_2	0.882	−0.454	0.257	−0.576	−0.605	−0.661	−0.562	−0.588	−0.694	−0.602	−0.73 5	−0.587	−0.471	1.000	1.000	−0.682	−0.682	0.727	0.718	−0.176	−0.074	−0.013	0.659
Z_3	−0.804	0.879	0.494	0.822	0.891	0.890	0.719	0.872	0.786	0.698	0.649	0.242	−0.028	−0.682	−0.682	1.000	1.000	−0.291	−0.285	0.838	0.326	0.690	−0.108
Z_4	−0.804	0.879	0.494	0.822	0.891	0.890	0.719	0.872	0.786	0.698	0.649	0.242	−0.028	−0.682	−0.682	1.000	1.000	−0.291	−0.285	0.838	0.326	0.690	−0.108
Z_5	0.557	0.046	0.614	−0.087	−0.135	−0.195	−0.078	−0.124	−0.265	−0.172	−0.289	−0.287	−0.319	0.727	0.727	−0.291	−0.291	1.000	0.999	0.188	0.259	0.480	0.768
Z_6	0.546	0.055	0.616	−0.086	−0.133	−0.189	−0.075	−0.122	−0.262	−0.175	−0.289	−0.287	−0.322	0.718	0.718	−0.285	−0.285	0.999	1.000	0.191	0.251	0.484	0.761
Z_7	−0.428	0.862	0.872	0.695	0.762	0.719	0.572	0.750	0.555	0.508	0.348	−0.088	−0.370	−0.176	−0.176	0.838	0.838	0.188	0.191	1.000	0.405	0.941	0.359
Z_8	−0.235	0.431	0.436	0.243	0.293	0.254	0.261	0.290	0.198	0.123	0.272	0.106	−0.014	−0.074	−0.074	0.326	0.326	0.259	0.251	0.405	1.000	0.473	0.150
Z_9	−0.274	0.822	0.949	0.648	0.685	0.634	0.549	0.677	0.474	0.460	0.306	−0.079	−0.349	−0.013	−0.013	0.690	0.690	0.480	0.484	0.941	0.473	1.000	0.519
Z_{10}	0.521	−0.059	0.666	−0.147	−0.160	−0.218	−0.285	−0.163	−0.305	−0.206	−0.423	−0.525	−0.537	0.659	0.659	−0.108	−0.108	0.768	0.761	0.359	0.150	0.519	1.000

Source: Own calculation based on FADN data [28].

References

1. Raport 2020. Polska Na Drodze Zrównoważonego Rozwoju. GUS, Warszawa, Poland. 2020. Available online: https://raportsdg.stat.gov.pl/2020/cel13.html (accessed on 15 January 2021).
2. Emisja Gazów Cieplarnianych. *Wybrane Zagadnienia Dotyczące Emisji CO2 w Polsce. Opracowanie Tematyczne OT-683*; Kancelaria Senatu: Warszawa, Poland, 2020. Available online: https://www.senat.gov.pl/gfx/senat/pl/senatopracowania/192/plik/ot-683.pdf (accessed on 15 January 2021).
3. Dziennik Ustaw. Protokół z Kioto do Ramowej Konwencji Narodów Zjednoczonych w Sprawie Zmian Klimatu Sporządzony w Kioto Dnia 11 Grudnia 1997 r. (Dz. U. 2005, nr 203, poz. 1683). Available online: http://isap.sejm.gov.pl/isap.nsf/DocDetails.xsp?id=wdu20052031684 (accessed on 15 January 2021).
4. Emmerling, C.; Krein, A.; Junk, J. Meta-Analysis of Strategies to Reduce NH_3 Emissions from Slurries in European Agriculture and Consequences from Greenhouse Gas Emissions. *Agronomy* **2020**, *10*, 1633. [CrossRef]
5. Nowakowicz-Dębek, B.; Wlazło, Ł.; Szymula, A.; Ossowski, M.; Kasela, M.; Chmielowiec-Korzeniowska, A.; Bis-Wencel, H. Estimating Methane Emissions from a Dairy Farm Using a Computer Program. *Atmosphere* **2020**, *11*, 803. [CrossRef]
6. Tongwane, M.I.; Moeletsi, M.E. A review of greenhouse gas emissions from the agriculture sector in Africa. *Agric. Syst.* **2018**, *166*, 124–134. [CrossRef]
7. Wiśniewski, P. Ocena wielkości emisji gazów cieplarnianych ze źródeł rolniczych na poziomie lokalnym w Polsce. *Rocz. Ochr. Sr.* **2018**, *20*, 1811–1829.
8. Gradziuk, P.; Gradziuk, B. Gospodarka niskoemisyjna–nowe wyzwanie dla gmin wiejskich. *Wieś I Rolnictwo* **2016**, *1*, 105–126. [CrossRef]
9. Zdeb, M. Redukcja Emisji Metanu i Węglowodorów Aromatycznych ze Składowisk Odpadów w Biofiltrze–Badania Polowe. *Rocz. Ochr. Srodowiska* **2015**, *17*, 1053–1073. Available online: https://ros.edu.pl/images/roczniki/2015/063_ROS_V17_R2015.pdf (accessed on 13 March 2021).
10. Rokicki, T.; Koszela, G.; Ochnio, I.; Golonko, M.; Zak, A.; Szczepaniuk, E.K.; Szczepaniuk, H.; Perkowska, A. Greenhouse Gas Emissions by Agriculture in EU Countries. *Rocz. Ochr. Srodowiska* **2020**, *22*, 809–824. Available online: http://ros.edu.pl/index.php?option=com_content&view=article&id=862:057-ros-v22-r2020&catid=62&lang=pl&Itemid=299 (accessed on 13 March 2021).
11. Kassenberg, A.; Świrkula, E. *Polska Niskoemisyjna, Od Idei Do Działania*; Instytut na rzecz Ekorozwoju: Warszawa, Poland, 2015.
12. Krajowy Raport Inwentaryzacyjny. *Inwentaryzacja Gazów Cieplarnianych w Polsce dla lat 1988–2018*; KOBiZE: Warszawa, Poland, 2020; Available online: https://www.kobize.pl/uploads/materialy/materialy_do_pobrania/krajowa_inwentaryzacja_emisji/NIR_2020_raport_syntetyczny_PL.pdf (accessed on 17 January 2021).
13. Krupin, V. Redukcja Emisji Gazów Cieplarnianych z Sektora Polskiego Rolnictwa: Misja Wykonalna? In *Go2′50 Klimat Społeczeństwo Gospodarka*; IOŚ–PIB: Warsaw, Poland, 2020; Volume 1, pp. 74–79. Available online: https://kobize.pl/uploads/materialy/materialy_do_pobrania/aktualnosci/2020/GO250/publikator_PL_finalna_web.pdf (accessed on 15 February 2021).
14. Rebolledo-Leiva, R.; Angulo-Meza, L.; Iriarte, A.; Gonzalez-Araya, M.C. Joint carbo footprint assessment and data envelopment analysis for the reduction of greenhouse gas emissions in agriculture production. *Sci. Total Environ.* **2017**, *594*, 346. [CrossRef]
15. Garnier, J.; Le Noe, J.; Marescaux, A.; Sanz-Cobena, A.; Lassaletta, L.; Silvestre, M.; Thieu, V.; Billen, G. Long-term changes in grennhouse gas emission from French agriculture and livestock (1852–2014): From traditional agriculture to conventional intensive systems. *Sci. Total Environ.* **2019**, *660*, 1486–1501. [CrossRef]
16. Sanz-Cobena, A.; Lassoletta, L.; Aguilera, E.; de Prado, A.; Garnier, J.; Billen, G.; Iglesias, A.; Sanchez, B.; Guardia, G.; Abalos, D. Strategies for greenhouse gas emissions mitigation in Mediterranean agriculture: A review. *Agric. Ecosyst. Environ.* **2017**, *238*, 5–24. [CrossRef]
17. Gao, J.; Hou, H.; Zhai, Y.; Woodward, A.; Vardoulakis, S.; Kovats, S.; Wilkinson, P.; Li, L.; Song, X.; Xu, L.; et al. Greenhouse gas emissions reduction in deifferent economic sectors: Mitigation measures, health co-benefits, knowledge gaps and policy implications. *Environ. Pollut.* **2018**, *240*, 683–698. [CrossRef] [PubMed]
18. Khan, M.T.; Ali, Q.; Ashfag, M. The nexus between greenhouse gas emission, electricity production, renewable energy and agriculture in Pakistan. *Renew. Energy* **2018**, *118*, 437–451. [CrossRef]
19. Zafeiriou, E.; Mallidis, I.; Galanopoulas, K.; Arabatzis, G. Greenhouse Gas Emissions and Economic Performance in EU Agriculture; AnEmpirical Study in a Non-Linear Framework. *Sustainability* **2018**, *10*, 3837. [CrossRef]
20. Kastratovic, R. Impact of foreign direct investment on greenhouse gas emissions in agriculture of developing countries. *Aust. J. Agric. Resour. Econ.* **2019**, *63*, 620–642. [CrossRef]
21. Wąs, A.; Kobus, P.; Krupin, V.; Witajewski-Baltvilks, J.; Cygiel, M. *Ocena Wpływu Polityki Klimatycznej Na Sector Polskich Gospodarstw Rolnych*; Centrum Analiz Klimatyczno-Energetycznych: Warszawa, Poland, 2020.
22. Castoldi, N.; Bechini, L. Intergated sustainability assesment of cropping systems with agro-ecological and economic indicators in northern Italy. *Eur. J. Agron.* **2010**, *32*, 59–72. [CrossRef]
23. Belanger, V.; Vanasse, A.; Parent, D.; Allard, G.; Pellerin, D. Delta: An integrated indicato–based self-assessment tool for the evaluation of dairy farms sustainability in Quebec Canada. *Agroecol. Sustain. Food Syst.* **2015**, *39*, 1022–1046. [CrossRef]

24. Escribano, A.J.; Gaspar, J.P.; Mesias, F.J.D.; Moreno, A.F.P.; Escribano, M. A sustainability assesment of organic and conventional beef cattle farms in agroforestry system: The case of the "dehesa" rangelands. *ITEA Inf. Tec. Econ. Agrar.* **2014**, *110*, 343–367. Available online: http://bazy.pb.edu.pl:2101/full_record.do?product=UA&search_mode=GeneralSearch&qid=11&SID=W2Wfvl6B (accessed on 30 January 2021).
25. Paracchini, M.L.; Bulgheroni, C.; Borreani, G.; Tabacco, E.; Banterle, A.; Bertoni, D.; Rossi, G.; Parolo, G.; Origgi, R.; De Paola, C. A diagnostics system to assess sustainability at a farm level: The SOSTARE model. *Agric. Syst.* **2015**, *133*, 35–53. [CrossRef]
26. Koloszko-Chomentowska, Z.; Zukovskis, J.; Gargasas, A. Ecological and economic sustainability of Polish and Lithuanian agricultural holdings specializing in animal production. In Proceedings of the 8th International Scientific Conference Rural Development Bioeconomy Challenges, Aleksandras Stulginskis University, Kaunas, Lithuania, 23–24 November 2017; pp. 1103–1107. [CrossRef]
27. Prus, P. Sustainable farming production and its impact on the natural environment–case study based on a selected group of farms. In Proceedings of the International Scientific Conference Rural Development Bioeconomy Challenges, Kaunas, Lithuania, 23–24 November 2017; pp. 1280–1285. [CrossRef]
28. Polski FADN. Available online: https://fadn.pl/publikacje/wyniki-standardowe-2/wyniki-standardowe-srednie-wazone/ (accessed on 20 January 2021).
29. Wilk, W. *Wykorzystanie Danych Statystycznych i Wyników Rachunkowości Rolnej do Oceny Wpływu Rolnictwa na Środowisko w Ujęciu Makro- i Mikroekonomicznym*; Studia I Raporty IUNG-PIB; Instytut Uprawy Nawożenia i Gleboznawstwa—Państwowy Instytut Badawcz: Pulawy, Poland, 2007; Volume 4, pp. 59–67. [CrossRef]
30. Piekut, K.; Machnacki, M. Ocena ekologiczno-ekonomiczna gospodarstw rolnych na podstawie danych FADN. *Woda-Środowisko-Obszary Wiejskie* **2011**, *11*, 203–219.
31. Syp, A.; Osuch, D. Szacowanie Emisji Gazów Cieplarnianych na Podstawie Danych FADN; Studia i Raporty IUNG-PIB. 2017, Volume 52, pp. 69–82. Available online: http://www.iung.pl/sir/zeszyt52_6.pdf (accessed on 21 January 2021).
32. IPCC 2006. Guidelines for National Greenhouse Gas Inventories. The Institute for Global Environmental Strategies (IGES) for the IPCC. Available online: https://www.ipcc.ch/report/2006-ipcc-guidelines-for-national-greenhouse-gas-inventories/ (accessed on 20 January 2021).
33. KOBiZE Krajowy Raport Inwentaryzacyjny. *Inwentaryzacja Gazów Cieplarnianych W Polsce Dla Lat 1988-2016*; IOŚ-PIB: Warszawa, Poland, 2018; Available online: https://www.kobize.pl/uploads/materialy/materialy_do_pobrania/krajowa_inwentaryzacja_emisji/NIR_2018_raport_syntetyczny_PL.pdf (accessed on 20 January 2021).
34. Ji, E.S.; Park, K.-H. Methane and nitrous oxide emissions from livestock agriculture in 16 local administrative districts of Korea. *Asian-Australas. J. Anim. Sci.* **2012**, *25*, 1768–1774. [CrossRef]
35. Yue, Q.; Xu, X.; Hiller, J.; Cheng, K.; Pan, G. Mitigating greenhouse gas emissions in agriculture: From farm production to food consumption. *J. Clean. Prod.* **2017**, *149*, 1011–1019. [CrossRef]
36. Wysocka-Czubaszek, A.; Czubaszek, R.; Roj-Rojewski, S.; Banaszuk, P. Methane and Nitrous Oxide Emissions from Agriculture on a Regional Scale. *J. Ecol. Eng.* **2018**, *19*, 206–217. [CrossRef]
37. Dick, J.; Smith, R.; Lilly, A.; Moxey, A.; Booth, J.; Campbell, C.; Coulter, D. *Calculating Farm Scale Greenhouse Gas Emissions*; University of Aberdeen, The Macaulay Institute, Carbon Plan, Pareto Consulting, SAOS Ltd.: Scotland, UK, 2008.
38. Fereniec, J. *Ekonomika I Organizacja Rolnictwa*; Key Text: Warszawa, Poland, 1999.
39. GUS. *Rocznik Statystyczny Rolnictwa Z Lat 2007–2019*; GUS: Warszawa, Poland, 2020.
40. Jasińska, E. *Indywidualne Gospodarstwa Rolne W Świetle Danych Rachunkowych Z Lat 1952-2001*; IERiGŻ-PIB: Warszawa, Poland, 2008.
41. Seber, W.G.A. *Multivariate Observations*; John Wiley & Sons: New York, NY, USA, 1984.
42. Morrison, D. *Wielowymiarowa Analiza Statystyczna*; PWN: Warszawa, Poland, 1990.
43. Carlson, K.M.; Gerber, J.S.; Mueller, N.D.; Herrero, M.; MacDonald, G.K.; Brauman, K.A.; Havlik, P.; O'Connell, C.S.; Johnson, J.A.; Saatchi, S.; et al. Grenhouse Gas Emissions Intensity of Global Croplands. *Nat. Clim. Chang.* **2017**, *7*, 63–68. [CrossRef]
44. Poore, J.; Nemecek, T. Reducing food's environment al imacts through producers and consumer. *Science* **2018**, *360*, 987–992. [CrossRef]
45. Wu, H.; Huang, H.; Tang, J.; Chen, W.; He, Y. Net greenhouse gas emissions from agriculture in China: Estimation, spatial correlation and convergence. *Sustainability* **2019**, *11*, 4817. [CrossRef]
46. Jarosz, Z.; Faber, A. Analiza przestrzennego zróżnicowania emisji podtlenku azotu z gruntów ornych w Polsce. Studia i Raporty IUNG-PIB. 2017, Volume 52, pp. 57–68. Available online: http://iung.pl/images/wyd/pib/zesz52.pdf (accessed on 20 March 2021).
47. Casey, J.W.; Holden, N.M. Analysis of greenhouse gas emissions from the average Irish milk production system. *Agric. Syst.* **2015**, *86*, 97–114. [CrossRef]
48. Dimitrov, D.D.; Wang, J. Geographic Inventory Framework for estimating spatial pattern of methane and nitrous oxide emissions from agriculture in Alberta, Canada. *Environ. Dev.* **2019**, *32*, 100461. [CrossRef]
49. Ren, C.; Liu, S.; van Grinsven, H.; Reis, S.; Jin, S.; Liu, H.; Gu, B. The impact of farm size on agricultural sustainability. *J. Clean. Prod.* **2019**, *220*, 357–367. [CrossRef]
50. Blandford, D.; Gaasland, I.; Vardal, E. Extensification versus Intensification in Reducing Greenhouse Gas Emissionsin Agriculture: Insights from Norway. *Eurochoices* **2013**, *12*, 4–9. [CrossRef]

51. Ariva, J.; Viira, A.H.; Poldaru, R.; Roots, J. Medium-run projections for greenhouse gas emissions: Arising from agriculture: The case of milk production in Estonia. *Agric. Food Sci.* **2015**, *24*, 300–312. [CrossRef]
52. Grainger, C.; Bauchemin, K.A. Can enteric methane emissions from ruminants be lowered without lowering their production? *Anim. Feed. Sci. Technol.* **2011**, *167*, 308–320. [CrossRef]
53. Hoglund-Isaksson, L.; Gomez-Sanabria, A.; Klimont, Z.; Rafaj, P.; Schopp, W. Technical potentials and costs for reducing global anthropogenic methane emissions in the 2050 timeframe–results from the GAINS model. *Environ. Res. Commun.* **2020**, *2*. [CrossRef]
54. Ocko, I.B.; Sun, T.; Shindell, D.; Oppenheimer, M.; Hristov, A.N.; Pacala, S.W.; Mauzerall, D.L.; Xu, Y.; Hamburg, S.P. Acting rapidly to deploy readily available methane mitigation measures by sector can immediately slow global warming. *Environ. Res. Lett.* **2021**, *16*, 054042. Available online: https://iopscience.iop.org/article/10.1088/1748-9326/abf9c8/pdf (accessed on 30 May 2021). [CrossRef]
55. Koloszko-Chomentowska, Z. Influence of the Common Agricultural Policy on the development of family agricultural farms in the Podlaskie Province. *Econ. Reg. Stud.* **2018**, *11*, 51–68. Available online: http://www.ers.edu.pl/OJS/index.php/erspl/article/view/340/335 (accessed on 10 March 2021). [CrossRef]
56. Some, T.E.; Mbaye, A.A.; Barbier, B. Greenhouse Gas Emission Reduction in Agriculture: Trade-off ot Win-Win Situation for Small Farmers in the Sudanian Area of Burkina Faso? *Afr. Dev. Rev. Rev. Afr. Dev.* **2017**, *29* (Suppl. 2), 163–178. [CrossRef]

energies

MDPI

Article

Economic and Energy Efficiency of Farms in Poland

Marcin Wysokiński [1], Bogdan Klepacki [1], Piotr Gradziuk [2], Magdalena Golonko [1], Piotr Gołasa [1], Wioletta Bieńkowska-Gołasa [1], Barbara Gradziuk [3], Paulina Trębska [1], Aleksandra Lubańska [1], Danuta Guzal-Dec [4], Arkadiusz Weremczuk [1] and Arkadiusz Gromada [1,*]

1 Institute of Economics and Finance, Warsaw University of Life Sciences, 02-787 Warsaw, Poland; marcin_wysokinski@sggw.edu.pl (M.W.); bogdan_klepacki@sggw.edu.pl (B.K.); magdalena_golonko@sggw.edu.pl (M.G.); piotr_golasa@sggw.edu.pl (P.G.); wioletta_bienkowska@sggw.edu.pl (W.B.-G.); paulina_trebska@sggw.edu.pl (P.T.); aleksandra_lubanska@sggw.edu.pl (A.L.); arkadiusz_weremczuk@sggw.edu.pl (A.W.)

2 Poland Economic Modelling Department, Institute of Rural and Agricultural Development, Polish Academy of Sciences, 00-330 Warsaw, Poland; pgradziuk@irwirpan.waw.pl

3 Poland Department of Management and Marketing, Faculty of Agrobioengineering, University of Life Sciences in Lublin, 22-033 Lublin, Poland; barbara.gradziuk@up.lublin.pl

4 Institute of Economics, Pope John Paul II State School of Higher Education in Biała Podlaska, 21-500 Biała Podlaska, Poland; danuta_guzal-dec@wp.pl

* Correspondence: arkadiusz_gromada@sggw.edu.pl

Abstract: Climate change and negative environmental effects are results of a simplified understanding of management processes, i.e., assuming economic effects as the basis for development, without taking into account external costs. Economically efficient facilities are not always environmentally efficient. Due to the existing conflict of economic and environmental goals, it seems necessary to look for measures that would include both economic and environmental elements in their structure. The above doubts were the main reasons for undertaking this research. One of the important sectors of the economy accepted for research, where energy is an essential factor of production, is agriculture. Agricultural production is very diversified both in terms of inputs and final products. Depending on the production direction, the processes of conversion of energy accumulated in inputs into energy accumulated in commodity products have different natures and relationships. Taking into account the importance of agriculture in the national economy and the current environmental needs of the world, the types of farms generating energy surplus and those in which the surplus is the least cost-consuming were indicated. The research used the economic and energy efficiency index, which makes it possible to jointly assess technical and economic efficiency. Assuming the need to produce food with low energy consumption and a positive energy balance, it is reasonable to develop a support system for those farms showing the highest economic and energy efficiency indicators.

Keywords: agriculture; energy consumption; efficiency; farms; FADN

Citation: Wysokiński, M.; Klepacki, B.; Gradziuk, P.; Golonko, M.; Gołasa, P.; Bieńkowska-Gołasa, W.; Gradziuk, B.; Trębska, P.; Lubańska, A.; Guzal-Dec, D.; et al. Economic and Energy Efficiency of Farms in Poland. *Energies* **2021**, *14*, 5586. https://doi.org/10.3390/en14175586

Academic Editors: Dalia Štreimikienė and Talal Yusaf

Received: 16 July 2021
Accepted: 27 August 2021
Published: 6 September 2021

1. Introduction

The current standard of living of mankind is possible thanks to the exploitation of natural capital on an unprecedented scale, which causes increasing interference in the state of the planet and uncertainty about its future [1]. Natural resource mismanagement leads to climate change and limitations in the biological productivity of the land [2–5]. In the history of mankind, there have been many cases of degradation of regional ecosystems as a result of human activity. One of them was "ecological suicide", the so-called Fertile Crescent that 12,000 years ago was the cradle of cities, empires, and great civilizations in the Middle East. The Fertile Crescent is a belt of more fertile lands, shaped like a great crescent, stretching from Egypt, through Palestine and Syria, to Mesopotamia. It extended from Memphis in the Nile Valley to Ur in southern Mesopotamia, including Syria and Canaan, the steppe between the mountain range of Asia Minor and the Syrian Desert. It is the geographic cradle of the great civilizations of the ancient Near East. Thanks to favorable conditions, the first agricultural areas were created here (around 10,000 BC). During the Neolithic revolution, wheat, millet, and barley were grown here. As man domesticated new species of plants and animals, legumes, figs, and grapevines began to be cultivated. Later, the civilizations of Mesopotamia and Ancient Egypt developed [6]. As a result of human activity, this area has become a dry, desert terrain, economically and socially backward. In the 21st century, humans are causing ecosystem destruction on a global scale, i.e., they will not be able to move to areas with favorable living conditions, as did the ancestors from the Fertile Crescent. According to Diamond [7], the societies that committed inadvertent "ecological suicide" were among the most developed and complex of their time. Currently, the most developed economies are experiencing trends based on the economic cult of economic growth and the consumption of goods and services, which are produced based on resources obtained from the environment. The appropriate counterbalance may be sustainable development [8–12], the condition of which is shaping the relations between the economy, society, and the environment in such a way that will not affect the ability of the environment to provide its services in the future. It is also important to treat environmental issues from a supranational and global perspective, treating the earth's ecosystem as a common good [13,14]. A contradiction of such an idea is, for example, the transfer of energy-intensive and "environmentally dirty" production by rich (pseudo-sustainable) countries to other parts of the globe. A significant problem is also the uneven distribution of natural resources, especially minerals, which are the main sources of energy. It is very dangerous to be in a situation where several countries have a good whose consumption can no longer be excluded. International raw material and energy dependencies are becoming an element of pressure and may be the cause of socio-economic crises.

The technological nature of human existence is dependent on external energy sources, which has become the condition of every civilization and the driving force behind every action. According to the Goban-Class [15], "without matter, there is nothing, without energy everything is stationary". This confirms the contemporary dependence of mankind on energy, which determines economic growth, living standards, and can also be a source of international conflicts [16,17]. One of the main problems is the limited energy sources, especially non-renewable ones. Therefore, there is a need for proper management, taking into account the needs of the present and future generations of the Earth's inhabitants [18] (Figure 1).

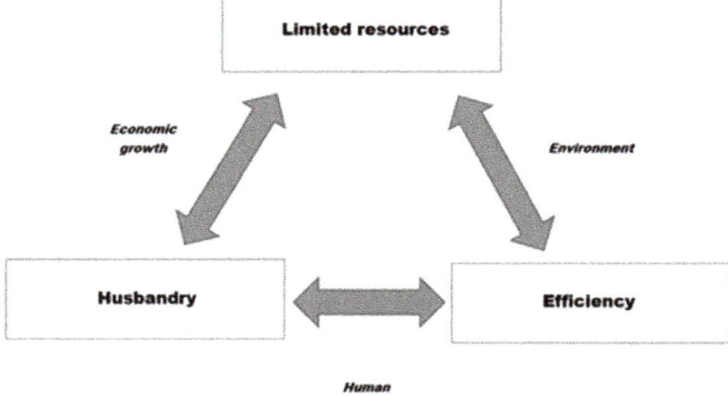

Figure 1. Energy management as an economic problem. Source: own study.

Another problem is the negative impact on the environment of the processes of obtaining energy from non-renewable sources. The main disadvantage is the high greenhouse gas emissions and interference with the ecosystem of conventional energy [19]. Climate change and any negative environmental effects are the results of a simplified understanding of management processes, i.e., assuming economic effects as the basis for development, without taking into account external costs. Performing an assessment solely using the classical measurement of economic efficiency turned out to be the wrong approach, providing inadequate information from the point of view of sustainable development. Economically efficient facilities are not always environmentally efficient. Due to the existing conflict of economic and environmental goals, it seems necessary to search for measures that would include both economic and environmental elements in their structure. The above doubts were one of the main reasons for researching the presented study. Another premise is the dependence of the world economy and its growth on limited natural resources and the growing energy demand. Moreover, there is a need to improve energy efficiency and reduce greenhouse gas emissions on a micro- and macro-scale [20]. Thus, the improvement of energy efficiency becomes the goal to which all activities aimed at reducing the energy consumption needed by the economy to produce products and services are subordinated.

A significant consumer of energy is modern agriculture, which, especially in developed countries, is fully dependent on external non-renewable energy sources. In the 20th century, when the world population increased 3.7 times and the inhabited area increased by about 40%, the energy input increased from 0.1 to almost 13 EJ. As a result, in 2000, on average, about 90 times more energy was used per hectare of arable land than in 1900. In 1900, gross global plant production (before losses in storage and distribution) was little more than the average human food demand, meaning that a large proportion of humanity had little or no nutrition, and the share of the harvest that could be used for feeding the animals was minimal. Increased energy inputs allowed the basic varieties to reach their full potential, which increased yields [21].

The use of means of production of industrial origin meant the introduction of a new source to agriculture—fossil raw materials, which was initiated by the use of solid fuels. Increasing energy resources increased production effects, in particular in plant production. Today, agriculture draws energy from two sources: biospheric resources and fossil resources, which correspond to two types of power—natural and industrial.

Agriculture using only natural sources of energy was a system with relatively high input processing efficiency. The production effects were not high, but the energy expenditure was also small. As the use of fossil fuel energy increases, the unit of energy expended yields less and less product revenue, which is a direct result of the law of diminishing returns. In addition, the increasing consumption of fossil fuels means an increase in greenhouse gas emissions from agriculture and an increasingly negative impact on the natural environment.

In agricultural activity, energy as a production input may determine the profitability of agricultural production, which in turn may affect the level of investments in farms aimed at improving production systems. It can be assumed that measures leading to the improvement of energy efficiency in agriculture and, consequently, to the reduction of production costs, are necessary both from an economic and environmental point of view by reducing GHG (greenhouse gas) emissions [22–24]. The discussion about energy use in agriculture most often focuses on direct energy consumption [25–29]. It is worth noting, however, that 50% or more of total energy consumption is due to the production of nitrogen fertilizers or other activities that indirectly affect the number of energy inputs [30,31]. Different agricultural production systems under different environmental conditions show different energy consumption and energy-saving potential. Therefore, the energy needs of agriculture depend on the nature of individual production processes, and agricultural production is very diverse both in terms of inputs and final products. Depending on the production direction, the processes of conversion of energy accumulated in inputs into energy accumulated in commodity products have different natures and relationships, hence the main objective of this research was to identify economic and energy efficiency in agriculture depending on the type and scale of production.

The research is an original contribution of the authors in the area of analyses of economic and energy efficiency. The proposed index is a measure that combines both economic data and technical data on energy consumption. The innovative approach consists in implementing the EROI (energy return on invested) method in agricultural research and treating the farm as a system that converts energy invested into commodity energy useful for humans in the form of food products. Recognizing that the most important task of agriculture is to feed humanity, a methodology was proposed to evaluate farms in terms of their efficiency and conversion of energy invested into commodity energy. Bearing in mind the negative impact of agriculture on the environment, inter alia through the consumption of non-renewable energy sources, the research results provide information on which farms generate energy surpluses and which of them do it at the lowest cost. The results of the research fill the gap in this respect because in the area of agriculture, the analyses conducted concern either only economic efficiency or only energy efficiency.

2. Materials and Methods

The research used the economic and energy efficiency index (*EEEI*, used for research on farms). The main theoretical assumption of the indicator is to treat a farm as a system that uses energy accumulated in inputs necessary for production, and on the other hand, a system that supplies energy contained in products sold, both of plant and animal origin (Figure 2). Based on Gołębiewska [32], a systemic approach was applied. The inputs "reaching" the farm are transformed in the production process into effects that "leave" the system. What "enters" the system (for example as raw material) is transformed within the system (farm) and leaves the system at the output (as products). The energy entering the system is energy that is purposefully invested by humans (the calculation does not include the energy provided by the sun and used by plants in photosynthesis), while the energy leaving the system is commodity energy (energy contained in animals and vegetable products) useful to consumers. This approach allows the assessment of the effectiveness, including economic, of conversion of invested energy into commodity energy.

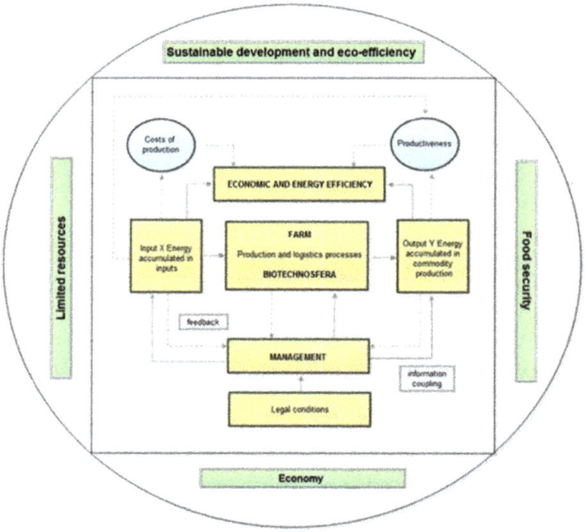

Figure 2. Generalized model of an agricultural farm as an energy system. Source: own study.

One of the goals of the work was, inter alia, diagnosing in which types of farming and on what scale of production the ratio of commodity energy to invested energy is the highest. Farms produce very diverse products that provide human living energy. While only a few dozen years ago the basis for the evaluation of a given production was the economic account, nowadays, it is necessary to include the environmental account. Humanity must very precisely produce food, paying attention to the lowest possible consumption of resources, including energy. Man needs a certain amount of protein and energy to live and work. The challenge is therefore to develop an optimal structure of food products produced with minimal energy inputs. The conducted research is the first step, where the purpose is to obtain information on which farms will provide more energy for the consumer than they use in the production process. The next step was to identify farms where the energy surplus is the least expensive. For this purpose, the following economic and energy efficiency index (*EEEI*) was constructed:

$$EEEI = \frac{\sum_{t=1}^{j=n}(G_p * Q_j) - \sum_{t=1}^{i=n}(G_r * Q_i)}{TPC} = \frac{CE - IE}{TPC} = \frac{SE}{TPC} \quad (1)$$

where:

EEEI—economic and energy efficiency index,
G_p—the weight or quantity of the individual products sold,
Q_j—the amount of energy contained in individual products sold (Table 1),
G_r—the weight or number of individual inputs,
Q_i—the amount of energy contained in individual inputs (Table 2),
TPC—total production costs (EUR).

$$SE = CE - IE \quad (2)$$

where:

SE—surplus energy (MJ),
CE—commodity energy (energy included in sold production) (MJ),
IE—invested energy (energy accumulated in direct and indirect inputs used in the production process) (MJ).

$$IE = E_{ec} + E_{ll} + E_{af} + E_{pf} + E_{pa} \quad (3)$$

where:

E_{ec}—energy from energy carriers (fuels, electricity) (MJ),
E_{ll}—energy equivalent to live labor (MJ),
E_{af}—energy contained in artificial fertilizers (MJ),
E_{pf}—energy contained in purchased feed (MJ),
E_{pa}—energy contained in animals from purchase (MJ).

Table 1. Energy value of products used to calculate the commodity energy.

Product Name	Energy Content per Unit (MJ)	Unit
Cereals for grain	13.70	kg
Peas	14.27	kg
Bean	13.18	kg
Broad bean (fresh matter)	3.18	kg
Lentin	14.69	kg
Soy	17.28	kg
Chickpeas	15.82	kg
Other edible legumes	14.00	kg
Forage peas (field peas)	14.27	kg
Horse bean	13.00	kg
Sweet lupine	15.53	kg
Vetch (fresh matter)	3.18	kg
Seradella (fresh matter)	3.18	kg
Other fodder legumes (fresh matter)	5.00	kg
Legume mixes with other plants	14.00	kg

Table 1. *Cont.*

Product Name	Energy Content per Unit (MJ)	Unit
Rape and turnip rape	27.00	kg
Sunflower	24.44	kg
Linen and linen cloth	10.59	kg
Oily soybeans	17.28	kg
Other oil plants	15.00	kg
Cattle	9.71	kg
Cow milk	2.68	kg
Pigs	6.49	kg
Hens	6.61	kg
Consumption chicken eggs	5.86	pieces
Geese	14.35	kg
Ducks	13.02	kg
Turkeys	5.40	kg
Ostriches	6.07	kg
Apples	2.09	kg
Pears	2.43	kg
Other crumb-pome fruits	2.26	kg
Plums	2.05	kg
Sour cherries	2.05	kg
Sweet cherries	2.64	kg
Peaches	2.09	kg
Apricots	2.09	kg
Other stone fruits	2.18	kg
Gooseberry	1.84	kg
Aronia	1.97	kg
Black currants	2.13	kg
White currants	2.03	kg
Red currants	1.93	kg
Raspberries	1.80	kg
Blackberries	1.84	kg
Blueberries	2.39	kg
Other berries	2.00	kg
Strawberries	1.38	kg

Reproduced from [33].

Table 2. Energy value of inputs adopted to calculate the invested energy.

<table>
<tr><th colspan="3">Description</th><th>Product Name</th><th>Energy Content per Unit (MJ)</th><th>Unit</th></tr>
<tr><td rowspan="9">IE—invested energy (energy accumulated in inputs for production)</td><td rowspan="9">E_d—direct energy</td><td rowspan="8">E_{ec}—energy from energy carriers (fuels, electricity)</td><td>Coal, coal dust, briquettes</td><td>30.00</td><td>kg</td></tr>
<tr><td>Fuel gas</td><td>25.19</td><td>l</td></tr>
<tr><td>Heating oil</td><td>36.55</td><td>l</td></tr>
<tr><td>Petrol fuel</td><td>32.25</td><td>l</td></tr>
<tr><td>Diesel fuel</td><td>45.22</td><td>l</td></tr>
<tr><td>Firewood</td><td>18.00</td><td>kg</td></tr>
<tr><td>Propellant</td><td>25.19</td><td>l</td></tr>
<tr><td>Electricity</td><td>13.60</td><td>kWh</td></tr>
<tr><td>E_{ll}—energy equivalent to live labor</td><td>Total working time</td><td>2.00</td><td>h</td></tr>
</table>

Table 2. *Cont.*

Description		Product Name	Energy Content per Unit (MJ)	Unit
	E_{af}—energy contained in artificial fertilizers	Nitrogen fertilizers	48.99	kg
		Phosphorus fertilizers	15.23	kg
		Potash fertilizers	9.68	kg
		Compound fertilizers	24.63	kg
		Other fertilizers (magnesium fertilizers)	6.70	kg
	E_{pf}—energy contained in purchased feed	Concentrated feed	3.80	kg
		Hay	1.90	kg
		Silage	0.90	kg
	E_{pa}—energy contained in animals from purchase	Cattle	9.71	kg
		Pigs	6.49	kg
		Hens	6.61	kg
		Geese	14.35	kg
		Ducks	13.02	kg
		Turkeys	5.40	kg
		Ostriches	6.07	kg
E_i—indirect energy	E_b—energy contained in buildings	Buildings and structures	153.00	m^2
	E_{md}—energy contained in machines and devices	Machines and devices	110.00	kg

Reproduced from [33–35].

For empirical research in the field of economic and energy efficiency of farms, data from the Polish FADN (Farm Accountancy Data Network) for 2016, from the entire territory of Poland, were used. The FADN operating in Poland is part of the European system, operating since 1965, based on the Regulation of the Council of 15 June 1965 setting up a network for the collection of accountancy data on the incomes and business operation of agricultural holdings in the European Economic Community [36]. Data in FADN are collected in the management accounting convention. The FADN database is economic and organizational. It is now the most complete source of information on the situation of agricultural holdings. The identical principles of operation of the FADN system throughout the EU make the results comparable for all EU countries. The obtained data are used both for decision-making by EU bodies, monitoring the effects of these activities, and scientists dealing with the economics and organization of agriculture. Participation in the FADN system is voluntary. Farmers participating in the research write down every economic event that took place on their farm, in a special book, then agricultural advisors transfer them to the system [24]. When selecting the research objects, the purposeful selection method was used—the results were adopted according to the set of classification coefficients "SO 2013". To eliminate the influence of the production structure and economic power on the results of the analyses, all calculations were performed using the division of farms into production types and economic size classes. The production type is defined as the share of standard outputs (SO) from particular production lines in the total value of standard production of a given farm. Two threshold values apply to the type of farming formula. Farms in which the share of one direction of plant or livestock production exceeds 2/3 of SO are called specialist farms. Farms where the share of any of the directions does not exceed 1/3 of the SO are defined as "mixed", i.e., combining animal and plant production (multidirectional) [37]. The economic size of farms is determined by the sum of standard outputs from all agricultural activities occurring on a given farm. Standard output (SO) is the five-year average value obtained from one hectare of a given type of crop production or, in the case of livestock production, from one head, in the production conditions average for a given region [38]. For analytical purposes, farms classified into 5 types of farming were selected following the FADN methodology:

- Specializing in the cultivation of cereals, oilseeds, and protein crops for seeds,
- Specializing in pig farming,
- Specializing in dairy cattle breeding,
- Specializing in slaughter cattle breeding,
- Specializing in the cultivation of fruit trees and shrubs.

The division of the researched farms into 3 economic size classes in each type was adopted, considering the economic size of the farm as a criterion for grouping:

- Small (8000 $\leq$ EUR < 25,000)—marked as economic size class I,
- Average (25,000 $\leq$ EUR < 50,000)—marked as economic size class II,
- Big ($\geq$50,000 EUR)—marked as economic size class III.

The surveyed population is 6261 farms. The most numerous group were farms specializing in dairy cattle breeding (2742 farms) and farms specializing in cereal cultivation (2036 farms). Taking into account the economic size classes, the share of farms in individual classes was very similar: 2143 in economic size class I, 2077 in economic size class II, and 2041 in economic size class III. Due to the type of farming and the economic size class of classified farms, the largest group in the study were medium-sized farms specializing in dairy cattle production (1165 farms).

The paper presents only selected production and economic indicators, which allowed for the characteristics of the examined objects in terms of assets involved, costs incurred, or effects of the activity. The selection of the presented data also resulted from their impact on economic and energy efficiency.

All the researched farms conducted their activities using land resources, which the greater they were, the higher the economic size (Table 3). The greatest amount of arable land was found on cereal farms—approximately 62 ha on average. However, the ownership structure of the land used is interesting, including the ratio of leased land to own use.

Among the analyzed types of farming, the largest amount of land was leased by farms specializing in cereal crops, which constituted approximately 32% of the total area of agricultural land in these farms. Farms with this type of production, working out by far the smallest direct surplus per hectare, are forced to increase their area more intensively to achieve acceptable income than farms with other types of production. Farms specializing in the cultivation of fruit trees and shrubs used the lease to the least extent, which is largely due to the specificity of production based on long-term plantings and significant related investments. The duration of the lease is usually limited to 5 years, which is a disadvantage in this case. It was found that with the increase in the scale of production, the share of leased agricultural land in each type of farm increases. This process was most dynamic in the pigs' type.

It was assumed that the number of tractors may also affect the efficiency considered in the study—fuel consumption is one of the main energy inputs in agriculture. It was found that farms specializing in the cultivation of fruit trees and shrubs were characterized by significantly higher than average equipment with tractors—on average almost 15 pieces per 100 ha of UAA (Utilized Agricultural Area). The reasons can be found in the large number of agrotechnical and agro logistic works carried out at the same time, which determines the need to have many low-power tractors. The use of large and efficient machines is also problematic, as in the case of cereal production, where on average 2.4 tractors are used per 100 ha of agricultural land. A negative correlation was observed between the production scale and the number of tractors per 100 ha of UAA.

Table 3. Characteristics of the researched farms.

		Number of Researched Farms		
Type of Farming	**I**	**II**	**III**	**Total**
Cereals	955	568	513	2036
Pigs	74	146	472	692
Dairy cattle	580	1 165	997	2742
Slaughter cattle	245	81	28	354
Fruit trees and shrubs	289	117	31	437
		Agricultural Land Area (ha)		
Type of Farming	**I**	**II**	**III**	**Average**
Cereals	22.95	56.02	139.81	61.57
Pigs	10.12	17.60	44.27	34.98
Dairy cattle	14.05	24.76	50.06	31.69
Slaughter cattle	19.25	40.17	83.41	29.11
Fruit trees and shrubs	8.61	17.98	37.30	13.16
		Share of Leased Space in the Total UAA (%)		
Type of Farming	**I**	**II**	**III**	**Average**
Cereals	22.94	32.34	34.81	32.11
Pigs	8.99	18.79	27.10	25.65
Dairy cattle	20.26	25.39	32.29	28.87
Slaughter cattle	17.80	28.52	38.55	25.88
Fruit trees and shrubs	6.47	9.56	15.66	9.45
		Number of Tractors (Pieces/100 ha of UAA)		
Type of Farming	**I**	**II**	**III**	**Average**
Cereals	8.13	4.76	2.40	4.00
Pigs	17.10	12.41	6.37	7.34
Dairy cattle	13.60	10.34	6.87	8.65
Slaughter cattle	9.73	6.58	3.81	7.39
Fruit trees and shrubs	19.08	13.12	9.34	14.94
		Labor Inputs per 100 ha of UAA (AWU/100 ha of UAA)		
Type of Farming	**I**	**II**	**III**	**Average**
Cereals	5.46	2.74	1.47	2.49
Pigs	13.85	9.14	4.75	5.50
Dairy cattle	11.96	7.75	4.54	6.30
Slaughter cattle	7.61	4.14	2.47	5.35
Fruit trees and shrubs	24.07	16.85	12.81	19.17
		Share of Hired Work Time in Total Work (%)		
Type of Farming	**I**	**II**	**III**	**Average**
Cereals	2.02	3.18	13.31	7.98
Pigs	0.93	0.40	11.85	9.01
Dairy cattle	0.56	1.41	8.10	4.04
Slaughter cattle	0.67	1.67	9.98	1.89
Fruit trees and shrubs	30.16	40.16	62.45	37.77

Source: own study.

One component of the invested energy is labor input. In the studied objects, they were the highest in fruit-growing farms—on average 19.17 AWU (Annual Work Unit) per 100 ha of UAA, and the smallest in farms specializing in the cultivation of cereals, oilseeds, and protein crops for seeds—on average 2.49 AWU per 100 ha of UAA. The level of labor inputs decreases with increasing economic size. Clear differences in the labor intensity of extreme types of farming are a consequence of the specificity of production and the possibility of using efficient machines and work automation, which should translate into savings in energy inputs.

Fixed assets include agricultural land, farm buildings, forest plantings, and machinery and equipment, as well as livestock animals (Table 4). For the calculation of the invested energy, energy accumulated in machines and devices as well as in buildings and structures was taken into account, as an indirect input. Taking these components into account gives grounds to believe that the conducted analyses have the features of a drawn calculus. Therefore, it was considered justified to present the significance of selected components of fixed assets and indicators of technical equipment for land and work in the researched farms.

Table 4. Structure of assets and technical infrastructure of land and work.

Share of Fixed Assets in Total Assets (%)				
Type of Farming	**I**	**II**	**III**	**Average**
Cereals	92.11	90.95	90.57	90.97
Pigs	90.67	88.39	84.88	85.47
Dairy cattle	89.95	88.90	88.44	88.72
Slaughter cattle	88.68	86.43	83.05	86.94
Fruit trees and shrubs	91.01	90.30	88.41	90.35
Share of Buildings in Fixed Assets (%)				
Type of Farming	**I**	**II**	**III**	**Average**
Cereals	15.31	10.87	7.82	10.11
Pigs	29.85	27.66	27.62	27.71
Dairy cattle	24.21	21.89	21.14	21.65
Slaughter cattle	19.91	15.96	16.47	18.07
Fruit trees and shrubs	24.69	22.35	18.37	22.86
Share of Machines and Devices in Fixed Assets (%)				
Type of Farming	**I**	**II**	**III**	**Average**
Cereals	14.55	19.79	20.97	19.38
Pigs	11.54	15.53	20.33	19.46
Dairy cattle	15.75	20.49	23.11	21.63
Slaughter cattle	13.84	17.46	18.07	15.71
Fruit trees and shrubs	15.58	18.70	25.36	18.22
Technical Equipment of the Land (EUR/ha of UAA)				
Type of Farming	**I**	**II**	**III**	**Average**
Cereals	2133.49	1917.56	1699.99	1830.94
Pigs	4800.46	4309.84	4495.07	4484.83
Dairy cattle	3420.42	3794.52	4203.87	3994.53
Slaughter cattle	2528.38	2219.60	1777.13	2260.64
Fruit trees and shrubs	6928.13	6863.86	5090.50	6535.08
Technical Equipment for Work (EUR/AWU)				
Type of Farming	**I**	**II**	**III**	**Average**
Cereals	39,103.95	70,096.93	115,269.42	73,520.55
Pigs	34,651.65	47,165.84	94,585.81	81,534.02
Dairy cattle	28,609.24	48,956.91	92,691.16	63,420.39
Slaughter cattle	33,232.65	53,650.36	71,922.32	42,270.70
Fruit trees and shrubs	28,777.41	40,729.39	39,734.99	34,095.51

Source: own study.

The share of fixed assets in total assets was at a similar level in all types. The highest level of the indicator was recorded in small horticultural farms (91.01%), and the lowest in large farms specializing in slaughter cattle (83.05%).

When analyzing the structure of fixed assets, clear differences in individual types were observed. In the case of buildings, their share in fixed assets ranged from 7.82% in the largest cereal farms to almost 30% in farms specialized in rearing pigs from economic size class I. It should be added that pig farms had the highest index in all economic size classes.

Therefore, these farms have the greatest negative impact of buildings and structures on the energy invested. This indicator decreases along with an increase in the economic size of farms. For buildings, it was assumed that the value of energy, which is the expenditure in a given year, constitutes 2.5% of the total energy accumulated in this fixed asset—following the principles of calculating depreciation for buildings and structures.

The researched farms were characterized by a very high share of machines and devices in the structure of fixed assets (18% on average). The differences between the individual types of farming were slight. The farms specializing in rearing cattle for slaughter were characterized by a lower index than the average. It was also found that the share of machines and devices in fixed assets increases with the increase in the scale of production. In the case of machines and devices, it was assumed that the value of energy, which is an input in a given year, constitutes 14% of the total energy accumulated in this fixed asset—following the principles of calculating depreciation for machines and devices.

A measure closely related to the value of buildings and machinery and equipment is the technical equipment of the land, which achieved the lowest value in large farms specializing in cereal cultivation (EUR 1699.99 per ha of UAA), while the highest value in small horticultural farms (EUR 6928.13 per ha of UAA), which is determined by the production technology appropriate for horticultural farms, where specialized buildings and structures (cold stores, etc.), as well as machines and devices, are required.

The study also counted the technical equipment of work, which in the studied group of farms is very diverse and ranges from EUR 28,609.24 per AWU in the smallest dairy farms, up to EUR 115,269.42 per AWU on cereal farms from economic size class III. Along with the increase in the economic size of farms, there is an increase in the technical equipment of work. The factor strongly affecting the level of this indicator is the number of people working on the farm, which is several times higher on dairy farms than on cereal farms.

3. Results

During the analyses, attention was also paid to economic effects (Table 5). One of the measures used for such calculations is economic labor productivity, which increases with the increase in the economic size of the researched farms, except farms specializing in the cultivation of fruit trees and shrubs. Average economic labor productivity for fruit farms is several times lower than in other types of production. The differences deepen with the increase in the scale of production. In every economy size class, cereal farms are the leader.

Table 5. Economic results of researched farms.

		Economic Labor Productivity (EUR/h)		
Type of Farming	**I**	**II**	**III**	**Average**
Cereals	2.84	6.17	11.78	6.87
Pigs	1.64	3.70	9.22	7.68
Dairy cattle	2.25	4.33	8.50	5.71
Slaughter cattle	2.49	4.57	8.85	3.67
Fruit trees and shrubs	1.73	2.18	1.94	1.90
		Land Profitability Index (EUR/ha of UAA)		
Type of Farming	**I**	**II**	**III**	**Average**
Cereals	342.72	384.32	402.03	387.17
Pigs	506.31	771.45	1009.66	968.74
Dairy cattle	610.59	784.48	901.17	835.19
Slaughter cattle	429.27	447.42	495.41	449.99
Fruit trees and shrubs	930.67	834.99	562.20	821.56

Table 5. *Cont.*

Charging Income with Energy Inputs (MJ/EUR)				
Type of Farming	**I**	**II**	**III**	**Average**
Cereals	82.06	87.55	89.16	87.68
Pigs	66.93	48.44	59.86	58.99
Dairy cattle	38.06	40.16	44.56	42.73
Slaughter cattle	35.05	40.68	31.00	35.80
Fruit trees and shrubs	38.15	46.56	65.31	45.04

Source: own study.

The land profitability index is the ratio of income from an agricultural holding to the UAA. It allows for the assessment of land use efficiency as one of the production factors. The highest profitability of land was characterized by large farms specializing in pig farming (EUR 1009.66 per ha of UAA) and small farms specializing in the cultivation of fruit trees and shrubs (EUR 930.67 per ha of UAA). The lowest values of this indicator were recorded for both small, medium-sized, and large farms specializing in the cultivation of cereals (approximately EUR 387.17 per ha of UAA), which use employees more effectively than the cultivated land. It is worth adding that cereal farms require relatively the largest amount of energy to be invested to earn EUR 1.00 of income—on average 87.68, which is a result 2.5 times worse than in farms of the slaughter cattle type.

The factor having a significant impact on the cost-intensity of the researched farms was energy costs (engine fuels, electricity, heating fuels). However, their impact on direct costs was varied (Table 6). By far the highest share of energy costs in direct costs was recorded in fruit-tree and shrub-type farms, which results from the specificity of production in these facilities. The dependence of this variable on the economic value was identified—with the increase in the scale of production, the share of energy costs in direct costs decreases. This relationship was most clearly visible in pig farms. Fruit farms were also characterized by the highest energy costs per hectare of UAA—on average EUR 210.89. The results of research on the structure of energy costs, which depended on the type of agricultural production, are interesting. For example, the cost of electricity was much more important for fruit and pig farms (on average over 30% share in energy costs) than for cereals (6.75%). Apart from the farm types of cereals and slaughter cattle, no clear correlation was found between the share and the economic size. In the case of the costs of propellants, their highest share in the energy costs is held by farms in the types of cereals and slaughter cattle—about 90%. The importance of individual energy sources depends on the needs of individual types of farming, resulting from the number of works and activities specific to a given production.

Table 6. Characteristics of energy costs in the researched farms.

Share of Energy Costs in Direct Costs (%)				
Type of Farming	**I**	**II**	**III**	**Average**
Cereals	27.87	21.71	21.12	22.20
Pigs	11.41	8.65	6.63	6.84
Dairy cattle	23.58	19.21	16.76	17.78
Slaughter cattle	33.64	24.91	24.38	28.25
Fruit trees and shrubs	37.78	38.92	36.03	37.87
Energy Costs per 1 ha of UAA (EUR)				
Type of Farming	**I**	**II**	**III**	**Average**
Cereals	79.22	76.37	81.74	79.94
Pigs	122.63	111.04	146.68	142.14
Dairy cattle	97.16	113.46	143.11	128.96
Slaughter cattle	75.53	75.61	64.70	73.10
Fruit trees and shrubs	200.55	227.89	202.21	210.89

Table 6. *Cont.*

Share of Electricity Costs in Energy Costs (%)				
Type of Farming	**I**	**II**	**III**	**Average**
Cereals	10.17	7.15	5.58	6.75
Pigs	28.23	25.33	31.62	31.01
Dairy cattle	24.16	24.06	21.27	22.29
Slaughter cattle	18.75	13.21	7.70	14.72
Fruit trees and shrubs	33.99	37.40	31.64	34.88
Propulsion Cost Share in Energy Costs (%)				
Type of Farming	**I**	**II**	**III**	**Average**
Cereals	87.93	90.06	87.00	87.90
Pigs	69.82	73.95	65.30	66.14
Dairy cattle	74.77	75.57	78.31	77.26
Slaughter cattle	79.69	86.38	92.03	84.35
Fruit trees and shrubs	64.70	61.50	67.85	64.04

Source: own study.

Invested energy is one of the key elements of the proposed economic and energy efficiency index. Therefore, it is important to recognize the impact of individual energy inputs on their amount. The structure of the energy invested is shown in Table 7. Mineral fertilizers, direct energy carriers (engine fuels, electricity, heating fuels), as well as machines and devices, had the greatest share. The energy inputs accumulated in buildings (this is a consequence of the adopted methodology of calculation—2.5% of the total expenditure, which corresponds to the methodology of depreciation) and the energy contained in the equivalent of live labor had a marginal impact. In the case of fertilizers, their dominant share was in the energy invested in cereal farms—75% on average. In this area, it is possible to seek efficiency improvement by reducing the most energy-consuming inputs. Direct energy carriers had the highest share in fruit farms, which results from the course and specificity of production in these facilities. Additionally, the share of machines and devices in shaping the invested energy was the highest in these farms.

Table 7. Invested energy structure.

Share of Direct Energy Carriers in Invested Energy (%)				
Type of Farming	**I**	**II**	**III**	**Average**
Cereals	15.55	12.07	11.94	12.49
Pigs	24.48	18.95	16.55	16.86
Dairy cattle	26.44	22.56	21.68	22.23
Slaughter cattle	30.12	23.53	22.16	26.05
Fruit trees and shrubs	38.88	41.29	36.82	39.40
Share of Energy Contained in Fertilizers in the Invested Energy (%)				
Type of Farming	**I**	**II**	**III**	**Average**
Cereals	70.98	75.02	76.15	75.12
Pigs	44.43	42.77	33.81	34.63
Dairy cattle	45.32	47.74	46.19	46.59
Slaughter cattle	38.66	47.44	42.64	42.65
Fruit trees and shrubs	32.32	29.58	34.50	31.70

Table 7. *Cont.*

Share of Energy Contained in Buildings in the Invested Energy (%)				
Type of Farming	**I**	**II**	**III**	**Average**
Cereals	0.06	0.03	0.02	0.03
Pigs	0.17	0.12	0.07	0.08
Dairy cattle	0.15	0.10	0.08	0.09
Slaughter cattle	0.17	0.10	0.09	0.13
Fruit trees and shrubs	0.20	0.16	0.10	0.16
The Share of Energy Contained in Machines and Devices in the Invested Energy (%)				
Type of Farming	**I**	**II**	**III**	**Average**
Cereals	12.39	12.36	11.52	11.86
Pigs	13.26	13.92	10.55	10.84
Dairy cattle	19.47	19.55	18.36	18.78
Slaughter cattle	23.13	21.39	20.32	21.90
Fruit trees and shrubs	25.34	26.99	26.99	26.30
Share of Energy Contained in the Equivalent of Live Labor in the Invested Energy (%)				
Type of Farming	**I**	**II**	**III**	**Average**
Cereals	0.86	0.37	0.19	0.33
Pigs	1.82	1.12	0.36	0.44
Dairy cattle	2.33	1.15	0.53	0.82
Slaughter cattle	2.29	1.07	0.73	1.52
Fruit trees and shrubs	3.03	1.97	1.58	2.33

Source: own study.

Concerning fertilizers, buildings, and live labor, it can be argued that along with the increase in the scale of production, the share of energy inputs in the invested energy decreases.

One of the objectives of the work was to calculate the EROI, i.e., the ratio of commodity energy to invested energy. Invested energy is energy accumulated in inputs used in the production process, while commodity energy is energy accumulated in sold products. The index should therefore be above 1, otherwise, it means that more energy has been invested than obtained in the production process. From an economic and environmental point of view, any activity should generate energy surpluses. In the researched farms, only the production of cereals and pigs generated such a surplus, regardless of the production scale (Table 8). Additionally, the smallest farms specialized in milk production recorded the indicator above 1. The greatest losses of energy were recorded in farms specialized in the production of slaughter cattle and fruit from economic size class III. It was found that with the increase in the production scale, the EROI index decreased. This is the result of a disproportionate increase in commodity energy in relation to the increasing energy inputs accumulated, among others in mineral fertilizers and larger, more advanced machines and devices used in farms with a larger production scale.

When analyzing the economic and energy efficiency separately and comparing their course, different relations between them depending on the type and scale of production were observed (Figure 3). It was found that farms with the highest energy efficiency (cereals) are characterized by the lowest economic efficiency, while the opposite was true for farms specialized in fruit production. It is worth adding that farms of the slaughter cattle type achieved the lowest values for both types of efficiency. The reaction of the examined efficiencies to changes in the production scale was also interesting. There was no common trend in this respect for the researched types of agricultural production. In pigs and dairy cattle farms, energy efficiency decreased and economic efficiency increased as the scale increased. The situation was quite different in fruit farms, where the growing production volume had negative effects on both economic and energy efficiency. In the case of cereals and slaughter cattle, the scale of production had a slightly positive impact on the economic effects per hectare of UAA, while in the case of energy efficiency, the direction of the trend cannot be clearly stated.

Table 8. Commodity energy to invested energy relations.

Value of Energy Contained in Production Inputs—Invested Energy (MJ)				
Type of Farming	**I**	**II**	**III**	**Average**
Cereals	646,251.09	1,882,479.47	5,039,707.20	2,098,127.66
Pigs	342,896.00	658,014.21	2,678,117.90	2,002,190.79
Dairy cattle	326,553.86	780,199.90	2,009,881.70	1,131,358.93
Slaughter cattle	289,784.55	731,121.38	1,281,315.20	469,194.55
Fruit trees and shrubs	306,019.67	699,495.59	1,369,634.84	486,817.73
Value of Energy Contained in Sold Products—Commodity Energy (MJ)				
Type of Farming	**I**	**II**	**III**	**Average**
Cereals	1,129,059.11	2,908,682.19	7,876,106.05	3,325,552.72
Pigs	625,481.55	1,117,827.97	3,133,211.11	2,439,832.60
Dairy cattle	365,025.17	734,011.13	1,689,047.87	1,003,216.01
Slaughter cattle	268,604.91	727,907.51	970,282.38	429,199.49
Fruit trees and shrubs	292,875.09	551,062.36	1,025,543.47	414,045.87
EROI—The Ratio of Commodity Energy to Invested Energy				
Type of Farming	**I**	**II**	**III**	**Average**
Cereals	1.75	1.55	1.56	1.59
Pigs	1.82	1.70	1.17	1.22
Dairy cattle	1.12	0.94	0.84	0.89
Slaughter cattle	0.93	1.00	0.76	0.91
Fruit trees and shrubs	0.96	0.79	0.75	0.85

Source: own study.

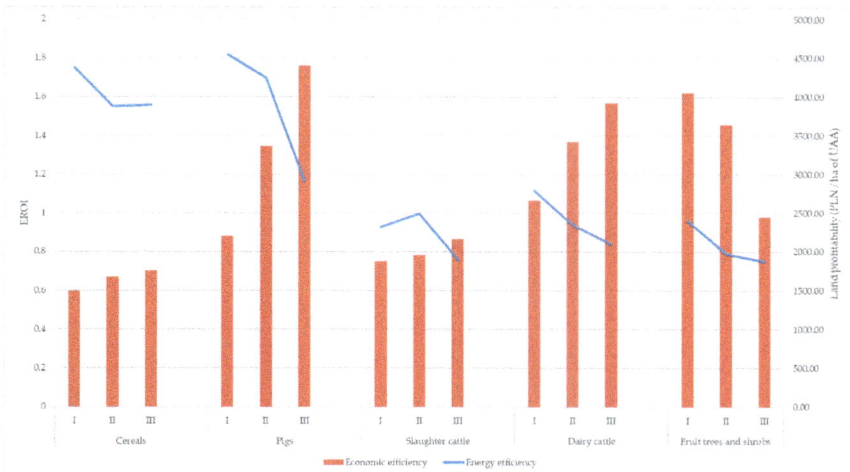

Figure 3. Economic and energy efficiency in the researched farms. Source: own study.

The last stage of the research was to calculate the economic and energy efficiency according to the proposed methodology (Figure 4). The highest ratio was achieved by farms specialized in cereal production—on average they generated 26.60 MJ of energy surplus per EUR 1.00 of costs. The result above zero was also achieved by pig producers and the smallest dairy farms.

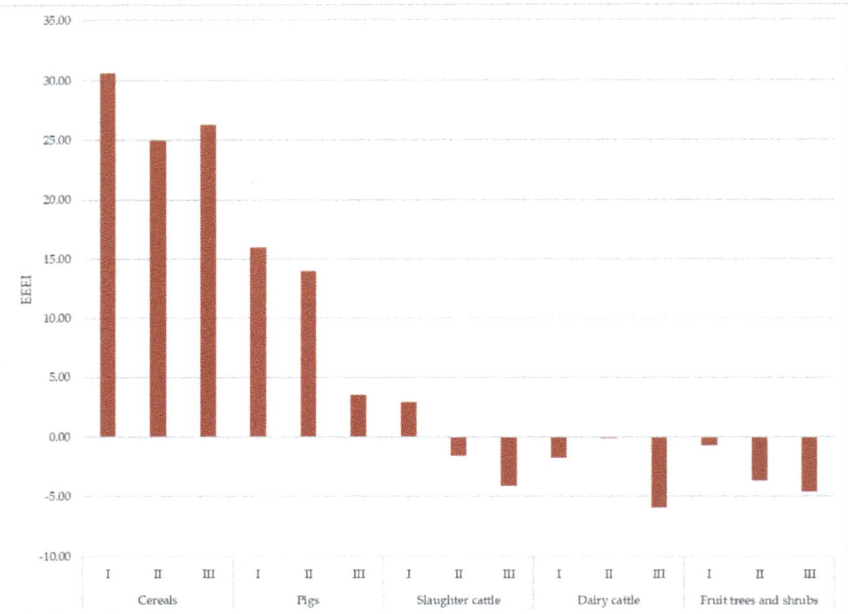

Figure 4. *EEEI*-economic and energy efficiency index (MJ/EUR). Source: own study.

In the group of effective farms (index above zero), the most effective farms were those with the smallest production scale. Therefore, it can be concluded that in the case of economic and energy efficiency, there are decreasing scale effects.

4. Discussion

Energy analysis, as an independent research approach, was first used in the early 1970s [39]. An impulse for research towards energy analyses was the work of Georgescu-Roegen "The entropy law and the economic process" [40] from 1971. Such analyses require combining biological and technical knowledge with economic knowledge [41–43]. It is not very easy, and therefore, no appropriate, uniform methodological foundations have yet been developed in energy analyses [44–46]. Previous studies usually focused on economic or energy efficiency, treating them separately. This approach was also most often used in agriculture. Energy efficiency indicators are used to evaluate various agricultural systems as well as production methods (ecological, conventional) [47–51]. According to Risoud [52] and the methodology developed in her work, the energy efficiency of a farm is defined as the following ratio: gross energy of useful products/non-renewable energies used to produce them. Research in the field of energy consumption and efficiency of its use was conducted in particular concerning selected crops or breeding. The energy and economic analysis of wheat cultivation in Bangladesh were carried out by Rahman and Hasan [53], and the rice production on farms in Iran by Pishgar-Komleh et al. [54]. In their opinion, mainly, large farms (more than 1 ha) had better management and were more successful in energy use and economic performance. Heidari et al. in their research determined the efficiency of energy use (EUE) for the production of broilers [55]. Energy consumption and energy efficiency for a representative crop (Flemish Farm Accountancy Data Network, FADN) of specialized dairy, arable, and pig farms in Flanders were determined by Meul et al. [56]. The most energy-efficient dairy and pig farms were intensive farms, which combined a high production with low energy use, and which possessed a gross value added per production unit comparable to, or even higher than the average.

Energy efficiency, yield efficiency, and labor requirements in the production of maize, wheat, potatoes, and apples were determined for organic (without synthetic fertilizers and pesticides) and conventional agricultural technologies in the studies of Pimentel et al. [57]. For all four crops, the labor input per unit of yield was higher for organic systems compared to conventional production. Similar studies for the comparison of cultivation systems (conventional, organic, and integrated) were carried out by the Italian research team of Falcone et al. [58]. The energy efficiency and economic effects of the main cultivation methods (conventional, organic, and integrated) of clementine's crops in Calabria (South Italy) were assessed by a combined use of the Life Cycle Energy Assessment (LCEA) approach and economic analysis. The economic efficiency of energy from clementine production was higher compared to the other two farming systems.

Keummel et al. [59] proposed, for example, a system of agricultural production combining food and energy production, which could be a step towards the development of sustainable agriculture. The purpose of introducing such a system would be to reduce the positive balance of carbon dioxide emissions by agriculture, which contributes to climate change. This goal would be achieved by replacing the use of energy from fossil fuels with energy from biofuels produced in mandatory separate areas within farms. In this way, the emission of carbon dioxide from fossil fuels would be significantly reduced and, additionally, the absorption of carbon dioxide from the atmosphere by crops could increase. These studies show that such a system would be economically acceptable both from the point of view of the farmer and the society. Introducing biofuel production on a local scale would have benefits not only in terms of energy and climate, but also reducing carbon dioxide emissions was estimated by the authors at the equivalent of EUR 300/ha of external benefits.

It is worth paying attention to the research of Alluvione and co-authors [60]. These researchers analyzed energy consumption and efficiency in three farming systems: low-cost, integrated, EU-compliant, and traditional-conventional. It was found that in the first two systems, the efficiency of energy use increases by 32.7% and 31.4% respectively, while maintaining similar results in terms of net energy. In the area of research on efficiency, the study by Uzal [61] deserves attention, where the energy efficiency of milk production was compared on two farms. In the first, dairy cattle were reared in a free-stall housing system, in the second—in a loose housing system. It was found that in both cases, the highest percentage of energy inputs came from feed and the electricity consumed. Total energy consumption per hectare was lower on loose housing system farms. In the research by Gronroos et al. [24], the energy consumption of traditional and organic milk and rye bread production in Finland was examined. Basic energy consumption in traditional milk production was 6.4 GJ per 1000 L of milk and 4.4 GJ in organic production. In the case of the production of rye bread, it was 15.3 and 13.3 GJ respectively, per 1000 kg of rye bread. Renewable energy use ranged from 7% to 16%, with a slightly higher percentage for organic farming.

An interesting approach to energy productivity in agriculture was presented by Uhlin [62], questioning the statements widely described in the earlier literature that to reverse the downward trend in energy productivity in Swedish agriculture, energy inputs from fossil fuels should be reduced. The author claims that the emphasis should be placed not on the reduction of the use of fossil fuels, but on the development of the use of energy from renewable sources, e.g., solar energy, as this approach offers many more benefits than just reducing energy inputs. In research and policymaking, technical development and modern technologies used in agriculture should not be overlooked

One of the methods of assessing effectiveness is Data Envelopment Analysis (DEA). Using this method, Ghali et al. [63] assessed the efficiency of the use of energy resources in French farms. Results show that disentangling energy resources from the rest of intermediate consumption highlights energy use excess, which is masked when considering intermediate consumption as a whole. Using DEA, Mohammadi et al. [64] assessed the energy efficiency of farmers, to find efficient and inefficient ones and to identify the wasteful uses of energy in kiwifruit production. Chemical fertilizers and chemical energy were the main inefficient consuming inputs.

An important research problem is also the relation between the energy obtained and the energy put into the production process. Many scientists in recent years have undertaken such research, among them Kuesters and Lammel [65], who in 1989–1997 analyzed the aforementioned relationship for winter wheat and sugar beet. It was found that the ratio of energy obtained to energy input was highest in the case of low-intensity crops, which means that extensive cultivation methods are preferred. However, with these production methods, low yields are obtained, and hence also low energy efficiency, therefore the authors additionally extended the analysis to include the net energy balance. The results were similar.

A similar aim of the research was adopted by Moitzi and his team [26], who verified energy consumption and energy efficiency in selected farms in Slovakia, Romania, Serbia, and Austria. It was found, inter alia, that the intensity of the use of production factors, i.e., fuel, seeds, fertilizers, and pesticides, affects the energy efficiency of plant production. The main analyzed index: energy generated for energy inputs, in the case of winter wheat cultivation was 5.6, with the range from 4.8 to 7.1.

This article proposed a combination of energy efficiency, economic efficiency, and EROI index, and the development of the *EEEI* economic and energy efficiency index (used for research on farms). The main theoretical assumption for the development of the indicator is to treat a farm as a system that uses energy accumulated in inputs necessary for production and as a system supplying energy contained in sold products of both plant and animal origin. Based on Gołębiewska [32], a systemic approach was applied. The inputs "reaching" the farm are transformed in the production process into effects that "leave" the system. What "enters" the system (for example as raw material) is transformed within the system (farm) and leaves the system at the output (as products). The energy entering the system is energy that is purposefully invested by humans (the calculation does not include the energy provided by the sun and used by plants in photosynthesis), while the energy leaving the system is commodity energy (energy contained in animals and vegetable products) useful to consumers. This approach allows the assessment of the effectiveness, including economic, of conversion of invested energy into commodity energy.

Recognizing that it is necessary to introduce a coherent environmental and energy policy in agriculture, the Common Agricultural Policy should be shaped differently, extending it with measures promoting the economical use of energy sources. Combining self-exclusive goals, i.e., economic and energy efficiency, requires regulation and support. Food production should use energy efficiently and carefully manage natural resources, and this requires a different policy than the current CAP of the EU. In the context of the current needs in the field of environmental protection and eco-efficiency, the obtained research results may be the basis for considering changes in the agricultural policy and its evolution towards supporting farms with the highest economic and energy efficiency. Using the proposed measure, it is possible to search for the best farms, and also within individual types and through the system of payments for these producers, encourage farmers to apply the most beneficial and energy-saving practices and activities. Farms that will be effective in terms of energy management will also emit relatively less GHG, which will have an impact on lower costs related to the planned fees for the GHG emissions.

To further develop research in the field of economic and energy efficiency, using the developed methodology, comparative analyses should be conducted between individual EU countries and a recommendation should be developed concerning in which regions of Europe particular production directions should be developed due to energy efficiency. Moreover, to deepen the analysis and identify the reasons for the differences in the indicator, a questionnaire survey among farmers is needed.

5. Conclusions

1. In the researched farms, mineral fertilizers, direct energy carriers (engine fuels, electricity, heating fuels), as well as machines and devices, had the largest share in the invested energy. In this area, it is possible to seek efficiency improvement by reducing the most energy-consuming inputs. Concerning fertilizers, buildings, and live labor, it can be argued that with the increase in the scale of production, the share of energy inputs accumulated in them in the energy invested.
2. Farms specialized in the production of cereals and pigs were characterized by the highest EROI index. The highest energy losses were found in farms specialized in the production of slaughter cattle and fruit from the groups with the highest economic size (class III). It was found that with the increase in the scale of production, the EROI index decreased. This is the result of the slower growth of commodity energy in relation to the increasing energy inputs accumulated, among others in mineral fertilizers and larger, more advanced machines and devices used in farms with a larger production scale.
3. The relationship between economic and energy efficiency varied depending on the type and scale of production. For example, in the pigs and dairy cattle farms, energy efficiency decreased and economic efficiency increased as the scale increased. The situation was completely different in fruit farms, where the growing production volume had negative effects on both economic and energy efficiency. It was found that each type of agricultural production had its specificity in this respect. It is therefore difficult to make decisions based on these two categories if you want to take into account both economic and environmental aspects. Therefore, the economic and energy efficiency index (*EEEI*) was proposed.
4. The highest economic and energy efficiency was achieved by farms specialized in the production of cereals. A positive result was also found in the case of pig producers and the smallest dairy farms. The factor differentiating economic and energy efficiency is the type of agricultural production. In the group of effective farms, the most favorable situation was in facilities with the smallest production scale. It can therefore be concluded that in the case of economic and energy efficiency, there are diminishing scale effects.

Author Contributions: Conceptualization, M.W. and B.K.; methodology, M.W. and P.G. (Piotr Gradziuk); software, P.T. and W.B.-G.; validation, M.W., B.K. and P.G. (Piotr Gradziuk); formal analysis, P.G. (Piotr Gradziuk) and B.K.; investigation, D.G.-D., A.L. and W.B.-G.; resources, A.L., D.G.-D. and B.G.; data curation, P.T., B.G. and A.G.; writing—original draft preparation, M.W. and P.G. (Piotr Gołasa); writing—review and editing, B.K. and P.G. (Piotr Gradziuk); visualization, A.W., A.G. and M.G.; supervision, M.W. and P.G. (Piotr Gołasa); project administration, M.W. and A.G.; funding acquisition, A.W. and M.G. All authors have read and agreed to the published version of the manuscript.

Funding: This research received no external funding.

Institutional Review Board Statement: Not applicable.

Informed Consent Statement: Not applicable.

Data Availability Statement: Not applicable.

Conflicts of Interest: The authors declare no conflict of interest.

References

1. Meadows, D.H.; Meadows, D.L.; Randers, J.; Behrens, W.J. *The Limits to Growth; a Report for the Club of Rome's Project on the Predicament of Mankind*; Universe Books: New York, NY, USA, 1972.
2. Balcerowicz, L. Wstęp do wydania polskiego. In *Dysponowanie Wspólnymi Zasobami*; Ostrom, E., Ed.; Wolters Kluwer Polska: Warsaw, Poland, 2013; pp. XI–XXVI.
3. Barbier, E.B. *Economics, Natural Resource Scarcity and Development*; Earthscan Publications: London, UK, 1989.
4. Barnes, P. *Capitalism 3.0: A Guide to Reclaiming the Commons*; Berrett-Koehler Publishers: San Francisco, CA, USA, 2006.

5. Barnett, J.; Morse, C. *Scarcity and Growth: The Economics of Natural Resource Availability*; Published for Resources for the Future, Inc. by The Johns Hopkins Press: Baltimore, MD, USA, 1967.
6. Tyloch, W. Bliski Wschód. In *Historia Starożytna*; Jaczynowska, M., Ed.; Wydawnictwa Szkolne i Pedagogiczne: Warsaw, Poland, 1974; p. 11.
7. Diamond, J. *Collapse: How Societies Choose to Fail or Succeed*; Penguin Putnam Inc.: New York, NY, USA, 2011.
8. Ayres, R.U. *The Bubble Economy: Is Sustainable Growth Possible?* MIT Press: Cambridge, MA, USA, 2014.
9. Czaja, S. Milenijne Cele Rozwoju—Wsparcie rozwoju zrównoważonego czy globalna fikcja. In *Rozwój—Godność Człowieka—Gospodarowanie—Poszanowanie Przyrody. Księga Pamiątkowa Prof. zw. dr hab. Franciszka Piontka*; Piontek, B., Piontek, W., Eds.; Polskie Wydawnictwa Ekonomiczne: Warsaw, Poland, 2007; p. 130.
10. Kiełczewski, D. Homo oeconomicus i homo sustinens jako wyzwania ekonomii zrównoważonego rozwoju. In *Teoretyczne Aspekty Ekonomii Zrównoważonego Rozwoju*; Poskrobko, B., Ed.; Wydawnictwo Wyższej Szkoły Ekonomicznej: Białystok, Poland, 2011.
11. Poskrobko, B. (Ed.) *Ekonomia Zrównoważonego Rozwoju w Świetle Kanonów Nauki*; Wyższa Szkoła Ekonomiczna: Białystok, Poland, 2011; p. 41.
12. Prandecki, K. Dobro wspólne a zrównoważony rozwój. *Optimum. Studia Ekon.* **2016**, *4*, 55–68. [CrossRef]
13. Ostrom, E. *Are Successful Efforts to Manage Common-Pool Problems a Challenge to the Theories of Garrett Hardin and Mancur Olson? Working Paper W85-31, Workshop in Political Theory and Policy Analysis*; Indiana University: Bloomington, IN, USA, 1985.
14. Ostrom, E. *Governing the Commons: The Evolution of Institutions for Collective Action*; Cambridge University Press: Cambridge, MA, USA, 1990.
15. Goban-Klas, T. Społeczeństwo niedoinformowane. Polityka 1988, No. 22 (Supplement to the Issue).
16. Stern, D.I. Economic Growth and Energy. In *Encyclopedia of Energy*; Cleveland, C., Ed.; Elsevier: Amsterdam, The Netherlands, 2004; Volume 2, pp. 35–51.
17. Stern, D.I. *The Role of Energy in Economic Growth*; Australian National University: Canberra, Australia, 2010.
18. Schefold, B. Ecological Problems as a Challenge to Classical and Keynesian Economics. *Metroeconomica* **1985**, *37*, 21–61. [CrossRef]
19. Ramanathan, R. A Multi-factor Efficiency Perspective to the Relationship Among World GDP, Energy Consumption and Carbon Dioxide Emissions. *Technol. Forecast. Soc. Change* **2006**, *73*, 483–494. [CrossRef]
20. Ilahi, S.; Wu, Y.; Raza, M.A.A.; Wei, W.; Imran, M.; Bayasgalankhuu, L. Optimization Approach for Improving Energy Efficiency and Evaluation of Greenhouse Gas Emission of Wheat Crop using Data Envelopment Analysis. *Sustainability* **2019**, *11*, 3409. [CrossRef]
21. Smil, V. *Energy in Nature and Society. General Energetics of Complex Systems*; MIT Press: Cambridge, MA, USA, 2008.
22. Yan, Q.; Yin, J.; Baležentis, T.; Makutėnienė, D.; Štreimikienė, D. Energy-related GHG emission in agriculture of the European countries: An application of the Generalized Divisia Index. *J. Clean. Prod.* **2017**, *164*, 686–694. [CrossRef]
23. Li, T.; Baležentis, T.; Makutėnienė, D.; Streimikiene, D.; Kriščiukaitienė, I. Energy-related CO_2 emission in European Union agriculture: Driving forces and possibilities for reduction. *Appl. Energy* **2016**, *180*, 682–694. [CrossRef]
24. Gołasa, P.; Wysokiński, M.; Bieńkowska-Gołasa, W.; Gradziuk, P.; Golonko, M.; Gradziuk, B.; Siedlecka, A.; Gromada, A. Sources of Greenhouse Gas Emissions in Agriculture, with Particular Emphasis on Emissions from Energy Used. *Energies* **2021**, *14*, 3784. [CrossRef]
25. Gronroos, J.; Seppala, J.; Voutilainen, P.; Seuri, P.; Koikkalainen, K. Energy Use in Conventional and Organic Milk and Rye Bread Production in Finland. *Agric. Ecosyst. Environ.* **2006**, *117*, 109–118. [CrossRef]
26. Moitzi, G.; Martinov, M.; Nozdrovicky, L.; Naghiu, A.; Gronauer, A. Energy Use and Energy Efficiency in Selected Arable Farms in Central and South Eastern Europe. *Agric. Conspec. Sci.* **2014**, *79*, 51–56.
27. Mousavi-Avval, S.H.; Rafiee, S.; Jafari, A.; Mohammadi, A. Energy Efficiency and Cost Analysis of Canola Production in Different Farm Sizes. *Int. J. Energy Environ.* **2011**, *2*, 845–852.
28. Martinho, V.D. Energy consumption across European Union farms: Efficiency in terms of farming output and utilized agricultural area. *Energy* **2016**, *103*, 543–556. [CrossRef]
29. Ramedani, Z.; Rafiee, S.; Heidari, M. An investigation on energy consumption and sensitivity analysis of soybean production farms. *Energy* **2011**, *36*, 6340–6344. [CrossRef]
30. Woods, J.; Williams, A.; Hughes, J.K.; Black, M.; Murphy, R. Energy and the food system. *Philos. Trans. R. Soc. B Biol. Sci.* **2010**, *365*, 2991–3006. [CrossRef]
31. Pelletier, N.; Audsley, E.; Brodt, S.; Garnett, T.; Henrikkson, P.; Kendall, A.; Kramer, K.; Murphy, D.; Nemecek, T.; Troell, M. Energy intensity of agriculture and food systems. *Annu. Rev. Environ. Resour.* **2011**, *36*, 233–246. [CrossRef]
32. Gołębiewska, B. *Organizacyjno-Ekonomiczne Skutki Zróżnicowania Powiązań Gospodarstw Rolniczych z Otoczeniem*; Wydawnictwo SGGW: Warsaw, Poland, 2010; pp. 60–61.
33. Kunachowicz, H.; Przygoda, B.; Nadolna, I.; Iwanow, K. *Tabele Składu i Wartości Odżywczej Żywności*, 2nd ed.; Wydawnictwo Lekarskie PZWL: Warsaw, Poland, 2017.
34. Gołaszewski, J. (Ed.) *Efektywność Energetyczna w Rolnictwie Europejskim—Studium Przypadków*; Wydawnictwo UWM: Olsztyn, Poland, 2013; p. 61.
35. Litwińczuk, Z. *Towaroznawstwo Surowców i Produktów Zwierzęcych*; Powszechne Wydawnictwo Rolnicze i Leśne: Warsaw, Poland, 2012.
36. Regulation No 79/65/EEC of the Council of 15 June 1965 Setting up a Network for the collection of Accountancy Data on the Incomes and Business Operation of Agricultural Holdings in the European Economic Community. Available online: https://eur-lex.europa.eu/legal-content/EN/TXT/PDF/?uri=CELEX:31965R0079&from=EN (accessed on 16 July 2021).

37. Goraj, L.; Olewnik, E. *FADN i Polski FADN (Sieć Danych Rachunkowych z Gospodarstwa Rolnych i System Zbierania Danych Rachunkowych z Gospodarstw Rolnych)*; Wydawnictwo IERiGŻ-PIB: Warsaw, Poland, 2010.
38. Wspólnotowa Typologia Gospodarstw Rolnych—Według Parametru Standardowej Produkcji. Available online: http://fadn.pl/metodyka/typologia/zasady-wtgr-wg-parametru-so/ (accessed on 16 July 2021).
39. Gradziuk, P. *Gospodarcze Znaczenie i Możliwości Wykorzystania Słomy na Cele Energetyczne w Polsce*; IUNG-PIB (Instytut Uprawy Nawożenia i Gleboznawstwa Państwowy Instytut Badawczy): Puławy, Poland, 2015; ISBN 978-83-7562-190-7.
40. Georgescu-Roegen, N. Energy and matter in mankind's technological circuit. *J. Bus. Adm.* **1979**, *10*, 107–127.
41. Cleveland, C.; Costanza, R.; Hall, C.; Kaufmann, R. Energy and the U.S. economy: A biophysical perspective. *Science* **1984**, *225*, 890–897. [CrossRef]
42. Cleveland, C.; Kaufmann, R.; Stern, D. Aggregation and the role of energy in the economy. *Ecol. Econ.* **2000**, *32*, 301–318. [CrossRef]
43. Cleveland, C.; Ruth, M. Capital humano, capital naturale limites biofísicos no processo economic. In *Meio Ambiente*; Cavalcanti, C., Ed.; Desenvolvimento Sustentávele Políticas Públicas: Sao Paulo, Brazil, 1997; pp. 131–164.
44. Czaja, S. Entropijno-energetyczna analiza funkcjonowania i dynamiki systemów społeczno-ekonomicznych. In *Ekologiczne Uwarunkowania Wzrostu Gospodarczego w Ujęciu Współczesnej Teorii Ekonomii*; Czaja, S., Fiedor, B., Jakubczy, Z., Eds.; Wydawnictwo Ekonomia i Środowisko: Białystok, Poland, 1993; p. 96.
45. Czaja, S.; Becla, A. *Ekologiczne Podstawy Procesów Gospodarowania*; Wydawnictwo Akademii Ekonomicznej im, Oskara Langego we Wrocławiu: Wrocław, Poland, 2002; pp. 72–109.
46. Czaja, S.; Jakubczyk, Z. Spory wokół przedmiotu badań ekonomii politycznej (sayowsko-langowskie a ekologiczne ujęcie związków człowiek-przyroda). *Pr. Nauk. Akad. Ekon. Wrocławiu* **1991**, *572*, 15–22.
47. Talukder, B.; van Loon, G.W.; Hipel, K.W. Energy efficiency of agricultural systems in the southwest coastal zone of Bangladesh. *Ecol. Indic.* **2019**, *98*, 641–648. [CrossRef]
48. Aydın, B.; Aktürk, D. Energy use efficiency and economic analysis of peach and cherry production regarding good agricultural practices in Turkey: A case study in Çanakkale province. *Energy* **2018**, *158*, 967–974. [CrossRef]
49. Lewandowska-Czarnecka, A.; Buller, L.S.; Nienartowicz, A.; Piernik, A. Energy and emergy analysis for assessing changes in Polish agriculture since the accession to the European Union. *Ecol. Model.* **2019**, *412*, 108819. [CrossRef]
50. Jat, H.S.; Jat, R.D.; Nanwal, R.K.; Lohan, S.K.; Yadav, A.K.; Poonia, T.; Sharma, P.C.; Jat, M.L. Energy use efficiency of crop residue management for sustainable energy and agriculture conservation in NW India. *Renew. Energy* **2020**, *155*, 1372–1382. [CrossRef]
51. Grassauer, F.; Herndl, M.; Nemecek, T.; Guggenberger, T.; Fritz, C.; Steinwidder, A.; Zollitsch, W. Eco-efficiency of farms considering multiple functions of agriculture: Concept and results from Austrian farms. *J. Clean. Prod.* **2021**, *297*, 126662. [CrossRef]
52. Risoud, B. Energy efficiency of various French farming systems: Questions to sustainability. In Proceedings of the International Conference Sustainable Energy: New Challenges for Agriculture and Implications for Land Use, Wageningen, The Netherlands, 18–20 May 2000.
53. Rahman, S.; Hasan, M.K. Energy productivity and efficiency of wheat farming in Bangladesh. *Energy* **2014**, *66*, 107–114. [CrossRef]
54. Pishgar-Komleh, S.H.; Sefeedpari, P.; Rafiee, S. Energy and economic analysis of rice production under different farm levels in Guilan province of Iran. *Energy* **2011**, *36*, 5824–5831. [CrossRef]
55. Heidari, M.D.; Omid, M.; Akram, A. Energy efficiency and econometric analysis of broiler production farms. *Energy* **2011**, *36*, 6536–6541. [CrossRef]
56. Meul, M.; Nevens, F.; Reheul, D.; Hofman, G. Energy use efficiency of specialised dairy, arable and pig farms in Flanders. *Agric. Ecosyst. Environ.* **2007**, *119*, 135–144. [CrossRef]
57. Pimentel, D.; Hepperly, P.; Hanson, J.; Douds, D.; Seidel, R. Environmental, Energetic, and Economic Comparisons of Organic and Conventional Farming Systems. *BioScience* **2005**, *55*, 573–582. [CrossRef]
58. Falcone, G.; Stillitano, T.; De Luca, A.I.; Di Vita, G.; Iofrida, N.; Strano, A.; Gulisano, G.; Pecorino, B.; D'Amico, M. Energetic and Economic Analyses for Agricultural Management Models: The Calabria PGI Clementine Case Study. *Energies* **2020**, *13*, 1289. [CrossRef]
59. Keummel, B.; Langer, V.; Maid, J.; De Neergaard, A.; Porter, J.R. Energetic, economic and ecological balances of a combined food and energy system. *Biomass Bioenergy* **1998**, *15*, 407–416. [CrossRef]
60. Alluvione, F.; Moretti, B.; Sacco, D.; Grignani, C. EUE (energy use efficiency) of cropping systems for a sustainable agriculture. *Energy* **2011**, *36*, 4468–4481. [CrossRef]
61. Uzal, S. Comparison of the energy efficiency of dairy production farms using different housing systems. *Environ. Prog. Sustain. Energy* **2012**, *32*, 1202–1209. [CrossRef]
62. Uhlin, H.-E. Energy productivity of technological Agriculture-lessons from transition of Swedish Agriculture. *Agric. Ecosyst. Environ.* **1999**, *73*, 63–81. [CrossRef]
63. Ghali, M.; Latruffe, L.; Daniel, K. Efficient Use of Energy Resources on French Farms: An Analysis through Technical Efficiency. *Energies* **2016**, *9*, 601. [CrossRef]
64. Mohammadi, A.; Rafiee, S.; Mohtasebi, S.S.; Mousavi Avval, S.H.; Rafiee, H. Energy efficiency improvement and input cost saving in kiwifruit production using Data Envelopment Analysis approach. *Renew. Energy* **2011**, *36*, 2573–2579. [CrossRef]
65. Kuesters, J.; Lammel, J. Investigations of the energy efficiency of the production of winter wheat and sugar beet in Europe. *Eur. J. Agron.* **1999**, *11*, 35–43. [CrossRef]

energies

MDPI

Article

Regional Diversification of Electricity Consumption in Rural Areas of Poland

Paweł Siemiński [1,*], Jakub Hadyński [1], Jarosław Lira [1] and Anna Rosa [2]

[1] Department of Economics and Economic Polity in Agribusiness, Poznań University of Life Sciences, Wojska Polskiego 28, 60-637 Poznań, Poland; jakub.hadynski@up.poznan.pl (J.H.); jaroslaw.lira@up.poznan.pl (J.L.)
[2] Institute of Rural and Agriculture Development, Polish Academy of Sciences, Nowy Świat 72, 00-330 Warsaw, Poland; arosa@irwirpan.waw.pl
* Correspondence: pawel.sieminski@up.poznan.pl

Abstract: Access to energy, including electricity, determines countries' socio-economic development. The growing demand for electricity translates into environmental problems. Energy is therefore a crucial element of the European Union's sustainable development strategy. This article aims to present the changes taking place in the electricity market in Poland considering the goals of the energy policy until 2040. This is the basis for the determination of the scale of processes taking place in the Polish energy sector from two perspectives, i.e., the production of electricity considering its level and energy carriers used, and the consumption of electricity in households depending on their location (rural vs. urban areas). The research was conducted at the regional level (NUTS 2 until 2017) in Poland. Secondary data from the Central Statistical Office (GUS) contained in the Local Data Bank were used, along with information from the European Commission and Eurostat websites. Results of the study made it possible to identify areas in which a greater environmental load is observed due to increasing electricity consumption. The coefficient of localization and concentration (by Florence) and the rate of change were applied. These results indicate that, in Poland, it is now the rural areas that have a greater negative environmental impact than urban areas, resulting from differences in unit energy consumption. Compared to the other provinces, rural areas of Podlaskie province had the highest rate of growth in energy consumption in the years 2004–2019, with an annual average of almost 20%.

Keywords: electricity; production; consumption; rural areas; energy carriers; Poland

Citation: Siemiński, P.; Hadyński, J.; Lira, J.; Rosa, A. Regional Diversification of Electricity Consumption in Rural Areas of Poland. *Energies* **2021**, *14*, 8532. https://doi.org/10.3390/en14248532

Academic Editor: Ashish Prakash Agalgaonkar

Received: 6 October 2021
Accepted: 14 December 2021
Published: 17 December 2021

1. Introduction

Due to development, more and more energy resources are necessary to satisfy social needs as well as production. There is a growth trend in electricity consumption all over the world. Abolhosseini et al. [1] indicate that electricity consumption will constitute an increasing share of global energy demand over the next two decades, contributing to climate change and environmental pollution, and constituting a serious threat to human health. Energy is therefore a crucial element of the European Union's sustainable development strategy.

Climate problems in EU countries are noticeable as issues that may significantly affect or limit future socio-economic development. The cause of climate problems is the increasing emission of greenhouse gases due to anthropogenic activities directly related to the combustion of fossil fuels for electricity, heat, and transport. However, it is primarily the combustion of fossil fuels that causes atmospheric pollutants that are harmful to the environment and human health. Fossil fuels play a dominant role in global energy systems [2]. They are responsible for more than 70% of world greenhouse gas emissions [3]. In 2019, the largest share of greenhouse gas emissions, 77%, was those related to energy production [4], while, in 2015, this share was 78% [5].

Coal is the most damaging fossil fuel for environmental concerns and carbon emissions. In many countries, it is increasingly being replaced by natural gas for electricity production. However, reserves of coal could last a long time and still play a role in meeting primary energy demand [6]. Poland is rich in energy resources, with hard coal and lignite, i.e., energy resources with a potentially high environmental impact, predominating [7].

The electricity market has a dominant position among other energy markets (heat, car fuel, etc.) in terms of the scale of production.

In the EU, the greatest role is played by the conventional energy sources, while a decrease in production has recently been observed (e.g., in the years 2017–2019 mean annual production was approximately 7% lower at 1.166 million GWh). Nuclear energy (0.729 GWh in 2019) is the second most important energy source in the EU.

Conventional energy production accounted for 42.8% of the total, while nuclear power rose to 26.7% of the total. In the EU countries, a considerable role is also played by wind power (13%) and hydro power (12%) (Figure 1).

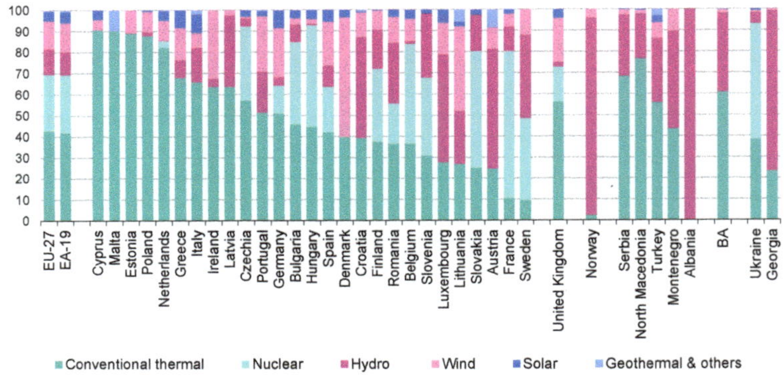

Figure 1. Energy production in the EU countries in 2019 (%). Source: Eurostat.

In the EU, a significant problem in energy production is related to limited and rapidly depleting energy resources, thus resulting in the increasing dependence on imported energy. Net imports cover over 50% of gross available energy in the EU, with the energy dependence rate exceeding 50.0% [8].

Electricity consumption in the EU has been characterized by considerable changes. In the years 2005–2007, it increased rapidly, reaching its highest level of over 76,000 tonnes of oil equivalent (TOE) in 2007. This led to decisive action aimed at reducing this trend. In the following years, a decrease was observed, which may be connected with the consequences of the economic crisis in Europe. Slight increases in the successive years did not cause an increase in energy consumption over the 2007 level, whereas, at present (Table 1), a marked economic slowdown triggered by the negative effects of the Covid-19 pandemic has led to a decrease in energy consumption and a general reduction in negative anthropogenic impacts on the natural environment.

The data in Table 1 indicate that Poland is one of the EU's highest electricity consumers. At the same time, it predominantly relies on conventional energy sources (Figure 1). Solutions and regulations are being sought to minimize electricity consumption, regardless of the current condition of the economy. However, it is crucial to ensure the country's energy security [9].

EU policies set objectives that are important for global priorities for the protection and conservation of the natural environment. The effects of these policies can be identified, among others, in Poland, where dynamic changes in the diversification of energy generation are taking place, which is confirmed by the research. It should also be noted that the energy market in Poland is undergoing dynamic changes which greatly affect rural areas. At the same time, the level of regional development determines the development of the electricity market.

Table 1. Electricity and heat generation in the EU countries (thousand tonnes of oil equivalent).

Country/Period	2017	2018	2019
EU-27	607,009.750	594,147.245	578,490.641
Belgium	18,552.910	15,172.833	19,293.727
Bulgaria	11,289.852	11,207.263	10,893.471
Czechia	21,816.784	22,035.561	22,233.952
Denmark	6704.290	6648.120	6148.094
Germany	125,613.205	121,608.696	112,069.343
Estonia	3679.360	3558.216	2293.173
Ireland	4801.722	4617.431	4472.922
Greece	10,139.682	9765.473	10,615.034
Spain	50,553.625	48,240.888	46,262.363
Francje	130,171.155	132,728.853	129,450.691
Croatia	1856.582	1930.725	1962.438
Italy	55,904.390	53,919.876	53,787.109
Cyprus	1059.351	1074.845	1062.102
Latvia	1588.248	1589.194	1531.975
Lithuania	1762.408	1654.311	1604.404
Luxsemborg	349.600	354.453	347.244
Hungary	8528.965	8251.994	8393.807
Malta	297.952	314.975	334.322
Holland	21,907.670	20,566.062	21,192.037
Austria	10,549.253	9975.852	10,341.393
Poland	41,250.397	40,852.733	38,536.183
Portugal	9798.661	9323.320	8047.017
Romonia	13,005.711	12,964.004	12,106.492
Slovenia	3192.512	3095.545	3052.134
Slovakia	6924.131	6451.268	6747.014
Finlandia	15,192.384	15,641.939	15,260.934
Switzerland	30,518.950	30,602.815	30,766.528
Descriptive statistic			
minimum	297.952	314.975	334.322
mean	22,481.843	22,005.454	21,437.256
median	10,139.682	9765.473	10,341.393
variation efficiency (%)	150.46	152.65	149.40
maximum	130,171.155	132,728.853	129,450.691

Source of data: Eurostat.

This article aims to present:

- Changes in the electricity production market in Poland in the period from Poland's accession to the EU (considering its level and energy sources used);
- Electricity consumption in households in Poland in terms of time and region (depending on the location—rural vs. urban areas).

The research was conducted at the regional level (NUTS 2 until 2017). Secondary data from the Central Statistical Office (GUS) in the Local Data Bank were used, along with information from the European Commission and Eurostat websites.

The research problem is important from the point of view of sustainable development and environmental protection as well as the economic security of electricity consumers. This paper does not discuss all the aspects of energy economics; however, the analysis covered two key issues, i.e., the production and consumption of electricity.

2. Background and Literature Review

2.1. EU Energy Policy

Energy policy must be long-term and beneficial to all member states. Accordingly, the EU is implementing an energy policy covering the full range of sources, from fossil fuels to nuclear and renewable energy. The goal is to transform economies into low-energy economies while ensuring greater security, competitiveness and sustainability of the energy used.

It may be assumed that the EU project, consisting in the integration of economies after WWII and peaceful cooperation between countries, was based, among other things, on energy policy initially related to the coal market and, subsequently, also to nuclear energy. The foundations for this process were laid by the Treaty of Paris, establishing the European Coal and Steel Community (ECSC) signed in 1951, which entered into force on 23 July 1952. At that time, coal was the main energy source worldwide, and it accounted for approximately 70% of the total. The energy balance of the six member states showed considerable differences between individual countries, since in France it was approximately 60%, whereas in Luxembourg it amounted to 95% [10]. In view of the current actions aiming at the establishment of a common energy market for the EU member states, cooperation within ECSC may be considered as the foundation for a common energy policy [11]. In turn, the Energy Card Treaty signed in 1991 by 46 countries is a document of significant importance, providing grounds for actions aimed at the improvement of energy efficiency and thus also the energy economy [11,12]. In their study, Grycan et al. indicated the following most important regulations in this respect [12]:

- The EU climate and energy package (September 2007), known for its 20-20-20 targets or the three "20 targets" by 2020, assuming a reduction in greenhouse gas emissions by 20% in relation to the levels of 1990, reduction in energy consumption by 20% compared to the EU forecasts for 2020, and an increase in renewable energy to 20% of the total EU energy consumption;
- The third energy package (September 2007), stipulating ownership unbundling, separating electricity and gas production from their transmission and a framework for enhanced mutual cooperation and aid in the case of threatened energy supplies;
- The climate and energy package (January 2008) comprising six proposals of legal acts concerning the promotion of renewable energy, vehicle CO_2 emission standards, fuel specifications, joint efforts to reduce greenhouse gas emissions and development of carbon capture and storage as well as a review of the European system of emissions trading;
- The Commission Communication "Energy 2020" (November 2010), defining five current priorities for the energy sector, i.e., energy saving in the transport and construction sectors, establishment of an internal energy market along with the respective infrastructure, execution of the common European energy policy and European leadership in energy generation technologies and innovations as well as guaranteeing access to secure, reliable and competitively priced energy for Europeans;
- The third energy package (March 2011), comprising two market directives, two transmission regulations and a regulation establishing the Agency for the Cooperation of Energy Regulators;
- The Communication from the European Commission on the plan to ensure energy efficiency and a roadmap for moving to a competitive low carbon economy in 2050 (March 2011), presenting a proposal to the EU member states for a joint strategy to convert to low carbon energy economies;
- Directive 2012/27/EU of the European Parliament and Council on energy efficiency, (10.2012) amending Directives 2009/125/EC and 2010/30/EU and repealing Directives 2004/8/EC and 2006/32/EC.

It seems that a milestone in the EU energy policy was the adoption of a strategy establishing the energy union, which reflected the new concept of energy security for the EU countries [13]. Its aim was to establish the energy union, which would provide secure, sustainable, competitive and reasonably priced energy supplies to EU households and enterprises.

A natural continuation of actions related to the development of the energy union was provided by the "clean energy for all Europeans" package [14]. Its objectives included, first of all, energy efficiency and increased energy production from renewable sources, with the simultaneous assurance of cheap energy accessible to all consumers. Thanks to the implementation of measures leading to modernization of the EU economies (Figure 2), it is assumed that the intensity of CO_2 emissions will be reduced by over 40%, while renewable sources will account for approximately 50% all electricity [15,16]. A significant component in this reform package for the energy market is related to the need for a 10-year integrated national energy and climate plan for the years 2021–2030.

The EU energy policy in the 2030 perspective—setting out national strategies—includes three primary goals [17]:

- Minimization of energy prices;
- Ensuring an appropriate level of energy security;
- Minimizing the consequences of energy technologies that are harmful to the environment.

Both for the Polish economy and the entire population, an important planning document is the Energy Policy of Poland by 2040 (PEP 2040) [18], which has been considerably modified by arrangements in the climate and energy policy adopted at the EU level. The essence of Poland' energy policy by the year 2040 is based on three pillars:

I. Just transformation;
II. A zero-emission power engineering system;
III. Good air quality.

The above-mentioned three pillars of the energy policy comprise eight specific goals, of which some directly concern the power sector, including green power engineering, which jointly constitute an energy supply chain, starting from the acquisition of raw materials and energy generation and supply, as well as energy use and sale, at the same time maintaining energy security for consumers.

2.2. Economic Transformation for Sustainable Development

The EU energy policy indicates the direction of activities oriented towards sustainable development, of which energy is an essential factor.

The concept of sustainable development [19–22] in the following decades was acknowledged and incorporated into various forms of socio-economic development.

The concept of sustainable development includes the following aspects: humans as subjects affecting the environment, our planet as an object of human activity and the mode of action, i.e., partnership, since only integrated measures will facilitate achieving the goal of the concept, i.e., sustainable prosperity [23]. Thus, sustainable development promotes environmental protection and, by preventing the over-exploitation of natural resources, it protects them for future generations.

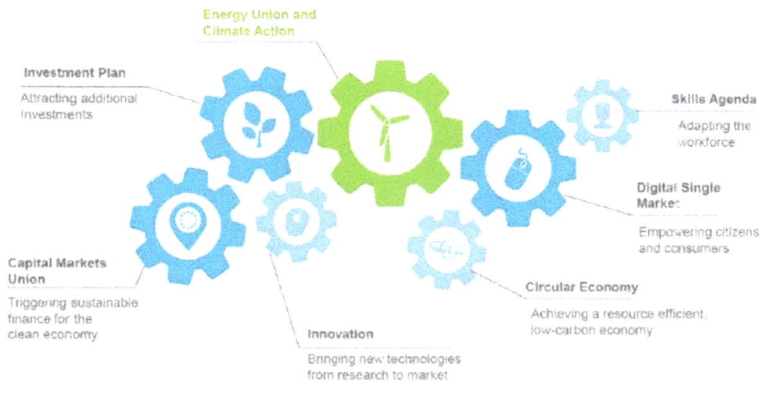

Figure 2. Role of the energy union and climate action. Source: [23].

In the EU countries, the concept of sustainable development has been an important part of economic development strategies for several decades. It is incorporated into the strategic socio-economic policies in all these countries, while being based on climate stability. When considering the implementation of sustainable development, we need to focus on the Europe 2020 strategy and the European Green Deal. One of the leading directions for the implementation of sustainable development within the Europe 2020 strategy has been connected with meeting the requirements of the climate and energy policy in the EU. In terms of the operations, it is intended to achieve goals such as a reduction in CO_2 emissions and reduced consumption of fossil fuels, particularly coal, since this exhibits the greatest emission loads. Such goals are to promote a low-emission economy, protecting sustainable resources both for the present and future generations. Figure 3 presents the primary environmental goals of Agenda 2030 and the Europe 2020 strategy as an intermediate stage in the execution of the ultimate goals. The goals established provide direction to national changes in the energy policy in the EU countries.

There are paths ("trajectories") to the RES target, i.e., presentations of the rate of implementation of the contribution in the period 2021–2030. These are agreed on individually with each country; however, they are not arbitrary. The regulations indicate the minimum levels of RES share in specific years [24]:

- In 2022—at least 18% of the planned (for 2030) RES growth share in the period 2021–2030;
- In 2025, at least 43%;
- In 2027, at least 65%.

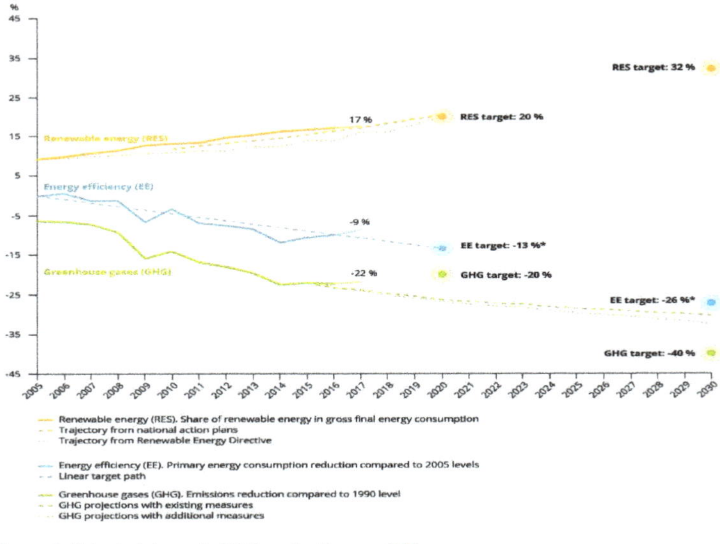

Figure 3. Selected Agenda 2030 goals. Source: [25].

Within the next three decades, more ambitious goals have been proposed for sustainable development in the EU countries by updating the list of climate and environmental problems which need to be solved, as specified in the European Green Deal strategy [26–29].

The European Green Deal (Figure 4) is a new strategy for growth, aimed at transforming the EU into a fair and prosperous society with a technologically advanced, resource-efficient and competitive economy, which will reach zero net greenhouse gas emissions by 2050 and within which economic growth will not be dependent on the consumption of natural resources [26].

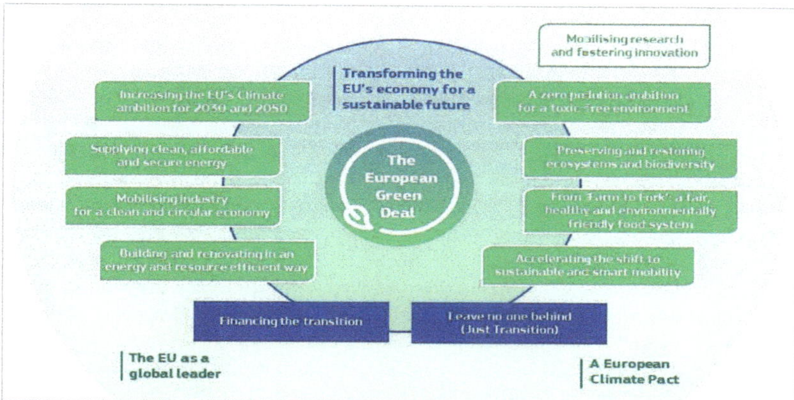

Figure 4. The European Green Deal. Source: [26].

Such a goal is a very ambitious developmental challenge both economically and socially. However, it is advisable to undertake such challenges considering potential benefits resulting from the reduction or possibly even halting of the progressive increase in global temperatures and environmental degradation, as well as the resulting climate change. The European Green Deal 2050 adopted by the EU is in line with the guidelines for the protection of climate and the natural environment within the UN 2030 sustainable development agenda, i.e., actions undertaken on the global scale in terms of environmental protection and responsible environmental management.

The essence of the EU economic transformation aimed at a sustainable future will be based on the joint funding of green investments and the simultaneous financial involvement of public and private stakeholders in the transition process. Moreover, it needs to be a just transformation, concentrating on the regions and sectors that will suffer most from its consequences due to their dependence on fossil fuels and high-emission processes. Within the cohesion policy, the Just Transformation Fund (JTF) is a new financial instrument covering the years 2021–2027, providing financial support to regions suffering serious socio-economic problems resulting from the transformation aimed at climate neutrality [27–29]. The Fund resources will curb the negative social, economic and environmental impacts of the energy transformation. Most probably, the funds will provide support to beneficiaries from the Śląskie, Dolnośląskie and the Wielkopolskie provinces, while the government is also trying to make the support available also to the Lubelskie, Łódzkie and Małopolskie provinces. Nevertheless, the national budgets will continue to play a key role in the green transformation processes, among other things using green budgeting tools, facilitating the transition of public investment and consumption and tax systems to further environmental priorities and reduce further degradation of the natural environment.

Realization of the economic transformation towards a sustainable future for the EU countries will be based on the following strategic tasks:

- Adoption of more ambitious EU climate goals for 2030 and 2050:
- Provision of clean, reasonably priced and secure energy;
- Mobilization of the secondary sector to implement the circular economy model;
- Building and renovating in an energy and resource-efficient manner;
- Zero emissions for a non-toxic environment;
- Conservation and restoration of ecosystems and biodiversity;
- The From Farm to Fork Strategy—a fair, healthy and environmentally friendly food production system;
- Accelerating the shift to sustainable and smart mobility;
- Supporting research and fostering innovations.

Gradual and successful implementation of the strategic tasks in the following years will result in positive changes, while, at the same time, the EU countries may become world leaders in preventing climate change and environmental degradation.

In view of these expected changes, the Executive Vice-President of the European Commission Frans Timmermans stated: *"We must show solidarity with the most affected regions in Europe, such as coal mining regions and others, to make sure the Green Deal gets everyone's full support and has a chance to become a reality"* [30].

As indicated by Jonker and Krukowska [31], the creation of "green economy" in all industrialized countries will establish a boundary between the two eras. A green economy is a circular economy. Its essence is manifested in the fact that changes progress from one phase to the next, followed by the return to the transitional phase, which reverses the previous linear economic system if changes follow the "make-take-dispose", step-by-step direction. The term green economy was first used by David Pearce [32], who observed that sustainable development is impossible in the current economy dependent on depleting resources such as oil and coal [33]. The green economy model involves low carbon dioxide emissions, efficient utilization of natural resources and the inclusion of all groups and individuals.

A green transformation both of the global and EU economies is becoming an increasingly accepted form of economic development. Such a socio-economic development is more and more universally accepted thanks to the potential positive effects of the green economy, with its environmentally friendly development pattern. Another crucial aspect is connected with the fact that the green transformation is to be a fair one, which means that inhabitants of the affected regions will not be disadvantaged and left alone but will be presented with a viable alternative. Such an approach seems to be highly beneficial for Poland, since our country may be the greatest beneficiary of this fund, with the allocated financial resources reaching two billion euro, i.e., approximately a quarter of the entire JTF. However, it needs to be remembered that the real absorption of funds is dependent on the acceptance of the Paris Agreement goals by the prospective beneficiary. For the time being, Poland is the only country which has refused to join in the climate neutrality goal, arguing that this stems from the specific character of its domestic energy sector, which needs much more time to adapt to changes, while automatic acceleration of energy transformation would involve huge social costs, greatly exceeding the amount allocated to Poland in the JTF. As a result, the real funds would be only 50% of those originally allocated.

The success of the green transformation in the EU countries, aimed at achieving climate neutrality by 2050 through the decarbonization of economies, will depend to a considerable extent on transformation programs such as the Regional Just Transformation Plans. In the opinion of officials from the Ministry of Development Funds and Regional Policy " . . . *the aim is to develop plans, which will ensure sustainable and just solutions for mining regions. The Regional Just Transformation Plans need to be compatible with the national energy and climate plans"* [34].

2.3. *Challenges in the Strategic Goal: Provision of Clean, Reasonably Priced, and Secure Energy*

In view of prevention of further climate change and progressive degradation of the natural environment throughout the EU, the primary issue is to implement the strategic goal ensuring the "provision of clean, reasonably priced and secure energy." EU countries' previous experience indicates that over 75% of greenhouse gas emissions are generated by the production and use of energy [35]. Such a high share of greenhouse gas emissions from the energy sector results from the predominant use of carbon energy carriers. To date, less than 18% of gross final energy consumption in the EU countries in 2017 has come from renewable energy sources [36]. The forecast objective in this respect in 2020 was 20%. In view of the current data in some EU countries, there is a risk of failing to reach the expected objective, with a serious risk of such failure indicated for such countries as Belgium, France and Poland, while a moderate risk is suggested for Luxembourg and the Netherlands. In turn, based on estimates for all the EU-28 countries, the probable share of renewable energy in 2020 was approximately 23% (For the group of EU-27 countries (excluding Great Britain) this level will amount to approximately 24%.); thus, it would be higher than expected [37]. This indicates the internal diversification between the EU-28 countries in terms of the development of renewable energy and achieving the adopted goals. The energy sector based on renewable energy needs to be further developed. To a considerable degree, this will facilitate the elimination of coal as the main source of energy, while at the same time reducing the emission levels in the economy.

The transition to clean energy is a long-term process, consisting of the transformation of the power engineering system; while ensuring the effectiveness of these changes, it will be necessary to involve all consumers and gain their acceptance thanks to the economic benefits offered and the awareness of the need for change. In the transition to clean energy, a key role will be played by renewable energy sources. The European Commission Communication [29] expressed an opinion that: "*Increasing offshore wind production will be essential, building on regional cooperation between Member States. The smart integration of renewables, energy efficiency and other sustainable solutions across sectors will help to achieve decarbonisation at the lowest possible cost. The rapid decrease in the cost of renewables, combined with improved design of support policies, has already reduced the impact on households' energy bills of renewables deployment*".

Nevertheless, some households still face the problem of energy poverty [38]. In view of the above, considering the poorest part of the population, special measures need to be introduced to protect financially stressed households when they cannot afford indispensable energy services to maintain a basic standard of living. The most important role is played by effective initiatives, e.g., those which may in the future reduce energy bills and which, through specific solutions, will have a positive and advantageous impact on the condition of the natural environment.

At present, individual countries are only starting the long process of transition to a low-emission power generation system and the low-emission economy. Profound changes related to energy transformation will require considerable public and political support. It is energy prices and the costs of energy transformation that should stimulate market transformation to achieve a climate-neutral economy within the next few decades. To ensure the success of the entire transition process, it is essential for energy consumers, both households and businesses, to have access to reasonably priced energy. In the last five years, an upward trend for wholesale electricity prices has been evident in the EU countries, followed by rising retail prices for end users. A culmination of the wholesale price increases was recorded in 2018, followed in 2019 by a reduction in prices mainly thanks to decreasing consumer demand as well as the rapid increase in the supply of renewable energy. In the EU countries, this phenomenon was far from universal; as a result, the diversification in price levels between the regional markets grew. In the first half of 2020 compared to the analogous period in 2019, prices dropped by 30% in some regional markets in southern Europe and up to 70% in certain northern regions [39]. This diversified reduction is explained by insufficient interconnection capacity, differences in the production of renewable energy on individual markets and the considerable growth of CO_2 prices, which had a considerable impact particularly in the EU countries with greater shares of fossil fuel in their energy basket. Thus, it stresses the need for additional investment in grid flexibility, transboundary transmission capacity and renewable energy sources, particularly in EU countries that are falling behind in this respect, which should, in the future, result in greater price integration of electricity between the regional markets and benefits for consumers. Taking the existing needs into consideration, achieving climate neutrality requires an intelligent infrastructure. Strengthening transboundary and regional cooperation between countries will benefit from the transition to affordable clean energy. Thus, it will be necessary to review the frameworks regulating the energy infrastructure, including the TEN-E regulation 12, in order to ensure cohesion aimed at climate neutrality. These frameworks need to promote innovative technologies and infrastructure, such as intelligent grids, hydrogen networks or the capture, storage and disposal of carbon dioxide as well as energy storage, while facilitating sector integration. Nevertheless, the general public has to realize that certain existing facilities and infrastructure will have to be modernized to further serve their role and resist climate change.

Energy consumers may be concerned about price levels on the retail electricity markets since, in the last decade, these have kept rising. In the years 2010–2019, electricity prices for households were increasing at 2.3% annually, while the general prices of consumer goods increased by 1.4% [39]. Over the same period, an increase was also recorded for electricity prices for business consumers; however, in this case, the annual mean growth rate was 1.1%. In turn, for energy consumers such as large industrial enterprises, energy prices decreased by 5%, in the 2010–2019 period; thus, they were advantageous for this group of consumers. Energy prices for the end users are determined by a variety of factors. These obviously include wholesale prices, but also grid charges as well as taxes and other fees, such as the current subsidies for renewable energy or the costs of energy-supply commercialization. At present, it is taxes and charges that are the most important cause of differences in retail prices at the regional level.

Results of the latest analyses of energy prices in the EU have confirmed considerable differences in taxation of electricity consumption between individual EU countries and, as a consequence, the impact of this element on retail energy prices. In 2019, environmental taxes paid by households ranged from 1 EUR/MWh in Luxembourg to 118 EUR/MWh in Denmark, while VAT rates ranged from 5% in Malta to 27% in Hungary. Fees charged on renewable energy range from 3 EUR/MWh in Sweden to 67 EUR/MWh in Germany. Moreover, in most countries, taxes and fees, as well as grid charges (i.e., the two price elements defined based on regulatory measures), considerably exceed the element imposed on energy and determined by market mechanisms.

3. Materials and Methods

This study is based on data from the Local Data Bank of the Central Statistical Office (Statistics Poland), titled *Electricity in households by consumer location* [40]. Moreover, both national and international reports were used along with numerous studies concerning electricity production and consumption in general, particularly in rural areas. The problems investigated were analyzed in terms of development conditions resulting from the EU development strategy by 2050 referred to as the European Green Deal and the Energy Policy of Poland by 2040.

The primary aim of this study was to identify homogeneous groups of provinces (województwa) characterized by a comparable rate of change in electricity consumption considering changes observed in rural areas. Moreover, the level, directions and rate of changes in electricity generation and consumption were also investigated.

The subject of studies on the energy economy presented in this paper are related to energy production, taking into consideration energy carriers used in the generation processes and electricity consumption in households by consumer location. Thus, energy consumption was analyzed separately for rural areas compared to urban areas or the overall consumption on the national level.

In the case of electricity production, its generation was analyzed for two periods, 2004 and 2019. In this way, the direction of change, the dynamics and the mean annual rate of change in electricity production in Poland were compared for these two periods, considering the environmental impact of energy generation processes, and assuming that an advantageous situation would be manifested in a situation considered to be desirable, i.e., the share of energy generated using fossil fuels will decrease in successive years, being replaced by renewable energy. In the case of electricity consumption in households, the time frame for the analyses covered the years 2004–2019.

A separate analysis was conducted for energy consumption in households in rural areas and in urban areas, thus identifying existing trends in this respect. It specified which of the areas contributes to a greater environmental load in absolute terms, resulting from higher electricity consumption. The rate of change in electricity consumption was also determined on a regional scale divided into rural and urban areas, which made it possible to identify which of the areas contributed to a greater environmental load related to growing electricity consumption in the period investigated. With the implementation of development assumptions stipulated by the EGD 2050 and a reduction in the negative anthropogenic impact on the environment, particularly by supplying clean and environmentally safe energy, it is crucial to have knowledge on the effect of these phenomena both in rural and urban areas. Insight into this problem will facilitate appropriate and adequate preventive or countermeasures addressing the needs identified. The degree of similarity was determined for the distribution of electricity consumption in the spatial unit system in rural and urban areas, applying Florence's coefficient [L_1] as presented by [41]:

$$L_1 = \frac{1}{200}\sum_{i=1}^{n}|u_{ir} - u_{is}| \tag{1}$$

where $u_{ir} = \frac{y_{ir}}{\sum_{j=1}^{n} y_{jr}} \cdot 100\%$ and $u_{is} = \frac{y_{is}}{\sum_{j=1}^{n} y_{js}} \cdot 100\%$ are percentages of the Y_r and Y_s features, respectively, and n denotes the number of objects ($i = 1, 2, \cdots, n$).

The total coefficient of localization L_1 assumes values in the range of <0, 1>, with the closer its value is to one, the greater the degree of discrepancy for the characteristic, while the closer the value is to zero, the greater the similarity.

Moreover, the degree of concentration of absolute electricity consumption in the system of spatial units in rural and urban areas made it possible to identify the degree of discrepancy in electricity consumption by provinces, while Florence's coefficient [K_1] was applied as proposed by [41]:

$$K_1 = \frac{1}{200}\sum_{i=1}^{n}\left|u_i - \frac{100}{n}\right| \tag{2}$$

where $u_i = \frac{y_i}{\sum_{j=1}^{n} y_j} \cdot 100\%$ is the percentage of the examined feature Y, and n denotes the number of objects ($i = 1, 2, \cdots, n$).

Values of the coefficient are found within the range of <0, 1>, with the coefficient value of 0 indicating a uniform distribution, i.e., lack of concentration, while the value of 1 denotes complete non-uniformity, i.e., complete concentration of the trait analyzed.

Analysis of the rate of changes made it possible to identify the existing trends in energy consumption in urban and rural areas. This was calculated on the basis of values throughout the entire period analyzed, which covered the years 2004–2019, applying the formula [41]:

$$rg = \frac{-3m + \sqrt{9m^2 + 24\mathrm{m}\ (\mathrm{n}-1)(\frac{1}{y_1}\ \sum_{t=1}^{n} y_t - n)}}{2m(n-1)} \tag{3}$$

where: y_t denoted the observation of the feature Y in the period t, $m = n(n+1)$ and n is the number of periods ($i = 1, 2, \cdots, n$).

Moreover, using the rate of change, the provinces were divided into homogeneous classes in terms of the scale of changes observed. This made it possible to distinguish regions of greater area, in relation to which similar instruments may be applied in the future to boost environmentally friendly actions. The classification of provinces from the high to the low rate of change was based, e.g., on an analysis of differences in the values of the rate of change. After the ordering of provinces according to the non-growing values of the rate of changes, differences were calculated between its values for neighbouring provinces, i.e., for the first and second, the second and the third, etc. Analyzing successive differences starting from the first (the difference between the second and the first province), a markedly higher value of this difference from the others will make it possible to distinguish classes of province with the highest rate of change, while the successive differences make it possible to identify the successive class.

4. Results

4.1. Electricity Production in Poland

Analysis of the national energy system in 2004 and 2019 indicates advantageous changes in Polish electric power engineering in view of the goals of the energy policy by 2040. Between 2004 and 2019, a simultaneous increase was recorded in installed capacity and electricity production. The growth dynamics of installed power in that period amounted to almost 35%, while electricity production increased by 3.5%; thus, it was ten times lower. In 2004, total installed power was almost 35 GW, while it grew in the successive years to reach almost 47 GW in 2019. Both in 2004 and 2019, the energy system was based primarily on carbon-based sources, with their share of the installed power types in power plants amounting to almost 60% and 50% for coal and 25% and 18% for lignite. Thus, in the next fifteen years, the share of carbon-based sources decreased by a total of approximately 17 p.p., including almost 10 p.p. for coal and by approximately 7 p.p. for lignite. In the analogous period, a gradual increase was observed in installed power based on gas-fired power plants, from less than 770 MW to almost 2800 MW, while their share in the structure amounted to 2.2% and 6%, respectively, i.e., it continued to be rather marginal despite the increase.

The analysis of data also showed that, in the following years, Poland realized that the process of gradual elimination of lignite as an energy source and a decreasing role of lignite-fired power plants in the national energy production system is indicated by the negative growth rate of installed power, the mean annual value of which was −0.33% in the years 2004–2019.

In view of the challenges resulting from the implementation of the Polish energy policy by 2040 in the successive decades, in line with the guidelines in the EGD 2050, particularly provision of clean and secure energy, in the last fifteen years advantageous changes were introduced in the national electric power engineering system in Poland. First, in the successive years since Poland's accession to the EU, interest in renewable energy has been increasing.

According Arıoğlu et al. [42], renewable energy is becoming the fastest growing energy source in the world. Gielen et al. [43] also note that renewable energy can meet two-thirds of the total global energy demand and, to a large extent, contribute to the reduction in greenhouse gas emissions responsible for climate change.

While, in 2004, wind power plants and renewable energy systems did not exist in Poland on a broader scale, by 2019 they had become a relatively important source of power in the national energy system. In 2019, wind power plants and other renewable energy sources accounted for almost 7500 MW, and their share in the total structure was 16%, i.e., slightly less than the share of lignite-fired power plants (17.9%), at the same time being over two-and-a-half times greater than gas-fired power plants (6%) (Table 2). Simultaneously, in 2019, the production of what is defined as clean and secure energy in the regulations adopted within the EGD 2050 amounted to more than 14,000 GWh in Poland, at 9% of energy generation. This share exceeded that of energy from gas-fired power plants by 1.5 p.p. and was as much as six times greater than energy from hydro power (1.5%). The conclusions provided by the observations of changes to hydro power in Poland between 2004 and 2019 are disturbing. This results from the drop in the energy generated from more than 3500 GWh in 2004 to less than 2500 GWh in 2019. The negative direction of change confirms the negative mean annual rate of change, which amounted to almost (−2.4%) and, at the same time, was the highest of all other energy sources. This is a greater reduction than in the case of coal-fired power plants (mean annual decrease of less than minus 0.6%) and lignite-fired power plants (mean annual decrease of approximately minus 1.5%). In view of the energy generation conditions in Poland, the most pressing need to reduce the energy sources in the national energy production system concerns coal- and lignite-fired power plants. Experience from the last fifteen years shows that this process is taking place in Poland; however, at a very slow rate. In view of the implementation of the objectives of the national energy plan PEP2040 and EGD 2050 in the coming years by providing consumers with clean and secure energy, further development of wind power and the use of other renewable energy sources needs to be promoted. together with the further development of gas-fired power plants, since this type of energy has a much less negative environmental impact. Experience in this respect obtained in the last fifteen years indicates that, in Poland, the use of gas in energy generation was developing dynamically, as indicated by the dynamics of change, amounting to almost 320%, as well as the high mean annual rate of change, which was positive and amounted to approximately 8% for this type of installation.

Table 2. The National Electric Power System in Poland in 2004 and 2019.

Specification	Installed Power 2004		Production 2004		Installed Power 2019		Production 2019		Dynamics of Change 2019 [2014 = 100]		Average Annual Rate of Change [%]	
	MW	%	GWh	%	MW	%	GWh	%	Installed Power	Production	Installed Power	Production
Professional power plants	32,162	92.6	144,821	94.4	36,675	78.4	134,245	84.6	114	92.7	0.88	−0.5
Professional hydropower plants	2168	6.7	3525	2.4	2346	6.4	2454	1.8	108.2	69.6	0.53	−2.38
Commercial thermal power plants:	29,994	93.3	141,296	97.6	34,329	93.6	131,791	98.2	114.7	93.3	0.92	−0.46
- hard coal	20,411	68.1	85,370	60.4	23,159	67.5	78,190	59.3	113.8	91.6	0.87	−0.58
- lignite	8856	29.5	52,136	36.9	8382	24.4	41,502	31.5	95.2	79.6	−0.33	−1.51
- gas	727	2.4	3791	2.7	2788	8.1	12,099	9.2	362.5	319.2	8.97	8.04
Wind farms and other renewable energy	-	-	-	-	7490	16.0	14,344	9.0	-	-	-	-
Industrial power plants	2553	7.4	8541	5.6	2634	5.6	10,178	6.4	103.2	119.2	0.21	1.18
Total	34,715	100	153,362	100	46,799	100	158,767	100	134.8	103.5	2.01	0.23

Source: [44,45].

4.2. Electricity Consumption in Poland Rural vs. Urban Areas

Since Poland became an EU member, the entire country, including rural areas, has received new development opportunities. Accession to the EU and the related development policies, particularly the cohesion policy and the CAP, as well as the trade and industrial policies, provided a new economic and social quality. An important impulse for development has related to the targeted support from the EU funds addressing rural development in Poland. For example, allocation and utilization of the EU CAP funds are typically almost twice as high in Poland as the EU mean (the Polish agri-food sector and rural areas after ten-year EU membership—a review of major changes in 2014). Thus, the transfer of funds, also including public funds, has contributed to a boost in economic activity, which has been manifested in increased electricity consumption. Economic activity in rural areas measured by the number of economic entities in the REGON registry has improved considerably, as indicated by the almost 25% increase in 2018 compared to 2010 (at that time there were 1.2 million out of the total 4.4 million entities), whereas, in towns and cities in the same period, the increase was as little as 8% [46].

Electricity consumption in Polish households in the last fifteen years has grown continuously. In 2004, it was 22.8 TWh, while, in 2019, it was 30.6 TWh, i.e., the absolute increase amounted to 7.8 TWh, or approximately a third in absolute terms (Figure 5). A particularly marked increase was recorded for energy consumption in rural areas, in 2004 consumers in those areas used 6.3 TWh electric energy, while in 2019 it was 12.7 TWh, an increase of 6.4 TWh. Rural areas were thus responsible for an over 80% increase in electricity consumption in Poland. This phenomenon was becoming even more pronounced in the successive years, since, while the share of rural areas in electricity consumption in Poland in 2004 was below 28%, in 2019 it reached over 41%. Rural areas in Poland are thus characterized by growing needs in terms of electricity supply. This is confirmed by the dynamics of changes in electricity consumption. While, in the extreme years, this amounted to slightly over 34%, in urban areas it was less than 10%, then in the same period in rural areas it was ten times higher, amounting to almost 100%, which shows a doubling of electricity consumption (Table 3). In the regional system, the dynamics of change in electricity consumption varied between individual provinces (Table 3). First of all, it may be observed that, in contrast to rural areas, electricity consumption in urban areas is generally characterized by minor changes. In provinces such as Pomorskie, Łódzkie and Warmińsko-mazurskie, a highly stable level of energy consumption was recorded, while the dynamics of change in 2019 compared to 2004 did not exceed 2.5%. Rural areas exhibited a much greater dynamic of electricity consumption in the years 2004–2019. In certain provinces, energy consumption increased several-fold. In the Podlaskie province, electricity consumption increased five-fold, while in the Łódzkie and Lubelskie provinces a minimum three-fold increase was recorded.

Table 3. Dynamics of change in electricity consumption in Poland (GWh).

Consumption	Total		Urban Areas		Rural Areas		Dynamics of Change 2019 [2004=100]		
Region	2004	2019	2004	2019	2004	2019	Total	Rural Areas	Urban Areas
Poland	22,804.4	30,613.2	6344.1	12,677.1	16,460.3	17,936.0	134.2	199.8	109.0
Dolnośląskie	1633.0	2373.4	403.2	839.7	1229.8	1533.7	145.3	208.2	124.7
Kujawsko-Pomorskie	1151.6	1551.9	377.8	697.4	773.7	854.5	134.8	184.6	110.4
Lubelskie	886.4	1448.9	266.1	803.9	620.3	644.9	163.5	302.1	104.0
Lubuskie	599.3	794.7	170.3	296.0	429.0	498.7	132.6	173.8	116.3
Łódzkie	1468.9	2041.5	276.2	830.5	1192.7	1211.0	139.0	300.7	101.5
Małopolskie	2398.7	2858.0	1006.0	1393.9	1392.6	1464.2	119.2	138.6	105.1
Mazowieckie	3508.4	5046.4	639.4	1820.2	2869.0	3226.2	143.8	284.7	112.5
Opolskie	674.8	823.6	282.4	421.9	392.4	401.7	122.1	149.4	102.4
Podkarpackie	786.9	1262.6	324.9	728.8	462.0	533.8	160.4	224.3	115.5
Podlaskie	503.4	937.7	94.0	476.4	409.4	461.3	186.3	507.0	112.7
Pomorskie	1517.8	1826.6	395.2	701.5	1122.6	1125.1	120.3	177.5	100.2
Śląskie	3322.6	3688.1	698.7	954.9	2623.9	2733.2	111.0	136.7	104.2
Świętokrzyskie	491.6	794.2	172.2	427.5	319.4	366.7	161.6	248.3	114.8
Warmińsko-Mazurskie	788.0	1029.5	258.5	488.5	529.5	541.0	130.7	189.0	102.2
Wielkopolskie	2061.0	2880.0	745.7	1394.3	1315.3	1485.6	139.7	187.0	113.0
Zachodniopomorskie	1012.1	1256.0	233.5	401.7	778.6	854.3	124.1	172.0	109.7
			Descriptive statistics						
minimum	491.6	794.2	94.0	296.0	319.4	366.7	111.0	136.7	100.2
mean	1425.3	1913.3	396.5	792.3	1028.8	1121.0			
median	1081.9	1500.4	303.7	715.2	776.2	854.4	136.9	188.0	110.1
Coefficient of variation (%)	67.0	62.8	62.9	53.5	74.2	74.3			
maximum	3508.4	5046.4	1006.0	1820.2	2869.0	3226.2	186.3	507.0	124.7

Source: [44]. https://bdl.stat.gov.pl/BDL/dane/podgrup/temat/11/57/1880, accessed on 12 September 2021.

Mean annual rate of change in electricity consumption in households (Table 3) in rural areas of Poland on average was almost nine times greater than in urban areas, 6.51% compared to 0.74%. The greatest growth rate for electricity consumption in rural areas was found in the Podlaskie and Lubelskie provinces, at almost 20% and 12.5%, respectively (Table 4).

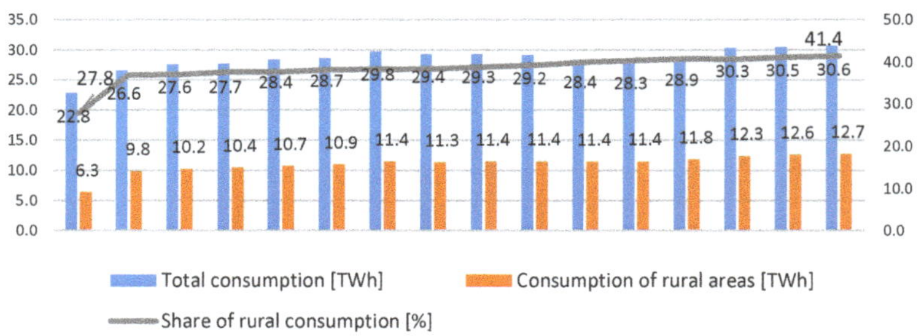

Figure 5. Electricity consumption in households in Poland in the years 2004–2019 (TWh). Source: [44].

Table 4. The rate of change in electricity consumption in households in Poland in the years 2004–2019 (%).

Region	Total	Urban Areas	Rural Areas
Poland	2.61	0.74	6.51
Dolnośląskie	3.04	2.06	5.66
Kujawsko-Pomorskie	2.91	0.95	6.19
Lubelskie	5.26	0.70	12.57
Lubuskie	2.14	0.75	5.09
Łódzkie	3.20	0.38	11.48
Małopolskie	1.20	0.27	2.38
Mazowieckie	3.04	0.65	10.58
Opolskie	2.25	0.53	4.30
Podkarpackie	4.67	1.31	8.42
Podlaskie	6.34	1.22	19.68
Pomorskie	1.77	0.51	4.78
Śląskie	0.87	0.47	2.25
Świętokrzyskie	4.81	1.04	10.01
Warmińsko-Mazurskie	2.57	0.44	6.07
Wielkopolskie	2.67	0.84	5.38
Zachodniopomorskie	1.69	0.62	4.72

Source: authors' calculations based on Local Data Bank [44], https://bdl.stat.gov.pl/BDL/dane/podgrup/temat/11/57/1880, accessed on 12 September 2021.

Analyzing unit electricity consumption in Poland in the years 2004–2019 (Figure 6), i.e., per capita, with the division into rural and urban areas, it may be indicated that a characteristic event took place in 2014. For the first time in Poland, unit electricity consumption was higher in rural areas than in urban areas—by 1.4%. In the following years, this phenomenon grew. As a result, in 2019, unit electricity consumption in rural areas was already 6.3% higher and amounted to 827 kWh compared to 778 kWh.

A comparison of the degree of concentration of electricity consumption by provinces between rural and urban areas suggests a low degree of concentration for electricity consumption by province both in rural and urban areas (Table 5). Analogously, a close similarity is also observed in electricity consumption between rural and urban areas, as evidenced by similar values of the coefficient of similarity according to Florence (Table 5). In Poland, only in the Podlaskie province in the years 2004–2019, was a high growth rate for electricity consumption observed (Figure 7). In the next five provinces, i.e., Mazowieckie, Lubelskie. Podkarpackie, Świętokrzyskie and Łódzkie, electricity consumption increased at a medium rate. In the other provinces, it exhibited a low growth rate.

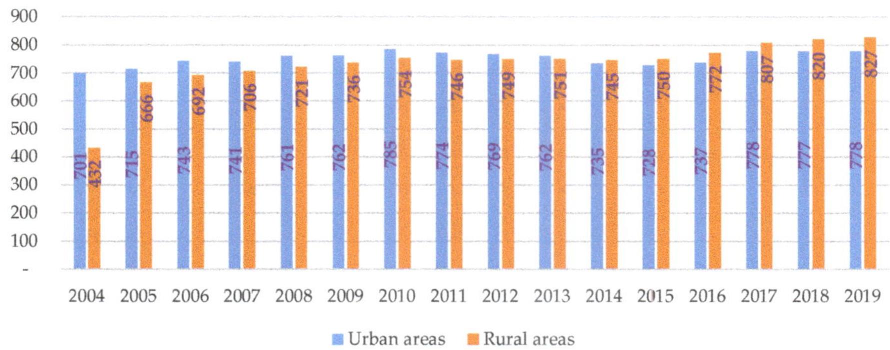

Figure 6. Unit electricity consumption in Poland (kWh/per capita). Source: authors' calculations based on Local Data Bank [44].

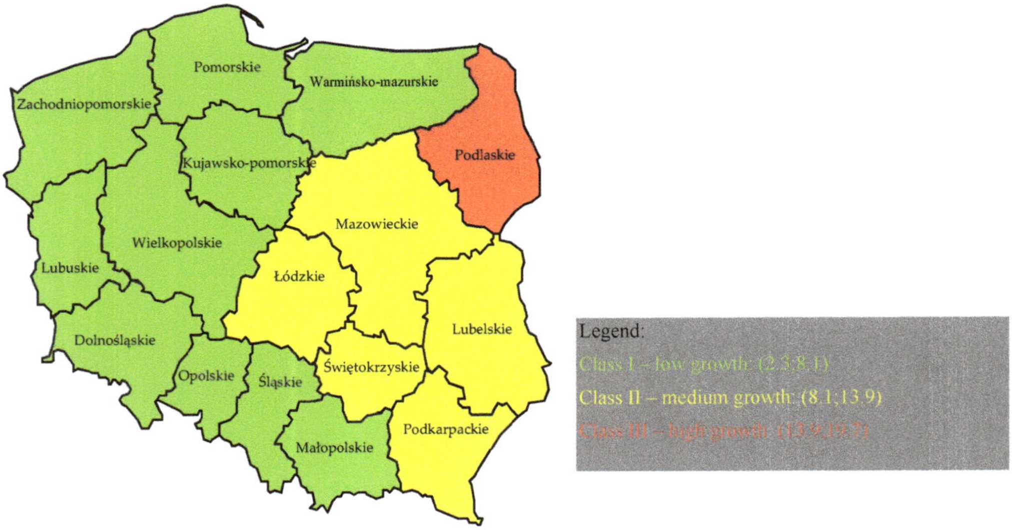

Figure 7. Regional variation in electricity consumption in rural areas of Poland in the years 2004–2019 according to mean annual rate of changes. Source: authors' study based on Table 3.

Table 5. The degree of similarity and concentration of electricity consumption.

Year	Concentration		Similarity	
	Rural Areas	Urban Areas	Rural Areas	Urban Areas
2004	0.24	0.28	0.04	0.04
2005	0.19	0.28	0.05	0.03
2006	0.18	0.27	0.06	0.03
2007	0.19	0.27	0.05	0.03
2008	0.19	0.27	0.05	0.03
2009	0.19	0.27	0.05	0.03
2010	0.19	0.27	0.06	0.03
2011	0.19	0.27	0.05	0.03
2012	0.19	0.28	0.05	0.03
2013	0.19	0.28	0.05	0.03
2014	0.19	0.27	0.05	0.03
2015	0.19	0.27	0.04	0.03
2016	0.19	0.27	0.04	0.03
2017	0.19	0.27	0.04	0.03
2018	0.19	0.27	0.04	0.03
2019	0.20	0.27	0.04	0.03

Source: authors' calculations based on Local Data Bank [44].

5. Discussion

The energy market in Poland is undergoing dynamic changes that affect rural areas. Since 2014, unit electricity consumption per capita in rural areas exceeded the level of consumption in urban areas. This relationship is permanent, since it has been evident in the following years, while, additionally, differences in the level of consumption also grew in subsequent years.

Rural areas in Poland are characterized by a growing need for electricity. The scope of the research made it possible to track the existing trends in electricity consumption. Moreover, the provinces of Poland were grouped based on the growth rate of their electricity consumption. Classes characterized by low, medium and high growth rates were distinguished. The analysis was carried out for rural areas in Poland, which show great territorial differences in the level of socioeconomic development. This is understood as changes taking place in a direction that meets the collective and individual needs and individual aspirations of residents and local communities to an ever-greater extent [47]. At the same time, the level of regional development determines the development of the electricity market. The voivodships with a high and medium growth rate of electricity consumption include rural areas, which are characterized by a very low and low socio-economic development.

The results indicate that, currently, rural areas contribute to the high adverse impact on the natural environment in Poland, which is the result of differences in specific energy consumption. In terms of the spatial arrangement, Podlaskie Voivodeship has a unique situation. This voivodeship had the highest increase in electricity consumption per capita in 2004–2019 (on average it was almost 20%).

The changes observed are related to the increase in economic activity in rural areas, the intensification of agriculture and the increase in the scale of livestock production on farms. According to Wójcicki [48], animal breeding is more energy-consuming than arable farming.

In Poland, Podlaskie Voivodeship has the highest cattle population per 100 ha of UAA. This creates excellent opportunities for biogas production due to the increased availability of the substrate. There are therefore possibilities to limit the negative impact of Podlaskie Voivodeship on the environment in the future. This reduces CO_2 emissions, particularly those from rural areas, and reduces the electricity consumption generated from fossil fuels. In the conditions of the energy transformation consisting of abandoning coal, a good state energy policy is needed [49]. Also, activities at the regional level support the existing solutions at the national level, for example, support from EU funds for investments in agricultural biogas plants.

Experience from the global energy market shows that it is possible to shift from fossil fuels to clean energy as the world moves towards decarbonizing economies and reducing greenhouse gas emissions worldwide [50]. As a result, environmental targets are no longer seen as an obstacle to economic development but as a solution to economic and social development.

According to Kaygusuz [51], renewable energy sources are a desirable form of practical solutions in the energy sector to develop clean and sustainable energy to reduce environmental pollution.

Over ten years ago, Marecki [52] and Jaczewski [53] indicated that, around 2020, the following may be expected:

- A moderate increase in the consumption of solid fuels, i.e., coal and lignite, in energy generation;
- A relatively low growth in oil consumption;
- A simultaneous high growth in natural-gas consumption;
- A very high increase in the production of renewable energy, whose share in the global energy consumption may increase to over 12 percent by 2020.

The results of the study confirm the accuracy of these forecasts; particularly in the Polish energy economy in the last fifteen years, a considerable increase has been recorded in natural gas consumption for electricity generation along with a dynamic development of renewable energy [54]. A gradual reduction in the use of coal for electricity production has also been observed, although the rate of changes in this respect is not adequate to meet the existing challenges. These gradual changes in the consumption of energy carriers are consistent with the strategic EU goals. Thanks to further actions resulting from the Polish national energy plan PEP 2040, in the coming years, the energy economy in Poland will be increasingly compatible with that of the EU.

Within approximately the last two decades, Poland has made progress in reducing the negative environmental impact of electricity production, and the reduction in coal and lignite consumption on average by 0.6% and 1.5% confirms this.

In the following years, the development of renewable energy generation may also be observed. In 2019, the share of electric energy produced by wind power as well as other renewable energy sources was 9%, while, together with hydro power plants, it amounted to more than 10%. Poland has reserves in the potential for clean energy generation, as indicated by the available production capacity which, for the two types together, amounts to 21% of total installed power in Poland.

Moreover, the realization of renewable energy investments with the support of EU funds in many local government units [54], also in rural communes, will, in the immediate future, bring an increase in the share of renewable energy.

Investments in renewable energy sources are desirable in rural areas. Sutherland et al. [55] highlight the countryside and the potential of agriculture to generate renewable energy. They indicate that, due to its historic commitment to managing essential resources, especially land and biomass, the agricultural sector plays, or at least can play, an important role in the transformation of renewable energy.

However, special care should be taken to ensure that the production of renewable energy in rural areas does not result in competition for food and energy resources, which could result in a deterioration of food security [56].

There are possible solutions that minimize the occurrence of such a phenomenon. According to Jasiulewicz [57], for example, to produce biomass, we should use lower quality soils that have been set aside, degraded or are at least not suitable for food production. The production of renewable energy in rural areas can also be based to a greater extent than before on biogas and the use of agriculture waste, which may be slurry or manure from the breeding of cattle, pigs or poultry [58–60].

6. Conclusions

Forecasted changes and the transformation of energy systems both in Europe and worldwide are taking place under the influence of the so-called development megatrends [61]. The most important of these in the European power engineering include decreasing costs of renewable, limitation of the environmental impact of the energy sector, the decreasing role of coal as an energy source and new business models in power engineering. On the international and EU scales, in the last five years, we have witnessed many crucial events and agreements, which, during the coming decades, will influence changes in the energy sector. These include the Paris Agreement on the global reduction of climate change, the adoption of the European Green Deal 2050, aimed at a zero-emission economy, and actions to introduce the Energy Union. National energy policy needs to adapt to such conditions on the macro scale, as new challenges have also appeared. The most important of these include the following issues: how to ensure energy security in a changing energy market? What might the future role of coal be in energy economy? In what areas should the development of renewable energy sources be promoted? What should the rate of integration in the national electric energy market with the EU market be and, as a consequence, what should the European compromise for the energy sector be? [62].

Over the last five years, the problems of the energy economy have gained importance for the public. At the same time, they are a constant and essential aspect in the work of the EU summit meetings. The research problem discussed is important from the point of view of sustainable development, environmental protection and economic security of electricity consumers.

Actions aimed at the implementation of sustainable development result from the fact that, at present, it is universally acknowledged that the previous paths of socio-economic development and a recreation of previously grounded trends for economic growth disregarding broadly understood environmental, social and economic consequences should not be maintained in the future.

Sustainable development requires the application of prevention and, first of all, foresight, preferably in all areas of socio-economic life [63]. A form of prevention seems to be enacting seeing changes in the consumption of electricity and the energy carriers used to generate it.

Based on experience, it is possible to reduce greenhouse gas emissions in the EU. This results mainly from reducing the use of hard coal and lignite in electricity production (in favor of increasing the importance of renewable energy). The effects of the 2005–2015 period showed that greenhouse gas emissions fell below the target set for 2020 [5].

Author Contributions: Conceptualization, P.S., J.H., J.L. and A.R.; methodology, P.S., J.L.; software, P.S., J.L.; validation P.S., J.H., J.L. and A.R.; formal analysis, P.S., J.H., J.L. and A.R., investigation, P.S., J.H., J.L. and A.R., resources, P.S., J.H., J.L. and A.R., data curation, P.S., J.H., J.L. and A.R.; writing—original draft preparation, P.S., J.H., J.L. and A.R., writing—review and editing, P.S., J.H., J.L. and A.R.; visualization, P.S., J.H. and R.A; supervision, P.S., J.H., J.L. and A.R.; project administration, P.S.; funding acquisition, P.S., J.H. and J.L. All authors have read and agreed to the published version of the manuscript.

Funding: This research was financed from resources of the Faculty of Economics Poznań University of Life Sciences.

Institutional Review Board Statement: Not applicable.

Informed Consent Statement: Not applicable.

Data Availability Statement: Not applicable.

Conflicts of Interest: Not applicable.

References

1. Abolhosseini, S.; Heszmati, A.; Altmann, J. *Review of Renewable Energy Supply and Energy Efficiency Technologies*; Discussion Paper Series; IZA: Bonn, Germany, 2014; Volume 8145, pp. 1–35.
2. Ritchie, H.; Roser, M.; Fossil Fuels. Our World Data. 2017. Available online: https://ourworldindata.org/fossil-fuels (accessed on 10 November 2021).
3. EPA. Greenhouse Gas Emissions. Global Emissions by Economic Sector, Washington: United States Environmental Protection Agency. Available online: https://www.epa.gov/ (accessed on 10 November 2021).
4. Emisje Gazów Cieplarnianych w Unii Europejskiej, Dostęp. Available online: https://www.europarl.europa.eu/news/pl/headlines/society/20180301STO98928/infografika-emisje-gazow-cieplarnianych-w-unii-europejskiej (accessed on 10 September 2021).
5. Energy and Climate Change. *Signals 2017—Shaping the Future of Energy in Europe, No 142*; ECC: Copenhagen, Denmark, 2017.
6. Altawell, N.; Milne, J.; Seowou, P.; Sykes, L. *Rural Electrification: Optimizing Economics, Planning and Policy in an Era of Climate Change and Energy Transition*, 1st ed.; Academic Press: Cambridge, MA, USA, 2020; pp. 19–38.
7. Bałamut, A. Polish energy security police in the years 2000–2015. (Polityka bezpieczeństwa energetycznego polski w latach 2000–2015). *Teoria Prakt.* **2017**, *29*, 49–64.
8. Energy Production and Imports: Eurostat 2020. Available online: https://ec.europa.eu/eurostat/statistics-explained/index.php?title=Energy_production_and_imports/pl (accessed on 25 March 2021).
9. Yergin, D. Ensuring Energy Security. In *Foregin Affairs;* Council on Foreign Relations: New York, NY, USA, 2006; Volume 85, pp. 69–82.
10. European Commission. *Twenty Five Years of the Common Market in Coal 1953–1978;* Office for Official Publications of the European Communities: Luxemburg, 1977; pp. 143–144.
11. Vallebona, M. The Genesis and Modification of the Common EU Energy Policy. *Econ. Political Thought* **2015**, *1*, 92–112.
12. Grycan, W.; Wnukowska, B.; Wróblewski, Z. Modelling of electric energy conditions of a region. *Electrotech. Rev.* **2014**, *90*, 230–233.
13. Trubalska, J. Towards the energy union. A new concept of energy security in the EU. *Zesz. Naukowe Inst. Gospod. Surow. Miner. Energią Pol. Akad. Nauk* **2017**, *97*, 21–32.
14. European Commission. *Final Communication from The Commission Clean Energy for All Europeans;* European Commission: Brussels, Belgium, 2019.
15. SWD. *Underlying Results of the Impact Assessment for the Amendment of the Energy Efficiency Directive;* SWD: Brussels, Belgium, 2016; p. 405.
16. SWD. *Impact Assessment for the Recast of the Renewables Directive;* SWD: Brussels, Belgium, 2016; p. 418.
17. European Commission. *European Energy and Transport Trends to 2030;* European Commission: Brussels, Belgium, 2010.
18. Poland's Energy Policy until 2040. Ministry of Climate and Environment. Available online: https://www.gov.pl/web/klimat/polityka-energetyczna-polski (accessed on 26 March 2021).
19. Trzepacz, P. Genesis and Essence of the Sustainable Development Concept. In *Sustainable Development—Global Challenges;* Instytut Geografii i Gospodarki Przestrzennej UJ: Cracow, Poland, 2012; pp. 11–37.

20. Sustainable Development and Its New Challenges. Available online: https://www.ceeol.com/search/article-detail?id=747496 (accessed on 15 September 2021).
21. Rokicka, E.; Woźniak, W. *Towards Sustainable Development. Concepts, Interpretations, Contexts*; University of Lodz: Łódź, Poland, 2016.
22. Latoszek, E. The Concept of Sustainable Development in Theory and Practice ONZ 25–65. In *Sustainable Development and Global Public Goods in Theory and Practice of International Organisations*; Latoszek, E., Proczek, M., Krukowska, M., Eds.; Elipsa: Warszawa, Poland, 2016; pp. 25–64.
23. European Commission. *Final Communication from The Commission Clean Energy for All Europeans*; European Commission: Brussels, Belgium, 2016.
24. Wróbel, P. Analyzes and dialogue. In *Small Steps to Big Changes: Impact of the Clean Energy for All Europeans Package on Energy*; Energy Forum: Warszawa, Poland, 2019; p. 9, [Forum energii. Analizy i dialog. Małymi krokami do wielkich zmian Wpływ pakietu „Czysta energia dla wszystkich Europejczyków" na energetykę, wrzesień 2019, p. 9].
25. Rising Energy Consumption Slows EU Progress on Renewables and Energy Efficiency Targets. Available online: https://www.eea.europa.eu/highlights/rising-energy-consumption-slows-eu (accessed on 25 March 2021).
26. The European Green Deal, Communication from The Commission COM (2019) 640 Final, Brussels, 11.12.2019. Available online: https://eur-lex.europa.eu/legal-content/EN/TXT/?uri=COM%3A2019%3A640%3AFIN (accessed on 14 September 2021).
27. Szpak, K. The EU Climate Policy for 2050. In *Climate Policy and Its Implementation in the First Half of the 21st Century*; Gajewski, G., Paprocki, W., Eds.; Centrum Myśli Strategicznych: Gdansk, Poland, 2020; pp. 34–53.
28. Mrowiec, M. Economic Aspects of the Climate Policy. In *Climate Policy and Its Implementation in the First Half of the 21st Century*; Gajewski, G., Paprocki, W., Eds.; Centrum Myśli Strategicznych: Gdansk, Poland, 2020; pp. 203–217.
29. Ceglarz, A.; Polish Energy Policy. Climate Change, Energy and the Environment. *Friedrich-Ebert-Stiftung*. 2020. Available online: http://library.fes.de/pdf-files/bueros/warschau/17099.pdf (accessed on 14 September 2021).
30. The Just Transition Mechanism: Making Sure No One is Left Behind. Available online: https://ec.europa.eu/info/strategy/priorities-2019--2024/european-green-deal/actions-being-taken-eu/just-transition-mechanism_en (accessed on 10 September 2021).
31. Jonker, J.; Krukowska, M. Transformation Towards Sustainable Development. In *Management and Business Administration*; Central Europe, 4/2012 (117); Akademia Leona Koźmińskiego: Warsaw, Poland, 2012; pp. 92–106. ISSN 2084–3356.
32. Pearce, D.; Markandya, A.; Barbier, E.B. *Blueprint for a Green Economy*; Earthscan: London, UK, 1989.
33. UNEP. Towards a Green Economy: Setting the Stage for a Green Economy Transition. 2011. Available online: http://www.unep.org/greeneconomy (accessed on 13 August 2011).
34. Regional Just Transformation Plans Will Promote Use of EU Funds. Available online: https://www.gov.pl/web/fundusze-regiony/terytorialne-plany-sprawiedliwej-transformacji-pomoga-korzystac-z-funduszy-unijnych (accessed on 25 March 2021).
35. European Commission. *A Clean Planet for All: A European Strategic Long-Term Vision for a Prosperous, Modern, Competitive and Climate Neutral Economy*; European Commission: Brussels, Belgium, 2018.
36. Report from the Commission to The European Parliament, The Council, The European Economic and Social Committee and The Committee of The Regions Renewable Energy Progress Report. Available online: https://eur-lex.europa.eu/legal-content/EN/TXT/?uri=CELEX%3A52017DC0057 (accessed on 30 September 2021).
37. Report from the Commission to The European Parliament, The Council, The European Economic and Social Committee and The Committee of The Regions Renewable Energy Progress Report. Com/2020/952 Final. Available online: https://ec.europa.eu/energy/sites/ener/files/renewable_energy_progress_report_com_2020_952.pdf (accessed on 30 September 2021).
38. Middlemiss, L.; Ambrosio-Albalá, P.; Ross, N.E.; Gilbertson, G.; Hargreaves, T.; Mullen, C.; Ryan, T.; Snell, C.; Tod, A. Energy poverty and social relations: A capabilities approach. *Energy Res. Soc. Sci.* **2019**, *55*, 227–235. [CrossRef]
39. Energy Prices and Costs in Europe. COM 951 Final. 2020. Available online: https://eur-lex.europa.eu/legal-content/PL/TXT/PDF/?uri=CELEX:52020DC0951&from=EN (accessed on 30 September 2021).
40. Electricity in Households by Recipient's Location, Local Data Bank of the Central Statistical Office. Available online: https://bdl.stat.gov.pl/BDL/metadane/cechy/1880 (accessed on 20 March 2021)(Energia elektryczna w gospodarstwach domowych wg lokalizacji odbiorcy, Bank Danych Lokalnych GUS, dostęp: 20 marzec 2021 r.).
41. Wysocki, F.; Lira, J. *Descriptive Statistics. Wydawnictwo Akademii Rolniczej Im*; Augusta Cieszkowskiego: Warszawa, Poland, 2003.
42. Arıoğlu Akan, M.; Selam, A.; Oktay Fırat, S.; Er Kara, M.; Özel, S. A Comparative Analysis of Renewable Energy Use and Policies: Global and Turkish Perspectives. *Sustainability* **2015**, *7*, 16379–16407. [CrossRef]
43. Gielen, D.; Boshell, F.; Saygin, D.; Bazilian, M.D.; Wagner, N.; Gorini, R. The role of renewable energy in the global energy transformation. *Energy Strategy Rev.* **2019**, *24*, 38–50. [CrossRef]
44. Report 2004 National Electricity Network—Compilation of Quantitative Data. Available online: www.pse.pl (accessed on 1 March 2021). (Raport 2004 Krajowa Sieć Elektroenergetyczna–zestawienie danych ilościowych. www.pse.pl–dostęp: 1 marca 2021 r.).
45. Report 2019 National Electricity Network—Compilation of Quantitative Data. Available online: www.pse.pl (accessed on 1 March 2021). (Raport 2004 Krajowa Sieć Elektroenergetyczna–zestawienie danych ilościowych. www.pse.pl–dostęp: 01.03.2021 r).
46. Statistical Analyses, Rural Areas in Poland in 2018, Statistics Poland, Warsaw, Olsztyn 2020. Available online: https://stat.gov.pl/obszary-tematyczne/rolnictwo-lesnictwo/rolnictwo/obszary-wiejskie-w-polsce-w-2018-roku,2,4.html (accessed on 30 August 2021).

47. Stanny, M.; Rosner, A.; Komorowski, Ł. *Monitoring of Rural Development. Stage III. Socio-Economic Structures, Their Spatial Diversification and Dynamics (Monitoring of Rural Development. Stage III. Socio-Economic Structures, Their Spatial Differentiation and Dynamics)*; EFRWP, IRWiR PAN: Warsaw, Poland, 2018; (Monitoring Rozwoju Obszarów Wiejskich. Etap III. Struktury Społeczno-Gospodarcze, Ich Przestrzenne Zróżnicowanie i Dynamika (Monitorowanie rozwoju wsi. Etap III Struktury społeczno-gospodarcze, ich zróżnicowanie przestrzenne i dynamika) EFRWP, IRWiR PAN: Warszawa, Polska, 2018.
48. Wójcicki, Z. Agricultural production energy consumption based on research. *Probl. Agric. Eng.* **2015**, *4*, 31–41, (Energochłonność produkcji rolniczej na podstawie badań. Problemy Inżynierii Rolniczej. Z. 4(90) s. 31–41).
49. Jedlińska, K.; Olkuski, T. The decision-making process in the EU in the field of energy policy. *Energy Policy J.* **2019**, *22*, 5–18. [CrossRef]
50. Altawell, N.; Milne, J.; Seowou, P.; Sykes, L. *Rural Electrification: Optimizing Economics, Planning and Policy in an Era of Climate Change and Energy Transition*; Academic Press: Cambridge, MA, USA, 2021; pp. 201–230.
51. Kaygusuz, K. Sustainable energy, environmental and agricultural policies in Turkey. *Energy Convers. Manag.* **2010**, *51*, 1075–1084. [CrossRef]
52. Marecki, J. Prospects for the development of the power industry in Poland until 2020. *Electrotech. Review.* **2002**, *78*, 210–214.
53. Jaczewski, M.; Marecki, J.; Sikora, W. *Programme for Coal and Electrictiy Demand by 2020*; Ekspertyza KPE PAN: Warszawa, Poland, 1999.
54. Standar, A.; Kozera, A.; Satoła, L. The Importance of Local Investments Co-Financed by the European Union in the Field of Renewable Energy Sources in Rural Areas of Poland. *Energies* **2021**, *14*, 450. [CrossRef]
55. Sutherlan, L.A.; Pete, S.; Zagata, L. Conceptualising multi-regime interactions: The role of the agriculture sector in renewable energy transitions. *Res. Policy* **2015**, *44*, 1543–1554. [CrossRef]
56. Popp, J.; Lakner, Z.; Harangi-Rákos, M.; Fári, M. The effect of bioenergy expansion: Food, energy, and environment. Renew. Sustain. *Energy Rev.* **2014**, *32*, 559–578.
57. Jasiulewicz, M. Production of energy from agribiomass in Poland in the background of selected countries of the European Union. *Rocz. Naukowe SERiA* **2015**, *8*, 94–99.
58. Owczuk, M.; Wardzińska, D.; Zamojska-Jaroszewicz, A.; Matuszewska, A. The use of biodegradable waste for the production of biogas as an alternative source of renewable energy. *Stud. Ecol. Bioet. UKSW* **2013**, *11*, 133–144.
59. Damyanova, S.; Beschkov, V. Biogas as a Source of Energy and Chemicals. Biorefinery Concepts, Energy and Products. In *Biorefinery Concepts*; IntechOpen: London, UK, 2020.
60. Nevzorova, T.; Kutcherov, V. Barriers to the wider implementation of biogas as a source of energy: A state-of-the-art review. *Energy Strat. Rev.* **2019**, *26*, 100414. [CrossRef]
61. Kamrat, W. Problems of Energy Sector Development in View of Increased Natural Gas Supply. Available online: https://www.cire.pl/ (accessed on 26 March 2021).
62. Maćkowiak-Pandera, J.; Rączka, J.; Bukowski, M. *Polish Energy Sector at Megatrends*; Forum Analiz Energetycznych: Warszawa, Poland, 2016.
63. Jeżowski, P. Sustainable Development and Its New Challenges. Available online: https://ssl-kolegia.sgh.waw.pl/pl/KES/czasopisma/kwartalnik/archiwum/Documents/PJezowski10a.pdf (accessed on 30 November 2021).

energies

MDPI

Article

Economic Feasibility of Agricultural Biogas Production by Farms in Ukraine

Galyna Trypolska [1], Sergii Kyryziuk [2], Vitaliy Krupin [3,*], Adam Wąs [4] and Roman Podolets [1,*]

1 State Institution Institute for Economics and Forecasting of the National Academy of Sciences of Ukraine, Panasa Myrnoho 26, 01011 Kyiv, Ukraine; g.trypolska@gmail.com
2 State Institution "Center for Evaluation of Activity of Research Institutions and Scientific Support of Regional Development of UkraineNAS of Ukraine", Volodymyrska 54, 01601 Kyiv, Ukraine; kyryzyuk.ief@gmail.com
3 Institute of Rural and Agricultural Development, Polish Academy of Sciences, Nowy Świat 72, 00-330 Warsaw, Poland
4 Institute of Economics and Finances, Warsaw University of Life Sciences (SGGW), Nowoursynowska 166, 02-787 Warsaw, Poland; adam_was@sggw.edu.pl
* Correspondence: vkrupin@irwirpan.waw.pl (V.K.); podolets@ief.org.ua (R.P.)

Abstract: Renewable energy generation in Ukraine is developing slower than state strategies and expectations, with the installations for energy generation based on biogas currently being among the lowest in terms of installed capacity. Most of those involved in energy generation from agricultural biogas are large enterprises, while the small and medium-sized farms are far less involved. Thus the article aims to assess the economic feasibility of biogas production from agricultural waste by specific farm types and sizes, with a special focus on small and medium-sized farms. The research results present findings in two dimensions, first defining the economic feasibility of biogas installations in Ukraine based on investment costs and the rate of return at both the current and potential feed-in tariff, and second, analyzing the influence of state regulation and support on the economic feasibility of agricultural biogas production in Ukraine. The results emphasize that the construction of small generation capacities does not provide sufficient funds under the current feed-in tariff to meet the simple return period expected by the domestic financing institutions. Except for the general support programs for agricultural activities, there are no support funds specifically for biogas producers, while there is tight competition with wind and solar energy due to diversified feed-in tariffs.

Keywords: agricultural biogas; farm; economic feasibility; investment; LCOE; state support; feed-in tariff; Ukraine

Citation: Trypolska, G.; Kyryziuk, S.; Krupin, V.; Wąs, A.; Podolets, R. Economic Feasibility of Agricultural Biogas Production by Farms in Ukraine. *Energies* **2022**, *15*, 87. https://doi.org/10.3390/en15010087

Academic Editor: Attilio Converti

Received: 1 November 2021
Accepted: 15 December 2021
Published: 23 December 2021

1. Introduction

Global climate change is increasing its tempo and impact [1] on humanity, while the origin of this change is primarily anthropogenic [2–5] due to the excessive negative influence of the intense use of fossil fuels and the consequent environmental pollution. A swift shift to renewable energy is among key solutions to this growing problem and needs to be implemented by all technically feasible means. These include energy generation from biogas, which is becoming an increasingly popular and important source of renewable energy from the standpoint of the circular economy [6–8], yet still falls behind the shares of solar and wind energy [9,10]. The global direct consumption of biogas in 2018 equaled ca. 35 Mtoe, including 16.1 Mtoe in Europe, 8.8 Mtoe in China and 4.0 Mtoe in North America [11].

While biogas can be generated from numerous types of organic waste, they can be aggregated into either of the two major ones: solid and agricultural. The latter includes the plant leftovers, weeds, leaf litter, sawdust, as well as the animal-originated solid, slurry and liquid waste. The utilization of biogas makes it possible to generate electricity and heat, reduce greenhouse gas emissions (methane and nitrous oxide from livestock waste), smooth overloads in the energy transfer grids and create new jobs in rural areas.

Ukraine is often referred to as a country with substantial agricultural development potential [12–14]. The same applies to bioenergy generation, yet in this case, a distinction needs to be made between different sources. While the biomass from crop production is increasingly available due to the subsector's growth [15,16], the livestock sector has been declining drastically in the past 30 years [17,18]. Thus the production of biogas from livestock feedstock could simultaneously be a significant chance to intensify renewable energy generation [19,20], but a difficult task due to specific conditions of the livestock subsector's development.

The integration of livestock production with the agricultural biogas installation could be especially beneficial for small and medium-sized farms, as it would allow them to use the livestock waste (utilization of which is currently an additional economic [21,22] and environmental burden [23]) in order to increase economic viability and achieve their own energy security, as well as aid the achievement of national goals of energy independence, regional energy generation diffusion and increasing the number and capacity of renewable energy sources.

The combined model of biogas production and use, in which biogas can be produced simultaneously from different types of feedstock, is a common approach in the global perspective [24]. While the average efficiency of electricity production (excluding heat) is ca. 34.6%, the simultaneous production of heat and electricity (i.e., co-generation) increases the efficiency of the production plant to 76.4% (provided there are heat consumers present on-site). Heat can be used for agricultural premises (e.g., farms, warehouses, greenhouses, processing), residential premises or the social infrastructure.

Despite the obvious advantages of biogas production for farm income diversification, the spread of biogas plants is still quite limited in Ukraine [20], especially compared to more economically developed European countries [25,26]. In particular, as of 1 January 2020, there were 49 biogas plants in Ukraine [27], of which only 21 were utilizing agricultural waste products as their feedstock, compared to almost 10,000 similar plants in nearby Germany [28]. Researchers say the limitations lie in the regulatory field [29–32], the lack of substantiated and efficient state support for energy generation from biogas [29,33] and a low level of understanding of the benefits of renewable energy generation by private entities and individuals [30,34]. In our opinion, there is still a gap within the research aimed at deepening the technological and economic feasibility of biogas production, taking into account national, regional and local conditions, and the suitability of installations to particular types of entities (according to their sizes and economic potential).

Until now, the typical approach in research articles regarding the development of biogas in Ukraine has been rather general [16,17,25,29,31] or technology-focused [21,35–38], still lacking a well-substantiated focus on farms, yet even more on their different sizes. Some research analyzed small energy generation installations that would be most suitable for small entities [37,39,40], but these focused rather on overview statistical and analytical data or technological issues, omitting the economic feasibility component. At the same time, work mentioning economic feasibility issues for various sizes of generation capacity does not tackle the state support system [34,41–43], or they take into account the feed-in tariff system for the energy generation from biogas [32,33,44] but do not view them from the perspective of farm size. This creates a research gap, especially for the small and medium-sized farms taking account of the economic elements of the support system, which would help understand the limitations and possibilities of development of small- and medium-scale bioenergy generation capacity by farming entities and individuals.

The study thus aims to assess the economic feasibility of biogas production from agricultural waste by farms, with a special focus on small and medium-sized farms involved in cattle, pig and poultry. The following research objectives were set for this:

- Investigate the state of the art of renewable energy in Ukraine;
- Assess the economic feasibility of biogas installations in Ukraine depending on their capacity, type and size of farms;

- Calculate the levelized costs of electricity and heat generation by biogas installations of various capacities and respective simple payback periods;
- Examine the impact of state regulation and support on the economic feasibility of agricultural biogas production in Ukraine, including the functioning of biogas installations under the feed-in tariff and the upcoming auction schemes.

The manuscript is divided into six sections. Following the introduction, Section 2 describes the current state, official plans and potential of renewable energy generation and biogas production in Ukraine. Section 3 presents the materials used and methods applied within the research. The results in Section 4 present findings of two dimensions: the assessment of the economic feasibility of biogas installations in Ukraine and the analysis of the influence of state regulation and support on the economic feasibility of agricultural biogas production in Ukraine. Section 5 discusses the results obtained and compares the current state of agricultural biogas development with selected EU experience. Section 6 summarizes the results and defines limitations and possible future directions of research within the topic.

2. Renewable Energy Generation and Biogas Production in Ukraine: Current State, Plans and Potential

The development of renewable energy in Ukraine is taking place in accordance with the National Renewable Energy Action Plan through 2020 (NREAP2020 [45,46]) and the Energy Strategy of Ukraine until 2035 (ESU2035 [47,48]). According to the strategic goals set in the NREAP2020, by 2020, it was expected that 11% of the final energy consumption would come from renewable sources, while the share of electricity from biomass should have reached 16.2% (of which 2.6% is biogas), and 85.5% in heating and cooling systems (of which 2.6% is biogas). By 2035, the share of energy from renewable sources is to reach 25%, in accordance with ESU2035. The total installed domestic capacity of bioenergy (including not only biogas-based installations but also other biomass plants) reached 0.98 GW in 2020, 1.3 GW in 2025, 1.67 GW in 2030 and 2.13 GW in 2035 [49].

Current achievements in renewable energy in Ukraine indicate it is falling behind the targets. In particular, in 2019, the share of energy generated by renewable energy sources in gross final energy consumption reached 8.1%. An increase in energy consumption from renewable energy sources in 2019 by 9.1% (0.36 Mtoe) compared to the previous year and the general reduction in energy consumption (by 2.6 Mtoe) made it possible to achieve some overall progress in the development of renewable energy (Figure 1), yet the rate of implementation is still not sufficient to meet the goals.

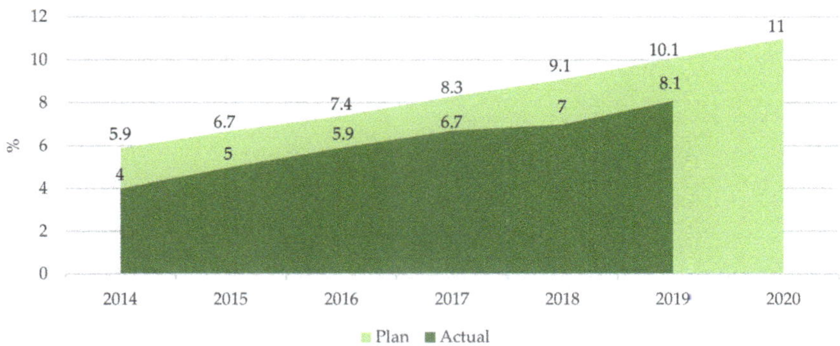

Figure 1. Renewable energy generation shares (actual vs. planned) in final energy consumption in 2014–2020 in Ukraine (in %). Source: own compilation based on [50].

During 2015–2019, the renewable electricity capacity in Ukraine (excluding the temporarily occupied territory of the Autonomous Republic of Crimea by the Russian Federation) performing under the set feed-in tariff increased by 5965 MW (from 967 MW to 6932

MW). The overall capacity of renewable energy generation in Ukraine is currently (as of 1 January 2020) dominated by solar installations (71%), while biogas made up only 1.24% (Figure 2).

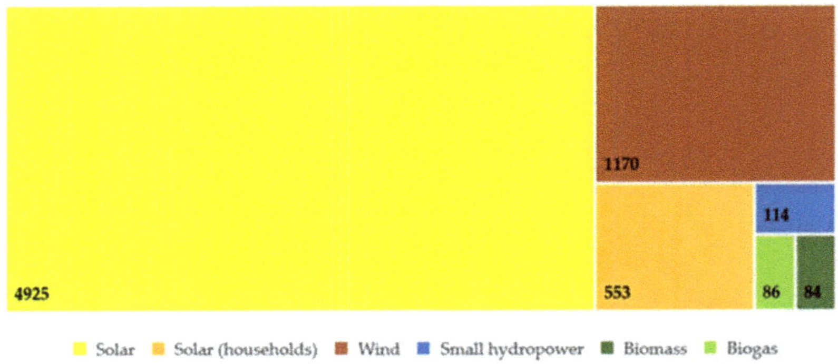

Figure 2. Renewable energy facilities functioning within the feed-in tariff as of 1 January 2020, by sources (in MW). Source: own compilation based on [50].

The development of renewable energy especially intensified in 2019. In that year alone, renewable energy facilities with a total capacity of 4642 MW were put into operation, which is five times the capacity installed in 2018. However, this was achieved mainly by expanding solar power plants: 76% of the total installed generation capacity in that year was from solar installations, while biomass and biogas grew by only 0.7% and 0.9%, respectively [50].

During 2019 the renewable energy facilities functioning under the feed-in tariff generated 5908 million kWh, of which: solar power plants generated 2932 million kWh, household photovoltaic stations generated 303 million kWh, wind power plants generated 2022 million kWh, small hydroelectric power plants generated 242 million kWh, biomass power plants generated 162 million kWh and biogas power plants generated 247 million kWh [50]. Thus the share of energy produced from biogas in 2019 amounted to 4.2% of the total from renewables. Of the total functioning 49 biogas plants, agricultural biomass is used as feedstock in 21 plants with a total installed capacity of 59 MW (Figure 3).

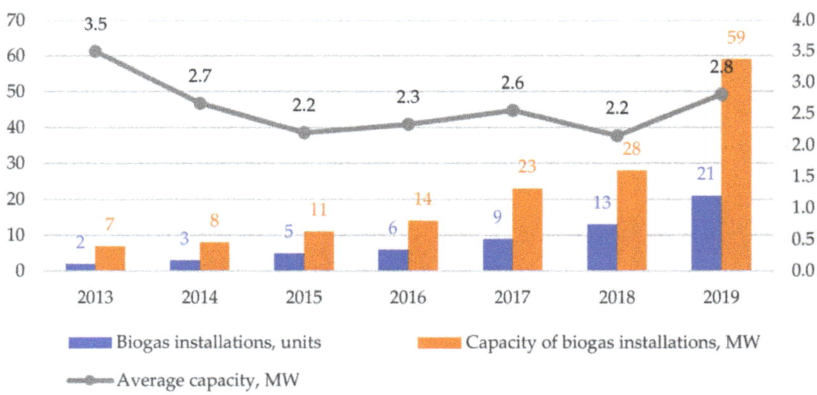

Figure 3. Characteristics of biogas installations running on agricultural feedstock in Ukraine within 2013–2019. Source: own compilation based on [51].

Energy generation based on agricultural biogas in Ukraine is currently carried out primarily by large agricultural or processing enterprises. Key factors of their intense involvement in this activity include large generation capacities, dependence on their own feedstock, a relatively high rate of return, state support and a greater ability to overcome bureaucratic obstacles.

Despite the rather limited overall development of biogas production, Ukraine has significant potential due to both the availability of feedstock and a developed gas distribution system, with ca. 70% of the population having direct access to the natural-gas grid. In 2018, the annual potential of biogas from agricultural waste, food industry and enterprises' wastewater was estimated at 7.8 billion m^3 of methane [52]. According to the Bioenergy Association of Ukraine, the development of biogas use in Ukraine could potentially replace 2.6–18 billion m^3 of natural gas annually [53], thus strengthening national energy security. The latter is a crucial issue in light of Ukraine's high energy dependence on imported resources [54,55].

The use of agricultural biogas for energy generation in Ukraine has numerous advantages compared to other renewable energy sources [56]:

- The production of biogas and thus generation of energy from this source does not depend on weather conditions;
- The combination of biomass from farms operating in various seasons with the biomass from processing plants (e.g., sugar producers) allows energy generation throughout the year;
- The creation of new jobs for workers employed in agricultural processing plants, which in Ukraine are typically located in small towns and are often core/sole employers in those areas;
- Organic fertilizers as by-products of biogas production [57,58] can be further used for organic farming, which is increasingly lacking stable supply (the use of organic fertilizers in Ukraine decreased from 6.2 t/ha in 1990 to 0.27 t/ha in 2019, while the area of organic fertilizer application decreased from 5.5 to 0.8 million ha [59]);
- The efficient management of a wide range of agricultural waste [19,42,60], especially manure, thus helping to reduce odor, restore soil quality and preserve agricultural land that otherwise would be used for waste storage;
- The possibility to facilitate energy generation closer to agricultural waste production sites, thus reducing transport costs and emissions, while optimizing the efficiency of waste use (which loses its energy generation capacity over time [61]);
- The stabilization of the peak loads in the energy transmission grids and covering the possible failures of power generation created by intermittent renewable energy sources such as wind and solar [62];
- A gradual transition to a model of decentralized energy supply beneficial both in a national perspective and in particular for local communities;
- The reduction in methane emissions, which have a higher Global Warming Potential [63] than CO_2, is important from the standpoint of mitigating climate change.

Studies by the Institute for Economics and Forecasting of the National Academy of Sciences of Ukraine [56] show that with the steady development of biogas production and use, its economically feasible potential could reach 9.9 Mtoe by 2030. The use of biogas replacing fossil fuels could result in greenhouse gas emission reduction within the range of 11.5–19.1 Mt CO_2eq. In order to achieve this, the additional consumption of corn silage could reach 13.9 million tonnes. In order to implement the necessary biogas development projects, ca. EUR 4 billion is necessary for the heat and electricity generation field. The implementation of such projects in Ukraine could lead to numerous positive macroeconomic consequences, such as additional GDP growth of 0.3% in 2025–2029, structural changes, including the increased output of machinery and construction sectors, and a slowdown in coal mining. Despite the fact that biogas projects have almost no effect on the level of real household income (from −0.1 to 0.3%), they can potentially aid households by reducing their expenses, including those for heat and electricity.

3. Materials and Methods

Livestock manure is the most suitable feedstock for agricultural biogas production, with more liquified options such as slurry making it possible to increase methane output. It is being used as feedstock for reactors, where the fermentation process is initiated under appropriate conditions [64]. Pig manure is also well suited for biogas production but requires dilution with water. Poultry manure is suitable only in the cage systems, as the floor system increases the risk of the appearance of solid minerals in the manure, which adversely affects the reactor. However, poultry manure can be efficiently combined with livestock manure. By using pure poultry manure as feedstock, there is a danger of high ammonia concentrations; thus, it is crucial to adhere to the proper composition of the feedstock in accordance with the technological solution selected [65]. For small feedstock volumes, it is recommended to use biogas installations with a mesophilic temperature range and a twentieth of the daily load of the total feedstock volume accompanied by a slow stirring process (every 4–6 h).

As in agriculture, there are both specialized and mixed farms; the type determines the availability of feedstock for biogas production in terms of volumes and composition. In order to estimate the amount of manure available for production within a farm, data were used to estimate the yield of excrements depending on the animal species and their age [66]. These data were obtained from a specialized statistical survey of agricultural enterprises, which contains detailed information about 2860 farms in Ukraine keeping specific types of animals (cattle, pigs and poultry). The age structure of farm livestock was estimated on the basis of averaged data provided by the State Statistics Service of Ukraine and extrapolated for the farms analyzed. Coefficients of feedstock conversion into biogas made it possible to calculate the theoretical yield of biogas for each type of farm.

The study focused on the Ukrainian agricultural sector and took into account the following farm types: (1) agricultural enterprises and (2) farming households, yet only those that are either legal entities or private entrepreneurs (referred to in the text as registered farming households), thus excluding the smallest subsistence farms.

For the purposes of this research, the farm sizes were assumed based on the farm livestock (cattle, pigs and poultry) population. This assumption is necessary as Ukrainian legislation does not provide precise definitions of farm sizes, yet these are needed to understand the energy generation potentials based on their own available feedstock and typical features in biogas generation by small, medium-sized and large farms. Thus for cattle farms, the assumed distribution is as follows: small farms with a population up to 100 head, medium in the range between 100 and 1000 head, and large farms of over 1000 head. Small pig farms are assumed to have up to 200 head, medium farms have between 200 and 1000 head and large farms have over 1000 head. For poultry farms, the small ones are assumed to keep under 5000 head, medium farms keep between 5000 and 50,000 head and large farms keep over 50,000 head.

Given the technical limitations of manure use, the study assumes that a biogas installation uses 20% (in terms of energy content) of corn silage and 80% of manure as its feedstock. The capacity of a biogas installation that could be theoretically installed is estimated based on the availability of feedstock (volume of manure + silage). Thus the installation's capacity equals the annual volume of biogas multiplied by two and divided by hours in a day multiplied by days in a year (capacity = annual volume $\times$ 2/(24 $\times$ 365)).

As emphasized in Section 2, the practical possibilities for the utilization of biogas potential (as is also the case for other renewable energy sources) are determined by profitability. The approach to its estimation in the research is based on the concept of the Levelized Cost of Energy (LCOE). In order to obtain the results, the following assumptions were made:

- A biogas installation's utilization lifetime is 25 years. Its construction and entry into service is estimated at 1–2 years (depending on capacity) and is expected to start generating energy in the second year of operation;

- Estimates of the necessary investments for different capacity installations were calculated based on a simple return approach, and detailed results are shown in Table 1;
- Annual operating costs are set at 15% of the investment costs;
- The costs of construction and entry into service of small biogas installations are set at 40% of the total investment costs [67] and at 80% for large installations;
- Annual decommissioning costs are 5% of the installation costs;
- The installed capacity utilization factor (ICUF) is 34.6% for electricity generation and 41.8% for heat production. The sale of heat produced by small installations is rather unlikely, but it could be used for farm purposes, such as heating buildings and greenhouses;
- Generated 1 kW × h = 0.000860 Gcal, 1 Gcal cost equals UAH 1650 or EUR 54, biogas installation consumes 8% of the generated electricity for its own needs;
- Manure transportation cost is 2 UAH/t (0.065 EUR/t);
- Purchased or produced corn silage costs equal 25 EUR/t;

Wherever applicable, the currency exchange rate from UAH to EUR was conducted based on the average rate of the National Bank of Ukraine for 2020, [68] at 30.79 UAH/EUR.

In order to calculate the cost of investment capital, it is assumed that the risk-free rate of return (e.g., on deposits) is set at 9% in Ukrainian currency (UAH), the annual interest rate on a loan is 19% in UAH, the share of equity capital is 30% (the remaining 70% is the loan capital) and the income tax rate is set at 18%.

In order to assess the feasibility of biogas installation, the authors used a simple payback period approach. The latter matters, as biogas technologies have relatively high investment costs, and the commercial banks in Ukraine tend to provide only medium-term loans.

The materials used include the openly accessible data provided by the State Statistics Service of Ukraine [59], as well as an additional purchased database prepared by the State Statistics Service of Ukraine for 2015, which included solely the farms registered as medium- and large-scale agricultural enterprises, covering 59% of all cattle in the sector, 86% of pigs and 100% of poultry.

4. Results

4.1. Assessment of the Economic Feasibility of Biogas Installations in Ukraine

In Ukraine, the group of farms studied holds 33.4% of the total livestock, 59.1% of pigs and 57.4% of poultry (as of 1 January 2020). However, the availability of feedstock for biogas production is diverse, as farms differ substantially according to their livestock quantity. In particular, most farms that keep cattle fall into the categories of either 100–500 head and up to 50 head. However, farms with the largest number of cattle hold over 1500 head (Figure 4).

According to estimates, a 1 MW biogas installation requires ca. 6000 head of cattle. If only cattle manure were used (i.e., without corn silage), then ca. 10,000 head of cattle may be needed [69]. There are examples of such farms in Ukraine, such as the farm in Bziv village (Kyivska region), which has a biogas installation with a generation capacity of 330 kW, fed by 950 cows [70]. For 75 kW installations, ca. 500 head of cattle is sufficient [69].

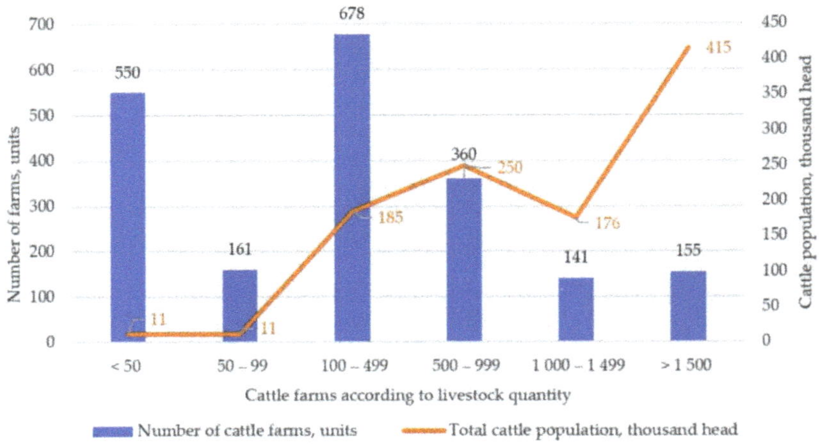

Figure 4. Cattle farms in Ukraine grouped according to livestock quantity as of 1 January 2020. Source: own compilation based on data of the State Statistics Service of Ukraine [59].

There is a similar distribution for pig farms. The most numerous are small farms (up to 100 pigs) and medium farms (ranging between 200 and 499 pigs). However, the largest pig population is held on large farms with over 10,000 head (Figure 5).

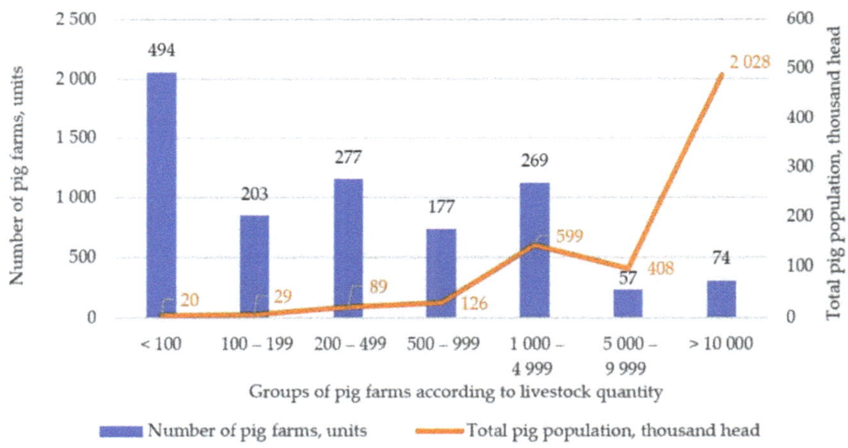

Figure 5. Pig farms in Ukraine grouped according to livestock quantity as of 1 January 2020. Source: own compilation based on data of the State Statistics Service of Ukraine [59].

Concentration is even more visible on poultry farms. Most small farms have fewer than 5000 head of poultry. However, the total number of poultry kept is significantly dominated by large enterprises with over 500,000 head (Figure 6).

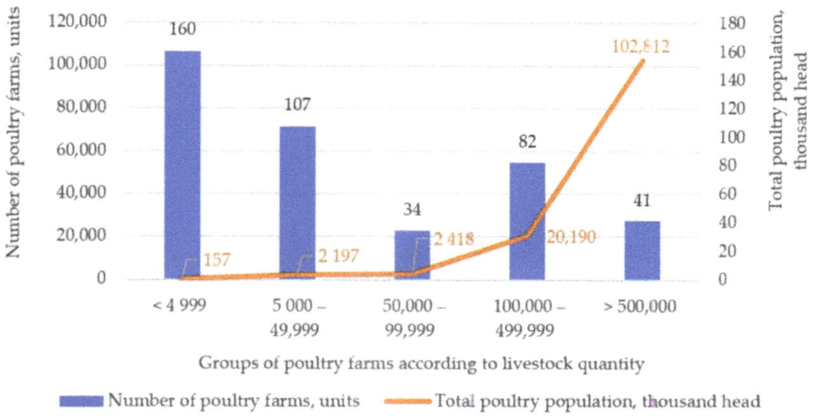

Figure 6. Poultry farms in Ukraine grouped according to livestock quantity as of 1 January 2020. Source: own compilation based on data of the State Statistics Service of Ukraine [59].

As already mentioned, the design of a biogas installation should take into account not only the amount of available feedstock (which directly depends on the number of animals on the farm) but also its composition. Based on the farm data, the share of farms within each livestock type that would be self-sufficient in terms of provision of feedstock for their potential biogas installations was estimated. These values are 33% for cattle farms, 30.8% for pig farms, 6.3% for poultry farms and 29.9% for mixed farms. The required biogas installation's capacity is calculated for each farm type based on the amount of its own available feedstock, maintaining a combination of 80% manure and 20% silage (Figure 7). The visualization shows the distribution of farms that can meet this constraint and makes it possible to see the shares of such farms from the smallest possible generation capacity.

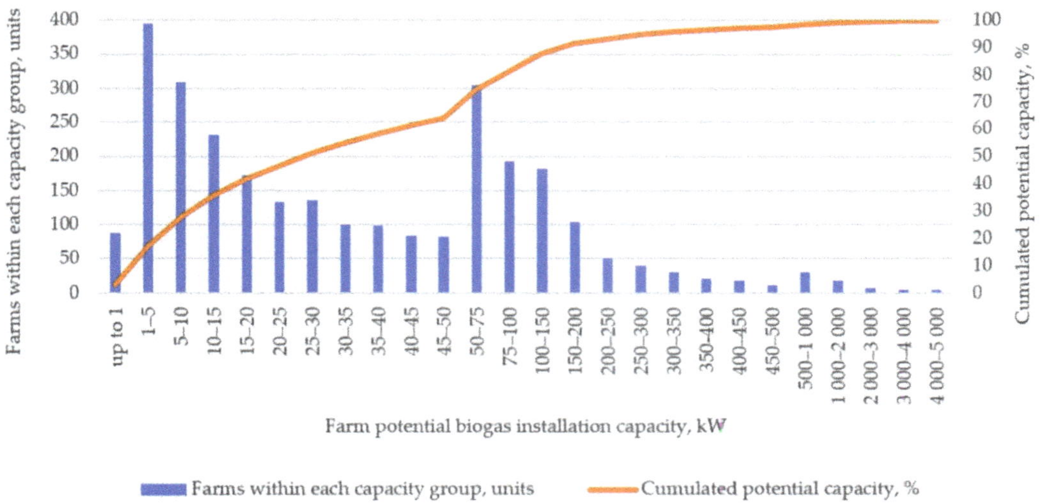

Figure 7. Distribution of farms in Ukraine by potential biogas installation's capacity based on availability of own feedstock. Source: own calculations based on data of the State Statistics Service of Ukraine [59].

The theoretical possibility of the construction of a biogas installation should be consistent with its economic feasibility. This assessment was carried out on the basis of LCOE.

Results of these calculations for different capacities of biogas installations are presented in Table 1.

Table 1. Levelized costs of electricity and heat generation by biogas installations of various capacities and corresponding investment return (simple payback) periods.

Parameters	Generation Capacity								
	25 kW	50 kW	100 kW	150 kW	350 kW	500 kW	1 MW	10 MW	20 MW
Investments, EUR/kW [69]	8400 [20]	8400	6000 [20]	5000 [71]	4500	4500	3500	1800	2000
Levelised generation costs, EUR/kWh	0.837	0.837	0.603	0.500	0.465	0.465	0.370	0.220	0.145
Simple return period, years	16.5	16.5	11.8	9.8	8.9	8.9	6.9	3.5	2.9
Simple return period (in case of potential feed-in tariff at 0.3 EUR/kWh), years	8.4	8.4	6.0	5	4.5	4.5	3.5	1.8	1.4

Source: own calculations.

As the results in Table 1 show, with the current renewable energy support system from biogas installations in Ukraine in the form of feed-in tariff at 0.1239 EUR/kWh, the construction of small-capacity installations (in our case, all types below 500 kW fall into this category) is not feasible, because:

- The current feed-in tariff for biogas installations is too low;
- Typical bank lending is provided for a period significantly shorter (usually up to 5 years) than the foreseeable investment's simple return period;
- Initial investment costs are relatively high.

Thus promoting the development of renewable energy generation from agricultural biogas by small and medium installations (from 100–500 kW) would require the introduction of additional stimulating measures for such farms.

Farms whose production of feedstock does not meet the necessary capacity of biogas installations of 100 kW and below could still be involved in the generation of renewable energy by cooperation with other farms. However, the economic feasibility of such cooperatives is limited by the transport costs; thus, the distance between them would need to be limited to ca. 20 km for liquid manure and ca. 50 km for dry manure. According to our estimates, in compliance with the above-mentioned criteria, it is possible to utilize up to 85.9% of feedstock from farms whose agricultural production volumes do not make it possible to meet the feasibility criteria for their own biogas installation.

4.2. *Impact of State Regulation and Support on the Economic Feasibility of Agricultural Biogas Production in Ukraine*

Currently, the main financial incentive for biogas projects in Ukraine is the feed-in tariff (called the "green tariff") introduced by the Law of Ukraine "On electricity" [72] and substantiated by the Law of Ukraine "On the electricity market" [73]. The feed-in tariff is set at 0.1239 EUR/kWh, and the state guarantees this until the end of 2029. In 2019, the opportunity to generate electricity from biogas was introduced for cooperatives with a capacity of up to 150 kW [74].

The downside is that the set feed-in tariff can be applied only to installations put into operation by January 2023 [75]. This latest legislation change introduced in July 2020 means that from 2023, all biogas installations, including small ones, will have to participate in state auctions to receive state support and qualify for the feed-in tariff. Participation in such auctions will be accompanied by an additional financial burden for producers, as each bidder for the right to participate in the auction must pay 5000 EUR/MW and an additional 15,000 EUR/MW if they win in the auction. This will limit the current financial benefits from biogas installations and will certainly demotivate potential investors. Thus possible investment return periods for biogas installations were assessed both according to the current conditions as well as those set to come into force from January 2023 (Table 2).

Table 2. Possible return periods of biogas installations according to selected generation capacity.

Parameters	Generation Capacity								
	25 kW	50 kW	100 kW	150 kW	350 kW	500 kW	1 MW	10 MW	20 MW
Simple return period under the current state support conditions, years	16.5	16.5	11.8	9.8	8.9	8.9	6.9	3.5	2.9
Simple return period under the state auction system requirements (planned to be initiated from 2023), years	16.6	16.6	11.9	9.9	8.9	8.9	6.9	3.6	3.0
Auction costs, thousand EUR	0.5	1	2	3	7	10	20	200	400

Source: own calculations.

Participation in auctions thus extends the simple return period of projects, yet the effect is not substantial. For large biogas installations, the fact of participation in the auction increases the cost of the project from EUR 20,000 to 400,000. For small installations, the increase ranges from EUR 0.5 to 10 thousand. Compared to the cost of biogas installations, this is not a large additional burden, but the question remains about the participation in such auctions, their clarity and level of bureaucracy. The key advantage of the auction system is that it guarantees the provision of state support for the next 20 years, while currently, the feed-in tariff is set to expire in 2029. However, the disadvantage is that according to [75], the only feed-in tariff for electricity generated besides wind and solar installations cannot exceed the above-mentioned 12 eurocents/kWh, which means that biogas projects are unprofitable for small producers due to the lack of a flexible economic support mechanism. It was estimated that the feed-in tariff would have to be increased to at least 0.3 EUR/kW if the state was aiming to support the development of small biogas projects based on livestock waste.

The legal basis for the introduction of feed-in tariff auctions in Ukraine was enacted in 2019 [74]. The pilot auction was to be held no later than October 2019, then it was postponed to April 2020, then to October 2021 [76] yet without success, and the new date is not yet known. For this purpose, annual quotas of support for particular types of renewable energy installations were to be determined. Given these implications and the need for significant investments, these auctions are obviously designed for large-capacity biogas installations, as the companies constructing them need to have access to large volumes of capital.

Another stimulating tool is also a premium on installations utilizing equipment produced domestically (Table 3). This premium is added on top of the feed-in tariff and depends on the share of domestic equipment used in the biogas project. At the same time, this premium is limited to 10% after six years of exploitation.

Table 3. Premiums for the use of domestic equipment in renewable energy generation projects.

Benefit Level, %	Shares of Domestic Equipment Utilized within the Energy Generation Project, %
5	between 30 and 50
10	between 50 and 70
20	over 70

Source: [75].

With the adoption of the Law of Ukraine "On the natural gas market" [77], biogas producers were granted the right to access gas transmission and distribution systems, gas storage facilities and LNG installations, as well as to connect to gas transmission and distribution systems, provided that technical and safety standards are met. The physical and technical characteristics of biogas should also meet the standards for natural gas.

In terms of electrical energy, according to the Law "On electricity market" [73], the taker (referred to in the legislation as the Guaranteed Buyer) is obligated to purchase electricity produced by households with installations of up to 50 kW capacity (the excess electricity above their monthly consumption volume), with the price set as the feed-in tariff. Similarly, the regional service provider is obligated to purchase electricity from producers (including energy cooperatives with a capacity of up to 150 kW) of all electricity supplied, reduced by the amount of electricity consumed for their operational needs. It is estimated that biogas installations typically consume ca. 5–8% of the generated electricity to ensure their operation. According to the Bioenergy Association of Ukraine [51], if biogas plants could sell 100% of the generated electricity to the grid, it would slightly increase the investment attractiveness of biogas production. Estimates were thus made to assess the simple return period of biogas installations with different capacities depending on the selected support scenarios and conditions: (1) at current feed-in tariff rate and without the purchase of electricity for the installation's operational needs; (2) at the current feed-in tariff rate and with the purchase of electricity for the installation's operational needs; (3) at a potential feed-in tariff rate of 0.3 EUR/kWh and with the purchase of electricity for the installation's operational needs (Figure 8).

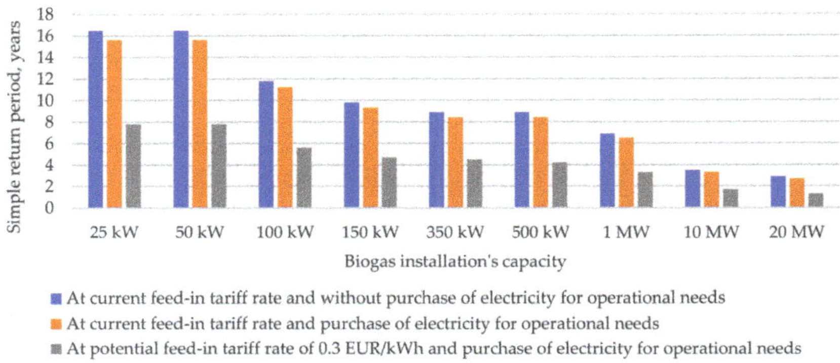

Figure 8. Simple return period of different-capacity biogas installations depending on the support scenarios. Source: own calculations.

The assessment (Figure 8) shows that a simultaneous increase in the feed-in tariff from 0.1239 EUR/kWh to 0.3 EUR/kWh together with permission to buy electricity from the grid for energy needs of a biogas installation allow an acceptable return period (up to 5 years [78]) of for installations with a capacity starting at 150 kW. In order to achieve an adequate return period for smaller biogas installations (below 150 kW), it would be recommended to additionally compensate their investment (equipment) costs if renewable energy from this source is to be stimulated.

Despite the adoption of amendments to the Law of Ukraine "On heat supply" [79] providing a financial mechanism for non-natural-gas boilers (i.e., biogas in co-generation units), this mechanism has not provided a significant impetus for the development of biogas projects. It is advisable for agro-industrial enterprises in Ukraine to continue developing the generation of electricity and heat with the subsequent sale of heat to neighboring households, as the available feed-in tariff means this option of biogas utilization is most favorable in Ukraine's economic conditions. However, farms are usually located at a distance from heat consumers (other farms and households), so the option of selling heat is rather an exception. Biogas projects can be located in areas where significant agricultural waste is produced, so the heat generated can be used in part to heat the farm itself. The key limitation here is that demand for heat is seasonal (at best half of a year in given climatic conditions), so it is impractical to focus solely on heat generation.

According to Ukrainian legislation [80], disposal of animal waste must be carried out exclusively by specialist companies and can not be performed by companies producing animal products for human consumption. Manure and animal residues belong to the second class of waste and can either be burned or converted into organic fertilizer after mandatory sterilization under pressure or converted into biogas by pressure sterilization. Processing facilities for animal waste must be separate from companies producing foodstuffs. Class 2 waste can be used to make organic fertilizers and soil improvers that can be put on the market. Waste disposal companies are market operators. The law does not define the minimum and maximum size of such enterprises, which means that it also applies to small enterprises. Market operators are required to report their activities to the central veterinary authority on a monthly basis. They dispose of animal residues at their own expense or at the expense of state or local budgets that provide subsidies to businesses to partially compensate for the costs associated with the disposal and removal of animal by-products.

The procedure for using state funds to finance measures related to the disposal of animal by-products is approved by the Cabinet of Ministers of Ukraine. The State Budget of Ukraine for 2021 does not provide for such expenditure. Market operators who dispose of or remove by-products without pressure sterilization or without processing into biogas under pressure after sterilization can be fined. For legal entities, the fine is 23–30 minimum wages (in 2020, the minimum monthly wage in Ukraine equaled EUR 153); for private entrepreneurs, it is 8–15 minimum wages. For using unsealed containers for the transport of livestock waste by market operators, legal entities are subject to a fine of 8–12 minimum wages, and individual entrepreneurs are subject to a fine of 5–8 minimum wages. Sometimes companies prefer to pay a fine without further measures to dispose of livestock waste. However, large agribusinesses are subject to inspections by the Ministry of Health, the Prosecutor's Office, the Sanitary and Epidemiological Service and the Environmental Inspectorate of the Ministry of Ecology and Natural Resources. In some cases, biogas installations using Ukrainian equipment are therefore not economically feasible even to cover the additional costs of waste disposal. For example, for a pig farm with 12,000 head, which produces 20,000 tonnes of waste, the manure disposal costs reach ca. EUR 10,000 annually [81].

In Ukraine, there are sectoral budget support programs for agricultural producers, with additional preferences for farms. In particular, the small farms are supported, which may be a synergy with the feed-in tariff for renewable energy generation utilizing biogas installations (Table 4).

Table 4. State farm-support programs of additional assistance to potential biogas producers.

Programs	Maximum Support Value	Specific Features
	Programs available to all types of agricultural producer	
	I. State support for livestock development	
Reimbursement of the cost of livestock holding facilities	(1) For facilities up to UAH 500 million (EUR 16.2 million) excluding VAT—30% of the total costs; (2) For facilities with a value of over UAH 500 million (EUR 16.2 million) excluding VAT—30% of UAH 500 million (EUR 16.2 million); (3) For facilities creating 500 or more workplaces, regardless of their cost (excluding VAT)—30% of the total costs.	Costs for livestock farms and complexes keeping cattle, pigs, poultry; for enterprises processing agricultural products and/or by-products of animal origin not intended for human consumption, belonging to II and III categories.
Reimbursement of the cost of breeding animals purchased	Up to 50% of the purchase cost (excluding VAT), but not more than: - For breeding heifers, heifers, cows—UAH 31,500 (EUR 1000) per head; - For breeding pigs and boars—UAH 10,000 (EUR 325) per head.	
Reimbursement of the value of facilities financed by bank loans	Reimbursement of 25% of the loan of up to 5 years for legal entities and private entrepreneurs.	Reimbursed is the share of loan for construction and/or reconstruction of livestock farms and complexes for keeping cattle, pigs and poultry; enterprises for processing agricultural products and / or by-products of animal origin, belonging to categories II and III, including the cost of equipment.

Table 4. *Cont.*

Programs	Maximum Support Value	Specific Features
	Programs available to all types of agricultural producer	
	I. State support for livestock development	
Partial reimbursement of costs for the purchase of agricultural machinery of domestic origin	Up to 40% of the costs of equipment purchased from domestic manufacturers, the list of which is determined by the Ministry of the Economy.	Twenty-five percent for all agricultural producers + 15% for registered farming households; the list includes equipment for the generation of renewable energy.
Loan reimbursements	UAH 15 million (EUR 487,000) for livestock producers, UAH 5 million (EUR 487,000) for other entities.	A 1.5 discount rate of the National Bank of Ukraine, but not higher than defined in the loan agreements, reduced by 5 percentage points.
	Programs dedicated to registered farming households	
Budget subsidy to registered farming households (except newly created)	UAH 40,000 (EUR 1300).	The head of the farm must be under 35 years old at the time of application.
Budget subsidy to newly created registered farming households	UAH 60,000 (EUR 1950).	Can be applied for in the first three years of the farm's existence.
Partial reimbursement of costs associated with receiving services from advisory services	90%, but not over UAH 10,000 (EUR 325).	Costs cannot be reimbursed under the "energy" section but can be reimbursed under "animal health" section.
Budget subsidy for keeping 5 or more cows	Up to UAH 250,000 (EUR 8120).	UAH 5000 (EUR 162) per head of cattle.
Reimbursement of Single Social Payments for registered farming households	-	Within 10 years in the range of 0.1–0.9 of the minimum single social payment premium if the remaining share is paid by the head or by members of the registered farming household.
State loans under the Ukrainian State Fund for Support of Farmers [82]	UAH 500,000 (EUR 16,240).	For acquisition of fixed assets and replenishment of working capital.

Source: compiled based on [83].

However, most of these programs are aimed at supporting agricultural production and could be treated as only supplementary support for the development of energy generation based on biogas. Moreover, the partial reimbursement of costs for the purchase of agricultural machinery of domestic origin includes a limited list of components for the generation of renewable energy (e.g., heat generators based on straw feedstock). At the same time, there should be dedicated state programs that would make it possible to take out a loan to finance biogas installations directly. It would also be advisable to extend the list of advisory services and include issues of renewable energy generation, as the lack of knowledge and advisory support to farms planning to engage and establish renewable energy installations is one of the key obstacles to their development.

In addition to government support, farmers and processors could benefit from a variety of sponsorships. In particular, the lending program through the Fund for Development of Entrepreneurship [84] was established jointly by the German state investment and development bank (KfW) [85], the Ukrainian government and the National Bank of Ukraine. Within this support, it is possible to apply for a loan of up to EUR 250,000 for five years to finance fixed assets of medium-sized enterprises (micro-crediting). Up to EUR 100,000 can be provided to enterprises, including for energy efficiency and energy saving, as well as job creation in depressed regions. The loan rate is calculated on the basis of the National Bank of Ukraine's discount rate plus 5%; the loan's timeframe is limited to six years [86]. Under the EU4Business program, it is possible to obtain a loan of up to EUR 5 million for up to ten years at a rate of 6–10% [87].

5. Discussion

By comparing the dynamics of energy generation based on biogas with the EU's experience, it can be stated that in the past three decades, Ukraine has shown relatively slow progress. In the EU, renewable energy generation based on biogas is developing at a much higher rate [88]. In 2018, 18,202 existing biogas plants had a total capacity of over 11 GW, producing 63,511 GWh of electricity [89]. In European countries, the main feedstock for biogas installations is crop residues and energy crops (almost half of the total feedstock) and livestock manure (ca. one-third) [11]. Germany is the European leader in the development of biogas energy, especially in terms of using the potential of agricultural feedstock. Thus, from 2014 to 2019, an annual average of 126 biogas installations were added, while the highest rate was from 2009 to 2011 (Figure 9). In addition, Germany is the European leader in the number of biomethane installations; as of 2018, the country had 195 units out of the total 540 units in the EU [90]. These installations are aimed at purifying biomethane and transferring it to the general distribution network.

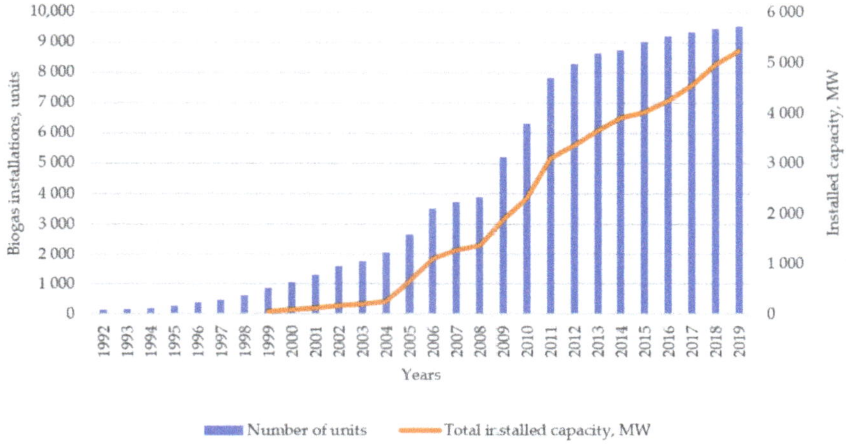

Figure 9. Biogas installations in Germany within 1992–2019. Source: own compilation based on [28].

The dynamics of average annual biogas installation capacity in Germany show a rising trend (Figure 10). While by 2005, the average capacity of the new biogas installations was in the range of 150–200 kW, in 2005–2013, it increased to an average of ca. 0.5 MW, and after 2014, a tendency of large installations began, stabilizing around 2016.

Figure 10. Average annual new biogas capacity in Germany (in MW). Source: own compilation based on [28].

This development of biogas installations in Germany became possible not only due to the stimulation of renewable energy generation but also through the reduction in the attractiveness of fossil fuel. Thus, in the early 2000s, a tax on fossil fuels was imposed at 0.47–0.67 EUR/L of petroleum products and 0.015 EUR/kWh on electricity generated from other fossil fuels. There was a feed-in tariff of 0.0616–0.27 EUR/kWh, as well as 15–35% of additional subsidies for mini biogas installations [91]. As of 2017, the feed-in tariff for biogas from biowaste oscillated around 0.134–0.237 EUR/kWh, the feed-in tariff for biogas from organic fertilizers (manure) was at 0.2314 EUR/kWh for installations with a capacity of less than 75 kW. Feed-in tariffs are ranked overall depending on the size of biogas installation and the type of feedstock used; the energy producers are guaranteed the level of the feed-in tariff for ten years. In order to participate in the auctions, the installation capacity needs to be over 150 kW. If the auctions are won, state support is provided for up to 20 years [92]. By comparing Germany and Ukraine in terms of energy generation from biogas, it can be stated that Ukraine's biogas market is in a similar condition and development level to Germany's in the early 1990s. Maintaining such favorable policies for renewable energy generation in Germany has ensured the development not only of large biogas investment projects but also stimulated the appearance of numerous small and medium-sized installations, which is of considerable interest for Ukraine.

Most studies in Ukraine [34,41–43] stress the importance of biogas production development with a later generation of bioenergy, but they fail to take into account the investment issues depending on the size and availability of feedstock in small and medium-sized farms. What is more important, studies [32,33,44] that go deeper into the feed-in tariff analysis do not search for alternatives in order to propose more detailed approaches helping to create more beneficial conditions for various types of entities that might engage in biogas production. Thus there is a gap between theory and practice, as many researches do not differentiate between the above-mentioned dimensions, therefore failing to analyze the key predisposition—economic feasibility—to the full extent. The assessment presented in Section 4 makes it possible to understand these issues and shows that economic feasibility is missing from current state support conditions for all small and most medium-sized farms.

Obtained results go in line with the conclusions from [34] in terms of Ukrainian high potential for biogas generation, especially by farming households. The constant rise of natural gas prices emphasized upon in the aforementioned study support the necessity to tackle the issues hindering economic feasibility of energy generation from biogas by small and medium farms, as these entities would be the first to lose due to increased energy costs in case of further exploitation of fossil fuels. Another study [44] states that "a necessity to implement such projects is the introduction of an economically substantiated feed-in tariff for generation of energy based on biogas", as with the current levels of the tariff, such feasibility is not reachable for most potential producers. These statements were proven true by the estimations conducted in the current study, as the small and most medium farms would not be able to return the investments in the expected timeframe based on the existing feed-in tariff for biogas installations.

The current state program aimed at supporting biogas energy generation does not differentiate the feed-in tariff, due to which there is a sharp polarization in investment—only large agricultural enterprises become involved in bioenergy based on biogas, and only 21 such installations have been established so far. Such a rate is not sufficient to make biogas energy a relevant renewable source, despite the existing agricultural potential defined in Sections 2 and 4.

Despite these implications, it is possible to seek solutions outside of solely the feed-in tariff. Farms whose feedstock production volumes do not allow profitable utilization of their biogas installations could potentially be merged into energy cooperatives based on the economic acceptability of feedstock transport costs. It is estimated that the economically feasible distance for the delivery of feedstock is up to 20 km for liquids and up to 50 km for solids [33]. Based on the previous administrative-territorial division [93,94], farms with biogas installations of up to 100 kW located in the same district as other farms could be united into cooperatives. While there are possible limitations, in many districts, the 20 km range criterion would be maintained, although this aspect would need to be studied further.

Additionally, the establishment of energy cooperatives allows the construction of energy facilities at the expense of local communities. Ukraine established its first municipal renewable energy cooperative, "Solar City" in Slavutych (Kyivska region), in February 2020, which is a 200-kWh-capacity solar power installation [95]. In Ukraine, the creation of energy cooperatives may be particularly appropriate in rural areas, as they are home to over a third of the population, while the costs and quality of energy services are not always satisfactory. There are examples of uniting nearby communities according to the principles of an energy cooperative when the community uses waste from the production/cultivation of basic agricultural products as energy feedstock (for example, the Yagidnyi Krai cooperative in the Ternopilska region) [96].

The definition of "energy cooperative" in Ukraine is established by the Law "On alternative energy sources" [97]. Further, in accordance with the provisions of the Law of Ukraine "On the electricity market" [73], energy cooperatives can sell their electricity either to the Guaranteed Buyer or to private households through the regional energy service provider. The creation of energy cooperatives is in line with a number of global trends, including distributed energy generation, the use of renewable energy sources and the concentration of energy production near places of direct consumption. The largest energy cooperatives are in the United States, Germany, Denmark, Sweden, the Netherlands and Austria [98–100]. As of 2015, there were 1000 cooperatives in Germany, which owned 47% of the renewable energy capacity [101].

Wider use of biogas requires changes in infrastructure, such as new roads for the supply of feedstock. Long-term contracts between suppliers of feedstock and enterprises that process it are necessary, which is especially relevant for small and medium-sized projects. Priority grid connection could be introduced for biogas projects, with the small installations (up to 500 kW) having these permits lifted. The construction of biogas pipelines is required to supply biomethane to the gas transportation systems. A Ukrainian corporation ("MHP Agro and Industrial Holding") has experience in building such biogas pipelines at ca. EUR 1 million per kilometer [102].

In the long run, it is possible to introduce the mandatory use of biogas by farms that produce the corresponding feedstock. This can be performed by introducing new national building standards for the construction of new agro-industrial companies whose activities are related to waste generation (farms, breweries) or the introduction of requirements for mandatory measures to reduce methane and carbon dioxide emissions. Legislation defining the need for sterilization of livestock waste under pressure should be repealed for biogas production, and existing fines for improper management of agricultural waste should be canceled.

Government loan guarantees would be beneficial. In order to enable the spread of biogas projects for small and medium-sized enterprises in Ukraine, interest rates on loans should be reduced through further cooperation of Ukrainian banks with international financial institutions such as the Global Environment Facility, the European Bank for Reconstruction and Development and the Clean Technology Fund.

There is poor dissemination of information and a lack of nationwide information campaigns on the use and construction of renewable energy sources in the agro-industrial complex. In our opinion, this barrier is no less important than financial barriers. Large agricultural corporations (referred to in Ukraine as agriholdings) already understand the benefits of using biogas and are launching large biogas projects. However, potential small and medium-sized producers need detailed information that not only provides information about the types of equipment but also about institutions providing financial support (state and private), as well as what would be the practical steps to construct a renewable energy installation and connect it to the energy grid or how to obtain the feed-in tariff for the energy. Changes should be made to requirements for obtaining state support, cutting bureaucracy and shortening the time lag between investments and actual support.

An important difference between small and large biogas projects (not to mention wind and solar energy) is access to development companies, which could be hired to prepare an investment feasibility study, change the land documentation to allow the placing of energy generation installations, obtain permits implementing the project, connect to the grid and start construction work [103]. Such functions are not widely available or affordable for small and medium-sized biogas projects, thus excluding them from investments in renewable energy generation. Detailed step-by-step information is needed to enable and speed up this process.

6. Conclusions

The study identified the economic feasibility of the development of renewable energy based on biogas projects in Ukraine, its key obstacles and legislation implications. Medium-sized and small farm capacities were focused upon as these types of farms in Ukraine are less economically viable. State support measures were analyzed to understand possible synergies between potential agricultural biogas projects and programs aimed at general or specific agricultural activities.

The analysis has shown there are relatively high initial investment costs, especially for small biogas installations. The smaller the installation, the higher the investment cost per unit of capacity. Loans and a special program for the implementation of small projects (up to 0.5 MW) are needed to aid small farms. This can be achieved in part through international financial institutions with energy efficiency and renewable energy generation programs, for example, by the International Finance Corporation or by the European Bank for Reconstruction and Development. The latter launched the EU4Business program in 2016 in cooperation with the European Commission, aimed at small- and medium-sized enterprises. It is also necessary to expand the Ukrainian state program to support farmers, including in terms of stimulating the purchase of certain types of machinery and equipment [104].

The feed-in tariff or the upper limit of the auction purchase price (currently the same as the feed-in tariff) is too low for biogas installations, which typically require high investment costs. The low feed-in tariff does not allow payback periods of biogas investment projects of less than seven years while, due to national currency volatility, inflation and the unstable political situation, bank institutions in Ukraine typically consider financing projects with a payback period of four to five years. In addition, it is advisable to differentiate feed-in tariff coefficients for biomass and biogas depending on the feedstock used so that the feed-in tariff coefficient or the upper limit of the auction price should be higher for biogas from agricultural waste than for biogas from by-products from alcoholic beverage production. In addition, the feed-in tariff or auction price should be differentiated depending on the installed capacity—the lower the capacity, the higher the tariff.

Currently, electricity from wind and solar energy installations is cheaper than that from biogas, while the feed-in tariffs in Ukraine for these sources of energy are higher. This is caused by significant differences in the costs of equipment, especially in the field of solar energy. It is likely that, with the introduction of liability for supply shortages, electricity from biogas would no longer be significantly more expensive, as upgrading meteorological stations to improve the quality of the forecast requires investment, and the question of who should invest in upgrading meteorological equipment in Ukraine has not been sufficiently considered. Moreover, wind and solar energy generation are dependent on seasonal and weather conditions, as well as daytime slots. At the same time, energy from agricultural biogas could be carried out either on a permanent basis or on-demand to cover gaps in the energy supply from more intermittent sources. Thus the lower feed-in tariffs for biogas are not well-grounded and do not take into account the importance and role of this source, yet this is a point for future research.

In conclusion, the state policy on biogas production and energy generation from agricultural sources is still fragmented and does not take into account the diversity of farms and their peculiarities, including the small and medium-sized ones. Within the existing legal framework, even after the launch of auctions, small farms have no real financial incentives to launch biogas projects. The current feed-in tariff is too low to provide reasonable payback periods. The marginal auction price is also too low. Small and medium-sized farms also need to prepare the investment projects on their own, as well as have limited access to financing. In order to overcome these problems, a specialist state biogas development program with government loan guarantees is needed, as well as advisory support and services to enable small and medium-sized farms to overcome the barriers of a lack of information. Such a complex approach could improve the conditions and make investments in bioenergy generation based on biogas more feasible, at least for a larger group of farms of various sizes.

Conclusions based on the studied Ukrainian experience could also serve as recommendations for other countries aiming to develop renewable energy generation. Financial (state or otherwise) support for diffusion of such innovations plays a crucial role, as its intensification is not possible if economic feasibility is not achieved and transfer to more "green" energy generation technologies are not incentivized, either in financial or organizational dimension. As for the latter, the experience of renewable energy development in Ukraine highlights the importance of institutional aspects, proving that transparency, consistency, stability and long-term predictability of state support and regulations are as crucial for the appropriate investment climate and transition to more environmentally-friendly energy generation solutions.

Author Contributions: Conceptualization, G.T., S.K. and V.K.; methodology, G.T. and S.K.; investigation, G.T., S.K., V.K., A.W. and R.P.; writing—original draft preparation, G.T., S.K., V.K. and A.W.; writing—review and editing, G.T., S.K., V.K., A.W. and R.P.; visualization, G.T., S.K. and V.K.; supervision, G.T. All authors have read and agreed to the published version of the manuscript.

Funding: This research received no external funding.

Institutional Review Board Statement: Not applicable.

Informed Consent Statement: Not applicable.

Data Availability Statement: Not applicable.

Conflicts of Interest: The authors declare no conflict of interest.

References

1. Gulev, S.K.; Thorne, P.W.; Ahn, J.; Dentener, F.J.; Domingues, C.M.; Gerland, S.; Gong, D.; Kaufman, D.S.; Nnamchi, H.C.; Quaas, J.; et al. Changing State of the Climate System. In *Climate Change 2021: The Physical Science Basis. Contribution of Working Group I to the Sixth Assessment Report of the Intergovernmental Panel on Climate Change*; Masson-Delmotte, V., Zhai, P., Pirani, A., Connors, S.L., Péan, C., Berger, S., Caud, N., Chen, Y., Goldfarb, L., Gomis, M.I., et al., Eds.; Cambridge University Press: Cambridge, UK, 2021.
2. Eyring, V.; Gillett, N.P.; Rao, K.M.A.; Barimalala, R.; Parrillo, M.B.; Bellouin, N.; Cassou, C.; Durack, P.J.; Kosaka, Y.; McGregor, S.; et al. Human Influence on the Climate System Supplementary Material. In *Climate Change 2021: The Physical Science Basis. Contribution of Working Group I to the Sixth Assessment Report of the Intergovernmental Panel on Climate Change*; Masson-Delmotte, V., Zhai, P., Pirani, A., Connors, S.L., Péan, C., Berger, S., Caud, N., Chen, Y., Goldfarb, L., Gomis, M.I., et al., Eds.; Cambridge University Press: Cambridge, UK, 2021.
3. NASA: Global Climate Change. Scientific Consensus: Earth's Climate Is Warming. 2021. Available online: https://climate.nasa.gov/scientific-consensus (accessed on 2 August 2021).
4. Berlie, A.B. Global warming: A review of the debates on the causes, consequences and politics of global response. *Ghana J. Geogr.* **2018**, *10*, 144–164.
5. Caserini, S.; Coyaud, S.; Persico, G.; Messori, G. Evaluating the scientific credentials of the supporters of public petitions denying anthropogenic climate change. *Tellus A Dyn. Meteorol. Oceanogr.* **2021**, *73*, 1–4. [CrossRef]
6. European Environment Agency. Emerging Waste Streams: Opportunities and Challenges of the Clean-Energy Transition from a Circular Economy Perspective. 2021. Available online: https://www.eea.europa.eu/publications/emerging-waste-streams-opportunities-and (accessed on 18 August 2021).
7. Chang, S.; Frauke, U. Circular economy for clean energy transitions: A new opportunity under the COVID-19 pandemic. *Appl. Energ.* **2021**, *289*, 116666. [CrossRef]
8. Heath, G.; Joint Institute for Strategic Energy Analysis. Moving toward a Circular Economy of Materials for Clean Manufacturing. 2019. Available online: https://www.nrel.gov/docs/fy19osti/73689.pdf (accessed on 7 September 2021).

9. Winquist, E.; Van Galen, M.; Zielonka, S.; Rikkonen, P.; Oudendag, D.; Zhou, L.; Greijdanus, A. Expert views on the future development of biogas business branch in Germany, The Netherlands, and Finland until 2030. *Sustainability* **2021**, *13*, 1148. [CrossRef]
10. IREA. *Global Energy Transformation: A Roadmap to 2050*; International Renewable Energy Agency: Abu Dhabi, United Arab Emirates, 2018; p. 76.
11. Outlook for Biogas and Biomethane: Prospects for Organic Growth. World Energy Outlook Special Report. 2020. Available online: https://www.iea.org/reports/outlook-for-biogas-and-biomethane-prospects-for-organic-growth/the-outlook-for-biogas-and-biomethane-to-2040 (accessed on 3 September 2021).
12. Strubenhoff, H.W. Unleashing the Potential of Agriculture in Ukraine. Brookings. 2016. Available online: https://www.brookings.edu/blog/future-development/2016/03/03/unleashing-the-potential-of-agriculture-in-ukraine (accessed on 18 August 2021).
13. FAO. Regional Office for Europe and Central Asia (REU). Assessment of the Agriculture and Rural Development Sectors in the Eastern Partnership Countries: Ukraine. 2021. Available online: https://www.eesc.europa.eu/sites/default/files/resources/docs/ukraine_assessment--2.pdf (accessed on 22 September 2021).
14. Deppermann, A.; Balkovič, J.; Bundle, S.; Di Fulvio, F.; Havlík, P.; Leclère, D.; Lesiv, M.; Schepaschenko, D. Crop production potentials in Russia and Ukraine—Intensification versus cropland expansion. Towards sustainable agri-food systems: Balancing between markets and society. In Proceedings of the XV EAAE Congress "Towards Sustainable Agri-Food Systems: Balancing between Markets and Society", Parma, Italy, 28 August–1 September 2017. [CrossRef]
15. Kyryzyuk, S.; Krupin, V.; Borodina, O.; Wąs, A. Crop Residue Removal: Assessment of Future Bioenergy Generation Potential and Agro-Environmental Limitations Based on a Case Study of Ukraine. *Energies* **2020**, *13*, 5343. [CrossRef]
16. Galchynska, J. Biomass energy potential forecasting as a preliminary development of bioenergy in Ukraine. *Bull. Chernivtsi Inst. Trade Econ.* **2019**, *I*, 12–27. [CrossRef]
17. Bondarenko, V.; Gontaruk, Y. Analysis of the state and prospects of the development of agriculture of Ukraine. *Probl. Prosp. Econ. Manag.* **2021**, *2*, 86–98.
18. Fursenko, K. Ways to Increase the Efficiency of Livestock Production. Kherson State University. 2020. Available online: http://ekhsuir.kspu.edu/handle/123456789/13170 (accessed on 1 October 2021).
19. Tokarchuk, D.; Prishlyak, N.; Palamarenko, Y. Prospects for use of crop waste for biogas production in Ukraine. *Agrosvit* **2020**, *22*, 51–57. [CrossRef]
20. Wąs, A.; Sulewski, P.; Krupin, V.; Popadynets, N.; Malak-Rawlikowska, A.; Szymańska, M.; Skorokhod, I.; Wysokiński, M. The potential of agricultural biogas production in Ukraine—Impact on GHG emissions and energy production. *Energies* **2020**, *13*, 5755. [CrossRef]
21. Gorobets, O.V. Classification of agricultural waste and selection of their utilisation technology. *Ecol. Sci.* **2020**, *4*, 225–229. [CrossRef]
22. Yurchenko, V.; Poltava State Agrarian Academy. Safe Neutralization of Organic Waste as an Important Prerequisite to the Environmental Stabilization of Poultry Enterprises Territories. 2018. Available online: http://repository.hdzva.edu.ua/handle/repoHDZVA/648 (accessed on 19 August 2021).
23. Sklyar, A.; Sklyar, R. *Directions for the Use of Organic Resources in Stock-Raising*; Tavria State Agrotechnological University: Oblast, Ukraine, 2011; Volume 11, Available online: http://elar.tsatu.edu.ua/handle/123456789/201 (accessed on 3 September 2021).
24. IEA Energy Technology Essentials. Biomass for Power Generation and CHP. 2007. Available online: https://www.nachhaltigwirtschaften.at/resources/pdf/iea_ete03.pdf (accessed on 2 September 2021).
25. Yevdokimov, Y.; Chygryn, O.; Pimonenko, T.; Lyulyov, O. Biogas as an alternative energy resource for Ukrainian companies: EU experience. *Innov. Mark.* **2018**, *14*, 7–15. [CrossRef]
26. SAF: Sustainable Agribusiness Forum. The Number and Total Capacity of Biogas Plants in Europe Continued to Grow. 2020. Available online: https://saf.org.ua/en/news/906 (accessed on 4 September 2021).
27. State Agency on Energy Efficiency and Energy Saving of Ukraine. Information about Capacity and Volumes of Electricity Generated by Renewable Energy Sources under Feed-In Tariff (as of 1 January 2020). 2020. Available online: https://saee.gov.ua/sites/default/files/VDE_2019.pdf (accessed on 2 September 2021).
28. Fachverband Biogas. Biogas Market Data in Germany 2020/2021. 2021. Available online: https://www.biogas.org/edcom/webfvb.nsf/id/EN-German-biogas-market-data/$file/21-10-14_Biogasindustryfigures_2020-2021_english.pdf (accessed on 19 October 2021).
29. Sakun, L.; Riznichenko, L.; Vielkin, B. Prospects of biogas market development in Ukraine and abroad. *Econ. Organ. Manag.* **2020**, *1*, 160–170. [CrossRef]
30. Bakulina, G. Wastes—A threat to the ecology or the wealth of the country? World practices and the prospects of biogas production for Ukraine. *Econ. Soc.* **2017**, *9*, 803–808. Available online: https://economyandsociety.in.ua/journals/9_ukr/140.pdf (accessed on 14 October 2021).
31. Okhota, Y. The main trends of efficient use of biogas in Ukraine. *Efektyvna Ekonomika* **2018**, *4*. Available online: http://www.economy.nayka.com.ua/pdf/4_2018/162.pdf (accessed on 15 October 2021).
32. Shuhalo, V. Ecological and Economic Efficiency of Biogas Production and Utilization in the Agricultural Sector of the Economy. Ph.D. Thesis, Lviv National Agrarian University, Dubliany, Ukraine, 2021. Available online: http://www.lnau.edu.ua/lnau/attachments/7068_dis.pdf (accessed on 1 November 2021).

33. Green Energy Technology Market—GETMARKET. Investments into Alternative Energy: How to Earn from Biogas in Ukraine. 2020. Available online: https://getmarket.com.ua/ua/news/investiciyi-v-al-ternativnu-energetiku-yak-v-ukrayini-zaroblyayut-na-biogazi (accessed on 10 August 2021).
34. Tokarchuk, D. Economic and ecological benefits of using biogas plants in households. *Econ. Financ. Manag. Top. Issues Sci. Pract. Act.* **2018**, *6*, 39–49. Available online: http://efm.vsau.org/storage/articles/February2020/t8QGzOvaTQ80lKBrnNpY.pdf (accessed on 5 October 2021).
35. Sidorov, Y. Modern biogas technologies. *Biotechnol. Acta* **2013**, *6*, 46–61. [CrossRef]
36. Kucher, I.; Adegov, A.; Yurchenko, Y.; Koval, O. Local biogas plants. *Constr. Mater. Sci. Mech. Eng.* **2017**, *99*, 107–112.
37. Tkachenko, S.; Stepanov, D.; Stepanova, N. Analysis of social and energy and sustainable efficiency of biogas technology implementation. *Visnyk VPI* **2020**, *2*, 34–41. [CrossRef]
38. Kucheruk, P. Improving the Efficiency of Biogas Production by Co-Digestion of Manure with Maize Silage. Ph.D. Thesis, Institute of Engineering Thermophysics of the National Academy of Sciences of Ukraine, Kyiv, Ukraine, 2016. Available online: http://ittf.kiev.ua/wp-content/uploads/2016/04/ar_kucheruk_sajt.pdf (accessed on 29 October 2021).
39. Zinchenko, A.; Kunbuttaeva, A. *Small Participants of RES-Market in Ukraine. Research of Generation Capacity Segment up to 1 MW*; The Heinrich Böll Foundation: Kyiv, Ukraine, 2020; Available online: https://ua.boell.org/sites/default/files/2020-08/%D0%9C%D0%B0%D0%BB%D1%96%20%D1%83%D1%87%D0%B0%D1%81%D0%BD%D0%B8%D0%BA%D0%B8%20%D0%92%D0%94%D0%95-%D1%80%D0%B8%D0%BD%D0%BA%D1%83%20%D0%B2%20%D0%A3%D0%BA%D1%80%D0%B0%D1%96%CC%88%D0%BD%D1%96.pdf (accessed on 28 October 2021).
40. Sydorchuk, O. Mini-biogas installations for private households. *Agrobiogas* **2020**, *5*, 104–107. Available online: https://agrobiogas.com.ua/mini-biogas-plants-for-private-households (accessed on 4 September 2021).
41. Gradovyi, V. Ecological and Economic Assessment and Provide Directions Biogas. Master's Thesis, Ternopil National Economic University, Ternopil, Ukraine, 2017. Available online: http://dspace.wunu.edu.ua/jspui/bitstream/316497/17253/1/%D0%93%D1%80%D0%B0%D0%B4%D0%BE%D0%B2%D0%B8%D0%B9.pdf (accessed on 10 October 2021).
42. Zapałowska, A.; Bashutska, U. The use of agricultural waste for the renewable energy production. *Proc. For. Acad. Sci. Ukr.* **2019**, *18*, 138–144. [CrossRef]
43. Doronin, A. *The Efficiency of Biogas Production in the Agricultural Enterprises of Husbandry Sector of Ukraine*; Scientific herald of Kherson State University: Kherson, Ukraine, 2015; Volume 11, pp. 52–55. Available online: http://www.ej.kherson.ua/journal/economic_11/96.pdf (accessed on 2 September 2021).
44. Palamarenko, Y. The current situation and prospects of development of the biogas industry of Ukraine. *Investytsiyi Praktyka Dosvid* **2019**, *21*, 54–62. [CrossRef]
45. Cabinet of Ministers of Ukraine. National Action Plan for Renewable Energy for the Period until 2020. 2014. Available online: https://zakon.rada.gov.ua/laws/show/902-2014-%D1%80#Text (accessed on 16 August 2021).
46. International Energy Agency. Ukraine Energy Profile: Country Report. 2020. Available online: https://www.iea.org/reports/ukraine-energy-profile (accessed on 8 October 2021).
47. Cabinet of Ministers of Ukraine. Energy Strategy of Ukraine until 2035: Security, Energy Efficiency, Competitiveness. 2017. Available online: https://zakon.rada.gov.ua/laws/show/605-2017-%D1%80#Text (accessed on 12 September 2021).
48. OECD. Monitoring the Energy Strategy of Ukraine 2035. 2020. Available online: https://www.oecd.org/eurasia/competitiveness-programme/eastern-partners/Monitoring-the-energy-strategy-Ukraine-2035-EN-.pdf (accessed on 29 August 2021).
49. Diachuk, O.; Podolets, R.; Yukhymets, R.; Pekkoiev, V.; Balyk, O.; Simonsen, M. Long-Term Energy Modelling and Forecasting in Ukraine: Scenarios for the Action Plan of Energy Strategy of Ukraine until 2035 (Final Report). 2019. Available online: https://ens.dk/sites/ens.dk/files/Globalcooperation/long-term_energy_modelling_and_forecasting_in_ukraine_english.pdf (accessed on 21 September 2021).
50. State Agency on Energy Efficiency and Energy Saving of Ukraine. Information about the Share of Energy Generated by Renewable Sources in the Final Energy Consumption in 2019. 2020. Available online: https://saee.gov.ua/sites/default/files/RE_SAEE_2019.pdf (accessed on 5 September 2021).
51. Mostova, M. *Biogas in Ukraine: Big Prospects and Reality*; Energy Transition: Abu Dhabi, United Arab Emirates, 2020; Available online: https://energytransition.in.ua/sfera-biohazu-v-ukraini-velyki-perspektyvy-ta-real-nist (accessed on 28 August 2021).
52. Geletuha, G.G. *Biogas Perspectives of Ukraine*; GreenPost: Freudenberg, Germany, 2020; Available online: https://greenpost.ua/news/biogazovi-perspektyvy-ukrayiny-i6724 (accessed on 16 August 2021).
53. Geletuha, G.G.; Kucheruk, P.P.; Y., B.M. Prospects of Biogas Generation and Use in Ukraine: Analytical Report No 4. 2013. Available online: https://saf.org.ua/wp-content/uploads/2019/04/position-paper-uabio-4-ua.pdf (accessed on 1 October 2021).
54. Kholod, N.; Evans, M.; Denysenko, A.; Roshchanka, V. *Improving Ukraine's Energy Security: The Role of Energy Efficiency*; U.S. Department of Energy: Washington, DC, USA, 2018. Available online: https://www.pnnl.gov/main/publications/external/technical_reports/PNNL-27447.pdf (accessed on 4 September 2021).
55. Sabishchenko, O.; Rębilas, R.; Sczygiol, N.; Urbański, M. Ukraine Energy Sector Management Using Hybrid Renewable Energy Systems. *Energies* **2020**, *13*, 1776. [CrossRef]
56. Trypolska, G.; Diachuk, O.; Podolets, R.; Chepeliev, M. Biogas projects in Ukraine: Prospects, consequences and regulatory policy. *Ekonomika Prognozuvannia* **2018**, *2*, 111–134. [CrossRef]

57. Shamanskyi, S. Scientifical and Technological Foundations of Improvement of Ecological Safety Processes of Sewerage. Ph.D. Thesis, National Aviation University, Kyiv, Ukraine, 2019. Available online: https://nau.edu.ua/site/variables/news/2019/5/disertation%20Shamanskii.pdf (accessed on 28 August 2021).
58. Livestock Keeping in Ukraine: Influence upon the Environment. Ecology, Law, Human. 2021. Available online: http://epl.org.ua/wp-content/uploads/2021/06/vidhody_tvarynnztva.pdf (accessed on 10 August 2021).
59. State Statistics Service of Ukraine. 2021. Available online: http://ukrstat.gov.ua (accessed on 1 October 2021).
60. Kremenchuk Mykhailo Ostrohradskyi National University. Perspectives of Biogas Installations' Use as Way to Cutting and Utilisation of Waste. 2020. Available online: http://www.kdu.edu.ua/Documents/KSNR_economica_2020/w9.pdf (accessed on 29 August 2021).
61. Production and Use Biogas in Ukraine. 2012. Available online: https://uabio.org/wp-content/uploads/2012/11/biogas-arzinger-handbook.pdf (accessed on 16 October 2021).
62. Findeisen, C. *Biogas—Trends on the German and the International Market*; German Biogas Association: Berlin, Germany, 2014; Available online: https://www.eclareon.com/sites/default/files/clemens_findeisen_-_biogas_-_trends_on_the_german_and_international_market.pdf (accessed on 10 October 2021).
63. Obtainement of Biogas and Organic Fertilisers through Anaerobic Fermentation. Available online: https://sites.google.com/site/otrimannabiogazu/otrimanna-biogazu (accessed on 2 September 2021).
64. Sklyar, A.; Sklyar, R. *Bases of Biogas Technologies and Parameters of Optimisation of Fermentation Process*; Tavria State Agrotechnological University: Melitopol, Ukraine, 2009; Available online: http://elar.tsatu.edu.ua/bitstream/123456789/3672/1/19.pdf (accessed on 7 September 2021).
65. Sydorchuk, O. *Raw Material Provision for Biogas Production, or What Is Best to Collect Energy Harvest from*; AgroBiogas: Kyiv, Ukraine, 2020; Available online: https://agrobiogas.com.ua/raw-materials-for-biogas-production-or-what-best-way-to-harvest-energy (accessed on 18 August 2021).
66. Kuznetsova, A.; Kutsenko, K. Biogas and "Green Tariffs" in Ukraine—Is Investment Beneficial? German-Ukrainian Agricultural Dialogue. 2010. Available online: https://www.apd-ukraine.de/images/AgPP_26_Ukr.pdf (accessed on 2 August 2021).
67. EnergyPedia. Costs of a Biogas Plant. 2020. Available online: https://energypedia.info/wiki/Costs_of_a_Biogas_Plant (accessed on 16 August 2021).
68. National Bank of Ukraine. Official UAH Average Exchange Rate to Foreign Currencies. 2021. Available online: https://bank.gov.ua/files/Exchange_r.xls (accessed on 12 October 2021).
69. Dykalenko, M. Biogas Technologies Can Substantially Improve the Economy of Agricultural Enterprises. 2018. Available online: https://landlord.ua/news/biogazovi-tehnologiyi-zdatni-znachno-polipshiti-ekonomiku-agrarnih-pidpriyemstv (accessed on 7 September 2021).
70. The Kyiv Chamber of Commerce and Industry. "ECODEVELOP" Company Launched Its Own Biogas Station. 2019. Available online: https://kiev-chamber.org.ua/uk/136/1749.html (accessed on 3 October 2021).
71. Ministry of Ecology and Natural Resources of Ukraine. Technology Needs Assessment Report Mitigation: Technology Prioritization. 2019. Available online: https://tech-action.unepdtu.org/wp-content/uploads/sites/2/2019/08/tna-01-mitigation-ua-final-190731.pdf (accessed on 16 August 2021).
72. Parliament of Ukraine. Law of Ukraine "On Electricity". 1998. Available online: https://zakon.rada.gov.ua/laws/show/575/97-%D0%B2%D1%80#Text (accessed on 5 September 2021).
73. Parliament of Ukraine. Law of Ukraine "On Electricity Market". 2017. Available online: https://zakon.rada.gov.ua/laws/show/2019-19#Text (accessed on 5 September 2021).
74. Parliament of Ukraine. Law of Ukraine "On Ammendments to Selected Laws of Ukraine on Ensuring Competetive Conditions of Electricity's Generation from Alternative Sources of Energy". 2019. Available online: https://zakon.rada.gov.ua/laws/show/2712-19#Text (accessed on 29 August 2021).
75. Parliament of Ukraine. Law of Ukraine "On Ammendments to Selected Laws of Ukraine on Improving Conditions Supporting Electricity Generation from Alternative Energy Sources". 2020. Available online: https://zakon.rada.gov.ua/laws/show/810-20#Text (accessed on 12 September 2021).
76. ExPro Consulting. Minenergo Discusses Options to Carry Out "Green" Auctions in 2021. 2021. Available online: https://expro.com.ua/novini/mnenergo-diskutu-nad-varantami-provedennya-zelenih-aukconv-u-2021r (accessed on 7 September 2021).
77. Parliament of Ukraine. Law of Ukraine "On the Natural Gas Market". 2015. Available online: https://zakon.rada.gov.ua/laws/show/329-19#Text (accessed on 11 September 2021).
78. Ecodevelop. Payback Periods of Biogas Complexes. 2019. Available online: https://ecodevelop.ua/termini-okupnosti-biogazovih-kompleksiv (accessed on 16 August 2021).
79. Parliament of Ukraine. Law of Ukraine "On Ammendments to the Law of Ukraine "On Heat Supply" on Improving Conditions Supporting Heat Generation from Alternative Energy Sources". 2017. Available online: https://zakon.rada.gov.ua/laws/show/1959-19#Text (accessed on 25 September 2021).
80. Parliament of Ukraine. Law of Ukraine "On Animal By-Products not Intended for Human Consumption". 2015. Available online: https://zakon.rada.gov.ua/laws/show/287-19#Text (accessed on 3 October 2021).

81. Agronews. Ukrainian Farmer Built a Completely Automatized Ecological Pig Farm Replying on Its Own Biogas. 2020. Available online: https://agronews.ua/news/ukrains-kyy-fermer-stvoryv-povnistiu-avtomatyzovanu-eko-svynofermu (accessed on 5 September 2021).
82. Ukrainian State Fund for Support of Farmers. Financial Aid. 2021. Available online: https://udf.gov.ua/index.php/finansova-pidtrymka/finansova-dopomoha (accessed on 10 October 2021).
83. Finsee. State Programs Supporting Registered Farming Households. 2021. Available online: https://finsee.com/%D1%84%D1%96%D0%BD%D0%B0%D0%BD%D1%81%D0%BE%D0%B2%D0%B0-%D0%BF%D1%96%D0%B4%D1%82%D1%80%D0%B8%D0%BC%D0%BA%D0%B0/%D1%84%D0%B5%D1%80%D0%BC%D0%B5%D1%80%D1%81%D1%82%D0%B2%D0%BE/ (accessed on 8 October 2021).
84. Entrepreneurship Development Fund. Available online: https://bdf.gov.ua/uk/pro-nuf (accessed on 29 August 2021).
85. KfW Development Bank. Available online: https://www.kfw.de/kfw.de-2.html (accessed on 16 August 2021).
86. Support Programs in Ukraine. Entrepreneurship Development Fund. Available online: https://bdf.gov.ua/uk/programi-1 (accessed on 29 August 2021).
87. Finsee. Financing of Small and Medium Enterprises within EU4Business. 2021. Available online: https://finsee.com/%D1%84%D1%96%D0%BD%D0%B0%D0%BD%D1%81%D0%BE%D0%B2%D0%B0-%D0%BF%D1%96%D0%B4%D1%82%D1%80%D0%B8%D0%BC%D0%BA%D0%B0/eu4business (accessed on 8 October 2021).
88. Eurostat. Renewable Energy Statistics. 2020. Available online: https://ec.europa.eu/eurostat/statistics-explained/index.php/Renewable_energy_statistics (accessed on 8 August 2021).
89. European Biogas Association. EBA Statistical Report 2019: European Overview. 2019. Available online: https://www.europeanbiogas.eu/eba-statistical-report-2019-european-overview (accessed on 29 August 2021).
90. Shafarenko, Y. *Ways to Overcome Barriers towards Successful Production and Use of Biomethane*; State Agency on Energy Efficiency and Energy Saving of Ukraine: Kyiv, Ukraine, 2020; Available online: https://uabio.org/wp-content/uploads/2020/06/shafarenko-online-seminar-biomethane-2020.pdf (accessed on 5 September 2021).
91. Naumenko, D. *Use of Biogas: Expirience of Germany, Conclusions for Ukraine*; Institute of Economic Research and Political Consultations: Kyiv, Ukraine, 2012; Available online: http://www.ier.com.ua/files/publications/News/2012/21.06.2012_discussion/Naumenko_Biogas_in_Germany_EU_Village_21062012.pdf (accessed on 29 August 2021).
92. Ivashchenko, A. *Foreing Experience in Stimulation of Renewable Energy Sources (Experience of Germany and Austria)*; Financial and Economic Analysis Office in the Parliament of Ukraine: Verkhovna Rada, Ukraine, 2017. Available online: http://kompek.rada.gov.ua/uploads/documents/29892.pdf (accessed on 10 October 2021).
93. Decentralisation. Monitoring of the Process of Power Decentralisation and Local Governance Reform. 2021. Available online: https://decentralization.gov.ua/en/mainmonitoring#main_info (accessed on 3 September 2021).
94. Kurilov, Y. Administrative reform in Ukraine: From its origins to the future. *Public Adm. Asp.* **2019**, *7*, 62–78. [CrossRef]
95. UNDP Ukraine. Turning Slavutych into Solar Town. 2020. Available online: https://undpukraine.medium.com/turning-slavutych-into-solar-town-8a5efc4c804c (accessed on 7 September 2021).
96. Fedosenko, N. Rural Community in Ternopilska Region Created Berry Cooperative Which Transformed into Energy Cooperative. 2017. Available online: https://ecotown.com.ua/news/Silska-hromada-na-Ternopilshchyni-stvoryla-yahidnyy-kooperatyv-yakyy-postupovo-peretvoryvsya-u-enerh (accessed on 2 August 2021).
97. Parliament of Ukraine. Law of Ukraine "On Alternative Energy Sources". 2003. Available online: https://zakon.rada.gov.ua/laws/show/555-15#Text (accessed on 5 September 2021).
98. Wierling, A.; Schwanitz, V.J.; Zeiß, J.P.; Bout, C.; Candelise, C.; Gilcrease, W.; Gregg, J.S. Statistical Evidence on the Role of Energy Cooperatives for the Energy Transition in European Countries. *Sustainability* **2018**, *10*, 3339. [CrossRef]
99. Schreuer, A.; Weismeier-Sammer, D. Energy Cooperatives and Local Ownership in the Field of Renewable Energy Technologies: A Literature Review. RiCC—Research Report 2010/4. Available online: https://base.socioeco.org/docs/schreuer-ea-2010-literature-review-energy-cooperatives.pdf (accessed on 7 September 2021).
100. Jasiński, J.; Kozakiewicz, M.; Sołtysik, M. The Effectiveness of Energy Cooperatives Operating on the Capacity Market. *Energies* **2021**, *14*, 3226. [CrossRef]
101. Zinchenko, A.; Skliarov, R.; Bodnarchuk, I. Energy of Communities. How Energy Cooperatives Aid Community Resilience across the World. Greencubator. 2016. Available online: https://www.dropbox.com/s/p5ac6ss2knwbbjx/kas_48094-1522-13-30.pdf?dl=0 (accessed on 1 October 2021).
102. Dombrovskyi, O. Prospects to Production of Biomethane in Agroindustrial Complex of Ukraine. Online Seminar "Perspectives of Production and Consumption of Renewable Gases in Ukraine". *UABIO*. 29 September 2020. Available online: https://uabio.org/events/8718 (accessed on 19 October 2021).
103. Podoliak, Y. What Is Lacking for Development of "Green" Bioenergy in Ukraine? 2020. Available online: https://getmarket.com.ua/ua/post/chogo-ne-vistachaye-dlya-rozvitku-zelenoyi-bioeneregetiki-v-ukrayini (accessed on 28 August 2021).
104. Geletuha, G.G.; Zheliezna, T.A.; Dragniev, S.V. Analysis of Barriers to Generation of Energy from Agrobiomass in Ukraine. Analytical Report UABIO No 21. 2019. Available online: https://uabio.org/wp-content/uploads/2020/01/position-paper-uabio-21-ua.pdf (accessed on 16 August 2021).

energies

MDPI

Article

Water Needs of Willow (*Salix* L.) in Western Poland

Daniel Liberacki [1,*], Joanna Kocięcka [1], Piotr Stachowski [1], Roman Rolbiecki [2], Stanisław Rolbiecki [2], Hicran A. Sadan [2], Anna Figas [3], Barbara Jagosz [4], Dorota Wichrowska [5], Wiesław Ptach [6], Piotr Prus [7], Ferenc Pal-Fam [8] and Ariel Łangowski [2]

1 Department of Land Improvement, Environmental Development and Spatial Management, Faculty of Environmental Engineering and Mechanical Engineering, Poznan University of Life Sciences, 60-649 Poznan, Poland; joanna.kociecka@up.poznan.pl (J.K.); piotr.stachowski@up.poznan.pl (P.S.)
2 Department of Agrometeorology, Plant Irrigation and Horticulture, Faculty of Agriculture and Biotechnology, Bydgoszcz University of Science and Technology, 85-029 Bydgoszcz, Poland; rolbr@pbs.edu.pl (R.R.); rolbs@pbs.edu.pl (S.R.); hicran_sadan_76@hotmail.com (H.A.S.); arilan000@pbs.edu.pl (A.Ł.)
3 Department of Agricultural Biotechnology, Faculty of Agriculture and Biotechnology, Bydgoszcz University of Science and Technology, 85-029 Bydgoszcz, Poland; figasanna@pbs.edu.pl
4 Department of Plant Biology and Biotechnology, Faculty of Biotechnology and Horticulture, University of Agriculture in Krakow, 31-120 Krakow, Poland; Barbara.Jagosz@urk.edu.pl
5 Department of Microbiology and Food Technology, Faculty of Agriculture and Biotechnology, Bydgoszcz University of Science and Technology, 85-029 Bydgoszcz, Poland; wichrowska@pbs.edu.pl
6 Department of Remote Sensing and Environmental Research, Institute of Environmental Engineering, Warsaw University of Life Sciences, 02-776 Warsaw, Poland; wieslaw_ptach@sggw.edu.pl
7 Department of Agronomy, Faculty of Agriculture and Biotechnology, Bydgoszcz University of Science and Technology, 85-029 Bydgoszcz, Poland; piotr.prus@pbs.edu.pl
8 Institute of Plant Production, Kaposvár Campus, Hungarian University of Agriculture and Life Sciences (MATE), H-7400 Kaposvar, Hungary; Pal-Fam.Ferenc.Istvan@szie.hu
* Correspondence: daniel.liberacki@up.poznan.pl

Abstract: Willows are one of the plants which can be used to produce biomass for energy purposes. Biomass production is classified as a renewable energy source. Increasing the share of renewable sources is one of the priority actions for European Union countries due to the need to reduce greenhouse gas emissions. To achieve the best possible growth of the willow and increase its biomass for fuel, it is crucial to provide optimal water conditions for its growth. The aim of the study was to determine the water requirements of willows under the conditions of the western Polish climate and to verify whether this area is potentially favourable for willow cultivation. The novelty of this paper lies in its multi-year climatic analysis in the context of willow water needs for the area of three voivodships: Lubusz, Lower Silesian, and West Pomeranian. This is one of the few willow water-needs analyses for this region which considers the potential for widespread willow cultivation and biomass production in western Poland. Reference evapotranspiration (ETo) was determined by the Blaney-Criddle equation and then, using plant coefficients, water needs for willow were determined. Calculations were carried out for the growing season lasting from 21 May to 31 October. The estimated water needs during the vegetation season amounted on average to 408 mm for the West Pomeranian Voivodeship, 405 mm for the Lubusz Voivodeship, and 402 mm for the Lower Silesian Voivodeship. The conducted analysis of variance (ANOVA) showed that these needs do not differ significantly between the voivodeships. Therefore, it can be concluded that the water requirements of willows in western Poland do not differ significantly, and the whole region shows similar water conditions for willow cultivation. Furthermore, it was found that water needs are increasing from decade to decade, making rational water management necessary. This is particularly important in countries with limited water resources, such as Poland. Correctly determining the water requirements of willow and applying them to the cultivation of this plant should increase the biomass obtained. With appropriate management, willow cultivation in Poland can provide an alternative energy source to coal.

Citation: Liberacki, D.; Kocięcka, J.; Stachowski, P.; Rolbiecki, R.; Rolbiecki, S.; Sadan, H.A.; Figas, A.; Jagosz, B.; Wichrowska, D.; Ptach, W.; et al. Water Needs of Willow (*Salix* L.) in Western Poland. *Energies* **2022**, *15*, 484. https://doi.org/10.3390/en15020484

Academic Editors: Vitaliy Krupin, Roman Podolets and Alban Kuriqi

Received: 1 December 2021
Accepted: 5 January 2022
Published: 11 January 2022

Keywords: willow; *Salix* L.; water needs; biomass; evapotranspiration; energy crops; precipitation deficit

1. Introduction

One of the priority strategies of the European Union is "Climate Neutral by 2050". This document assumes reducing greenhouse gas emissions by the Member States and striving for climate neutrality (i.e., zero emissions). The implementation of this task focuses, among other things, on the use of alternative sources of energy to eliminate conventional fuels. This is particularly important in Poland, where most energy comes from burning fossil fuels. In 2019, the share of renewable energy in gross final energy consumption in this country was only 12.2% [1]. This is one of the lower results concerning the European Union countries. One of the ways to improve the current situation is to replace classical energy sources with biofuels created from biomass. Biomass can be obtained from energy crops. The energy plants cultivated in the conditions of Polish climate include poplar (*Populus* L.), Robinia acacia (*Robinia pseudoacacia* L.), Virginia fanpetals (*Sida hermaphrodita*), Jerusalem artichoke (*Helianthus tuberosus*), Giant miscanthus (*Miscanthus* × *giganteus*), and willows (*Salix* L.).

The willow is a plant that is easy to reproduce and overgrows. The willow comprises more than 300 species that occur as trees, shrubs, or dwarf shrubs. It has low soil requirements but broadly responds to habitat conditions, especially water conditions and organic and mineral fertilization [2]. *Salix* tolerates moist habitats, and its cultivation is not complicated. It usually occupies clay soils with poor permeability and difficult groundwater recharge [3]. Researchers highlight that the willow is a cleaner energy source than fossil fuels and one of the most promising biomass fuels [4]. According to Heinsoo [5], under moderate climate conditions, annual woody biomass production of *Salix* species can approach 20 Mg of woody dry matter per hectare. Once planted, it provides yields for about 25–30 years. Furthermore, it was estimated that the mean yield of dry willow biomass achieved in Poland is approx. 8.5 Mg ha^{-1} y^{-1} d.m. [6]. It was also found that 0.5 ha of willow (*Salix viminalis*) cultivation can supply any farm with fuel throughout the year [2]. In biomass production from willow for energy purposes, an important role is played by the density of plants per area unit. Research results conducted by different authors [7,8] indicate that the optimum plant density should be from 15,000 to 25,000 stem cuttings per ha and depend on habitat conditions and the willow variety. Based on the review of previous research, it was concluded that in Poland, the cultivation of new willow clones in the SRC system (short rotation coppice) on former agricultural land should be at a density of about 20,000 cuttings ha^{-1} and harvested in three-year cycles [6]. Moreover, the willow has higher calorific values (19 MJ·kg^{-1}) than other energy crops under the conditions of the Polish climate (Table 1). Although this value does not match the value of hard coal, which is 21 MJ·kg^{-1}, the willow is nevertheless a viable alternative energy source for heating purposes [9]. Furthermore, when assessing the calorific value, the percentage of ash after biomass combustion is an important aspect. It has been estimated that this should not exceed 1.5 percent. This value for the willow is around 1.10 percent and thus meets the requirements. It should be noted that for other energy crops such as giant miscanthus or Jerusalem artichoke, this value is exceeded up to four times (Table 1).

Table 1. Energy crop parameters [9].

Energy Crop	Calorific Value [MJ·kg^{-1}]	Ash Content [%]
Energy willow	19.23	1.1
Tuberous sunflower	15.31	5.36
Miscanthus giganteus	16.28	6.89

Willows have a high demand for nutrients, and adequate soil moisture promotes the tree's movement and uptake of these nutrients. One way of meeting willow requirements is to provide nutrients by fertilising with municipal sewage after pre-treatment. This is particularly applicable on land that does not directly include food or fodder crops [10]. It should be mentioned that willow stands significantly reduce nitrogen and phosphorus concentrations in wastewater. According to Börjesson [11], this value is even 75–90%. Using wastewater to irrigate willow crops can significantly increase yield and biomass. Furthermore, it also reduces the risk of groundwater pollution and eutrophication of surface waters through the trees' partial uptake of nitrogen and phosphorus. The use of willow as a natural filter for wastewater treatment is an excellent way to clean the environment while increasing biomass production without using additional costs associated with (e.g., fertilization). Another factor in preferring this type of management is the reduction of natural water resources under climate change conditions in favour of using water from municipal [12] or agricultural wastewater [13].

The main factor limiting the growth of energy crops, including the willow, is water availability. This is especially true in regions with temperate climatic conditions, where insufficient rainfall limits biomass growth even with high nitrogen fertilization [14]. In the period of the maximum increase of plant mass (from June to August), the willow reacts particularly to the course of weather conditions. Precipitation and moderately high temperature in this time have a positive effect on biomass yields, while drought may cause a decrease in yields even by 50% [15]. Plant water consumption depends mainly on the species, the yield obtained and the meteorological conditions, and the length of the growing season [16]. To quantify trees water use (including willow), methods such as Bowen ratio energy balance system, Eddy covariance, and plant water flow (SF) evaluation techniques are used. SF of willow shows diurnal and seasonal variability. Air temperature is the main factor controlling the seasonal variation of SF. The highest water use by willow occurs in May, June, July, August, and September [17]. The willow's groundwater use also depends on the soil's depth, the plant's root system structure, and the soil type. The willow has a deep and well-developed root system, allowing it to use shallow groundwater, unlike field crops usually supplied with water stored in the aeration zone [18]. The range of optimum groundwater table under different soil conditions for willow cultivation is wide and between 1 and 3 m. The increase in willow yields is mainly related to the enhancement of transpiration and the correct ratio between water and air in the soil [19]. The measured transpiration rates for willow are among the higher values compared to other cultivated trees. This is partly due to the fact that the willow is a highly hydrophilic plant that requires high transpiration for biomass production [20]. Despite willow's high water use efficiency (6.3 g dry biomass per kg transpired water), researchers note that water availability is a critical factor shaping willow short-rotation forestry [21]. Therefore, it is crucial to carry out research on willow water management and the possibilities of meeting its water needs.

This study aims to determine the water requirements of willow in western Poland. To check whether the water needs of willow are fulfilled in this region, an analysis of the course of climatic conditions and precipitation deficit values was carried out. The hypothesis that the water needs of willow differ between the three analyzed voivodships (Lubusz, Lower Silesian, and West Pomeranian) was considered in this study. Also, trends in changes in water needs were determined. Estimating water needs based on current climatic conditions is essential in appropriate crop management of this plant in this part of the country. This is one of the few studies on the water needs of willow cultivation in western Poland. The conducted research will be a valuable practical guideline for cultivating this plant for farmers and growers.

2. Materials and Methods

The assessment of water needs of willow (*Salix* L.) was carried out for three voivodships located in western Poland, namely the Lower Silesian, Lubusz, and West Pomeranian voivodships. Calculations were based on data obtained from meteorological stations located in the largest cities of each province (i.e., for the Lower Silesian voivodship from Wrocław, for the Lubusz voivodship from Zielona Góra, and the West Pomeranian voivodship from Szczecin (Figure 1)).

Figure 1. The location of the analyzed voivodships and meteorological stations.

The analysis was carried out for the years from 1981 to 2010. The calculations were made for the growing season lasting in the studied area from the third decade of May (21 May) to the end of October. Based on meteorological data, the reference evapotranspiration (ETo) was determined. ETo is an agrometeorological parameter essential in irrigation planning and management [22]. ETo was calculated using the Blaney-Criddle (B-C) Formula (1) modified by Żakowicz [23] for the conditions of Poland. In this study, equation B-C was chosen to estimate ETo due to the limited availability of meteorological data (only monthly temperature values for the period from 1981 to 2010 are accessible). FAO Irrigation and Drainage Paper No. 24 'Crop water requirements' [24] suggests using the Blaney-Ciddle formula when only air temperature data are available [25]. The B-C formula is commonly used to estimate evapotranspiration with a limited number of available meteorological parameters. This is confirmed by several studies in various world areas [26–28].

$$\mathrm{ETo} = \mathrm{n} \times [\mathrm{p} \times (0.437 \times \mathrm{t} + 7.6) - 1.5] \quad (1)$$

where:

ETo = reference evapotranspiration (mm);
n = number of days in the month;

p = evaporation coefficients according to Doorenbos and Pruitt [24] for months and latitude determined from the tables;
t = monthly mean air temperature (°C).

Crop (potential) evapotranspiration was then estimated with equation (2). This method is widely used in scientific research [29–32]. Moreover, this equation is also used to calculate crop transpiration in the AquaCrop model developed by the Land and Water Division of FAO [33].

$$ETp = ETo \times kc \quad (2)$$

where:

ETp = crop (potential) evapotranspiration (mm);
ETo = reference evapotranspiration (mm);
kc = crop coefficient defined as the quotient of evapotranspiration measured in conditions of sufficient soil moisture and reference evapotranspiration (Figure 2) [34].

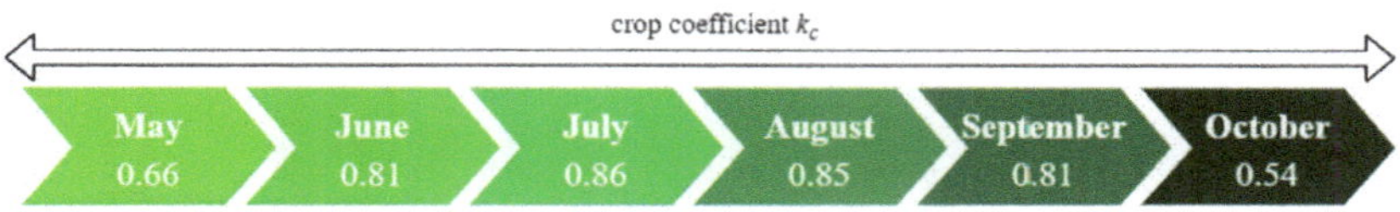

Figure 2. Crop coefficient for the Blaney-Criddle equation for willow depending on the month [34].

The final calculation stage determined the precipitation deficit using Ostromecki's formula (3) [35,36]. The rainfall deficit determines the difference between the sum of evapotranspiration of plants and the sum of precipitation. Therefore, by estimating the precipitation deficit, it is possible to determine to what extent plants' water needs are met by precipitation, how much water is lacking, and what amount should be supplied to crops for adequate growth. Rainfall deficit assessments were made for the occurrence probability of the normal years (N50%), medium dry years (N25%), and very dry years (N10%).

$$Np\% = Ap\% \times ETp - Bp\% \times P \quad (3)$$

where:

Np%= precipitation deficit at the probability occurrence p% (mm period^{-1});
ETp = average multi-year amount of evapotranspiration in the analyzed period (mm period^{-1});
P = multi-year average amount of precipitation in the analyzed period (mm period^{-1});
Ap% and Bp% = numerical factors characterizing the variability of evapotranspiration and precipitation for a given meteorological station.

The obtained results were statistically analyzed in R and Microsoft Excel environments. They aimed to determine the tendency of changes in the water needs of willow and significant differences in the results for individual voivodeships. ANOVA analysis of variance was used to find significant differences in water needs between the three voivodeships, preceded by Shapiro-Wilk tests and Bartlett's test for equality of variance. The analysis considered the following hypothesis:

Hypothesis 1 (H1): the water needs of willow do not differ between voivodships.

The linear correlation coefficient (r) method, widely used and proven in many studies, was used to determine the trend of changes in water needs.

3. Results

The results of willow ETp for individual growing seasons (third decade of May to the end of October) in the years from 1981 to 2010 were analyzed. The calculated values for each of the three provinces were compared to precipitation in the growing season. The analysis for the West Pomeranian voivodship showed that only in two years, 1996 and 2007, precipitation values (*P*) were higher than water needs (ETp). Therefore, it can be concluded that for this region, the course of climatic conditions in the studied period is unfavorable concerning the water needs of willow (Figure 3). The highest ETp values were estimated for 2006 and amounted to 442 mm. At the same time, the precipitation reached 253 mm. The lowest rainfall values were recorded in 1982–1983 and were about 162–174 mm. However, it should be noted that this was at the same time also in the period when the greatest differences between ETp and *P* occurred, amounting to 262 and 247 mm, respectively. Large differences were observed in 1992 (221 mm) and 1994 (209 mm). The standard deviation (SD) value for precipitation in the West Pomeranian province in 1981–2010 was 75.1. For the same period, the SD of water needs reached 13.6. Other statistical characteristics are presented in Table 2.

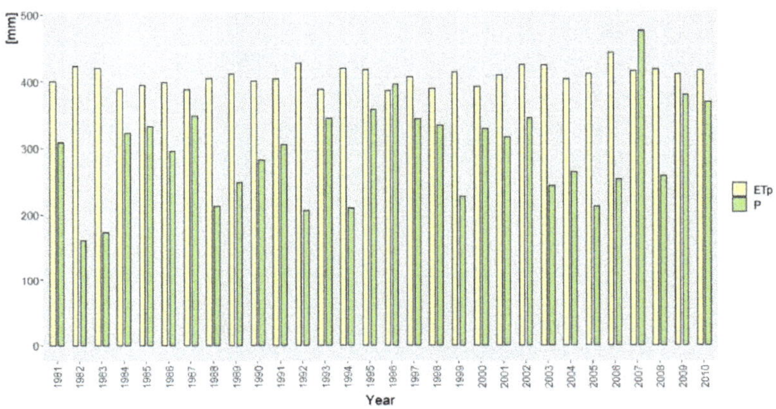

Figure 3. Precipitation (*P*) and water needs (ETp) in the growing season (third decade of May—October) in 1981–2010 in the West Pomeranian voivodeship (SD of *P* = 75.1, SD of ETp = 13.6).

Table 2. Statistical characteristics of willow water needs (ETp) and precipitation during 1981–2010 for each province.

Specification	West Pomeranian		Lubusz		Lower Silesian	
	P	ETP	*P*	ETP	*P*	ETP
Maximum [mm]	538	442	544	444	522	429
Minimum [mm]	207	385	189	377	251	378
Mean [mm]	365	408	381	405	384	402
Standard Deviation SD	75.1	13.6	93.0	17.4	85.7	12.6

The situation was slightly better in Lubusz voivodship, where precipitation was greater than water needs in the growing seasons in the analyzed 5 out of 30 years (Figure 4). This group includes 1981, 1985, 1993, 1996, and 1998. The highest precipitation values occurred during the growing season in 1981 and amounted to 463 mm, while the lowest of 150 mm was recorded a year later (1982). The standard deviation of precipitation for the years analyzed reached 93. The difference in the driest year between *P* and ETp was 249 mm. However, it should be noted that this was not the highest value for the analyzed multi-year period. The greatest difference between precipitation and water needs in the West Pomeranian Voivodship was recorded in 1992, and it amounted to 278 mm. The maximum value of water needs was in 2006 and amounted to 444 mm. The standard deviation of the ETp in the analyzed years was 17.4 (Table 2).

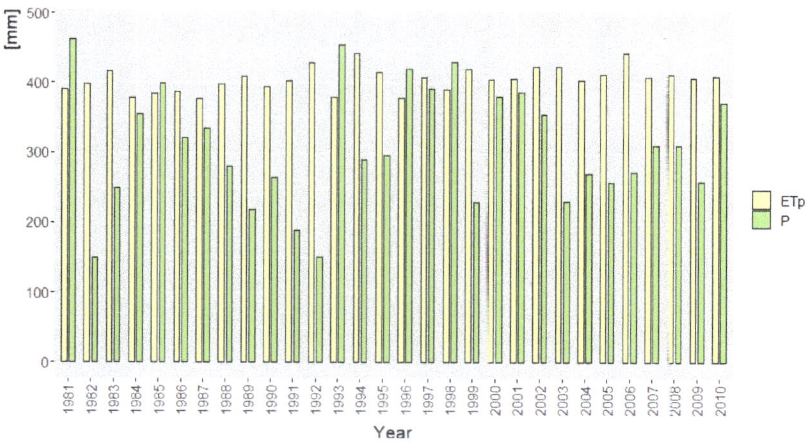

Figure 4. Precipitation (*P*) and water needs (ETp) in the growing season (third decade of May—October) in 1981–2010 in the Lubusz voivodeship (SD of *P* = 93, SD of ETp = 17.4).

Among the analyzed regions, Lower Silesian voivodship had the highest number of vegetation seasons in which precipitation exceeded water needs. In this case, positive values were recorded in 6 out of 30 years, more precisely in 1981, 1986, 1995, 1997, 2001, and 2009 (Figure 5). The year 1981 was also in this region characterized by the lowest precipitation in the vegetation season, amounting to 179 mm. It was also the year when the difference between precipitation and ETp was the greatest and amounted to 233 mm. A relatively large difference (220 mm) was re-recorded in 1994. The highest precipitation values were reached in 1997 with 522 mm. For the 30 years analyzed, the SD of precipitation was 85.7, while for water needs, the SD was 12.6 (Table 2).

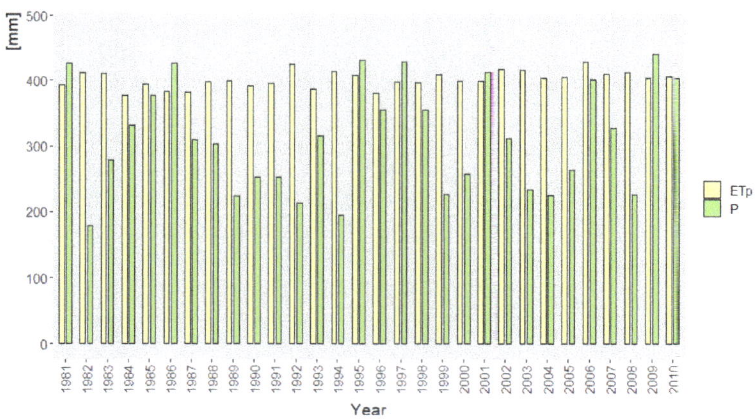

Figure 5. Precipitation (*P*) and water needs (ETp) in the growing season (third decade of May—October) in 1981–2010 in the Lower Silesian voivodeship (SD of *P* = 85.7, SD of ETp = 12.6).

Analyzing the climatic conditions during 30 years (1981–2010) in the growing seasons for the three provinces, it can be seen that in most years in all voivodships the ETp values were higher than *P*. Only in a few years the situation was reversed, and interestingly for each province, there were different years in most cases. Looking at the graphs (Figures 3–5), one can conclude that conditions were unfavorable for willow cultivation in most years. A potential solution to the problem of large differences between precipitation and ETp values for *Salix* cultivation could be the application of an adequate irrigation rate with treated wastewater. Many scientific studies underline that willow is a plant for which this type of irrigation can benefit and contribute to plant development [10]. Wastewater irrigation could also be a favourable solution in the case of Poland due to the country's limited water resources.

As part of the analyses, mean values of monthly water needs were determined for each of the studied voivodships (Lower Silesian, Lubusz, and West Pomeranian) for the measurement periods in 1981–2010 (Figure 6). It was observed that the beginning of the growing season (3rd decade of May) is characterized by relatively low water needs resulting from lower temperature and limited evaporation. These values are 18 to 25 mm for the Lower Silesian and the West Pomeranian voivodships. In the case of Lubusz Voivodeship, the maximum value of water requirements in the third decade of May is 47 mm. Low values of water needs are also noted at the end of the growing season (October). Moreover, the highest water needs are observed in June, July, and August. They resulted from plant growth enhancement and increased evaporation.

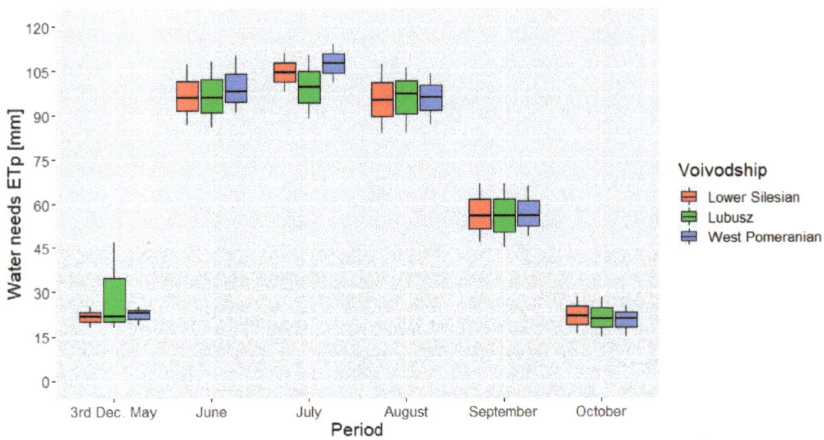

Figure 6. Water requirements (ETp) in the growing season calculated for individual voivodeships in 1981–2010.

The highest values of standard deviation (SD) characterizing the variability of water demand also occurred in the summer months (i.e., June, July, August, and September (Table 3)). Even though the value of the average potential evapotranspiration of willow ETp for the three voivodeships was similar in the whole growing season and amounted to 402–408 mm, there were differences in the value of the SD index. Among the studied voivodeships, Lubusz had the highest monthly SD index values ranging from 3.0 to 9.3. In the two other voivodeships, the values were similar. The highest SD value was reached in July for Lubusz, and it was 9.3.

Table 3. Statistical characteristics of willow water needs in the period 1981–2010 determined as ETp [mm].

Specification		Period						
		3rd Dec.V	VI	VII	VIII	IX	X	3rd Dec.V-X
Mean	WP	22	98	114	96	57	21	408
	L	23	97	111	97	57	22	405
	LS	22	97	111	94	56	22	402
Standard Deviation (SD)	WP	1.5	4.8	7.4	4.3	3.9	2.7	13.6
	L	4.8	5.8	9.3	5.3	5.2	3.0	17.4
	LS	1.4	5.1	6.7	4.4	4.3	2.8	12.6
Coefficient of Variation (VC [%])	WP	6.8	4.9	6.5	4.5	7.0	13.1	3.3
	L	21.0	6.0	8.4	5.5	9.2	14.1	4.3
	LS	6.2	5.3	6.1	4.6	7.6	12.9	3.1

Voivodeship: WP, West Pomeranian; L, Lubusz; LS, Lower Silesian.

Based on meteorological data from 1981–2010 for each analyzed voivodship, the average water needs in the growing season (from the 3rd decade of May to the end of October) were calculated (Figure 7). For the Lubusz Voivodship, average water needs from 30 years were 405 mm, while the median was 407 mm. The maximum value was 444 mm, and the minimum was 377 mm. For the Lower Silesian province, the mean value for the growing season was 3 mm lower than for the Lubusz vhoivodship and amounted to 402 mm. The minimum value of water needs of this voivodship in the growing season was 378 mm and a maximum 429 mm. In West Pomeranian voivodship, the average value of water needs was the highest of all three studied voivodships and amounted to 408 mm. The range of values, in this case, was from 385 mm to 442 mm.

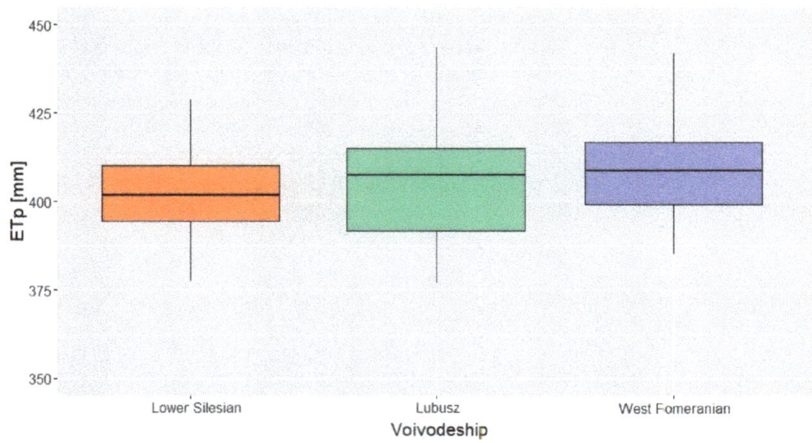

Figure 7. Average water requirements of willow for the analyzed voivodeships during the growing season (3rd decade of May to the end of October) calculated based on data from 1981 to 2010. (SD equaled 12.7, 17.4 and 13.6 for Lower Silesian, Lubusz and West Pomeranian, respectively).

The calculated Etp values for the growing season for each year in the years 1981–2010 in the studied three voivodeships were also subjected to statistical analyses in R to check whether there were significant differences between the provinces. First, for a data sequence for each voivodship, the hypothesis of normality of distribution was verified using the Shapiro-Wilk test. A significance level of $\alpha = 0.05$ was assumed. The obtained results for the probability level (p-value) were greater than 0.05. Based on this (p-value $> \alpha$), it was concluded that the assumption of normality of distribution is fulfilled (Table 4). Next,

Bartlett's test for equality of variance was performed. Also, a significance level was taken as $\alpha = 0.05$. A *p*-value of 0.1811 was obtained, greater than 0.05 (p-value > α). Therefore, it was concluded that at a significance level of $\alpha = 0.05$, there are no grounds to reject the hypothesis stating the equality of variance of water needs in the growing season for the three provinces. Therefore, further analyses can be carried out.

Table 4. *P*-value obtained after Shapiro-Wilk test.

Voivodeship	*p*-Value
West Pomeranian (WP)	0.5853
Lubusz (L)	0.4485
Lower Silesian (LS)	0.8892

Another analysis performed in the R environment to determine whether there were significant differences between water needs in the individual voivodships was an ANOVA analysis of variance. It included hypothesis H0: the water needs of willow in the analysed provinces do not differ. The significance level was taken as $\alpha = 0.05$. The ANOVA analysis resulted in a *p*-value of 0.33 (i.e., greater than 0.05 (p-value > α)). Therefore, it was concluded that at a significance level of $\alpha = 0.05$, there were no grounds to reject the H0 hypothesis. Thus, the values of water requirements in 1981–2010 calculated for the growing season for the three voivodships do not differ statistically significantly. As there was no basis for rejecting the H0 hypothesis, no further multiple tests were conducted to test the significance of the differences. Post-hoc tests were not performed since, based on ANOVA analysis, they would not show differences between water needs in the analyzed provinces. Therefore, it can be concluded that the whole area of western Poland is characterized by similar water needs in the case of willow.

Analyzing the obtained values of water needs in the studied period compared to precipitation for Lubusz, West Pomeranian, and Lower Silesian voivodships, it can be noticed that water requirements in western Poland are not fulfilled. The precipitation values were lower than the needs in all three voivodships (Figure 8). The most remarkable difference is visible in West Pomeranian voivodship, where the average amount of rainfall in the growing season is 365 mm, and the calculated water needs are 408 mm. In Lubusz and Lower Silesian voivodeships, this difference was lower and ranged from 18–24 mm, almost half as much as in West Pomeranian voivodeships. The standard deviation values for the data are shown in Table 2.

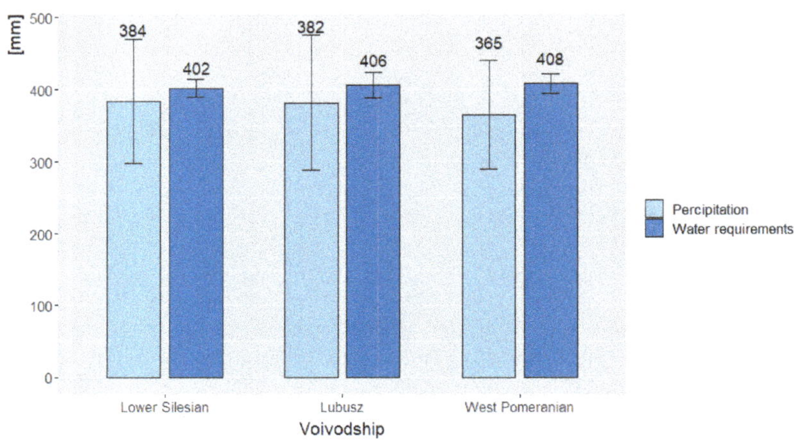

Figure 8. Water needs of willow and average precipitation during the growing season for individual provinces in 1981–2010.

The estimated water requirements of willow in individual decades (1981–1990, 1991–2000, and 2001–2010) of the 1981–2010 period showed an increasing trend (Figure 9). In Lubusz province in 1981–1990, the water needs were 394 mm and in 2001–2010 as much as 416 mm. Thus, the needs in this province increased by 22 mm. It was the highest increase in comparison to all three analyzed voivodeships. In Lower Silesian, this increase was 16 mm, and in West Pomeranian 14 mm. In the first analyzed decade of 1981–1990, the highest water requirements for willow were in the West Pomeranian voivodship. They amounted to 403 mm, while the lowest (394 mm) were in the Lubusz voivodship. On the other hand, the Lower Silesian province's lowest value in 2001–2010 of 411 mm. Analysing standard deviation (SD), the highest values in all three regions were recorded for data from the decade 1991–2000. The greatest SD was for Lubusz province and reached 20.53. The lowest SD was estimated for data from 2001 to 2010 in Lower Silesian and earned 8.51 (Table 5).

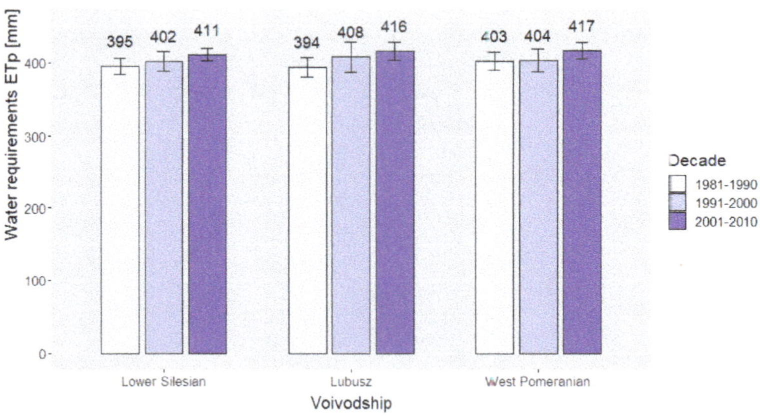

Figure 9. Average water needs of willow in each decade (10-year periods) from 1981 to 2010 for the analyzed voivodeships.

Table 5. Values of standard deviation (SD) for decadal data for each voivodship.

Voivodeship	Decade		
	1981–1990	1991–2000	2001–2010
West Pomeranian	11.96	14.93	11.26
Lubusz	12.87	20.53	11.95
Lower Silesian	11.29	13.30	8.51

4. Discussion

The impact of water resources on crops is particularly important in Poland, as Polish water resources are relatively small compared to other European countries. The average total precipitation is about 630 mm (i.e., 196 km^3 per year) [37]. Additionally, water resources in Poland are characterized by high spatial and temporal variability. The recommended way of rational management of limited water resources is retention. It consists in storing water when there is an excess and giving it back to users and the environment in times of shortage. Climate change scenarios indicate that we will increasingly be dealing with extreme weather events that will cause droughts or floods, and therefore appropriate water management is extremely important.

The assessment of climatic water balance for western Poland in the six-month growing season (April-September) indicates the occurrence of large negative differences between precipitation and evapotranspiration values. The increasing likelihood of heatwaves in summer may result in short- or long-lasting droughts that significantly impact crops [38].

Also, in this study on willow, large differences between precipitation values and potential evapotranspiration are noticeable (Figures 3–5). It is predicted that the values of the P-E index will change unfavorably in the future, leading to more frequent and more severe summer water stress. Therefore, western and central Poland areas require the necessary protection of agriculture against the negative effects of water shortage during the growing season [39]. Moreover, studies conducted in this paper indicate that plants' water needs will also increase. The determined time trend of the variability of water needs and the linear correlation coefficient (r) showed that in all of the three studied voivodeships, there was a significant tendency for the water needs of willow to increase during the growing season (Table 6). Significant differences were also observed in the summer months (June, July) for Lubusz province. In the Lower Silesian voivodeship, significant differences were also observed in August besides these two months. However, at the same time, no differences were noticed in West Pomeranian.

Table 6. Water needs (mm) of willow in the years between 1981 and 2010 in the provinces of western Poland.

Periods	Provinces		
	West Pomeranian	**Lubusz**	**Lower Silesian**
	Linear correlation coefficient (r)		
3rd Decade of May	0.01 ns	0.020 ns	0.118 ns
June	0.255 ns	0.460 **	0.450 **
July	0.282 ns	0.425 **	0.382 **
August	0.176 ns	0.240 ns	0.334 *
September	0.153 ns	0.104 ns	0.099 ns
October	0.128 ns	0.123 ns	0.089 ns
May 21–October 31	0.316 *	0.469 ***	0.461 **
	The tendency of water needs [mm·decade^{-1}]		
May 21–October 31	5.0	9.4	6.7

***—significant at $p = 0.01$; **—significant at $p = 0.05$; *— significant at $p = 0.1$; n.s.—not significant (r = 0.46398, r = 0.362 and r = 0.30692 for $p = 0.01$, $p = 0.05$, and $p = 0.1$, respectively (i.e., for probability 99%, 95%, and 90% accordingly)).

Analyzing the trend of changes in water needs in the thirty years 1981–2010, it can be observed that for the whole growing season in Lubusz, West Pomeranian, and Lower Silesian voivodeships, the trend is positive (Table 4). The highest is in Lubusz, and it is 9.4 mm-decade^{-1}. The values are similar in the following two provinces, and they are 6.7 mm·decade^{-1} for Lower Silesian and 5.0 mm·decade^{-1} for West Pomeranian. Analyzing monthly data, the highest increasing tendency of water needs of willow was 4.6 mm·decade^{-1} in July. On the other hand, slight negative tendencies for all voivodeships were observed in October (Figure 10).

Poland's western and central areas require necessary protection of agriculture against the adverse effects of water shortage during the growing season. The water deficit is evident on light soils with low water retention capacity. Security of crop production against drought is ensured by irrigation. Therefore, irrigation becomes an indispensable element of cultivation in large parts of the country, especially where the climatic water balance is negative (e.g., Wielkopolska, Kujawy), and the soils are characterized by low water retention [39]. The problem of water scarcity is also analyzed in this publication in the context of willow plantations. The precipitation deficits for three voivodeships in the growing season determined by Ostromęcki's method [35] are presented in Figure 11. The results of the calculations were shown for three categories of years: very dry years (once per ten years, N10%), medium dry (once per four years, N25%), and normal years (once per two years, N50%). The highest rainfall deficits occurred in the West Pomeranian Province in the very dry year and amounted to 286 mm. Among the three analyzed voivodships, in West Pomeranian voivodship, the deficits were also the highest in average dry years

(222 mm) and normal years (132 mm). Moreover, the lowest precipitation deficits in the studied period were recorded in Lower Silesian Province. They amounted to 263 mm for very dry years, 193 mm for medium dry years, and 99 mm for normal years.

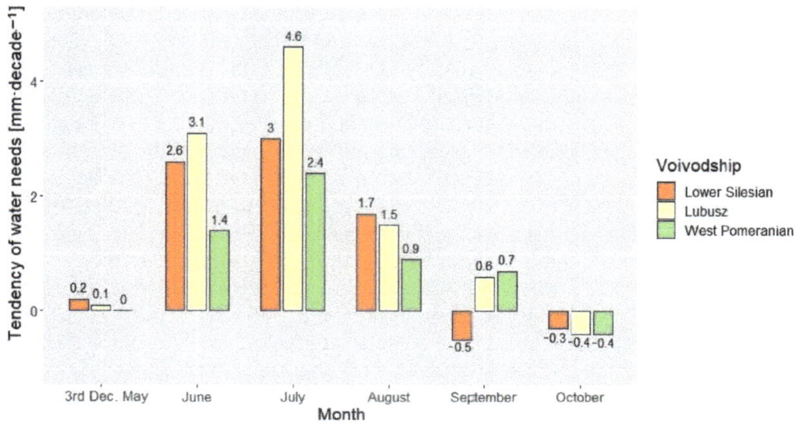

Figure 10. The trend of water needs of willow in each month for the analyzed voivodships.

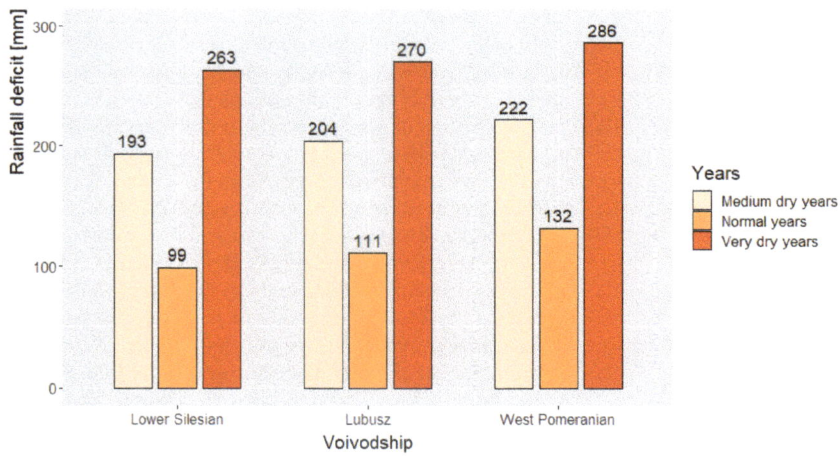

Figure 11. Rainfall deficit for willow cultivation in the analyzed provinces in 1981–2010.

The results confirm a problem of precipitation shortage in western Poland. In combination with its low water resources, it creates a particular threat to agriculture and the production of productive crops. Researchers emphasize that the decision on the location of energy crops with a high share in the catchment should take into account the water needs of energy crops, as their cultivation can significantly affect water balance parameters [40]. The analyses carried out on water need help to determine the potential of a region for willow cultivation. Published scientific articles note the potential of relatively large areas of marginal land and fallow grounds in Poland as suitable sites for willow growth. It should be noted that fallow grounds constitute as much as 10% of shares of arable lands [41]. Research carried out by scientists shows that in the management of fallow and uncultivated land in Poland for willow weed cultivation, the amount of energy obtained would be 3083 TJ. Moreover, it was shown that one of the voivodeships with the largest potential energy resource is the Lower Silesian voivodeship analyzed in this paper (with a volume of 386 TJ) [42]. Subsequent estimations carried out assuming willow cultivation on the total fallow area in Poland in 2014 showed that the energy value of willow wood dry mass would be 138,285,528 GJ·yr^{-1}. Moreover, the dry wood mass for the whole country would be 7,128,120 t·yr^{-1}. The ecological effects of obtaining energy willow biomass for heating purposes were also estimated in these analyses. It was found that as a result of this practice, nitrogen oxide emissions could potentially be reduced in Poland by 26,274 tons per year, carbon dioxide emissions by 13,828,553 tons and sulphur dioxide emissions by about 103,714 tons per year [43]. Therefore, it can be concluded that growing willow on fallow land and harvesting it for use would be a potentially positive measure in the face of climate change and the need to reduce greenhouse gases. Moreover, it would allow meeting the objectives of the European Union policy on greenhouse gas reduction.

Moreover, studies conducted in Poland on willow cultivation in a short rotation woody crops (SRWC) system (of three- to four-year rotation) and Eko-Salix systems (five-year rotation) have shown that the energy gain obtained is even 20 times higher than the inputs needed to run the plantation and harvest the willow biomass [44]. Therefore, this shows that willow cultivation benefits the climate, environment, and economic context. All of these studies support Poland's high potential for this type of cultivation. As Jadczyszyn et al. [45] estimated, the potential area of willow cultivation for energy purposes in Poland amounts to 9541 km^2 or 4.6% of agricultural land. However, this analysis did not cover soils with the highest production potential belonging to the wheat and very good rye complexes and the weakest, too dry soils of the very weak rye complex. It also excluded mountain areas, protected areas and areas with annual precipitation <550 mm. For the West Pomeranian voivodship, the estimated potential area for willow cultivation amounted to 1094 km^2 (6.5% of agricultural land), for the Lubusz voivodship 534 km^2 (6.5% of agricultural land) and the Lower Silesian voivodship 883 km^2 (6.8% of agricultural land) [45]. Thus, it can be concluded that these provinces not only do not differ concerning the water needs of willow (as shown in this paper) but also the potential area under willow cultivation is similar for the percentage of agricultural land in the given voivodship (it is about 6.5–6.8% of the agricultural land of the voivodship). The research carried out in this study has shown that in the area of the three voivodships of western Poland, there are no significant differences in the water needs of willow estimated for the growing season between 1981 and 2010. Lack of significant differences in the obtained ETp values results most probably from the course of air temperatures in the analyzed period in the voivodships. The used Blaney-Criddle formula is based on air temperature, and hence the data strongly affect the obtained results. However, it should be noted that the use of the B-C formula in this study captured an extremely important trend, namely the increase of water needs of willow. This is due to the fact that the predicted climatic changes under Polish conditions include an increase in air temperature, which affects the growth of plants' water needs.

The water requirements of willow estimated in the paper showed the current unfavorable state of conditions for the cultivation of this plant in western Poland. To obtain the largest possible biomass for energy purposes, providing the willow plant with optimal water conditions is crucial. World research papers have increasingly focused on water management in crops and estimated the necessary amount of water required for adequate irrigation. It is also essential to carry out studies considering the impact of climate change on plants' water needs [46]. Increasingly, research work is being conducted to model crop water requirements and the necessary amount of irrigation water under different climate change scenarios [30,47]. Previous analyses show that climate change has an impact on irrigation water requirements. Furthermore, irrigation demand will increase for many crops due to climate change [48].

Worldwide research indicates the need for precise estimation of plant water requirements. When making such calculations, it is crucial to use appropriate kc coefficients. Measurements conducted for Peach-leaf willow (*Salix amygdaloides*) in the Platte River basin in central Nebraska, USA, contributed to the development of crop evapotranspiration coefficient (KcET) curves for this cultivar [49]. However, it should be remembered that the water needs of willow depend on climatic conditions. In a study conducted on *Salix gooddingii* grown in restoration plots in three irrigation districts on the Lower Colorado River, reference crop evapotranspiration (ETo) values ranged from 1890 to 1969 mm·yr^{-1}. For the same sites, irrigation requirement was estimated from 1817 to 1962 mm·yr^{-1} [50]. Evapotranspiration (ET) values were also evaluated for wetlands and the willow variety *Salix miyabeana*. From May to October, the average evapotranspiration rate in eastern Canada was 22.7 mm·day^{-1} [51]. Therefore, it can be concluded that the estimation of ETp for willow in this study is fundamental in the context of its proper cultivation, and these analyses fit into the trend of global research. Moreover, this study fills a gap in science concerning the determination of the water needs of this plant for the conditions of western Poland. This is one of the few studies on this subject for this region.

Scientists emphasize that the willow is not a demanding plant in cultivation conditions. Furthermore, it has been noted that willow also shows salt tolerance, which has been defined as sensitive to moderately tolerant [52]. Moreover, it has been demonstrated that irrigation of willow with stormwater up to 1625 mg Cl had no short-term effect on biomass accumulation and evapotranspiration [53]. All of these measurements indicate that willow has a high tolerance to different growing conditions. However, it should be remembered that it is a plant that needs an adequate amount of water for optimal growth. Soil water availability is one of the determinants of willow growth in montane riparian communities in the USA [54]. High available water content (AWC) values were also the most critical determinant of willow yield in the Danish area. AWC had a much greater effect on yield than precipitation, radiation sum, and region [55]. The Swedish researchers found that water is critical for the excellent profitability of willow short-rotation forestry [56]. In addition, studies have shown that the Carolina Willow (*Salix caroliniana*) seeds in saturated soils kept moist by capillarity had the highest germination capacity [57]. The research also included estimating factors affecting aboveground biomass allocation and water storage ratio in alpine willow shrubs. It was observed that relative water storage allocation was significantly affected by species types [58].

The analyses carried out in this paper have shown that water needs of willow in Poland have an increasing tendency year by year. Due to the ongoing climatic changes, the occurrence of drought periods, and thus precipitation deficits, it will be more and more challenging to fulfill them. It may not be possible to supply the appropriate amount of rainwater necessary for irrigation during drought periods. However, it should be remembered that willow is a plant that can also be successfully irrigated with wastewater. Worldwide research shows that using willow for energy production is an opportunity to reduce greenhouse gas emissions. It has been found that the biomass of this plant can be a carbon negative or low-carbon energy source with high emissions and energy return on investment. This applies to regions with similar conditions for the plant's growth, transport distances, and infrastructure [59]. Therefore, it is crucial to continue research into willow cultivation to optimize its cultivation and widespread use. Research should include field experiments on different willow cultivation practices and varying water availability. In view of climate change and the need for crop adaptation, all kinds of experiments simulating stress conditions such as drought are desirable. Increasing research and knowledge is extremely important, especially in countries such as Poland, which soon must change their energy policy and drastically reduce coal burning in favour of other alternative energy sources, including biomass.

5. Conclusions

The calculations and analyses carried out in this paper determined the water needs of willow and evaluated the current conditions of its cultivation in three voivodships of western Poland: West Pomeranian, Lubusz, and Lower Silesian. The main conclusions of the study are as follows:

1. Estimated water needs for the years between 1981 and 2010 in the growing season (from 3rd decade of May to the end of October) amount on average to 408 mm for West Pomeranian Voivodeship, 405 mm for Lubusz Voivodeship, and 402 mm for Lower

Silesian Voivodeship. The highest values of water needs can be found in June, July, and August, while the lowest can be found in the third decade of May and October.
2. The analysis of variance (ANOVA) of the willow water needs values for the growing season from 1981 to 2010 did not show significant differences between the voivodeships (p-value = 0.33 > 0.05). The hypothesis H0 stating that the water needs of willow in the analyzed provinces do not differ was not rejected. Therefore, it can be concluded that the whole region shows similar conditions for willow cultivation.
3. Precipitation in the studied area of all three voivodeships was lower than the calculated ETp values in most years. However, in only a few years, precipitation fulfilled the water requirements of willow. Concerning this aspect, the best conditions prevailed in the Lower Silesian voivodship, where rainfall was higher than ETp in 6 years out of the analyzed 30 years.
4. Water needs of willows in the analyzed thirty-year period from 1981 to 2010 generally show an increasing trend. This trend forces growers to manage water appropriately, particularly important given Poland's limited water resources.
5. The highest rainfall deficits are found in the West Pomeranian voivodship and range from 132 to 288 mm, depending on the year, while the lowest values are found in the Lower Silesian. The problem of precipitation shortage is currently one of the major challenges for agriculture in Poland.

The analyses carried out show that similar conditions for willow cultivation characterize West Poland, and its water needs mainly were not fulfilled by precipitation in the period from 1981 to 2010. Based on previous worldwide studies, one should consider trying to apply adequate irrigation with water or irrigation with treated wastewater, which would provide an appropriate amount of water to this plant. At the same time, this measure could contribute to obtaining an adequate or even higher yield. The use of wastewater in Poland for this purpose would also be potentially beneficial due to the relatively small water resources. However, more research should be conducted in Poland to verify this hypothesis. Nevertheless, without estimating the water requirements, it is not possible to use adequate irrigation and carry out further experiments correctly. Therefore, this study provides guidance and encouragement for further research and valuable practical advice for farmers and growers.

Currently, scientists emphasize that due to Poland's limited water resources, the selection of a suitable energy crops should be based on the water needs of the plants [40], which were estimated in this paper for the willow. This research project is one of the few attempts to estimate willow ETp for the climate of western Poland. With the accurate estimation of the water needs of the willow, it will be possible to optimize the cultivation of this plant and thus increase the biomass obtained. With increased biomass, this plant could potentially be a source of renewable energy for Poland, thus speeding up the country's transition away from coal mining. Such an action fits into the climate neutrality policy of the European Union, which is currently one of the priorities to be implemented by member states.

Author Contributions: Conceptualization, R.R., S.R., B.J., W.P. and P.P.; methodology, S.R., R.R., P.P., F.P.-F., H.A.S., J.K., A.F., W.P. and D.W.; validation, A.F., D.W., R.R., P.P., S.R., P.S. and H.A.S.; formal analysis, S.R., J.K., R.R. and B.J.; investigation, R.R., S.R., B.J., W.P., D.L., P.S., F.P.-F. and A.Ł.; writing—original draft preparation, J.K., D.L., P.S., R.R. and S.R.; writing—review and editing, J.K., D.L., P.S., R.R. and S.R; visualization, J.K. supervision, R.R., S.R. and D.L. All authors have read and agreed to the published version of the manuscript.

Funding: This research received no external funding.

Institutional Review Board Statement: Not applicable.

Informed Consent Statement: Not applicable.

Data Availability Statement: Data sharing not applicable.

Conflicts of Interest: The authors declare no conflict of interest.

References

1. GUS 2021. Energy, Warszawa. 2021. Available online: https://stat.gov.pl/en/topics/environment-energy/ (accessed on 30 November 2021).
2. Sobczyk, W. Evaluation of harvest of energetic basket willow. *Teka Kom. Motoryz. Energetyki Rol.* **2011**, *11*, 343–353.
3. Volk, T.A.; Abrahamson, L.P.; Nowak, C.A.; Smart, L.B.; Tharakan, P.J.; White, E.H. The development of short-rotation willow in the Northeastern United States for bioenergy and bioproducts, agroforestry and phytoremediation. *Biomass Bioenergy* **2006**, *30*, 715–727. [CrossRef]
4. Kuzovkina, Y.A.; Quigley, M.F. Willows beyond Wetlands: Uses of *Salix* L. Species for Environmental Projects. *Water Air Soil Pollut.* **2005**, *162*, 183–204. [CrossRef]
5. Heinsoo, K.; Sild, E.; Koppel, A. Estimation of shoot biomass productivity in Estonian Salix plantations. *For. Ecol. Manag.* **2002**, *170*, 67–74. [CrossRef]
6. Stolarski, M.J.; Niksa, D.; Krzyżaniak, M.; Tworkowski, J.; Szczukowski, S. Willow productivity from small-and large-scale experimental plantations in Poland from 2000 to 2017. *Renew. Sustain. Energy Rev.* **2019**, *101*, 461–475. [CrossRef]
7. Bullard, M.J.; Mustill, S.J.; McMillan, S.D.; Nixon, P.M.I.; Carver, P.; Britt, C.P. Yield improvements through modification of planting density and harvest frequency in short rotation coppice *Salix* spp.—1. Yield response in two morphologically diverse varieties. *Biomass Bioenergy* **2002**, *22*, 15–25. [CrossRef]
8. Wilkinson, J.M.; Evans, E.J.; Bilsborrow, P.E.; Wright, C.; Hewison, W.O.; Pillbeam, D.J. Yield of willow cultivars at different planting densities in a commercial short rotation coppice in the north of England. *Biomass Bioenergy* **2007**, *31*, 469–474. [CrossRef]
9. Magazyn Biomasa. Porównanie Roślin Energetycznych Uprawianych w Polsce. 2017. Available online: https://magazynbiomasa.pl/porownanie-roslin-energetycznych-uprawianych-polsce (accessed on 14 October 2021).
10. Hasselgren, K. Use of municipal waste products in energy forestry: Highlights from 15years of experience. *Biomass Bioenergy* **1998**, *15*, 71–74. [CrossRef]
11. Börjesson, P. Environmental effects of energy crop cultivation in Sweden—I: Identification and quantification. *Biomass Bioenergy* **1999**, *16*, 137–154. [CrossRef]
12. Perttu, K.L.; Kowalik, P.J. *Salix* vegetation filters for purification of waters and soils. *Biomass Bioenergy* **1997**, *12*, 9–19. [CrossRef]
13. Elowson, S. Willow as a vegetation filter for cleaning of polluted drainage water from agricultural land. *Biomass Bioenergy* **1999**, *16*, 281–290. [CrossRef]
14. Mitchell, C.P.; Stevens, E.A.; Watters, M.P. Short-rotation forestry—operations, productivity and costs based on experience gained in the UK. *For. Ecol. Manag.* **1999**, *121*, 123–136. [CrossRef]
15. Juliszewski, T.; Kwaśniewski, D.; Baran, D. Wpływ wybranych czynników na przyrosty wierzby energetycznej. *Inżynieria Rol.* **2006**, *10*, 225–232.
16. Larsen, S.U.; Jørgensen, U.; Lærke, P.E. Willow Yield Is Highly Dependent on Clone and Site. *BioEnergy Res.* **2014**, *7*, 1280–1292. [CrossRef]
17. Yin, L.; Zhou, Y.; Huang, J.; Wenninger, J.; Hou, G.; Zhang, E.; Wang, X.; Dong, J.; Zhang, J.; Uhlenbrook, S. Dynamics of willow tree (*Salix matsudana*) water use and its response to environmental factors in the semi-arid Hailiutu River catchment, Northwest China. *Environ. Earth Sci.* **2014**, *71*, 4997–5006. [CrossRef]
18. Kanecka-Geszke, E. Zużycie wody i plonowanie wierzby wiciowej (*Salix viminalis* L.) na glebach mineralnych o gospodarce opadowo-retencyjnej. *Maszynopis. Praca doktorska* **2015**, 162.
19. Rydłowski, M. Wpływ poziomu wody gruntowej oraz warunków roku badań nawielkość ewapotransoiracji wierzby wiciowej (*Salix viminalis* L.) w świetle badań lizymetrycznych. *Woda-Sr -Obsz. Wiej.* **2016**, *16*, 73–84.
20. Hall, R.L.; Allen, S.J.; Rosier, P.T.W.; Hopkins, R. Transpiration from coppiced poplar and willow measured using sap-flow methods. *Agric. For. Meteorol.* **1998**, *90*, 275–290. [CrossRef]
21. Lindroth, A.; Cienciala, E. Water use efficiency of short-rotation *Salix viminalis* at leaf, tree and stand scales. *Tree Physiol.* **1996**, *16*, 257–262. [CrossRef]
22. Łabędzki, L.; Kanecka-Geszke, E.; Bąk, B.; Słowińska, S. *Estimation of Reference Evapotranspiration Using the FAO Penman-Monteith Method for Climatic Conditions of Poland*; Institute of Technology and Life Sciences: Falenty, Poland, 2011. Available online: https://www.intechopen.com/chapters/14190 (accessed on 26 December 2021).
23. Żakowicz, S. *Podstawy Technologii Nawadniania Rekultywowanych Składowisk Odpadów Komunalnych*; SGGW: Warszawa, Poland, 2010.
24. Doorenbos, J.; Pruitt, W.O. *Guidelines for Predicting Crop Water Requirements*; FAO Irrigation and Drainage Paper 24; Food and Agriculture Organization: Rome, Italy, 1977.
25. Allen, R.G.; Pereira, L.S.; Raes, D.; Smith, M. *Crop Evapotranspiration. Guidelines for Computing Crop Water Requirements*; FAO Irrigation and Drainage No. 56; Food and Agriculture Organization: Rome, Italy, 1998.
26. Hafeez, M.; Khan, A.A. Assessment of Hargreaves and Blaney-Criddle Methods to Estimate Reference Evapotranspiration Under Coastal Conditions. *Am. J. Sci. Eng. Technol.* **2018**, *3*, 65–72. [CrossRef]
27. Heydari, M.M.; Tajamoli, A.; Ghoreishi, S.H. Evaluation and calibration of Blaney–Criddle equation for estimating reference evapotranspiration in semiarid and arid regions. *Environ. Earth Sci.* **2015**, *74*, 4053–4063. [CrossRef]
28. Xiong, Y.; Luo, Y.; Wang, Y.; Traore, S.; Xu, J.; Jiao, X.; Fipps, G. Forecasting daily reference evapotranspiration using the Blaney–Criddle model and temperature forecasts. *Arch. Agron. Soil Sci.* **2016**, *62*, 790–805. [CrossRef]

29. Mendez-Costabel, M.; Morgan, A.; Dokoozlian, N.; Thoreson, B.; Clark, B. Remote sensing of irrigation requirements in wine grapes; validation of an energy balance model and potential application of vegetation indices. *Acta Hortic.* **2014**, *1038*, 249–254. [CrossRef]
30. Ewaid, S.H.; Abed, S.A.; Al-Ansari, N. Crop Water Requirements and Irrigation Schedules for Some Major Crops in Southern Iraq. *Water* **2019**, *11*, 756. [CrossRef]
31. Łabędzki, L. Susze rolnicze. Zarys problematyki oraz metody monitorowania i klasyfikacji. *Woda. Sr. Obsz. Wiej. Rozpr. Nauk. Monogr.* **2006**, *17*, 1–107.
32. Doorenbos, J.; Kassam, A. *Yield Response to Water*; FAO Irrigation and Drainage Paper 33; Food and Agriculture Organization: Rome, Italy, 1979.
33. Durodola, O.S.; Mourad, K.A. Modelling Maize Yield and Water Requirements under Different Climate Change Scenarios. *Climate* **2020**, *8*, 127. [CrossRef]
34. Żyromski, A.; Biniak-Pieróg, M.; Szulczewski, W.; Kordas, L.; Kabała, C.; Gałka, B. *Mathematical Modeling of Evapotranspiration of Selected Energy Crops*; Wydawnictwo Uniwersytetu Przyrodniczego we Wrocławiu: Wrocław, Poland, 2016; ISBN 978-83-7717-253-7.
35. Rolbiecki, S.; Rolbiecki, R.; Jagosz, B.; Figas, A.; Ptach, W.; Stachowski, P. Irrigation effects of red beet as affected by rainfall in different regions of Poland. In *Infrastructure and Environment*; Krakowiak-Bal, A., Vaverkova, M., Eds.; Springer: Cham, Switzerland, 2020. [CrossRef]
36. Łabędzki, L.; Bąk, B.; Kanecka-Geszke, E.; Kasperska-Wołowicz, W.; Smarzyńska, K. Związek między suszą meteorologiczną i rolniczą w różnych regionach agroklimatycznych Polski. *Woda Sr. Obsz. Wiej.* **2008**, *25*, 137.
37. Ziernicka-Wojtaszek, A.; Kopcińska, J. Variation in atmospheric precipitation in Poland in the years 2001–2018. *Atmosphere* **2020**, *11*, 794. [CrossRef]
38. Szwed, M.; Karg, G.; Pinskwar, I.; Radziejewski, M.; Graczyk, D.; Kędziora, A.; Kundzewicz, Z.W. Climate change and its effect on agriculture, water resources and human health sectors in Poland. *Nat. Hazards Earth Syst. Sci.* **2010**, *10*, 1725–1737. [CrossRef]
39. Kundzewicz, Z.; Banasik, K.; Błażejewski, R.; Januchta-Szostak, A.; Jokiel, P.; Kochanek, K.; Konieczny, R.; Kutek, K.; Majewski, W.; Nachlik, E.; et al. Alert wodny nr 2—Konieczna jest redukcja ryzyka powodzi i suszy. *Gospodarka Wodna* **2020**, *8*, 4–6. [CrossRef]
40. Pierzgalski, E.; Mańk, K. Energy crops in aspect of water resources in Poland. In *Agro-Energy for Sustainable Agriculture and Rural Development: Good Practices from Slovakia-Serbia Bilateral Cooperation*; Singidunum University, Faculty of Applied Ecology Futura: Belgrade, Serbia, 2016.
41. Jasiulewicz, M. Biomass from willow plantation as a source of renewable energy and reduction of smog emissions. *Studia Obsz. Wiej.* **2020**, *57*, 145–153. [CrossRef]
42. Igliński, B.; Cichosz, M.; Skrzatek, M.; Buczkowski, R. Potencjał energetyczny biomasy na gruntach ugorowanych i nieużytkach w Polsce [Energy Potential of Biomass on Uncultivated Lands and Fallows in Poland]. *Inżynieria Ochr. Sr.* **2018**, *21*, 79–87. [CrossRef]
43. Gawrońska, G.; Gawroński, K. Metoda szacunku potencjalnych efektów ekologicznych pozyskania energii biomasy drewna wierzby energetycznej na przykładzie Polski. *Acta Sci. Pol. Form. Circumiectus* **2016**, *15*, 5–16. [CrossRef]
44. Szczukowski, S.; Tworkowski, J.; Stolarski, M.J.; Krzyzaniak, M. The energy efficiency of willow biomass production in Poland-a comparative study. *Pap. Glob. Chang.* **2015**, *22*, 123–130. [CrossRef]
45. Jadczyszyn, J.; Faber, A.; Zaliwski, A. Wyznaczanie obszarów potencjalnie przydatnych do uprawy wierzby i ślazowca pensylwańskiego na cele energetyczne w Polsce. *Studia Rap. IUNG-PIB* **2008**, *11*, 55–65.
46. Acharjee, T.K.; Ludwig, F.; van Halsema, G.; Hellegers, P.; Supit, I. Future changes in water requirements of Boro rice in the face of climate change in North-West Bangladesh. *Agric. Water Manag.* **2017**, *194*, 172–183. [CrossRef]
47. Zhang, Y.; Wang, Y.; Niu, H. Effects of temperature, precipitation and carbon dioxide concentrations on the requirements for crop irrigation water in China under future climate scenarios. *Sci. Total Environ.* **2019**, *656*, 373–387. [CrossRef]
48. Masia, S.; Trabucco, A.; Spano, D.; Snyder, R.L.; Sušnik, J.; Marras, S. A modelling platform for climate change impact on local and regional crop water requirements. *Agric. Water Manag.* **2021**, *255*, 107005. [CrossRef]
49. Irmak, S.; Kabenge, I.; Rudnick, D.; Knezevic, S.; Woodward, D.; Moravek, M. Evapotranspiration crop coefficients for mixed riparian plant community and transpiration crop coefficients for Common reed, Cottonwood and Peach-leaf willow in the Platte River Basin, Nebraska-USA. *J. Hydrol.* **2013**, *481*, 177–190. [CrossRef]
50. Hartwell, S.; Morino, K.; Nagler, P.L.; Glenn, E.P. On the irrigation requirements of cottonwood (*Populus fremontii* and *Populus deltoides* var. *wislizenii*) and willow (*Salix gooddingii*) grown in a desert environment. *J. Arid. Environ.* **2010**, *74*, 667–674. [CrossRef]
51. Frédette, C.; Grebenshchykova, Z.; Comeau, Y.; Brisson, J. Evapotranspiration of a willow cultivar (*Salix miyabeana* SX67) grown in a full-scale treatment wetland. *Ecol. Eng.* **2019**, *127*, 254–262. [CrossRef]
52. Mirck, J.; Zalesny, R.S. Mini-Review of Knowledge Gaps in Salt Tolerance of Plants Applied to Willows and Poplars. *Int. J. Phytoremediation* **2015**, *17*, 640–650. [CrossRef] [PubMed]
53. Mirck, J.; Volk, T.A. Response of three shrub willow varieties (Salix spp.) to storm water treatments with different concentrations of salts. *Bioresour. Technol.* **2010**, *101*, 3484–3492. [CrossRef]
54. Gage, E.A.; Cooper, D.J. Constraints on willow seedling survival in a Rocky Mountain Montane floodplain. *Wetlands* **2004**, *24*, 908–911. [CrossRef]
55. Larsen, S.; Jaiswal, D.; Bentsen, N.S.; Wang, D.; Long, S.P. Comparing predicted yield and yield stability of willow and Miscanthus across Denmark. *GCB Bioenergy* **2016**, *8*, 1061–1070. [CrossRef]

56. Lindroth, A.; Båth, A. Assessment of regional willow coppice yield in Sweden on basis of water availability. *For. Ecol. Manag.* **1999**, *121*, 57–65. [CrossRef]
57. Castro-Morales, L.M.; Quintana-Ascencio, P.F.; Fauth, J.E.; Ponzio, K.J.; Hall, D.L. Environmental Factors Affecting Germination and Seedling Survival of Carolina Willow (*Salix Caroliniana*). *Wetlands* **2014**, *34*, 469–478. [CrossRef]
58. Liu, Z.W.; Chen, R.S.; Song, Y.X.; Han, C.T. Aboveground biomass and water storage allocation in alpine willow shrubs in the Qilian Mountains in China. *J. Mt. Sci.* **2015**, *12*, 207–217. [CrossRef]
59. Yang, S.; Volk, T.A.; Fortier, M.-O.P. Willow Biomass Crops Are a Carbon Negative or Low-Carbon Feedstock Depending on Prior Land Use and Transportation Distances to End Users. *Energies* **2020**, *13*, 4251. [CrossRef]

energies

MDPI

Article

Influence of the Fertilization Method on the *Silphium perfoliatum* Biomass Composition and Methane Fermentation Efficiency

Dumitru Peni [1,*], Marcin Dębowski [2] and Mariusz Jerzy Stolarski [1]

[1] Department of Genetics, Plant Breeding and Bioresource Engineering, Faculty of Agriculture and Forestry, University of Warmia and Mazury in Olsztyn, Plac Łódzki 3, 10-724 Olsztyn, Poland; mariusz.stolarski@uwm.edu.pl

[2] Department of Environmental Engineering, Faculty of Geoengineering, University of Warmia and Mazury in Olsztyn, Warszawska 117, 10-719 Olsztyn, Poland; marcin.debowski@uwm.edu.pl

* Correspondence: dumitru.peni@uwm.edu.pl

Abstract: Biogas production is one of the solutions for replacing fossil fuels, which promotes the widespread use of green energy. The aim of this study was to determine the potential of *Silphium perfoliatum* as an energy crop for biogas production, as well as the effect of different fertilization doses (0, 85 and 170 kg N ha^{-1}) on the production potential (NL CH_4 kg^{-1} VS) of *Silphium perfoliatum*. The study investigated the use of different feedstocks, such as raw and ensiled *Silphium perfoliatum* biomass. The methane production ranged between 193.59 and 243.61 NL CH_4 kg^{-1} VS. The highest biogas production potential was achieved with the biomasses which were cultivated with the highest fertilization dose (170 kg N ha^{-1}), both for raw and ensiled crop biomasses, although the difference from the other fertilization doses was not significant. The feedstock (biomass and silage) and digestate parameters were investigated as well. The use of *Silphium perfoliatum* for biogas production seems very promising since its methane production potential was found to be similar to that of the most common energy crop, such as maize, indicating that *Silphium perfoliatum* can compete in the future with maize.

Keywords: anaerobic digestion; fertilization; *Silphium perfoliatum*; biogas; biomass characteristics; digestate

Citation: Peni, D.; Dębowski, M.; Stolarski, M.J. Influence of the Fertilization Method on the *Silphium perfoliatum* Biomass Composition and Methane Fermentation Efficiency. *Energies* **2022**, *15*, 927. https://doi.org/10.3390/en15030927

Academic Editors: Vitaliy Krupin and Roman Podolets

Received: 11 January 2022
Accepted: 26 January 2022
Published: 27 January 2022

1. Introduction

In the last few decades of the 20th century, fossil fuels were the most important source of energy used worldwide, having a huge impact on the technological and economic development of many countries [1]. However, their widespread use has had detrimental consequences for the environment by increasing the level of CO_2 released into the atmosphere, which highly contributes to global warming and climate change [2,3]. Now, fossil fuels are continuing to be used widely in energy production. But facing the depletion of fossil fuels and their constantly rising prices, new and considerably less polluting renewable energy sources are needed [4]. Although European countries have set out to reach a certain percentage/target of renewable energy share to reach by 2020 [5], some of them, including Poland, have not been able to achieve the proposed target. Therefore, it is very important to investigate and implement the use of renewable energy sources on a large scale.

Anaerobic digestion is a biological process that converts biomass into energy/methane [6]. Digestate is the main anaerobic digestion byproduct from which methane has been obtained that can be used as a substitute for mineral fertilizers [7,8]. Nowadays, this process is used to convert biomass to obtain methane as a source of green renewable energy. The production and use of biogas contribute in many aspects to economic, environmental, and social factors [9–11]. Biogas can be obtained from a manifold of substrates. All that is needed is a biomass that contains carbon, carbohydrates, proteins, fats, cellulose and hemicellulose. Currently, biogas is produced from various types of biomass, such as residues from fruits and vegetables, slurry, maize, silage, manure, agricultural and industrial wastes, municipal organic wastes, sewages and sludges [12]. Poland is one of the most important biomass exporters in Europe [13,14] and at the national level, the biomass represents the highest raw material (79%) to obtain energy from renewable energy sources (RES) [15] and could influence the EU market towards bio-based economy [16]. At the end of 2019, there were 103 biogas agricultural plants in Poland [17], while for example, Germany has more than 9000 [6]. The most popular agricultural feedstocks for biogas production in Poland in the last year were distillery stillage (20.6%), residues from fruits and vegetables (19.4%), slurry (18.5%). Maize silage (10.6%) is widely used as well [17]. Mono–digestion of, for example, manure releases low methane yields, and hence co-digestion with energy crops is more preferable [18], and mixing substrates can also have positive effects [19].

Some recent studies [20,21] describe a possible and promising use of *Silphium perfoliatum* for biogas production after an anaerobic digestion process [22]. In this context, it is highly possible that the use of *Silphium perfoliatum* for co-digestion could improve and contribute to the higher yield of methane. *Silphium perfoliatum* belongs to the sunflower family *Asteraceae*. It is used and investigated for its properties as feed [23–27], as well as for its content of active compounds which are used for pharmaceutical and cosmetic purposes [28–30]. *Silphium perfoliatum* is a species that in the future may provide a source of renewable energy [10,20]. *Silphium perfoliatum* plantations can be exploited for a period of over 20 years with one or two harvests during the growing/vegetation season. It contributes to the improvement of the soil quality due to the reduced agricultural operations, as well as being beneficial to biodiversity [31–33]. It is less competitive with food and feed, compared with the most common agricultural crop used for biogas production, i.e., maize [33,34]. Its productivity of biogas is currently investigated [31–33] as the plant is considered a promising energy crop that could replace maize [35] in regions where lands are highly affected by intensive agriculture, as well as on marginal soils that are not suitable for the cultivation of other agricultural crops [36]. Another product following the production of biogas is digestate, which—depending on its characteristics and origin—can be used as a bio-fertilizer or for the production of solid biofuel [37,38].

Therefore, the novel perennial energy crop *Silphium perfoliatum*, which is more often proposed as an alternative crop for biogas production, has been investigated in the present study to gain new knowledge regarding its suitability as a biogas substrate in order to expand its use to a much wider scale. Particular attention was paid to its use as raw material or as silage, as well as the influence of fertilization on biogas and methane production. Withal, the digestate after the end of biogas fermentation was investigated to allow for a more comprehensive evaluation of its properties. The aim of this research was to investigate the potential of biogas production from *Silphium perfoliatum* as raw biomass, and silage as a feedstock for agricultural biogas plants depending on the form (organic and mineral) and doses of fertilization applied, as well as the characteristics of the digestate.

2. Materials and Methods

2.1. Organization of Experimental Works

The research was composed of two different experimental stages focused on the type of feedstock: raw biomass *Silphium perfoliatum* (stage 1) and silage of *Silphium perfoliatum* (stage 2), and three experimental series, analyzing the form of fertilization (organic, mineral and control without fertilizer). In each series, there were three variants, i.e., doses of fertilization (0, 85 and 170 kg ha^{-1} N) used every year after the onset of vegetative growth. The research design is presented in Table 1.

2.2. Feedstock Origin

The substrate used in the present study was raw and ensiled *Silphium perfoliatum*. The biomass collected was green forage from whole plants. The field experiment was conducted on land owned by the Research Station of the University of Warmia and Mazury in Olsztyn (UWM) in the village of Łężany (Poland). Plants of *Silphium perfoliatum* were harvested at the maturity stage in the third and fourth years of growth. They were cut with a rotary mower in the first ten days of September 2019 and 2020. Chopping was performed with a device for cutting and grinding (Viking Ge220).

Table 1. The research design.

Stage 1				
Raw biomass *Silphium perfoliatum*				Control
Series				
Organic fertilizer		Mineral fertilizer		Without fertilization
Variant		Variant		Variant
85	170	85	170	0
Symbol				
RO85(A)	RO170(B)	RM85(C)	RM170(D)	RC0(E)
Stage 2				
Silage *Silphium perfoliatum*				Control
Series				
Organic fertilizer		Mineral fertilizer		Without fertilization
Variant		Variant		Variant
85	170	85	170	0
Symbol				
SO85(A)	SO170(B)	SM85(C)	SM170(D)	SC0(E)

(A) RO85—raw biomass with organic fertilizer 85 kg ha^{-1} N; (B) RO170—raw biomass with organic fertilizer 170 kg ha^{-1} N; (C) RM85—raw biomass with mineral fertilizer 85 kg ha^{-1} N; (D) RM170—raw biomass with mineral fertilizer 170 kg ha^{-1} N; (E) RC0—raw biomass without fertilization; (A) SO85—silage with organic fertilizer 85 kg ha^{-1} N; (B) SO170—silage with organic fertilizer 170 kg ha^{-1} N; (C) SM85—silage with mineral fertilizer 85 kg ha^{-1} N; (D) SM170—silage with mineral fertilizer 170 kg ha^{-1} N; (E) SC0—silage without fertilization.

Anaerobic sludge, which served as the inoculum for the fermentation process in bioreactors, came from a closed bioreactor with a capacity of 7300 m^3 operated at 36 °C. The characteristics of anaerobic sludge are provided in Table 2.

Table 2. Characteristics of anaerobic sludge used as the inoculum for the fermentation process in bioreactors.

Item	Inoculum
DM	7.1 ± 0.3
ODM	70.6 ± 2.9
Ash	29.4 ± 2.9
Carbon	40.4 ± 1.3
Nitrogen	3.9 ± 0.1
C/N ratio	10.2 ± 0.1

DM—dry mass (% d.m.); ODM—organic dry mass (% d.m.); Ash (% d.m.); Carbon (% d.m.); Nitrogen (% d.m.); C/N—carbon to nitrogen ratio.

2.3. Silage Preparation

Silage investigated in the present study was preserved by ensiling *Silphium perfoliatum* in 1000 mL plastic silos on the day of harvesting, immediately after chopping. Ensiling began by filling in every silo manually by compaction. All silos were filled completely and subsequently sealed, obtaining an average density of 867 kg m^{-3}. The silos were stored at a temperature of 10–15 °C, for a period of 7 months before starting the measurements. Moreover, fresh silage samples were dried at 105 °C for 24 h and then crushed in a fiber mill (Retsch SM 200, Retsch GmbH, Haan, Germany) to a particle size of 1 mm for chemical analyzes. Silage was prepared in triplicate, without silage additives or preservatives.

2.4. Anaerobic Digestion Test

The biogas generation from raw material and silage was carried out for 25 days under mesophilic conditions (37 °C) in Automatic Methane Potential Test System II (AMPTS II) reactors coupled with a system recording changes in partial pressure. The amount of methane produced in AMPTS II was measured every three hours throughout the process. Tests were performed on a laboratory scale in 500 cm^3 reactors (glass vessels) filled with approximately 190 g of the inoculum and then assumed amounts of substrate were added (depending on the content of organic matter). Moreover, 200 g of inoculum substrate-free was used as a negative control sample. In the technological repetitions, the initial load varied between 4.5 and 5.5 g VS/L, respectively. Was recalculated amounts of substrate. Anaerobic conditions were achieved by removing oxygen from the reactors (the feedstock and gas phase of the reactors), which was purged with compressed nitrogen. Reactors were equipped with automated stirrers (mixing the content at 100 rpm for 30 s every 10 min), a stabilizing system, and temperature control. The pressure (biomethane production) report was automatically recorded daily, in already normalized data (1.0 standard atmospheric pressure, 0 °C and zero moisture content), using the bioprocess control software. The composition of biogas was measured at the end of the process using a 10 mL injection volume syringe probing into the gas chromatograph connected to a thermal conductivity detector (TCD) (GC Agilent 7890 A–Agilent Technologies, Santa Clara, CA, USA). Helium (He) and argon (Ar) were used as the carrier gases at a flow of 15 mL/min. The temperatures of the injection and detector ports were 150 °C and 250 °C, respectively. Methane yields were calculated as the methane volume produced over a period of 25 days. The perfect gas equation was the basis for computing the volume of produced methane. The endogenous production of the anaerobic sludge was excluded from the calculations of methane production of the tests.

2.5. Analytical Methods

At the beginning of the trials, substrates (raw and silage), inoculum and digestate were analyzed for dry matter content, organic dry matter content, ash content, carbon content and total nitrogen. For the substrates, inoculum and digestate samples, moisture, dry matter content and organic dry matter were determined with the gravimetric method by drying in an oven at 105 °C for 24 h (EN ISO 18134–1:2015 using the FD series laboratory dryer (FD BINDER series, Tuttlingen, Germany)). Ash content was determined using an automatic ELTRA TGA–THERMOSTEP analyzer (ELTRA GmbH, Neuss, Germany) according to the PN–EN ISO 18122:2016–01. The carbon content was determined using an automatic ELTRA CHS–500 analyzer (ELTRA GmbH, Neuss, Germany) PN–EN ISO 16948:2015–07. The total nitrogen was determined by the Kjeldahl method with the use of a K–435 mineralizer and B–324 BUCHI distiller (Büchi Labortechnik AG, Flawil, Switzerland).

2.6. Statistical Analysis

All experimental variants were conducted in triplicate. Statistical analysis of the results was supported by a Statistical 13.3 PL package. Thus, the reaction rate constants (k) based on experimental data were determined by non-linear regression. The rate of biogas production (r) could be determined for each experimental variant. The iterative method was applied, in which the function is replaced in each iterative step with a linear differential in relation to the determined parameters. The coefficient of convergence φ2 was adopted as the measure of the curve's fit (with determined parameters) to experimental data. This coefficient is the ratio of the sum square of deviations of experimental values to the sum square of deviations of experimental values from the mean value. A three-way analysis of variance (ANOVA) was carried out to determine the significance of differences between the variables. To determine the significance of differences between the analyzed variables, Tukey's HSD test was used. In all tests, differences were considered significant at $p < 0.05$. The Pearson correlation coefficient between the analyzed trials was also determined.

3. Results and Discussion

3.1. Biomass Characteristics

The characteristics of the *Silphium perfoliatum* (raw biomass and silage) used in the study are presented in Table 3. The DM had values between 22.5–25.3% for raw material and 20.6–21.8% for silage. In the studied biomass, the ODM had values between 90.3–92.0% for raw material and 89.5–91.9% for silage. The analysis showed that the substrate type (raw biomass and silage) influenced the ODM, ash content, C, N content and the C/N ratio (Table 4). In turn, the fertilization type significantly influenced ODM, ash content, dry matter content, N content and C/N ratio. On the other hand, the fertilization dose had a significant effect only on the N content and C/N ratio. By analyzing the influence of the interaction of the main factors, it was found that only the fertilization type × N dose had an effect on DM content. A positive correlation between DM and methane and biogas production was also observed (Table 5). Previous studies on other perennial crops as well show that the N fertilization dose influences the N content and C/N ratio and improved the biomass quality [39,40].

Table 3. Characteristics of raw material and silage of *Silphium perfoliatum* used for the preparation of feedstock. (O85—organic fertilization 85 kg ha^{-1} N; O170—organic fertilization 170 kg ha^{-1} N; M85—mineral fertilization 85 kg ha^{-1} N; M170—mineral fertilization 170 kg ha^{-1} N; C—without fertilization).

		O85	O170	M85	M170	C0
DM	Raw material	22.9 ± 0.8	23.1 ± 0.7	25.3 ± 0.8	22.5 ± 0.6	22.9 ± 0.1
	Silage	21.4 ± 0.0	21.8 ± 1.1	21.1 ± 2.6	20.6 ± 1.9	20.6 ± 1.5
ODM	Raw material	90.7 ± 0.2	90.3 ± 0.8	91.0 ± 1.5	92.0 ± 0.1	91.4 ± 0.8
	Silage	89.5 ± 0.1	89.7 ± 0.6	90.5 ± 1.1	91.9 ± 0.4	89.7 ± 0.1
Ash	Raw material	9.3 ± 0.2	9.7 ± 0.8	9.0 ± 1.5	7.9 ± 0.1	8.6 ± 0.8
	Silage	10.5 ± 0.1	10.3 ± 0.6	9.5 ± 1.1	8.1 ± 0.4	10.3 ± 0.1
Carbon	Raw material	43.7 ± 0.5	44.4 ± 1.1	43.6 ± 1.1	44.8 ± 1.0	43.7 ± 0.7
	Silage	45.1 ± 0.9	45.5 ± 0.0	44.9 ± 2.5	46.8 ± 1.1	45.4 ± 1.0
Nitrogen	Raw material	0.8 ± 0.0	0.8 ± 0.0	0.9 ± 0.0	1.1 ± 0.3	0.7 ± 0.0
	Silage	0.9 ± 0.0	0.9 ± 0.0	1.0 ± 0.0	1.1 ± 0.1	0.8 ± 0.1
C/N ratio	Raw material	62.3 ± 4.4	53.0 ± 1.2	47.6 ± 1.1	41.7 ± 10.6	61.4 ± 3.1
	Silage	53.3 ± 1.4	49.4 ± 0.4	46.4 ± 1.4	43.8 ± 3.9	54.5 ± 4.5

DM—dry mass (% d.m.); ODM—organic dry mass (% d.m.); Ash (% d.m.); Carbon (% d.m.); Nitrogen (% d.m.); C/N—carbon to nitrogen ratio.

Table 4. Analysis of variance (p values) for the analyzed features.

Source of Variation	Organic Dry Matter	Ash Content	Dry Matter	C Content	N Content	C/N Ratio	CO_2%	CH_4%	Methane Production	Biogas Production
Substrate type	0.005 *	0.005 *	0.115	0.000 *	0.027 *	0.005 *	0.164	0.164	0.021 *	0.006 *
Fertilization type	0.002 *	0.002 *	0.019 *	0.560	0.000 *	0.000 *	0.603	0.603	0.801	0.659
Nitrogen (N) dose	0.208	0.208	0.232	0.068	0.021 *	0.017 *	0.704	0.704	0.584	0.645
Substrate type × Fertilization type	0.058	0.058	0.719	0.879	0.205	0.078	0.936	0.936	0.894	0.830
Substrate type × N dose	0.692	0.692	0.381	0.877	0.455	0.307	0.984	0.984	0.715	0.736
Fertilization type × N dose	0.067	0.067	0.031 *	0.238	0.089	0.175	0.949	0.949	0.763	0.767
Substrate type × Fertilization type × N dose	0.956	0.956	0.801	0.829	0.719	0.732	0.965	0.965	0.466	0.368

* Significant values ($p < 0.05$).

Table 5. The Pearson correlation coefficients for the analyzed trials.

Item	Organic Dry Matter	Ash Content	Dry Matter	C Content	N Content	C/N Ratio	Methane Production	Biogas Production	CO_2 %	CH_4 %
Organic dry matter	1.00									
Ash content	−1.00 *	1.00								
Dry matter	0.03	−0.03	1.00							
C content	0.08	−0.08	−0.58 *	1.00						
N content	0.27	−0.27	−0.02	0.33 *	1.00					
C/N ratio	−0.20	0.20	−0.20	−0.23	−0.93 *	1.00				
Methane production	−0.03	0.03	0.40 *	−0.48 *	0.20	−0.16	1.00			
Biogas production	0.11	−0.11	0.41 *	−0.54 *	0.26	−0.22	0.96 *	1.00		
CO_2%	0.49 *	−0.49 *	0.08	−0.27	0.15	−0.18	−0.08	0.20	1.00	
CH_4%	−0.49 *	0.49 *	−0.08	0.27	−0.15	0.18	0.08	−0.20	−1.00 *	1.00

* Significant values ($p < 0.05$).

The results of our experiments showed that the highest C/N ratio was found in *Silphium perfoliatum* raw biomass with the organic fertilization dose of 85 kg ha^{-1} N (62.3); meanwhile, the lowest C/N ratio was found in *Silphium perfoliatum* raw biomass with the mineral fertilization dose of 170 kg ha^{-1} N (41.7) (Table 3). However, the fertilization type and N dose had a significant impact on the C/N ratio in the biomass (Table 4), but the values of this parameter were not correlated with the production of biogas (Figure 1). The results of our investigation showed the opposite of the observation regarding other energy crops which show that there was a correlation between the C/N ratio and biogas yield; this is not the rule, but the inappropriate C/N ratio is unfavorable for AD [40].

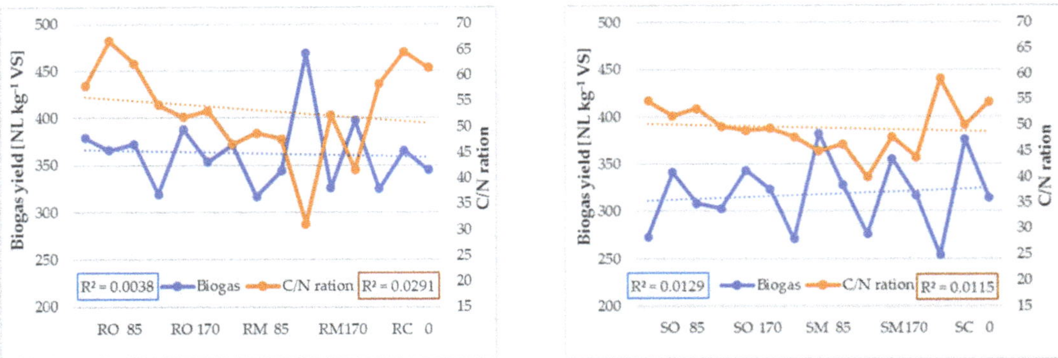

(**a**) substrates as raw biomass (**b**) substrates as silage

Figure 1. Correlation of biogas yields of substrates with a C/N ratio of raw (**a**) and silage (**b**) from *Silphium perfoliatum*.

The characteristics of the mixture of anaerobic sludge and substrate are provided in Table 6. The carbon and nitrogen (C/N) ratio in the substrate is one of the most important factors that influence biogas production [41]. In the present study, it was found that the use of *Silphium perfoliatum* biomass as a feedstock for anaerobic digestion significantly improved the value of the C/N ratio. However, the literature review provides the optimal ranges of the C/N ratio for an undisturbed course of anaerobic digestion in the range of 10 to 30 [42] or even in a narrower range from 20 to 30 [43].

Table 6. Characteristics of the mixture of anaerobic sludge and substrate raw material/silage of *Silphium perfoliatum* used for the anaerobic digestion test.

Item	Substrate Type	O85	O170	M85	M170	C0
DM (%)	Raw material	8.1 ± 0.7	8.2 ± 0.7	8.2 ± 0.7	8.1 ± 0.6	8.1 ± 0.6
	Silage	7.5 ± 0.0	7.6 ± 0.0	7.6 ± 0.1	7.6 ± 0.1	7.6 ± 0.1
ODM (% d.m.)	Raw material	68.9 ± 0.0	68.9 ± 0.0	68.8 ± 0.1	68.9 ± 0.0	68.9 ± 0.0
	Silage	74.3 ± 2.0	74.3 ± 2.0	74.2 ± 2.1	74.4 ± 2.0	74.3 ± 2.0
Ash (% d.m.)	Raw material	31.1 ± 0.0	31.1 ± 0.0	31.2 ± 0.1	31.1 ± 0.0	31.1 ± 0.0
	Silage	25.7 ± 2.0	25.7 ± 2.0	25.8 ± 2.1	25.6 ± 2.0	25.7 ± 2.0
Carbon (%d.m.)	Raw material	39.3 ± 1.1	39.3 ± 1.1	39.2 ± 1.1	39.2 ± 1.1	39.3 ± 0.1
	Silage	41.8 ± 2.0	41.9 ± 1.9	41.8 ± 2.1	41.9 ± 2.0	41.9 ± 2.0
Nitrogen (% d.m.)	Raw material	3.7 ± 0.2	3.7 ± 0.2	3.7 ± 0.2	3.7 ± 0.2	3.7 ± 0.2
	Silage	3.9 ± 0.0	3.9 ± 0.0	3.9 ± 0.0	3.9 ± 0.0	3.9 ± 0.0
C/N ratio	Raw material	12.8 ± 0.5	12.3 ± 0.3	11.9 ± 0.3	11.7 ± 0.8	12.7 ± 0.4
	Silage	12.5 ± 0.4	12.3 ± 0.4	11.9 ± 0.4	12.0 ± 0.6	12.5 ± 0.6

DM—dry mass (% d.m.); ODM—organic dry mass (% d.m.); Ash (% d.m.); Carbon (% d.m.); Nitrogen (% d.m.); C/N—carbon to nitrogen ratio.

3.2. Methane and Biogas Production

Daily biogas and methane production from the raw substrates and silages of *Silphium perfoliatum* is presented in Figure 2. The methane and biogas yields of *Silphium perfoliatum* averaged between 222.82 and 361.18 NL kg^{-1} VS, respectively, for raw biomass, and 200.09 and 317.59 NL kg^{-1} VS for silage. Methane yields are evaluated on a DM–basis, but the results are presented and discussed on a VS–basis. Higher methane yields were obtained in Germany, 232–321 NL kg^{-1} VS [32,44–47], the Republic of Moldova, 275 NL kg^{-1} VS [48] and the Czech Republic, 276 NL kg^{-1} VS [33]. In the present study, methane and biogas production showed differences between substrate type—raw material and silage. The methane production differed significantly in the case of substrate type, ranging from 193.59 to 243.61 NL kg^{-1} VS (Figure 2). The highest methane and biogas yield: 243.61 and 395.15 NL kg^{-1} VS, respectively, was achieved from raw material with the mineral fertilization dose of 170 kg ha^{-1} N (Figure 2a), as well as 204.26 and 327.45 NL kg^{-1} VS, respectively from silage with the mineral fertilization dose 170 kg ha^{-1} N (Figure 2b). The highest effectiveness was achieved with the mineral fertilization dose of 170 kg ha^{-1} N and silage (Figure 2b) at the production rate of r = 90.0 cm^3 d^{-1} and methane content of 63.2 ± 0.8% (Table 7) and with the mineral fertilization dose of 170 kg ha^{-1} N and raw biomass (Figure 2a), at the production rate of r = 75.1 cm^3 d^{-1} and methane content of 62.2 ± 3.2% (Table 7). Nonetheless, there were not any significant statistical differences regarding biogas and methane production between fertilization type, and the N dose and between all interactions (Table 4). In a similar study where different energy crops were investigated, it was found that the N dose significantly influenced the methane and biogas production (maize and sunflower) but was not a rule for all energy crops (sorghum and triticale) [40].

Table 7. Biogas production rate (r), the reaction rate constant (k) and methane (CH_4), carbon dioxide (CO_2) content in biogas of the analyzed raw material, and silage of *Silphium perfoliatum*.

Substrate Type		O85	O170	M85	M170	C0
Raw material	r [cm^3 d^{-1}]	66.6	70.4	72.2	75.1	65.5
	k [l d^{-1}]	0.18	0.20	0.21	0.19	0.19
Silage	r [cm^3 d^{-1}]	89.1	80.8	81.7	90.0	88.2
	k [l d^{-1}]	0.29	0.25	0.25	0.29	0.28
Raw material	CH_4 [%]	60.9 ± 2.5	61.5 ± 3.5	61.8 ± 4.1	62.2 ± 3.2	62.4.6 ± 2.9
	CO_2 [%]	39.1 ± 2.5	38.5 ± 3.5	38.2 ± 4.1	37.8 ± 3.2	37.6 ± 2.9
Silage	CH_4 [%]	62.7 ± 2.3	62.7 ± 0.2	62.4± 1.0	63.2 ± 0.8	63.6 ± 1.0
	CO_2 [%]	37.3 ± 2.3	37.3 ± 0.2	37.6 ± 1.0	36.8 ± 0.8	36.4 ± 1.0

3.3. Methane Content

The biogas production rate, reaction rate constant, methane and carbon dioxide content in biogas from raw material and silage is presented in Table 7. It was found that the *Silphium perfoliatum* silage was easier and faster biodegradable in anaerobic conditions. In the case of silage, r values ranged from 80.8 to 90.0 cm^3 d^{-1}. In the case of raw biomass, they ranged from 65.5 to 75.1 cm^3 d^{-1}. The highest content of methane was achieved in silage and raw biomass substrates without fertilization, 63.6% and 62.4%, respectively. But, there were no significant statistical differences regarding the content of methane and carbon dioxide between analyzed features (substrate type, fertilization type and N dose) and between their interaction (Table 4). In a study conducted in Poland, the methane content of maize straws was found between 48.97 and 50.26% CH_4 [49]. Of course, this value (of corn straw) is much lower compared to the maize silage, which has a value corresponding to 56–59% CH_4 [40], or even up to 65% CH_4 at the beginning of the process [50].

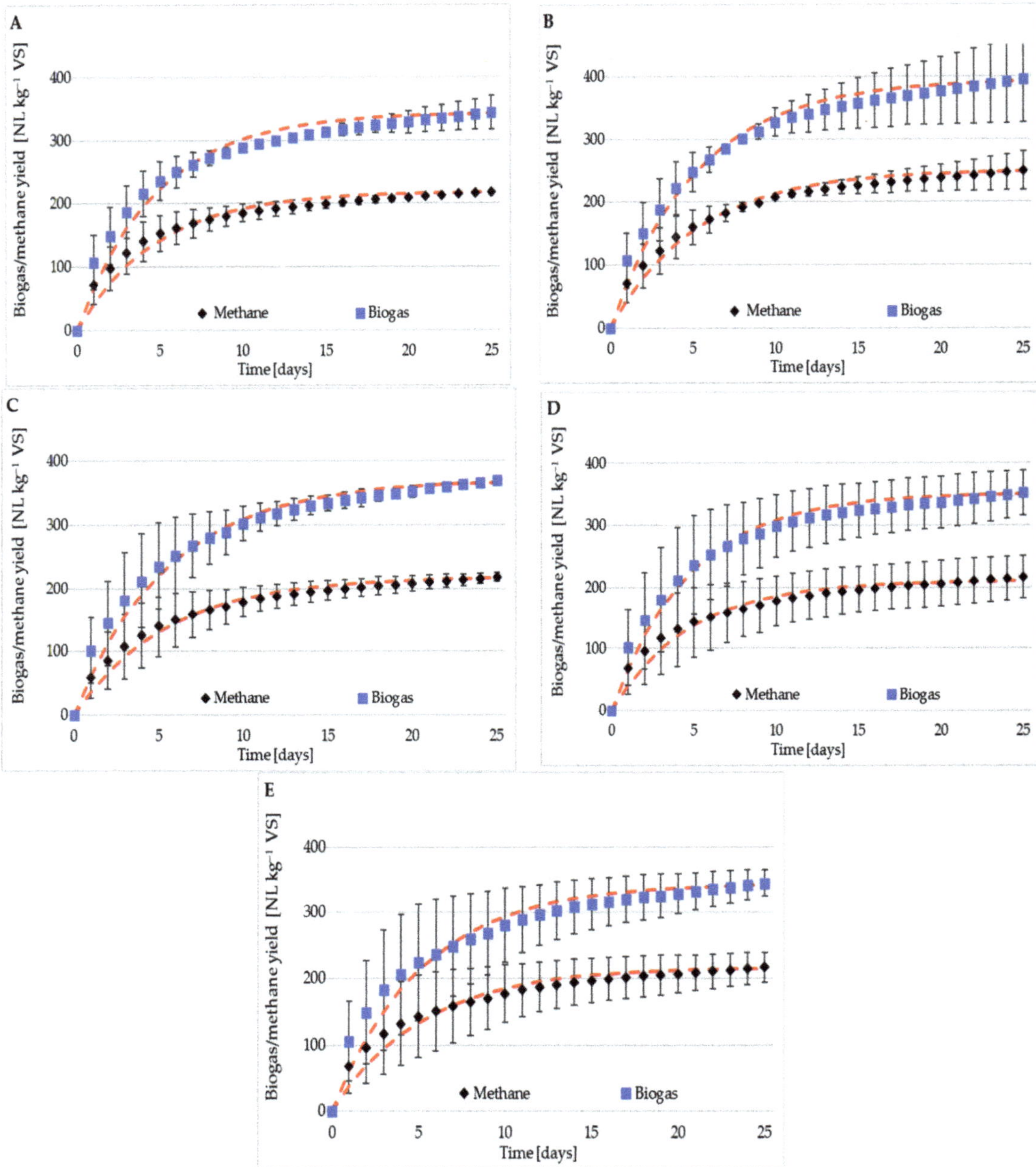

(a) Silphium perfoliatum as raw biomass

Figure 2. *Cont.*

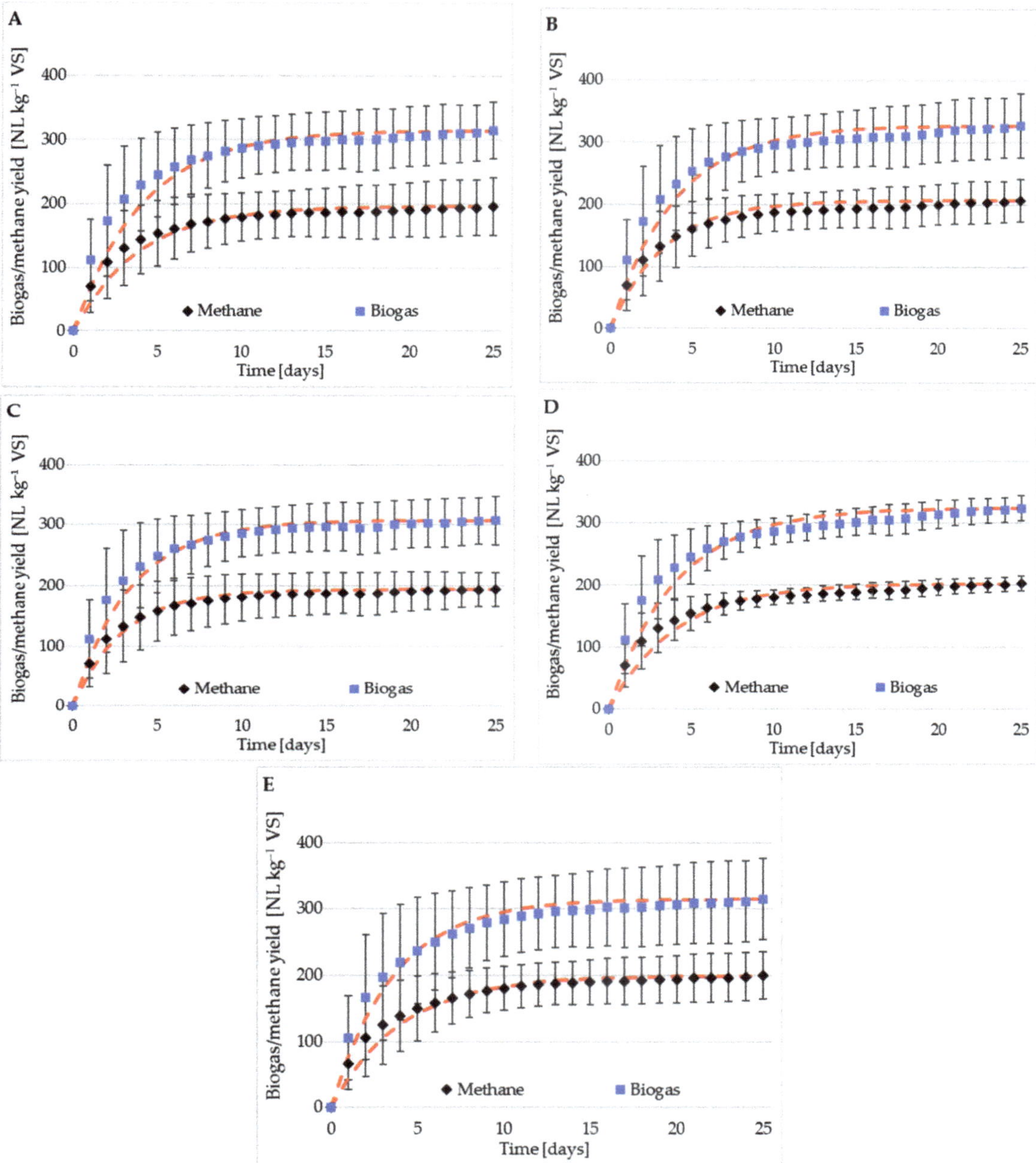

(**b**) Silphium perfoliatum as silage

Figure 2. Cumulative methane and biogas yield (NL kg^{-1} VS) of *Silphium perfcliatum* as raw biomass (**a**) and as silage (**b**) in the 25-day test, depending on the type of fertilization (kg ha^{-1} N): (**A**) mineral fertilization 85, (**B**) mineral fertilization 170, (**C**) organic fertilization 85, (**D**) organic fertilization 170, (**E**) without fertilization.

3.4. Digestate Characteristic

Digestate is the substance that remains after the end of the biogas generation process. It is rich in various active substances that can be extracted or reused as fertilizer, depending on the content of micro and macronutrients [38].

Its composition and the amounts of certain elements largely depend on the raw material used to produce biogas. The content of the DM, ODM, ash content, C/N ratio, and nitrogen (N) of the digestate and inoculum after anaerobic digestion (I.A.A.D) is presented in Table 8. The measurements were carried out in three replication and the results were averaged. The studied digestate DM had values between 6.7 and 6.9% depending on the fertilization dose supplied to *Silphium perfoliatum*, and was slightly lower for the digestate obtained after AD (anaerobic digestion). The carbon and nitrogen content of digestate is very important for the C/N ratio. To be used safely as fertilizer in agriculture, it is recommended the C/N ratio be between 15 and 20 without some pretreatment operations [51]. This ratio was much lower in the present study (between 11.5–12.3 from silage and 12.1–12.8 from raw material).

Table 8. Characteristic of digestate from raw material and ensiled *Silphium perfoliatum*.

Item	Substrate Type	O85	O170	M85	M170	C0	I.A.A.D.
DM	Raw material	6.8 ± 0.6	6.7 ± 0.4	6.9 ± 0.5	6.8 ± 0.5	6.8 ± 0.4	6.7 ± 0.6
	Silage	6.3 ± 0.3	6.2 ± 0.3	6.4 ± 0.4	6.2 ± 0.2	6.3 ± 0.2	5.7 ± 0.0
ODM	Raw material	71.2 ± 1.0	69.8 ± 1.8	72.1 ± 0.7	71.5 ± 0.8	71.7 ± 0.8	70.4 ± 1.4
	Silage	69.7 ± 2.3	69.9 ± 2.5	71.1 ± 1.5	70.8 ± 2.4	70.5 ± 1.9	70.0 ± 2.1
Ash	Raw material	28.8 ± 1.0	30.2 ± 1.8	27.9 ± 0.7	28.6 ± 0.8	28.4 ± 0.8	29.6 ± 1.4
	Silage	30.3 ± 2.3	30.1 ± 2.5	28.9 ± 1.5	29.2 ± 2.4	29.5 ± 1.9	30.0 ±2.1
Carbon	Raw material	39.9 ± 0.6	38.9 ± 1.3	40.2 ± 0.9	39.7 ± 0.9	39.5 ± 1.0	39.4 ± 1.4
	Silage	40.8 ± 1.6	41.1 ± 2.1	40.6 ± 0.7	41.2 ± 1.4	41.5 ± 1.9	40.1 ± 2.0
Nitrogen	Raw material	3.2 ± 0.2	3.1 ± 0.3	3.3 ± 0.1	3.2 ± 0.2	3.1 ± 0.3	3.5 ± 0.2
	Silage	3.1 ± 0.2	3.2 ± 0.1	3.2 ± 0.3	3.3 ± 0.1	3.1 ± 0.1	3.4 ± 0.0
C/N ratio	Raw material	12.7 ± 0.7	12.1 ± 0.7	12.4 ± 0.1	12.6 ± 0.6	12.8 ± 0.9	11.3 ± 0.1
	Silage	11.9 ± 1.1	11.7 ± 1.0	11.5 ± 1.1	11.6 ± 0.8	12.3 ± 1.2	11.1 ± 0.4

DM—dry mass (% d.m.); ODM—organic dry mass (% d.m.); Ash (% d.m.); Carbon (% d.m.); Nitrogen (% d.m.); C/N—carbon to nitrogen ratio; I.A.A.D—inoculum after anaerobic digestion.

4. Conclusions

In the present study, the influence of different fertilization types, nitrogen dose and substrate types (row biomass and silage) of the Silphium perfoliatum for biogas production was investigated. It was found that the substrate type has a significant influence on most of the analyzed features. This study indicates that *Silphium perfoliatum* can be used to produce biogas. However, the yield of biogas and methane may differ under the effect of different types and doses of fertilizers, although the differences were small. On the other hand, the yield of methane and biogas significantly depends only on the substrate type. It is noteworthy that the C/N ratio is an important factor that influences biogas production but, in our study, there was no correlation between these two parameters. Future studies will require an investigation of the amount of biomass per hectare, to determine what amount/yield of biogas and methane can be obtained from a given area depending on the amount of fertilizer used per hectare.

Author Contributions: Conceptualization, D.P., M.D. and M.J.S.; methodology, M.D. and D.P.; formal analysis, M.D. and M.J.S.; validation, M.D. and M.J.S.; investigation, D.P. and M.D.; resources, M.D. and M.J.S.; data curation, D.P., M.D. and M.J.S.; writing—original draft preparation, D.P.; writing—review and editing, D.P., M.D. and M.J.S.; visualization, D.P., M.D. and M.J.S.; supervision M.D. and M.J.S.; funding acquisition, M.D. and M.J.S. All authors have read and agreed to the published version of the manuscript.

Funding: The results presented in this paper were obtained as part of a comprehensive study financed by the University of Warmia and Mazury in Olsztyn, Faculty of Agriculture and Forestry, Department of Genetics, Plant Breeding and Bioresource Engineering (granted by Ministry Education and Science No. 30.610.007-110).

Institutional Review Board Statement: Not applicable.

Informed Consent Statement: Not applicable.

Data Availability Statement: Not applicable.

Acknowledgments: Dumitru Peni is a recipient of a scholarship from the program of Interdisciplinary Doctoral Studies in Bioeconomy (POWR. 03.02.00-00-I034/16-00), which is funded by the European Social Fund. We would also like to thank the technical staff of the Department of Genetics, Plant Breeding and Bioresource Engineering and the Department of Environmental Engineering for their technical support during the experiment.

Conflicts of Interest: The authors declare no conflict of interest.

References

1. Kazimierowicz, J.; Dzienis, L. Giant miscanthus as a substrate for biogas production. *J. Ecol. Eng.* **2015**, *16*, 139–142. [CrossRef]
2. Paolini, V.; Petracchini, F.; Segreto, M.; Tomassetti, L.; Naja, N.; Cecinato, A. Environmental impact of biogas: A short review of current knowledge. *J. Environ. Sci. Health A* **2018**, *53*, 899–906. [CrossRef] [PubMed]
3. Masson-Delmotte, V.; Zhai, P.; Pörtner, H.-O.; Roberts, D.; Skea, J.; Shukla, P.R.; Pirani, A.; Moufouma-Okia, W.; Péan, C.; Pidcock, R.; et al. *IPCC Special Report: Global Warming of 1.5 °C—Summary for Policymakers*; World Meteorological Organization: Geneva, Switzerland, 2018; pp. 1–32.
4. Zieliński, M.; Dębowski, M.; Kazimierowicz, J. The Effect of Electromagnetic Microwave Radiation on Methane Fermentation of Selected Energy Crop Species. *Processes* **2022**, *10*, 45. [CrossRef]
5. EU Commission. Directive 2009/28/EC of the European Parliament and of the Council of 23 April 2009 on the Promotion of the Use of Energy from Renewable Sources and Amending and Subsequently Repealing Directives 2001/77/EC and 2003/30/EC. *Off. J. Eur. Union* **2009**, *11*, 39–85.
6. Scarlat, N.; Dallemand, J.-F.; Fahl, F. Biogas: Developments and perspectives in Europe. *Renew. Energy* **2018**, *129*, 457–472. [CrossRef]
7. Kazimierowicz, J. Organic waste used in agricultural biogas plants. *J. Ecol. Eng.* **2014**, *15*, 88–92. [CrossRef]
8. Lukehurst, C.T.; Frost, P.; Al Seadi, T. *Utilisation of Digestate from Biogas Plants as Biofertiliser*; IEA Bioenergy: Paris, France, 2010; pp. 1–36.
9. Thrän, D.; Schaubach, K.; Majer, S.; Horschig, T. Governance of sustainability in the German biogas sector—Adaptive management of the Renewable Energy Act between agriculture and the energy sector. *Energy Sustain. Soc.* **2020**, *10*, 3. [CrossRef]
10. Stolarski, M.J.; Krzyżaniak, M.; Śnieg, M.; Słomińska, E.; Piórkowski, M.; Filipkowski, R. Thermophysical and chemical properties of perennial energy crops depending on harvest period. *Int. Agrophys.* **2014**, *28*, 201–211. [CrossRef]
11. Hijazi, O.; Munro, S.; Zerhusen, B.; Effenberger, M. Review of life cycle assessment for biogas production in Europe. *Renew. Sustain. Energy Rev.* **2016**, *54*, 1291–1300. [CrossRef]
12. Kazimierowicz, J.; Zieliński, M.; Dębowski, M. Influence of the Heating Method on the Efficiency of Biomethane Production from Expired Food Products. *Fermentation* **2021**, *7*, 12. [CrossRef]
13. Gołuchowska, B.; Sławiński, J.; Markowski, G. Biomass utilization as a renewable energy source in polish power industry–current status and perspectives. *J. Ecol. Eng.* **2015**, *16*, 143–154. [CrossRef]
14. Stolarski, M.J.; Warmiński, K.; Krzyżaniak, M.; Olba–Zięty, E.; Akincza, M. Bioenergy technologies and biomass potential vary in Northern European countries. *Renew. Sustain. Energy Rev.* **2020**, *133*, 110238. [CrossRef]
15. Główny Urząd Statystyczny. Available online: https://stat.gov.pl/obszary-tematyczne/srodowisko-energia/energia/energia-2021-folder,1,9.html (accessed on 30 September 2021).
16. Jankowski, K.J.; Dubis, B.; Budzyński, W.S.; Bórawski, P.; Bułkowska, K. Energy efficiency of crops grown for biogas production in a large-scale farm in Poland. *Energy* **2016**, *109*, 277–286. [CrossRef]
17. Biuletyn Informacji Publicznej Krajowego Ośrodka Wsparcia Rolnictwa. Available online: https://bip.kowr.gov.pl/informacje-publiczne/odnawialne-zrodla-energii/biogaz-rolniczy/dane-dotyczace-dzialalnosci-wytworcow-biogazu-rolniczego-w-latach-2011--2019 (accessed on 29 July 2021).
18. Zieliński, M.; Rusanowska, P.; Zielińska, M.; Dudek, M.; Nowicka, A.; Purwin, C.; Fijałkowska, M.; Dębowski, M. Influence of preparation of Sida hermaphrodita silages on its conversion to methane. *Renew. Energy* **2021**, *163*, 437–444. [CrossRef]
19. Kazimierowicz, J.; Dzienis, L.; Dębowski, M.; Zieliński, M. Optimisation of methane fermentation as a valorisation method for food waste products. *Biomass Bioenergy* **2021**, *144*, 105913. [CrossRef]
20. Peni, D.; Stolarski, M.J.; Bordiean, A.; Krzyżaniak, M.; Dębowski, M. Silphium perfoliatum—A Herbaceous Crop with Increased Interest in Recent Years for Multi-Purpose Use. *Agriculture* **2020**, *10*, 640. [CrossRef]

21. Siwek, H.; Włodarczyk, M.; Możdżer, E.; Bury, M.; Kitczak, T. Chemical Composition and Biogas Formation potential of *Sida hermaphrodita* and *Silphium perfoliatum*. *Appl. Sci.* **2019**, *9*, 4016. [CrossRef]
22. Von Cossel, M.; Amarysti, C.; Wilhelm, H.; Priya, N.; Winkler, B.; Hoerner, L. The replacement of maize (*Zea mays* L.) by cup plant (*Silphium perfoliatum* L.) as biogas substrate and its implications for the energy and material flows of a large biogas plant. *Biofuels Bioprod. Biorefin.* **2020**, *14*, 152–179. [CrossRef]
23. Pastukhova, M.A.; Sheliuto, B.V. Cultivation of *Silphium perfoaliatum* under agricultural crops cover. *Bull. Belarus. State Agric. Acad.* **2019**, *3*, 83–87.
24. Pichard, G. Management, production, and nutritional characteristics of cup-plant (*Silphium perfoliatum*) in temperate climates of southern Chile. *Cienc. Investig. Agrar.* **2012**, *39*, 61–77. [CrossRef]
25. Guoyan, P.; Ouyang, Z.; Qunying, L.; Qiang, Y.; Jishun, W. Water consumption of seven forage cultivars under different climatic conditions in the North China plain. *J. Resour. Ecol.* **2011**, *2*, 74–82. [CrossRef]
26. Gimbatov, A.S.; Alimirzaeva, G.A. Optimization of technology for the cultivation of new feed crops in the irrigated conditions of the lowland zone of Dagestan. *Probl. Dev. Agric. Ind. APK Reg.* **2011**, *5*, 8–11.
27. Stanford, G. Silphium perfoliatum (cup-plant) as a new forage. In *Recapturing a Vanishing Heritage, Proceedings of the 12th North American Prairie Conference, Cedar Falls, IA, USA, 5–9 August 1990*; University of Northern Iowa: Cedar Falls, IA, USA, 1990; pp. 33–37.
28. Kowalski, R. Analysis of lipophilic fraction from leaves, inflorescences and rhizomes of *Silphium perfoliatum* L. *Acta Soc. Bot. Pol.* **2005**, *74*, 5–10. [CrossRef]
29. Kowalski, R.; Kędzia, B. Antibacterial activity of *Silphium perfoliatum*. Extracts. *Pharm. Biol.* **2007**, *45*, 494–500. [CrossRef]
30. Feng, W.-S.; Pei, Y.-Y.; Zheng, X.-K.; Li, C.-G.; Ke, Y.-Y.; Lv, Y.-Y.; Zhang, Y.-L. A new kaempferol trioside from *Silphium perfoliatum*. *J. Asian Nat. Prod. Res.* **2014**, *16*, 393–399. [CrossRef] [PubMed]
31. Gansberger, M.; Montgomery, L.F.R.; Liebhard, P. Botanical characteristics, crop management and potential of *Silphium perfoliatum* L. as a renewable resource for biogas production: A review. *Ind. Crop. Prod.* **2015**, *63*, 362–372. [CrossRef]
32. Mast, B.; Lemmer, A.; Oechsner, H.; Reinhardt-Hanisch, A.; Claupein, W.; Graeff-Hönninger, S. Methane yield potential of novel perennial biogas crops influenced by harvest date. *Ind. Crop. Prod.* **2014**, *58*, 194–203. [CrossRef]
33. Ustak, S.; Munoz, J. Cup-plant potential for biogas production compared to reference maize in relation to the balance needs of nutrients and some microelements for their cultivation. *J. Environ. Manag.* **2018**, *228*, 260–266. [CrossRef]
34. Thompson, W.; Meyer, S. Second generation biofuels and food crops: Co-products or competitors? *Glob. Food Secur.* **2013**, *2*, 89–96. [CrossRef]
35. Gerstberger, P.; Asen, F.; Hartmann, C. Zur Ökonomie und Ökologie der Becherpflanze (*Silphium perfoliatum* L.) im Vergleich zum Silomais [Economy and ecology of cup plant (*Silphium perfoliatum* L.) compared with silage maize]. *J. Kult.* **2016**, *68*, 372–377. [CrossRef]
36. Stolarski, M.J.; Niksa, D.; Krzyżaniak, M.; Tworkowski, J.; Szczukowski, S. Willow productivity from small-and large-scale experimental plantations in Poland from 2000 to 2017. *Renew. Sustain Energy Rev.* **2019**, *101*, 461–475. [CrossRef]
37. Ogwang, I.; Kasedde, H.; Nabuuma, B.; Kirabira, J.B.; Lwanyaga, J.D. Characterization of Biogas Digestate for Solid Biofuel Production in Uganda. *Sci. Afr.* **2021**, *12*, e00735. [CrossRef]
38. Drosg, B.; Fuchs, W.; Al Seadi, T.; Madsen, M.; Linke, B. *Nutrient Recovery by Biogas Digestate Processing*; IEA Bioenergy: Paris, France, 2015.
39. Oleszek, M.; Matyka, M. Nitrogen fertilization level and cutting affected lignocellulosic crops properties important for biogas production. *BioResources* **2017**, *12*, 8565–8580. [CrossRef]
40. Oleszek, M.; Matyka, M. Determination of the efficiency and kinetics of biogas production from energy crops through nitrogen fertilization levels and cutting frequency. *BioResources* **2018**, *13*, 8505–8528. [CrossRef]
41. Dębowski, M.; Kisielewska, M.; Kazimierowicz, J.; Rudnicka, A.; Dudek, M.; Romanowska-Duda, Z.; Zieliński, M. The effects of Microalgae Biomass Co-Substrate on Biogas Production from the Common Agricultural Biogas Plants Feedstock. *Energies* **2020**, *13*, 2186. [CrossRef]
42. Schnurer, A.; Jarvis, A. *Microbiological Handbook for Biogas Plants. Swedish Waste Management U2009:03 Wedish Gas Centre Report 207*; Avfall Sverige and Svenskt Gastekniskt Center AB: Malmö, Sweden, 2009; pp. 1–74.
43. Drosg, B.; Braun, R.; Bochmann, G.; Al Saedi, T. Analysis and characterisation of biogas feedstocks. In *The Biogas Handbook: Science, Production and Applications*; Wellinger, A., Murphy, J., David, B., Eds.; Woodhead Publishing: Sawston, UK, 2013; pp. 52–84. [CrossRef]
44. Schoo, B.; Kage, H.; Schittenhelm, S. Radiation use efficiency, chemical composition, and methane yield of biogas crops under rainfed and irrigated conditions. *Eur. J. Agron.* **2017**, *87*, 8–18. [CrossRef]
45. Wever, C.; Höller, M.; Becker, L.; Biertümpfel, A.; Köhler, J.; van Inghelandt, D.; Westhoff, P.; Pude, R.; Pestsova, E. Towards high-biomass yielding bioenergy crop *Silphium perfoliatum* L.: Phenotypic and genotypic evaluation of five cultivated populations. *Biomass Bioenergy* **2019**, *124*, 102–113. [CrossRef]
46. Ruf, T.; Emmerling, C. Site-adapted production of bioenergy feedstocks on poorly drained cropland through the cultivation of perennial crops. A feasibility study on biomass yield and biochemical methane potential. *Biomass Bioenergy* **2018**, *119*, 429–435. [CrossRef]

47. Von Cossel, M.; Steberl, K.; Hartung, J.; Pereira, L.A.; Kiesel, A.; Lewandowski, I. Methane yield and species diversity dynamics of perennial wild plant mixtures established alone, under cover crop maize (*Zea mays* L.), and after spring barley (*Hordeum vulgare* L.). *GCB Bioenergy* **2019**, *11*, 1376–1391. [CrossRef]
48. Ţîţei, V. The evaluation of biomass of the *Sida hermaphrodita* and *Silphium perfoliatum* for renewable energy in Moldova. *Sci. Pap. Ser. A Agron.* **2017**, *60*, 534–540.
49. Mazurkiewicz, J.; Marczuk, A.; Pochwatka, P.; Kujawa, S. Maize Straw as a Valuable Energetic Material for Biogas Plant Feeding. *Materials* **2019**, *12*, 3848. [CrossRef] [PubMed]
50. Dinuccio, E.; Balsari, P.; Gioelli, F.; Menardo, S. Evaluation of the biogas productivity potential of some Italian agro-industrial biomasses. *Bioresour. Technol.* **2010**, *101*, 3780–3783. [CrossRef] [PubMed]
51. Visvanathan, Z.; Visvanathan, C. Evaluation of anaerobic digestate for greenhouse gas emissions at various stages of its management. *Int. Biodeterior. Biodegrad.* **2014**, *95*, 167–175. [CrossRef]

energies

MDPI

Article

Influence of Growing *Miscanthus x giganteus* on Ecosystem Services of Chernozem

Yana Vodiak [1], Yurii Tsapko [1], Anatolii Kucher [2], Vitaliy Krupin [3,*] and Iryna Skorokhod [4]

1 National Scientific Center "Institute for Soil Science and Agrochemistry Research named after O.N. Sokolovsky", Chaikovska Str. 4, 61024 Kharkiv, Ukraine; 1994demidova@gmail.com (Y.V.); tsapkoul@i.ua (Y.T.)

2 Department of Ecology and Environmental Management, V.N. Karazin Kharkiv National University, Svobody Sq. 4, 61022 Kharkiv, Ukraine; kucher@karazin.ua

3 Institute of Rural and Agricultural Development, Polish Academy of Sciences, Nowy Świat 72, 00-330 Warsaw, Poland

4 Department of International Economic Relations and Project Management, Lesya Ukrainka Volyn National University, Volia Ave. 13, 43025 Lutsk, Ukraine; skorokhodiryna1@gmail.com

* Correspondence: vkrupin@irwirpan.waw.pl

Abstract: The paper investigates the optimization of ecosystem services of podzolized heavy loamy chernozem (black soil) as a result of the cultivation of the perennial energy culture of *Miscanthus x giganteus*. The research was conducted on an experimental land plot during 2016–2021. No fertilization was applied to the soil during the experiments, and over the years of research, the growing seasons were accompanied by abnormal droughts, but even under such conditions, the plants of *Miscanthus x giganteus* gradually increased their yield. At the initial stage of research, in the third year of cultivation, dry biomass of *Miscanthus x giganteus* was obtained at 14.3 t/ha, in the fourth year–18.6 t/ha, and already in the fifth and sixth years, 21.7 and 24.5 t/ha, respectively. That is, energy-wise, the harvest for the last year was equivalent to 15.9 tons of coal or 12,618 m^3 of natural gas. Cultivation of *Miscanthus x giganteus* on black soil for six years has improved the provision of its ecosystem services, regulation, and ecosystem maintenance services. The possibility of growing perennial energy crops on agricultural soils has been proven by obtaining a significant amount of biomass and a positive phytoremediation effect on the soil by reducing erosion, preserving biodiversity, sequestering carbon, and sustainably improving the ecological situation.

Keywords: *Miscanthus x giganteus*; biomass; energy crops; soil; ecosystem services; carbon sequestration; podzolized chernozem; black soil

Citation: Vodiak, Y.; Tsapko, Y.; Kucher, A.; Krupin, V.; Skorokhod, I. Influence of Growing *Miscanthus x giganteus* on Ecosystem Services of Chernozem. *Energies* **2022**, *15*, 4157. https://doi.org/10.3390/en15114157

Academic Editor: Dino Musmarra

Received: 14 May 2022
Accepted: 4 June 2022
Published: 6 June 2022

Publisher's Note: MDPI stays neutral with regard to jurisdictional claims in published maps and institutional affiliations.

1. Introduction

The importance of a stable energy supply is increasing in the global perspective, as the energy demand is expected to grow at a fast pace in the next decades [1–3] along with population and economic growth [4,5]. Recent geopolitical events associated with the Russian aggression against Ukraine [6] revealed the vulnerability of the current energy supply structure [7], where dependency not only on fossil fuels but also on its particular unstable suppliers, has the potential to distort the global energy security in case of unforeseen political shocks, thus undermining the feasibility of substantiated and set development paths [8] worldwide.

In the past years, key factors influencing the energy policies of the developed countries have been arising mainly from the climate change agenda [9,10], thus targeting to increase the share of renewable energy generation [11,12] and search for ways to limit the greenhouse gas emissions [13,14] from the economic sectors. In current conditions, the role of renewable energy generation representing decentralized and local sources [15,16] of sustainable energy is gaining additional importance and puts the energy transformation agenda on top of development priorities.

Energy based on the use of biomass is among the key sustainable approaches [17] to its generation, allowing us to obtain clean energy in terms of associated greenhouse gas emissions. Additionally, the cultivation of energy crops has numerous economic, environmental, and even social effects, which combined show the importance of substantiated development of this approach in theory and its implementation in practice.

These effects can also be assessed within the concept of ecosystem services, the benefits of which are within reach due to their integrative and interdisciplinary nature as they include economic, environmental, and social dimensions [18]. The services provided by such ecosystems could have a positive effect on climate regulation (through the level of carbon sequestration), structure and quality of soils, as well as water availability and cycling [19,20].

Applying and valuing ecosystem services is seen as an innovative step toward sustainable land use. Assessing these services can reveal the impact of energy crops and add objectivity to the renewable energy debate. The concept of sustainable land management should take into account the current and possible future impacts of energy crop production as well as people's preferences for achieving long-term sustainable solutions [18]. Increased demands on renewable energy are likely to result in the allocation of more land for the production of bioenergy plants. Therefore, land-use change is being increasingly verified through environmental impact assessments, which [21] propose to include a more complete study of ecosystem services.

For a comprehensive assessment of the sustainability of energy production from biomass, it is necessary to take into account the entire life cycle of production and combustion of crops for a wide range of indicators of ecosystem services. Therefore, Lovett et al. [22] in their article present the basis for such an assessment based on the synthesis of a large amount of data on the impact on ecosystem services of bioenergy crops (for example, low vegetation and *Miscanthus x giganteus*), which makes it possible to compare the impact on ecosystems between energy systems.

Ecosystem services are the bridge that exists between nature and people [23], as they can be defined as the direct or indirect contribution of ecosystems to human well-being [24]. Soils provide and regulate a large number of ecosystem services, yet imbalanced or non-sustainable practices can as well produce disservices and lead to soil degradation [25]. Thus, it is important to utilize experimental approaches to verify agricultural practices in set conditions, which allow for substantiated results and grounded policy and practical recommendations.

The Ukrainian background is especially important in this research, as for this country, the latest geopolitical developments further limit the availability of traditional energy sources based primarily on fossil fuels. While the renewable energy generation in Ukraine took place in past decades in a rather slow manner [26], current conditions further amplify the need to diversify the energy sources, yet also continue transformation towards more sustainable energy generation. Thus, the current deficit of fossil fuels in Ukraine contributes to the even more urgent necessity for the development of "green energy", among others, through the cultivation of energy crops with a low-carbon footprint, which opens up opportunities for sustainable biofuel production and, together with the development of wind and solar energy production, steadily reduces the release of greenhouse gases into the atmosphere. Increasing the supply volume of such fuels is extremely important and relatively easily achievable due to the assimilation of large areas not only of sown plantations but also of low-productive "problem" lands, as alternatives to traditional ones, for growing energy crops. In addition, the cultivation of energy crops and biofuels serves as a very important compromise between the development of the energy sector and the "environmental friendliness" of industrial production.

It is this strategy for the development of "green energy" that allows solving the problem of balance between meeting the social, economic, and environmental issues arising in the bioenergy generation industry, which are considered by the concept of ecosystem services. The main advantages of energy crops from the point of view of soil science and the provision of ecosystem services are the suitability for growing on low-productive lands, the possibility of minimal application of fertilizers (or even abandonment of them), no need for weed control (except for one-time treatment in the first year of planting), and phytoremediation ability. This is especially noticeable when compared to permanent fertilization and pest control when growing traditional food and forage crops, which have significantly higher costs than those that cannot function normally and grow on non-agricultural land. Thus, the cultivation of energy crops invariably meets the social and economic needs of society, increasing the profits of entrepreneurs or farmers, meeting energy needs, and, in the long term, increasing the value of ecosystem services through effective management and potential restoration of soil quality. In this work, we draw attention to the fact that in the realities of Ukraine, the cultivation of energy crops with rational use is quite possible on agricultural land [27,28].

Therefore, the purpose of this publication is to detect and reveal the possibilities for optimization of ecosystem services of podzolized heavy loamy chernozem due to the cultivation of *Miscanthus x giganteus*.

2. Materials and Methods

The research was carried out during 2016–2021 aimed to optimize the ecosystem services of podzolized heavy loamy chernozem while growing the "Zvezdotsvetosenniy" *Miscanthus x giganteus* (referred to later on as *Mischanthus*) variety in the stationary field experiment of the National Scientific Center "Institute for Soil Science and Agrochemistry Research named after O.N. Sokolovsky", enterprise "DG "Grakivske", the village of Novyi Korotychin the Kharkiv region of Ukraine (Figure 1).

Figure 1. Experimental research area in Ukraine in Kharkiv region. Source: own elaboration.

The relief of the experimental field is leveled, has a gentle 2–3° slope of northern exposure. The field is bounded on all four sides by protective strips. Podzolized heavy loamy chernozem in the experimental area is characterized by the following parameters of the arable layer: pH aq. 5.9–6.0; the carbon content of organic matter is 1.89%; physical clay content is 43%. A one-factor experiment, which did not provide for the application of fertilizers and the use of plant protection products to assess the direct impact of growing *Mischanthus* on the studied soil, it was planted twice: in 2016 (*Mischanthus x giganteus* I) and 2019 (*Mischanthus x giganteus* II). Soil sampling was carried out from layers 0–20, 20–40, and 40–60 cm directly under the plants in triplicate according to Ukrainian state standards DSTU 4287: 2004 and DSTU ISO 11464: 2007. The number of ground invertebrates-microarthropods (the method of eclectation according to Berlese in Tullgren's modification). Counting the yield of *Miscanthus* was carried out by the method of test plots (sheaves) in triplicate followed by weighing. The carbon content in organic matter was determined by the oxidometric method–DSTU 4289:2004.

The number of microarthropods was determined by the selective Tullgren method, which is based on the use of a trait common to all soil inhabitants-the desire to penetrate deep into the soil when the upper layers of the soil dry out. Brief description of the measurement method: soil samples (from a layer of 0–20 cm) of a fixed volume (150 cm^3) were placed in a sieve inserted into a funnel of a slightly larger diameter, under which a vessel with a fixing solution (70% alcohol) was placed. Natural light was used to dry the surface of the soil sample. The number of microartopods that moved down and, sliding along the walls of the funnel, moved into the fixative liquid was counted after distillation using a magnifying glass, previously filtered on filter paper. The results were statistically processed using Microsoft Excel.

3. Results

The idea of the study is based on obtaining new scientific knowledge about the influence of constant (for five years) cultivation of a perennial energy crop of *Miscanthus* on the optimization of ecosystem services in podzolized heavy loamy chernozem. It is especially important to receive up-to-date scientific information on improving the provision of ecosystem services by soil, which contributes to the solution of the tasks under the UN Sustainable Development Goals (UN SDGs) related to food security, water scarcity, climate change, loss of biodiversity, and threats to public health [29].

The spread of degradation processes due to irrational use of land—excessive plowing, short crop rotations, and rapid climatic changes—forces the scientific community and specialists to intensively search for fundamentally new ways to restore soil fertility, which are used in agricultural production. In this context, it is important to understand that the guarantee of agroecological stability of soils, the promotion of the development of self-reproduction of their fertility, and buffering capacity is the preservation of biological diversity [30,31].

Based on the positive phytoremediation experience of growing energy crops [32,33], we have established a positive effect of growing *Miscanthus* on podzolized heavy loamy chernozem in relation to the optimization of ecosystem services.

3.1. Weather Conditions at the Research Site

The climatic changes observed in recent years [34] are confirmed by the fact that on the experimental site during the research period from 2016 to 2021, there was an increase in average monthly temperatures (Figure 2) as well as a noticeable decrease in the amount of precipitation (Figure 3).

Assessing the weather data, we note that in the Kharkiv region, even before the beginning of the second decade of the 21st century, the average monthly rainfall was at the level of 43 mm, which means, about 520 mm came to the earth's surface annually, and about 260 mm per year during the growing season. Since 2011, almost every year the amount of precipitation has dropped significantly, and the average value of the air temperature has increased.

Aridization, or signs of desertification, is especially noticeable in September, which is the very month in Ukraine when agricultural enterprises plant winter crops. However, now the realities of the weather conditions in September are as follows: The air temperatures are quite high, and there is practically no natural moisture in the soil (see Figure 2), which means the problems of agricultural production caused by climatic changes are clearly observed.

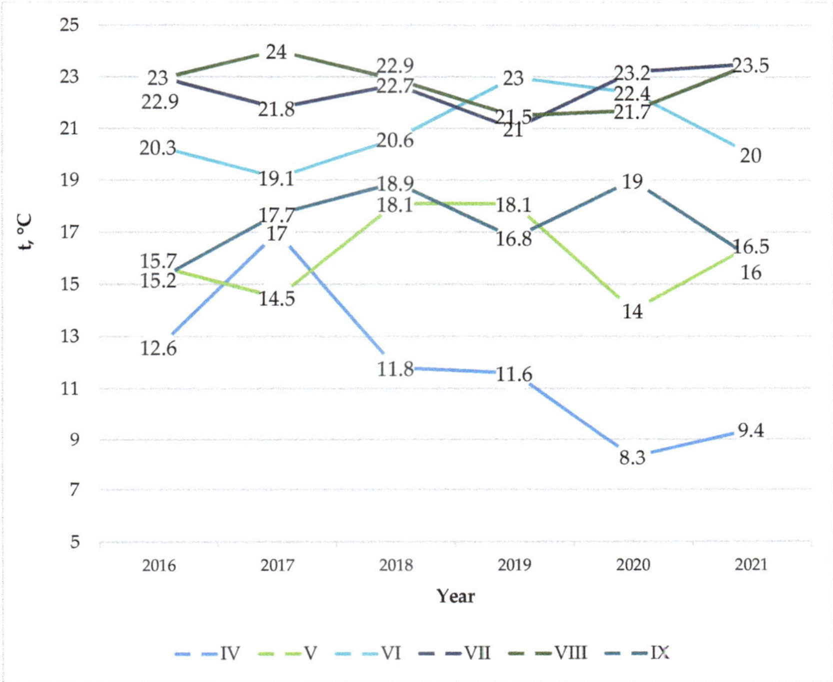

Figure 2. Average monthly temperature within the periods April–September in 2016–2021. Source: own elaboration.

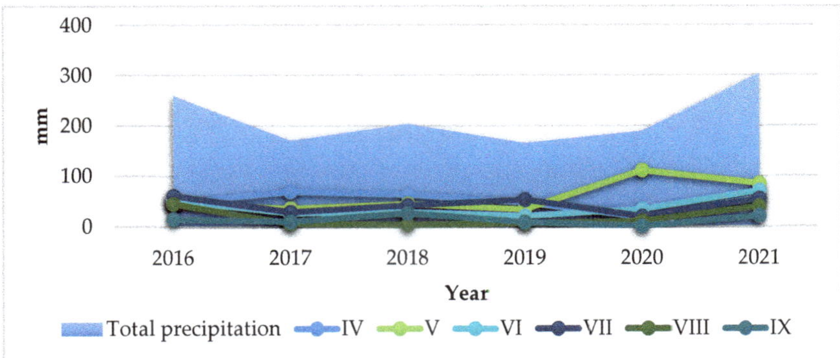

Figure 3. The amount of precipitation within the periods of April–September in 2016–2021. Source: own research results.

However, growing *Miscanthus* on black soil, even in such challenging weather conditions, has not prevented the soil from improving ecosystem services, as evidenced by the gradual increase in yields of this energy crop.

3.2. *Harvest of Miscanthus x giganteus*

The world practice of growing *Miscanthus* involves accounting for the harvest in the third year after planting, so we have provided data on the harvest since 2018 (Table 1).

Table 1. Yield of *Miscanthus x giganteus* on black soil in 2018–2021, dry weight t/ha.

Energy Culture	2018	2019	2020	2021
Miscanthus x giganteus	14.3	18.6	21.7	24.5

Source: own research results.

The peculiarities of *Miscanthus* cultivation are that the harvest is usually not taken into account for the first two years, and the harvest is recorded from the third year. It was found that even in relatively dry conditions, *Miscanthus* plants produce significant volumes of biomass without reducing soil productivity, even without fertilization. So, in the third year of cultivation in 2018, the yield of dry biomass of *Miscanthus* was 14.3 t/ha, in 2019–18.6 t/ha, in 2020–21.7 t/ha, and already in September 2021, it was 24.5 t/ha. The harvest for the last year is equivalent to 15.9 tons of coal, 9.8 tons of crude oil, 41.7 tons of timber, or 12,618 m^3 of natural gas [35].

Due to the fact that in our studies, podzolized chernozem during the cultivation of *Miscanthus* was not subjected to agrotechnological processing, starting from the second year of cultivation, and the complete rejection of any fertilizers and plant protection products, we received a significant amount of biomass of *Miscanthus*, which indicates a high ecological value and the profitability of growing it. According to scientists [36], starting from the third year, the profitability of *Miscanthus* cultivation is 726% and can remain almost at this level for many years until its complete elimination.

Energy raw materials are referred to as an ecosystem supply service that is the easiest to understand and quantify. At the same time, the resulting harvest is a material benefit that has a specific price in monetary terms and a guaranteed energy supply. On the Ukrainian market, the cost of 1 ton of *Miscanthus* pellets is about UAH 4500, and the cost of 1 ton of straw briquettes is only UAH 2700.

3.3. Carbon Content in Soil Organic Matter

Studies have shown that under the influence of growing *Miscanthus* on podzolized heavy loamy chernozem, the amount of organic matter carbon in the arable and subsoil layers increases (Figure 4).

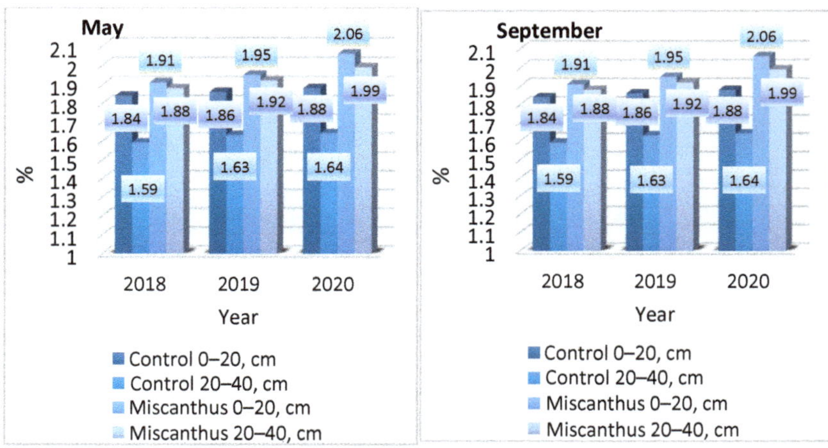

(a) (b)

Figure 4. Carbon content in organic matter of chernozem, May (**a**)–September (**b**) 2018–2020, in %. Source: own research results.

It was found that over three years in the 0–20 cm layer in May, the organic carbon content increased from 1.91% in 2018 to 2.06% in 2020, while in September this indicator changed over the same years, respectively, from 1.92% to 2.11%.

In the subsurface layer (20–40 cm) of the studied chernozem, a similar tendency is observed with respect to a gradual increase in the carbon content of organic matter. The established pattern is extremely important for the development of measures to reduce greenhouse gas emissions into the atmosphere, which is a powerful argument for fulfilling Ukraine's obligations, which are reflected in a number of state documents, in particular: "Concept for the implementation of state policy in the field of climate change for the period up to 2030" (Order of the Cabinet of Ministers of Ukraine dated 6 December 2016 No. 932-r); "National Action Plan to Combat Land Degradation and Desertification" (Order of the Cabinet of Ministers of Ukraine dated 30 March 30 2016 No. 271-r). A gradual increase in the organic carbon content in the studied chernozem under the *Miscanthus* indicates an improvement in the supporting ecosystem service, which, together with a regulatory service (habitat formation, soil formation), provides a significant improvement in the ecological state of this chernozem. The ability of energy crops to store carbon in soil can be attributed to several ecosystem services. Firstly, this is a regulation service—that is, improving soil quality by increasing carbon as the main humus-forming element and improving air quality due to a decrease in carbon dioxide near plantations with energy crops; secondly, the service of maintaining ecosystems, because the content of this element in the soil is part of the process of the biogeochemical carbon cycle, and, consequently, a decrease in the release into the atmosphere, and therefore, counteraction to global warming.

3.4. The Number of Microarthropods in the Experiment

The biodiversity of soil microfauna, numerous representatives of which are microarthropods–invertebrate oribatids (Oribatida, Acarina carapace mites) and colembola (Springtail Collembola), plays an important role in the destruction and transformation of organic matter. It should also be noted that these soil microorganisms are extremely sensitive and are often used as bioindicators of environmental changes. Since their number clearly reacts to air temperature, moisture, and soil chemical composition, our results vary somewhat depending on the month and year of the study. The number of oribatids on the studied soil under the *Miscanthus* is shown in Figure 5.

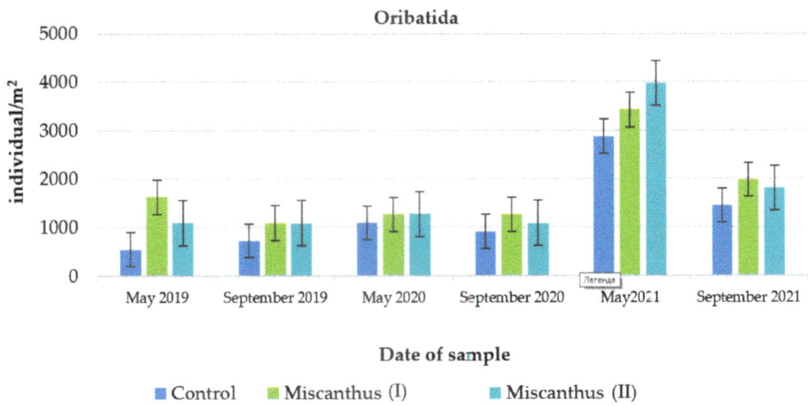

Figure 5. The number of oribatids on the studied soil under the *Miscanthus*: May 2019–HIP05 = 225; September 2019–HIP05 = 146; May 2020–HIP05 = 130; September 2020–HIP05 = 192; May 2021–HIP05 = 354; September 2021–HIP05 = 142. Source: own research results.

The figure clearly shows that in 2020 the number of oribatids in May and September was 1086 and 908 specimens/m^2, respectively, and 1260 and 1264 specimens/m^2 under *Miscanthus*. However, in 2021, the number of oribatids almost tripled compared to previous years because May 2021 was characterized by moderate temperatures and relatively high rainfall of 84 mm.

Our studies, carried out on podzolized heavy loamy chernozem in 2020, found that in the control, the number of colemboles (Figure 6) in May was 1622, and in September, 1986 ind./m^2, and under *Miscanthus* plants, respectively, 1628 and 3240 ind./m^2.

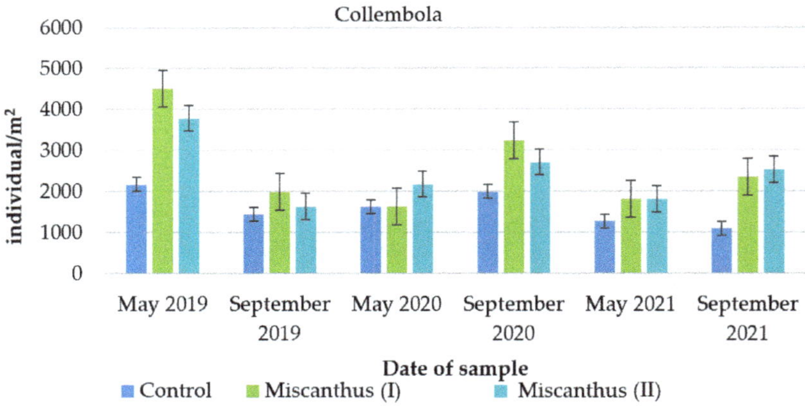

Figure 6. The number of colemboles on chernozem under *Miscanthus*: May 2019–HIP05 = 362; September 2019–HIP05 = 232; May 2020–HIP05 = 284; September 2020–HIP05 = 188; May 2021–HIP05 = 204; September 2019–HIP05 = 192. Source: own research results.

The number of colembol specimens differs from oribatids in variants and decreases somewhat over time, although it remains much higher than in the control. This indicator is due, to a large extent, to the increase in the number of ticks (oribatids) because some of them are predators that feed on colemboles. Thus, microarthropods sensitively react not only to weather conditions but also to the species composition of plants in the ecosystem and to the species composition of soil microorganisms.

In general, the number of microarthropods indicates that under the plants of the *Miscanthus* there are more favorable conditions for their habitation and development and, consequently, for the biodiversity of the ecosystem as a whole, which refers to a supporting ecosystem service. Under such conditions, the activation of the biological factor (microarthropods) enhances the course of the soil-forming process towards self-adaptation and self-reproduction, which will certainly lead to an improvement in soil fertility.

4. Discussion

Based on the achieved results, it is assumed that this research has obtained positive evidence regarding the impact of the cultivation of *Miscanthus* on the optimization of ecosystem services in podzolized loamy chernozem, which is usually used for growing traditional crops of wheat, corn, sunflower, and beet.

The results are consistent with the findings of other researchers on the cultivation of individual bioenergy crops on marginal lands [37,38]. The results of the study are of great practical importance in the context of achieving climate neutrality by 2050, in particular, by increasing the area of cultivation of bioenergy crops, including those on marginal lands [39]. It should be noted that the use of biomass for energy crops in combination with other alternative energy sources [40,41] is one of the priority areas for ensuring low-carbon development of the agricultural sector and the economy as a whole.

Integration of energy crops into agricultural landscapes can foster permanence and maintain sustainability if they are placed in such a way as to stimulate multiple ecosystem services and mitigate harmful ecosystem effects from existing crops [42], as well as promote balanced land use [43].

For example, energy crops in the coastal regions of the midwest United States have a positive effect on ecosystem services while the benefits-costs ratio has fluctuated significantly. At the same time, the overall monetary value of the improved ecosystem services associated with the introduction of perennial energy crops was much lower than the opportunity cost. The mismatch between recoverable costs and social value is a fundamental challenge for the expansion of perennial energy crops and sustainable agricultural landscapes [42] and the potential for biomass supply [44]. Analyzing the dynamics and uncertainty of land-use transformation for the production of perennial energy crops, [45] examined the effects of payment for ecosystem services policies. It has been found that the current expected profit from growing perennial energy crops (including switchgrass) is insufficient for these crops to be widely adopted by American farmers due to relatively unstable yields, volatile incomes, and high costs of growing crops. At the same time, switchgrass has the potential to provide energy while reducing greenhouse gas emissions [46].

In this context, the results of a survey of farmers and non-experts on the perception of energy crop production in Germany turned out to be interesting. In particular, it was found that many farmers consider themselves responsible for the provision of many ecosystem services while they prefer the regional scale of growing energy crops based on conventional crops. Most of the non-specialists interviewed noted the ambiguity of energy crops as a source of energy without side effects. In layman's opinion, the use of biomass for renewable energy production is not an important ecosystem service. Biomass production should be limited to fields that do not require food production and the use of crop residues or materials for landscape management [46].

Global scientists and practitioners are mainly exploring the possibilities and cultivation of energy crops on marginal lands. For example, [37] notes in the article that: (i) ecosystem services differ depending on the type of marginal land; (ii) special bioenergy crops can improve ecosystem services on marginal lands; (iii) there is a need to intensify research in this direction, as there is currently a lack of field data on the productivity of energy crops on marginal lands, and ecosystem services are hardly discussed in the literature. This was among the reasons why the current research was conducted, aiming to fill in the existing gap in experimental data, which would be beneficial for further substantiation of strategies to expand the cultivation of *Miscanthus* in particular soil conditions, as well as to take these findings into account in the policies being implemented towards the protection of the environment, achieving climate-neutrality and improving biodiversity, as well as supplying clean energy from a renewable source.

Another direction to increase the impact of the cultivation of specialized energy crops is to do this on marginal lands, which can provide, in particular, such ecosystem services as biomass production, control of water and wind erosion of soil, sequestration of carbon in the soil, absorption or content of pollutants or metals, stabilization or reclamation of disturbed soils, and improvement of properties. It is summarized in [37] that growing energy crops on marginal lands can increase soil carbon sequestration, restore contaminated or compacted soils, and improve biodiversity. Fertilizing or adding organic improvers increases biomass yield and carbon sequestration on marginal lands [37].

Growing energy crops on marginal lands is considered a useful opportunity for farmers against the progressive risk of underutilization or non-use of these lands. Scenario modeling results indicate the positive impact of energy crops on ecosystem services in terms of environmental quality and biodiversity value [38]. At the same time, other studies show that increased production of bioenergy crops leads to increased soil-use and land-use conflicts and also decreases the supply of several ecosystem services, such as regulation of soil erosion, carbon sequestration, environmental value, and landscape aesthetic value [38]. Therefore, this indicates the need to continue experimental research to answer the question of the impact of energy crops on soil ecosystem services. Among the new and promising areas of research is also the evaluation of the efficiency of growing bioenergy crops using alternative fertilizer systems, including green manure [47].

A substantiated approach needs to be taken with each agricultural practice, as particular ones are especially influential on sustainable development and its goals (SDGs). The authors argue [48], in this context, that such ecosystem service as soil conservation service can be among those, as it contributes simultaneously to SDG 15 (Life on land), SDG 13 (Climate action), and SDG 6 (Clean water and sanitation), as well as several others to a lesser extent. In our opinion, the cultivation of *Miscanthus* on black soils under the conditions verified within the experimental research proves that such an approach is highly beneficial for the soil conservation service and thus is of high importance in light of ensuring sustainable development.

5. Conclusions

Based on experimental studies conducted in 2016–2021 on the optimization of ecosystem services of podzolic heavy loamy chernozem by growing *Miscanthus*, its positive impact on the analyzed soil ecosystem services—supply and regulation—was established. The cultivation of *Miscanthus* on chernozem even in relatively dry conditions over the years of research has not prevented the improvement of ecosystem services provided by this soil, as evidenced by the gradual annual increase in the yield of this energy crop.

It is established that under the influence of the growth of *Miscanthus* on chernozem, the amount of carbon organic matter in the soil increases both in the arable and in the underlying layer. The gradual increase in the organic carbon content of chernozem under *Miscanthus* indicates improved support for ecosystem services, which together with regulatory services (habitat formation, soil formation), provides a significant improvement in the agro-ecological condition of the soil and environment. Growing perennial energy crops on agricultural soils provides a significant amount of biomass and a positive phytoremediation effect on the soil by reducing erosion, conserving biodiversity, carbon sequestration and improving the agri-environmental situation.

Further research on the impacts of *Miscanthus* cultivation on chernozem ecosystem services should focus on (i) the economic valuation of possible ecosystem services and an analysis of the cost-benefit ratio of growing energy crops; (ii) strategies for sustainable management of energy crops in specific areas, climatic and socio-economic criteria; (iii) the development of innovative bioenergy projects for the cultivation of energy crops and an assessment of their economic efficiency and investment attractiveness.

Author Contributions: Conceptualization, Y.V., Y.T. and A.K.; methodology, Y.V., Y.T. and A.K.; software, Y.V. and Y.T.; validation, Y.V., Y.T. and A.K.; formal analysis, Y.V., Y.T. and A.K.; investigation, Y.V. and Y.T.; resources, Y.V. and Y.T.; data curation, Y.V., Y.T. and V.K.; writing—original draft preparation, Y.V., Y.T., V.K., A.K. and I.S.; writing—review and editing, V.K., A.K. and I.S.; visualization, Y.V. and A.K.; supervision, Y.T. and A.K.; project administration, A.K. and V.K.; funding acquisition, V.K. All authors have read and agreed to the published version of the manuscript.

Funding: This research received no external funding.

Institutional Review Board Statement: Not applicable.

Informed Consent Statement: Not applicable.

Data Availability Statement: Not applicable.

Conflicts of Interest: The authors declare no conflict of interest.

References

1. Electricity Market Report (July 2021). International Energy Agency. Available online: https://www.iea.org/reports/electricity-market-report-july-2021 (accessed on 2 February 2022).
2. Jiang, P.; Fan, Y.V.; Klemeš, J.J. Impacts of COVID-19 on energy demand and consumption: Challenges, lessons and emerging opportunities. *Appl. Energy* **2021**, *285*, 116441. [CrossRef] [PubMed]
3. Doğan, B.; Ghosh, S.; Shahzadi, I.; Balsalobre-Lorente, D.; Nguyen, C.P. The relevance of economic complexity and economic globalization as determinants of energy demand for different stages of development. *Renew. Energy* **2022**, *190*, 371–384. [CrossRef]
4. Gu, D.; Andreev, K.; Dupre, M.E. Major Trends in Population Growth Around the World. *China CDC Wkly.* **2021**, *3*, 604–613. [CrossRef]
5. The Long View: How Will the Global Economic order Change by 2050? The World in 2050. PricewaterhouseCoopers. 2017. Available online: https://www.pwc.com/gx/en/world-2050/assets/pwc-the-world-in-2050-full-report-feb-2017.pdf (accessed on 3 March 2022).

6. Eleventh Emergency Special Session, 5th & 6th Meetings (2 March 2022). General Assembly of the United Nations. Available online: https://www.un.org/press/en/2022/ga12407.doc.htm (accessed on 19 April 2022).
7. Joint European Action for More Affordable, Secure Energy. European Commission. 2021. Available online: https://ec.europa.eu/commission/presscorner/detail/en/ip_22_1511 (accessed on 9 May 2022).
8. The Sustainable Development Goals Report 2020. United Nations. Available online: https://unstats.un.org/sdgs/report/2020/The-Sustainable-Development-Goals-Report-2020.pdf (accessed on 1 April 2022).
9. Biesenbender, S. The EU's energy policy agenda: Directions and developments. In *Energy Policy Making in the EU*; Springer: London, UK, 2015; pp. 21–40.
10. Hughes, S. The politics of urban climate change policy: Toward a research agenda. *Urban Aff. Rev.* **2017**, *53*, 362–380. [CrossRef]
11. Directive (EU) 2018/2001 of the European Parliament and of the Council of 11 December 2018 on the Promotion of the Use of Energy from Renewable Sources, 11 December 2018. Available online: http://data.europa.eu/eli/dir/2018/2001/oj (accessed on 4 April 2022).
12. Maniatis, K.; Chiaramonti, D.; van den Heuvel, E. Post COVID-19 Recovery and 2050 Climate Change Targets: Changing the Emphasis from Promotion of Renewables to Mandated Curtailment of Fossil Fuels in the EU Policies. *Energies* **2021**, *14*, 1347. [CrossRef]
13. Eskander, S.M.; Fankhauser, S. Reduction in greenhouse gas emissions from national climate legislation. *Nat. Clim. Chang.* **2020**, *10*, 750–756. [CrossRef]
14. Meyer-Ohlendorf, N.; Voß, P.; Velten, E.; Görlach, B. EU Greenhouse Gas Emission Budget: Implications for EU Climate Policies. What Does 2050 Mean for 2030? 2018. Available online: https://www.ecologic.eu/sites/default/files/publication/2018/2120_eu_emission_budgets_ecologic_report20180124_final.pdf (accessed on 17 March 2022).
15. O'Sullivan, M.; Overland, I.; Sandalow, D. The Geopolitics of Renewable Energy. Working Paper, June 2017. Available online: https://energypolicy.columbia.edu/sites/default/files/CGEPTheGeopoliticsOfRenewables.pdf (accessed on 10 April 2022).
16. McKenna, R. The double-edged sword of decentralized energy autonomy. *Energy Policy* **2018**, *113*, 747–750. [CrossRef]
17. Güney, T.; Kantar, K. Biomass energy consumption and sustainable development. *Int. J. Sustain. Dev. World Ecol.* **2020**, *27*, 762–767. [CrossRef]
18. Bastian, O.; Lupp, G.; Syrbe, R.; Steinháußer, R. Ecosystem services and energy crops—spatial differentiation of risks. *Ekol. Bratisl.* **2013**, *32*, 13–29. [CrossRef]
19. MacDicken, K.G. Global Forest Resources Assessment 2015: What, why and how? *For. Ecol. Manag.* **2015**, *352*, 3–8. [CrossRef]
20. Castro-Díez, P.; Vaz, A.S.; Silva, J.S.; Van Loo, M.; Alonso, Á.; Aponte, C.; Bayón, Á.; Bellingham, P.J.; Chiuffo, M.C.; Di Manno, N.; et al. Global effects of non-native tree species on multiple ecosystem services. *Biol. Rev.* **2019**, *94*, 1477–1501. [CrossRef] [PubMed]
21. Coleby, A.M.; van der Horst, D.; Hubacek, K.; Goodier, C.; Burgess, P.J.; Graves, A.; Howard, D. Environmental impact assessment, ecosystems services and the case of energy crops in England. *J. Environ. Plan. Manag.* **2012**, *55*, 369–385. [CrossRef]
22. Lovett, A.A.; Dockerty, T.L.; Papathanasopoulou, E.; Beaumont, N.J.; Smith, P. A framework for assessing the impacts on ecosystem services of energy provision in the UK: An example relating to the production and combustion life cycle of UK produced biomass crops (short rotation coppice and miscanthus). *Biomass Bioenergy* **2015**, *83*, 311–321. [CrossRef]
23. Kadykalo, A.N.; López-Rodriguez, M.D.; Ainscough, J.; Droste, N.; Ryu, H.; Ávila-Flores, G.; Le Clec'h, S.; Muñoz, M.C.; Nilsson, L.; Rana, S.; et al. Disentangling 'ecosystem services' and 'nature's contributions to people'. *Ecosyst. People* **2019**, *15*, 269–287. [CrossRef]
24. Haines-Young, R.; Potschin, M. Common International Classification of Ecosystem Services V5.1. Guidance on the Application of the Revised Structure. Fabis Consult. 2018. Available online: https://cices.eu/content/uploads/sites/8/2018/01/Guidance-V51-01012018.pdf (accessed on 29 April 2022).
25. Pereira, P.; Bogunovic, I.; Muñoz-Rojas, M.; Brevik, E.C. Soil ecosystem services, sustainability, valuation and management. *Curr. Opin. Environ. Sci. Health* **2018**, *5*, 7–13. [CrossRef]
26. Trypolska, G.; Kyryziuk, S.; Krupin, V.; Wąs, A.; Podolets, R. Economic Feasibility of Agricultural Biogas Production by Farms in Ukraine. *Energies* **2022**, *15*, 87. [CrossRef]
27. Kaletnik, H.M. Biopalyvo: Prodovolcha, Enerhetychna ta EkolohichnaBezpekaUkrainy. *Bioenerhetyka* **2013**, *2*, 12–14. Available online: http://nbuv.gov.ua/UJRN/Bioen_2013_2_6 (accessed on 29 April 2022).
28. IRENA. *REmap 2030: Prospects for the Development of Renewable Energy in Ukraine*; IRENA: Abu Dhabi, United Arab Emirates, 2015.
29. Keesstra, S.D.; Bouma, J.; Wallinga, J.; Tittonell, P.; Smith, P.; Cerdà, A.; Montanarella, L.; Quinton, J.N.; Pachepsky, Y.; van der Putten, W.H.; et al. The significance of soils and soil science towards realization of the United Nations Sustainable Development Goals. *SOIL* **2016**, *2*, 111–128. [CrossRef]
30. Bender, S.F.; Bowles, T.M.; Scow, K.M.; van der Heijden, M.G.A. Soil Biodiversity for Agricultural Production and Environmental Integrity. In Proceedings of the Keep Soil Alive, Protect Soil Biodiversity: Global Symposium on Soil Biodiversity, Rome, Italy, 19–22 April 2021; FAO: Zürich, Switzerland, 2021.
31. Frouz, J. Effects of soil macro- and mesofauna on litter decomposition and soil organic matter stabilization. *Geoderma* **2017**, *332*, 161–172. [CrossRef]
32. Kulyk, M.I.; Galytska, M.A.; Samoylik, M.S.; Zhornyk, I.I. Phytoremediation aspects of energy crops use in Ukraine. *Agrology* **2018**, *1*, 373–381. [CrossRef]

33. Vodiak, Y.N. Impact of energy crops on the quality of ecosystem services of Luvic Cherniphaozem. Agrochemistry and Soil Science. *Collect. Papers* **2021**, *91*, 64–71. [CrossRef]
34. Monitoring of Weather for Farms. Meteo.farm. Available online: https://www.meteo.farm/dashboard (accessed on 12 November 2021).
35. Bioenergy and Biotechnology. The Institute of Bioenergy Crops and Sugar Beet NAAS of Ukraine. 2020. Available online: https://bio.gov.ua (accessed on 5 March 2022).
36. Humentyk, M.Y. Agrotechnological Fundamentals of the Productivity Formation of Perennial Grasses for Biofuel Production in the Ukrainian Forest-Steppe. Ph.D. Thesis, Institute of Bioenergy Crops and Sugar Beet NAAS of Ukraine, Kyiv, Ukraine, 2021.
37. Blanco-Canqui, H. Growing dedicated energy crops on marginal lands and ecosystem services. *Soil Sci. Soc. Am. J.* **2016**, *80*, 845–858. [CrossRef]
38. Cervelli, E.; Scotto di Perta, E.; Pindozzi, S. Energy crops in marginal areas: Scenario-based assessment through ecosystem services, as support to sustainable development. *Ecol. Indic.* **2020**, *113*, 106180. [CrossRef]
39. Wąs, A.; Krupin, V.; Kobus, P.; Witajewski-Baltvilks, J.; Jeszke, R.; Szczepański, K. Towards Climate Neutrality in Poland by 2050: Assessment of Policy Implications in the Farm Sector. *Energies* **2021**, *14*, 7595. [CrossRef]
40. Bazaluk, O.; Havrysh, V.; Nitsenko, V. Energy and Environmental Assessment of Straw Production for Power Generation. *E3S Web Conf.* **2021**, *228*, 01010. [CrossRef]
41. Nitsenko, V.; Mardani, A.; Streimikis, J.; Shkrabak, I.; Klopov, I.; Novomlynets, O.; Podolska, O. Criteria for Evaluation of Efficiency of Energy Transformation Based on Renewable Energy Sources. *Monten. J. Econ.* **2018**, *14*, 253–263. [CrossRef]
42. Meehan, T.D.; Gratton, C.; Diehl, E.; Hunt, N.D.; Mooney, D.F.; Ventura, S.J.; Barham, B.L.; Jackson, R.D. Ecosystem-service tradeoffs associated with switching from annual to perennial energy crops in riparian zones of the US midwest. *PLoS ONE* **2013**, *8*, e80093. [CrossRef]
43. Furdychko, O.; Drebot, O.; Palianychko, N.; Dankevych, S.; Okabe, Y. On the way to balance of forestry land use of Ukraine: Ecological-and-economic aspect. *Agric. Resour. Econ.* **2021**, *7*, 218–244. [CrossRef]
44. Zalesny, R.; Headlee, W. Poplar energy crop development in the north central United State: Biomass supply potential and ecosystem services. In Proceedings of the International Bioenergy and Bioproducts Conference 2013, IBBC 2013, Green Bay, WI, USA, 18–20 September 2013; pp. 33–53.
45. Li, X.; Zipp, K.Y. Dynamics and uncertainty in land use conversion for perennial energy crop production: Exploring effects of payments for ecosystem services policies. *Agric. Resour. Econ. Rev.* **2019**, *48*, 328–358. [CrossRef]
46. Lupp, G.; Bastian, O.; Syrbe, R.; Steinhäußer, R. Perceptions of energy crop production by lay people and farmers using the ecosystem services approach. *Morav. Geogr. Rep.* **2014**, *22*, 15–25. [CrossRef]
47. Khalep, Y.; Moskalenko, A.; Shapovaliuk, M.; Arabska, E. Ecological and economical value of green manuring for Polissia organic agrocenoses. *Agric. Resour. Econ.* **2021**, *7*, 211–231. [CrossRef]
48. Yin, C.; Zhao, W.; Pereira, P. Soil conservation service underpins sustainable development goals. *Glob. Ecol. Conserv.* **2022**, *33*, e01974. [CrossRef]

energies

Article

The Potential of Ukrainian Agriculture's Biomass to Generate Renewable Energy in the Context of Climate and Political Challenges—The Case of the Kyiv Region

Adam Wąs [1,*], Piotr Sulewski [1], Nataliia Gerasymchuk [2], Ludmila Stepasyuk [3], Vitaliy Krupin [4], Zoia Titenko [5] and Kinga Pogodzińska [1]

1 Institute of Economics and Finance, Warsaw University of Life Sciences, Nowoursynowska 166, 02-787 Warsaw, Poland
2 Department of Marketing, Poltava State Agrarian University, 36003 Poltava, Ukraine
3 Department of Economic, National University of Life and Environmental Sciences of Ukraine, 03041 Kyiv, Ukraine
4 Institute of Rural and Agricultural Development, Polish Academy of Sciences, Nowy Świat 72, 00-330 Warsaw, Poland
5 Department of Finance, National University of Life and Environmental Sciences of Ukraine, 03041 Kyiv, Ukraine
* Correspondence: adam_was@sggw.edu.pl

Abstract: Increasing the share of renewable energy in the final energy consumption is a way to ensure independence from external supplies of fossil fuels, which is a fundamental political and economic challenge for many countries nowadays. One such country is Ukraine, which depended on Russian gas supplies and energy (electricity) from nuclear power plants. Russian gas is not delivered anymore to Ukraine, and Russians have recently taken over some of the nuclear power plants. The changes in the political situation force Ukraine to search for alternative energy sources. In countries with high agricultural production potential, one of the basic options seems to be popularization of modern methods of obtaining energy from biomass (bioenergy), which so far has played a minor role in the country's energy mix (less than 2% in the case of Ukraine). The analysis carried out on the case of the Kyiv Region indicates that the annual economic potential of biomass in the region is equivalent to 1743 thousand toe (tonnes of oil), and its use allows them to save about 43% of fossil fuel annually.

Keywords: biomass; energy potential; alternative energy sources; resources; enterprises; fuel

Citation: Wąs, A.; Sulewski, P.; Gerasymchuk, N.; Stepasyuk, L.; Krupin, V.; Titenko, Z.; Pogodzińska, K. The Potential of Ukrainian Agriculture's Biomass to Generate Renewable Energy in the Context of Climate and Political Challenges—The Case of the Kyiv Region. *Energies* **2022**, *15*, 6547. https://doi.org/10.3390/en15186547

Academic Editor: Alberto-Jesus Perea-Moreno

Received: 30 July 2022
Accepted: 4 September 2022
Published: 7 September 2022

1. Introduction

Due to globally observed climate challenges, the energy issue has become one of humanity's most important problems to solve in the near future [1–3]. Global energy production is still dominated by fossil fuels, accounting for 80% of the global energy mix [4]. Simultaneously combustion of fossil fuels (coal, oil, and gas for electricity, heat, and transformation) is the main contributor to global climate change, accounting for over 75% of global GHG emissions [5] and almost 90% of all carbon dioxide emissions [6]. Climate scientists' position is clear—moving away from fossil fuels is essential to stop further climate change [7,8].

The current level of renewable energy development differs significantly between different regions of the world and even neighboring countries [9,10]. On average, less than 11% of global primary energy consumption came from renewable sources in 2019, of which 6.4% was traditional biomass combustion [9]. These statistics show that biomass, particularly modern methods of its use, such as processing into biogas or biomethane, remains a relatively underused renewable energy source. Nevertheless, the production of agricultural biogas and other forms of biomass is an advantageous option in countries with significant agricultural production potential [11,12]. One of them is Ukraine, one of the most important agricultural producers in Europe and the world [13].

In the case of Ukraine (and also in many other countries), increasing the degree of biomass use (including the production of agricultural biogas) is a way to increase energy independence [14,15]. Awareness of this challenge significantly increased after Russia invaded Ukraine. Although Ukraine has not imported natural gas directly from Russia since 2015, it still remains dependent on gas imports from Western countries [16]. However, the invasion has significant consequences for the global energy sector. In particular, it relates to natural gas, which can, at least partially, be replaced by gas from biomass.

In the face of the Russian invasion of Ukraine, the challenge of increasing energy independence has become much more critical than ever before. Searching for the possibilities of using biomass for energy purposes is, therefore, a task justified for environmental reasons (replacement of fossil fuels and reduction of GHG emissions) and political reasons (increasing energy independence). There is an urgent need to search for new, alternative sources of energy, and biomass use for energy production is the most attractive option [17]. In this context, the study's goal was to assess the potential of using agricultural biomass for renewable energy production in the Kyiv region.

2. Background Information

2.1. Biomass and Ukrainian Energy Sector—General Information

Biomass is a renewable organic material of plant and animal origin and can be helpful in substitution fossil fuels. Energy from biomass can be obtained in processes such as direct combustion, thermochemical conversion, and biological conversion [18]. In practice, the most frequently applied solutions include co-firing in coal power plants, combustion of biomass in dedicated power and CHP plants, or biomass conversion into biogas in anaerobic fermentation or thermo-chemical processes (pyrolysis) [19]. One of the essential advantages of generating energy from biomass is that it is a carbon-free process because emitted CO_2 was previously assimilated by plants [19]. Moreover, among the benefits of biomass can be mentioned wide availability, reduced overreliance on fossil fuels, usually lower prices compared to fossil fuels, and reduced waste in landfills [20]. Besides, biomass is a local fuel, and its use increases the regional added value by minimizing fossil fuel imports. Moreover, biomass production and supply contribute to creating new workplaces, mainly in rural areas, which is vital for the local economy.

Shortcomings of biomass use are reflected in the lower efficiency of some biofuels compared to fossil fuels. Although burning biomass is carbon neutral, it still generates air pollution. Overuse of wood can lead to deforestation, and biomass plants usually require a lot of space [20]. Biomass used for energy production may include wood and wood processing wastes, agricultural products, and food wastes as well as municipal, solid, and liquid wastes [18]. Particular hopes for the development of biomass are connected to the production of biogas [12,21].

Ukraine is one of Europe's largest energy consumers, with a primary energy consumption of 93 Mtoe (million tonnes of oil equivalent) in 2018. Domestic energy production is insufficient to meet total energy demand; it covers about 65% of energy needs [22]. In recent years, there has been a significant decrease in domestic energy production compared to 2007 by over 30% (Figure 1). Energy exports have also dropped significantly, and to a lesser extent, imports. As a result, Ukraine is most dependent on imported oil (83% of consumption) and to a lesser extent on coal (50% of consumption) and natural gas (33%), which meant that in 2018 it was necessary to import 8.5 Mtoe of natural gas, 13.5 Mtoe of coal, and 10.4 Mtoe of oil products [22].

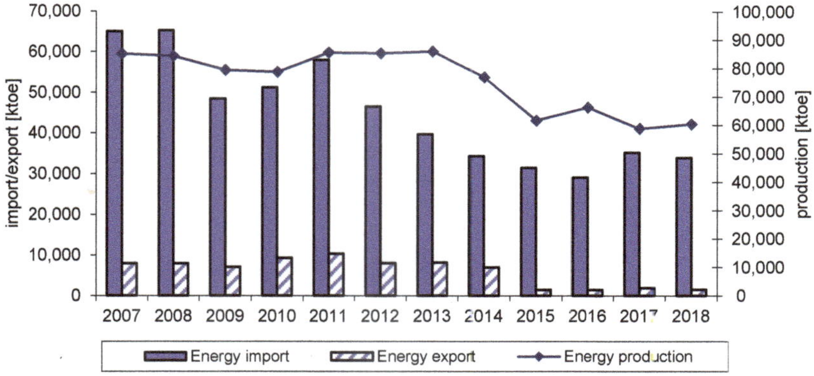

Figure 1. Production, export, and import of energy in Ukraine, in thousands toe. Source: The author's calculations are based on statistics from the State Statistics Service of Ukraine [23].

The Ukrainian energy sector is mainly based on fossil fuels (natural gas, oil, and coal) and nuclear energy [22]. In the structure of primary energy consumption dominates coal (28.3%), followed by natural gas (28.2%), nuclear energy (23.4%), and oil (13.8%) [4]. The share of other energy sources, including hydropower, wind, solar, and other renewables, is less than 6% [4] (Figure 2).

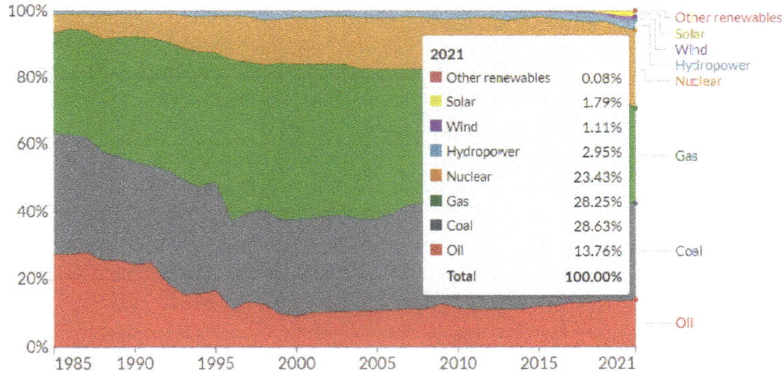

Figure 2. Structure of primary energy consumption in Ukraine by source. Source: [24].

The importance of nuclear energy for Ukraine should be emphasized—although it accounts for less than 25% of the total primary energy, it meets half of the country's electricity needs [22]. Nuclear power plants increase the country's energy independence, but they also pose a severe threat to the whole world in the face of the Russian invasion of Ukraine. An example is the Zaporizhzhia Nuclear Power Plant (the largest nuclear plant in Europe), occupied by the Russians, over which the Ukrainian authorities lost control [25]. This case, as well as other negative experiences with the safety of nuclear power plants, forces us to ask about the further development of the energy sector in Ukraine. It is also worth paying attention to the changes in the share of renewable energy. Although it is still relatively small, in the last few years, it has increased from less than 2% in 2015 to about 6% in 2021 (excluding traditional biofuel use) [4]. The progress in the development of RES observed in recent years is connected with increasing awareness that renewable energy sources have a high potential to reduce natural gas dependency and enhance energy security. The government's decision in 2016 to withdraw from subsidizing the production of heat from natural gas turned out to be particularly important for the development of RES and made the production of heat from renewables (including biomass) comparatively competitive (in comparison to fossil fuels) [22]. The share of renewable energy in the heating and cooling sector in 2020 was 9.3%; in the electricity sector, it was 13.9%; and in the transport sector, it was 2.5% [26]. The total installed capacity of active renewable energy projects (excluding large scale hydro generation >10 MW) was around 7.7 MW, of which 72% belong to industrial solar, 8% solar in a private household, 15.7% wind, 1.5% small hydro, and 2.3% to biomass and biogas [27]. These data indicate the relatively low importance of biomass in energy production in Ukraine, although the analyses of Lakyda et al. [28] show that the technical potential of forest biomass can be estimated at the level of 2.1 Mtoe and that of agricultural waste at the level of 12 Mtoe. Assuming the demand for primary energy is at the level of 86.4 Mtoe (in 2020), this would meet approximately 16.3% of the country's energy needs.

Many authors underline the need to diversify the Ukrainian energy mix and improve energy efficiency. For example, Lewicki [29] stressed the need for diversification of supplies, differentiation of energy balance through increased use of renewable energy sources, and increasing the energy efficiency in the historical aspect, while Gerasymchuk [30] outlined the background of using renewable energy sources in order to ensure the energy efficiency of Ukraine, given the statistic and existing situation in the energy market, and analyzed the resource base for renewable energy sources and local fuels for the energy efficiency and the reliability of Ukraine's energy supply, which became a start for this research.

The use of biomass seems particularly justified in the case of heat production, because sometimes it seems to be the only feasible option to replace fossil fuels to provide heating for buildings without easy access to other supply options [31]. Ukraine's total thermal energy consumption in 2012 was estimated at 14.03 Mtoe, of which only about 6% was covered by biomass (solid biomass and biogas) [31]. Currently, the share of biomass in Ukraine's total heat production is estimated at 9% [32].

The growth of energy production from renewable sources is an important area for replacing natural gas, as there is a large reserve for reorientating biomass exports to the domestic market. Energy security in the face of the Russian military aggression against Ukraine is another perspective that needs to be assessed and considered in the energy and bioenergy development plans. Energy generation from biomass in this regard seems to be not only sustainable, but also highly dependable [33–35]. Local generation of energy based on locally available sources allows them to sustain the needs of particular farms or even communities and ensures maintaining of their functions regardless of the exogenous shocks and national or regional grid malfunctions [36,37]. In this case, a tight connection between food and energy security strengthens the sustainability and resilience of the local food systems. It allows them to carry on with the provision of essential system functions [38].

Although the current contribution of biomass to energy generation in Ukraine remains small, it can be expected that this situation will change in the future. Geletukha et al. [39], in their complex assessment of the future bioenergy developments in Ukraine, assume the country would follow the European Green Deal and align its climate neutrality achievement and environmental development along the current European priorities. Ukraine is also a member of the European Energy Community, which has declared its conscious participation in a global policy aimed at sustainable development and reduction of harmful effects on the environment. As Ukraine is already committed to the Paris Agreement to work on the reduction of greenhouse gas emissions and the Energy Community Treaty to work towards transformation to clean energy, the development of the bioenergy sector to fulfil its green transitions is crucial.

2.2. Ukrainian Agricultural Sector—Supplier of Biomass for Energy Generation

As a country with a large agricultural sector, Ukraine has significant development potential in bioenergy. The development of the bioenergy sector is eased by vast areas of fertile croplands and less productive lands suitable for growing undemanding energy plants, a favorable climate for plant and livestock production, and the availability of the necessary human and material resources. In addition, high yields of major crops provide a sustainable resource base, which has not been exploited so far. In this regard, plant biomass plays one of the key roles in the development of bioenergy.

In the European Union, biomass for energy generation reached a share of ca. 60% among renewable energy sources, which directly contributes to the EU's energy security, as most of the demand (about 96%) is covered by domestically produced biomass [40]. Already, as of 2020, the volume of biomass consumption for energy production in the European Union is more than 120 million tonnes of oil equivalent per year [41]. As Ukraine has been granted EU candidate status, compliance with EU legislation and principles will be increasingly important. According to the EU Energy Security Strategy [42], members need to become more energy "independent" by saving energy and producing more local (RES) energy.

According to Geletukha et al. [43], Ukraine has considerable potential for renewable energy sources, one of the most extensive being biomass. Despite some fluctuations, Ukraine's volume of agricultural biomass increases almost every year due to the general trend of growth in the production and yield of major crops. Thus, in 2019, the country harvested a record amount for the last 20 years of sunflower, corn for grain, and some other cereals. Since 2000, the energy potential of straw of cereal eared crops, byproducts and waste of grain, corn and sunflower production in Ukraine has tripled, from 2.8 Mtoe in 2000 to 8.5 Mtoe in 2020. As the abovementioned authors state, agro-biomass (agricultural residues and energy crops) will remain Ukraine's primary type of bioenergy potential. Expanding the use of agricultural residues requires working out technologies for baling corn and sunflower stalks. On the other hand, energy crops for solid biofuels will continue to grow on unused (low-yield) agricultural lands.

Given that the agro-industrial resource is becoming a leading strategic bioresource, biomass from products produced in the agricultural sector can give Ukraine new opportunities for sustainable development through the production of cheap, environmentally friendly bioenergy products through efficient use of agricultural biomass. However, analysis of the use of agricultural biomass for energy purposes showed that the current level of use of energy potential of biomass in the country is very low—from 0 to 2–3% depending on the specific species, and only sunflower husk shows the level of 73.1% [44]. As the authors state, biomass's leading destination is thermal energy production, which is used for heating and hot water supply. Between 2014 and 2018, biomass's share of thermal energy was within 97% of all renewable thermal energy.

Kulyk [45] emphasizes that bioenergy development in Ukraine requires searching for ways to reduce the cost of various types of bio-raw materials in the economic justification of their production. Currently, the main components of the bioenergy production potential in Ukraine are primary agricultural residues (cereal straw, corn, and sunflower residues) and industrial cultivation of energy crops. However, the biomass production of renewable plant material from energy crops depends on many factors determining their cultivation's feasibility. Environmental influences on energy crop cultivation are mainly reflected in its effect on seed germination and the initial stages of plant growth. In order to achieve balanced cultivation and use of energy crops as plant material, ecological aspects must be taken into account. Reducing the pressure on the environment requires establishing energy plantations and growing energy crops on marginal lands with low fertility, showing signs of degradation and requiring reclamation [45,46].

Other studies [47] show that the estimated biological yield of plant biomass in Ukraine could be 64.3 million tonnes. They have also established that an increase in the use of straw for energy needs can be ensured only by increasing green manure crops in crop rotation (in particular, cultivations of cover crops within crop rotation cycles). At the same time, Hutsol [41] states that at this time, Ukraine has not approved a standardized system for measuring and accounting for solid biomass resources of forest and agricultural origin. Lack of such information, especially on energy crops, hinders the development and implementation of sustainable energy policies and projects in a particular area and the country as a whole. They also emphasize that it can be argued that there is the active use of renewable energy and increased energy efficiency in the regions of Ukraine. However, many certain regions are cautiously implementing renewable energy production systems. The authors stress, among others, that the results of their research on solid biomass (mainly from agricultural residue) in Ukraine have a high potential for fuels that can be quickly applied.

The potential of biogas production from the livestock sector of Ukraine has been assessed previously [21]. It was found that, in absolute terms, Ukraine has a significant potential for the production of agricultural biogas from animal manure, reaching nearly 3 billion m^3. However, the practical possibilities of using this potential are severely limited by the dual structure of agriculture. More than half of the available manure is produced on small livestock farms that are too small scale to consider investing in biogas plants.

It should be kept in mind that the ongoing war is also affecting the agricultural sector. Reduced production (reduced sown area) of primary agricultural commodities reduces the potential for food production and limits the amount of biomass used to generate energy (even if only agri-food waste was included). Therefore, looking for biomass sources other than agriculture is worthwhile.

3. Material and Methods

3.1. Case Study Area Description

Sustainable development of bioenergy requires, first of all, a careful assessment of the available biomass potential. Therefore, we assessed, as an example, the potential of biomass used as an energy source in the Kyiv region.

The Kyiv region is one of the largest regions of Ukraine, with an area of 28.1 thousand square kilometers (without the city of Kyiv), which is 4.7% of Ukraine's territory (Figure 3). By size, the Kyiv region ranks eighth among other regions of Ukraine. Kyiv region is a metropolitan region, in the center of which is located Kyiv, the capital of Ukraine, a powerful political, business, industrial, scientific, technical, transport, and cultural center of the country, connected with the region with close commercial and social ties. The distance from Kyiv to the region's northern border is 118 km, to the southern border is 128 km, to the western border is 76 km, and to the eastern border is 112 km. A feature of the Kyiv region is the absence of a regional center. Kyiv city, where the central administrative bodies of the region are located, is the autonomic region and does not count in Kyiv region statistics. Another feature of the region is the presence of a Slavutych city, which belongs to the Chernihiv region.

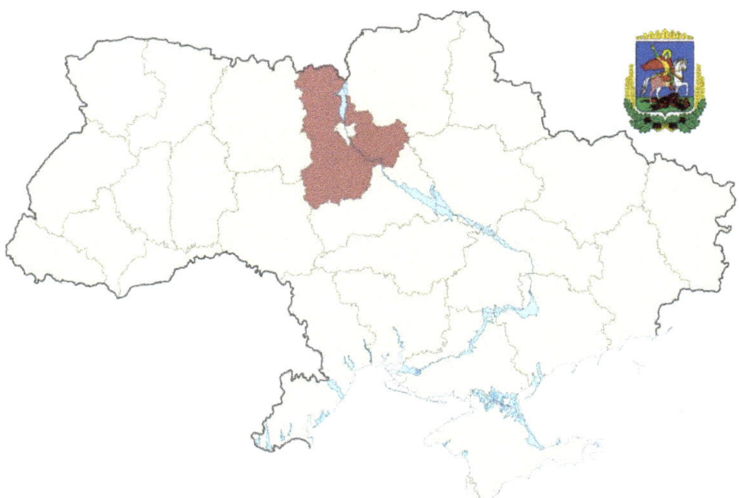

Figure 3. Geographical location and herb of the Kyiv region in Ukraine. Source: Strategy of Kyiv region for 2021–2027 years [48].

Kyiv Region has favorable conditions for agriculture, namely the region's climate, the structure of agricultural lands, availability of the capital city Kyiv as a sales market, and a robust scientific base for implementing innovative technologies in the production and processing of agricultural products. As a result, by volume of gross agricultural production, the region ranks second among all other regions of Ukraine.

Crop production in the total agricultural production of the region takes a significant share with 62.4%. The main crops grown are grain crops, potatoes, sugar beets, and sunflowers, with a large share of perennial gardens. By zone division, the north part of the Kyiv region is located in the Polissia zone, which is characterized by a larger share of forests over the fields with the developed forest harvesting, and the south is in the forest-steppe zone. The Kyiv region is one of the leaders among grain and oil storage market operators in Ukraine, with 53 elevators operating in the region with a total capacity of 2.6 million tonnes storage of the specified crops and 45 fruit storage refrigerators. The region ranks fourth in Ukraine for egg production and takes first place by volume of livestock meat and poultry.

The Kyiv region belongs to the energy-rich regions. Energy enterprises located on its territory have a total capacity of 3200 MW, namely the Trypilska thermal power station, Kyiv hydroelectric power station, Kyiv Hydroaccumulating Electric Power Station, Bilotserkivska Thermal Power Station, and small hydroelectric power stations, as well as the Dymerska solar power plant that is among the ten most powerful solar power plants of Ukraine [48].

The Kyiv region is an agrarian region in which large volumes of byproducts and waste suitable for energy use are generated. Large areas of agricultural land create significant potential for growing energy crops. The main source of biomass in the Kyiv region is primary crop waste.

3.2. General Assumptions—Methods of Calculating

Typically, three main types of biomass potential are considered when energy production possibilities are considered, i.e., theoretical, technical, and economic potentials.

The theoretical potential is the maximum amount of terrestrial biomass theoretically available for energy production. For example, the theoretical potential of waste and residues of various types is the total volume from which energy can be extracted. The technical potential limits the theoretical to the amount of biomass that is available for processing, which is available at a specific moment under certain structural, technical, and technological conditions. When calculating it, it is essential to consider spatial restrictions caused by competition between land users and environmental factors [49].

Economic potential is an even narrower concept because it includes only that share of technical potential that provides the desired level of profitability. Therefore, within this study's framework, the cost estimation method was used instead of profitability analysis for its evaluation based on the resource-consuming concept. This approach better represents the importance of the amount of available biomass when planning the production of energy products [50]. Furthermore, the used technique allows, in addition to planning the volume of energy production, to forecast financial results from the activity.

Wood biomass, which can serve as fuel, is produced due to general and sanitary felling. Firewood, wood chips, branches, stumps and crowns, and secondary processing products—shavings and sawdust—are involved in energy production. We used the approach of determining the energy potential of wood waste P_w according to the formula:

$$P_w = (V_w * K_1 + (V_w - V_{com}) * K_2) * Q_w$$

where:

V_w—a volume of wood logging, m^3;

K_1 = 0.1—waste ratio;

V_{com}—the volume of round timber, density m^3;

K_2 =1 − (0.2 . . . 0.25) = 0.8 . . . 0.75—the total coefficient of waste of wood's main and secondary processing. Considering standard losses during wood processing of 5–10%, we accept K_2 = 0.70;

Q_w = 0.186 toe/dense m^3—calorific value of dense wood during logging [51].

We suggest calculating the energy potential of biogas (toe) from organic waste using the formula:

$$E_{LS} = \sum_{i=1}^{n} \frac{365 * N_i * q_{mi} * \frac{TS_i}{100} * \frac{VS_i}{100} * q_i^{bg} Q_{LHV}^{bq}}{Q_{LHV}^{oe}}$$

where:

N_i—the total number of animals of the i species, heads;

q_{mi}—yield of organic waste of the i-th type, kg/(hour-day);

TS_i—share of dry substance in organic waste of the i-th type, %;

VS_i—the proportion of organic substance in the dry residue, %;

$q_i{}^{bg}$—expected yield of biogas from an organic waste of the i-th type, m^3/kg DOM (dry organic matter);

Q_{LHV}^{bq}—expected lower heat of combustion of biogas (LHV), generation from an organic waste of the i type, MJ/nm^3;

Q_{LHV}^{oe} = 41.868 MJ/kg—lower heat of combustion of oil equivalent [46].

The formula determines the economic potential of biomass from pruning of fruit trees:

$$P_e = \sum Spac_i * Pr_i * Kt_i * Koe_i$$

where:

$Spac_i$—land, the area of which is occupied by fruit trees of the i-th species at the fruit-bearing age, thousand hectares;

Pr_i (2.4 for pome fruit, 3.0—for stone fruit trees)—specific productivity of pruning fruit trees of the II species at the fruit-bearing age for calculating the theoretical potential of biomass, t/ha;

Kt_i = (2.4 for seed trees, 3.0—for stone trees)—the criterion of the technical possibility of pruning for calculating the technical potential of biomass;

Koe_i (0.406 for pome fruit, 0.400—for stone fruit trees)—the potential biomass coefficient in oil equivalent: the calorific value of plant waste in oil equivalent [51].

The following formula is used to determine the economic potential of processing waste:

$$P_e = \sum_{i=1}^{n} Cpr_i * Kr_i * Koe_i$$

where:

Cpr_i—the volume of processed raw materials of the i-th type (for example, sunflower seeds);

Kr_i—the coefficient of waste generated during the processing of raw materials (Kr = 0.15 for sunflower seeds shows that 1 tonne of processed seeds yields 150 kg of husk, i.e., 15% of the total volume);

Koe_i—coefficient of conversion of biomass potential into oil equivalent: (for sunflower husk, it is 0.358) [51].

4. Results

Based on the results of the analysis of sunflower seed processing by agricultural enterprises in the Kyiv region, it is clear that from 2014 to 2019, its volume remained practically unchanged. In the reporting year (i.e., 2019), it amounted to 370,000 tonnes. Therefore, the economic potential of such energy-oriented production, calculated according to the above formulas, is 19.9 thousand toe (Figure 4).

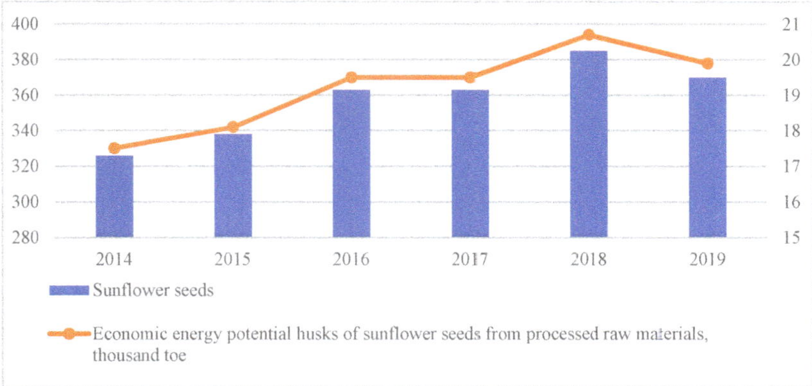

Figure 4. Volumes of processed raw materials, thousand tonnes. Source: author's calculations are based on statistics from the State Statistics Service of Ukraine [23].

At the same time, it was observed that timber felling in the region increased significantly: from 1608.7 thousand m^3 of wood in 2015 to 2015.2 thousand m^3 of wood in 2019. Thus, the difference between the reporting and base years exceeds 25%. At the same time, the volume of wood harvesting increased by 2.1 times. This increased the economic potential of wood by 1.6 times (Table 1).

Table 1. The economic potential of wood waste.

Years	Wood Logging, Thous. m^3	Round Wood (Commercial Wood), Thous. m^3	The Volume of Wood Waste		Firewood Logging, Thous. m^3	The Economic Potential of Wood Waste, Thous. Toe
			Logging, Thous. m^3	Processing, Thous. m^3		
2015	1608.7	563.5	161	732	518.7	262.5
2016	1785.3	547.1	179	867	551.6	297.0
2016	1912.2	522.7	191	973	1139.0	428.3
2017	1916.0	440.0	192	1033	1219.2	454.6
2018	2077.5	589.2	208	1042	1196.5	455.0
2019	2015.2	559.7	202	1019	1100.8	431.7

Source: author's calculations based on statistics from Kyiv Oblast Statistical State service [52].

A representative sample is key to objective analysis and reliable results. Based on this postulate, we took for the study of economic energy potential only agricultural enterprises of the Kyiv region with a significant number of livestock, namely: cattle—from 2000 heads, pigs—from 9000 heads, and poultry—from more than 400,000 heads. Small enterprises that do not have a centralized collection of organic waste in animal husbandry were not taken into account.

The results of the calculations show significant dynamics in the size of livestock. Thus, from 2014 to 2019, the poultry number increased by 6%, while the number of cows and pigs decreased by 15% and 1.3% (Table 2). In animal husbandry, poultry farming is the largest source of organic waste for obtaining biogas. The bird population in 2019 was the largest and amounted to about 17 million. From this volume, it is possible to get 40.6 thousand toe. Therefore, the total livestock that was analyzed to calculate the economic potential of biogas production can provide a sound output of 62 thousand toe.

Table 2. The economic potential of biogas from organic waste.

	2014	2015	2016	2017	2018	2019
Cattle total, thousand heads	40.0	38.0	37.0	35.0	34.0	34.0
Biogas from organic waste cattle total a thousand toe	10.2	9.7	9.4.0	8.9	8.6	8.6
Pigs, thousand heads	228.0	226.5	225.0	224.0	226.0	225.0
Biogas from organic waste pigs, thousand toe	13.0	12.9	12.8	12.8	12.9	12.8
Poultry, thousand heads	15,958.0	16,005.0	16,402.0	16,454.0	16,589.0	16,921.0
Biogas from organic waste poultry, thousand toe	38.3	38.4	39.3	39.5	39.8	40.6

Source: author's calculations based on statistics from Kyiv Oblast Statistical State Service [52].

Corn and sunflower stalks, wheat, rye, barley, buckwheat, pea, soybean, rapeseed, and millet straw can be used in crop production for energy needs. Almost all grain and oil subsectors have significant potential for biogas production. The byproduct output rate determines the available amounts of straw in accordance with the agricultural crop yield. Table 3 presents the results of the analysis of the economic potential of crop production in the Kyiv region. Enterprises in the region specialize in growing corn, sunflower, soybeans, rapeseed, barley, and wheat. Byproducts are used as fertilizer, as well as for livestock maintenance, especially barley straw. This trend confirms the structure of the use of byproducts adopted in our methodology, namely that 40% of oil crops and 30% of cereals are free for biogas production.

Table 3. The economic potential of grain and industrial waste, thousand toe.

Crops	2014	2015	2016	2017	2018	2019
Wheat	123.6	143.3	167.1	91.8	125.1	153.8
Rye	3.7	3.4	3.4	3.7	4.0	3.5
Barley	22.9	22.6	27.6	13.3	18.6	27.8
Oat	1.0	0.7	0.6	0.4	0.4	0.4
Millet	0.1	0.1	0.1	0.0	0.1	0.5
Buckwheat	1.0	1.0	1.0	0.9	1.0	0.8
Legumes	2.5	2.4	2.7	2.3	2.2	2.6
Corn stalks	592.0	413.0	525.0	456.0	819.0	795.0
Sunflower stalks	60.0	59.0	91.0	80.0	114.0	100.0
Soybean straw	104.0	87.0	98.0	73.0	86.0	61.0
Rapeseed straw	30.0	26.0	13.0	20.0	34.0	36.0
Total	**940.8**	**758.5**	**929.5**	**741.4**	**1204.4**	**1181.4**

Source: Own calculations based on statistics from the State Service of Statistics [53].

In addition, the methodology uses the coefficient of losses and the volume of slaughtered products for fertilizers and animal husbandry needs (up to 50%).

The distribution of straw in the enterprises of the Kyiv region by areas of use shows that in 2019 the available amount of straw was 1538.4 thousand tonnes, 769 thousand tonnes of which are applied as fertilizers, 189 thousand tonnes were used for litter, and 547.8 thousand tonnes can be used for energy production.

The results show that the area under grain fruit plantations for 2014–2019 decreased by 2.6 times, and fruit trees remained unchanged under the stones. Therefore, according to our calculations, the economic energy potential of perennial plantations in the Kyiv region is 0.9 thousand toe (Table 4). However, its potential is somewhat irrelevant due to the relatively small scale of plantations.

Table 4. Areas of perennial plantations in fruiting age in agriculture enterprises of Kyiv region, thous. ha.

Plantations	2014	2015	2016	2017	2018	2019
			Thousand ha			
Areas of perennial plantations in fruiting age in agriculture enterprises						
Pome fruits trees	2.6	1.8	1.5	1.3	1.0	1.0
Stone fruits trees	0.1	0.1	0.1	0.1	0.1	0.1
The economic potential of energy from pruning of fruit trees and vineyards waste wood						
			Thousand toe			
Pome fruits	2.0	1.4	1.2	1.0	0.8	0.8
Stone fruits	0.1	0.1	0.1	0.1	0.1	0.1
Total	**2.1**	**1.5**	**1.3**	**1.1**	**0.9**	**0.9**

Source: Own calculations based on statistics from State Service of Statistics [53,54].

Considering the direct relationship between the yield of crops and the economic potential of byproducts, the results of the analysis showed that, in 2019, the most significant potential in oil equivalent was as follows: straw waste in the amount of 1182 thousand tonnes, wood at 432 thousand tonnes, and manure at 62 thousand tonnes. On the other hand, the husk has the lowest energy potential at only 19.9 thousand toe (Table 5). Thus, the total economic energy potential of agricultural enterprises of the Kyiv region in the amount of 1697 thousand tonnes is distributed by sources in the following ratio: stalks and straw occupy 70% of the structure, wood accounts for 25%, manure for 4%, and sunflower husks produce only 1% (Figure 5).

Table 5. The economic energy potential of waste in the Kyiv region, thousand toe.

Source of Energy	2014	2015	2016	2017	2018	2019
Straw and waste	941.0	758.0	929.0	741.0	1205.0	1182.0
Sunflower husk	18.0	18.0	20.0	20.0	21.0	19.9
Pruning trees	2.0	1.0	1.0	1.0	1.0	1.0
Wood	262.0	297.0	428.0	455.0	455.0	432.0
Manure	61.4	61.0	61.6	61.1	61.3	62.0
Total	**1284.4**	**1135.0**	**1439.6**	**1278.1**	**1743.3**	**1696.9**

Source: Own calculations based on statistics from the State Service of Statistics [53].

In 2018, 4019 thousand toe was used in the Kyiv region. At the same time, according to the results (Table 5), the economic energy potential of crop, livestock, and horticulture waste amounted to 1743 thousand toe. This amount could provide about 43 3% of the fuel needs at the expense of alternative types of biofuel. Such optimization will contribute to reducing the destructive impact of harmful emissions from petroleum fuel on the environment, increase the self-sufficiency of enterprises and their organizational and financial independence from external conditions, and reduce the production process cost.

There are no specific data about current renewable energy usage in Kyiv oblast. However, the share of the total capacity of boiler plants operating on alternative fuel types to the total number of boiler plants is 16.9% (against 16.594 in 2019), which is 0.4 percentage points more than in the same period of the 2019 year.

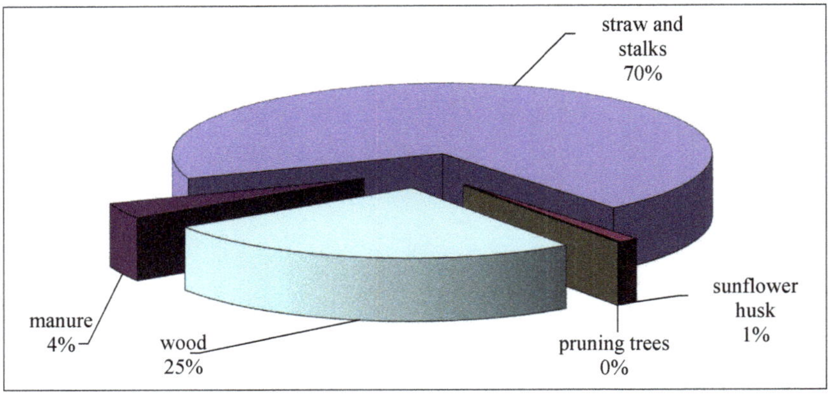

Figure 5. Structure of energy potential of the agricultural production in the Kyiv region, 2019. Source: Own calculations based on statistics from the State Service of Statistics [53].

As of January 1, 2021, 378 boiler stations for communal purposes were converted to alternative fuels, which is 23 units more than in the same period in the previous year. The number of boiler plants producing energy from installations converted to alternative fuel is constantly increasing and is 27.3% now against 26.7% for the same period in 2019. There is a positive trend in implementing measures to replace natural gas consumption [55].

Referring to one of the best practices of usage of renewable energy could be mentioned company "UMK" in the Zguriv district of the Kyiv region, which implemented the biogas project of the "Ukrainian Dairy Company" with a capacity of 1 MW, which processes manure from 4000 cows and corn silage. The energy produced is enough for a dairy farm and the village of Velikiy Krupil.

Another example of the use of alternative energy sources is a large biogas plant located in the village of Rokytne, Kyiv region, with a capacity of 2.38 MW. The enterprise's output to the design capacity made it possible to provide energy to about 800 individual households. Furthermore, the project initiator—the group of companies "Silhospproduct"—plans to use the mentioned technologies to construct similar factories [39].

Research on the economic efficiency of production and implementation of granules from agricultural raw materials on the domestic market shows that the average payback period of these projects is 2.8 years for the production of sunflower husk pellets and four years for the production of pellets from grain straw and corn stalks. In addition, the population uses straw fuel briquettes to heat their buildings in solid fuel boilers as a substitute for coal. [56].

5. Discussion and Conclusions

International studies of the energy security of humanity indicate trends in the increase in the price of energy sources. This issue is particularly tough for Ukraine, as the country depends on oil supplies from abroad. The civilian population and producers in various areas of the economy are sensitive to price fluctuations. The fuel shortage endangers the operation of vehicles, machines, and equipment involved in the production. National authors Geletukha and Zheliezna [39], in their Roadmap for Bioenergy Development in Ukraine, until 2050, have forecasted the growth of renewable energy generation and its implications. While the authors of the current work can highly relate to the trends for the next two to three decades described in the referenced article, it is doubtful that it would be possible to achieve the complete transition to renewables in energy generation by 2050. While it is a welcomed scenario, the current situation triggered by the Russian war against Ukraine will definitely set back the development of Ukraine, either economic or agricultural development, as well as development in energy transformation. Geletukha and Zheliezna [39] assumed the consumption of biomass for energy production in 2050 at

20 Mtoe/year, ehich seems to be quite substantiated and relevant to the expected share of renewables (63%) in the total primary energy supply in the same year. The achievement of the targets defined in the abovementioned roadmap requires numerous legislative improvements in Ukraine and considers the needed investments to neutralize the harmful effects of the Russian aggression and accompanied damages.

Modelling results based on the TIMES-Ukraine energy system model [57] prove the need to develop and implement the national strategy to increase energy generation from biomass. The highly valuable contribution proposed by the referenced authors includes the analysis of the current policy environment in the context of future biomass development and concluded with a set of policy recommendations for utilization of the biomass potential in Ukraine. However, as Kaletnik and Larina [17,58] point out, there are numerous issues hindering bioenergy development at the level of the legislative framework; methodological approaches to the economic, environmental, and social efficiency of production; and the use of biological types of energy. The Kyiv region has all the necessary conditions for biofuel production regarding available land resources and plant potential. Already today, the potential in the region of biomass, which is suitable for the cost-effective production of liquid biofuels (bioethanol and biodiesel), gives grounds to argue about the prospects of this area. According to our calculations (Figure 6), the energy from biomass produced in the Kyiv region can annually replace 43.3% of fossil fuels.

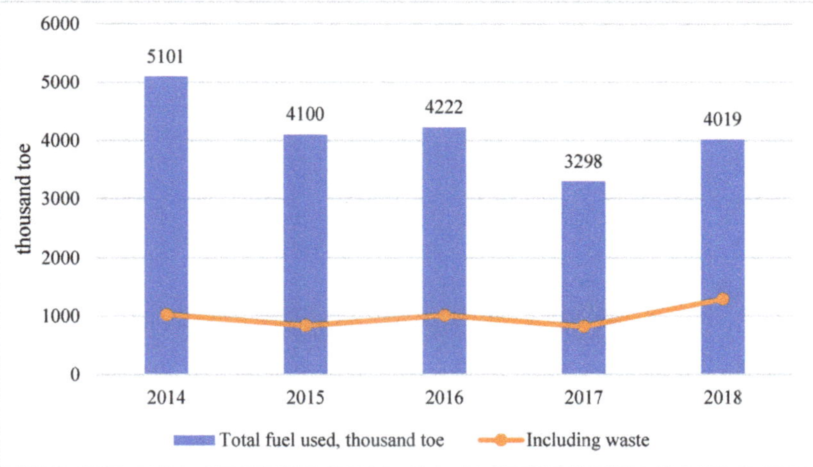

Figure 6. The total amount of fuel used in the Kyiv region (thousand toe). Source: own calculations based on statistics from [52].

It can be assumed that covering more than 40% of energy needs with locally produced biomass would represent considerable progress in the energy transformation of the region. However, this requires investments, which must be assessed regarding their profitability, energy security, and state security.

The findings of the current article support the abovementioned findings and show the urgency to implement legislative support for renewable energy generation, in particular forming a transparent and understandable regulatory environment for investments in bioenergy projects. These must be verified against the current conditions due to numerous changes in the local environment due to Russian atrocities and inflicted material and human damages on Ukrainian land. Approaches must be taken to intensify the restoration of the local economic infrastructure and farm property. Transport and energy infrastructure are of the utmost importance to ensure local development and the possibility of further improvements.

Author Contributions: Conceptualization, A.W., N.G., L.S., P.S. and Z.T.; methodology, N.G., L.S., Z.T. and P.S.; validation, A.W., P.S. and V.K.; formal analysis, N.G., L.S. and Z.T.; investigation, P.S., N.G., L.S. and Z.T.; resources, N.G., L.S., Z.T. and V.K.; data curation, N.G., L.S., Z.T., P.S. and V.K.; writing—original draft preparation, A.W., P.S., V.K., N.G., L.S. and Z.T.; writing—review and editing, A.W., P.S., V.K. and K.P.; visualization, N.G. and K.P.; supervision, A.W. and P.S.; project administration, A.W.; funding acquisition, A.W. and P.S. All authors have read and agreed to the published version of the manuscript.

Funding: This research was funded in part by National Science Centre, Poland 2021/43/B/HS4/02367. For the purpose of Open Access, the author has applied a CC-BY public copyright licence to any Author Accepted Manuscript (AAM) version arising from this submission.

Data Availability Statement: Not applicable.

Conflicts of Interest: The authors declare no conflict of interest.

References

1. Owusu, P.A.; Asumadu-Sarkodie, S. A review of renewable energy sources, sustainability issues and climate change mitigation. *Cogent Eng.* **2016**, *3*, 1167990. [CrossRef]
2. IPCC. *Climate Change 2022: Mitigation of Climate Change*; IPCC: Geneva, Switzerland, 2022. Available online: https://www.ipcc.ch/report/ar6/wg3/downloads/report/IPCC_AR6_WGIII_Full_Report.pdf (accessed on 14 July 2022).
3. Lamb, W.F.; Wiedmann, T.; Pongratz, J.; Andrew, R.; Crippa, M.; Olivier, J.G.J.; Wiedenhofer, D.; Mattioli, G.; Al Khourdajie, A.; House, J.; et al. A review of trends and drivers of greenhouse gas emissions by sector from 1990 to 2018. *Environ. Res. Lett.* **2021**, *16*, 073005. [CrossRef]
4. BP Plc. *Statistical Review of World Energy*, 70th ed.; BP Plc: London, UK, 2021.
5. Center for Sustainable Systems. *Greenhouse Gases Factsheet*; Center for Sustainable Systems: Ann Arbor, MI, USA, 2021. Available online: https://css.umich.edu/publications/factsheets/climate-change/greenhouse-gases-factsheet (accessed on 14 July 2022).
6. Olivier, J.G.J.; Peters, J.A.H.W. *Trends in Global CO_2 and Total Greenhouse Gas Emissions*; PBL Netherlands Environmental Assessment Agency: The Hague, The Netherlands, 2020.
7. United Nations (UN). *Renewable Energy–Powering a Safer Future*; United Nations: New York, NY, USA, 2022. Available online: https://www.un.org/en/climatechange/raising-ambition/renewable-energy (accessed on 14 July 2022).
8. IRENA. *World Energy Transitions Outlook 2022*; IRENA: Bonn, Germany, 2022. Available online: https://www.irena.org/-/media/files/irena/agency/publication/2022/mar/irena_weto_summary_2022.pdf?la=en&hash=1da99d3c3334c84668f5caae029bd9a076c10079 (accessed on 14 July 2022).
9. BP Plc. *Statistical Review of World Energy*, 71st ed.; BP Plc: London, UK, 2022.
10. Ember. Global Electricity Review. 2022. Available online: https://ember-climate.org/app/uploads/2022/03/Report-GER22.pdf (accessed on 20 July 2022).
11. Saleem, M. Possibility of utilizing agriculture biomass as a renewable and sustainable future energy source. *Heliyon* **2022**, *8*, e08905. [CrossRef] [PubMed]
12. Sulewski, P.; Majewski, E.; Wąs, A. The importance of agriculture in the renewable energy production in Poland and the EU. *Probl. Agric. Econ.* **2017**, *350*, 50–74. [CrossRef]
13. Foreign Agricultural Service U.S. Department of Agriculture. USDA Ukraine Agricultural Production and Trade. 2022. Available online: https://www.fas.usda.gov/sites/default/files/2022-07/Ukraine-Factsheet-July2022.pdf (accessed on 20 July 2022).
14. Perrio, M. Community Biomass: For Energy Independence, Stable Prices and Local Control. 2022. Available online: https://energypost.eu/community-biomass-for-energy-independence-stable-prices-and-local-control/ (accessed on 22 July 2022).
15. Moussu, N. Europe Rediscovers Biogas in Search for Energy Independence. Available online: https://www.euractiv.com/section/energy/news/europe-rediscovers-biogas-in-search-for-energy-independence/ (accessed on 21 July 2022).
16. Polityuk, P.; Chestney, N. Factbox: Ukraine's Energy Options Limited in Event of Russian Gas Disruption. Available online: https://www.reuters.com/business/energy/ukraines-energy-options-limited-event-russian-gas-disruption-2022-02-24/ (accessed on 21 July 2022).
17. Larina, Y.; Galchynska, J.; Kucheruk, P.; Zghurska, O.; Ortina, G.; Al-Nadzhar, F.; Marusei, T.; Kuboń, M.; Dzieniszewski, G. Estimation of the Domestic Agricultural Sector Potential for the Growth of Energy Cultures for Bioenergy Fuel Production. *Agric. Eng.* **2021**, *25*, 73–82. [CrossRef]
18. EIA Biomass Explained. Available online: https://www.eia.gov/energyexplained/biomass/ (accessed on 15 July 2022).
19. OECD/IEA. *Biomass for Power Generation and CHP*; OECD/IEA: Paris, France, 2007. Available online: https://www.iea.org/reports/biomass-for-power-generation-and-chp (accessed on 16 July 2022).
20. Energypedia Advantages and Disadvantages of Biomass. Available online: https://energypedia.info/wiki/Advantages_and_Disadvantages_of_Biomass (accessed on 16 July 2022).

21. Wąs, A.; Sulewski, P.; Krupin, V.; Popadynets, N.; Malak-Rawlikowska, A.; Szymańska, M.; Skorokhod, I.; Wysokiński, M. The Potential of Agricultural Biogas Production in Ukraine—Impact on GHG Emissions and Energy Production. *Energies* **2020**, *13*, 5755. [CrossRef]
22. IEA Ukraine Energy Profile. International Energy Agency. Available online: https://www.iea.org/reports/ukraine-energy-profile (accessed on 16 July 2022).
23. State Statistics Service of Ukraine. *Statistical Yearbook of Ukraine for 2019*; State Statistics Service of Ukraine: Kyiv, Ukraine, 2019.
24. Ritchie, H.; Roser, M. Ukraine: Energy Country Profile. Available online: https://ourworldindata.org/energy/country/ukraine (accessed on 10 July 2022).
25. Lederer, E.M. UN Nuclear Chief: Ukraine Nuclear Plant Is 'Out of Control'. *AP News*, 3 August 2022. Available online: https://apnews.com/article/russia-ukraine-science-accidents-d2e0077af104f2692b76f737c58e1984(accessed on 6 August 2022).
26. SAEE. Звіт про Результати Стимулювання та Використання Енергії, Виробленої з Відновлюваних Джерел, в Україні за 2019–2020. 2020. Available online: https://saee.gov.ua/sites/default/files/blocks/Report%20RES%202019-2020_eng%2013.12.2021_0.pdf (accessed on 15 July 2022).
27. SAEE. Інформація щодо Потужності та Обсягів Виробництва Електроенергії Об'єктами Відновлюваної Електроенергетики, Яким Встановлено «Зелений» Тариф. 2020. Available online: https://saee.gov.ua/sites/default/files/VDE_2019.pdf (accessed on 15 July 2022).
28. Lakyda, P.; Vasylyshyn, R.; Zibtsev, S.; Lakyda, I.; Geletukha, G.G.; Zhelyezna, T.; Bottcher, H. *Energy Potential of Biomass in Ukraine*; Center of NUBiP of Ukraine: Kyiv, Ukraine, 2011; ISBN 978-9-66800-683-8
29. Lewicki, W. Energy integration as an element of EU security in a historical outline. *Mod. Manag. Rev.* **2019**, *3*, 41–49. [CrossRef]
30. Gerasymchuk, N. Background of using renewable energy sources in order to ensure energy efficiency of Ukraine. *Humanit. Soc. Sci.* **2018**, *25*, 71–78. [CrossRef]
31. Panoutsou, C.; Zheliezna, T. Strategic Guide for Biomass Heat Policy in Ukraine. Available online: https://www.s2biom.eu/images/Publications/D8.3d_S2Biom_strategic_guide_for_biomass_heat_policy_in_Ukraine_Final.pdf (accessed on 10 July 2022).
32. Prince, T.; Solonyna, Y. As Gas Prices Surge, Bioenergy Could Help Heat Ukraine. So Why Is It on the Back Burner? Available online: https://www.rferl.org/a/ukraine-bioenergy-dependence-russian-fuel/31572075.html (accessed on 16 July 2022).
33. Johansson, B. Security aspects of future renewable energy systems—A short overview. *Energy* **2013**, *61*, 598–605. [CrossRef]
34. Freiberg, A.; Scharfe, J.; Murta, V.C.; Seidler, A. The Use of Biomass for Electricity Generation: A Scoping Review of Health Effects on Humans in Residential and Occupational Settings. *Int. J. Environ. Res. Public Health* **2018**, *15*, 354. [CrossRef]
35. Purkus, A.; Gawel, E.; Szarka, N.; Lauer, M.; Lenz, V.; Ortwein, A.; Tafarte, P.; Eichhorn, M.; Thrän, D. Contributions of flexible power generation from biomass to a secure and cost-effective electricity supply—A review of potentials, incentives and obstacles in Germany. *Energy Sustain. Soc.* **2018**, *8*, 18. [CrossRef]
36. Rosas-Casals, M.; Marzo, M.; Salas-Prat, P. Sovereignty, robustness, and short-term energy security levels. The Catalonia case study. *Front. Energy Res.* **2014**, *2*, 16. [CrossRef]
37. Khawaja, C.; Janssen, R.; Mergner, R.; Rutz, D.; Colangeli, M.; Traverso, L.; Morese, M.M.; Hirschmugl, M.; Sobe, C.; Calera, A.; et al. Viability and Sustainability Assessment of Bioenergy Value Chains on Underutilised Lands in the EU and Ukraine. *Energies* **2021**, *14*, 1566. [CrossRef]
38. Paas, W.; Accatino, F.; Bijttebier, J.; Black, J.E.; Gavrilescu, C.; Krupin, V.; Manevska-Tasevska, G.; Ollendorf, F.; Peneva, M.; San Martin, C.; et al. Participatory assessment of critical thresholds for resilient and sustainable European farming systems. *J. Rural Stud.* **2021**, *88*, 214–226. [CrossRef]
39. Geletukha, G.; Zheliezna, T. Prospects for bioenergy development in Ukraine: Roadmap until 2050. *Ecol. Eng. Environ. Technol.* **2021**, *22*, 73–81. [CrossRef]
40. Scarlat, N.; Dallemand, J.; Taylor, N.; Banja, M. *Brief on Biomass for Energy in the European Union*; Sanchez Lopez, J., Avraamides, M., Eds.; JRC109354; Publications Office of the European Union: Luxembourg, 2019.
41. Hutsol, T.; Glowacki, S.; Mudryk, K.; Yermakov, S.; Kucher, O.; Knapczyk, A.; Muliarchuk, O.; Koberniuk, O.; Kovalenko, N.; Kovalenko, V.; et al. *Agrobiomass of Ukraine—Energy Potential of Central and Eastern Europe (Engineering, Technology, Innovation, Economics)*; International Visegrad Fund: Bratislava, Slovakia, 2021; pp. 7–22. Available online: http://dglib.nubip.edu.ua/bitstream/123456789/8102/1/Agrobiomass%20of%20Ukraine_druk.pdf (accessed on 16 July 2022).
42. European Commission (EC). *Communication from the Commission to the European Parliament and the Council European Energy Security Strategy*; COM/2014/0330 Final; European Commission (EC): Brussels, Belgium, 2014. Available online: https://eur-lex.europa.eu/legal-content/EN/ALL/?uri=CELEX%3A52014DC0330 (accessed on 16 July 2022).
43. Geletukha, G.; Zheliezna, T.; Drahniev, S.; Haidai, O. Analysis of Actions for Ukraine to Replace Russian Natural Gas. *Ecol. Eng. Environ. Technol.* **2022**, *23*, 1–9. [CrossRef]
44. Kucher, O.; Pustova, Z.; Pustova, N.; Tkach, O.; Butorina, V. The role of biomass in the bioeconomic policy of Ukraine and its legal regulation. In Proceedings of the 2020 International Scientific Conference 'Economic Sciences for Agribusiness and Rural Economy' No 4, Warsaw, Poland, 21–22 September 2020; pp. 84–91. [CrossRef]
45. Kulyk, M.; Taranenko, A.; D'omin, D.; Rozhko, I. Agroecological aspects of rare energy crops growing in order to produce sustainable plant biomass. In *Global Trends and Prospects of Socio-Economic Development of Ukraine*; Baltija Publishing: Riga, Latvia, 2022; pp. 132–160.

46. Taranenko, A.; Kulyk, M.; Galytska, M.; Taranenko, S. Dynamics of soil organic matter in Panicum virgatum sole crops and intercrops. *Zemdirb.-Agric.* **2021**, *108*, 255–262. [CrossRef]
47. Golub, G.; Chuba, V.; Tsyvenkova, N.; Marus, O.; Yarosh, Y. Bioenergy potential of Ukrainian agriculture. *Int. J. Renew. Energy Res.* **2021**, *11*, 1223–1229. [CrossRef]
48. *Kyiv Region Council Strategy of Kyiv Region for 2021–2027 Years;* Decision of the Kyiv Regional Council Dated 19.12.2019 No. 789-32-VII; Kyiv Region Council: Kyiv, Ukraine, 2019; pp. 1–158. Available online: https://www.minregion.gov.ua/wp-content/uploads/2020/05/strategiya-rozvytku-kyyivskoyi-oblasti-na-2021-2027-roky.pdf (accessed on 16 July 2022).
49. Stepasyuk, L.; Nahorna, O. Переваги біомаси в порівнянні з традиційними видами біопалива. *Біоекономіка І Аграрний Бізнес* **2021**, *11*, 96–104. [CrossRef]
50. Gerasymchuk, N.; Kyrlyk, N.; Shtuler, I.; Stepasyuk, L.; Titenko, Z. Efficiency of the agricultural industry export potential as the foundation of the state financial stability. *Financ. Credit Act. Probl. Theory Pract.* **2020**, *2*, 434–440. [CrossRef]
51. Klius, S.V. Determination of straw and plant waste during independence period of Ukraine. *Renew. Energy* **2012**, *3*, 71–79.
52. State Statistics Service of Ukraine. *Statistical Yearbook of the Kyiv Region for 2020 Year;* Main Administration of Statistics in Kyiv Region: Kyiv, Ukraine, 2021; 390p.
53. Statistical Compendium. *Agriculture of Ukraine 2019;* State Statistics Service of Ukraine: Kyiv, Ukraine, 2020. Available online: https://ukrstat.gov.ua/druk/publicat/kat_u/2020/zb/11/zb_yearbook_2019_e.pdf (accessed on 10 July 2022).
54. Statistical Compendium. *Agriculture of Ukraine 2016;* State Statistics Service of Ukraine: Kyiv, Ukraine, 2017. Available online: https://ukrstat.gov.ua/druk/publicat/kat_u/2017/zb/09/zb_agriculture_2016pdf.zip (accessed on 16 July 2022).
55. Department of Ecology and Natural Resources of Kyiv Regional State Administration. *Regional Report on the State of the Natural Environment in the Kyiv Region in 2020 Year;* Kyiv Regional State Administration: Kyiv, Ukraine, 2021; 212p.
56. Heletukha, G.; Dragnev, S.; Kucheruk, P.; Matveev, Y. A practical guide to the use of biomass as fuel in the municipal sector of Ukraine (for representatives of the agro-industrial complex).
57. Chepeliev, M.; Diachuk, O.; Podolets, R.; Trypolska, G. The role of bioenergy in Ukraine's climate mitigation policy by 2050. *Renew. Sustain. Energy Rev.* **2021**, *152*, 111714. [CrossRef]
58. Kaletnik, G.; Pryshliak, N.; Tokarchuk, D. Potential of Production of Energy Crops in Ukraine and their Processing on Solid Biofuels. *Ecol. Eng. Environ. Technol.* **2021**, *22*, 59–70. [CrossRef]

MDPI
St. Alban-Anlage 66
4052 Basel
Switzerland
Tel. +41 61 683 77 34
Fax +41 61 302 89 18
www.mdpi.com

Energies Editorial Office
E-mail: energies@mdpi.com
www.mdpi.com/journal/energies

www.ingramcontent.com/pod-product-compliance
Lightning Source LLC
Chambersburg PA
CBHW061608220926
51850CB00005B/453

Machinability and Tribological Performance of Advanced Alloys

Machinability and Tribological Performance of Advanced Alloys

Editor

George A. Pantazopoulos

MDPI • Basel • Beijing • Wuhan • Barcelona • Belgrade • Manchester • Tokyo • Cluj • Tianjin

Editor
George A. Pantazopoulos
ELKEME Hellenic Research
Centre for Metals S.A
Oinofyta Viotias, Greece

Editorial Office
MDPI
St. Alban-Anlage 66
4052 Basel, Switzerland

This is a reprint of articles from the Special Issue published online in the open access journal *Metals* (ISSN 2075-4701) (available at: https://www.mdpi.com/journal/metals/special_issues/Machin_Tribo_Perf).

For citation purposes, cite each article independently as indicated on the article page online and as indicated below:

LastName, A.A.; LastName, B.B.; LastName, C.C. Article Title. *Journal Name* **Year**, *Volume Number*, Page Range.

ISBN 978-3-0365-8280-1 (Hbk)
ISBN 978-3-0365-8281-8 (PDF)

Contents

About the Editor

George A. Pantazopoulos

George A. Pantazopoulos graduated from the Chemical Engineering Department of the National Technical University of Athens (NTUA) in 1992 (order of graduation considered equivalent to "summa cum laude") with specialization in Materials Science. He received his Ph.D. from the Mechanical Engineering Department of NTUA in the field of Manufacturing Technology and Processing of Advanced Materials in 1999. He has served in the metals industry for more than 23 years as a quality control engineer, as a senior R&D project leader, and more recently, as a quality and technology manager. His principal scientific research interests are focused on the field of failure and fracture analysis, materials processing, manufacturing process technology, mechanical testing, and tribology. He is the author or co-author of more than 160 papers in the above areas, presented at international journals and conferences, as well as two books and various monographs/book chapters. He was an invited speaker at various conferences and academic institutions. He serves as a subject editor/editorial board member and reviewer for journals in the field of materials science and failure analysis. He has been the recipient of various awards and distinctions, including a best paper prize and a dissertation thesis prize. He was also included among the top-cited scientist career-long list, based on a statistical analysis of bibliometric indicators from the Scopus database (based on calculations using all Scopus author profiles as of 1 September 2022).

metals

MDPI

Editorial

Machinability and Tribological Performance of Advanced Alloys

George A. Pantazopoulos

ELKEME Hellenic Research Centre for Metals S.A., 61st km Athens—Lamia National Road, 32011 Oinofyta, Viotias, Greece; gpantaz@elkeme.vionet.gr; Tel.: +30-2262-60-4463

1. Introduction and Scope

Machining is specially utilized to manufacture special, precision parts and difficult-to-form materials. The most challenging aspects of this topic are the continuously evolving quality and productivity requirements in modern industries and creating new and smart materials that meet energy and environmental (green/clean) regulations. In general, among the conventional machining operations, high-end, modern material removal processes are addressed in this Special Issue (e.g., using high-energy beams, such as laser machining).

The tribological performance of alloys plays an important role in the production or service environment. Because it is closely related to their materials and manufacturing characteristics, it is considered to be a combined topic in the frame of this Special Issue. The study of tribological behaviour suitably addresses the surface engineering aspects of modern and conventional alloys, either in bulk or in coating form, and receives special attention in the generic context of industrial component production and/or service in aggressive working environments.

The core subject of this Special Issue is the elaboration and presentation of studies on alloy design, manufacturing, testing and characterization in order to provide a clear insight into their machinability and/or tribological behaviour. The improvement of these properties has led to the development of new alloy chemistries, novel engineered microstructures and the application of coatings that produce better machinability and/or tribological endurance properties under poor service conditions. Environmental and health and safety regulations demand the use of ecofriendly components. The manufacture of anti-microbial copper alloys in healthcare facilities and the elimination of lead in brass components for drinking water applications constitute examples of this modern industrial trend.

Altering the manufacturing and service conditions is also essential for achieving the optimization of their machinability and tribological performance. The application of surface processes to enhance the tribological performance of alloys is be addressed in this Special Issue.

Briefly, the main machinability parameters and outcomes are schematically depicted in Figure 1.

Citation: Pantazopoulos, G.A. Machinability and Tribological Performance of Advanced Alloys. *Metals* **2023**, *13*, 1190. https://doi.org/10.3390/met13071190

Received: 9 June 2023
Accepted: 26 June 2023
Published: 27 June 2023

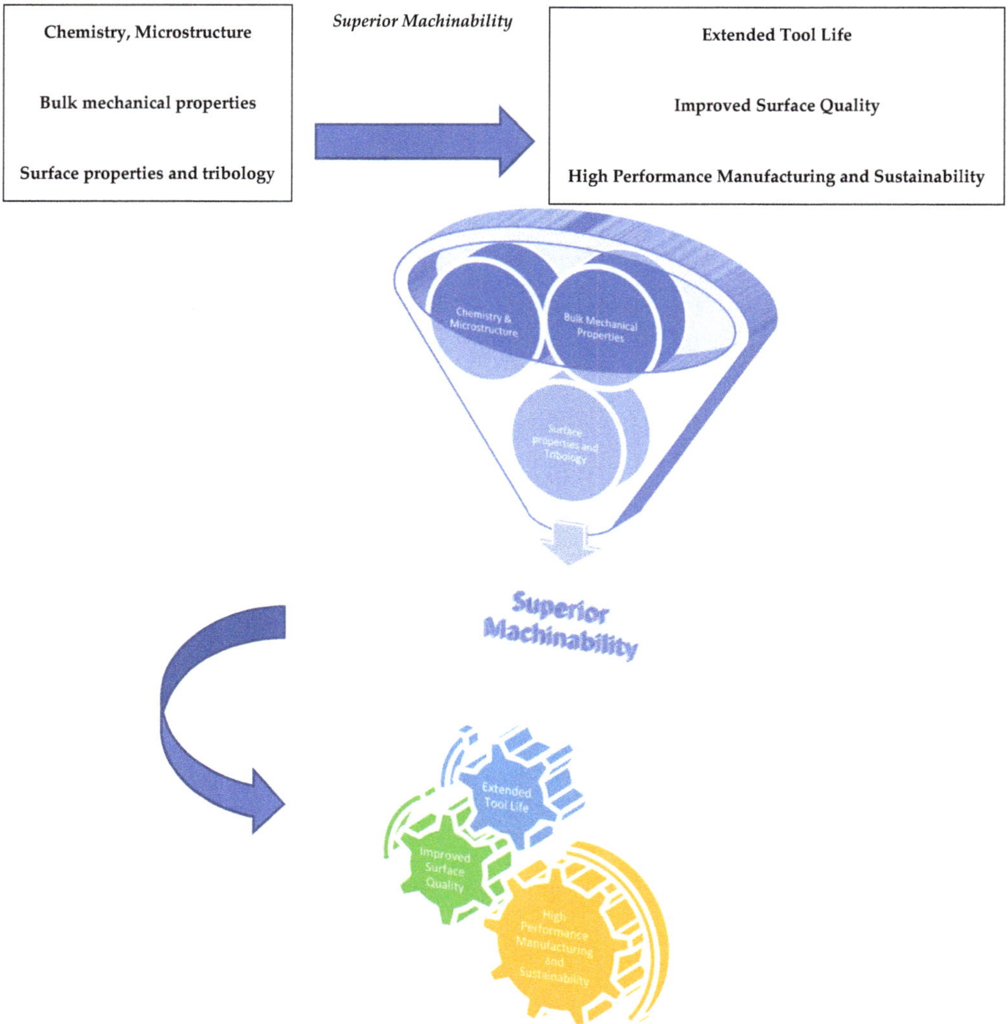

Figure 1. Schematic diagram illustrating that the special design of basic and generic material parameters could control level of the machinability and the major outcomes.

2. Contributions

The Special Issue "Machinability and Tribological Performance of Advanced Alloys" contains fifteen research articles [1–15], including one review paper [5]. The contents of this collection cover a wide spectrum of cross-disciplinary fields related to the machinability and tribological behaviour of materials with a variety of industrial applications. In addition, there are a few papers that, although they are not directly relevant to the core-subject area, provide valuable insight into the broader area of manufacturing technology [2,8] and the advanced characterization methodology of W-Cu nanocrystalline materials using FEM nanoindentation [4].

The main papers take a materials-focus perspective utilizing various manufacturing processes, such as turning [5,13], planing [10] and electric discharge machining [3,15]. A review paper reported on the basic categories of leaded and lead-free machinable brass alloys [5], while a specialized crystallographic investigation using Electron Backscatter

Diffraction (EBSD) analysis revealed the texture and fracture behaviour relationships of lead-free machinable brass alloys (such as CW511L and CW510L) [7]. Texture and phase analyses were performed before and after a heat treatment, which was applied for machinability improvement [7]. The combination of LAGBs and Σ3 boundaries together with β-phase content play a significant role in energy absorption and fracture resistance development [7]. The influence of induction heating on the machining of an SDK11 (high C, Cr-Mo) alloy steel is presented [12]. According to this study, the steel microstructure was not changed after heating it at a temperature between 200 and 400 °C, while a significant cutting force reduction was observed [12]. In [13], a comparative study on vibration monitoring among three brass alloys, two lead-free ones (CW510L and CW724R) and one leaded one (CW614N) was performed. According to this research, on average, CW510L and CW614N brass alloys demonstrated three times lower vibration damping values compared to that of CW724R brass alloy. In [15], the surface quality of an EDM-treated nickel based superalloy (Inconel 617) using cryogenically and non-cryogenically treated electrodes and various modified dielectrics of transformer oil was investigated.

Tribological behaviour research that aimed to improve the lifetime of materials/components by increasing the wear and/or corrosion resistance was undertaken in [1,6,9]. A comparative study concerning the abrasive wear resistance of different plughshare steels is presented in [9]. Furthermore, a study on the lubrication efficiency of Inconel 718 and the WC tribosystem is presented in [11]. The sulfurized fatty acid ester results in excellent anti-wear and anti-friction properties [11]. According to [14], a reverse thermal model and parametric modelling were applied to predict the temperature history of surface clads (carbide/steel systems) produced via a Concentrated Solar Energy surface treatment (CSE). This could result in the correlation between CSE processing parameters and the performance of created surface layers.

The content of the Special Issue is also presented to facilitate the readership and for taxonomy purposes (Table 1).

Table 1. List of topics and materials elaborated in the contents of the current S.I.

Reference Number	Relevant Topic			Materials
	Machinability	Tribology/Surface Properties	Miscellaneous	
[1]		■		X210CrW12 steel
[2]			■	AA5083 Al-alloy
[3]	■			DC53 steel
[4]			■	W-Cu (nanocrystalline)
[5]	■			Copper alloys (brasses)
[6]		■		Fe-Bulk metallic glasses
[7]	■			Copper alloys (brasses)
[8]			■	Mild steel
[9]		■		Steels (mainly S355J2G3, 37MnSi5, Hardox 450, UTP 690, OK 84.58)
[10]	■			EN C45 carbon steel
[11]		■		Inconel 718/WC
[12]	■			SKD11 Alloy Steel
[13]	■			Brass alloys (CW510L, CW614N, CW724R)
[14]		■		TiC/carbon steel, Cr_3C_2/carbon steel
[15]	■			Inconel 617

3. Conclusions and Outlook

The current Special Issue aims to place emphasis on this critical area of manufacturing, shedding light onto relationships governing the mutually influenced and interrelated properties of the machinability and tribological behaviour of metallic materials.

Since the industry demands more sustainable, energy efficient- and environmental-/health and safety (E&HS)-friendly solutions, more research will be conducted on innovative and "green" materials (such as eco-friendly or lead-free copper alloys, environmentally friendly coatings and lubricants) and cost-efficient manufacturing processes. Industrial digitalization and modernization (in the frame of Industry 4.0) may further contribute to high-performance lean production, leading to the highest quality products and components, displaying minimum variation in their quality characteristics.

Funding: This research received no external funding.

Acknowledgments: The Guest Editor expresses his deep gratitude to the contributing authors who shared their valuable research works in this collection of papers, sharing research in the field of Machinability and Tribological Behaviour of Materials and shedding light on significant industrial applications and challenges. Moreover, we acknowledge the voluntary contributions of peer reviewers, together with the assistance and support provided generously by the MDPI *Metals* Editorial Team. Finally, the continuous support of ELKEME colleagues and management team is highly appreciated.

Conflicts of Interest: The author declares no conflict of interest.

Profound Appreciation and Dedication : The launch of the current Special Issue occurred almost three years ago and coincided with the outbreak of COVID-19, which was an unprecedented, dystopic period during which millions of lives were lost, people were fearful, strict measures were applied, including social distancing and the abrupt interruption of communication. At that time, Stoic philosophical/thinking principles provided me with outstanding mental support, courage, discipline and the vision to try to do the best I could within my power and my control. During this sorrowful and "dark" period, I received plentifully special and dedicated support from "close and skillful contributors and friends" who gave me courage and strength to continue and keep persisting towards my targets and goals. Because I feel sincerely indebted to them, I would like to express my deepest gratitude to these talented and Amazing Supporters (abbreviated as A.S.) who showed the path to find joy and the flowers of life. A.S. contributed to building an "inner citadel", which together with the catalytic influence and wise insight amply offered to me, will remain unforgettable.

References

1. Duan, Y.; Qu, S.; Jia, S.; Li, X. Evolution of the Fretting Wear Damage of a Complex Phase Compound Layer for a Nitrided High-Carbon High-Chromium Steel. *Metals* **2020**, *10*, 1391. [CrossRef]
2. Du, Z.; Wang, G.; Wang, H. The Process Design and Rapid Superplastic Forming of Industrial AA5083 for a Fender with a Negative Angle in a Small Batch. *Metals* **2021**, *11*, 497. [CrossRef]
3. Khan, S.A.; Rehman, M.; Farooq, M.U.; Ali, M.A.; Naveed, R.; Pruncu, C.I.; Ahmad, W. A Detailed Machinability Assessment of DC53 Steel for Die and Mold Industry through Wire Electric Discharge Machining. *Metals* **2021**, *11*, 816. [CrossRef]
4. Tserpes, K.; Bazios, P.; Pantelakis, S.G.; Pappa, M.; Michailidis, N. Mechanical Characterization of Nanocrystalline Materials via a Finite Element Nanoindentation Model. *Metals* **2021**, *11*, 1827. [CrossRef]
5. Stavroulakis, P.; Toulfatzis, A.I.; Pantazopoulos, G.A.; Paipetis, A.S. Machinable Leaded and Eco-Friendly Brass Alloys for High Performance Manufacturing Processes: A Critical Review. *Metals* **2022**, *12*, 246. [CrossRef]
6. Zhang, G.; Sun, W.; Xie, L.; Zhang, C.; Tan, J.; Peng, X.; Li, Q.; Ma, X.; Zhao, D.; Yu, J. Multicomponent Fe-Based Bulk Metallic Glasses with Excellent Corrosion and Wear Resistances. *Metals* **2022**, *12*, 564. [CrossRef]
7. Vazdirvanidis, A.; Rikos, A.; Toulfatzis, A.I.; Pantazopoulos, G.A. Electron Backscatter Diffraction (EBSD) Analysis of Machinable Lead-Free Brass Alloys: Connecting Texture with Fracture. *Metals* **2022**, *12*, 569. [CrossRef]
8. Kuo, S.-K.; Ou, Y.-L.; Wang, D.-A. An Analytical Model for Stress and Curvature Prediction of a Strip Leveling Process. *Metals* **2022**, *12*, 757. [CrossRef]
9. Vargova, M.; Tavodova, M.; Monkova, K.; Dzupon, M. Research of Resistance of Selected Materials to Abrasive Wear to Increase the Ploughshare Lifetime. *Metals* **2022**, *12*, 940. [CrossRef]
10. Monka, P.P.; Monkova, K.; Vasina, M.; Kubisova, M.; Korol, M.; Sekerakova, A. Effect of Machining Conditions on Temperature and Vickers Microhardness of Chips during Planing. *Metals* **2022**, *12*, 1605. [CrossRef]
11. Yang, Y.; Luan, H.; Guo, S.; Liu, F.; Dai, Y.; Zhang, C.; Zhang, D.; Zhou, G. Tribological Behaviors of Inconel 718–Tungsten Carbide Friction Pair with Sulfur Additive Lubrication. *Metals* **2022**, *12*, 1841. [CrossRef]
12. Mac, T.-B.; Luyen, T.-T.; Nguyen, D.-T. Assessment of the Effect of Thermal-Assisted Machining on the Machinability of SKD11 Alloy Steel. *Metals* **2023**, *13*, 699. [CrossRef]
13. Monka, P.P.; Monkova, K.; Pantazopoulos, G.A.; Toulfatzis, A.I. Effect of Wear on Vibration Amplitude and Chip Shape Characteristics during Machining of Eco-Friendly and Leaded Brass Alloys. *Metals* **2023**, *13*, 828. [CrossRef]

14. Zervaki, A.D.; Lambrakos, S.G.; Mourlas, A.G.; Papantoniou, I.G.; Rodríguez, J.; Psyllaki, P.P. Inverse Thermal Analysis as a Tool for Optimizing Concentrated Solar Energy Elaboration of Wear Resistant Surface Layers. *Metals* **2023**, *13*, 942. [CrossRef]
15. Ishfaq, K.; Sana, M.; Mahmood, M.A.; Anwar, S.; Waseem, M.U. Evaluating Surface Quality of Inconel 617 by Employing Deep Cryogenically Treated Electrodes in Surfactant-Added Dielectrics of Transformer Oil. *Metals* **2023**, *13*, 1092. [CrossRef]

metals

MDPI

Article

Evolution of the Fretting Wear Damage of a Complex Phase Compound Layer for a Nitrided High-Carbon High-Chromium Steel

Yong Duan, Shengguan Qu *, Siyu Jia and Xiaoqiang Li

School of Mechanical and Automotive Engineering, South China University of Technology, Guangzhou 510640, China; yong_duan_scut@outlook.com (Y.D.); J15635223374@163.com (S.J.); lixq@scut.edu.cn (X.L.)
* Correspondence: qusg@scut.edu.cn

Received: 29 September 2020; Accepted: 14 October 2020; Published: 19 October 2020

Abstract: In this paper, the X210CrW12 steel was subjected to gas nitriding to obtain a complex phase compound layer with limited porosity. The nitrided layer was characterized by scanning electron microscopy (SEM), energy dispersive spectroscopy (EDS) and X-ray diffraction (XRD). The fretting wear behavior and the evolution of fretting wear damage of the compound layer were studied, and the worn surfaces were characterized by SEM/EDS and 3D optical profilometry. The results indicated that the compound layer showed superior fretting wear resistance and sufficient load-carrying capacity in the low loading case of 35 N, but the fracture of coarse nitrides (transformed primary carbides) was obviously detrimental to wear resistance. For the high loading case of 70 N, the low toughness of the compound layer led to the occurrence of brittle cracks, and the decrease in the thickness of the compound layer due to wear resulted in the cracking and spalling of the compound layer.

Keywords: compound layer; brittle delamination; oxidative mechanism; fretting wear

1. Introduction

Gas nitriding, as a typical thermo-chemical surface treatment, is one of the most efficient methods to enhance the surface properties, such as surface hardness, corrosion and wear resistance [1–8]. The typical nitrided layer consists of two different structures, known as the compound layer and diffusion layer. Generally, the compound layer is composed of γ'-Fe_4N, ε-$Fe_{2-3}N$, or a mixed phase ($\varepsilon + \gamma'$). The diffusion layer brings about an improvement of the fatigue properties, whereas the improvement of wear and corrosion properties is ascribed to the compound layer [3].

There are three main types of compound layer. One type of layer consists of a predominant percentage of ε phase. Such a layer is very porous and its wear resistance is worsened [9,10]. However, the γ' phase is more ductile than ε phase, but the compound layer consisting of monophase γ' is usually thin and prone to spalling in heavy conditions due to the extremely hardness differences occurring between the compound layer and diffusion zone [11]. The next layer consists of $\gamma' + \varepsilon$ iron nitrides with a limited percentage of ε phase. Investigations show that if the porosity in this type of compound layer can be well controlled, the compound layer with a relative reasonable thickness is very beneficial to the wear and corrosion resistance in many cases [12–15]. In practice, the complex phase compound layers (with limited porosity) are widely applied in non-impact conditions to improve the resistance to damage affecting the component life such as wear and corrosion. Further, the sliding wear characteristics of this type of compound layer have been extensively studied [4–7,11–13,16–18]. As reported by Rad et al. [17], the specimens with a good thickness compound layer had a lower wear rate, but the thin compound layer was broken in small sliding distance, resulting in a high wear rate. In the paper [18], the sliding wear test of the nitrided specimens was carried out on a ball-on-disc machine under loads of 100 and 300 N, and the GCr15 steel ball (with a hardness of 63–65 HRC) was used as the counterbody material. The results showed that the compound layer consisting of $\gamma' + \varepsilon$ phases showed excellent wear resistance under the light load of 100 N, but poor wear properties at the heavy load of 300 N.

Fretting wear, which is caused by the vibration of the mechanical system or the variation in loads, is found to be different from the sliding wear [19] due to the small displacement amplitude, and the material loss caused by it is much less than that caused by sliding wear. However, fretting wear can produce a loss of fitting and degrade the functionality of the mechanical system. The degradation of the component surface caused by fretting wear is the result of several wear mechanisms, such as abrasion, adhesion, crack and oxidation [20]. Although many experimental works have been carried out to study the sliding wear behavior of the compound layer, there is a lack of knowledge relating to the fretting wear behavior. Identification of the evolution of fretting wear damage of the compound layer is crucial to ensure stable service of the nitrided components.

In this study, the X210CrW12 steel was selected as the experiment material. It is a very hard material with high wear-resistance and can be widely used in tribological applications [21]. The material was subjected to gas nitriding to obtain a complex phase compound layer with limited porosity, and the fretting wear behavior and the evolution of fretting wear damage of the compound layer were analyzed in detail.

2. Materials and Methods

2.1. Material Preparation and Nitriding Treatment

The chemical composition (wt.%) of X210CrW12 steel used in this study is shown in Table 1. The discs with a diameter of 40 mm and thickness of 11 mm were cut from a bar received in the annealed condition. Before nitriding, specimens were austenitized at 1080 °C for 30 min followed by oil quenching, then triple tempered at the temperature of 580 °C for 2 h, leading to a hardness of 495 ± 15 $HV_{0.2}$. After the heat treatment, the thickness of all specimens was machined to 8 mm to remove the decarburized layer. Prior to the gas nitriding process, all specimens were ground with SiC abrasive papers of 600, 1200, 1500 and 2000 grits followed by mirror polishing, and ultrasonically cleaned in acetone for 10 min. Then, all specimens were hung in a gas nitriding furnace. Gas nitriding was performed using an industrial unit at 500 °C for nearly 70 h, and the high-purity NH_3 gas was used as the nitriding medium.

Table 1. Compositions of X210CrW12 steel specimen (wt.%).

C	Si	Mn	P	Cr	V	W	Fe
2.16	0.2	0.26	0.27	12.2	0.12	0.81	Bal.

2.2. Fretting Wear Test and Microstructure Characterization

Fretting wear tests were performed on an Optimol SRV IV (DIN51834, SRV, Optimol, Munchen, Germany) reciprocating friction and wear test machine under the dry condition at a temperature of 25 °C. The ball-on-flat method was adopted to perform the test, as illustrated in Figure 1. The upper ball specimen (SiC ball with a hardness of about 2600 HV) with a diameter of 10 mm was mounted in a holder attached to an oscillating electro-mechanical drive, and pressed against the stationary lower specimen (nitrided specimen). In addition, load, frequency and amplitude conditions were chosen for best test stability. The fretting wear tests were carried out under the load of 35 N and 70 N, because the fretting wear behavior of the nitrided samples under these two load conditions was representative. In order to study the evolution of fretting wear damage, the fretting cycle ranged from 24×10^3 to 72×10^3. Each test was repeated at least three times to check the repeatability. The main test parameters are listed in Table 2. After each fretting wear test, the specimens were ultrasonically cleaned for 15 min, thus leaving any debris that was more firmly adhered to the worn surface.

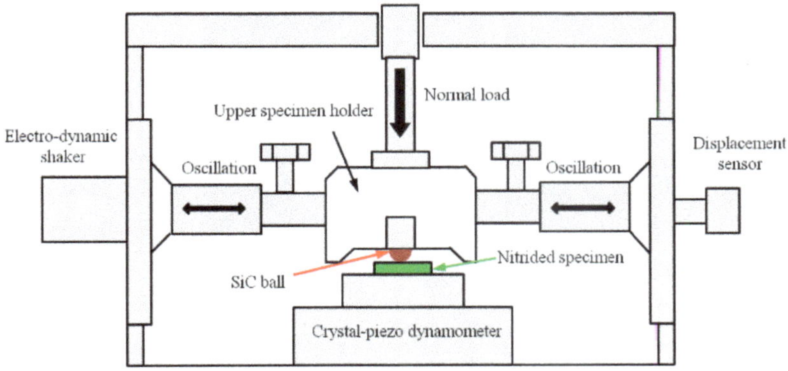

Figure 1. Schematic diagram of ball-on-disk fretting tribometer.

Table 2. Main test parameters.

Load (N)	Frequency (Hz)	Stroke (μm)	Duration, N (Number of Cycles)
35, 70	20	200	24×10^3, 48×10^3, 72×10^3

The cross-sectional microstructure of nitrided specimen and the fretting wear mechanism was studied by scanning electron microscopy using backscattered electron mode (SEM/BSE; Quanta 200 & Nova nanosem 430, FEI, Hillsboro, OR, USA). The microhardness of the nitrided specimen was measured using a Vickers microhardness tester (HDV-10002, SCTMC, Shanghai, China) with a load of 1.96 N and a holding time of 15 s. Phase composition of the un-nitrided and nitrided specimens was studied by X-ray diffraction (XRD; D8 Advance, Bruker-axs, Karlsruhe, Germany). The surface topography of wear scar was characterized by mean of a 3D optical profiler (America RTEC Up Dual-Mode), and the depth and volume of the wear scar were obtained by the instrument software.

3. Results and Discussion

3.1. Microstructural Characteristics

Figure 2 shows the cross-sectional microstructure of the nitrided layer. It can be found that the primary and secondary alloy carbides are dispersed throughout the base material. These phases represent the main microstructural characteristics of X210Crw12 steel. During nitriding, these initial carbides will be transformed into nitrides [22,23], as confirmed by the results of EDS analysis (Figure 2c), and the released carbon atoms will diffuse (i) toward the nitrogen diffusion front, forming the cementite precipitates along grain boundaries, and (ii) toward the outer surface, leading to decarburization of the surface adjacent region [23]. Figure 2b clearly shows that the compound layer is a compact layer, and the thickness of the compound layer was measured to be in the range of about 8–10 μm. However, small pores formed on the top surface of the compound layer, as shown with arrows. Figure 3 shows the cross-sectional microhardness profile of the nitrided specimen. After nitriding, the surface hardness is 960 ± 30 $HV_{0.2}$, which was measured after the removal of pores on the top surface of the compound layer. The decrease in the surface hardness may be caused by the occurrence of decarburization in the surface adjacent regions [23] or the long nitriding time [12,23].

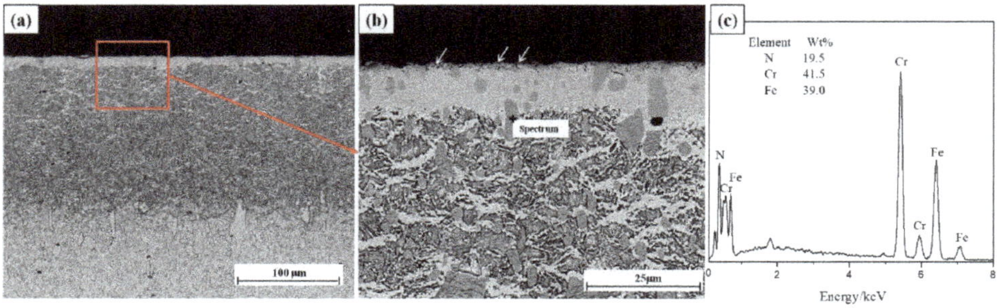

Figure 2. Cross-sectional microstructure of the nitrided layer: (**a**) overall features, (**b**) a close view of the rectangle in (**a**) and (**c**) energy dispersive spectroscopy (EDS) analysis of the point in (**b**).

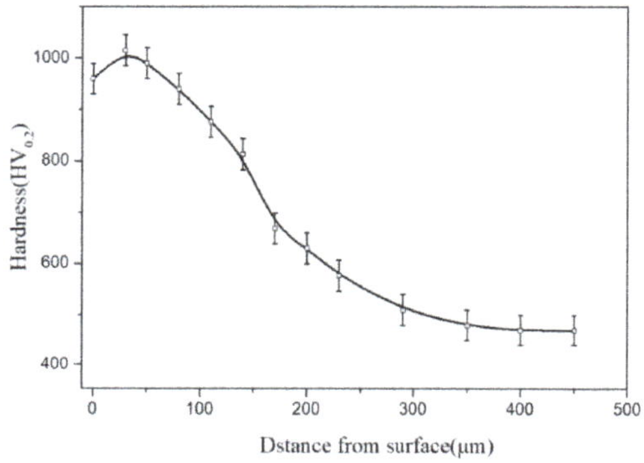

Figure 3. Cross-sectional hardness–depth profile for the nitrided specimen.

Figure 4 shows the X-ray diffraction patterns on the surface of the un-nitrided and nitrided specimens. Before nitriding, the substrate is characterized by ferrite (tempered martensite) and (Cr, M)$_7$C$_3$ carbide diffraction patterns (Figure 4a). After nitriding, the presence of the γ' (Fe_4N) and ε (Fe_3N) phases, together with the CrN phase can be identified (Figure 4b), confirming the formation of the compound layer constituted by these phases. Then, the nitrided specimen was polished to remove the pores on the top surface of the compound layer. Figure 4c shows the XRD phase analysis of the compound layer after polishing. It can be found that after polishing, the peak intensity of ε phase becomes stronger, while the intensity of γ' phase peak becomes weak. Therefore, the existence of ε phase should not be responsible for the formation of pores on the top surface of the compound layer. The reason for the formation of pores and the increase in the amount of γ' phase at the top surface of the compound layer may be due to the occurrence of decarburization in this region [24]. As estimated from the intensities of the respective diffraction lines (Figure 4c), the ratio of ε/γ' for the compound layer is about 1.3, which provides high hardness for the compound layer and makes its porosity be well controlled.

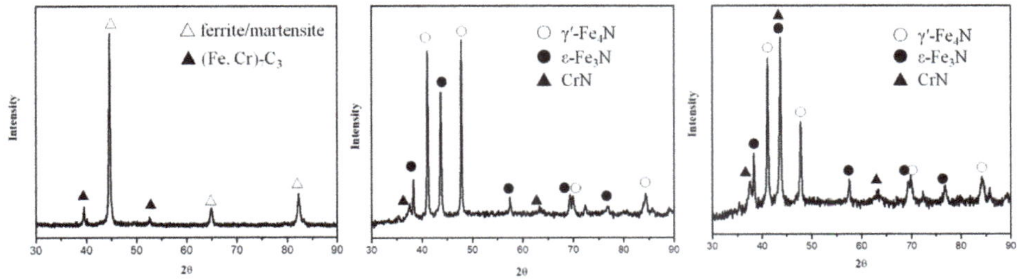

(**a**) (**b**) (**c**)

Figure 4. X-ray diffraction from the surface of (**a**) un-nitrided specimen, (**b**) nitrided specimen before polishing and (**c**) nitrided specimen after polishing.

3.2. Fretting Wear Behavior

Figure 5 plots the evolution of the friction coefficient for 72×10^3 fretting cycles under the two loading conditions. In the loading case of 35 N, the friction coefficient varies very much in the initial 24×10^3 cycles, and then gets to a steady state with a small value of around 0.32. However, it is worth noting that the friction coefficient can increase rapidly at a certain time, and then decreases gradually to the stable value again. Under the normal load of 70 N, the frictional coefficient curve shows different characteristics from that in the loading case of 35 N. The friction coefficient varies very much throughout the whole wear process, and the average friction coefficient is 0.45, which is higher than that in the loading case of 35 N. This finding is in contrast with the results reported by many other investigators [25,26]. In their work, the friction coefficient decreased with increasing the normal load, which was proposed to be due to the fact that the surface contact was the elastic contact which resulted in the asperities interlock in the low loading case [26].

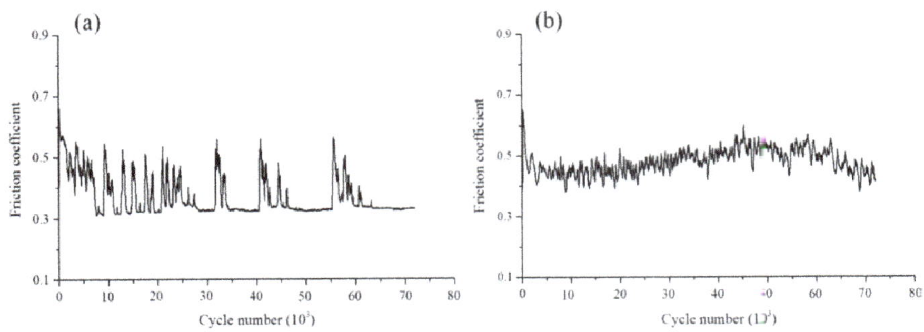

Figure 5. Evolution of friction coefficient under the load of (**a**) 35 N and (**b**) 70 N.

Figure 6 illustrates the wear volume of the nitrided specimens as a function of cycle number. For the two loading conditions, the wear volume increases with the number of cycles as expected. However, for the loading case of 35 N, it can be seen that the wear rate rapidly decreases to a very low value after the initial 24×10^3 cycles. When the fretting cycle increases from 24×10^3 to 72×10^3 the wear volume increases slowly from 0.62×10^{-3} mm^3 to 0.82×10^{-3} mm^3. Under the normal load of 70 N, the wear volume shows a different trend compared to that under the load of 35 N, it increases almost linearly with increasing the fretting cycle.

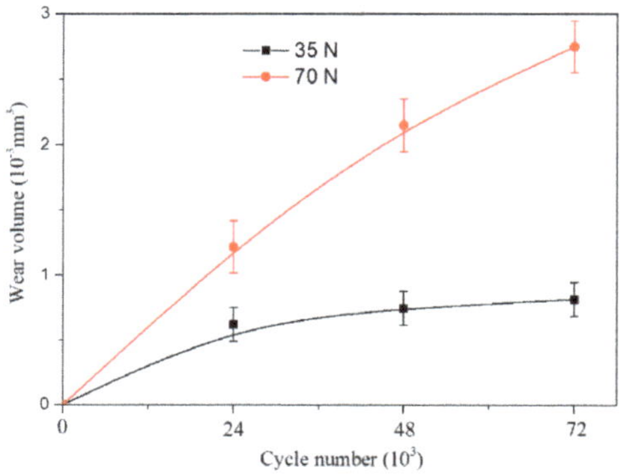

Figure 6. Wear volume of the nitrided specimens under the load of 35 N and 70 N.

Figure 7 shows the 3D surface morphologies and BSE images of the wear scars after different fretting cycles under the normal load of 35 N. The 3D surface morphologies show that the worn surfaces are dominated by the deep grooves parallel to the sliding direction, suggesting that the main wear mechanism is two-body abrasive wear. Furthermore, it can be seen that wear depth in the central region of the wear scars is maximum, which is proposed to be caused by the Hertzian contact pressure distribution. However, the wear area and depth increase slowly with the increasing number of cycles. From the SEM-BSE images of the wear scars, it can be seen that the worn surfaces for different cycle's tests show similar wear characteristics with some dark areas. EDS analysis of the dark area (point in Figure 7h) is shown in Figure 8, revealing that the dark area is oxidized wear debris consisting mainly of O, Fe and Si. These indicate that the abrasive particles generated from the wear of the nitrided specimens and SiC balls were oxidized, and material transfer from the balls to the worn surfaces occurred.

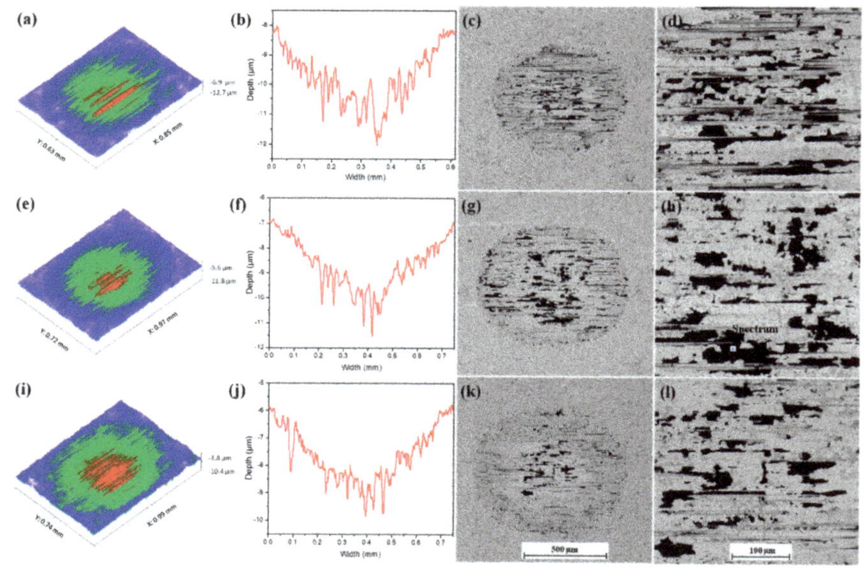

Figure 7. 3D surface morphologies, cross-sectional profiles and scanning electron microscopy using backscattered electron mode (SEM-BSE) images of the wear scars after fretting cycles of 24 × 10^3 (**a–d**), 48 × 10^3 (**e–h**) and 72 × 10^3 (**i–l**) under the normal load of 35 N.

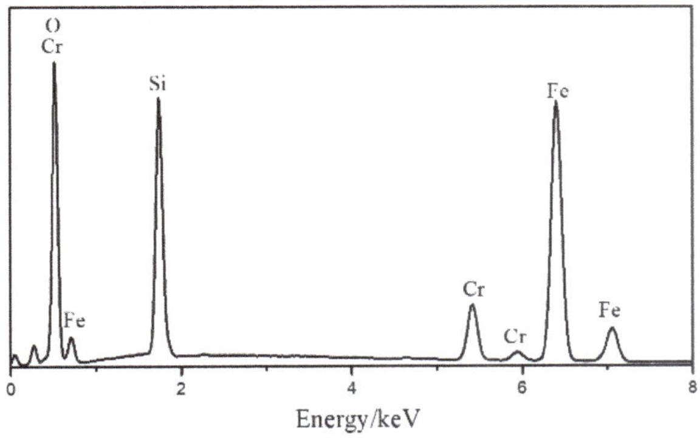

Element	Wt%
O	31.6
Si	13.6
Cr	5.9
Fe	48.9
Totals	100.00

Figure 8. EDS analysis of the dark area (the point in Figure 7h).

Figure 9 shows the worn surface at the center of the wear scar after 72 × 10^3 cycles. Some cracks can be observed on the worn surface, resulting from the fracture of coarse nitrides (transformed primary carbides) (Figure 9a). Under the high-frequency reciprocating sliding conditions, the cyclic stresses and local concentrated stresses were imposed, leading to the cracking of coarse nitrides due to their low toughness, and to their removal as fragments, leaving deep cavities on the worn surface (Figure 9a). Figure 9b clearly shows that cracks are observed around the broken coarse nitride, suggesting that the interfaces of the transformed primary carbides and matrix are the potential sites for the initiation as well as the propagation of cracks. It becomes clear that these transformed primary carbides are obviously detrimental to wear resistance in this situation. These observations are consistent with the work of Singh et al. [27] who studied the wear behavior of AISI D2 steel in two-body abrasive wear and stated that when the induced stress exceeded the fracture strength of primary carbides, the carbides fractured due to their low ductility, and the fracture and exfoliation of the primary carbides were the dominating factors affecting the wear resistance of the material. It can be observed that apart from the fracture of the coarse nitrides, brittle cracks or metallic ribbons ready to detach can be hardly observed in the matrix (Figure 9a), indicating that this type of compound layer shows good toughness and wear resistance in this loading case.

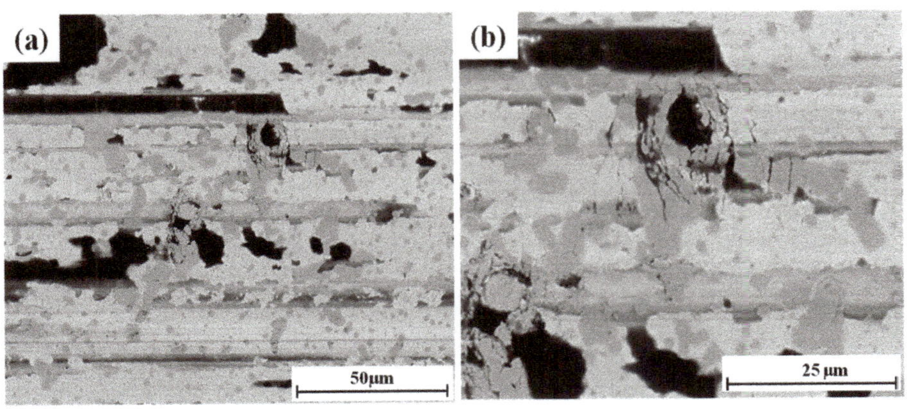

Figure 9. (**a**) Worn surface at the center of the wear scar after 72 × 10^3 cycles under the normal load of 35 N. (**b**) Higher magnification BSE image of (**a**).

In all cases, the wear process itself can lead to a reduction of contact pressure due to the increased area of contact, subsequently resulting in the decrease in wear rate, as predicted by the literature [28] which presented the macroscopic wear process using the finite element analysis (FEA). On the other hand, the oxide layer formed on the worn surfaces should be responsible for the very low wear rate in this loading case, because it acted as a lubricating film during the wear process [29,30], which favoured the wear resistance and decreased the wear rate and friction coefficient.

The friction coefficient shows that there are two main wear mechanisms involved in the wear of the nitrided specimens during the wear progress (Figure 5a). During the early fretting cycles of about 24×10^3, the friction coefficient varies very much. At this stage, the dominant wear mechanism is abrasive wear, accompanied by the rapid removal of material, which is confirmed by the deep grooves on the worn surface (Figure 7a) and the high wear rate (Figure 6). The oxidation mechanism is also active (confirmed by the oxide layer on the worn surface as shown in Figure 7c and d), but it is not the primary wear mechanism. Then, as the fretting cycle continues to increase, the friction coefficient decreases to a small value of around 0.32, but it is worth noting that the friction coefficient can increase rapidly at a certain time, and then decreases gradually to the stable value again. At this stage, the low friction coefficient suggests that oxidation becomes the primary wear mechanism, which explains the rapid decrease in the wear rate in this stage (Figure 6). It is reported that the changes in friction coefficient are controlled by the wear particles produced by microcutting and spalling, and the size of the wear particles at the contact surface plays a more important in raising friction, rather than the number of particles [31]. Therefore, it can be inferred that the rapid increase in the friction coefficient at the steady state should be caused by the large and hard particles resulted from the fracture of primary carbides. These particles were comminuted by repeated plastic deformation and fracture. Once they were reduced to a sufficiently small size, the friction coefficient was reduced to the stable value again. On the other hand, the accumulation and cohesion of the wear debris in the form of the fine oxide particles and their subsequent de-cohesion after reaching a critical size could also cause the appearance of peaks in the friction coefficient curve. Table 3 shows the wear mechanisms of nitrided specimens at different stages of testing.

Table 3. The wear mechanisms of nitrided specimens at different stages of testing under the load of 35 N.

Number of Cycles	Wear Mechanism	Coefficient of Friction	Wear Rate
24×10^3	Abrasive wear * Oxidation mechanism	High	High
48×10^3	Abrasive wear Oxidation mechanism *	Low	Low
72×10^3	Abrasive wear Oxidation mechanism *	Low	Low

* represents the main wear mechanism.

Figure 10 shows the SEM-BSE images of the wear scars and their corresponding 3D surface topographies and cross-sectional profiles under the normal load of 70 N. The cross-section profiles of the wear scars were measured vertical to the sliding direction, passing through the center of the wear scars. It can be seen that the worn surfaces exhibit similar characteristics to that from the tests conducted under the normal load of 35 N, showing many grooves parallel to the sliding direction and darker-coloured oxides. In contrast to the loading case of 35 N, the wear scar area and depth have an obvious increase with an increasing number of cycles. However, the average wear depth at the central region of the wear scar after the initial 24×10^3 fretting cycles is about 6 μm. When the fretting cycle continues to increase, the average wear depth at the central region gets to about 8 μm at 48×10^3 cycles and about 11 μm at 72×10^3 cycles. It is clear that the compound layer at the central region of the wear scar after the cycles of 72×10^3 is worn through. On the other hand, a large number of cracks perpendicular to the sliding direction can be clearly observed in the central region of the wear scars following 48×10^3 and 72×10^3 cycles.

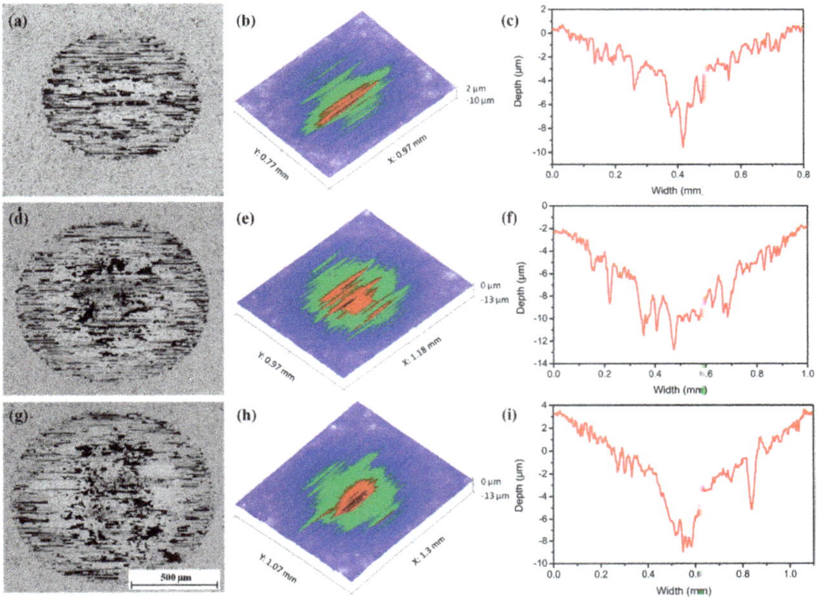

Figure 10. Plan view SEM-BSE images of the wear scars and their corresponding 3D surface morphologies and cross-sectional profiles under the normal load of 70 N after fretting cycles of 24×10^3 (**a–c**), 48×10^3 (**d–f**), and 72×10^3 (**g–i**).

Figure 11 shows the SEM-BSE images of the central region of the wear scars and their corresponding 3D surface morphologies under the normal load of 70 N. It can be seen that there are few visible cracks that can be observed at the central region of the wear scar after the initial fretting cycles of 24×10^3 (Figure 11a). However, the higher magnification image corresponding to the rectangular areas in Figure 11a and b clearly reveals that some brittle cracks perpendicular to the sliding direction do occur at the edge of the groove. Under the high normal load, the high contact pressure and shear stress can promote the rapid removal of the material and the generation of deep grooves, especially at the contact center (Figure 10). As shown in Figure 10b and c, the micro-cutting depth is quite deep in the early stage of wear, inducing great stresses around the grooves. Consequently, the brittle cracks emerged on the edge of the deep grooves due to the low toughness of the compound layer. It should be noted that the obvious brittle delamination of the compound layer cannot be observed on the worn surface, indicating that the compound layer still maintains a good load carrying capacity in this situation.

However, when the fretting cycle increases to 48×10^3, a large number of cracks perpendicular to the sliding direction can be clearly seen on the worn surface (Figure 11c), and the higher magnification image of rectangular areas in Figure 11c and d shows clear evidence that the propagation and interconnection of the cracks result in severe delamination type wear in the zone between two deep grooves. As the fretting cycle continues to increase to 72×10^3, there are only a few visible cracks in the contact center corresponding to the maximum depth, but large amounts of cracks are concentrated at the edge of the contact center, leading to severe delamination wear in these areas.

As predicted in the literature [28], the maximum contact pressure goes down with wear and the contact pressure becomes more uniform. Therefore, the high contact pressure should not be the main factor that leads to the severe brittle delamination of the compound layer at the central region of the wear scars. Studies have shown that the load-carrying capacity of the coated system increased with increasing substrate stiffness and coating thickness, and the very thin coatings were detrimental to the load-carrying capacity [17]. The brittle delamination mechanism for the compound layer in the central region of the wear scar can be explained as follows. The rapid wear at the contact center reduced the thickness of the compound layer, which reduced the strength against the external stress resulting in the cracking or spalling of the compound layer. In general, the compound layer containing both γ' phase and ε phase led to weak interface boundaries resulting in an increase in brittleness [16]. For the area between two grooves, highly concentrated stresses were imposed during abrasion, promoting the brittle fracture propagation, as shown in the higher magnification image of the rectangular areas in Figure 11c. The delaminated layer then turned into bulk and hard abrasive particles, promoting the wear in the contact region. Therefore, the primary wear mechanism throughout the wear process was abrasive wear, which led to the high friction coefficient and wear rate. The wear mechanisms of nitrided specimens at different stages of testing are shown in Table 4.

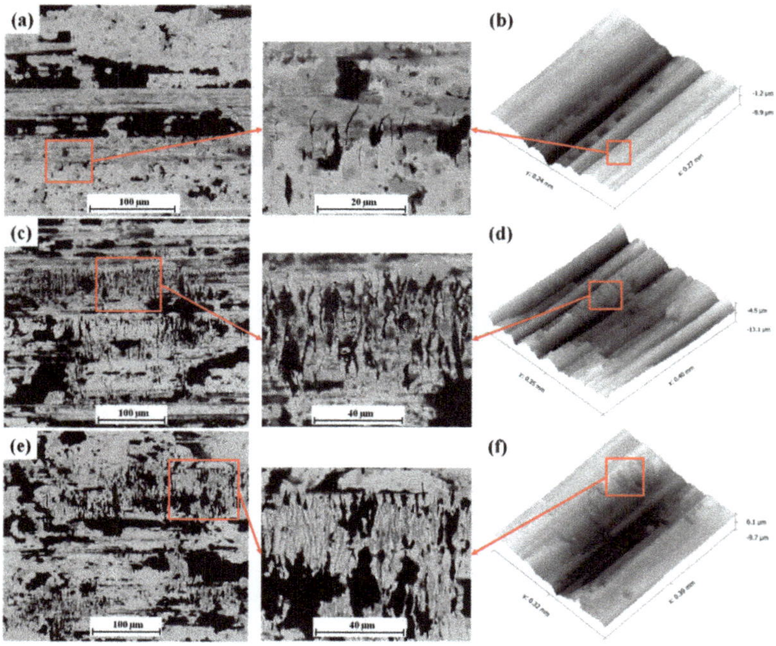

Figure 11. SEM-BSE images of the central region of the wear scars and their corresponding 3D surface morphologies under the normal load of 70 N after fretting cycles of 24 × 10^3 (**a**,**b**), 48 × 10^3 (**c**,**d**), and 72 × 10^3 (**e**,**f**), and the higher magnification SEM-BSE image corresponding to the rectangular area.

Table 4. The wear mechanisms of nitrided specimens at different stages of testing under the load of 70 N.

Number of Cycles	Wear Mechanism	Coefficient of Friction	Wear Rate
24 × 10^3	Abrasive wear * Oxidation mechanism	High	High
48 × 10^3	Abrasive wear * Brittle delamination Oxidation mechanism	High	High
72 × 10^3	Abrasive wear * Brittle delamination Oxidation mechanism	High	High

* represents the main wear mechanism.

4. Conclusions

In this paper, the X210CrW12 steel was selected and subjected to gas nitriding to obtain the complex phase compound layer. The fretting wear behavior and the evolution of fretting wear damage of the compound layer were studied in detail. Some conclusions were summarized as follows:

(1) The obtained compound layer was a compact layer with limited porosity, and the ratio of ε/γ' as estimated from XED peaks is about 1.3.
(2) The compound layer showed superior fretting wear resistance and sufficient load-carrying capacity in a low loading case of 35 N, and provided a rigid substrate to support the oxide layer, leading to the oxidative mechanism.
(3) In the low loading case of 35 N, the fracture of coarse nitrides (transformed primary carbides) was obviously detrimental to wear resistance, leading to crack initiation and propagation in the matrix, and the generated hard particles controlled the friction coefficient.
(4) For the high loading case of 70 N, the compound layer still maintained good load carrying capacity and the brittle delamination of the compound layer was suppressed in the early stage of wear. However, some brittle cracks can be observed on the edge of the deep grooves due to the induced great stresses around the grooves and the low toughness of the compound layer.
(5) Under the high loading case of 70 N, the rapid wear at the contact center reduced the thickness of the compound layer, which reduced the strength against the external stress resulting in the cracking and spalling of the compound layer.

Author Contributions: Conceptualization, Y.D.; methodology, Y.D. and S.Q.; software, Y.D. and S.J.; validation, X.L., Y.D. and S.J.; formal analysis, Y.D.; investigation, Y.D.; resources, S.Q.; data curation, X.L. and S.Q.; writing—original draft preparation, Y.D.; writing—review and editing, Y.D. and S.Q.; visualization, X.L.; supervision, S.Q.; project administration, S.Q.; funding acquisition, S.Q. All authors have read and agreed to the published version of the manuscript.

Funding: This research received no external funding.

Conflicts of Interest: The authors declare no conflict of interest.

References

1. Birol, Y. Analysis of wear of a gas nitrided H13 tool steel die in aluminium extrusion. *Eng. Fail. Anal.* **2012**, *26*, 203–210. [CrossRef]
2. Wang, B.; Sun, S.; Guo, M.; Jin, G.; Zhou, Z.; Fu, W. Study on pressurized gas nitriding characteristics for steel 38CrMoAlA. *Surf. Coat. Technol.* **2015**, *279*, 60–64. [CrossRef]
3. Akhtar, S.S.; Arif, A.F.M.; Yilbas, B. Evaluation of gas nitriding process with in-process variation of nitriding potential for AISI H13 tool steel. *Int. J. Adv. Manuf. Technol.* **2009**, *47*, 687–698. [CrossRef]
4. Yang, J.; Liu, Y.; Ye, Z.; Yang, D.; He, S. Microstructural and tribological characterization of plasma- and gas-nitrided 2Cr13 steel in vacuum. *Mater. Des.* **2011**, *32*, 808–814. [CrossRef]
5. Nolan, D.; Leskovsek, V.; Jenko, M. Estimation of fracture toughness of nitride compound layers on tool steel by application of the Vickers indentation method. *Surf. Coat. Technol.* **2006**, *201*, 182–188. [CrossRef]
6. Kundalkar, D.; Mavalankar, M.; Tewari, A. Effect of gas nitriding on the thermal fatigue behavior of martensitic chromium hot-work tool steel. *Mater. Sci. Eng. A* **2016**, *651*, 391–398. [CrossRef]
7. Yang, C.; Liu, J. Intermittent vacuum gas nitriding of TB8 titanium alloy. *Vacuum* **2019**, *163*, 52–58. [CrossRef]
8. Aghajani, H.; Torshizi, M.; Soltanieh, M. A new model for growth mechanism of nitride layers in plasma nitriding of AISI H11 hot work tool steel. *Vacuum* **2017**, *141*, 97–102. [CrossRef]
9. Schwarz, B.; Göhring, H.; Meka, S.; Schacherl, R.E.; Mittemeijer, E.J. Pore Formation Upon Nitriding Iron and Iron-Based Alloys: The Role of Alloying Elements and Grain Boundaries. *Met. Mater. Trans. A* **2014**, *45*, 6173–6186. [CrossRef]
10. Hernández, M.; Staia, M.; Cabrera, E.P. Evaluation of microstructure and mechanical properties of nitrided steels. *Surf. Coat. Technol.* **2008**, *202*, 1935–1943. [CrossRef]

11. Karamiş, M.; Gerçekcioğlu, E. Wear behaviour of plasma nitrided steels at ambient and elevated temperatures. *Wear* **2000**, *243*, 76–84. [CrossRef]
12. Karamis, M. An investigation of the properties and wear behaviour of plasma-nitrided hot-working steel (H13). *Wear* **1991**, *150*, 331–342. [CrossRef]
13. Xi, Y.-T.; Liu, D.-X.; Han, D. Improvement of corrosion and wear resistances of AISI 420 martensitic stainless steel using plasma nitriding at low temperature. *Surf. Coat. Technol.* **2008**, *202*, 2577–2583. [CrossRef]
14. Terčelj, M.; Smolej, A.; Fajfar, P.; Turk, R. Laboratory assessment of wear on nitrided surfaces of dies for hot extrusion of aluminium. *Tribol. Int.* **2007**, *40*, 374–384. [CrossRef]
15. Bombac, D.; Terčelj, M.; Peruš, I.; Fajfar, P. The progress of degradation on the bearing surfaces of nitrided dies for aluminium hot extrusion with two different relative lengths of bearing surface. *Wear* **2013**, *307*, 10–21. [CrossRef]
16. Castro, G.; Fernández-Vicente, A.; Cid, J. Influence of the nitriding time in the wear behaviour of an AISI H13 steel during a crankshaft forging process. *Wear* **2007**, *263*, 1375–1385. [CrossRef]
17. Rad, H.F.; Amadeh, A.; Moradi, H. Wear assessment of plasma nitrided AISI H11 steel. *Mater. Des.* **2011**, *32*, 2635–2643. [CrossRef]
18. Wang, B.; Zhao, X.; Li, W.; Qin, M.; Gu, J. Effect of nitrided-layer microstructure control on wear behavior of AISI H13 hot work die steel. *Appl. Surf. Sci.* **2018**, *431*, 39–43. [CrossRef]
19. Chen, G.; Zhou, Z. Study on transition between fretting and reciprocating sliding wear. *Wear* **2001**, *250*, 665–672. [CrossRef]
20. Varenberg, M.; Halperin, G.; Etsion, I. Different aspects of the role of wear debris in fretting wear. *Wear* **2002**, *252*, 902–910. [CrossRef]
21. Knauf, F.; Baadjou, R.; Hirt, G. Analysis of semi-solid extrusion products made of steel alloy X210CRW12. *Int. J. Mater. Form.* **2009**, *2*, 733–736. [CrossRef]
22. Jégou, S.; Barrallier, L.; Kübler, R. Phase transformations and induced volume changes in a nitrided ternary Fe–3%Cr–0.345%C alloy. *Acta Mater.* **2010**, *58*, 2666–2676. [CrossRef]
23. Van Wiggen, P.C.; Rozendaal, H.C.F.; Mittemeijer, E.J. The nitriding behaviour of iron-chromium-carbon alloys. *J. Mater. Sci.* **1985**, *20*, 4561–4582. [CrossRef]
24. Ruset, C.; Ciuca, S.; Grigore, E. The influence of the sputtering process on the constitution of the compound layers obtained by plasma nitriding. *Surf. Coat. Technol.* **2003**, *174*, 1201–1205. [CrossRef]
25. Shima, M.; Okado, J.; Mccoll, I.R. The influence of substrate material and hardness on the fretting behavior of TiN. *Wear* **1999**, *225*, 38–45. [CrossRef]
26. Zhang, D.K.; Ge, S.R.; Qiang, Y.H. Research on the fatigue and fracture behaviour due to the fretting wear of steel wire in hosting rope. *Wear* **2003**, *255*, 1233–1237. [CrossRef]
27. Singh, K.; Khatirkar, R.K.; Sapate, S.G. Microstructure evolution and abrasive wear behavior of D2 steel. *Wear* **2015**, 206–216. [CrossRef]
28. Ding, J.; Leen, S.B.; McColl, I. The effect of slip regime on fretting wear-induced stress evolution. *Int. J. Fatig.* **2004**, *26*, 521–531. [CrossRef]
29. Stott, F. The role of oxidation in the wear of alloys. *Tribol. Int.* **1998**, *31*, 61–71. [CrossRef]
30. So, H. The mechanism of oxidational wear. *Wear* **1995**, *184*, 161–167. [CrossRef]
31. Hwang, D.; Kim, D.; Lee, S. Influence of wear particle interaction in the sliding interface on friction of metals. *Wear* **1999**, *225*, 427–439. [CrossRef]

metals

MDPI

Article

The Process Design and Rapid Superplastic Forming of Industrial AA5083 for a Fender with a Negative Angle in a Small Batch

Zhihao Du [1,*], Guofeng Wang [2] and Hailun Wang [2,3]

[1] School of Mechanical and Electric Engineering, Nanyang Normal University, Nanyang 473061, China
[2] School of Material Science and Engineering, Harbin Institute of Technology, Harbin 150001, China; gfwang@hit.edu.cn (G.W.); hlwyhzs@163.com (H.W.)
[3] Capital Aerospace Machinery Corporation Limited, Beijing 100071, China
* Correspondence: duzhihaohit@126.com; Tel.: +86-377-6351-3563

Abstract: A front automobile fender with a negative angle was trial produced via rapid superplastic forming (SPF) technology. The tensile test of industrial AA5083 was carried out at elevated temperatures, and the results showed that the maximum elongation was 242% at 480 °C/0.001 s^{-1}. A rigid-plastic constitutive model of the SPF process was established. Initial dies of preforming and final forming were designed. The finite element method (FEM) was used to simulate the forming process and predict the thickness distribution of different areas. Furthermore, the dies were optimized to make the thickness distribution uniform. In the final structure, the maximum thinning ratio decreased from 83.2% to 63% due to the optimized design of the forming dies. The front automobile fender was then successfully fabricated by the preforming process and final forming process at 480 °C. A thickness measurement was carried out, and the minimum thickness of the preforming structure was 2.17 mm at the transverse tank, while that of the final structure was 2.49 mm near the edge of the lamp orifice. The average grain size grew from 20 to 35 μm. The grain growth led to the reduction of mechanical properties. Compared with the mechanical properties of the initial material, the maximum decrease in tensile strength for the material after superplastic forming was 5.78%, and that of elongation was 18.5%.

Keywords: AA5083; rapid SPF; FEM; mechanical property; microstructure evolution

Citation: Du, Z.; Wang, G.; Wang, H. The Process Design and Rapid Superplastic Forming of Industrial AA5083 for a Fender with a Negative Angle in a Small Batch. *Metals* **2021**, *11*, 497. https://doi.org/10.3390/met11030497

Academic Editors: Diego Celentano and Anders E. W. Jarfors

Received: 26 January 2021
Accepted: 14 March 2021
Published: 17 March 2021

1. Introduction

Due to its high specific strength, high specific stiffness, excellent corrosion resistance, and mechanical properties, AA5083 has been in the spotlight of railway vehicles, shipbuilding, and other industries [1–4]. Many investigations have been conducted to determine the influencing factors such as superplasticity and the superplastic forming (SPF) of AA5083 [5–11]. The chemical composition optimization of AA5083 has been attempted to reduce cost [5], and the superplastic deformation behavior was analyzed [6]. The SPF of AA5083 has been widely used to fabricate automotive liftgates and decklids [7], "high-end" automobiles [10], and components for aircraft applications [11].

The front fender is considered an essential part needed to protect nearby components and support other exteriors; however, it is the most challenging part to manufacture due to its complex shape and requirement for high machining accuracy [12–14]. The conventional forming process of a fender was the cold stamping of steel [15] and plastic injection molding [16]. Cold stamping has the disadvantages of spring-back and a high die cost, while injection molding has complex forming processes [17]. SPF has the advantages of low die costs, eliminating spring-back, and producing complex components [18]. Moreover, SPF is an effective forming process for lightweight alloys for automotive components [19]. Thus, SPF, in theory, could be used to fabricate the automotive fender of AA5083.

Jarrar et al. [20] investigated the superplastic forming of triangular channels with sharp radii, and the forming process was simulated. Zhang et al. [21] researched the controlling of the thickness distribution in superplastic forming. In comparison, a large thickness deviation existed in the produced part. As such, direct-reverse SPF was adopted to make the thickness more uniform [22–24]. Xing et al. [22] proposed a simple and efficient preform design method to improve the uniformity of wall thickness. Lan et al. [24] investigated the two-stage superplastic forming of a V-shaped aluminum sheet into a trough with a deep and irregular contour, and a superior thickness profile was provided. Another limitation of SPF was the slow forming rate, which made it inefficient. Thus, it was necessary to investigate the rapid direct-reverse SPF process for an automobile front fender with a negative angle.

This manuscript aimed to fabricate a complex-shaped front automobile fender by the rapid direct-reverse SPF technology. The tensile test of an industrial AA5083 was carried out at elevated temperatures, and a constitutive equation in SPF was established. With the help of FEM, the forming process and the dies were established. The front automobile fender was then successfully manufactured by rapid SPF technology. A thickness measurement was made, and both the macroscopic and the microstructure behaviors were investigated.

2. Experimental

An industrial AA5083 sheet was used in this study, and the thickness of the sheet was 4 mm. The sheet was provided by the Northern Light Alloy Company Ltd. (Harbin, China), and the chemical composition is shown in Table 1. Figure 1 shows the initial microstructure of AA5083, and it can be seen that the grain size was about 20 μm. The grain size of less than 10 μm is typically one of the requirements for superplasticity [18], and this grain size was larger than that in other research papers investigating the superplasticity and SPF of AA5083 [25–28].

Table 1. Chemical composition of AA5083.

Elements	Si	Fe	Cu	Mn	Mg	Cr	Al
Content (wt %)	0.073	0.18	0.020	0.68	4.66	0.074	Bal.

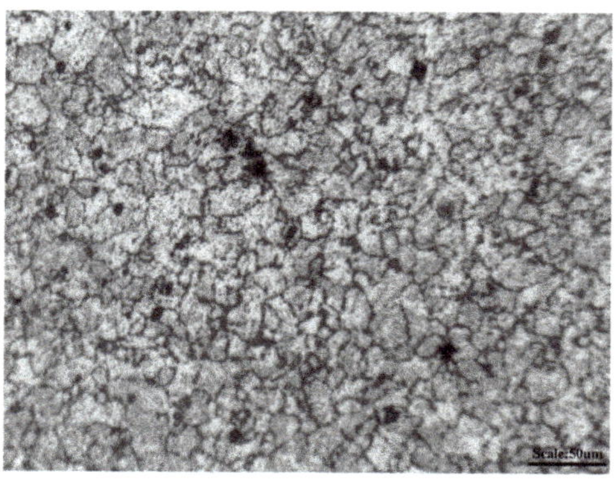

Figure 1. Initial microstructure of AA5083.

Tensile test specimens were cut parallel to the rolling direction, and the gauge size was 15 mm × 5 mm × 4 mm. Tensile tests at elevated temperatures of 400–560 °C with an interval of 40 °C were carried out at the strain rate of 5×10^{-3} s^{-1}–5×10^{-4} s^{-1}, and the tests were conducted on an Instron 3343 testing machine (Instron Corporation, Norwood, MA, USA). The temperature diversity was less than 3 °C along the gauge length. The tensile tests at room temperature were conducted on the AG-X Plus testing machine, and the tensile strain rate was 1.67×10^{-3} s^{-1}. These specimens were cut from the initial sheet and the sheet after the forming process. To ensure the accuracy of tensile data, at least three tensile tests were performed for each condition, and the intermediate data at each condition were selected as the final result. The stress–strain data were captured at 10 Hz to guarantee the continuity of the data.

Rapid SPF of the fender was carried out on a 10,000 kN SPF machine. This rapid SPF process was divided into two stages: the preforming process and the final forming process. Each stage was composed of drawing and SPF, and the forming temperature was 480 °C. The target strain rate of the SPF process was 1×10^{-3} s^{-1}.

The microstructural sample was cut into small pieces of 5 mm × 4 mm × T mm (where T is the thickness of different regions) by electron discharge wire cutting, then the plane of 5 mm × 4 mm was mechanically ground using SiC paper (from 400# to 2000#). Subsequently, the sample was electrolytically polished. The samples were corroded with Keller reagent for 60 s then cleaned with alcohol and distilled water. Scanning electron microscope (SEM) analysis of the microstructure of the as-received material, the preforming structure, and the final forming structure was performed on the Quanta 200 FEG-SEM machine (FEI company, Hillsboro, OR, USA).

3. Results and Discussion

3.1. Flow Stress of AA5083 at Elevated Temperatures

The flow stress curves of AA5083 at the strain of 5×10^{-3} s^{-1}–5×10^{-4} s^{-1} are shown in Figure 2. The flow stress was dependent on the temperature, strain rate, and true strain. The tensile strength decreased with increasing temperatures and decreasing strain rates, while the elongation increased firstly then reduced with the increasing temperature.

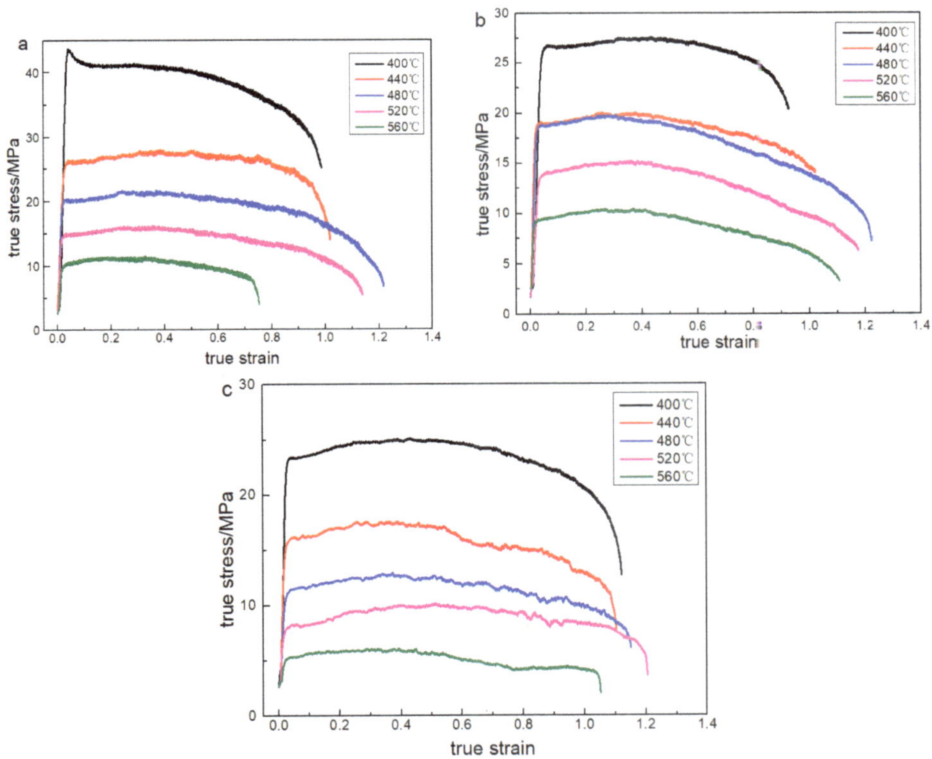

Figure 2. Flow stress curves of AA5083 at: (**a**) 0.005 s^{-1}; (**b**) 0.001 s^{-1} (**c**) 0.0005 s^{-1}.

It can be seen that the elongation at some deformation parameters was more than 200% from Figure 2, and the maximum elongation obtained at 480 °C and 1×10^{-3} s^{-1} was 242%. Thus, it can be concluded that the industrial AA5083 exhibited superplastic-like large elongation. This superplastic-like elongation of coarse-grained aluminum alloys with a grain size of more than 10 μm has been reported in other investigations [29–31].

In the SPF process, the strain hardening was neglected. Thus, the constitutive model used to describe the SPF process was set as:

$$\sigma = K\dot{\varepsilon}^{m} \tag{1}$$

where m is the strain rate sensitivity coefficient indicated the effect of strain rate on the flow stress. K is the material constant related to the deformation temperature and the microstructure. Equation (1) was taken for the logarithms, then it was expressed as:

$$\ln \sigma = \ln K + m \ln \dot{\varepsilon} \tag{2}$$

For Equation (2), after being differential, the equation for m can be obtained as:

$$m = \frac{\partial \ln \sigma}{\partial \ln \dot{\varepsilon}} \tag{3}$$

The value of m was then calculated to be about 0.35, and K was estimated to be 155 MPa·s on account of the flow stress curves at 480 °C with different strain rates. When the strain rate sensitivity exponent was 0.35, it was the value of the solute drag creep, which is different from the value indicating grain boundary sliding [32]. This value was lower than in the case of the fine-grained materials [33]. Furthermore, superplastic forming by solute drag creep has been reported by Honda Motor Co., Ltd. (Tokyo, Japan) and UACJ as high-temperature, high-speed blow forming. Thus, the material in this paper could be applied to the SPF process.

3.2. Structural Design and Process Optimization

The forming process of the fender is complex. Moreover, it is difficult to observe the deformation and the variation of thickness in the manufacturing process. Thus, the forming process was simulated by FEM, and the thickness distribution of the forming process was predicted, providing a theoretical reference for the final manufacturing process. The conventional SPF cannot achieve the final shape in one process due to the complex shape; thus the whole process included two steps. The first step was the preforming process, which comprised the drawing and the reverse superplastic bulging. The second step was the final process, consisting of the drawing and the direct superplastic bulging.

3.2.1. Initial Design

The initial designs of the male die for the preforming process and the male die for the final forming process are shown in Figure 3. To ensure the thickness uniformity of the final product, appropriate material storage was prepared on the preformed male die. Storage tanks were mainly set in A, B and C parts, as shown in Figure 3a). The male die of the final forming shown in Figure 3b was mainly used for further stretching and as supplementing material for the superplastic bulging of the final process.

The numerical simulation of the forming process was carried out by the MSC.MARC software (2010, MSC software Corporation, LA, USA). The initial model of the FEM analysis is shown in Figure 4. The color scales represent the plate thickness, and the units of the legend in each figure are millimeters. The dies were rigid, and the sheet was deformable in this model. The size of the sheet was 1900 mm × 2500 mm. The four-node shell element was selected to represent the sheet. The size of the shell element was 20 mm × 20 mm, and the number of the four-node shell element was 11,875. The thickness was set to be 3 mm. The friction coefficient between the die and the face sheet was set as 0.3, and the Coulomb friction model was selected to be the friction model.

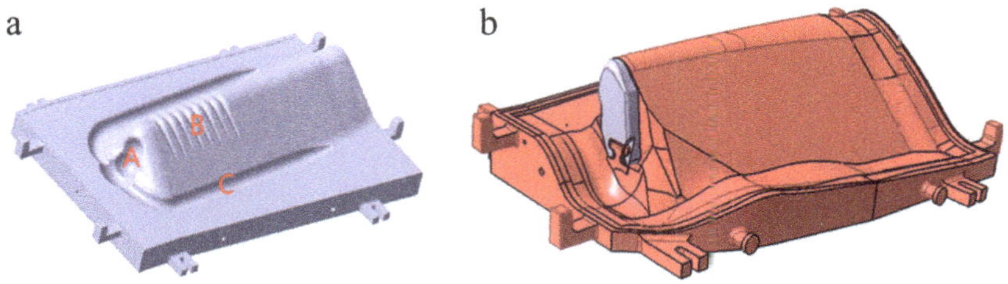

Figure 3. Male dies of (**a**) the preforming process; (**b**) final forming process.

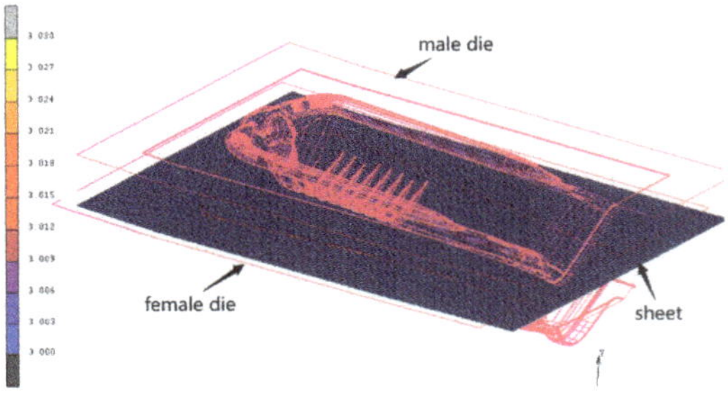

Figure 4. Initial FEM model.

(1) Drawing of the preforming process

The drawing depth was 250 mm. The thickness distribution of the drawing during the preforming process is shown in Figure 5. It can be seen that the minimum thickness at the lowest position of the drawing was 2.78 mm. The radial tensile stress of the material at this position was greater than the tangential compressive stress. Therefore, this resulted in thickness reduction, but the plate thickness reduction rate was only 7.33%.

Figure 5. Thickness distribution of the drawing during the preforming process.

(2) Reverse superplastic bulging of the preforming process

The thickness distribution of the reverse superplastic bulging is shown in Figure 6. The minimum thickness was 1.631 mm at Region A, and the thinning ratio reached was 45.6%. The serious thickness reduction at this region resulted from the thickness reduction in the drawing of the preforming process and the further reduction in bulging.

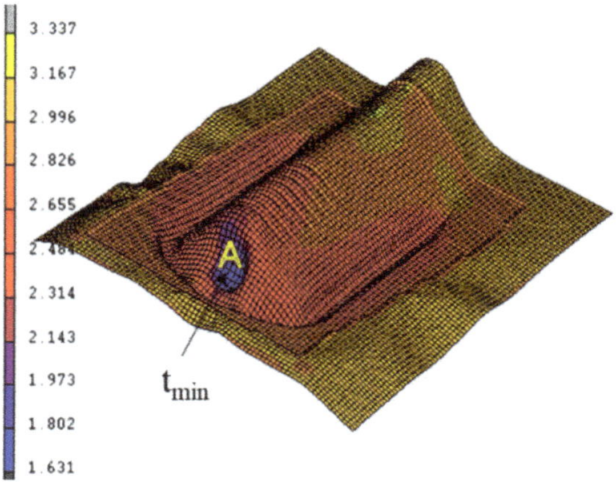

Figure 6. Thickness distribution of the drawing during the preforming process.

(3) Drawing of the final forming process

A finite element analysis of the drawing was carried out, and the results are shown in Figure 7. The hot drawing depth of the final forming was 380 mm. The minimum thickness was reduced from 1.631 to 1.608 mm due to the stretching of the storage tank. In addition, the material of the storage tank at Region C in the preform process was provided to supply the drawing, and it decreased the thickness reduction. Hence, the thickness at the lowest point of the drawing was 2.106 mm.

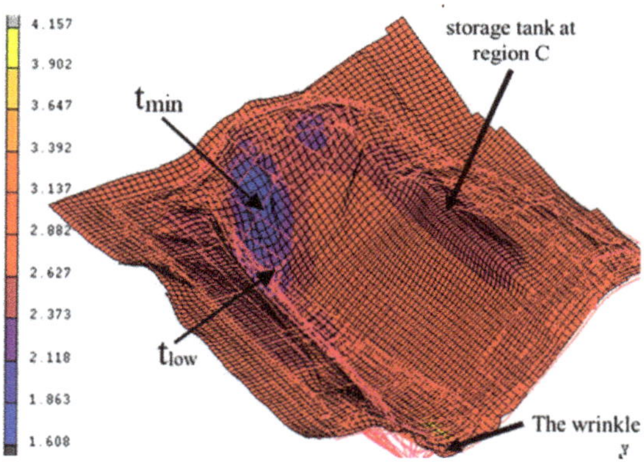

Figure 7. Thickness distribution of the drawing during the final forming process.

(4) Superplastic bulging of the final forming process

The thickness distribution of the SPF process and the contour of the fender are shown in Figure 8. It can be found that the complex surfaces such as the negative angle at the lamp mouth, the wheel mouth plane, and the steps were formed at the last stage. The minimum thickness occurred at the edge of the lamp mouth. The maximum thickness reduction ratio was 83.2%, and it was so severe that the material could not meet this requirement.

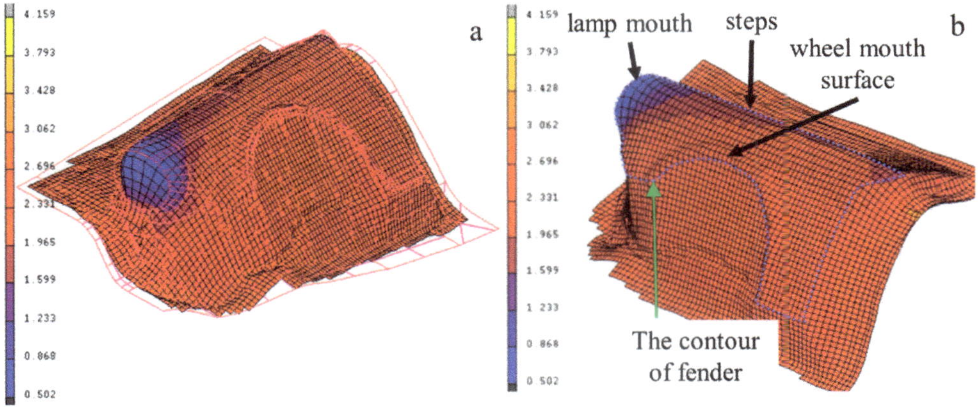

Figure 8. (**a**) Thickness distribution of the direct superplastic bulging; (**b**) contour of the fender.

3.2.2. Optimized Design

In order to optimize the process and make the thickness more uniform, the preformed die was redesigned. The stamping depth of the preformed die was increased, and the preformed material storage tanks were also added to provide sufficient material for the final product. The optimized preformed die is shown in Figure 9. The drawing depth was increased to 300 mm, material storage tanks near Region B were added, and the width of Region A was also increased.

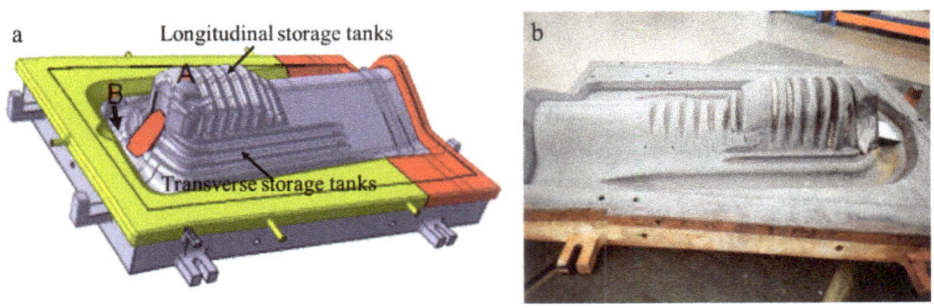

Figure 9. Male die of the preforming process: (**a**) optimized model; (**b**) machined die.

(1) Drawing of the preforming process

The thickness distribution resulting from the preforming process drawing after the optimized design is shown in Figure 10. During the stamping process, it was found that the minimum thickness was 2.736 mm, and it occurred at the lowest point of the drawing. The maximum thickness reduction ratio was 8.8%, and no cracks and other defects occurred. Simultaneously, a wrinkle was observed at Region A. Thus, the drawing rate needed to be controlled strictly to reduce wrinkle defects.

Figure 10. Thickness distribution of the drawing in the preforming process by the optimized design.

(2) Reverse bulging of the preforming process

The thickness distribution after reverse bulging is shown in Figure 11. It can be seen that the thickness diversity increased due to the complex shape of the female die and the improved storage tanks. The minimum thickness was 1.937 mm in Region A, and the minimum thickness was 2.23 mm at the longitudinal storage tanks and 2.26 mm at the transverse storage tanks, respectively. Though the thickness at different regions was nonuniform, it was still greater than 1.9 mm. In addition, there are wrinkles at Regions B and E. The wrinkle at Region B could be flattened in the final forming process, while the deformation at Region E had little effect on the forming process.

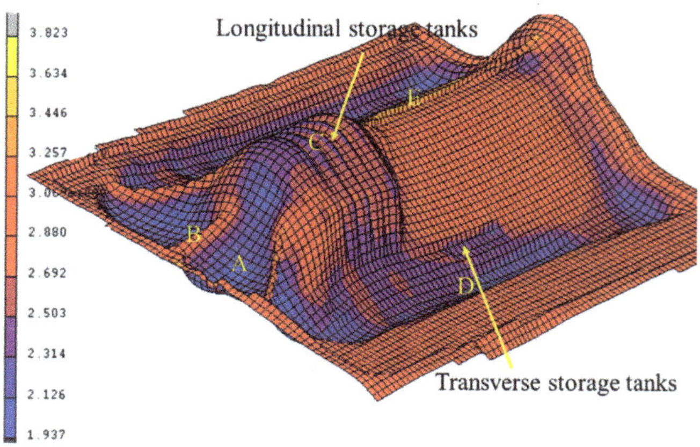

Figure 11. Thickness distribution of reverse bulging during the preforming process.

(3) Drawing of the final forming process

The maximum drawing depth of the final forming was 380 mm, and the minimum thickness at the initial stage of the drawing was 1.94 mm. The thickness distribution of the drawing at the beginning and at the end of the drawing is shown in Figure 12. The storage tanks expanded to supply more material for the next forming process, and the minimum thickness increased slightly to 1.98 mm. It can be seen in Figure 12b that the preformed storage tanks still retained a specific shape.

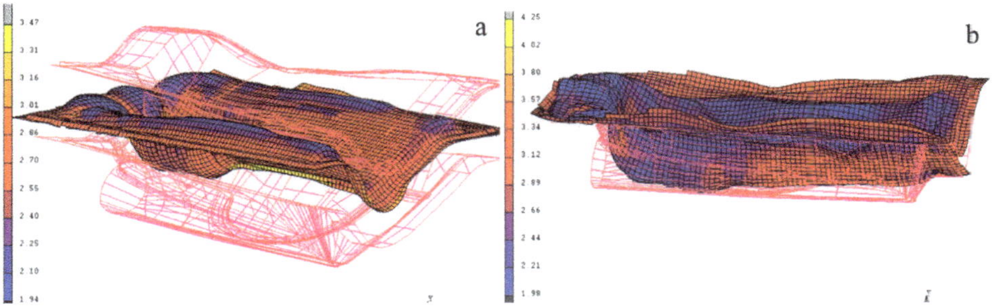

Figure 12. Thickness distribution of the drawing during the final forming process: (**a**) beginning; (**b**) end.

(4) SPF of the final forming process

SPF of the final forming process is shown in Figure 13. The wrinkled area at the end of the preformed structure had been flattened by the air bulging. When the fender was fully formed, the minimum thickness occurred at the lamp orifice and the thickness reduction ratio was 63%. Compared to the thickness distribution of the initial design, the minimum thickness increased by 0.6 mm, indicating that the optimization of the preforming male die was effective. It was also observed that small fillet thickness was more than 2.25 mm, which could guarantee the uniformity of the thickness and part performance.

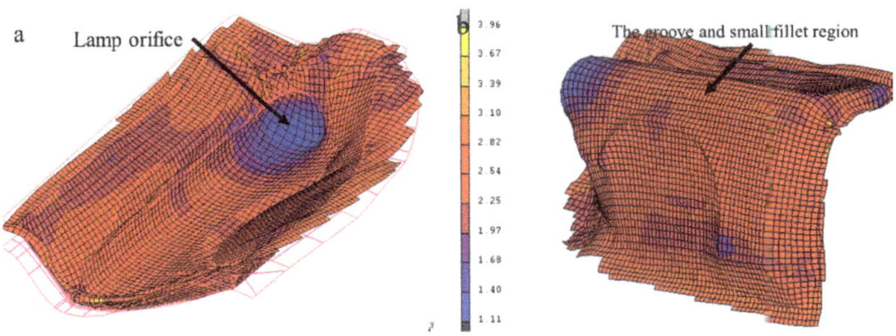

Figure 13. Thickness distribution of superplastic forming (SPF) during the final forming process. (**a**)interior perspective; (**b**) lateral perspective.

It was concluded that the optimized design could increase the minimum thickness of the final structure, and the optimized design had more feasibility than the original design in regards to fabricating the fender.

3.3. Rapid SPF of the Front Fender

Based on the FEM results of the optimized design, the dies required to fabricate the fender were machined consistently with the dies of the optimized design. The front fender was then trial produced on the SPF machine, with the forming temperature set to be 480 °C.

3.3.1. Thickness Distribution

(1) Preforming process

Figure 14 illustrates the measured points and the thickness distribution of the preforming structure. The structure after reverse superplastic bulging is shown in Figure 14a. Although wrinkles were observed, no crack appeared, indicating that superfluous material was supplied to some regions during the forming process. The thicknesses of three different areas were measured and are shown in Figure 14b–d. The results showed that the minimum thickness of the end was 2.34 mm, and the "W" shape of the thickness distribution was presented. Furthermore, the thickness measurement of the storage tanks showed that the thickness was more than 3.0 mm at the longitudinal tanks, and the minimum thickness of the transverse tanks was 2.17 mm.

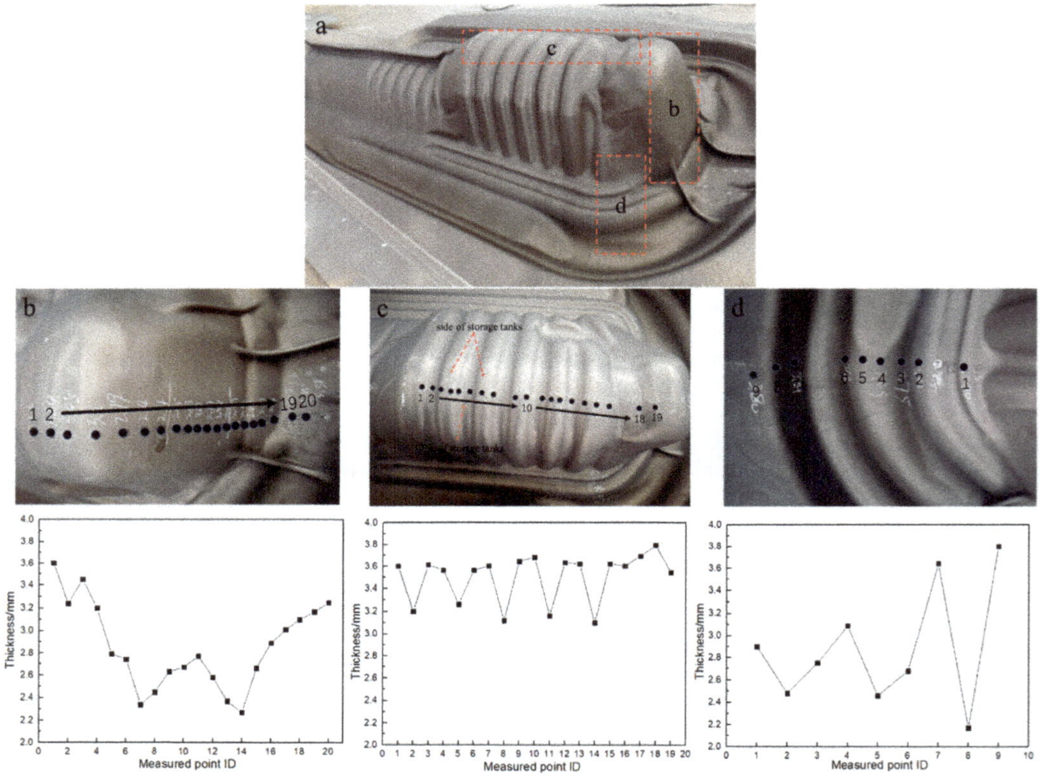

Figure 14. Measurement point ID and thickness: (**a**) preforming structure; (**b**) end; (**c**) longitudinal storage tanks; (**d**) transverse storage tanks.

(2) Final forming process

Figure 15 illustrates the measured points of the structure obtained after the final forming and the thickness of each point. The thickness measurement showed that the thickness was almost more than 2.5 mm, and the thickness gradually reduced as the measured points approached the lamp orifice. The thickness reduction resulted from the large deformation at the lamp orifice, and the thickness distribution meets the demand for service. It can also be seen that the facet of the wheel opening contacted the female mold of the final forming process. Furthermore, the long groove and small fillet were formed completely.

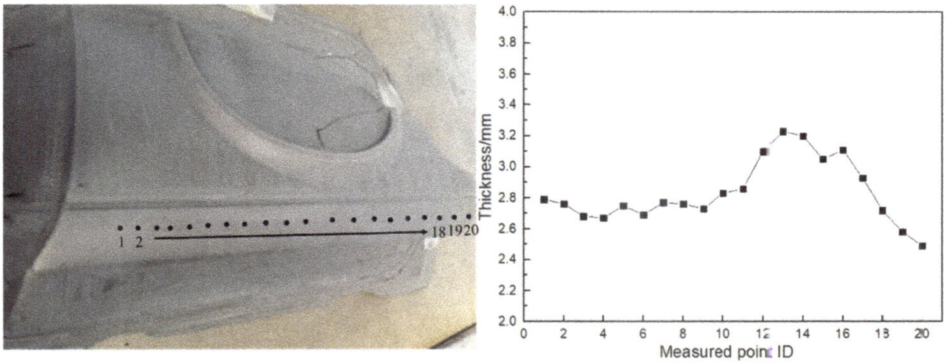

Figure 15. Final structure and thickness measurement points.

The final structure was wire cut and etched, as shown in Figure 16. The front automobile fender was successfully trial produced.

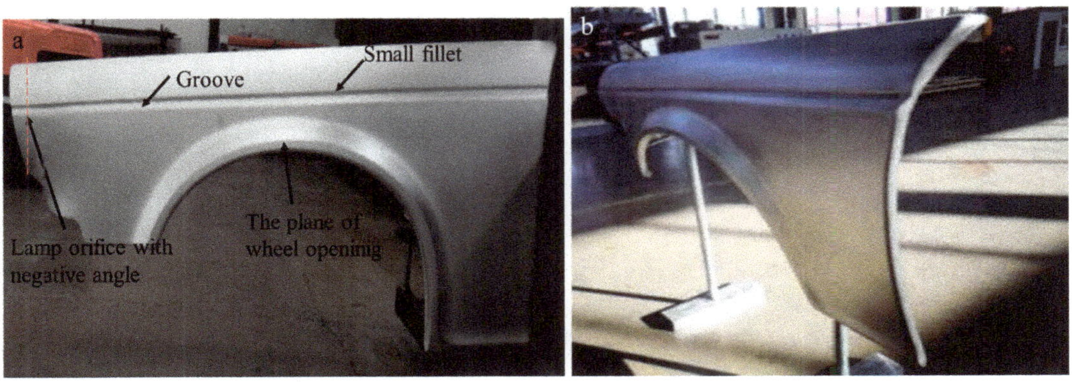

Figure 16. Front automobile fender: (**a**) elevation view; (**b**) side view.

3.3.2. Macro/Micro Performance of the Postmaterial

The elongation, tensile strength, and yield stress of the postforming material at different regions are shown in Figure 17. The result showed that the tensile strength of the original material was 294 MPa, the yield stress was 143 MPa, and the elongation was about 31%. The minimum tensile strength of the post-formed material was 277 MPa, and the minimum yield stress was 115 MPa at the transverse tank, while the minimum elongation was 24.9% at the lamp orifice. The measured tensile strength yield stress, and elongation of the postformed material ranged between 270 and 284 MPa, 115 and 121 MPa, and 24.9 and 26.6%, respectively. The mechanical performance decreased after the rapid SPF process due to the grain growth resulted from the long exposure to elevated temperature and deformation.

The microstructures of the material at two stages are shown in Figure 18. The initial grain size was about 20 μm in Figure 1. Hence, based on the comparison between Figures 1 and 18, it can be concluded that the grain size increased by 5 μm after the preforming process and then further increased by 10 μm after the final forming process. There is grain size growth resulting from the long time exposure at elevated temperature in the preforming process and the final forming process.

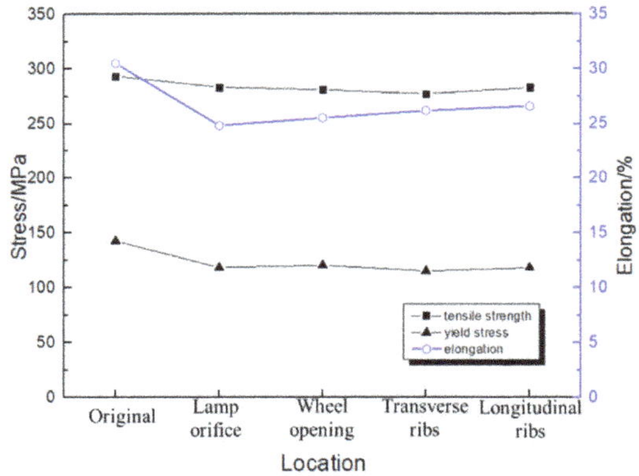

Figure 17. Mechanical properties of the material after the forming process.

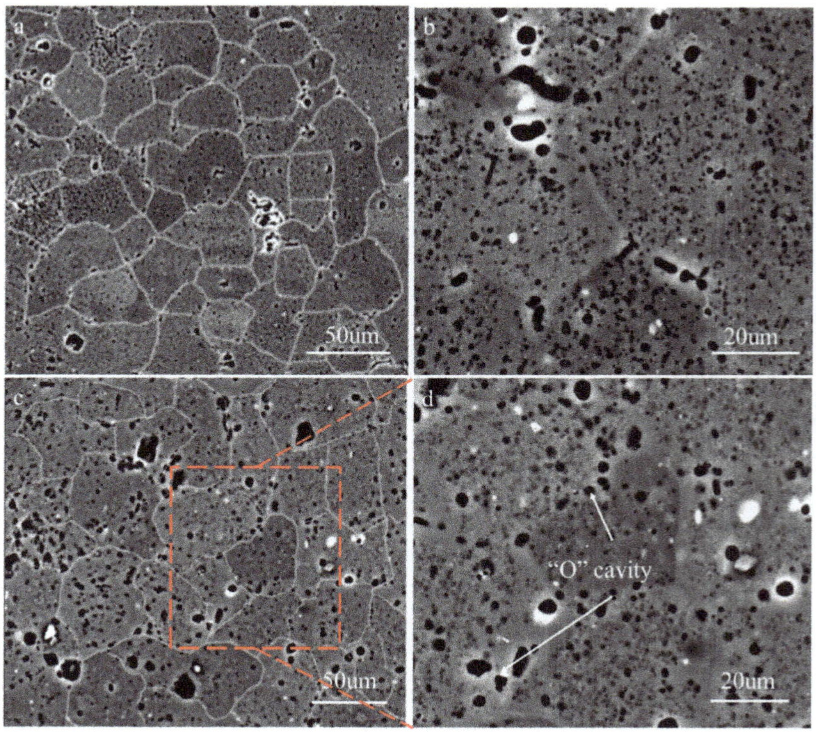

Figure 18. Microstructure of (**a**,**b**) the preforming structure; (**c**,**d**) final structure.

The cavity size grew with the increase of the deformation and holding time. It has been reported that most cavities nucleated at the interfaces between the particle and the matrix at both the grain boundary and the grain interior [34]. With the increase of deformation and insulation time, the size of the cavity increased gradually. As seen in Figure 18c,d, some cavities appeared at the grain boundary; however, these cavities were not interconnected. Therefore, the mechanical properties of the post-material decreased slightly.

4. Conclusions

The mechanical properties of AA5083 were investigated at elevated temperatures, and the FEM established the structural design of the dies. Furthermore, the front fender was trial produced by a rapid SPF process. The following conclusions are drawn:

(1) Industrial AA5083 with the grain size of 20 μm exhibited superplasticity at 480 °C/0.001 s^{-1}, and the maximum elongation was about 242% with the strain rate sensitivity value of 0.35.

(2) The optimized design of the forming dies and the forming process could fabricate the fender with a complex shape.

(3) The front fender was successfully trial produced by the rapid SPF process, and the minimum thickness was 2.49 mm at the edge of the lamp orifice.

(4) The grain size increased by 15 μm after the forming process. The tensile strength reduced by 5.78%, and the elongation reduced by 18.5%. The inharmonious grain boundaries sliding resulted in the appearance of cavities at the grain boundaries.

Author Contributions: Z.D., data collection, writing; G.W., study desigh; H.W., data analysis, figures. All authors have read and agreed to the published version of the manuscript.

Funding: This reseach was funded by the Henan Provincial Department of Science and Technology Research Project (Grant No. 182102210468), and National Natural Science Foundation of China (Grant No. 51875122). And the APC was funded by the Henan Provincial Department of Science and Technology Research Project (Grant No. 182102210468).

Institutional Review Board Statement: Not applicable.

Informed Consent Statement: Not applicable.

Data Availability Statement: Data sharing not applicable.

Acknowledgments: This work was supported by the Henan Provincial Department of Science and Technology Research Project (Grant No. 182102210468), and National Natural Science Foundation of China (Grant No. 51875122).

Conflicts of Interest: The authors declare no conflict of interest.

References

1. Ma, M.Y.; Wang, B.; Liu, H.Q.; Yi, D.Q.; Shen, F.H.; Zhai, T.G. Investigation of fatigue crack propagation behavior of 5083 aluminum alloy under various stress ratios: Role of grain boundary and schmid factor. *Mater. Sci. Eng. A* **2020**, *773*, 138871. [CrossRef]
2. Holroyd, N.J.H.; Burnett, T.L.; Seifi, M.; Lewandowski, J.J. Improved understanding of environment-induced cracking (EIC) of sensitized 5XXX series aluminum alloys. *Mater. Sci. Eng. A* **2017**, *682*, 613–621. [CrossRef]
3. Magee, A.C.; Ladani, L. Representation of a microstructure with bimodal grain size distribution through crystal plasticity and cohesive interface modeling. *Mech. Mater.* **2015**, *82*, 1–12. [CrossRef]
4. Magee, A.; Ladani, L.; Topping, T.D.; Lavernia, E.J. Effects of tensile test parameters on the mechanical properties of a bimodal Al-Mg alloy. *Acta Mater.* **2012**, *60*, 5838–5849. [CrossRef]
5. Jin, H. Optimization of aluminum alloy AA5083 for superplastic and quick plastic forming. *Metall. Mater. Trans. A* **2019**, *50*, 3868–3890. [CrossRef]
6. Giuliano, G.; Samani, F. Comparison between superplastic and non-superplastic grade AA 5083. *J. Test. Eval.* **2016**, *44*, 2114–2119. [CrossRef]
7. Krajewski, P.E.; Schroth, J.G. Overview of quick plastic forming technology. *Mater. Sci. Forum* **2007**, *551*, 3–12. [CrossRef]
8. Liu, J.; Tan, M.J.; Lim, C.V.S.; Chua, B.W. Process optimization and microstructural development during superplastic-like forming of AA5083. *Int. J. Adv. Manuf. Technol.* **2013**, *69*, 2415–2422. [CrossRef]
9. Liu, J.; Tan, M.J.; Aue-u-lan, Y.; Jarfors, A.E.W.; Fong, K.S.; Castagne, S. Superplastic-like forming of non-superplastic AA5083 combined with mechanical pre-forming. *Int. J. Adv. Manuf. Technol.* **2011**, *52*, 123–129. [CrossRef]
10. Barnes, A.J.; Raman, H.; Lowerson, A.; Edwards, D. Recent application of superformed 5083 aluminum alloy in the aerospace industry. *Mater. Sci. Forum* **2012**, *735*, 361–371. [CrossRef]
11. Hefti, L.D. Co mmercial airplane applications of superplastically formed AA5083 aluminum sheet. *J. Mater. Eng. Perform.* **2007**, *16*, 136–141. [CrossRef]
12. Park, H.S.; Dang, X.P.; Roderburg, A.; Nau, B. Development of plastic front side panels for green cars. *CIRP J. Manuf. Sci. Technol.* **2013**, *6*, 44–52. [CrossRef]

13. Tao, W.; Liu, Z.; Zhu, P.; Zhu, C.; Chen, W. Multi-scale design of three dimensional woven composite automobile fender using modified particle swarm optimization algorithm. *Compos. Struct.* **2017**, *181*, 73–83. [CrossRef]
14. Hu, W.; Yao, L.G.; Hua, Z.Z. Optimization of sheet metal forming processes by adaptive response surface based on intelligent sampling method. *J. Mater. Process. Technol.* **2008**, *197*, 77–88. [CrossRef]
15. Fan, Z.J.; Gui, L.J.; Su, R.Y. Research and development of automotive lightweight technology. *J. Automo. Saf. Energy* **2014**, *5*, 1–16.
16. Park, H.S.; Nguyen, T.T. Optimization of injection molding process for car fender in consideration of energy efficiency and product quality. *J. Comput. Des. Eng.* **2014**, *1*, 256–265. [CrossRef]
17. Roth, R.; Clark, J.; Kelkar, A. Automobile bodies: Can aluminum be an economical alternative to steel? *JOM* **2001**, *53*, 28–32. [CrossRef]
18. Zhang, K.F.; Jiang, S.S. Superplastic forming. In *Comprehensive Materials Processing*, 1st ed.; Hashmi, S., Batalha, G.F., van Tyne, C.J., Yilbas, B., McGeough, J., Eds.; Elsevier: Amsterdam, The Netherlands, 2014; Volume 5, pp. 371–392.
19. Jafar, R.A.; Jarrar, F.S.; Al-Huniti, N.S. Two-stage approach for improving the thickness distribution in superplastic forming. *J. Mater. Sci. Res.* **2015**, *4*, 12–27. [CrossRef]
20. Jarrar, F.S.; Liewald, M.; Schmid, P.; Fortanier, A. Superplastic forming of triangular channels with sharp radii. *J. Mater. Eng. Perform.* **2014**, *23*, 1313–1320. [CrossRef]
21. Zhang, K.F.; Wang, G.F.; Wu, D.Z.; Wang, Z.R. Research on the controlling of the thickness distribution in superplastic forming. *J. Mater. Process. Technol.* **2004**, *151*, 54–57. [CrossRef]
22. Xing, H.L.; Zhang, K.F.; Wang, Z.R. A preform design method for sheet superplastic bulging with finite element modeling. *J. Mater. Process. Technol.* **2004**, *151*, 284–288. [CrossRef]
23. Jarrar, F.S.; Jr, L.G.H.; Khraisheh, M.K.; Bower, A.F. New approach to gas pressure profile prediction for high temperature AA5083 sheet forming. *J. Mater. Process. Technol.* **2010**, *210*, 825–834. [CrossRef]
24. Lan, H.C.; Fuh, Y.K.; Lee, S.; Chu, C.L.; Chang, T.C. Two-stage superplastic forming of a V-shaped aluminum sheet into a trough with deep and irregular contour. *J. Mater. Eng. Perform.* **2013**, *22*, 2241–2246. [CrossRef]
25. Xu, C.; Langdon, T.G.; Horita, Z.; Furukawa, M. Using equal-channel angular pressing for the production of superplastic aluminum and magnesium alloys. *J. Mater. Eng. Perform.* **2004**, *13*, 683–690. [CrossRef]
26. Hojjati, M.H.; Zoorabadi, M.; Hosseinipour, S.J. Optimization of superplastic hydroforming process of Aluminium alloy 5083. *J. Mater. Process. Technol.* **2008**, *205*, 482–488. [CrossRef]
27. Ko, Y.G.; Shin, D.H.; Park, K.T.; Lee, C.S. Superplastic deformation behavior of ultra-fine-grained 5083 Al alloy using load-relaxation tests. *Mater. Sci. Eng. A* **2007**, *449*, 756–760. [CrossRef]
28. Balasubramanian, M.; Stalin, B.; Ramanathan, K.; Ravichandran, M. Hot tensile test for determining the material constant on superplastic 5083Al alloy sheet. *Mater. Today Proc.* **2020**, *21*, 324–328. [CrossRef]
29. Taleff, E.M.; Wadsworth, L.J. Enhanced ductility in coarse-grained Al-Mg alloys. *Metall. Mater. Trans. A* **1996**, *27*, 343–352. [CrossRef]
30. Chezan, A.R.; de Hosson, J.T.M. Superplastic behavior of coarse-grained aluminum alloys. *Mater. Sci. Eng. A* **2005**, *410*, 120–123. [CrossRef]
31. Qiao, J.; Taleff, E.M. Superplasticity-like creep behavior of coarse grained ternary Al alloys. *Trans. Nonferrous Met. Soc. China* **2010**, *20*, 564–571. [CrossRef]
32. Sherby, O.D.; Taleff, E.M. Influence of grain size, solute atoms and second-phase particles on creep behavior of polycrystalline solids. *Mater. Sci. Eng. A* **2002**, *322*, 89–99. [CrossRef]
33. Bae, D.H.; Ghosh, A.K. Grain size and temperature dependence of superplastic deformation in an Al–Mg alloy under isostructural condition. *Acta. Mater.* **2000**, *48*, 1207–1224. [CrossRef]
34. Iwasaki, H.; Mori, T.; Hosokawa, H.; Tagata, T.; Mabuchi, M.; Higashi, K. Cavitation behavior of coarse-grained Al-4.5Mg alloy exhibiting superplastic-like elongation. *Mrs Proc.* **1999**, *601*, 67–72. [CrossRef]

metals

Article

A Detailed Machinability Assessment of DC53 Steel for Die and Mold Industry through Wire Electric Discharge Machining

Sarmad Ali Khan [1], Mudassar Rehman [2], Muhammad Umar Farooq [1,3,*], Muhammad Asad Ali [1,*], Rakhshanda Naveed [1], Catalin I. Pruncu [4,5,*] and Waheed Ahmad [1]

1 Department of Industrial and Manufacturing Engineering, University of Engineering and Technology Lahore, Lahore 54890, Pakistan; drsarmad@uet.edu.pk (S.A.K.); rakhshanda@uet.edu.pk (R.N.); 2012msime15@student.uet.edu.pk (W.A.)
2 Department of Industrial Engineering, School of Mechanical Engineering, Northwestern Polytechnical University, Xi'an 710072, China; mudassar@mail.nwpu.edu.cn
3 Department of Industrial and Systems Engineering, Korea Advanced Institute of Science and Technology, Daejeon 34141, Korea
4 Department of Mechanical Engineering, Imperial College London, Exhibition Rd., London SW7 2AZ, UK
5 Design, Manufacturing & Engineering Management, University of Strathclyde, Glasgow G1 1XJ, UK
* Correspondence: umarmuf0@gmail.com (M.U.F.); asad.ali@uet.edu.pk (M.A.A.); c.pruncu@imperial.ac.uk or catalin.pruncu@strath.ac.uk (C.I.P.)

Abstract: Recently, DC53 die steel was introduced to the die and mold industry because of its excellent characteristics i.e., very good machinability and better engineering properties. DC53 demonstrates a strong capability to retain a near-net shape profile of the die, which is a very challenging process with materials. To produce complex and accurate die features, the use of the wire electric discharge machining (WEDM) process takes the lead in the manufacturing industry. However, the challenge is to understand the physical science of the process to improve surface features and service properties. In this study, a detailed yet systematic evaluation of process parameters investigation is made on the influence of a wire feed, pulse on duration, open voltage, and servo voltage on the productivity (material removal rate) and material quality (surface roughness, recast layer thickness, kerf width) against the requirements of mechanical-tooling industry. Based on parametric exploration, wire feed was found the most influential parameter on kerf width: KW (45.64%), pulse on time on surface roughness: SR (84.83%), open voltage on material removal rate: MRR (49.07%) and recast layer thickness: RLT (52.06%). Also, the optimized process parameters resulted in 1.710 μm SR, 10.367 mm^3/min MRR, 0.327 mm KW, and 10.443 μm RLT. Moreover, the evolution of surface features and process complexities are thoroughly discussed based on the involved physical science. The recast layer, often considered as a process limitation, was explored with the aim of minimizing the layers' depth, as well as the recast layer and heat-affected zone. The research provides regression models based on thorough investigation to support machinists for achieving required features.

Keywords: WEDM; DC53 steel; recast layer thickness; material removal rate; Kerf

Citation: Khan, S.A.; Rehman, M.; Farooq, M.U.; Ali, M.A.; Naveed, R.; Pruncu, C.I.; Ahmad, W. A Detailed Machinability Assessment of DC53 Steel for Die and Mold Industry through Wire Electric Discharge Machining. *Metals* **2021**, *11*, 816. https://doi.org/10.3390/met11050816

Academic Editor: George A. Pantazopoulos

Received: 7 April 2021
Accepted: 14 May 2021
Published: 17 May 2021

1. Introduction

Die-making industries employ a wide variety of engineering materials, such as variants of steel, for the fabrication of tools, dies, and molds based on their considerable application performance and increasing demand. Conventional steels involving D2 and D3 steel have been engaged for over a decade in the manufacturing of dies and molds. Among these steel families, DC53 steel is considered as an advancement over D2 and D3 steel in terms of its high hardness (64 HRC), better toughness, improved fatigue strength and wear resistance. Various engineering applications of DC53 steel involve the manufacturing of different rolling, forging, injection molding, extrusion, and stamping dies along with molds, cutting tools, along with a wide variety of high speed and wear resistant parts [1].

The processing of such materials also becomes challenging as a result of their improved properties. Therefore, advanced machining setups are employed to fulfil the requirements. These involve electrochemical machining, electric discharge machining, plasma arc cutting, ultrasonic machining and abrasive water jet machining. These setups are considered an improvement in machining hard materials and generating a variety of machining profiles. Processing of hard materials such as titanium alloys and various other hardened steel grades—for instance, DC53 steel—may be difficult in a conventional machining scenario because of their excessive tool wear during machining. Therefore, non-conventional machining processes may be employed to serve the same purpose with little/no tool wear. Wire electric discharge machining (WEDM) is one of the non-traditional machining processes in which spark-based discharge energy is produced among two electrodes (wire electrode and workpiece) and responsible for material erosion during machining with no mechanical stresses generated unlike traditional machining processes [2].

WEDM is employed to process D2, D3 and different tool and die steel materials. Several investigations are made on the performance attributes like kerf width, material removal rate, and surface morphology, including roughness and microstructural examination involving recast layer [3–5]. The aspects of DC53 steel machining have not been thoroughly explored, therefore, are mainly investigated herein. Thiagarajan et al. [6] reported the combined influence of machining variables like pulse on duration (P_{on}), pulse off time (P_{off}), wire feed (WF) and wire tension (WT) at the MRR and SR in electric discharge machining of AISI D3 steel. A detailed study as multi-objective optimization was made by Manajaiah et al. [7] for AISI D2 steel in WEDM and inferred that the MRR, SR and recast layer thickness (RLT) are found in direct proportion with P_{on}. Ikram et al. [8] carried an insight investigation in WEDM of AISI D2 steel. The study reports that P_{on} is influenced more on SR (52.14%), MRR (49.62%), and KW (47.06%). In the study, results proved that P_{on} is two times more significant for SR compared to open voltage (OV), the combined influence of (WT, P_{on} and OV) found significant for KW, and the combination of servo voltage (SV), OV and P_{on} found significant for MRR respectively.

The experimental investigation performed by Vaghela et al. [9] for AISI D3 steel in WEDM inferred that the set of process variables peak current, P_{on} and P_{off} to maximize MRR, minimize KW and SR, respectively. Zhang et al. [10] performed mathematical modelling using back-propagation neural-network integrated with genetic algorithm and response surface methodology (RSM) to obtain optimal values of MRR and SR in WEDM of SKD11(D2 steel). Rupajati et al. [11] determined the machining performance regarding SR and recast layer thickness (RLT) for AISI H13 tool steel in electric discharge machining using Taguchi methodology along with the fuzzy-logic technique.

Significant improvements were obtained in RLT (22.04% reduction) and SR (13.15% reduction) from design of experiments results. It was also inferred that the significant variables for SR and RLT included P_{on} (40.08%) and O_V (30.38%), respectively. Hasriadi et al. [12] machined two different tool steel grades in WEDM for the assessment of surface morphology in terms of SR, RLT, and density of microcracks and resulted in the previously stated response variables forming a direct linear relation with both P_{on} and the arc on duration. Castillo et al. [13] explored the parametric effects on AISI 304 stainless steel.

Saini et al. [14] achieved multi-objective optimization in WEDM of 16MnCr5 alloy steel. Obwald et al. [15] conducted an analysis of different pulses types generated in high-speed WEDM. A detailed study was made by Solomon et al. [16] to comprehend the effect of variation in process variables for obtaining optimized parameters in WEDM of the steel family. Multi-objective optimization performed by Shivade and Shinde [17] in WEDM of AISI D3 steel resulted that gap current, and MRR varied directly with P_{on} and peak current, and inversely with the machining time (tm). Azam et al. [18] optimized cutting speed (C.S) by the variation in P_{on}, P_{off}, and pulse frequency in WEDM of HSLA steel. The machined surface characterization was conducted by Dhobe et al. [19] in the electric discharge of AISI D2 steel. It was reported that RLT is found directly proportional with both SR and fatigue life. Based upon scanning electron microscopy (SEM) and X-ray diffraction results, it was reported that triple tempering (a heat treatment process) is better than single tempering as it improves the machined surface quality by reducing RLT and SR. Khanna and Singh [20] performed a comparative study in WEDM of cryogenic-treated and normal D3 tool steel for the cutting rate, MRR and SR. It was reported that the cryogenic treatment significantly influenced both the MRR (decreased 5.6%) and SR (surface finish increased 10.6%) compared to untreated D3 steel. Sharma et al. [21] endorsed similar findings on AISI D2 steel.

Ramaswamy et al. [22] predicted an optimal setting of machining parameters while performing optimization using a desirability function approach. Payla et al. [23] inferred that P_{on} directly influenced the MRR and power consumption during electric discharge machining of EN-31 die steel. It was also concluded that the MRR is reduced with an increase in SV because the energy density decreased as the spark gap expanded. Chen et al. [24] found that discharge capacitance has a direct and significant influence on MRR and the machining gap in microreciprocated WEDM of SKD11 (D2 steel). Singh et al. [25] supported the phenomenon for EN-31 tool steel. Abdulkareem et al. [26] worked to enhance the SR because of more adhesion of the debris particles and surface irregularities on the processed surface. Mouralova et al. [27] concluded that the frequency of discharge occurrence of the recast layer is highly influenced by cutting speed. Negrete [28] made a similar finding on AISI 01 tool steel. Kanlayasiri and Boonmung [1] investigated the influence of P_{on}, P_{off}, peak current and WT on SR in WEDM of DC53 steel. Nawaz et al. [29] used molybdenum to machine steel with the aim of achieving higher material removal rate and lesser surface roughness. The study quantified the influence of process in enhancing the pulse discharge energy because of the deeper craters produced on the machined surface. Similarly, Rehman et al. [30] evaluated gamma wire and resulted the significant influence of P_{on} for C.S (80.21%), KW (48.25%), and MRR (45.21%) during WEDM of DC53 steel.

The literature review reveals that a significant amount of work has already been done on various tools and die steels, especially D2, and D3 steel, to improve machining accuracy and productivity in wire electric discharge machining. However, we have noted a lack of systematic evaluation for the main parameters affecting the output responses in the WEDM of DC53 steel (having superior mechanical properties and machining performance over D2 and D3 steel), which was the key objective of this research. Moreover, the supremacy of zinc-coated brass wire (as recommended by literature) is not thoroughly studied, which may provide superior results as compared to previous established methodologies. To resolve this problem and further improve the state of the art, the following issue are discussed in this research:

- Investigating the influence of machining parameters on recast layer thickness (process limitation), surface roughness and kerf width (quality), and material removal rate (productivity).
- Optimizing input parameters such as wire feed (WF), pulse on duration (P_{on}), open voltage (OV), and servo voltage (SV) controlling process productivity and work quality of DC53 die steel processes with zinc-coated brass wire
- Exploring the statistical significance of machining process science on physical changes on the material surface.
- Examining microstructural surface evolution using optical microscopy and scanning electron microscopy.

The results gathered in this work are of paramount importance for the forging industry (i.e., metal forming, flow forming, spinning) in order to create tools with higher integrity and superior surface characteristics.

2. Materials and Methods

DC53 steel has numerous applications in die- and mold-making industries and is the workpiece material for investigation. DC53 is a high chromium steel with high strength, good toughness, wear-resistance and excellent machinability over that of D2 and D3 steel. The composition details and other characteristics of DC53 are shown in Table 1. A DC53 steel flat bar was used with density of 7.85×10^3 kg/m^3. Experimental trials were performed in deionized water dielectric medium using zinc-coated brass wire of 0.25-mm diameter as an electrode on a CNC wire EDM setup (G43S CHMER, KNB Technologies Sdn Bhd, Malaysia). The wire electrode is recommended for fast-roughing and fine-trimming with low SR features providing less frequent breakage. Wire breakage is limiting phenomena that minimizes the productivity and upsurges the tooling cost. A detailed demonstration to illustrate the machining setup and 3D CAD view of the machined specimen accompanied by its dimensions are presented in Figure 1.

The effects of four machining variables, namely wire feed (WF), pulse on duration (P_{on}), open voltage (OV), and servo voltage (SV) have been carefully analyzed on the kerf width (KW), material removal rate (MRR), surface roughness (SR), and recast layer thickness (RLT). The choice of the said input parameters was made based on a comprehensive literature survey [1,15–17,19–25]. Additionally, preliminary experimentation (six trial experiments) was also performed before the mature experimentation to express the process parameter levels. In addition, the dielectric resistivity was carefully monitored and maintained at same level throughout the entire experimentation phase. Variables, in addition to the machining variables, were considered as the fixed factors. The particulars relating to the input variables and their ranges along with constant factors are reported in Table 2.

The use of full factorial design as an experimental process involves the consumption of time and cost [32] compared to the Taguchi experimental design. The Taguchi methodology is considered useful because of its precise and robust analysis by a reduction in experimental runs [30]. Consequently, Taguchi's fractional factorial design orthogonal array (OA) has been used for performing the experimentation. In total, 18 experiments under mature experimentation were successfully completed using L_{18} OA and were followed by the measurement of the established output responses.

Table 1. Key features of DC53 steel.

DC53 Chemical Specifications		DC53 Characteristics [29–31]		
Element	**Wt (%)**	**Name**	**Unit**	**Value**
Carbon (C)	1.10	Modulus of elasticity	GPa	150
Chromium (Cr)	8.50	Modulus of rigidity	GPa	58.5
Molybdenum (Mo)	2.00	Rockwell hardness	HRC	64
Silicon (Si)	0.90	Poisson's ratio	-	0.28
Vanadium (V)	0.30	Thermal conductivity (at room temperature)	W/m-K	23.86
Manganese (Mn)	0.35	Coefficient of thermal expansion	1/°C	13×10^{-6}
Phosphorous (P)	0.03	Density	kg/m^3	7.85×10^3
Sulphur (S)	0.03	-	-	-
Iron (Fe)	BAL	-	-	-

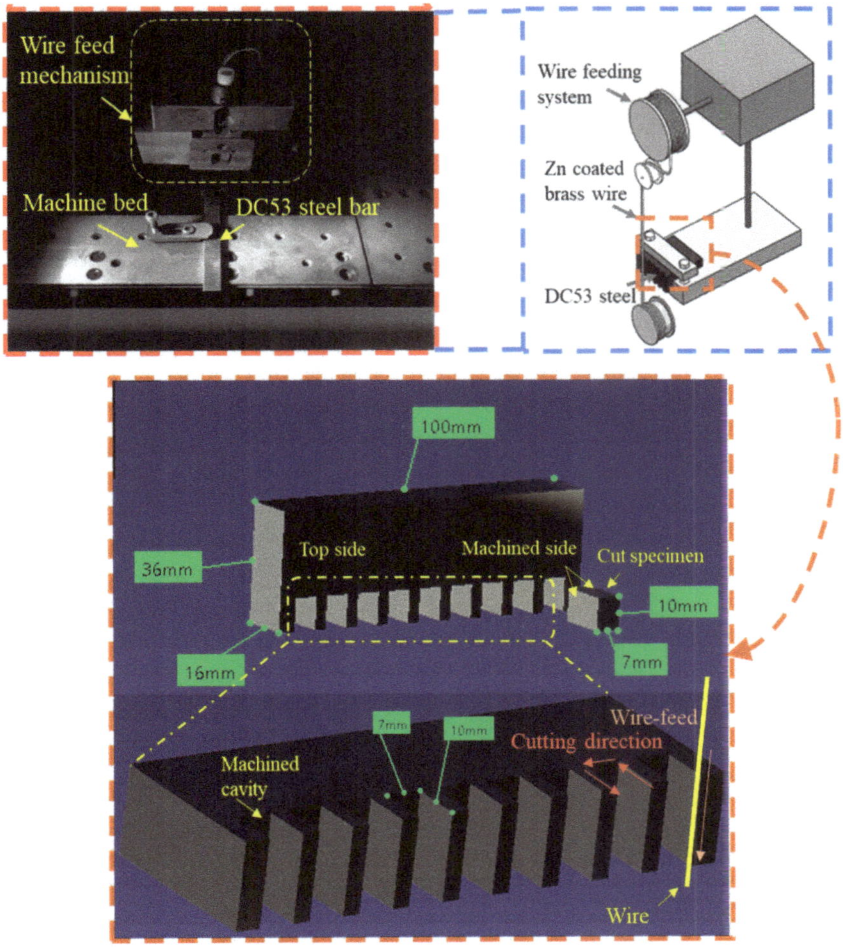

Figure 1. Illustration of experimental setup along with the workpiece.

Table 2. Input machining variables levels along with response measures.

Process Variables	Symbol	Unit	Level			Output Responses	Symbol	Unit	Constant Parameters	Unit	Value
			1	2	3						
Wire feed	WF	mm/s	5	8		Kerf width	KW	µm	Pulse off-duration	µs	25
Pulse on-duration	P_{on}	µs	4	5	6	Surface roughness	SR	µm	Resistivity of dielectric	kg-Ω-cm	50–70
Open voltage	OV	volt	80	90	100	Material removal rate	MRR	mm^3/min	Wire tension	gms-f	1390
Servo voltage	SV	volt	40	50	60	Recast layer thickness	RLT	µm	Arc on time	µs	2

Kerf-width is the measurement of extra material removed from the processed surface of the workpiece during machining. KW was examined by measuring the cutting slot distance using a calibrated video probe of a coordinate-measuring-machine (CMM) (CE-450DV, Chien Wei Precise Technology Co., Ltd., Fengshan District, Taiwan), as illustrated in Figure 2. MRR calculation methodology is reported in Equation (1):

$$MRR = \left[\frac{W_b - W_a}{t_m * \rho_w}\right] \quad (1)$$

where W_b and W_a denoted the weights of the workpiece before and after processing; t_m symbolized the machining time for each machining experiment, and ρ_w showed the workpiece density, respectively. The surface texture meter (Surtronic S128, Taylor Hobson Leicester, United Kingdom) measured the surface roughness SR of the machined surface. The SR measurements were performed at three different points on the machined work surface by calibrating at an evaluation length of 4 mm and a cut-off length of 0.8 mm. The averages of measurements were used for analytical purposes. A scanning electron microscope (Vega 3-TESCAN, TESCAN, Brno, Czechia) was employed for microstructural examination of machined surface of the workpiece and recast layer thickness was measured by using optical microscope (Olympus MM6C-PS2, Olympus Corporation, Tokyo, Japan).

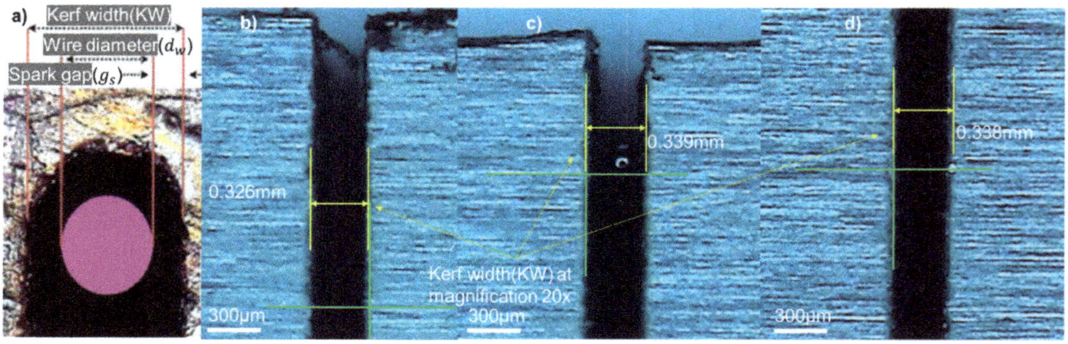

Figure 2. Schematic and physical representation of (**a**) kerf width along with physical evidence of its measurement using CMM, (**b**) For Exp. 1 (WF 5 mm/s, P_{on} 4 µs, OV 80 volt, SV 40 volt), (**c**) For Exp. 10 (WF 8 mm/s, P_{on} 4 µs, OV 80 volt, SV 60 volt), and (**d**) Exp. 18 (WF 8 mm/s, P_{on} 6 µs, OV 100 volt, SV 50 volt).

To carry out the microscopic examination, samples were prepared by mounting, etching, grinding and polishing. SEM-based morphological analysis was used to explain the process science and physical phenomenon. The statistical analysis was completed by using Minitab 19 software (Minitab, LLC, State College, Pennsylvania, USA). Confirmatory experimentation was performed based on parametric settings for the validation of research results.

3. Results and Discussion

The experimentation on the DC53 steel work part was successfully performed according to Taguchi L_{18} to examine and investigate the behaviour of input process variables for output responses. The results of the examination are more profoundly observed and analysed using statistical analysis techniques, optical microscopy and the scanning electron microscopy. These analyses are conducted to enlighten the machining effects along with process physics involved on the microstructure of machined surface. The design of experiments (DOE) results revealed that minimum kerf width (KW) of 0.318 mm, least (average) surface roughness (SR) of 1.69 µm, higher material removal rate (MRR) of 17.82 mm^3/min, and a least average recast layer thickness (RLT) of 9.63 µm were obtained. The obtained MRR results have shown maximum values as pulse on duration (P_{on}) was increased. In addition to this, KW, SR, and RLT values were also decreased by reducing P_{on} because it influenced directly on the number of sparks produced resulting in adequate melting of the workpiece during machining. The facts explanation pertaining to the analysis are stated in the upcoming segments of discussion in variance and trend analyses.

3.1. Parametric Significance Analysis

Variance analysis is considered a decisive tool to determine each process parameter's influence for a certain output response [8,30,33,34]. The insights of this analysis have been reported in Table 3, containing the p-value and percent contribution (PCR). The p-value describes the significance in such a way that the process parameter with lower (<0.05) value is considered highly influential for a specific output response compared to other control variables [35]. The is because variables with a p-value $\leq 5\%$ in statistical output data, indicating a 95% confidence interval, were regarded as significant with respect to other input parameters [33]. Percentage contribution is another tool to determine the more detailed control of the machining parameters on responses. In this study, PCR was estimated to evaluate and prioritize their significance in accordance with output responses as described by formula reported in Equation (2).

$$\text{Percentage contribution (PCR)} = \left[\frac{\text{Adj SS}}{\text{Total SS}} \times 100\right]\% \quad (2)$$

where Adj SS = Adjusted sum of squares, and Total SS = Total sum of squares are used from analysis of variance to calculate the contribution. The detailed results of variance analysis from Table 3 inferred that WF (45.648%), P_{on} (41.56%), and SV (3.77%) are the significant factors for KW. However, the detailed influential variables for SR include P_{on} (84.83%), WF (8.32%), OV (2.10%), and SV (1.15%). Open voltage, pulse on duration, and wire feed with percentage-contribution of 49.07%, 32.81%, and 10.51% respectively, influenced on material removal. Continually, the significant variables for recast layer thickness involved OV (52.06%) and P_{on} (40%). Thorough analysis of the statistical data reported in Table 3 resulted in the finding that the pulse on duration: P_{on} was the common and most influential input parameter for all defined output responses in this study.

Table 3. Summary of analysis of variance.

Sr No.	Output Responses	p Value Variables				Percentage Contribution (PCR)				R-Sq (%)	R-Sq (Adj) (%)
		WF	P_{on}	OV	SV	WF	P_{on}	OV	SV		
1	Kerf width (KW)	**<0.001**	**<0.001**	0.08	**0.02**	45.64	41.56	1.95	3.77	92.95%	90.78%
2	Surface roughness (SR)	**<0.001**	**<0.001**	**0.016**	0.06	8.32	84.83	2.10	1.15	96.42%	95.31%
3	Material removal rate (MRR)	**0.001**	**<0.001**	**<0.001**	0.41	10.51	32.81	49.07	0.39	92.80%	90.58%
4	Recast layer thickness (RLT)	0.593	**<0.001**	**<0.001**	0.56	0.17	40.00	52.06	0.20	92.45%	90.13%

3.2. Process-Parametric Effect Analysis

The parametric-effect analysis was done to study the behaviour of process variables through graphical trends of output responses and the results based on statistical analysis. The analysis is shown in Figures 3–7.

3.2.1. Effects of the Wire Feed

The main effects result for WF, presented in Figure 3, has inferred that kerf width and surface roughness were decreased by increasing WF. The key reason behind this reduction is the rapid transport of wire, which resulted in less workpiece–wire electrode interaction, and produced a small heat-affected zone (HAZ). Moreover, it resulted in minor thermal damages, therefore, less molten material/debris was redeposited on the workpiece surface (will be revealed in Section 3.4). The machining efficiency of the wire electrode in WEDM was increased by increasing the WF. The effect was experienced irrespective of considering wire consumption because of a larger amount of the material removed in a shorter machining cycle was due to a rapid traverse of wire. Therefore, the material removal rate (MRR) increased with an increasing level of WF. Similar results on the wire feed were reported by Mussada et al. [36] and Negrete and Carmita [28]. It was confirmed that the increase in the WF assisted in the stable spark discharges produced and thus resulted in more melting and a redeposit of work material.

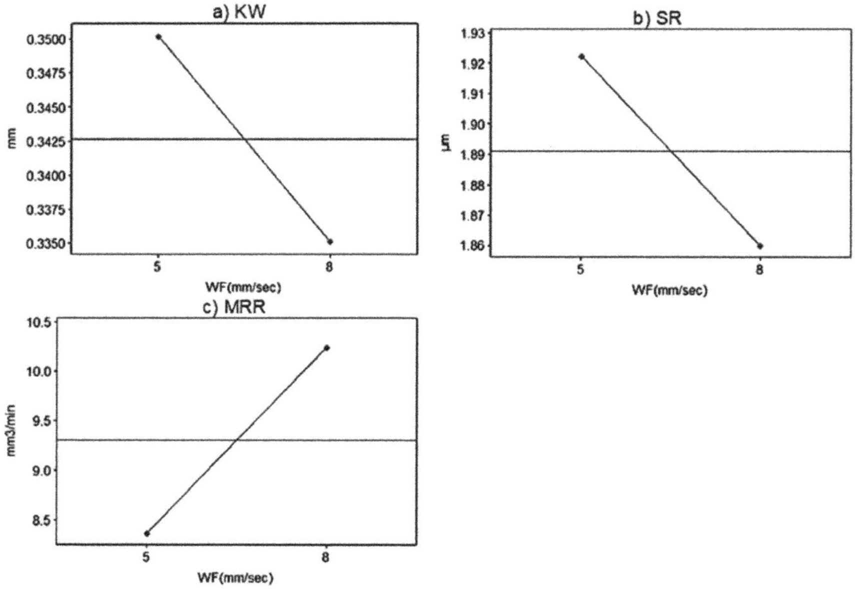

Figure 3. Parametric effects of wire feed (WF) for (**a**) kerf width (KW), (**b**) surface roughness (SR), and (**c**) material removal rate (MRR).

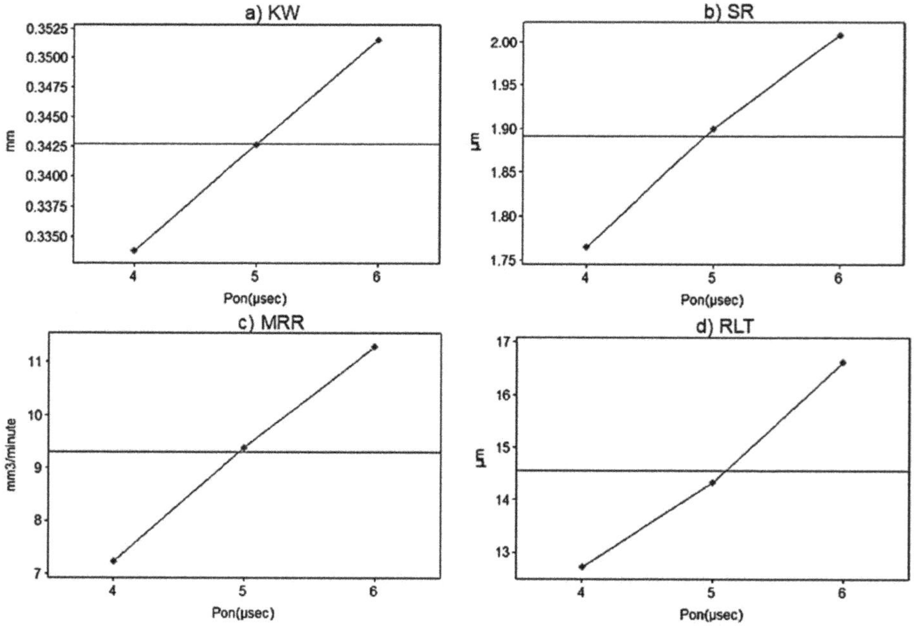

Figure 4. Parametric effects of pulse on duration (P_{on}) for (**a**) kerf width (KW), (**b**) surface roughness (SR), (**c**) material removal rate (MRR), and (**d**) recast layer thickness (RLT).

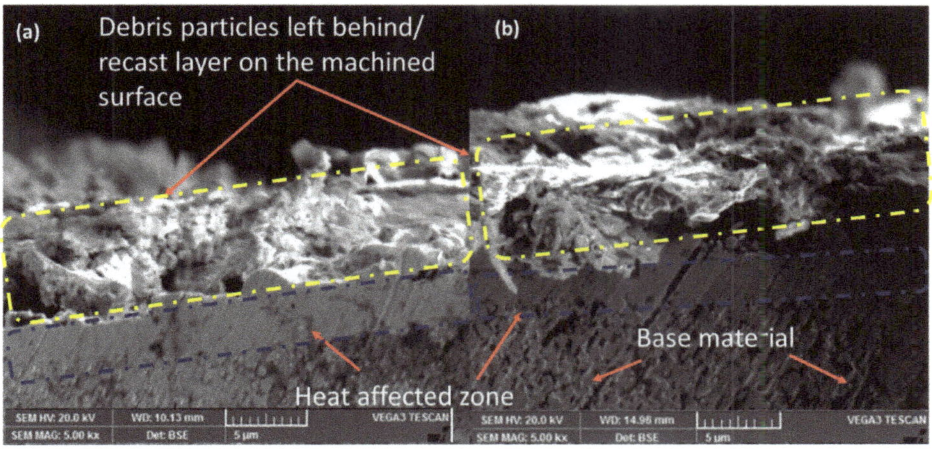

Figure 5. Representation of the recast layer formed on the machined work surface at random experimental settings (**a**) P_{on} 4 µsec (**b**) P_{on} 5 µsec.

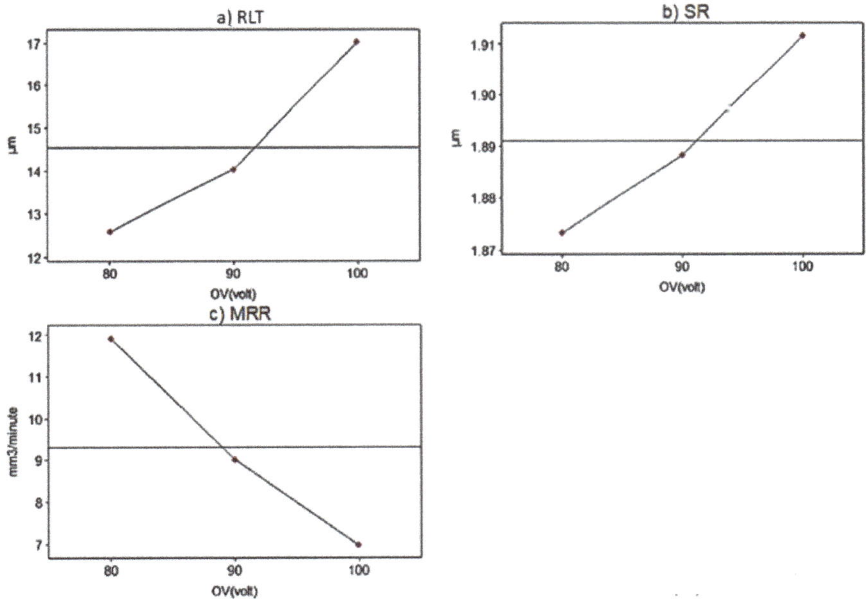

Figure 6. Parametric effects of open voltage (OV) for (**a**) recast layer thickness (RLT), (**b**) surface roughness (SR), (**c**) material removal rate (MRR).

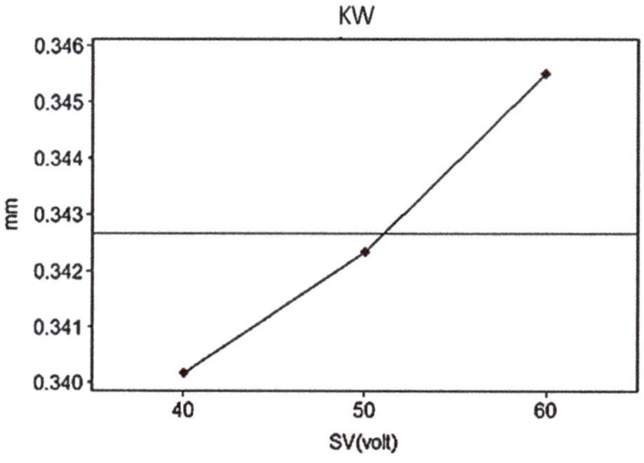

Figure 7. Parametric effects of servo voltage (SV) for kerf width (KW).

Kerf-width is defined as the measurement of excessive material removed during machining and its measurement demonstration is indicated in Figure 2. The analysis of means is shown as main effect plots in Figure 3a. It indicated an inverse relationship of WF with KW. The decreasing trend is observed for KW from 0.350 to 0.335 mm by increasing WF level from 5 to 8 mm/s. The rationale behind this effect lies in the minimal spark gap produced because of enhanced WF. The continual incoming fresh wire assisted in the minimizing the workpiece–wire electrode interaction time, and thus reduced the removal of excessive material from the work surface during machining.

However, the excessive increase in the WF produces irregularities on the machined surface, as reported by Somashekhar et al. [37]. The intensity of the current is enhanced because of consistent incoming fresh wire. This is because of the availability of increasingly present fresh ions and interacting surface. Furthermore, the wire breakage frequency may also be reduced because of adequate increase in WF. The decrease in KW happened at low P_{on} values. Hence, the kerf width is decreased by increasing the wire feed. Based on variance analysis from Table 3, the wire feed was also found to be top significant factor for kerf width (45.64%) compared to other process variables.

Surface roughness is the measurement of surface asperities on the machined surface [38]. The graphical trends of WF for SR from Figure 3b have resulted that the surface roughness was declined from 1.92 to 1.86 μm because of increased levels of wire feed. This is because of less contact duration between the wire electrode and the workpiece. Therefore, a greater amount of debris is flushed away by the dielectric and less discrepancies occurred on the machined surface thus resulted in small SR. Similar findings were identified by Tilekar et al. [39] and it was also inferred that the incoming fresh wire in the machining zone assisted in reducing the heat energy (function of heat transfer and plasma generation) because of instability of discharges resulting from higher wire feed which produced less irregularities on the machined surface. In addition, increased wire feed decreases the spark stability time due to the speedy motion of the wire and increases the tooling cost (similar science is discussed by Farooq et al. [40]). ANOVA observations also indicated that WF was identified as the influential factor for SR (8.20%).

The material removal rate is the measurement of amount of material-removed in a particular machining time [38]. Higher MRR is usually desired to enhance the overall machining productivity of the machine and the process. The results data have indicated that the MRR increased from 8.5 to ~10.3 mm^3/min by increasing the WF values (Figure 3c). The reason is the improved stability of thermal energy produced in the cutting zone with the help of constantly moving electrode wire. This has caused an increase in the sparks density in a shorter duration of machining cycle, which leads to enhance material melting and vaporization and thus resulted in more MRR. The comparison of the MRR trends was made with the existing literature and similar findings were obtained [37]. Wire feed was found to be significant variable for MRR (10.51%) as viewed from the variance analysis in Table 3.

3.2.2. Effects of Pulse-On Duration

The analysis of the parametric effects of pulse-on duration on the selected output responses is indicated in Figure 4. It is generally said that P_{on} plays its key role in improving the machining performance/machinability of WEDM. The statistical data results have elaborated that all the response measures namely KW, SR, MRR and RLT were linearly boosted up by increasing P_{on}. The reason is that by increasing this variable leads to increase in the discharge energy among two electrodes and creates a plasma energy channel among the workpiece and wire electrode gap. This plasma channel consists of pool of electrons and ions melted material because of the generation of an adequate heat-affected zone on the machined surface. Therefore, it usually plays an important role in influencing on the machining quality. Similar insights were reported by Rehman et al. [30] and coherent results were obtained in their experimental investigation.

The graphical trend of P_{on} for KW is similar as of WF for MRR. The parametric plots from Figure 4a have shown the direct effect of pulse-on duration (P_{on}) on KW, which means that kerf width is alleviated by varying the duration of pulses. This was because the number of sparks was amplified by increasing the time interval, resulting in the enhancement of the sparks energy and thus, a larger heat-affected zone was produced. The kerf width was increased by increasing the pulse duration. However, the analysis of trends also inferred that a small increase in KW was obtained when the pulse on duration was increased to one upper level from 4 to 5 μs, whereas a sudden increase in the KW was observed as P_{on} was promoted from 5 to 6 μs. However, 5 μs is a threshold level of the pulse on duration, after which relatively more thermal energy is produced to melt the extra material from the sides of the workpiece because of consistently increasing sparks erosion. The variance analysis findings have also proved the most influential of P_{on} (41.56%) for KW.

The trend analysis of P_{on} is shown in Figure 4b,c for surface roughness (SR) and material removal rate (MRR), respectively. The melt material is highly depended on increased number of discharges and the discharge energy generates a large pool of electrons and ions. Thus, with the increase in the amount of material removed and surface asperities on the machined surface of the workpiece also increase. Pulse duration was found as the most remarkable parameter for SR (84.83%) and also influential variable for MRR (32.81%).

The white layer is produced during machining when the leftover melt material re-solidifies on the machined surface of the workpiece [41]. Therefore, the recast layer thickness plays a vital role in evaluating the surface integrity of the machined specimen. The number of leftover debris and unflushed material on the surface has increased linearly with the increase of spark on duration. The reason is thermally affected region due to enhanced thermal energy and stability of sparks because of more spark's duration. Therefore, the number of unflushed particles were increased as the P_{on} is increased from 4 to 6 μs, resulting in a greater recast layer thickness, as shown in Figure 4d. A microscopic overview of the recast layer formation is shown in Figure 5. The craters produced are shallow and wider whereas the density of microcracks formation is also increased by increasing the P_{on}.

3.2.3. Effects of Open Voltage

The role of the open voltage is to perform a thermal breakdown to generate plasma channel [33]. SR was enhanced greatly by increasing OV as can be seen in Figure 6. SR possessed a direct relationship with OV similar to RLT. The thorough analysis of observations indicated that the increase in OV resulted in increasing SR, and RLT. The key reason involves discharge energy phenomenon because of thermal breakdown during machining.

The plasma channel width is increased because of the enhanced open voltage levels resulting in a larger thermal affected zone. The surface irregularities were increased at higher values of OV as shown in Figure 6a. The surface asperities are enhanced because of increased discharge energy. This enlarged the width of the plasma channel, creating deeper craters. The volume of the material removed was decreased by increasing the OV from 80 to 100 V because of the instability of discharge energy (Figure 6b). The phenomenon involves inadequate melt generation due to a small amount of thermal energy and poor flushing conditions, resulting in the insufficient removal of melt. OV was found to have significant influence (49.07%) in controlling MRR.

The thickness of the recast layer increased linearly by enhancing the OV, as presented in Figure 6c. In this case, from OV from 80 to 100 V, a linearly increasing trend of RLT is observed similar to SR was obtained due to plasma energy channel. It was produced because the material was melted and redeposited in the surface. Therefore, discharge energy was enhanced by increasing OV leaving more material behind on the work surface. The open voltage was found as foremost significant variable (52.06%) compared to other variables affecting the recast layer thickness.

3.2.4. Effects of Servo Voltage

Servo voltage is usually responsible for the transfer of heat flux on the dedicated workpiece surface through controlling the workpiece and wire electrode gap. All the output responses possessed quadratic behaviour by increasing the SV. A detailed discussion of SV provided herein, and trends are shown in Figure 7.

Kerf width was increased gradually by increasing the SV from 40 to 50 V. The rationale of the gradual increase in KW at the first two levels of SV is the relatively small amount of erosion of the material from both edges of the workpiece. However, a relatively greater increase in KW with increased servo voltage levels from 50 to 60 V is due to increased spark gap between the wire and workpiece electrodes, which resulted in a wider kerf. The phenomenon behind the increasing trend of KW was endorsed by Rehman et al. [30].

3.3. *Mathematical Modelling and Parametric Optimization*

To determine the predictability of responses (productivity and quality), the general regression model is developed through converting variable parameters to continuous predictive variables [42]. The general algebraic expression of model is given as following.

$$y = b_0 + b_1x_1 + b_2x_2 \ldots\ldots\ldots\ldots + b_kx_k \tag{3}$$

In the general model, y represents the value of the response while b_0 is the constant value and $b_1, b_2, b_3, \ldots, b_k$ represents the estimated variation in mean response for each unit change in the variable parameters. Experimental data was used to generate regression models using variable parameters as continuous predictors. The relationship of WEDM variable parameters on productivity quality measures using multiple regression analysis.

$$\text{RLT} = -14.03 - 0.070\ \text{WF} + 1.947\ \text{Pon} + 0.2222\ \text{OV} - 0.0139\ \text{SV} \tag{4}$$

$$\text{W} = 0.3007 - 0.005037\ \text{WF} + 0.00883\ \text{Pon} + 0.000192\ \text{OV} + 0.000267\ \text{SV} \tag{5}$$

$$\text{SR} = 1.3159 - 0.02074\ \text{WF} + 0.12167\ \text{Pon} + 0.001917\ \text{OV} - 0.001417\ \text{SV} \tag{6}$$

$$\text{MRR} = 18.53 + 0.624\ \text{WF} + 2.027\ \text{Pon} - 0.2478\ \text{OV} - 0.0223\ \text{SV} \tag{7}$$

Where "WF", "P_{on}", "OV" and "SV" are representing wire feed (mm/s), pulse on time (µs), open voltage (V), and servo voltage (V), respectively.

Utilizing the experimental data, Minitab 19 was used to develop multiple regression models associated with recast layer thickness, kerf width, surface roughness and material removal rate presented in Equations (4)–(7), respectively. The output values of the response measures were predicted after obtaining the optimal parameters for each response, and the optimization results were validated by performing confirmatory experiments. The results of the experimentation (three experiments) as a result of this optimization produced a very minute standard deviation from the design of experiments results.

Moreover, the desirability function was employed to optimize process settings, such as the minimum, which was better for recast layer thickness, kerf width, surface roughness, and the maximum, which was better for the material removal rate. The results of the experimentation provided a major process improvement. The optimized parameters obtained are "WF 8 mm/s", "P_{on} 4 µs", "OV 80 V" and "SV 56V". In Table 4, the close agreement is shown between the predicted values and experimental trials on the optimized process settings ensures the adequacy of empirical models.

Table 4. Measured responses on optimized process parameters.

Sr. No.	Process Parameters				Response Indicators			
	WF (mm/s)	P_{on} (µs)	OV (V)	SV (V)	SR (µm)	MRR (mm^3/min)	KW (mm)	RLT (µm)
1	8	4	80	56	1.709	10.307	0.327	10.093
2	8	4	80	56	1.703	10.299	0.326	10.256
3	8	4	80	56	1.719	10.496	0.330	10.980
		Avg. experimental value			1.710	10.367	0.327	10.443
		Standard deviation			0.006	0.091	0.001	0.385
		Predicted value			1.711	10.554	0.326	10.191
		Error %			0.64	1.77	0.31	2.47

3.4. *Recast Layer Measurement and Microstructural Evaluation*

The microscopic analysis is usually performed to analyse the physical phenomena on the surface morphology of the workpiece. The optical and scanning electron microscopy-based analyses are explored herein.

3.4.1. Optical-Based Microscopic Analysis

The detailed recast layer thickness analysis was carried out through optical microscope, (Figure 8A,B) to investigate the physics involved on the morphological alterations on the workpiece surface. The recast layer was measured at three distinct positions on the processed surface and average of the measurements was chosen for the analysis purpose.

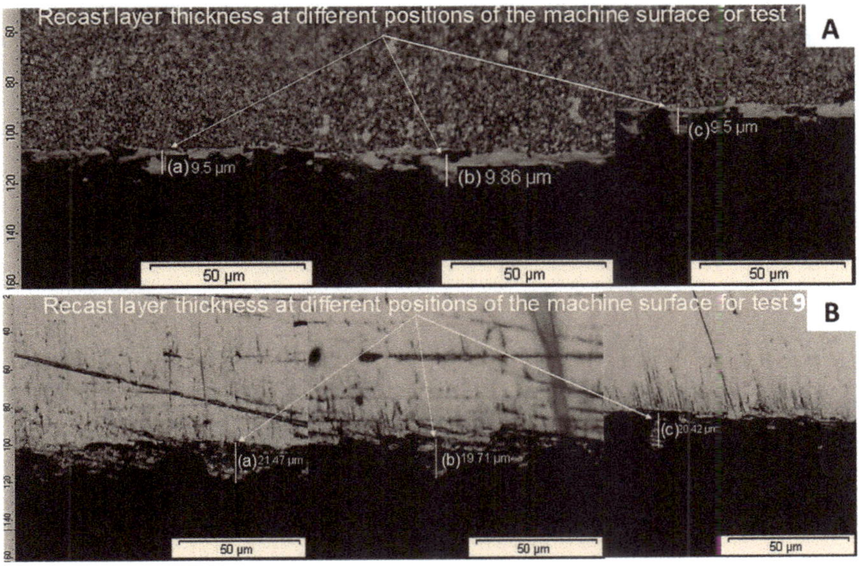

Figure 8. Optical microscopic examination (at three different points for one experiment) for recast layer measurement at different parametric settings for (**A**) WF 5 mm/s, Pon 4 µs, OV 80 V, and SV 40 V (**B**) WF 5 mm/s, Pon 6 µs, OV 100 V, and SV 40 V.

It can be extracted from Figure 8A,B that the significant thickness is deposited on the base metal layer. Its microstructure is different from the unaffected surface consisting of primary carbides. It was inferred that the maximum recast layer thickness of 19.52 μm was obtained at the minimum WF (5 mm/s), high level of P_{on} (6 μs), higher value of OV (100 V), and the minimum SV (40 V). Similarly, the higher control is responsible for the thermal breakdown to improve the sparks' stability and smaller control level means that the sparks energy is focused on the specific region where material melting and redeposition occurred. Therefore, the recast layer thickness was increased at these parametric settings. Similarly, a minimum RLT was observed at WF 5 mm/s, P_{on} (4 μs), OV (80 V), and SV (40 V). The reason is that the debris redeposition is decreased as the sparks on duration is decreased because P_{on} is highly influential parameter for the recast layer thickness compared to other variables.

3.4.2. Scanning Electron Microscopic Analysis

The detailed investigation for the microstructure of the processed surface is demonstrated in Figure 9a,b. The evidence shows that microcracks appeared on the machined surface. Moreover, the properties of melt material redeposit on machined surface varied because of bonding and cracks generated in heat effected region. The recast layer properties are primarily different from the base metal because of heat treatment and quenching, as shown in Figure 9. The thermal energy generated between wire electrodes results in enlarged heat effected zone produced on the machined surface, providing opportunity to redeposit onto the machined layer.

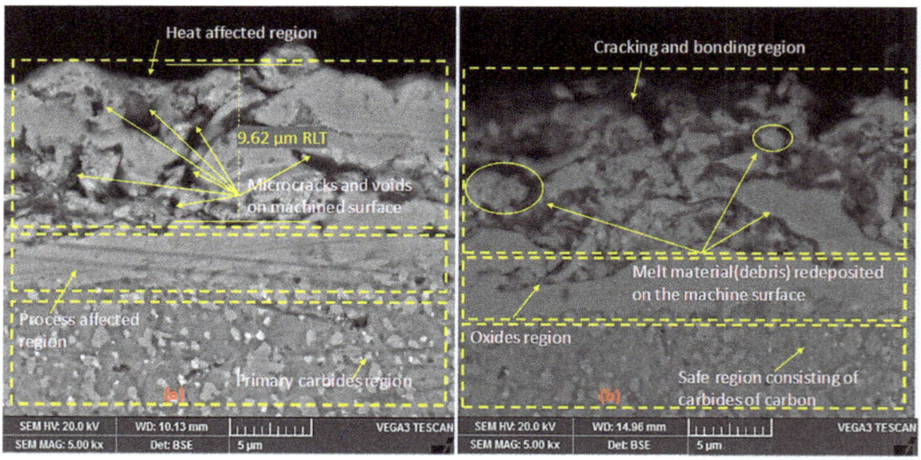

Figure 9. Scanning electron microscopy results of machined specimens (**a**) WF1, Pon1, OV2, SV2, (**b**) WF1, Pon3, OV1, and SV2.

It can be seen from Figure 9 that the craters of different sizes and shapes, microcracks and thermally redeposit melt material/debris were evident on machined surface. The density of crack formation and the size of the craters increased by increasing the sparks-on duration (the discharge energy is linearly dependent on it). Likewise, the process is influenced by process variables on the microstructural evaluation, as evident from Table 3. Similarly, pulse on duration influenced its maximum on controlling the craters size, density of cracks formation, and size of debris globules formed on the processed surface. The measured RLT on the machined surface was found 9.62 μm. The analysis of SEM images from Figure 9 inferred that an oxide layer formed just below the heat-affected layer. This process-affected region was generated because of large plasma channel width, resulting in more irregularities on the machined surface. The layer below the process affected zone was considered as the safe region, consisting of primary carbides of carbon.

3.5. Comparison with Previous Studies

The tabular comparison of current output results was also compared with the published literature for DC53 steel in WEDM and summary is reported in Table 5. The research findings were found almost in good agreement with the existing results which also validated the experimental work. In addition, machinability of DC53 steel with zinc-coated wire produced superior results in comparison to established practices [29,30].

Table 5. Comparison of the obtained results with the existing literature.

Output Response	Unit	Current Work	Previous Work on Steel	References
SR	μm	1.71	1.72, 1.89, 1.68	[1,29,30]
MRR	mm^3/min	10.367	5.443, 10.205	[29,30]
KW	mm	0.327 (at 0.25 mm wire Φ)	0.220, 0.287 (at 0.20 mm wire Φ)	[29,30]
RLT	μm	10.443	-	-

4. Conclusions

This work was dedicated to providing experimental insight into the machining of DC53 die steel, and considering the most important process variables, namely wire feed-WF, pulse on duration-P_{on}, open voltage-OV, and servo voltage-SV. The influence of these variables on kerf width-KW, surface roughness-SR, material removal rate-MRR, and recast layer thickness-RLT during wire electric discharge machining was studied. Through rigorous analysis based on physical phenomenon on surface morphology and material characteristics, the following conclusions are extracted:

- RLT was directly influenced by varying the P_{on} and it was observed that SEM examination revealed three surface regions; (i) thermally affected region consisting of debris particles of different sizes and shapes, cracks and craters, (ii) process-affected zone consisting of oxides, and (iii) safe region consisting of the primary carbides of carbon. Minimum average RLT of 9.63 μm was measured and observed after the examination of machined surface morphology.
- The variance analysis results inferred that P_{on} was found to be the most significant machining variable for all output responses with higher contribution percentages such as SR (84.83%), KW (41.56%), RLT (40.00%), and MRR (32.81%), respectively. High P_{on} results ina high plasma channel consisting the pool of electrons and ions melted material because of the generation of an adequate heat-affected zone on the machined surface.
- WF (45.64%), P_{on} (41.56%), and the SV (3.774%) are the significant factors for KW. However, the detailed influential variables for SR include P_{on} (84.83%), WF (8.320%), OV (2.10%), and SV (1.15%). The significant variables for recast layer thickness involved OV (52.06%) and P_{on} (40.00%). Open voltage, pulse on duration, and wire feed with percentage contributions of 49.07%, 32.81%, and 10.51%, respectively, are significant for MRR. The rapid transport of wire resulted in less workpiece–wire electrode interaction, and produced a smaller heat-affected zone because of minor thermal damages. Therefore, less molten material was redeposited on the workpiece surface.
- The experimental results reveal that WF influences prominently in reducing KW, SR, and RLT and increasing MRR. All response measures were directly increased by increasing P_{on} due to the amplified discharge energy of the sparks produced.
- The optimized parameters obtained are "WF 8 mm/s", "P_{on} 4 μs", "OV 80 V" and "SV 56 V" resulting 1.710 μm SR, 10.367 mm^3/min MRR, 0.327 mm KW, and 10.443 μm RLT. The regression models (showing less than a 3% error from the physical experimentation) are developed based on a thorough investigation to support the machinists for achieving the required features.

Future work that could further improve the quality of dies tools should be focused on experimental investigation using other control variables like flushing pressure, wire tension, and corner accuracy (both top and bottom) in WEDM of DC53 steel using different wire electrodes.

Author Contributions: Conceptualization, S.A.K. and W.A.; Data curation, W.A.; Formal analysis, R.N. and M.R.; Funding acquisition, M.U.F., M.A.A. and C.I.P.; Investigation, W.A., M.R. and R.N.; Methodology, S.A.K. and M.R.; Resources, R.N.; Software, W.A. and M.R.; Supervision, S.A.K.; Validation, S.A.K., R.N. and C.I.P.; Visualization, M.U.F., M.A.A. and C.I.P.; Writing—original draft, M.U.F., M.R. and M.A.A.; Writing—review & editing, M.U.F., M.A.A., C.I.P. All authors have read and agreed to the published version of the manuscript.

Funding: This study did not receive any funding.

Institutional Review Board Statement: Not applicable.

Informed Consent Statement: Not applicable.

Data Availability Statement: The data is available as request from corresponding authors.

Conflicts of Interest: The authors declare no conflict of interest.

Abbreviations

ANOVA	Analysis of variance
CMM	Coordinate measuring machine
DOE	Design of experiments
HAZ	Heat affected zone
KW	Kerf width (μm)
MRR	Material removal rate (mm^3/min)
OV	Open voltage (volt)
OA	Orthogonal array
ρ	Density (kg/m^3)
PCR	Percent contribution (%)
Pon	Pulse on duration (μs)
Poff	Pulse off time (μs)
RLT	Recast layer thickness (μm)
SR	Surface roughness (μm)
SV	Servo voltage (volt)
SEM	Scanning electron microscopy
WEDM	Wire electric discharge machining
WF	Wire feed (mm/s)
WT	Wire tension (gms-f)
W_b	Weight of the workpiece prior machining (g)
W_a	Weight of the workpiece prior after machining (g)

References

1. Kanlayasiri, K.; Boonmung, S. Effects of wire-EDM machining variables on surface roughness of newly developed DC 53 die steel: Design of experiments and regression model. *J. Mater. Process. Technol.* **2007**, *192–193*, 459–464. [CrossRef]
2. Singh, V.; Bhandari, R.; Yadav, V.K. An experimental investigation on machining parameters of AISI D2 steel using WEDM. *Int. J. Adv. Manuf. Technol.* **2017**, *93*, 203–214. [CrossRef]
3. Durairaj, M.; Sudharsun, D.; Swamynathan, N. Analysis of process parameters in wire EDM with stainless steel using single objective Taguchi method and multi objective grey relational grade. *Procedia Eng.* **2013**, *64*, 868–877. [CrossRef]
4. Zhang, G.; Zhang, Z.; Guo, J.; Ming, W.; Li, M.; Huang, Y. Modeling and optimization of medium-speed WEDM process parameters for machining SKD11. *Mater. Manuf. Process.* **2013**, *28*, 1124–1132. [CrossRef]
5. Lodhi, B.K.; Agarwal, S. Optimization of machining parameters in WEDM of AISI D3 steel using taguchi technique. *Procedia CIRP* **2014**, *14*, 194–199. [CrossRef]
6. Thiagarajan, C.; Sararvanan, M.; Senthil, J. Performance Evaluation of Wire Electro Discharge Machining on D3-Tool Steel. *Int. J. Pure Appl. Math.* **2018**, *118*, 943–949.

7. Manjaiah, M.; Laubscher, R.F.; Kumar, A.; Basavarajappa, S. Parametric optimization of MRR and surface roughness in wire electro discharge machining (WEDM) of D2 steel using Taguchi-based utility approach. *Int. J. Mech. Mater. Eng.* **2016**, *11*, 7. [CrossRef]
8. Ikram, A.; Mufti, N.A.; Saleem, M.Q.; Khan, A.R. Parametric optimization for surface roughness, kerf and MRR in wire electrical discharge machining (WEDM) using Taguchi design of experiment. *J. Mech. Sci. Technol.* **2013**, *27*, 2133–2141. [CrossRef]
9. Vaghela, J.R.; Valaki, J.B.; Pandit, J.H. Parametric Optimization of Machining Parameters of AISI D3 Tool Steel Using Wire Cut Electric Discharge Machining—A Taguchi Based Approach. *Natl. Conf. Emerg. Res. Trends Eng.* **2016**. Available online: https://www.researchgate.net/publication/300262070 (accessed on 22 March 2021).
10. Zhang, G.; Zhang, Z.; Guo, J.; Ming, W.; Li, M.; Huang, Y.; Shao, X. The multi-objective optimization of medium-speed WEDM process parameters for machining SKD11 steel by the hybrid method of RSM and NSGA-II. *Int. J. Adv. Manuf.* **2014**, *70*, 2097–2109. [CrossRef]
11. Rupajati, P.; Soepangkat, B.O.P.; Pramujati, B.; Agustin, H.C.K. Optimization of Recast Layer Thickness and Surface Roughness in the Wire EDM Process of AISI H13 Tool Steel Using Taguchi and Fuzzy Logic. *Appl. Mech. Mater.* **2014**, *493*, 529–534. [CrossRef]
12. Hasriadi, B.O.P.; Soepangkat, H.S. The Effects of Pulse on Time and Arc on Time on Surface Quality in Wire-EDM of ASSAB XW-42 and ASSAB 8407 2M Tool Steels. *Appl. Mech. Mater.* **2016**, *836*, 173–178. [CrossRef]
13. Hernández-Castillo, I.; Sánchez-López, O.; Lancho-Romero, G.A.; Castañeda-Roldán, C.H. An experimental study of surface roughness in electrical discharge machining of AISI 304 stainless steel. *Eng. Res.* **2018**, *38*, 90–96. [CrossRef]
14. Saini, T.; Goyal, K.; Bhandari, D. Multi-response optimization of *WEDM* parameters on machining 16MnCr5 alloy steel using Taguchi technique. *Multiscale Multidiscip. Model. Exp. Des.* **2018**. [CrossRef]
15. Oßwald, K.; Lochmahr, I.; Schulze, H.P.; Kröning, O. Automated Analysis of Pulse Types in High Speed Wire EDM. *Procedia CIRP* **2018**, *68*, 796–801. [CrossRef]
16. Solomon, R.S.; Sevvel, P.; Jaiganesh, V.; Satheesh, C. An Investigational Review on Influence of Performance Parameters during Wire Electrical Discharge Machining of Various Grades of Steels. *J. Adv. Res. Dyn. Cont. Sys.* **2017**, *9*, 101–109.
17. Shivade, A.S.; Shinde, V.D. Multi-objective optimization in WEDM of D3 tool steel using integrated approach of Taguchi method & Grey relational analysis. *J. Ind. Eng. Int.* **2014**, *10*, 149–162. [CrossRef]
18. Azam, M.; Jahanzaib, M.; Abbasi, J.A.; Wasim, A. Modeling of cutting speed (CS) for HSLA steel in wire electrical discharge machining (WEDM) using moly wire. *J. Chin. Inst. Engr.* **2016**, *39*, 802–808. [CrossRef]
19. Dhobe, M.M.; Chopde, I.K.; Gogte, C.L. Investigations on Surface Characteristics of Heat Treated Tool Steel after Wire Electro-Discharge Machining Investigations on Surface Characteristics of Heat Treated Tool Steel after Wire Electro-Discharge Machining. *Mater. Manuf. Process.* **2013**, *28*, 1143–1146. [CrossRef]
20. Khanna, R.; Singh, H. Comparison of optimized settings for cryogenic-treated and normal D-3 steel on WEDM using grey relational theory. *Proc. Inst. Mech. Eng. Part L J. Mater. Des. Appl.* **2016**, *230*, 219–232. [CrossRef]
21. Sharma, S.; Kumar Vates, U.; Bansal, A. Parametric optimization in wire EDM of D2 tool steel using Taguchi method. *Mater. Today Proc* **2020**. [CrossRef]
22. Ramaswamy, A.; Perumal, A.V.; Jagadeesan, J.; Nagarajan, H.V. Optimization of WEDM process parameters for D3 die steel using RSM. *Mater. Today Proc.* **2020**, *37*, 2063–2069. [CrossRef]
23. Pavla, A.; Chopra, K.; Mussada, E.K. Investigations on power consumption in WEDM of EN31 steel for sustainable production. *Mater. Manuf. Process.* **2019**, *34*, 1855–1865. [CrossRef]
24. Chen, X.; Wang, Z.; Wang, Y.; Chi, G. Investigation on MRR and Machining Gap of Micro Reciprocated Wire-EDM for SKD11. *Int. J. Precis. Eng. Manuf.* **2020**, *21*, 11–22. [CrossRef]
25. Singh, S.; Maheshwari, S.; Pandey, P.C. Some investigations into the electric discharge machining of hardened tool steel using different electrode materials. *J. Mater. Process. Technol.* **2004**, *149*, 272–277. [CrossRef]
26. Abdulkareem, S.; Khan, A.A.; Zain, Z.M. Experimental investigation of machining parameters on surface roughness in dry and wet wire-electrical discharge machining. *Adv. Mater. Res.* **2011**, *264–265*, 831–836. [CrossRef]
27. Mouralova, K.; Benes, L.; Bednar, J.; Zahradnicek, R.; Prokes, T.; Matousek, R.; Hrabec, P.; Fiserova, Z.; Otoupalik, J. Using a DoE for a comprehensive analysis of the surface quality and cutting speed in WED-machined hadfield steel. *J. Mech. Sci. Technol.* **2019**, *33*, 2371–2386. [CrossRef]
28. Camposeco-Negrete, C. Prediction and optimization of machining time and surface roughness of AISI O1 tool steel in wire-cut EDM using robust design and desirability approach. *Int. J. Adv. Manuf. Technol.* **2019**, *103*, 2411–2422. [CrossRef]
29. Nawaz, Y.; Maqsood, S.; Naeem, K.; Nawaz, R.; Omair, M.; Habib, T. Parametric optimization of material removal rate, surface roughness, and kerf width in high-speed wire electric discharge machining (HS-WEDM) of DC53 die steel. *Int. J. Adv. Manuf. Technol.* **2020**, *107*, 3231–3245. [CrossRef]
30. Rehman, M.; Khan, S.A.; Naveed, R. Parametric optimization in wire electric discharge machining of DC53 steel using gamma phase coated wire. *J. Mech. Sci Technol.* **2020**, *34*, 2767–2773. [CrossRef]
31. Çardaklı, İ.S. *Thin Section High Cooling Rate Solidification, Thermomechanical Processing and Characterization of AISI DC53 Cold Work Tool Steel*; Middle East Technical University: Ankara, Turkey, 2019; Available online: https://open.metu.edu.tr/bitstream/handle/11511/45197/index.pdf (accessed on 31 March 2021).
32. Ishfaq, K.; Anwar, S.; Ali, M.A.; Raza, M.H.; Farooq, M.U.; Ahmad, S.; Salah, B. Optimization of WEDM for precise machining of novel developed Al6061-7.5% SiC squeeze casted composite. *Int. J. Adv. Manuf. Technol.* **2020**. [CrossRef]

33. Mughal, M.P.; Farooq, M.U.; Mumtaz, J.; Mia, M.; Shareef, M.; Javed, M.; Pruncu, C.I. Surface modification for osseointegration of Ti6Al4V ELI using powder mixed sinking EDM. *J. Mech. Behav. Biomed. Mater.* **2020**, 104145. [CrossRef] [PubMed]
34. Umar Farooq, M.; Pervez Mughal, M.; Ahmed, N.; Ahmad Mufti, N.; Al-Ahmari, A.M.; He, Y. On the Investigation of Surface Integrity of Ti6Al4V ELI Using Si-Mixed Electric Discharge Machining. *Materials* **2020**, *13*, 1549. [CrossRef] [PubMed]
35. Tosun, N.; Cogun, C.; Tosun, G. A study on kerf and material removal rate in wire electrical discharge machining based on Taguchi method. *J. Mater. Process. Technol.* **2004**, *152*, 316–322. [CrossRef]
36. Mussada, E.K.; Hua, C.C. Rao AKP Surface hardenability studies of the die steel machined by WEDM. *Mater. Manuf. Process.* **2018**, *33*, 1745–1750. [CrossRef]
37. Somashekhar, K.P.; Ramachandran, N.; Mathew, J. Material removal characteristics of microslot (kerf) geometry in μ-WEDM on aluminum. *Int. J. Adv. Manuf. Technol.* **2010**, *51*, 611–626. [CrossRef]
38. Groover, M. *Fundamentals of Modern Manufacturing: Materials, Processes, and Systems*, 4th ed.; John Wiley & Sons, Inc.: Hoboken, NJ, USA, 2013; ISBN 978-0470467008.
39. Tilekar, S.; Das, S.S.; Patowari, P.K. Process Parameter Optimization of Wire EDM on Aluminum and Mild Steel by Using Taguchi Method. *Procedia Mater. Sci.* **2014**, *5*, 2577–2584. [CrossRef]
40. Farooq, M.U.; Ali, M.A.; He, Y.; Khan, A.M.; Pruncu, C.I.; Kashif, M.; Asif, N. Curved profiles machining of Ti6Al4V alloy through WEDM: Investigations on geometrical errors. *J. Mater. Res. Technol.* **2020**, *9*, 16186–16201. [CrossRef]
41. Ishfaq, K.; Farooq, M.U.; Anwar, S.; Ali, M.A.; Ahmad, S.; El-Sherbeeny, A.M. A comprehensive investigation of geometrical accuracy errors during WEDM of Al6061-7. *5% SiC composite. Mater. Manuf. Process.* **2020**, *36*, 362–372. [CrossRef]
42. Asad, A.M.; Kashif, I.; Raza, M.H.; Umar, F.M.; Mufti, N.A.; Pruncu, C.I. Mechanical characterization of aged AA2026-AA2026 overcast joints fabricated by squeeze casting. *Int. J. Adv. Manuf. Technol.* **2020**, *107*, 3277–3297.

metals

MDPI

Article

Mechanical Characterization of Nanocrystalline Materials via a Finite Element Nanoindentation Model

Konstantinos Tserpes [1,*], Panagiotis Bazios [1], Spiros G. Pantelakis [1], Maria Pappa [2] and Nikolaos Michailidis [2]

1 Laboratory of Technology & Strength of Materials, Department of Mechanical Engineering & Aeronautics, University of Patras, 26500 Patras, Greece; pbazios@upatras.gr (P.B.); pantelak@mech.upatras.gr (S.G.P.)
2 Physical Metallurgy Laboratory, Mechanical Engineering Department, School of Engineering, Aristotle University of Thessaloniki, 54124 Thessaloniki, Greece; pappmari@auth.gr (M.P.); nmichail@auth.gr (N.M.)
* Correspondence: kitserpes@upatras.gr; Tel.: +30-261-096-9498

Abstract: The difficulty of producing sufficient quantities of nanocrystalline materials for test specimens has led to an effort to explore alternative means for the mechanical characterization of small material volumes. In the present work, a numerical model simulating a nanoindentation test was developed using Abaqus software. In order to implement the model, the principal material properties were used. The numerical nanoindentation results were converted to stress–strain curves through an inverse algorithm in order to obtain the macroscopic mechanical properties. For the validation of the developed model, nanoindentation tests were carried out in accordance with the ISO 14577. The composition of 75% wt. tungsten and 25% wt. copper was investigated by producing two batches of specimens with a coarse-grain microstructure with an average grain size of 150 nm and a nanocrystalline microstructure with a grain diameter of 100 nm, respectively. The porosity of both batches was derived to range between 9% and 10% based on X-ray diffraction analyses. The experimental nanoidentation results in terms of load–displacement curves show a good agreement with the numerical nanoindentation results. The proposed numerical technique combined with the inverse algorithm predicts the material properties of a fully dense, nanocrystalline material with very good accuracy, but it shows an appreciable deviation with the corresponding compression results, leading to the finding that the porosity effect is a crucial parameter which needs to be taken into account in the multiscale numerical methodology.

Keywords: nanocrystalline materials; finite element analysis; inverse algorithm

Citation: Tserpes, K.; Bazios, P.; Pantelakis, S.G.; Pappa, M.; Michailidis, N. Mechanical Characterization of Nanocrystalline Materials via a Finite Element Nanoindentation Model. *Metals* **2021**, *11*, 1827. https://doi.org/10.3390/met11111827

Academic Editor: George A. Pantazopoulos

Received: 11 October 2021
Accepted: 10 November 2021
Published: 13 November 2021

1. Introduction

Notable attempts have been made, in recent years, at the production of materials with a nanocrystalline microstructure, due to their improved mechanical properties [1,2]. The enhancement in terms of strength by decreasing the grain size has been described by the Hall-Petch equation effect [3].

Various manufacturing methods are applied to obtain nanocrystalline alloys (i.e., a grain size less than 100 nm), including electrodeposition [1,4], powder metallurgy [5,6], magnetron sputtering [7–10], and inert gas condensation followed by the consolidation of powders [1] and severe plastic deformation (SPD) [11,12]. The vast majority of nanocrystalline materials are produced by the mechanical alloying method or electrodeposition technique. The electrodeposition technique has some inherent limitations in terms of the quality of the produced specimens because they show extensive porosity or impurities. On the other hand, the fabrication technique of mechanical alloying is able to produce materials with a nanocrystalline microstructure at a high manufacturing speed and with complete control of the fabrication parameters, so as to obtain defect-free bulk materials [13]. However, the main obstacle of the mechanical alloying method is the production of specimens without a uniform grain size, which in turn influences the macroscopic mechanical response of the materials, especially in a harsh environment. The material microstructure changes rapidly at elevated temperatures. This is the commonly-known coarsening effect. The mechanical response of said materials is completely influenced by their different interphases into the microstructure. This kind of material is the so-called graded material [14,15].

Taking for granted said facts, all of the proposed production techniques are cost-effective, but the produced materials are influenced by several imperfections (like flaws and pores) which drive to the degradation of the material properties. With respect to limitations in the manufacturing methods of nano-crystalline materials, the mass production of fully dense, bulk forms with a nanocrystalline microstructure at sufficient quantities has hardly been achievable up to now, and it has not been industrialized.

With the aforementioned increasing interest in the production of innovative metallic materials, mechanical testing procedures using small material volumes are more than necessary nowadays. Moreover, the conventional testing techniques like tensile or compression testing become noticeably challenging with specimens of small volume. Furthermore, limited results are shown in journals concerning the correlation between the microstructure and the resulting material properties. Some of the desirable material properties in the frame of nanocrystalline materials are the yield stress, the ultimate strength at fracture, the enhanced wear resistance and the improved plasticity compared to that of their microcrystalline counterparts [16–20]. A more detailed and extensive mechanical characterization is needed so as to obtain more reliable results.

Based on the obstacles of the mechanical characterization utilizing conventional experimental tests, scientific progress has resulted in the extensive use of nanoindentation experiments [21–34]. Berkovich indenters allow precise measurements by varying the force of the nanoindentation (P) in relation to the depth of penetration (h). Observations through experimental nanoindentation tests were performed on many materials so as to determine material features such as residual stresses and hardness [23–25,29,33–37].

At the same time, a noticeable effort has been made to use analytical and numerical investigations in order to determine the deformation mechanisms and the contact mechanics of the nanoindentation effect. The main purpose is to define the mechanical response to the applied load of nanoindentation (P) versus penetration depth (h) diagrams acquired from instrumented nanoindentation [23,25,31,32,36–41]. More specifically, researchers [23,25] suggested analytical approaches in which the modulus of elasticity and hardness could be extracted from the P_{max} and the unloading incline at the starting stages. Additionally, Giannakopoulos and Suresh [40] proposed a procedure in which all of elastoplastic properties may be extracted through a P-h diagram. Based on [40], Suresh and Giannakopoulos [42] developed an up-to-date methodology which was able to export the residual stresses. In terms of numerical methods, several numerical attempts were conducted on the case studies of thin-film systems [43–45].

To conclude, a methodology was developed by Hill et al. [46] for the case study of materials with power law plastic behavior under the indentation load of a spherical indenter. The calculations of a sharp indenter, such as a Berkovich indenter, for the case of elastoplastic materials were numerically extracted by Larsson et al. [47] and Giannakopoulos et al. [38]. Moreover, scale factors were carried out so as to investigate bulk materials [31,32,39] and coated material [45]. To this end, researchers [32] have developed a comprehensive analytical approach, and several investigations were conducted on the basis of the dimensional analysis of sharp indentation. The results from the aforementioned approach showed that Kick's Law is a constitutive factor of the dimensional analysis of sharp indentation.

The scope of the present paper is to develop an innovative multiscale finite element methodology simulating the nanoindentation test which accounts for the different material phases, as they are described in [48], in order to define the macroscopic mechanical response of nanocrystalline materials. Through the aforementioned model, the nanoindentation behavior of nanocrystalline materials was extracted, and the corresponding macroscopic mechanical behavior was evaluated via the implementation of an inverse analytical algorithm. By utilizing a homogenized RVE for the last simulation in conjunction with the inverse algorithm, we are able to predict the macroscopic mechanical response.

2. Modelling

2.1. Analytical Approach

As referred to above, the nanoindentation method is a very attractive testing procedure to obtain the mechanical properties of such kinds of materials as nanocrystalline metals, which are difficult to produce in large quantities, leading to the need for the mechanical characterization of small-volume specimens. The uniqueness of the nanoindentation technique relies on the extracted results which come from the *P-h* diagram. More specifically, the estimation of the contact pressure between the nano-indenter and the target material—and the resulting hardness—comes from the measurement of the contact area between said parts in terms of the indentation load and penetration depth. A continuous data acquisition in the loading phase of a typical indentation test is involved. In the following unloading part of the *P-h* diagram, the Young's modulus and Poisson ratio of the indented material can be exported at the early stages of the unloading phase because of the strain relaxation within the material [49,50]. Furthermore, material features like residual stress [46,51], initial plasticity [51–53], creep [54], and hardening [55] are very important parameters for property extraction in the nanoindentation testing method. Additionally, the surface deformation around the plastic deformation induced by the nano-indenter creates sink-in and pile-up residual strains, which are very important for the mechanical characterization of the target material [56]. The aforementioned alteration of the surface plays an important role in determining the contact area, and is also an effective factor for the definition of material properties like the strain-hardening behavior [57,58].

According to the literature for the indentation testing method, a nanoindentation test is commonly known for the creation of a non-uniform stress field at the surface of the target material, which leads to the average calculation of induced stresses [59]. The mean value of stresses in terms of von Mises stress, according to Meyer's hardness [49,59], is defined utilizing the Tabor Relation [59–61]:

$$\sigma \approx \frac{H}{3}, \quad (1)$$

$$H = \frac{P}{A}, \quad (2)$$

where H is the resulting hardness, P is the applied force of the nano-indenter on the target material, and A is the contact area of the Berkovich nanoindenter. The indentation area is calculated taking into consideration the geometry of the nanoindenter's tip and the pileup or sink-in that take place around the tip, as shown in Figure 1.

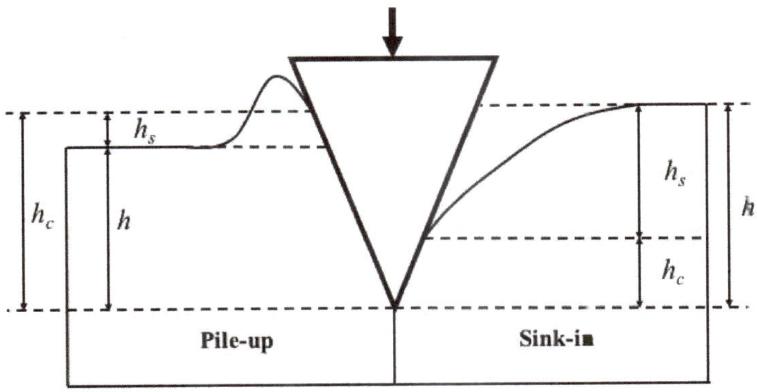

Figure 1. Indentation geometry showing the pile-up and sink-in.

The mathematical expressions followed for the calculation of the projected contact area of the nanoindenter are presented in Equation (3) for the case study of a Berkovich indenter's tip [62], and in Equation (4) for the case study of the spherical indenter's tip:

$$A = 24.5{h_c}^2 + Ch_c, \tag{3}$$

$$A = \pi\left(2Rh_c - {h_c}^2\right), \tag{4}$$

where h_c is the penetration depth of the indenter tip into the target specimen, C is a constant for the case of Berkovich indenters (150 nm) referring to the geometrical features of the specific indenter [62], and R refers to the curvature of the nanoindenter. The penetration depth of the indenter's tip was calculated according to the contact mechanics' analytical expressions (Figure 1) [49,59,63]:

$$h_c = h - h_s = h - \delta\left(\frac{P_{max}}{S}\right), \tag{5}$$

$$S = \frac{dP}{dh}, \tag{6}$$

where h is the overall displacement of the nanoindenter; δ is a geometric constant parameter referring to the geometrical characteristics of the indenter's tip, which is equal to 0.75 for Berkovich indenters; the variable h_s describes the displacement of the target material's surface during the nanoindentation load, taking into account the pile-up and sink-in deformation mechanisms; and S is the stiffness at the unloading phase. At this point, it is of paramount importance to mention that the stiffness at the unloading step is defined by the incline of the *P-h* diagram (an indicative plot is shown in Figure 2).

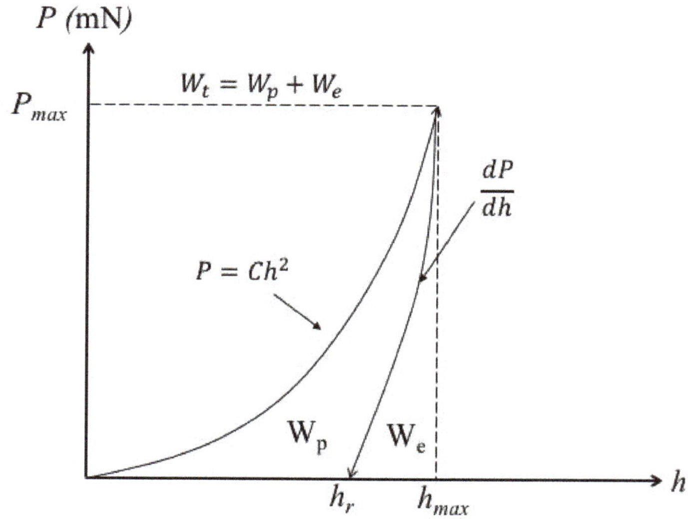

Figure 2. Indentation load–penetration depth curve from an indentation test.

It is noteworthy to mention that Equations (2)–(6) were from research papers in the frame of homogeneous materials. Based on this aspect, extensive work needs to be carried out so as to verify the effectiveness of the proposed mathematical procedure on heterogeneous materials like the nanocrystalline tungsten–copper materials produced and tested in the present paper. The selection of the tungsten–copper alloy was made in the frame of the European FET-Open project ICARUS [64] due to its inherent thermodynamic stability for aerospace applications. Moreover, the strain rate of the indentation at the surface of the target material is also non-uniform, and it leads to the creation of the corresponding non-uniform stress field, as referred to above. It is of paramount importance to define the mean deformation rate as the rate of the complete penetration depth into the target material over the current total displacement of the indenter [60]:

$$\dot{\varepsilon} = \frac{\dot{h}}{h} \quad (7)$$

2.2. Numerical Model

For the purpose of the nanoindentation simulation of a nanostructured material, a FE-based model was utilized. The recommended numerical methodology is utilized on RVEs with a nanocrystalline morphology. The numerical methodology is presented in the flowchart of Figure 3. The methodology begins with the computer tomograph YXLON FF35 CT (YXLON International GmbH, Hamburg, Germany) and X-ray diffraction images (Bruker D8 Discover, Bruker Corp., Billerica, MA, USA), and evolves with the image analysis and the multi-level simulation [48].

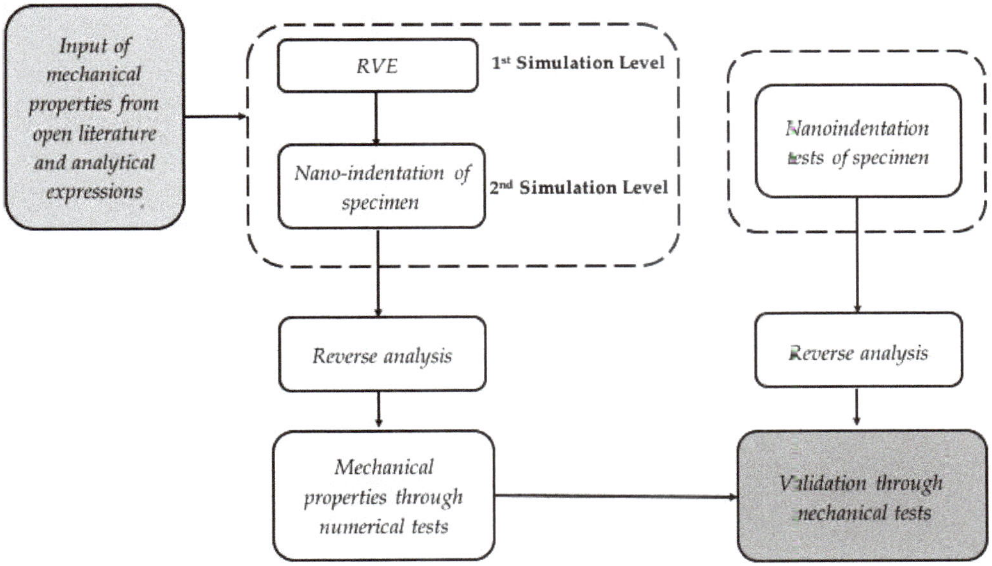

Figure 3. Flowchart of the methodology.

As it is commonly known, RVE is the smallest numerical volume for any material in which a macro-mechanical property can be defined through a multi-scale modelling approach. Due to the microscopic size of representative-volume elements, the detailed morphology of the materials' microstructural characteristics can be modelled. The analysis' objective is focused on the development of a numerical procedure via the parametric interaction representing the geometrical characteristics of nanostructured metals. The presumptions and the extensive numerical methodology used in the present work are described extensively in [48,65], on account of the completeness of the paper. The nanocrystalline materials are heterogeneous materials due to their discrete areas showing different physico-chemical and mechanical properties at the microscale level. Based on this paradox, the mechanical response, taken by the proposed numerical methodology shown in [48,65], is the homogenized average medium of the equivalent heterogenous nanocrystalline material. More specifically, in [48,65], the presumptions were applied for the investigation of mechanical behavior of a fully dense, nanocrystalline material under uniaxial compressive loading, so as to extract the stress–strain response of an equivalent homogenized material from the heterogenous RVE (see Figure 4) at the nanoscale level under compression loading. In the present publication, we follow strictly the above-said numerical methodology by extending it to the application of a nanoindentation testing method, so as to define the mechanical response of nanocrystalline metals at the microscale level.

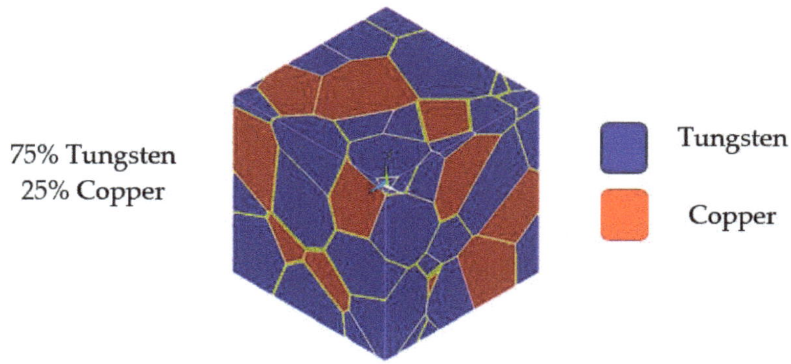

Figure 4. The representative volume element of a tungsten–copper nanocrystalline alloy.

The finite element modeling of the nanoindentation test was developed utilizing Abaqus CAE (Computer Aided Engineering) (Dassault Systemes Simulia Corp., Providence, RI, USA). The model was developed in three dimensions consisting of 2 objects, a representative pyramidal-like indenter with the real dimensions of Berkovich indenter and a cube representing the specimen with dimensions of 5 μm × 5 μm × 3 μm (Figure 5). The indenter was modeled to be rigid. The definition of elastoplastic response of cubic specimen was achieved by using the innovative nanoscale numerical model which was described above [48]. The material properties applied to the cubic specimen were extracted from the RVE as the homogenized average medium. The aforementioned mechanical properties were achieved by applying a compressive loading on the RVE. Additionally, a surface-to-surface frictional contact with a coefficient of 0.1 was implemented for the indentation model. Several numerical investigations have shown that the frictional coefficient between the indenter and target material is an insignificant parameter regarding the experimental nanoindentation outcomes. In the present work, the influence of the frictional contact was taken into account in the numerical model in order to develop numerically the same experimental conditions, with the ultimate goal of the maximum similarity between the numerical model and the experiments. Regarding the numerical parameters, the target material was meshed with 62,000 hexahedral elements of type C3D8R (eight node), with a dense mesh at the area of the target material directly beneath Berkovich's tip. Moreover, the indenter had a constant penetration speed of 5.5 nm/s until it reached the load of 15 mN, in the loading phase, which is a pre-defined setup parameter. After this step, the indenter was removed from the target material with a constant speed of 16 nm/s in the unloading phase. At this point, it is important to mention that the ratio of the cube's height to the maximum penetration depth is approximately equal to 8 in the current model, so as to reduce computational time. From the scope of parametric study, higher ratios of 10 and 20 were also investigated, and the results presented an insignificant divergence in the resulting *P-h* diagram compared to that of the ratio of 8.

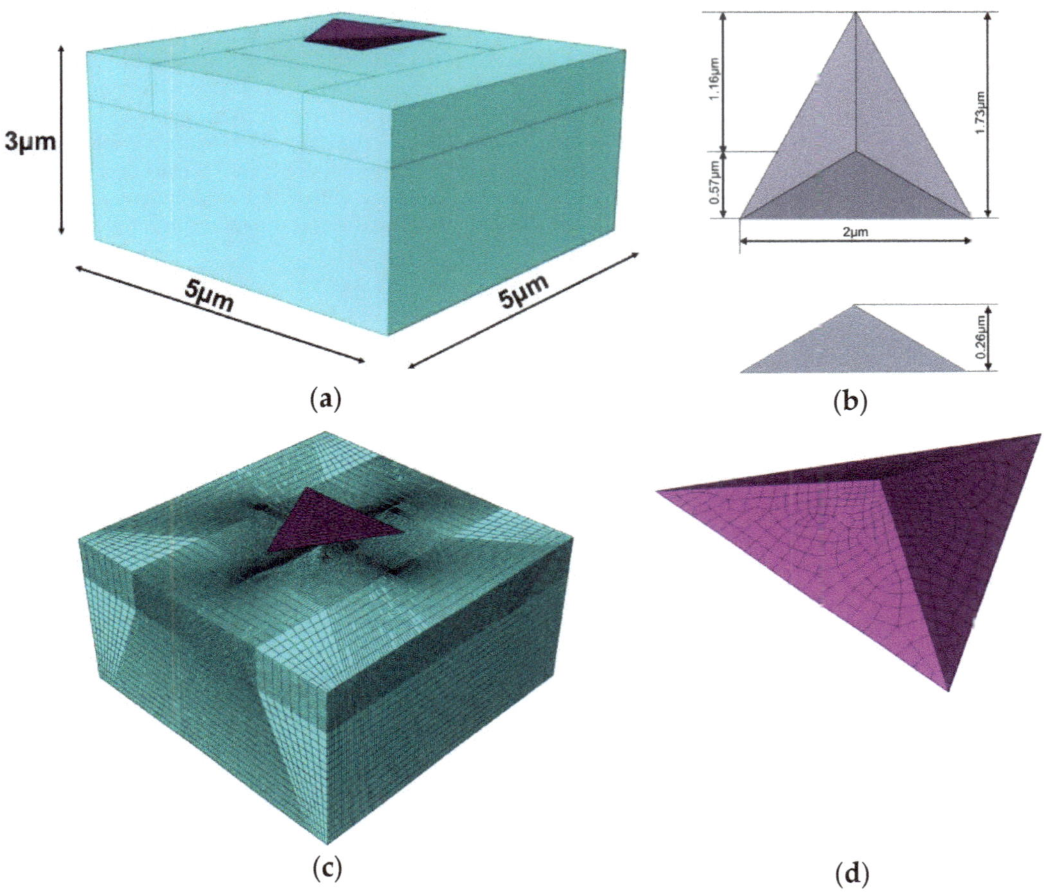

Figure 5. Three-dimensional finite element model of the nanoindentation: (**a**) the dimensions of the target material, (**b**) the dimensions of the Berkovich indenter, (**c**) the meshed model and (**d**) the meshed Berkovich indenter.

The main advantage of the numerical nanoindentation models is that they are able to acquire the load-versus-depth curve of a specified reference point by continuous data acquisition throughout the duration of the simulation, just as during an experiment in situ. Furthermore, another important aspect that should be mentioned is that the proposed numerical methodology has taken into account the importance of the dimensional definition of the indented cube. In this way, the boundary effects are eliminated on the nanoindentation model, and the assumption of a continuous medium was validated. During the setup of the FE parameters, one of the most important boundary conditions applied in this model is the fixation of the target material's base so as to ensure an ideal indentation test. A second important boundary condition applied is the constraint of the nanoindenter in such a way that the nanoindenter is fixed to allow motion in the vertical direction.

2.3. Inverse Analysis Algorithm

A comprehensive analytical methodology for instrumented indentation was identified in this paper, which allows us to define the mechanical properties of materials by using a Berkovich nanoindenter. The current paper is focused on the development of an inverse analysis algorithm so as to obtain the material properties via nanoindentation tests and numerical nanoindentation models. Figure 2 shows the *P*-*h* curve for a Berkovich indenter.

In the loading phase, it seems that the curve is in consistency with the relation $P = Ch^2$, where parameter C is responsible for the indentation curvature. This curvature represents the resistance of the target material to the indentation effect. Additionally, the calculated pressure of contact, $p_{av} = P_{max}/A_{max}$, could be determined with the hardness of the target material, where P_{max} is the maximum applied force for the indentation. The P_{max} provides the necessary energy to the nanoindenter to penetrate the target material by a depth of h_{max}, by that means creating the A_{max} (the maximum projected contact area on the surface of the target material). Elastoplastic FE models with a Berkovich indenter, conducted on the present paper utilizing computational analyses comparable with those presented in reference [38], also show that

$$\frac{h_r}{h_{max}} = 1 - d^*S, \tag{8}$$

where d^* = 4.678 for the case of a Berkovich indenter. Equation (8) is in rational accordance with the experimental observations of Breval and MacMillan [66]. Taking into consideration the true contact area on three-dimensional numerical models and the strain hardening effects imposed on pile-up and sink-in strains, the subsequent mathematical expression, which correlates A_{max} with h_{max}, was obtained for the case study of elastic–plastic materials [30,38]:

$$\frac{A_{max}}{h^2{}_{max}} = 9.96 - 12.64 * (1-S) + 105.42(1-S)^2 - 229.57(1-S)^3 + 157.67(1-S)^4,\ with\ S = \frac{p_{av}}{E^*} \tag{9}$$

The aforementioned mathematical expression is a polynomial regression to the calculated values of $A_{max}/h^2{}_{max}$. In Equation (9), the resulting Young's modulus of the Berkovich nanoindenter–target material system, E^*, is derived as:

$$E^* = \frac{1}{c^*\sqrt{A_{max}}}\left(\frac{dP}{dh}\right), \tag{10}$$

where dP/dh is the incline of the P-h diagram at the unloading phase from P_{max}, while the c^* parameter is equal to 1.167 for the case study of the Berkovich nanoindenter. Furthermore, the ratio of the residual penetration depth h_r to the maximum depth of indentation, h_{max}, is typical of the magnitude of strain hardening and plastic deformation [41,67,68]:

$$\frac{\sigma_{0.29} - \sigma_y}{0.29E^*} = 1 - 0.142\frac{h_r}{h_{max}} - 0.957\left(\frac{h_r}{h_{max}}\right)^2 \tag{11}$$

In the aforementioned Equation, σ_y refers to the yield strength, while $\sigma_{0.29}$ refers on the indicative plastic strain of 0.29 for the target material in uniaxial loading.

At this point, it is noteworthy to mention that the combination of Equations (8) and (9) gives an important relationship which correlates A_{max} with h_{max}. Using the aforementioned combined mathematical expression, we are able to extract the true contact area from the P-h diagram without the need of any visual examination. Knowing that

$$P_{max} \approx C h_{max}{}^2, \tag{12}$$

we can calculate the parameter C so as to use it into the following equation, which comes from 3D numerical simulations of elastic–plastic nanoindentation, in addition to the corresponding experiments [36,38,42,67,68]:

$$C = \frac{P}{h^2} = M_1\sigma_{0.29}\left\{1 + \frac{\sigma_y}{\sigma_{0.29}}\right\}\left\{M_2 + ln\frac{E^*}{\sigma_y}\right\} \tag{13}$$

The constants in this mathematical expression are M_1 and M_2, where M_1 is equal to 6.618 and M_2 is equal to -0.875 for the case study of a Berkovich indenter [40].

With all of the values calculated using the above-said analytical expressions, the strain hardening exponent can be easily calculated by applying the subsequent equation:

$$n \approx ln\{\sigma_{0.29} - \ln(\sigma_y)\}/5 \quad (14)$$

3. Experimental

3.1. Materials

To begin with, the tungsten–copper alloy system was selected for the present investigation. As mentioned above, the selection of the present alloying element combination was made in the frame of the European project ICARUS [64], which was focused on the investigation of coarsening-resistant alloys for aerospace applications. The alloy consisted of 75% tungsten and 25% copper in weight. For this material composition, two different types of tungsten–copper alloy were produced in terms of the resulting microstructure. Both types were manufactured using the same powders originating from the same powder batch.

The coarse-grained tungsten–copper specimens (cW-Cu) were fabricated through the simple mixture of the as-received commercial powders of tungsten and copper elements without any grain refinement process being involved. Additionally, the well-known manufacturing approach consisting of cold pressing, hot isostatic pressing (HIP) (University of Miskolc, Miskolc, Hungary) and heat treatment was implemented so as to obtain the consolidated specimens. The coarse-grained tungsten–copper specimen has an average grain size of 150 nm.

On the other hand, the high energy ball milling (HEBM) (MBN Nanomaterialia S.p.A., Treviso, Italy) method was utilized in order to mill tungsten and copper powders, achieving a nanocrystalline morphology with a grain diameter of 100 nm. The nanocrystalline tungsten–copper powders used for the production of the alloy will be referred in the following as W-Cu. Following the same consolidation technique of the cW-Cu specimens for the production of the corresponding W-Cu specimens, the fabrication method of cold pressing, hot isostatic pressing (HIP) and heat treatment was implemented here again. An extensive description of the produced materials' microstructures and their resulting imperfections (pores and impurities) can be found in [48]. The above-mentioned production of the two types of tungsten–copper alloys was performed by the company MBN Nanomaterialia.

3.2. Nanoindentation Test

Nanoindentation refers to a variety of hardness tests which are applied on a small scale. Penetration is perhaps the most common way of checking the mechanical properties of materials facilitated by high-precision instrumentation in the nanometer scale. In 2008, the ISO/TR 29381 was published, allowing for the evaluation of the tensile properties of metallic materials by instrumented indentation [69].

A nanoindentation device was developed at the Physical Metallurgy Laboratory, and is presented in Figures 6 and 7. The device performs testing using a Berkovich, Vickers, or Knoop diamond, or 0.5–2 mm indenters. The travel distance of the workbench can reach 50 by 140 mm, guided by specially developed computer software. The height of the optics is measured by means of a laser distance meter coupled with a rotary encoder attached to the fine adjustment dial of the microscope. The accuracy of the device in the XY direction is in the order of 1 µm. The applied load can vary from 1 to 1000 mN with a resolution of 0.1 mN, strictly following the ISO 14577 [70]. The measurement consists of two stages: the loading phase and the unloading phase. During the loading, force is applied gradually to the diamond indenter, and as it penetrates the test piece, the depth measurement is recorded. During the unloading, a residual depth remains due to the plastic deformation of the sample, which depends on the properties of the material, the size of the applied force, and the geometry of the indenter. The maximum indentation depth fluctuations measured on the same sample are mainly induced by different contact conditions between the indenter tip and the tested surface due to roughness [71]. The roughness effect can be confronted with the execution of an appreciable number of measurements to attain the stabilization of the maximum indentation depth's mean value [71]. The proper preparation of the examined specimens is a meticulous and time-consuming process that ensures the repeatability and robustness of the results, while eliminating those parameters that could tamper with the outcome due to mishandling or human error. The mounted samples were gently, mechanically ground and polished in several sequential steps employing a BUEHLER Vanguard^{TM} 2000 (Lake Bluff, IL, USA) automatic sample preparation system [72]. The mechanical grinding took place in several sequential steps, starting with coarse grinding discs (60 grit) and ending with finer ones (1600 grit). The polishing was obtained by employing an aqueous suspension of fine alumina powder with a particle size of 0.3 μm applied on the respective polishing cloths. In this work, 50 nanoindentation tests were performed on each specimen to eliminate roughness effects.

Figure 6. Nanoindentation test of nanocrystalline specimens according to ISO 14577.

Figure 7. Specimen's fixation for the nanoindentation test.

The evolution of FEM-based algorithms in the evaluation of nanoindentation results offers advanced capabilities in the determination of the exact contact between the indenter and the test piece, thus allowing the accurate calculation of the material's hardness and the stress–strain curve. The FEM model simulating the nanoindentation procedure and the actual indenter tip geometry, introduced in [72], can be applied to calculate the contact geometry during the loading and the shape of the surface impression after unloading. The

"SSCUBONI" algorithm employed for the continuous simulation of the nanoindentation enables the extraction of materials' stress–strain laws. FEM-supported methods offer advanced capabilities in the determination of mechanical properties, such as the Young's modulus, yield and rupture strength. Stress–strain curves of various materials—such as cemented carbide, ceramics and hardened steels—determined by a FEM-based algorithm of nanoindentation can be applied in FEM-supported simulations of micro- and macro-scale indentation procedures to allow the capture of the material response at various scales [73].

4. Results

The indentation force (*P*)–penetration depth (*h*) curves obtained using a Berkovich indenter tip, after a set of 50 measurements that were statistically evaluated to present the averages for the cW-Cu and the W-Cu type of alloy are presented in Figure 8, as well as the scatter of the measurement at the maximum indentation depth. As shown in Figure 8, the coarse-grained tungsten–copper alloy (cW-Cu) exhibits a higher penetration of about 400nm depth at an indentation load of 15 mN [73,74] compared to that of nanocrystalline the tungsten–copper alloy (W-Cu), which presents a penetration depth of about 250 nm at the applied penetration load of 15 mN, correspondingly. The aforementioned outcomes are reasonable due to the fact that the Hall–Petch effect governs this type of material, and the yield stress is increased by decreasing the grain size. Thus, the nanocrystalline specimen shows a higher "resistance" to the indentation effect and to plastic deformation.

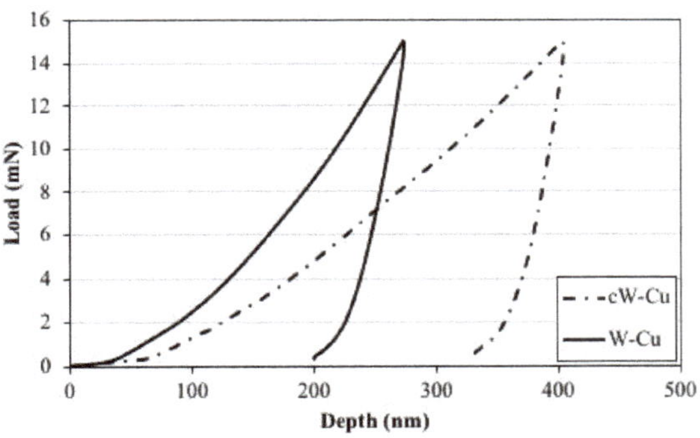

(a)

	cW-cu	W-cu
Min. Depth (nm)	352	224
Aver. Depth (nm)	407	258
Max. depth (nm)	522	284

(b)

Figure 8. (**a**) The experimental results of cW-Cu and WCu specimens in accordance with ISO 14577, and (**b**) the scatter of the nanoindentation results.

4.1. Validation of the Numerical Nanoindentation Model

Using the numerical multiscale model, the homogenized nanoindentation response of the fully dense, indented material was calculated for the case of a coarse-grained W-Cu (cW-Cu) material, and is shown in Figure 9. In this Figure, the computed *P-h* curves of a fully dense, nanocrystalline material using two Berkovich indenters—one with a sharp tip and one with a rounded tip with a roundness of 100 nm—are displayed and compared to the respective experimental curve. The geometrical features and the material properties applied on the cubic specimen are described extensively in [48]. The divergence of the numerical outcomes from the experimental result is negligible because the nanoindentation model does not take into account any defect (like pores) which might be located in the microstructure of the specimens. The aforementioned assumption of the absence of defects on the numerical model leads to the simulation of a fully dense, homogeneous-like material which may be different from that reality. On the other hand, the existing impurities or pores are mostly located at the lower layers of specimens, and as a result the nanoindenter cannot reach it through the nanoindentation test due to the very small indentation depth.

This phenomenon leads to the prediction of a nanoindentation response of a fully dense material without microstructure irregularities.

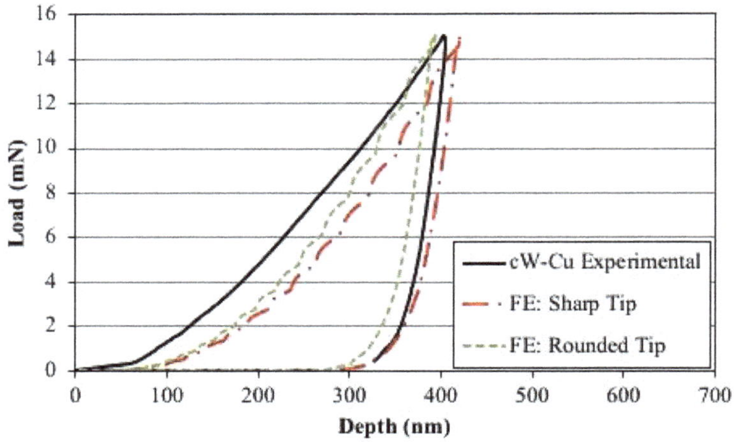

Figure 9. Comparison of the experimental result of the cW-Cu specimen with the numerical results utilizing sharp and rounded Berkovich indenters.

Taking for granted the above assumption, the only parameter which plays an important role on the definition of the nanoindentation response is the sharpness of the nanoindenter. In terms of the tip sharpness, both the tungsten–copper alloys and their corresponding numerical results (Figures 9 and 10) present the same tendency. More specifically, the numerical results of the sharp Berkovich indenter present the higher penetration depth of both materials. This phenomenon is reasonable due to the extensive stress concentration beneath the Berkovich tip. The sharper the tip, the higher the stress concentration factor is. On the other hand, the implementation of a rounded tip of the Berkovich indenter is a more realistic approach due to the unavoidable degeneration of the nanoindentation apparatus through its extensive usage. The above fact is logical, and the effect of the indenter's sharpness on the mechanical response in nanoindentation experiment was reported in [75]. Taking this aspect for granted, the numerical results of the rounded Berkovich indenter tend to show the most reliable nanoindentation behavior.

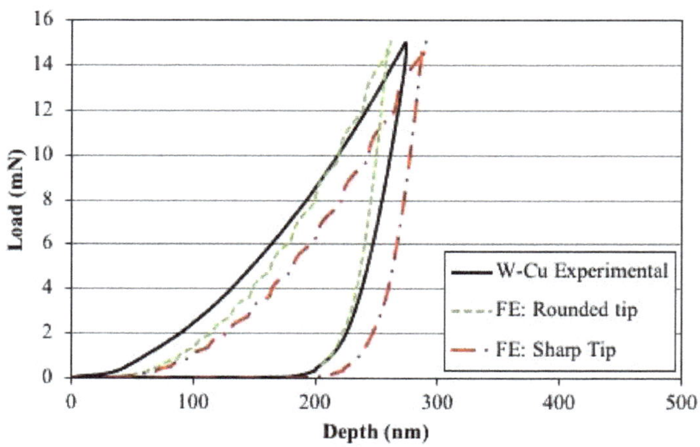

Figure 10. Comparison of the experimental result of a W-Cu specimen with the numerical results of a fully dense model utilizing sharp and rounded Berkovich indenters.

To sum up, the experimental *P-h* curve better reflects the nanoindentation behavior of a fully dense material than of a material with an extensive porosity like the examined specimens in this paper. The proposed numerical methodology shows a very good convergence with the nanoindentation test. On the other hand, utilizing the inverse algorithm as it is shown in Section 4.2, the macroscopic compression behavior obtained by the inverse algorithm of the numerical nanonindentation results will differ from the experimental compression tests examined in [48] because it does not take into account the inherent defects (impurities and pores) of the microstructure.

4.2. Comparison of Nanoindentation Experimental Results Utilizing the SSCUBONI Algorithm

Implementing the SSCUBONI algorithm, the resulting representative von Mises stress–strain curves of the cW-Cu and W-Cu specimens from the nanoindentation tests are shown in Figures 11 and 12, respectively. The aforementioned curves are compared to those of the compression tests for the same batches.

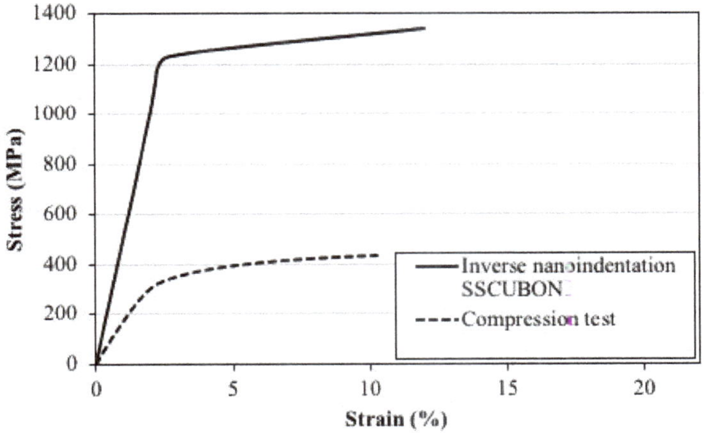

Figure 11. Comparison of the S-S curves of the experimental compression test with the corresponding inverse SSCUBONI-based experimental test for the cW-Cu specimens.

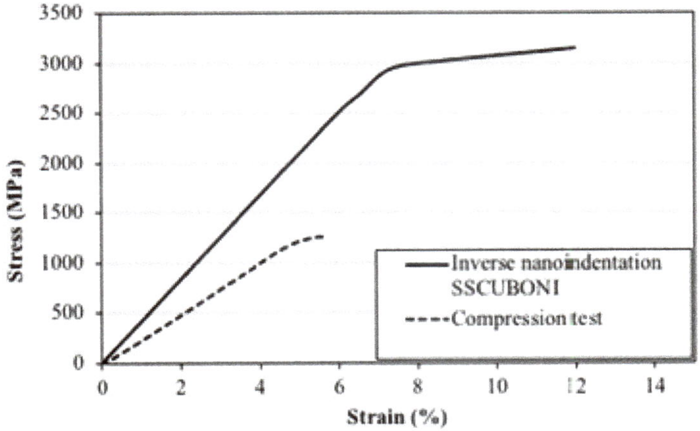

Figure 12. Comparison of the S-S curves of the experimental compression test with the corresponding inverse SSCUBONI-based experimental test for the W-Cu specimens.

As can be seen from both Figures, the inverse experimental nanoindentation results present a remarkable deviation from the compression test of the cW-Cu and W-Cu counterparts. More specifically, in Figure 11, the inverse SSCUBONI nanoindentation curve exhibits a Young's modulus of 200 GPa and a yield stress of 1340 MPa, while the corresponding compression curve shows a Young's modulus of 180 GPa and a yield stress of 450 MPa. Moreover, in the case of W-Cu specimens (Figure 12), the inverse SSCUBONI curve from the experimental nanoindentation curve shows a Young's modulus of 250 GPa and a yield stress of 2.8 GPa, while the respective compression curve for the case of the W-Cu specimen exhibits a Young's modulus of 220 GPa and a yield stress of 1150 MPa.

It is evident that the deviation of the inverse experimental nanoindentation outcomes from the experimental compression results is appreciable but reasonable. The porosity effect is the main characteristic which relies on the aforementioned deviation. As was already explained above, the nanoindenter penetrates the target material at a contact area of a few nanometers for an indentation depth of some nanometers. The contact area of the nanoindentation is extremely small compared to the volume of the specimen, and as a result the obtained experimental results reflect a nanoindentation behavior of a homogeneous-like material excluding all of the present heterogeneities into its microstructure by extracting a nanoindentation behavior of an ideal material. This procedure leads to unreliable results in the case of a material with several defects or impurities in its microstructure. The experimental test is not able to perceive the extensive presence of pores which are located in the lower layers. In our case, the experimental *P-h* curve better reflects the nanoindentation behaviour of a fully dense material than that of a material with an extensive porosity like the examined specimens of the current project.

4.3. Comparison of Numerical Nanoindentation Results Utilizing the Inverse Algorithm

In an effort to predict the macroscopic material properties via the nanoindentation testing method, an inverse algorithm was proposed and utilized in this paper. The macroscopic mechanical responses of the experimental results and the numerical outcomes of nanoindentation models were calculated for the case of a cW-Cu material and a W-Cu material, and they are shown in Figures 13 and 14, respectively. As can be seen, the experimental nanoindentation results obtained from the SSCUBONI algorithm show an appreciable convergence with the corresponding numerical nanoindentation results obtained by the proposed inverse algorithm.

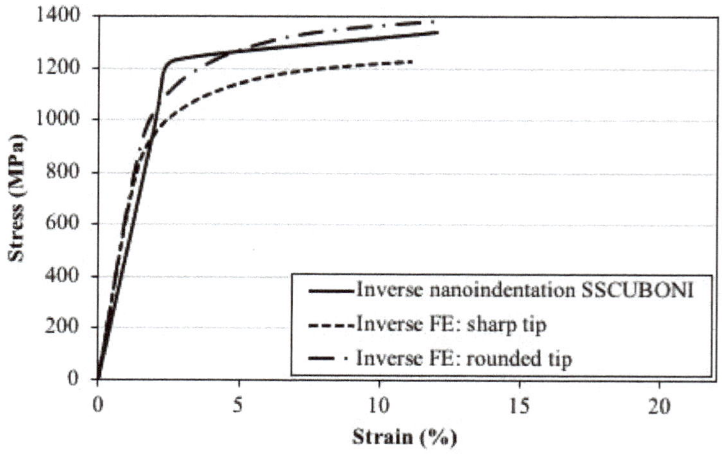

Figure 13. Comparison of the S-S curves of the experimental compression test with the corresponding inverse numerical nanoindentation results and the inverse SSCUBONI-based experimental test for the cW-Cu specimens.

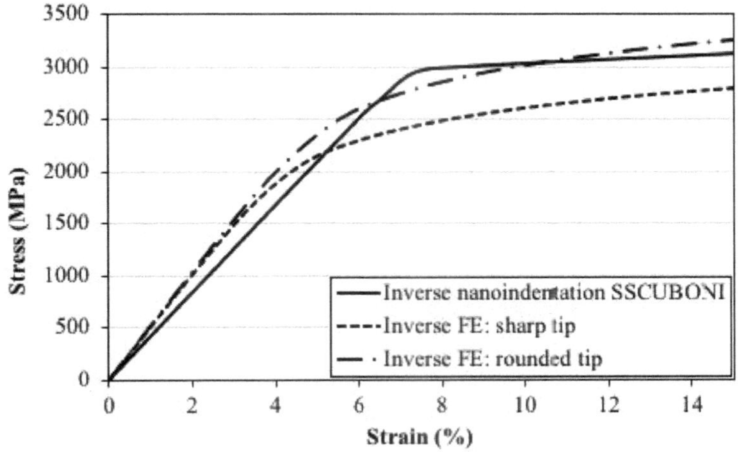

Figure 14. Comparison of the S-S curves of the experimental compression test with the corresponding inverse numerical nanoindentation results and the inverse SSCUBONI-based experimental test for the W-Cu specimens.

Both of the inverse analysis algorithms seem to be efficient in converting the penetration load–indentation depth curve to stress–strain curves, but the proposed multiscale nanoindentation model in conjunction with the inverse algorithm provides an advantage to the scientific community because it is able to predict the macroscopic mechanical response of any nanocrystalline material without executing any experimental test. However, it is noteworthy to mentioned at this point that the porosity effect is a crucial parameter for the mechanical behavior. For the aforementioned reason, the numerical multiscale nanoindentation methodology implementing the porosity effect for the case study of porous nanocrystalline alloys in combination with the proposed inverse analysis algorithm needs to be developed so as to predict the macroscopic mechanical response more efficiently.

5. Conclusions

Herein, a methodology was developed for the simulation of the nanoindentation testing of fully dense nanocrystalline materials utilizing a combination of numerical and analytical approaches. The numerical nanoindentation outcomes correlate very well with the experimental nanoindentation results, which validates the proposed methodology in terms of simulating the nanoindentation test. However, in the scope of the prediction of mechanical properties at the macroscale level, the proposed methodology shows an appreciable divergence with the experimental compression tests due to the porosity effect. For the aforementioned reason, it is of paramount importance that the proposed multiscale numerical methodology be modified so as to take into account the porosity effect. Additionally, a thorough investigation of the proposed methodology has to be carried out by applying different indentation loads, through numerical simulations and experimental tests, and by testing various specimens with different alloying elements, compositions and grain sizes so as to be validated for any alloy. Considering the above, it can be stated that the proposed numerical approach in conjunction with the inverse algorithm represents a preliminary contribution towards the development of a numerical nanoindentation model, and may serve as the basis for the development of large-scale models to be applied in the quality control of the mass production systems of the aforementioned promising materials in the near future.

Author Contributions: Conceptualization, S.G.P., K.T. and N.M.; methodology, S.G.P. and K.T.; software, P.B.; validation, P.B. and M.P.; formal analysis, P.B.; investigation, P.B.; data curation, P.B. and M.P.; writing—original draft preparation, P.B.; writing—review and editing, P.B. and M.P.; visualization, P.B.; supervision, S.G.P., K.T. and N.M.; project administration, S.G.P. All authors have read and agreed to the published version of the manuscript.

Funding: This research was funded by ICARUS, grant agreement No. 713514, of the European Union's Horizon 2020 research and innovation programme.

Data Availability Statement: Data sharing not applicable.

Acknowledgments: The work described in this paper received funding from the European Union's Horizon 2020-FETOPEN research and innovation programme under Grant Agreement no. 713514: ICARUS project (Innovative Coarsening-resistant Alloys with enhanced Radiation tolerance and Ultrafine-grained Structure for aerospace application). This paper originated from a presentation made in the frame of the sixth International Conference of Engineering Against Failure-ICEAF VI.

Conflicts of Interest: The authors declare no conflict of interest.

References

1. Gleiter, H. Nanostructured materials: Basic concepts and microstructure. *Acta Mater.* **2000**, *48*, 1–29. [CrossRef]
2. Kumar, K.; Van Swygenhoven, H.; Suresh, S. Mechanical behavior of nanocrystalline metals and alloys The Golden Jubilee Issue—Selected topics in Materials Science and Engineering: Past, Present and Future, edited by S. Suresh. *Acta Mater.* **2003**, *51*, 5743–5774. [CrossRef]
3. Naik, S.; Walley, S. The Hall–Petch and inverse Hall–Petch relations and the hardness of nanocrystalline metals. *J. Mater. Sci.* **2019**, *55*, 2661–2681. [CrossRef]
4. Hibbard, G.; Erb, U.; Aust, K.; Klement, U.; Palumbo, G. Thermal Stability of Nanostructured Electrodeposits. *J. Metastable Nanocryst. Mater.* **2002**, *13*, 387–396. [CrossRef]
5. Suryanarayana, C. Nanocrystalline materials. *Int. Mater. Rev.* **1995**, *40*, 41–64. [CrossRef]
6. Groza, J. Nanostructured Materials. *Mater. Today* **2002**, *5*, 55.
7. Liao, W.; Lan, S.; Gao, L.; Zhang, H.; Xu, S.; Song, J.; Wang, X.; Lu, Y. Nanocrystalline high-entropy alloy (CoCrFeNiAl0.3) thin-film coating by magnetron sputtering. *Thin Solid Films* **2017**, *638*, 383–388. [CrossRef]
8. Liao, W.; Zhang, H.; Liu, Z.; Li, P.; Huang, J.; Yu, C.; Lu, Y. High Strength and Deformation Mechanisms of Al0.3CoCrFeNi High-Entropy Alloy Thin Films Fabricated by Magnetron Sputtering. *Entropy* **2019**, *21*, 146. [CrossRef] [PubMed]
9. Xing, Q.; Ma, J.; Zhang, Y. Phase thermal stability and mechanical properties analyses of (Cr,Fe,V)-(Ta,W) multiple-based elemental system using a compositional gradient film. *Int. J. Miner. Metall. Mater.* **2020**, *27*, 1379–1387. [CrossRef]
10. Xing, Q.; Ma, J.; Wang, C.; Zhang, Y. High-Throughput Screening Solar-Thermal Conversion Films in a Pseudobinary (Cr, Fe, V)-(Ta, W) System. *ACS Comb. Sci.* **2018**, *20*, 602–610. [CrossRef]
11. Lowe, T.; Valiev, R. Producing nanoscale microstructures through severe plastic deformation. *JOM* **2000**, *52*, 27–28. [CrossRef]
12. Valiev, R.; Islamgaliev, R.; Alexandrov, I. Bulk nanostructured materials from severe plastic deformation. *Prog. Mater. Sci.* **2000**, *45*, 103–189. [CrossRef]
13. Zhou, H.; Hu, L.; Sun, Y.; Zhang, H.; Duan, C.; Yu, H. Synthesis of nanocrystalline AZ31 magnesium alloy with titanium addition by mechanical milling. *Mater. Charact.* **2016**, *113*, 108–116. [CrossRef]
14. Suresh, S. Graded Materials for Resistance to Contact Deformation and Damage. *Science* **2001**, *292*, 2447–2451. [CrossRef]
15. Delfosse, D. Fundamentals of Functionally Graded Materials. *Mater. Today* **1998**, *1*, 18. [CrossRef]
16. Masumura, R.; Hazzledine, P.; Pande, C. Yield stress of fine grained materials. *Acta Mater.* **1998**, *46*, 4527–4534. [CrossRef]
17. Jeong, D.; Gonzalez, F.; Palumbo, G.; Aust, K.; Erb, U. The effect of grain size on the wear properties of electrodeposited nanocrystalline nickel coatings. *Scr. Mater.* **2001**, *44*, 493–499. [CrossRef]
18. Hayes, R.; Witkin, D.; Zhou, F.; Lavernia, E. Deformation and activation volumes of cryomilled ultrafine-grained aluminum. *Acta Mater.* **2004**, *52*, 4259–4271. [CrossRef]
19. Jia, D.; Ramesh, K.; Ma, E. Effects of nanocrystalline and ultrafine grain sizes on constitutive behavior and shear bands in iron. *Acta Mater.* **2003**, *51*, 3495–3509. [CrossRef]
20. Cavaliere, P. Strain Rate Sensitivity and Fatigue Properties of an Al-fe Nanocrystalline Alloy Produced by Cryogenic Ball Milling. *Multidiscip. Model. Mater. Struct.* **2007**, *3*, 225–234. [CrossRef]
21. Tabor, D. *The Hardness of Metals*; Oxford University Press: New York, NY, USA, 1951.
22. Tabor, D. The hardness of solids. *Rev. Phys. Technol.* **1970**, *1*, 145–179. [CrossRef]
23. Doerner, M.; Nix, W. A method for interpreting the data from depth-sensing indentation instruments. *J. Mater. Res.* **1986**, *1*, 601–609. [CrossRef]
24. Pharr, G.; Cook, R. Instrumentation of a conventional hardness tester for load-displacement measurement during indentation. *J. Mater. Res.* **1990**, *5*, 847–851. [CrossRef]

25. Oliver, W.; Pharr, G. An improved technique for determining hardness and elastic modulus using load and displacement sensing indentation experiments. *J. Mater. Res.* **1992**, *7*, 1564–1583. [CrossRef]
26. Field, J.; Swain, M. A simple predictive model for spherical indentation. *J. Mater. Res.* **1993**, *8*, 297–306. [CrossRef]
27. Field, J.; Swain, M. Determining the mechanical properties of small volumes of material from submicrometer spherical indentations. *J. Mater. Res.* **1995**, *10*, 101–112. [CrossRef]
28. Gerberich, W.; Nelson, J.; Lilleodden, E.; Anderson, P.; Wyrobek, J. Indentation induced dislocation nucleation: The initial yield point. *Acta Mater.* **1996**, *44*, 3585–3598. [CrossRef]
29. Bolshakov, A.; Oliver, W.; Pharr, G. Influences of stress on the measurement of mechanical properties using nanoindentation: Part II. Finite element simulations. *J. Mater. Res.* **1996**, *11*, 760–768. [CrossRef]
30. Alcalá, J.; Giannakopoulos, A.; Suresh, S. Continuous measurements of load-penetration curves with spherical microindenters and the estimation of mechanical properties. *J. Mater. Res.* **1998**, *13*, 1390–1400. [CrossRef]
31. Cheng, Y.; Cheng, C. Scaling approach to conical indentation in elastic-plastic solids with work hardening. *J. Appl. Phys.* **1998**, *84*, 1284–1291. [CrossRef]
32. Cheng, Y.; Cheng, C. Relationships between hardness, elastic modulus, and the work of indentation. *Appl. Phys. Lett.* **1998**, *73*, 614–616. [CrossRef]
33. Suresh, S.; Nieh, T.; Choi, B. Nano-indentation of copper thin films on silicon substrates. *Scr. Mater.* **1999**, *41*, 951–957. [CrossRef]
34. Gouldstone, A.; Koh, H.; Zeng, K.; Giannakopoulos, A.; Suresh, S. Discrete and continuous deformation during nanoindentation of thin films. *Acta Mater.* **2000**, *48*, 2277–2295. [CrossRef]
35. Johnson, K. The correlation of indentation experiments. *J. Mech. Phys. Solids* **1970**, *18*, 115–126. [CrossRef]
36. Suresh, S.; Alcala, J.; Giannakopoulos, A. Depth Sensing Indentation and Methodology for Mechanical Property Measurements. U.S. Patent 6134954A, 19 June 2020.
37. Dao, M.; Chollacoop, N.; Van Vliet, K.; Venkatesh, T.; Suresh, S. Computational modeling of the forward and reverse problems in instrumented sharp indentation. *Acta Mater.* **2001**, *49*, 3899–3918. [CrossRef]
38. Giannakopoulos, A.; Larsson, P.; Vestergaard, R. Analysis of Vickers indentation. *Int. J. Solids Struct.* **1994**, *31*, 2679–2708. [CrossRef]
39. Cheng, Y.; Cheng, C. Can stress–strain relationships be obtained from indentation curves using conical and pyramidal indenters? *J. Mater. Res.* **1999**, *14*, 3493–3496. [CrossRef]
40. Giannakopoulos, A.; Suresh, S. Determination of elastoplastic properties by instrumented sharp indentation. *Scr. Mater.* **1999**, *40*, 1191–1198. [CrossRef]
41. Venkatesh, T. Determination of elasto-plastic properties by instrumented sharp indentation: Guidelines for property extraction. *Scr. Mater.* **2000**, *42*, 833–839. [CrossRef]
42. Suresh, S.; Giannakopoulos, A. A new method for estimating residual stresses by instrumented sharp indentation. *Acta Mater.* **1998**, *46*, 5755–5767. [CrossRef]
43. Bhattacharya, A.; Nix, W. Finite element simulation of indentation experiments. *Int. J. Solids Struct.* **1988**, *24*, 881–891. [CrossRef]
44. Laursen, T.; Simo, J. A study of the mechanics of microindentation using finite elements. *J. Mater. Res.* **1992**, *7*, 618–626. [CrossRef]
45. Tunvisut, K.; O'Dowd, N.; Busso, E. Use of scaling functions to determine mechanical properties of thin coatings from microindentation tests. *Int. J. Solids Struct.* **2001**, *38*, 335–351. [CrossRef]
46. Hill, R.; Storakers, B.; Zdunek, A. A theoretical study of the Brinell hardness test. *Proc. R. Soc. Lond. A Math. Phys. Sci.* **1989**, *423*, 301–330.
47. Larsson, P.; Giannakopoulos, A.; SÖderlund, E.; Rowcliffe, D.; Vestergaard, R. Analysis of Berkovich indentation. *Int. J. Solids Struct.* **1996**, *33*, 221–248. [CrossRef]
48. Bazios, P.; Tserpes, K.; Pantelakis, S. Modelling and Experimental Validation of the Porosity Effect on the Behaviour of Nano-Crystalline Materials. *Metals* **2020**, *10*, 821. [CrossRef]
49. Oliver, W.; Pharr, G. Measurement of hardness and elastic modulus by instrumented indentation: Advances in understanding and refinements to methodology. *J. Mater. Res.* **2004**, *19*, 3–20. [CrossRef]
50. Chrobak, D.; Kim, K.; Kurzydłowski, K.; Nowak, R. Nanoindentation experiments with different loading rate distinguish the mechanism of incipient plasticity. *Appl. Phys. Lett.* **2013**, *103*, 072101. [CrossRef]
51. Wang, Y.; Raabe, D.; Klüber, C.; Roters, F. Orientation dependence of nanoindentation pile-up patterns and of nanoindentation microtextures in copper single crystals. *Acta Mater.* **2004**, *52*, 2229–2238. [CrossRef]
52. Tadmor, E.; Miller, R.; Phillips, R.; Ortiz, M. Nanoindentation and incipient plasticity. *J. Mater. Res.* **1999**, *14*, 2233–2250. [CrossRef]
53. Zhang, J.; Zhang, J.; Wang, Z.; Hartmaier, A.; Yan, Y.; Sun, T. Interaction between phase transformations and dislocations at incipient plasticity of monocrystalline silicon under nanoindentation. *Comput. Mater. Sci.* **2017**, *131*, 55–61. [CrossRef]
54. Fischer-Cripps, A. A simple phenomenological approach to nanoindentation creep. *Mater. Sci. Eng A* **2004**, *385*, 74–82. [CrossRef]
55. Elmustafa, A.; Stone, D. Nanoindentation and the indentation size effect: Kinetics of deformation and strain gradient plasticity. *J. Mech. Phys. Solids* **2003**, *51*, 357–381. [CrossRef]
56. Karthik, V.; Visweswaran, P.; Bhushan, A.; Pawaskar, D.; Kasiviswanathan, K.; Jayakumar, T.; Raj, B. Finite element analysis of spherical indentation to study pile-up/sink-in phenomena in steels and experimental validation. *Int. J. Mech. Sci.* **2012**, *54*, 74–83. [CrossRef]

57. Biener, M.; Biener, J.; Hodge, A.; Hamza, A. Dislocation nucleation in bcc Ta single crystals studied by nanoindentation. *Phys. Rev. B* **2007**, *76*, 165422. [CrossRef]
58. Morris, J.; Bei, H.; Pharr, G.; George, E. Size Effects and Stochastic Behavior of Nanoindentation Pop In. *Phys. Rev. Lett.* **2011**, *106*. [CrossRef]
59. Hasnine, M.; Mustafa, M.; Suhling, J.; Prorok, B.; Bozack, M.; Lall, P. Characterization of aging effects in lead free solder joints using nanoindentation. In Proceedings of the 2013 IEEE 63rd Electronic Components and Technology Conference, Las Vegas, NV, USA, 28–31 May 2013; pp. 166–178. [CrossRef]
60. Lucas, B.; Oliver, W. Indentation power-law creep of high-purity indium. *Metall. Mater. Trans. A* **1999**, *30*, 601–610. [CrossRef]
61. Zhang, P.; Li, S.; Zhang, Z. General relationship between strength and hardness. *Mater. Sci. Eng. A* **2011**, *529*, 62–73. [CrossRef]
62. Hay, J.; Agee, P.; Herbert, E. Continuous stiffness measurement during instrumented indentation testing. *Exp. Tech.* **2010**, *34*, 86–94. [CrossRef]
63. Chen, X.; Xiang, Y.; Vlassak, J. Novel technique for measuring the mechanical properties of porous materials by nanoindentation. *J. Mater. Res.* **2006**, *21*, 715–724. [CrossRef]
64. Innovative Coarsening-Resistant Alloys with Enhanced Radiation Tolerance and Ultrafine-Grained Structure for Aerospace Application | ICARUS. Available online: http://icarus-alloys.eu (accessed on 14 June 2020).
65. Bazios, P.; Tserpes, K.; Pantelakis, S. Numerical Computation of Material Properties of Nanocrystalline Materials Utilizing Three-Dimensional Voronoi Models. *Metals* **2019**, *9*, 202. [CrossRef]
66. Breval, E.; Macmillan, N. Elastic recovery at Vickers hardness impressions. *J. Mater. Sci. Lett.* **1985**, *4*, 741–742. [CrossRef]
67. Suresh, S.; Giannakopoulos, A.; Alcalá, J. Spherical indentation of compositionally graded materials: Theory and experiments. *Acta Mater.* **1997**, *45*, 1307–1321. [CrossRef]
68. Suresh, S.A.; Giannakopoulos, A. *Report Inst-2/98*; Massachusetts Institute of Technology: Cambridge, MA, USA, 1998.
69. ISO/TR 29381. *Metallic Materials—Measurement of Mechanical Properties by an Instrumented Indentation Test—Indentation Tensile Properties*; ISO: Geneva, Switzerland, 2008.
70. ISO 14577. *Metallic Materials–Instrumented Indentation Test for Hardness and Materials Parameters, Part 1: Test Method, 2002, Part 2: Verification and Calibration of Testing Machines, 2002, Part 3: Calibration of Reference Blocks, 2002, Part 4: Test Method for Metallic and Non-Metallic Coatings*; ISO: Geneva, Switzerland, 2007.
71. Bouzakis, K.; Michailidis, N.; Hadjiyiannis, S.; Skordaris, G.; Erkens, G. The effect of specimen roughness and indenter tip geometry on the determination accuracy of thin hard coatings stress–strain laws by nanoindentation. *Mater. Charact.* **2002**, *49*, 149–156. [CrossRef]
72. Available online: http://www.horizonindia.in/buehler/grinders-polishers/3-BUEHLERVANGUARD.pdf (accessed on 31 October 2021).
73. Michailidis, N.; Pappa, M. Application of strength properties determined by nanoindentations to describe the material response in micro- and macro-indentation. *CIRP Ann.* **2009**, *58*, 511–514. [CrossRef]
74. Michailidis, N.; Bouzakis, K.; Koenders, L.; Herrmann, K. *Nanoindentation, CIRP Encyclopedia of Production Engineering*; Springer: Berlin/Heidelberg, Germany, 2014; pp. 1–9.
75. Lichinchi, M.; Lenardi, C.; Haupt, J.; Vitali, R. Simulation of Berkovich nanoindentation experiments on thin films using finite element method. *Thin Solid Films* **1998**, *312*, 240–248. [CrossRef]

metals

MDPI

Review

Machinable Leaded and Eco-Friendly Brass Alloys for High Performance Manufacturing Processes: A Critical Review

Paul Stavroulakis [1], Anagnostis I. Toulfatzis [2], George A. Pantazopoulos [2,*] and Alkiviadis S. Paipetis [3]

1. Sir Robert Hadfield Building, Department of Materials Science and Engineering, The University of Sheffield, Mappin St., Sheffield S1 3JD, UK; ppstavroulakis1@sheffield.ac.uk
2. ELKEME Hellenic Research Centre for Metals S.A., 61st km Athens—Lamia National Road, 32011 Oinofyta, Greece; atoulfatzis@elkeme.vionet.gr
3. Department of Materials Science and Engineering, University of Ioannina, 45110 Ioannina, Greece; paipetis@cc.uoi.gr

* Correspondence: gpantaz@elkeme.vionet.gr; Tel.: +30-2262-60-4463

Abstract: The recent environmental/health and safety regulations placed restrictions of use of hazardous substances on critical manufacturing sectors and consumers' products. Brass alloys specifically face a challenging issue concerning the elimination of lead (Pb) which has been a critical element affecting both the machinability and overall quality and efficiency of their manufacturing process. The adaptation of novel materials and processing routes in the green economy constitutes a crucial decision for competitive business and industry growth as a worldwide perspective with substantial industrial and social impact. This paper aims to review the emergent innovative and sustainable material solutions in the manufacturing industry, in line with environmental regulations, by highlighting smart alloy design practices and promoting new and innovative approaches for material selection and manufacturing process optimisation. In this review we analyse the processing, structure and machinability aspects of leaded brasses and underline the major guidelines and research methodologies required to overcome this technical challenge and further improve the mechanical properties and machinability of lead-free brass alloys. Various alloying and processing strategies were reviewed together with the most important failure types, as they were extracted from the existing industrial and technological experience, covering more than 20 years of research in this field.

Keywords: leaded brasses; lead-free brasses; eco-friendly; machinability; brass failures; design-of-experiments

Citation: Stavroulakis, P.; Toulfatzis, A.I.; Pantazopoulos, G.A.; Paipetis, A.S. Machinable Leaded and Eco-Friendly Brass Alloys for High Performance Manufacturing Processes: A Critical Review. *Metals* **2022**, *12*, 246. https://doi.org/10.3390/met12020246

Academic Editor: Joao Paulo Davim

Received: 13 December 2021
Accepted: 24 January 2022
Published: 27 January 2022

1. Introduction

Cu–Zn alloys (brasses) are widely used industrial materials because of their superior properties such as high corrosion resistance, low friction coefficient, non-magnetism, good plastic deformability (forgeability) and machinability [1–4]. Lead (Pb) has been commonly added to brass alloys to ensure pressure tightness and improve the machinability of the alloy, directly affecting cutting-tool life [5–7]. Owing to its low solidification temperature, lead acts as a filler for micropores that form during casting. Subsequently, these lead particles improve the machinability of brasses through their dual functionality as a lubricant and a stress concentration point, promoting the formation of small discontinuous chips and minimising tool wear [8]. As such, leaded brasses have seen use in a wide range of applications such as the base screw machine material for architectural hardware, cosmetics, musical instruments, screw machine parts, valves, fitting and general fasteners, while also being extensively used for plumbing applications [9]. Nevertheless, the health hazards associated with lead resulted in the implementation of stricter regulations for permissible lead content levels in brass products. More specifically, as leaded brasses have been extensively used for plumbing applications, particulate lead could be traced in drinking water due to the gradual corrosion of the brass components. Additionally, particles originating from solders used in plumbing led to the further contamination of the potable water supply. The gradual intake of particulate lead would elevate the blood lead levels, eventually leading to poisoning with devastating effects to human health. These include slower cognitive development in children, hypertension and even renal failure [10–16]. Thus, the development of lead-free brasses became a focal research point to ensure compliance with the regulatory standards and improve quality of life [17,18].

From a purely manufacturing–economic standpoint, substitution of lead-containing brass with a lead-free alternative does not appear to be an economically viable option, but it is technically possible, satisfying the requirements of the enforced legislation [19]. As such, efforts were made into incorporating soft second phase particles (SPPs) in the microstructure of copper–base alloys to attain the positive effects of Pb. Soft SPPs reduce shear ductility allowing for the formation of small chips, resulting in lesser machining power demands and a temperature reduction in the cutting tool, extending tool life. Such particles also reduce the shear strength of the brass at the secondary shear zone through the provision of internal lubrication, improving surface finish and further reducing the forces applied on the cutting tool [20]. In turn, this challenge sparked significant interest for the development of brass systems that enable the formation of said particles through the incorporation of various alloying elements such as bismuth (Bi), graphite, magnesium (Mg) silicon (Si), tin (Sn) and titanium (Ti) [21–23]. C. Yang et al. proposed the development of novel brass compositions by including beta (β) phase stabilisers and solid-solution strengtheners while simultaneously reducing the Zn content to enable the manifestation of high-strength and proper machinability [24]. The main drawback of these design routes however is the deviation from standardised compositions, requiring for additional approval requirements by International Organizations and Regulatory Authorities to be adopted in the supply chain. An alternative pathway towards improving the machinability of lead-free brasses through optimising the heat-treatment process of standardised compositions was proposed and investigated by Toulfatzis et al. [25]. This is achieved through a partial or complete suppression of the beta (β) to alpha (α) phase transformation by water quenching post heat-treating, resulting in beta-dominant microstructures. This microstructural modification allows for a shift in the fracture mechanism enabling intergranular fracture, thus promoting easier chip breaking and lower power consumption [26]. However, although interesting in nature, both approaches to this technical challenge are still in their infancy and require further development before they meet industrial implementation. It is thus equally as important to investigate both novel alloy design approaches that yield suitable material properties and their viability for incorporation in an industrial manufacturing process.

The primary microstructural features of conventional and modern brass rods have been extensively highlighted and discussed in relation to behaviour during machining [27]. Factors such as the chip size, cutting tool type and wear, cutting force, surface roughness and testing methodology play an important role in the optimisation of cutting processes in terms of time and cost [28–30]. In this direction, optimisation algorithms have been applied to investigate the optimal number of passes and the corresponding parameters such as speed, feed rate and depth of cut in each pass for multi-pass and multi-criterion plain turning problems [31]. This optimisation could be achieved by fractional factorial designs, preventing any potential complications arising from the use of analysis of variance (ANOVA) even in small samples [32,33]. The machinability of various alloys can also be reviewed, evaluated, and optimised using neural networks; however, noise factors generated during the machining process are known to cause errors in such computational efforts [34,35]. Therefore, it becomes clear that a precise data extraction process needs to be facilitated, minimising the relevant noise factors to enable more advanced computational tools. As such, the introduction of automated image processing tools would be beneficial to aid the data acquisition process especially in cases where segmented chip formation is observed [36]. The methodology of Taguchi Design of Experiments (DOE) applied using orthogonal arrays is very efficient in statistics for small sampling rates, constituting an optimisation technique for improving reliability [37,38]. Fractional factorials in robust design methodology designs have been extensively reviewed by Arvidsson and Gremyr [39]. The fundamental principle of robust design is to improve the quality of a product through minimising the property variation causes without their exclusive elimination [40]. This can be achieved by a joint product-process optimisation to disassociate the end-properties from the various causes of variation. As reported by Manna and Bhattacharyya, these causes were affected due to the noise factors associated with parameter design [40]. In their work, the signal to noise factor was used as a measure of the quality, and orthogonal arrays were used to study many design parameters simultaneously. Various examples of similar scientific work based on the practical design of experiments (DOE) methodology have been published in recent years [41]. These investigations considered the size of the experiments, the applied design, the number of factors that influenced the response variable and the sector of application of the experimental design. Viles et al.. presented a real case of design of experiments focusing on the preliminary stages of experimentation, e.g., how to choose the best response analysis, how to evaluate the crucial factors and what the data sample collection should be [42]. Engineers may encounter key barriers in their trial to use DOE methodologies, so an exhaustive literature review was performed to classify these barriers [43]. Numerous scientific articles have verified that there is a gap between theoretical development of DOE and its effective application in industry [44]. Therefore, a simple methodology was proposed to facilitate the implementation of DOE in companies, through a framework that follows the traditional DMAIC (Define–Measure–Analyse–Improve–Control) steps of 'Six Sigma' as a generic problem-solving methodology.

In this paper, a brief and concise review of conventional leaded (Pb-ed) as well as modern lead-free (Pb-free) brass alloys was realised, evaluating the alloys in terms of their microstructure, mechanical properties and machinability. The principles governing the manifestation of an optimal combination of properties in leaded brasses are laid out in detail to be used as a foundation for novel lead-free brass development, evaluation and optimisation. Furthermore, the state of the art concerning novel lead-free brass design is presented and summarised to facilitate the benefits and drawbacks associated with each material design approach. We believe that highlighting the efforts of various research groups around the globe concerning brass design, thermomechanical processing and machining is a crucial step for tackling this critical technical challenge. The introduction of innovative and sustainable material solutions in the manufacturing industry that are aligned with environmental/health and safety regulations. To the best of the authors' knowledge, no such updated and comprehensive review focused on machinable brass alloys has been published so far. The authors' team incorporate long term expertise in the field of research of copper and copper alloys processing and properties counting more than 20 years, where a considerable amount of papers have been published and reviewed in the current study.

2. Chemical Composition of Leaded and Lead-Free Brass Alloys

The elemental chemical analysis of the alloys presented within our work and the related references are listed in Tables A1–A9 (Appendix A).

3. Microstructure, Mechanical Properties and Processing Relationships

3.1. Leaded Brasses

3.1.1. Microstructure

The phase structure of leaded (Pb) brass alloys is governed by a variety of mechanisms. As mentioned previously, Pb addition in conventional brass alloys serves to improve their machinability. Pb preferred accumulation sites are mainly the α/β interphase boundaries (Figure 1). Depending upon the casting and thermomechanical process (TMP) parameters, Pb as a liquid phase during metal forming tends to reduce its surface energy by attaining a spherical shape [27]. To further minimize their interfacial energy. the Pb globules coalesce, forming coarse islands arranged at the α/β interfaces. Pb coalescence increases mean particle size and reduces the Pb particle distribution density. with an adverse effect on machinability [27]. On the contrary, fine and uniform lead distribution enables superior machining behaviour through better chip fracturing and minimises tool wear, while an increased volume fraction of Pb or a non-uniform particle distribution may cause surface cracking during hot working due to hot-shortness [45]. However, hot shortness can also be induced from the excess heat generated in the brass billet from the friction between the brass billet and the container during the extrusion process [25]. As mentioned, the α/β interphase boundaries constitute high interfacial energy sites and thus potential Pb accumulation areas [27,46,47]. When machined, an $(\alpha + \beta)$-brass has better chip breaking properties than a pure α-alloy. Further machinability improvements are attained by increasing the length of interphase boundaries, which in turn depends on the β-phase content and the size of the α-phase crystals. Higher β-phase volume fraction and finer α-crystals (for a constant α-phase content) tend to create longer interphase boundaries acting as Pb concentration sites [27,48]. Therefore, lower temperatures would be used during an extrusion process to enable the formation of extended α/β interfaces, an increase in strength and improve the machinability [47,48]. As Pb forms near the end of the solidification process, its particle size and distribution are primarily affected by the casting and solidification conditions, while the phase structure (volume fraction, grain size) is mainly influenced by the extrusion conditions (e.g., temperature, extrusion ratio, speed and cooling rate) [49].

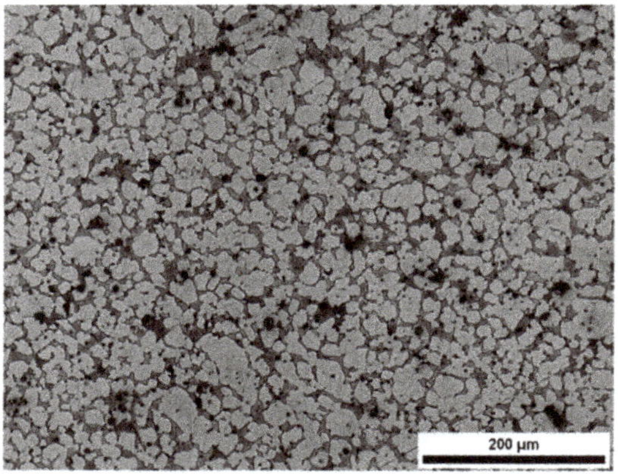

(**a**)

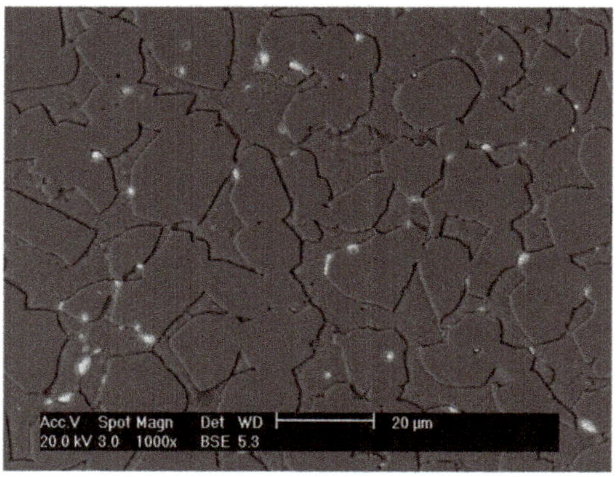

(**b**)

Figure 1. (**a**) Optical micrograph and (**b**) SEM micrograph showing the typical microstructure of CuZn39Pb3 (CW614N) brass [50].

3.1.2. Thermomechanical Processing

The sawing performance of brass bars (CuZn39Pb3) and its relationship to Pb particle size, distribution and the phase structure was reported in a previous study [49]. In this work the presence of β-phase reduced the overall ductility of the alloy and therefore enhanced the segmentation of the machining chips. Holler et al. examined the influence of the extrusion parameters on the phase structure in CuZn40Pb2 [51]. They showed that the percentage of β-phase decreased with an increasing billet temperature and extrusion ratio, while the grain size decreased with decreasing billet temperature and increasing extrusion ratio. Both the grain size and β-phase fraction were less influenced by the extrusion speed. Nevertheless, a study by Loginov et al. concluded that brasses with compositions similar to those mentioned above could benefit greatly from increasing the extrusion temperature to 780–800 °C as the TMP would be taking place entirely within the beta single phase region ensuring better property uniformity [52]. Kunčická et al. further highlighted the importance of maintaining controlled manufacturing processes by studying the failure of CuZn40Pb2 brass fittings from a computational and experimental standpoint. They unveiled that an inhomogeneous temperature profile during the manufacturing process may produce residual stresses in the cooler regions and initiate undesired phase transformations which eventually lead to the manifestation of forming defects [53]. In another case, they investigated a hot-die forged component [53,54]. There, failure originated from a region characterised by plastic flow instability and inhomogeneous stress distribution deeming the area prone to the generation of defects. The cracked region was also characterised by a localised depletion of β phase in favour of the formation of the α phase, thus decreasing the ductility. Sub-micron oxide particles were also present within the crack which were detrimental for the ductility and the cohesive toughness of the material [55].

According to Pantazopoulos and Vazdirvanidis, the beta (β) phase formed in the CuZn39Pb3 alloy may exhibit a characteristic plate-like type morphology possessing different orientations among the various grains [27]. Such morphological characteristics may be considered beneficial for chip-breaking during the machining process. Chandra et al. investigated the intergranular and interphase cavitation in binary alpha/beta brass at 600 °C during superplastic deformation [56]. The process of cavitation was associated with grain boundary sliding. Cavity nucleation occurred at points of stress concentration and such as the sliding interfaces between α–β boundaries rather than on α–α and β–β interfaces [56]. Hentati et al. investigated an incidence of cracking of CuZn39Pb2 during hot forging [57]. The forgeability of the alloy was significantly affected by the work hardening capacity. A decrease in ductility was observed as a result of work hardening and α-phase size reduction during a hot stamping process at 750 °C. Microhardness tests confirmed that stamping increased the brass hardness, while the size of α-phase also decreased. The reduction in α-crystal size and subsequent hardness increase degraded the ductility of the material, causing cracks during the forming process. Such cracks could have been caused from residual stresses generated during TMP. On the other hand, a heat treatment at 300 °C for 1 h had a positive contribution in both tackling the cracking problem and enabling an increase in elongation to fracture up from 22% to 28% [57]. Blaz et al. showed that the reduction in the flow stress during hot deformation of a CuZn39Pb3 alloy resulted from dynamic recovery and dynamic recrystallisation occurring during operation above ~677 °C [58]. Fine Pb particles did not fully inhibit the grain coarsening during dynamic recrystallisation at higher deformation temperatures. During hot deformation above ~727 °C, $\alpha \rightarrow \beta$ phase transformation was found to overlap with the structural softening process, resulting in effective grain coarsening followed by effective flow stress reduction. All of the cases presented here further strengthen the fact that attaining a proper microstructure, appropriate mechanical properties and avoid in-process defects or failures heavily depends on the careful control of the TMP.

3.1.3. Mechanical Behaviour

Pantazopoulos and Toulfatzis studied the mechanical behaviour and fracture mechanisms of conventional leaded brasses [59]. A comparison between the CuZn39Pb3 and CuZn36Pb2As brasses showed that the former contained a significantly higher volume fraction of the β-phase than the latter. The beta (β)-phase (CuZn) is an ordered intermetallic phase possessing a body-centred cubic (BCC) crystal structure. This phase is characterised by a higher hardness but less ductility compared to the α-phase which refers to a Cu-based solid solution with the face-centred cubic (FCC) crystal structure. Therefore, the CuZn39Pb3 brass which is a dual phase (α + β) brass containing ~35% β-phase exhibits higher strength, hardness and lower total elongation than the near single α-phase CuZn36Pb2As brass. As expected, the CuZn39Pb3 brass had higher yield strength and tensile strength, but lower elongation compared to the CuZn36Pb2As brass [46]. Upon investigating the side views of the fractured specimens in tension, a ductile fracture mechanism was uncovered in both compositions. The reduction in area was also higher in the CuZn36Pb2As brass compared to CuZn39Pb3, indicating that post-uniform elongation was more pronounced for the CuZn36Pb2As brass [46]. This deviation between the two specimens was attributed to a more extended region of non-uniform elongation in the case of the CuZn36Pb2As brass and was indicative of pronounced post-necking deformation during tensile testing. As the microstructure of the aforementioned specimen consisted primarily of α-phase grains with a uniform distribution of a small percentage of β-phase, it was found to be more resistant to micro-void growth and coalescence under tri-axial stress leading to substantially higher post-uniform elongation and higher reduction of area during necking [59]. Hardness measurements indicated a more homogeneous hardness profile (lower hardness gradient) to CuZn39Pb3 brass compared to the CuZn36Pb2As brass [46]. In this research, the CuZn39Pb3 alloy was characterized by higher hardness values (core ~136 to surface ~149 HV) as compared to CuZn36Pb2As (core ~104 to surface ~136 HV). The topographic features of tensile fracture surfaces of CuZn39Pb3 and CuZn36Pb2As alloys were studied using Scanning Electron Microscopy [59]. In this study, a mixed-mode fracture combining dimple formation and planar faceted fracture induced by a quasi-cleavage process was identified in the case of CuZn39Pb3 alloy. On the other hand, the CuZn36Pb2As alloy exhibited a typical ductile fracture consisting of fine equiaxed dimples. Pantazopoulos investigated the mechanical properties of the CuZn39Pb3 leaded brass [8]. Appreciable ductility was observed prior to fracture through the formation of a localised neck. The subsequent fracture was primarily ductile, and the resulting matched surfaces consisted of interconnected shear cracks. The final metal forming process, cold drawing, led to hardness inhomogeneities due to heterogeneous plastic deformation between the surface and core [8]. Tensile testing at higher temperatures during quasi-static and static loading revealed that the flow stress decreased proportionally to the testing temperature which was expected. The presence of brittle phases (such as β′) together with their morphology is also related to better machinability [60]. Nevertheless, a significant decrease in the ductility between 200 and 400 °C was observed, owing to the partial or complete melting of Pb [46].Moreover, the strain rate sensitivity index increased proportionally to the testing temperature without enabling relevant increases in ductility to take place due to the hot-shortness mechanism [61].

Generally, copper alloys tend to exhibit a reduction in ductility between 300 and 600 °C through a phenomenon called intermediate temperature embrittlement (ITE). This is due to a change in the involved fracture mechanism from transgranular to intergranular fracture as exposure to intermediate temperatures enables grain boundary voiding. Eventually, the ductility of the alloy is restored once the intermediate temperature range is exceeded, enabling dynamic recrystallisation and an operative grain boundary mobility. This effect is enhanced by the presence of alloying elements in the grain boundaries that reduce their structural integrity by encouraging void generation and growth along grain boundaries [62]. In the case of leaded brasses, however, this effect is further amplified due to the synergistic effect of liquid metal embrittlement (LME) where, as previously seen, the partial or complete melting of lead may introduce a liquid phase between crack faces. As such, the local cohesion of the alloy is modified and the kinetics of dynamic embrittlement related to ITE are accelerated as the presence of the liquid phase promotes the rapid diffusion of impurities and void mobility within the grain boundaries [62,63].

The mechanical behaviour of CC763S brass during cyclic loading was investigated by Flegeau et al. under torsional stress. They pinpointed that the fracture mechanism consisted of both ductile and brittle characteristics, with crack growth taking place along α/β grain boundaries where the concentration of intermetallic compounds and pores was the highest. Generally, a higher dislocation density was identified in the interphase α/β boundaries, while α grains remained relatively vacant of dislocations. Subsequently, the interphase boundaries were deemed as the crack nucleation points [64]. In their investigation, Johansson et al. also highlighted the interesting mechanical properties of the CW724R brass compared to the CW614N. The alloy was characterised by $R_{p0.2}$ and R_m values of 441.4 MPa and 811.9 MPa, respectively, compared to 254.9 MPa and 561.5 MPa for the high-lead counterpart while retaining similar levels of ductility.

3.2. *Eco-Friendly Brasses (Pb-Free)*

3.2.1. Novel Lead-Free Brasses

The absence of Pb and potentially other alloying elements directly affects the microstructure of Pb-free brasses. As such, the introduction of soft intermetallic phases that could imitate the effects caused by Pb particles in leaded brasses became a focal research point. The addition of iron (Fe) alongside silicon (Si) was investigated by Johansson et al. [65], attempting to formulate a low-Pb variant of the CW614N brass. Although, the Fe_4Si intermetallics were formulated, their intergranular precipitation points did not enable similar effects as in the case of Pb to come into play [65]. Atsumi et al. investigated a high-strength brass (Cu-40% Zn) alloy with magnesium (Mg) fabricated through a powder technology process to uncover the effect of the Mg addition on the microstructural properties of extruded α–β duplex brass alloys [23]. In this work, pre-mixed Cu-40% Zn alloy powder with 0.5–1.5 mass% pure Mg powder (Cu-40% Zn + Mg) was consolidated using spark plasma sintering (SPS). The SPSed Cu-40% Zn + Mg specimens consisted of α–β duplex microstructure containing $Mg(Cu1\text{-}xZnx)_2$ intermetallic compounds (IMCs) with a mean particle size of 10–30 µm. Initially, Mg was completely dissolved in the α–β duplex phases through a heat-treatment at 700 °C for 15 min. Thus, to disperse fine IMCs on α–β duplex phase matrix, the SPSed Cu-40% Zn + Mg specimens were pre-heated to achieve a solid solution condition and were afterwards immediately extruded. They studied the tensile properties of the extruded Cu-40% Zn + 1.0% Mg brass and reported an average yield strength (YS) of 328 MPa, ultimate tensile strength (UTS) of 553 MPa and elongation of 25% [66]. The extruded Cu-40% Zn + 1.0% Mg alloy outperformed the conventional binary brass alloy, exhibiting 229 MPa yield strength (YS) and 464 MPa ultimate tensile strength (UTS). The strengthening mechanism of this wrought brass alloy was attributed to the grain refinement due to the pinning effect produced by the fine $Mg(Cu1\text{-}xZnx)_2$ precipitates forming at interphase boundaries. The extruded specimen exhibited fine α–β duplex phases containing very fine precipitates of the above $Mg(Cu1\text{-}xZnx)_2$ IMCs with 0.5–3.0 µm in diameter [66]. The fracture surfaces of both extruded specimens consisted of fine dimples owing to the evident extremely fine α–β duplex phase structures. The hills and valleys of the extruded Cu-40% Zn fracture surface indicated that the crack paths preferentially followed the primary powder particle interface region. On the other hand, such hills and valleys were never observed in the extruded Cu-40% Zn + 1.5% Mg specimen; however, some fine particles were observed on the fracture surface [66]. Atsumi et al. also investigated the microstructural properties of the Cu-40Zn-0.5Ti ternary alloys having undergone a solid solution treatment [23]. The specimen was pre-heated at 700 °C for 15 min and subsequently extruded consisting of a fine and uniform α–β duplex phase structure with an average grain size of 2.14 µm. Furthermore, fine Ti-based particles with a mean size of 0.5 µm were densely dispersed in the matrix [23]. Atsumi et al. reported that the extruded Cu-40%Zn-0.5 Ti alloy had a mean yield strength of 304 MPa, ultimate tensile strength of 543 MPa and elongation of 44% [23]. Cu-40%Zn-0.5Ti brasses revealed satisfactory strength and good ductility. The high strength of the wrought brass alloy was mainly due to the fine grain size of α and β phases caused by the precipitation of fine Ti-based particles at the grain boundaries [23]. In a previous research Li et al. reported that remarkable grain-refinement and strengthening effects were achieved by the addition of Ti to 60/40 brass through a powder metallurgy process route [67]. However, Ti segregated to the powder particle interface at elevated temperatures, significantly deteriorating the mechanical properties of the BS40–1.0Ti. To inhibit this segregation behaviour of the Ti in the BS40–1.0Ti, the addition of Sn was investigated. Ti was incorporated in the formation of CuSn3Ti5 particles at the grain boundaries, rather than segregating at the powder particle interface, leading to significant grain refinement and enabling strengthening effects on the BS40–0.6Sn1.0Ti brass. These results showed that the introduction of Sn via powder metallurgy to BS40–1.0Ti brass can effectively impede the segregation of Ti and stabilise the mechanical properties at high processing temperatures [67].

An investigation attempting to introduce both Ti and graphite (Gr) to the Cu-40Zn brass showed that the sole addition of Gr to the alloy would yield a better cutting performance alongside a progressive deterioration of the mechanical properties. Nevertheless, the combination of both additives should enable precipitation and dispersion strengthening mechanisms to manifest through the formation of nano Ti clusters and Cu_2Ti_4O particles, thus maintaining reasonable mechanical properties [68]. The concept of Gr globules incorporation as a microstructural constituent was previously studied by Zhuo et al. through the introduction of cementite (Fe_3C) particles in the alloy and a subsequent heat treatment to enable the Fe_3C to undergo a graphitisation process [69]. Si has also been extensively investigated as an alloying addition to replace Pb. It acts as a β and β′ phase stabiliser and enables the formation of Widmanstätten-type α phase nucleating from the β interface. Moreover, conducting a solution heat treatment followed by water quenching would further suppress the diffusional nature of the $\beta \rightarrow \beta'$ transformation, resulting in a martensitic transformation instead. The ease of this process is directly related to the Si content. This concept was explored by Alirezaei and Doostmohammadi and used to design a high-hardness Cu30Zn-Si brass [70]. For the design of novel Pb-free brasses the Zn equivalent rule has been used extensively to predict the formulating phases in said alloys (Equation (1)) [24,71].

$$X(\%) = 100\% \frac{(C_{Zn} + \sum C_i \cdot K_k)}{(C_{Zn} + C_{Cu} + \sum C_i \cdot K_k)} \quad (1)$$

where $X\%$ is the fictitious zinc equivalent, C_{Zn} and C_{Cu} the Zn and Cu contents in the specific alloy of interest and $\Sigma C_i \cdot K_i$ the sum of the products of content C_i and the fictitious Zn equivalent K_i of each alloying element. As such, dual phase or single-phase brasses can be designed and compositionally fine-tuned to achieve an ideal combination of properties. Yang et al. used this approach to design an α + β Cu-39Zn-Si-Al brass and manufacture it through low-pressure die casting (LPDC). By altering the processing parameters, they were able to significantly alter the average grain size of the β phase and the morphology of the α phase. Their investigation led to the design of a microstructure consisting of fine β grains and a higher content of rod-like or granular α phase allowing for the manifestation of superior mechanical properties, namely a YS of 359.8 MPa, UTS of 598.9 MPa and an elongation of 15%. Nevertheless, the present composition has not been evaluated in terms of machinability [72]. Expanding upon the concept of Al and Si-containing brasses, D. Suksongram et al. sought to create a high-hardness dual-phase β + γ CuZnSiAlSn brass through the addition of antimony (Sb). According to their research, increasing amounts of Sb up to 1.5 wt.% would enable the inter and intragranular formation of increased amounts of γ phase to the β phase resulting in a significant hardness increase. This could be attributed to an increased solid solution strengthening effect taking place due to the incorporation of Sb within the γ phase. Nonetheless, upon reaching 2.0 wt.% Sb content, a high-Sb intermetallic phase would form, reducing the hardness of the investigated alloying system by roughly 17% [73].

Chandra et al. found that both the ductility as well as the flow stress of α–β brass is significantly affected by the addition of Cerium (Ce) [74]. Ce decreased the solubility of Zn in the α (FCC) phase between 475 °C and 675 °C, consequently shifting the α/(α + β) region of the phase diagram towards lower Zn concentrations. More specifically, the addition of 0.07 wt.% of Ce decreased the α/β ratio from 70:30 to 55:45, improving the ductility and decreasing the flow stress up to a factor of three, at the tested temperature range. Moreover, the hindering effect of Ce addition on Zn solubility indirectly contributed to an expected increase in stacking fault energy (SFE) of the α-phase, improving dislocation mobility. Atsumi et al. investigated the mechanical properties of the high-strength, Pb-free machinable α–β duplex phase Cu–40Zn–Cr–Fe–Sn–Bi alloys [75]. The extruded Cu–40Zn–Cr–Fe–Sn–Bi alloys consisted of α–β phases containing the fine, uniform Cr–Fe IMCs and Bi particles. The average yield strength (YS) and ultimate tensile strength (UTS) of the extruded Cu–40Zn–Cr–Fe–Sn–Bi alloys were 288 MPa and 601 MPa, respectively. These alloys exhibited YS and UTS that were 29% and 40% higher than that of conventional machinable brass Cu–40Zn–Pb alloy. The extruded Cu–40Zn–Cr–Fe–Sn–Bi alloys also maintained 75% of the machinability of the Cu–40Zn–Pb alloy [23]. Fontaine and Keast, compared a lead-free Bi-containing brass (EnviroBrass C89520) to a conventional brass C84400 [5]. In this work, the large difference in grain size was attributed to the dependence of grain growth on alloy chemistry, while the excess concentration of Sn around the particles and at the grain boundaries was due to 'coring', a well-documented phenomenon. However, the presence of Sn prevented the segregation of Bi to the grain boundaries and inhibited hot-tearing phenomena during processing involving Bi segregation. It was not clear at this stage how the presence of Sn prevented the segregation of Bi. Either the wetting behaviour of Bi was different in a Cu–Sn alloy than in pure copper or an excess of Sn at the boundaries prevented the segregation of Bi. Scanning transmission electron microscopy (STEM) coupled with electron dispersive X-ray spectroscopy (EDS) did uncover traces of Bi segregation at the boundaries in the brass, even though Bi segregates strongly in pure copper alloys [5]. Further analysis regarding the Bi addition as a possible replacement for Pb was conducted by Thomson et al., who modified the C93700 to replace Pb with Bi. Bi was noted to accomplish a similar role as that of Pb in the present alloying system. Indeed, Bi proved to be a possible replacement, achieving similar mechanical properties as the Pb case. Increased amounts of Bi would be detrimental for the mechanical properties instead. Nevertheless, the morphological characteristics of Bi deem it inferior to Pb by comparison, specifically for bearing applications. The Pb-containing alloy also exhibited greater fatigue resistance compared to the Bi cases. Lastly, tribological testing showed that the Bi-containing compositions induced higher friction forces than those encountered in the Pb-containing alloy. However, the friction forces of the Bi-containing compositions showed less fluctuation compared to that encountered in the Pb-containing alloy [76].

Basori et al. modified the Cu-29Zn brass through the addition of manganese (Mn). The progressive addition of Mn increased the volume fraction of the β-phase in the microstructure as Mn acts as a β-phase stabiliser. Furthermore, the increased Mn content had a minimal effect on twin formation during deformation, while dislocation motion through slip was inhibited due to the pronounced solid solution strengthening effect of Mn [77]. Quan Li et al. showed that the microstructure of the hot extruded Sb-Mg lead-free brass was composed of α, β phases and (Sb, Mg)-containing intermetallic compounds [78]. In this research, the developed Sb-Mg brass was compared to the HPb59-1 brass (CuZn40Pb2) in terms of its mechanical properties and was found to be a suitable replacement, reducing the cost and environment impact of the material. Basori et al. noted that in the Cu-29ZnMn system, the progressive addition of Mn was directly associated with an increase in hardness, $R_{p0.2}$ and R_m. An unexpected trend was obtained regarding the maximum elongation of the samples where Mn additions over 3 wt.% would reduce negatively impact ductility.

Nevertheless, an addition of up to 3 wt.% Mn had a positive effect instead. This was attributed to the increased β-phase presence at higher Mn contents [77].

3.2.2. Conventional Lead-Free Brasses—Microstructure and Mechanical Properties Optimisation

Toulfatzis et al. studied the fracture behaviour of lead-free brass alloys for machining applications [79]. In this work, the mechanical behaviour under static and dynamic loading of three lead-free brass alloys CuZn42 (CW510L), CuZn38As (CW511L) and CuZn36 (C27450) lead-free brass alloys in comparison to the conventional CuZn39Pb3 (CW614N) leaded brass alloy was studied. The fractographic evaluation was performed on impact tested brass samples to identify the involved fracture mechanisms and their relation to the alloy microstructure. The action of the dimpled fracture mechanism was evident to all the studied alloys; however, the dimple size and distribution exhibited a considerable variation signifying differences in the evolved plastic deformation prior to failure. The largest dimples were detected in the case of CW511L and C27450 alloys, which is consistent with their higher values of impact energy and fracture toughness. In this study, it was also shown that the CuZn42 (CW510L) lead-free brass alloy is a potential candidate in replacing the conventional CuZn39Pb3 (CW614N) leaded brass, combining high tensile strength and tolerable fracture toughness, due to the high presence of the β-phase in the alloy microstructure. The fracture toughness of CW510L and CW511L lead-free alloys, after post-processing heat treatment, was assessed via impact and Crack-Tip-Opening-Displacement (CTOD) testing and presented in Ref. [80]. Moderate improvement in impact energy and CTOD values were obtained for heat treated CW510L brass, showing a salient fracture mode, exhibiting a dominant intergranular crack propagation [80]. Similarly, Kozana et al. manufactured a CuZn39 alloy using various casting techniques. They observed significant differences in the hardness and R_m values of all samples while the only condition altered was the casting mould [81]. G. Iecks et al. carried out investment casting trials on the C35ZA and C30ZA alloys with the addition of aluminium (Al). Due to the segregation of the alloying elements, a finer microstructure was formed with greater hardness compared to other regions of the casting. A thinner cast component exhibited a fine β-phase Widmanstätten structure because of the higher cooling rate. The fineness of Widmanstätten structure is also directly affected by the Al content resulting in variations of hardness [82].

Microstructural evolution studies during severe plastic deformation (SPD) conditions are also important to discern the characteristics and evolution of the phases evident in dual phase brasses throughout the thermomechanical processing of the material [83]. Dhinwal et al. investigated the microstructural changes of a dual phase brass during severe plastic deformation (SPD). The α phase formed mechanical twins with each successive pass resulting in a more random crystallographic texture. The β-phase grains formed an interconnected structure. The overall grain size of the material was also significantly reduced [84]. Kozana et al. attempted to manipulate the microstructural characteristics of the CW511L brass using different casting moulds. Due to the differences between the moulds in terms of the heat transfer coefficients, various microstructures were obtained, the point of variation being the average grain size of the constituent microstructural features [81]. Toulfatzis et al. attempted to modify the microstructure of the CW510L, CW511L and C27450 brasses to improve their machinability by retaining an increased metastable β-phase fraction. This was accomplished by heat treating the alloys at temperatures between 700 and 850 °C for 60–120 min and then immediately subjecting the material to a water quench to suppress the β to α phase transformation. In turn, the increased volume fraction of β phase allowed further strengthening of the alloy at the cost of ductility. Specifically, in the case of the CW510L the amount of β phase present in the microstructure increased from 60% to 100%, whilst reaching 35% and 20% from an initial 5% and 2% for the CW511L and C27450 brasses, respectively, under the ideal heat treatment conditions (Figures 2 and 3, see Ref. [25]). As previously encountered in research investigating similar para-equilibrium conditions, Widmanstätten α-phase was present in the case of CW510L [85].

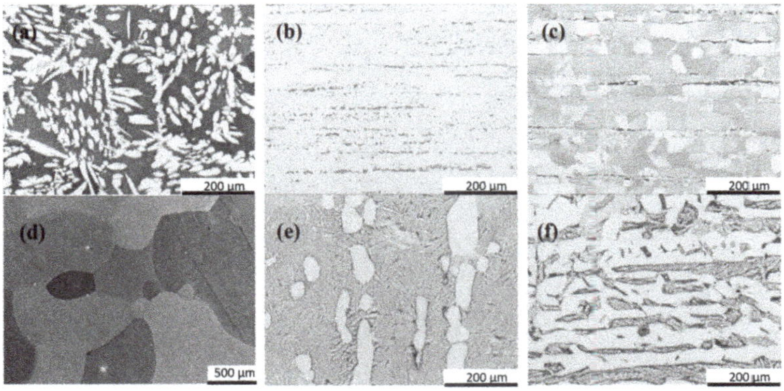

Figure 2. Optical micrographs of the (**a**) CuZn42, (**b**) CuZn38As, (**c**) CuZn36 alloys before heat-treating and (**d**) CuZn42, (**e**) CuZn38As, (**f**) CuZn36 alloys after the heat-treatment process [25].

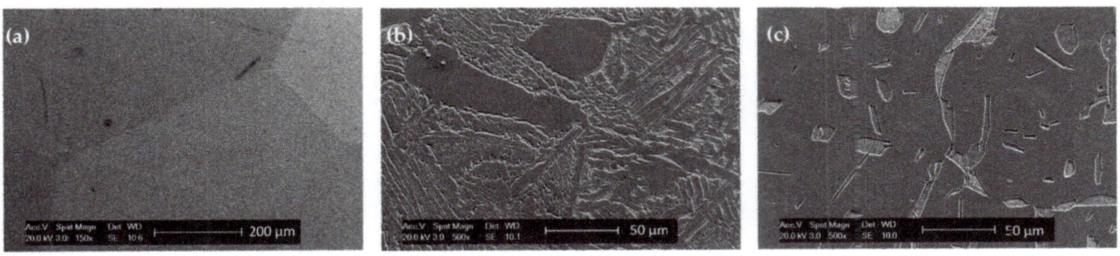

Figure 3. SEM micrographs on longitudinal sections showing the microstructures of (**a**) CuZn42, (**b**) CuZn38As and (**c**) CuZn36, after the selected heat treatment condition [25].

3.3. Failure Types Encountered in Leaded and Lead-Free Brass Components

As a wider range of brasses becomes integrated into the production line, the matter of component failure and the underlying principles dictating the manifestation of such incidents need to be highlighted. Generally, different types of brasses are more susceptible to different modes of failure owing to their processing route or service life and are extensively analysed in the work by Pantazopoulos [86] and Gupta et al. [87]. However, failure modes that are particularly related to the chemical composition of brasses are further expanded below to provide a potential alloy design direction.

3.3.1. In-Process Failure Modes

The failure of brasses during their thermomechanical processing can be attributed to factors owing mostly to the parameters involved in the casting, extrusion or cold-drawing process [86]. Nevertheless, there are also cases that are dependent exclusively on the alloy composition, with the most prominent one being hot-shortness.

Hot-Shortness

Hot-shortness occurs during the extrusion process in cases where the extrusion temperature is amplified due to the friction between the billet and the container. In turn, this causes SPPs with a low melting point such as Pb particles and regions exhibiting notable chemical segregation to partially melt or become hot short resulting in the possible formation of tears in close proximity to the metal surface [45,86,88]. In cases where novel brass compositions are designed, aiming to replace the Pb particles with a different type of SPPs, this failure mode becomes particularly important. This becomes apparent in cases where Pb has been replaced by Bi. As Bi migrates to the grain boundaries and has a low melting point overheating the brass component can yet again cause hot-shortness [89–92]. Therefore, a suitable SPP melting point needs to be one of the main points of alloy design to pre-emptively address hot-shortness. This can also be entirely circumvented through the complete elimination of SPPs in cases where the machinability of the brass would not be compromised.

3.3.2. In-Service Failure Modes

Brass components tend to fail during service mostly due to corrosion-related phenomena that are evident in most alloying systems. However, the increased Zn content of brasses can lead to the manifestation of mechanisms almost exclusive to brasses such as dezincification or enhance certain types of corrosion leading to failure such as stress corrosion cracking (SCC) [93,94].

Dezincification

Dezincification may occur during the exposure of brass components to saline and water environments (Figure 4). The mechanism is based upon the selective leaching of Zn due to the electrochemical potential formulating through the interaction between Zn and Cu atoms. This process is facilitated either through the exclusive interaction of Zn with oxygen or chloride ions present in water or through the combined oxidation of both Cu and Zn leading to the eventual establishment of an equilibrium between Cu ions and metallic water and the subsequent deposition of metallic Cu on the brass surface. Therefore, the base material is depleted of Zn leading to the formation a porous Cu-rich layer in the form of red patches on the surface of the material. Dezincification has devastating effects on the mechanical properties of the brass component. This corrosive mechanism can be inhibited from an alloy design perspective through the introduction of tin and arsenic as alloying elements or by controlling the Zn content as this effect is most prominent in brasses containing between 15 and 30 wt.% Zn [86,87].

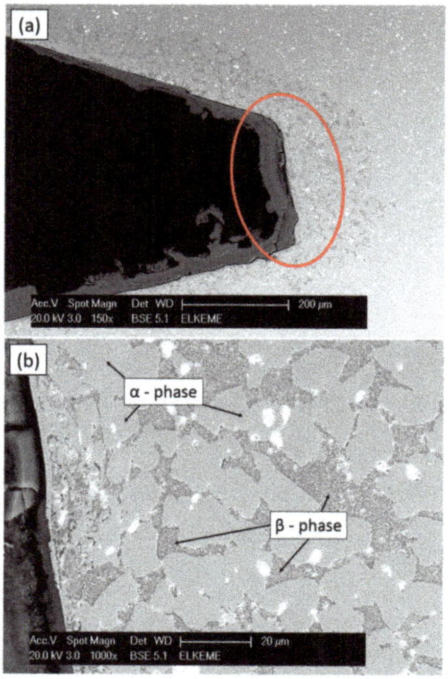

Figure 4. (**a**) BSE SEM Micrograph showing the dezincification area of a brass component. (**b**) Detail of (**a**) [93].

Stress Corrosion Cracking (SCC)

Stress corrosion cracking consists of the formation of intergranular cracks propagating through a metallic material due to the combinatory effect of high residual stresses alongside exposure to a specific corrosive medium and alkalinity. SCC manifests through the selective corrosion of the base material along the more chemically active paths such as interphase and grain boundaries compromising the material integrity in said points. Upon exposure to additional external stresses, cracks form along the compromised regions leading to the formation of additional free surfaces where the corrosion products are deposited and further accelerate the process. SCC occurs mostly in dual-phase brasses with a Zn content around 20–40 wt.%. This failure mode can be avoided by subjecting the component to a low-temperature heat-treatment to erase residual stresses caused by the manufacturing process in the component. In addition, laser shock peening has been proven to improve the SCC resistance of the material by introducing compressive stresses. Furthermore, protective coatings can be employed to further enhance the corrosion resistance of the material. As the Zn content range where this effect occurs is extensive, there are limited alloy design strategies that can be employed as of yet to alleviate SCC in brasses [86,87,95,96]. The fracture caused by SCC usually follows an intergranular path with multiple branching patterns. Typical SCC fracture surfaces and the corresponding EDS spectrum of a machinable brass component are presented in Figure 5 [93].

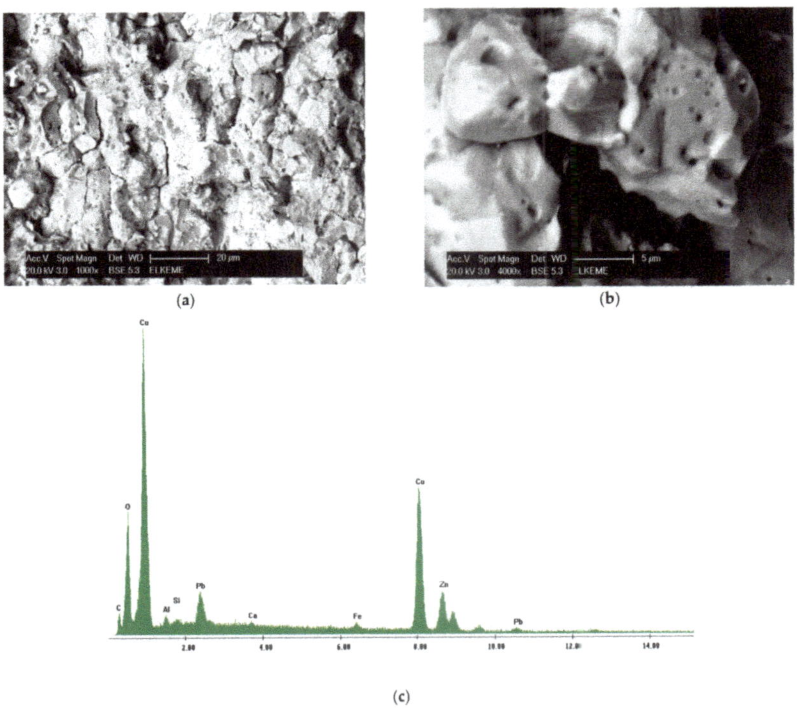

Figure 5. (**a**,**b**) Backscattered Electron (BSE) micrographs showing the fracture surface of a brass component showing intergranular morphology, (**c**) EDS spectrum of the fracture surface showing the presence of oxides. See Reference [93].

4. Machinability Evaluation

4.1. Chip Formation and Tool Wear

4.1.1. Leaded Brasses

Toulfatzis et al. investigated the machinability of CuZn39Pb3 and CuZn36Pb2As brasses [46]. In this research, machinability was assessed qualitatively and quantitatively by evaluating the chip size and morphology, and the corresponding cutting tool wear land, employing a single point turning technique according to ISO 3685 [97]. Experimental results indicated that the most critical factors affecting the chip formation quality during brass bar machining were the type of copper alloy used (CuZn39Pb3 and CuZn36Pb2As alloys) and the cutting speed, while it was not found to be sensitive to depth of cut and feed rate variations [46]. It was evident that a significant variety of chip morphology was produced by changing the turning conditions. The single needle, feathery and continuous helical were representative forms that constituted a qualitative indicator of the machining process performance [46]. The formation of shorter chips resulted in better surface finish, lower tool wear and ease of operation, allowing higher cutting speeds, while chip fracturing and removal led to the reduction in tool–chip contact, minimising the risk of built-up edge formation [46].

Gane has shown that the addition of a small quantity of Pb to brass was very effective in reducing friction, both in cutting experiments and in simple sliding-friction experiments [98]. In addition, the ductility of brass was substantially reduced when Pb was present. Both these factors contributed to the considerable improvement in machining properties observed in lead-containing alloys. The reduced friction on the rake face of the tool decreased the specific cutting energy by a factor of about two. As a result, the tool temperature decreased and the tool life improved [98]. As the Pb particles became finer (less than 5 μm) and their distribution more homogeneous, chip-breaking during cutting became more pronounced, and, as a result, machinability improved. Chip fracture in short pieces allows for ease of machining operation; maximum productivity, lower cutting forces, better surface finish of the workpiece, closer tolerances of the final component and less tool wear, thus longer tool life [98]. Preliminary machinability tests were performed by Pantazopoulos in turning operation [8]. The turning speed was up to 395 rpm (the linear speed was up to 0.165 m/s) using a cemented tungsten carbide insert as a cutting tool. The chip size tended to increase by increasing the depth of cut, while no appreciable changes were recorded by changing the feed rate. The depth of cut was constant within the turning cycle (i.e., 0.4, 1.0, or 3.0 mm). The horizontal advancement of the tool per revolution was defined as a function of the feed rate (0.25 or 0.50 mm/rev) [8]. Morphological observations of the produced chips, using optical microscopy, indicated the absence of a flow-zone. Segmented chips with intense shear fractures were produced during turning as a result of the dual phase microstructure (α–β) and the distribution of Pb particles at interphase boundaries. The presence of Pb decreased friction and temperature at the tool–workpiece interface, therefore increasing tool life [8]. Cutting tool wear was principally controlled mainly by the cutting speed and depth of cut used during machining experiments while it appeared to be less sensitive to feed rate or the type of copper alloy variations [46]. Adhesion and sticking of the workpiece debris on the flank face intensified at higher cutting speeds. Moreover, the flank zone revealed several typical wear features, such as abrasive marks and chipping (edge fracturing), of increasing severity as the cutting speed increased (from 2000 to 4000 rpm) [46].

As such, chipping caused the edge fracture of the machining tool, impairing cutting edge sharpness and leading to tool blunting. This type of tool failure increased cutting forces, resulting therefore in further degradation of the machining performance (longer chips, higher fluctuations and chattering, part distortion), causing serious defects on the machined surface quality. Severe damage phenomena such as thermal cracking or flaking were not observed in this study [46]. Toulfatzis et al. determined the optimum cutting conditions using non-parametric design of experiments methods applied in turning processes of leaded brass bars (CuZn39Pb3 and CuZn36Pb2As) for industrial applications [46].

Non-linear fractional factorial designs for the nine-run, three-level, four-factor saturated orthogonal array (L9 = 3^4) were applied to collect sufficient data to investigate the multi-response optimisation of two main machinability quality characteristics such as the chip morphology and cutting tool wear. The study suggested that concurrent multi-response optimisation gave rise to no specific effect. These optimal settings of the cutting parameters for chip morphology could also be used for cutting tool wear optimization. Confirmation experiments seem to verify the conclusion of this work [46].

Hassan et al. evaluated the influence of cutting parameters such as cutting speed, feed rate and depth of cut in C34000 leaded brass using an analysis of variance (ANOVA) methodology [99]. The purpose of ANOVA was to identify the important parameters for predicting material removal rate. Feed rate was found to be the most significant factor and its contribution to material removal rate was 42.77% within a 95% confidence interval. As such, the feed rate was the most significant factor affecting the material removal rate. The interaction between cutting speed and feed rate was found to be significant with a contribution of 7.75%. The best results for optimal material removal rate were achieved when C34000 leaded brass was machined at a cutting speed of 55 m/min, using a depth of cut of 0.2 mm and a feed rate of 0.35 mm/rev. Johansson et al. investigated brasses with both higher and lower Pb content. It was observed that the machinability of the conventional leaded brass exceeded that of its lower Pb content counterparts by a significant margin. It was however mentioned that the machinability of the CW724R brass was tolerable compared to the original CW614N, while also exhibiting interesting mechanical properties [65]. Thus, the importance of Pb as an addition cannot be overlooked as it directly affects the machinability of the material to a significant extent.

4.1.2. Eco-Friendly Brasses (Pb-Free)

The design criterion for the Pb-free brasses typically concerns the introduction of a soft intermetallic phase uniformly dispersed within the material. Upon contact with the cutting tool, the intermetallic particles should act as a stress concentration point, leading to the segmentation of the formulating chip. Li et al. studied the influence of Ti addition on lead-free brasses [67]. They showed that BS40-2.2Bi alloy exhibited an excellent machinability compared to the Cu40Zn3.2Pb conventional leaded brass. The average drilling time increased with increasing Ti content from 0.3 wt.% to 1.0 wt.%. BS40-2.2Bi with 1.0 wt.% Ti addition could not be penetrated by drilling for more than 180 sec. The different chip morphologies obtained during drilling operations were used to evaluate the machinability of the alloys produced. Compared with the chip shape of the conventional leaded brass (Cu40Zn3.2Pb), BS40-2.2Bi exhibited a different morphology due to the substantial brittleness of Bi affecting the fracture behaviour of the chip and the Ti addition to BS40-2.2Bi, affecting the chip geometry considerably, which transformed into a continuous spiral shape when the Ti content increased to 0.5 wt.% [67].

As the phase diagram indicates, both antimony (Sb) and magnesium can be dissolved partly in copper forming intermetallic compounds with copper [78]. If these intermetallic compounds were dispersedly distributed in the Cu-Zn system, such particles or strips would disturb the metallic matrix continuity and make the chips rupture easily. Thus, the cutting performance of the brass would be improved without the addition of Pb. Additionally, the particles in Sb-Mg brass are not expected to produce stress concentration during forming as they deformed plastically during extrusion, while improving the cutting performance and mechanical properties of the alloy [78].

Toulfatzis et al. evaluated the machinability of the CuZn42, CuZn38As and CuZn36 lead-free brasses, comparing them to the conventional leaded CuZn39Pb3 brass [25]. Their results reinforced the perception that an increased volume fraction of β-phase would enable the generation of micro-cracks during the machining process owing to the formation of shear bands. Nevertheless, the chip segmentation process was predominantly controlled by the distribution of Pb particles. Expanding upon this concept, by subjecting the Pb-free alloys to a series of heat treatments (at 775 °C for 60 min for CuZn42 and at 850 °C for 2 h for

CuZn38As), a higher volume fraction of β-phase would be stabilised in room temperature as previously discussed. As such, a mild improvement in machinability was to be expected. Indeed, the alloys showed an improvement in machinability relevant to the increased volume fraction of the β-phase and, specifically in the case of the CuZn38As alloy, surpass the conventional leaded counterpart in terms of surface quality [20,21,25,35]. The chip morphologies of the heat-treated lead-free brass alloys compared to the as received ones are depicted in Figures 6 and 7, while the typical characteristic topographic patterns obtained by optical profilometry is presented in Figure 8.

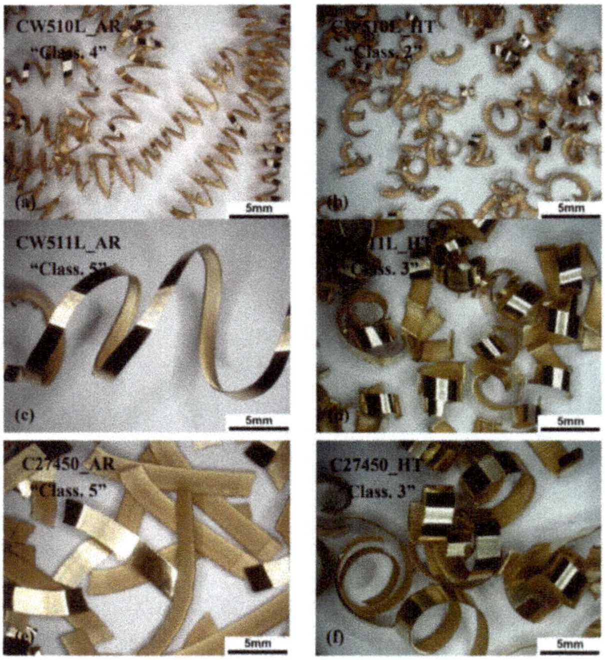

Figure 6. Chip morphology of the as-received (AR) and heat-treated (HT) CW510L-CuZn42 (**a**,**b**), CW511L-CuZn38As (**c**,**d**) and C27450-CuZn36 (**e**,**f**) brass alloys [25]. Note: the decrease in Class number denotes improvement in chip morphology [25].

Nobel et al. also studied the machinability of various lead-free alloys, i.e., CW510L, CW511L and CW724R, correlating their machinability to the use of different cutting tool materials [100]. CW724R lead-free silicon brass, in comparison to other lead-free brasses, such as CuZn42 (CW510L) and CuZn38As (CW511L), showed improved machinability concerning chip formation. However, CW724R caused higher tool wear due to the abrasive nature of the Si-rich κ-phase when used alongside a TiAlN-coated carbide tool. In this study the CW510L machining performance using various cutting tools with different coatings (TiAlN, TiB2, DLC, CVD-diamond) was investigated. Adhesive and abrasive wear was reduced by using DLC-coating among the tested PVD-coated tool systems. An additional improvement of cutting performance was achieved using CVD-diamond coating which exhibited exceptional abrasive wear resistance, while adhesion on the rake and flank face was minimised. Polycrystalline diamond tools (PCD) provide another alternative with significant abrasive wear resistance and low adhesion properties. However, the production demand for long tubular chips restricts the use of PCD tools and poses the need for the manufacture of such tools with favourable chip breaking geometry.

Experiments and simulations were implemented by Laakso et al. in order to investigate the effects of different cutting parameters on chip morphology of a low-lead brass [101]. This low-lead brass is not standardised but is compositionally close to the standard CuZn38As (CW511L) lead-free brass. Based on these simulations and experiments it was found that the use of a positive rake angle (+5°), high cutting speed (300 m/min) and low cutting feed rate (0.1 mm/rev) improved chip breakage from a continuous chip to a chip of an average length of 4 mm. High cutting speeds favour chip breakage by increasing chip brittleness due to the suppression of softening effects at high strain rate conditions. Minimisation of cutting force and friction force is achieved by using a positive rake angle (e.g., +5°). Reduction in friction could be realised by decreasing of tool–chip contact area by decreasing feed rate and/or depth of cut. Finite Element Modelling (FEM) simulations agreed with the above results. Furthermore, in a very recent work, the influence of Zn equivalent on the evaluation of chip morphology and machining surface quality of free cutting silicon brasses were investigated by adding Pb replacers (Si and Al elements) [24].

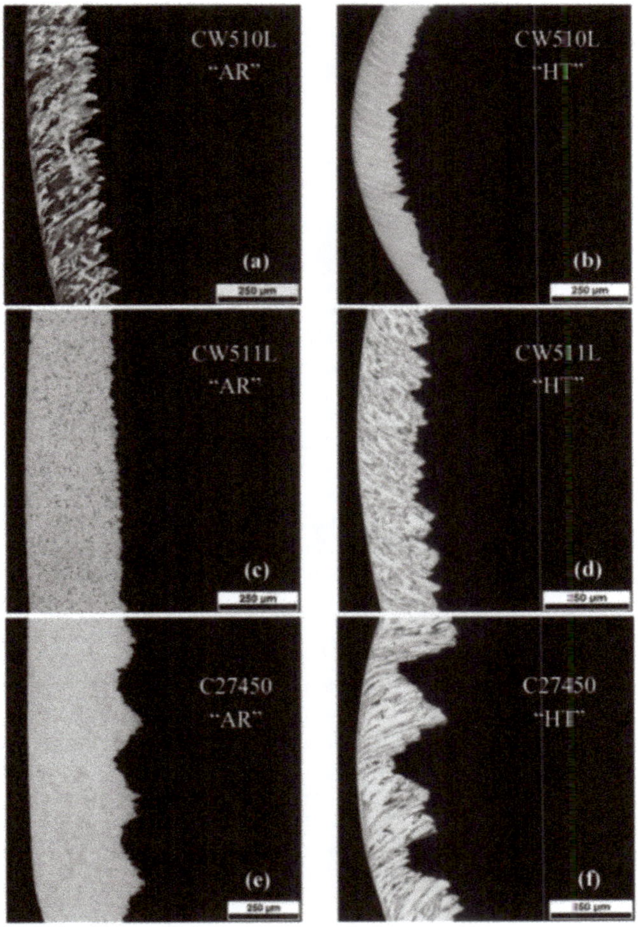

Figure 7. Optical micrographs showing the morphology of chips generated during the machining process of (**a**) CuZn42, (**c**) CuZn38As, (**e**) CuZn36 alloys before heat-treating (AR) and (**b**) CuZn42, (**d**) CuZn38As, (**f**) CuZn36 alloys after the heat-treatment process, HT [25].

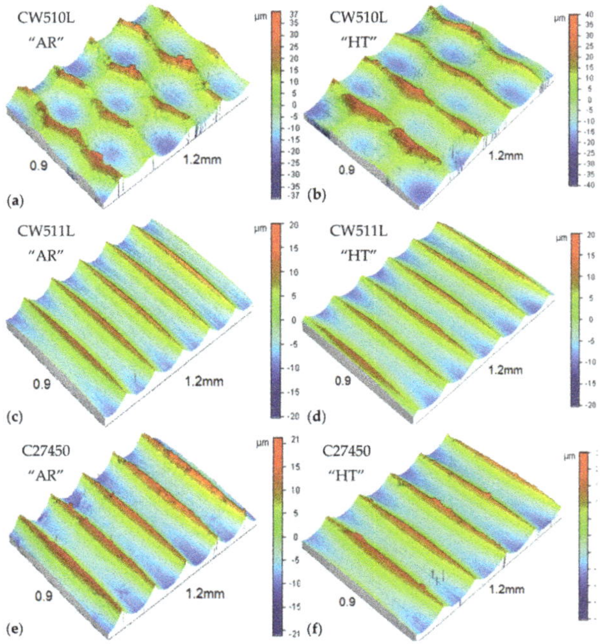

Figure 8. Surface topography derived from profilometric measurements: (**a**) CW510L, as received, (**b**) CW510L, heat-treated, (**c**) CW511L, as received, (**d**) CW511L, heat-treated, (**e**) C27450, as received and (**f**) C27450, heat-treated [25].

The introduction of additional alloying elements may also enable the stabilisation of various primary phases (i.e., the presence of α, β or γ phases in the Cu-Zn system) at equilibrium. This effect can be expanded further through TMP, directly affecting the fracture behaviour and consequently, the chip-breaking properties and machinability performance of novel compositions [102,103]. Taha et al. showed that, as the Si content increased from 0 wt.% to 1 wt.% in modern lead-free brasses (CuZn40), tool wear slightly increased as well [104]. This was attributed to the decreased volume fractions of the softer phase α and the respective increase in the harder phase β. In Si-containing brass alloys, the detrimental effect of β phase on tool wear has been confirmed. The increase in Si content, from 1 wt.% to 4 wt.% Si, led to a moderate increase in tool wear due to the increased volume fraction of brittle λ, η, and χ intermetallic phases, albeit no significant effects of hardness and microstructure changes were observed. It was also seen that chip formation changed from continuous to discontinuous type with decreasing cutting speed [104]. However, decreasing the cutting speed is recommended as a method to achieve the desirable discontinuous chip type as it results in lower productivity. By adding 4 wt.% Si to the unleaded CuZn40 brass, the chip formation process changed from the undesirable continuous type to an acceptable long bevel and cylindrical long types. It was clearly shown that the chip type was affected by the formation of hard secondary λ-phase precipitates rather than the variation in mechanical properties [104]. Scientific research revealed the benefits of the recycled bismuth–tin (Bi-Sn) solder addition in the lead-free brass alloy Cu-38Zn-0.5Si by reducing the chip-size morphology [105]. The improvement of chip-breaking efficiency was attributed to the presence of the κ-phase in CuZn21Si3P, as opposed to the effect of the α-phase in CuZn38As brass. Likewise, the high percentage of β-phase in the microstructure of CuZn41.5 resulted in the reduction in chip morphology. The influence of the coating type (TiN, TiAlN, TiB2 and DLC on carbide tools) as well as the use of polycrystalline diamond (PCD) tools on chip formation and workpiece quality was analysed for the evaluation of machinability in three low-lead brass alloys (CuZn38As, CuZn42 and CuZn21Si3P). The machining problems were diminished using a diamond-like carbon coating, especially through the reduction in the friction in the secondary shear zone [106]. A machinability comparison between leaded (CuZn39Pb3) and lead-free brass (CuZn21Si3P) alloys was implemented in a relevant work concerning tool wear during machining. Machining of the lead-free brass alloy (CuZn21Si3P) resulted in longer chip size and, eventually, higher tool-wear rates using cemented carbides. The use of coating on carbide tools, e.g., (Ti, V, Zr, Hf, Nb, Ta), was recommended as a possible solution for overcoming excessive tool-wear rates [107,108]. In terms of the chip morphology specifically, high-Si brasses are generally found to produce discontinuous chip types while a lower Si content is associated with better tool interaction and superior surface quality with the added benefit of greater mechanical properties. Nevertheless, the produced chip is of the discontinuous type for the low-Si brasses investigated in the present study [104].

4.2. Cutting Force and Surface Roughness

Vilarinho et al. uncovered several relationships concerning the influence of the chemical composition and the hardness on the different components of cutting forces [109]. The suggested empirical relationships considered both the well-established effect of Pb on the machinability behaviour already shown by several authors and the effect of various alloying elements usually present in brass composition. Conventional alloys (CuZn36Pb3, CuZn39Pb3), with and without additional alloying elements, showed lower values of cutting forces (Fz) than the modern brass alloys of the studied ternary systems, Cu-Sn-Zn, Cu-Al-Zn and Cu-Fe-Zn [109]. However, in some cases conventional alloys exhibited similar or higher values of feed and depth forces in relation to modern brasses. The cutting forces used for machining modern brass alloys of the ternary systems studied in this work were affected by the type and content of each alloying element. Comparing the development of residual stresses on the CuZn21Si3P and CuZn39Pb3 brasses showed the retainment of lower residual stresses in the Si-containing alloy and achieving better abrasion resistance, although higher cutting forces were required. This could also point towards better fatigue and stress-corrosion cracking resistance, but additional investigations are required [110]. When comparing the machinability conventional leaded brasses to their Pb-free counterparts, it becomes apparent that in both cases the depth of cut and feed rate are the most vital parameters influencing the resulting cutting force and surface roughness. The optimisation of said parameters enables the control of both machinability criteria and thus their investigation is of the utmost importance [111]. Scientific research revealed the benefits of the recycled Bi-Sn solder addition in lead-free brass alloy (Cu-38Zn-0.5Si) by reducing the required cutting forces in machining [105]. The influence of the coating type (TiN, TiAlN, TiB2 and DLC on carbide tools), as well as the use of polycrystalline diamond (PCD) tools on machining forces, was analysed to evaluate the machinability in three low-lead brass alloys (CuZn38As, CuZn42, and CuZn21Si3P). The machining problems were diminished using a diamond-like carbon coating, especially through the reduction in the friction in the secondary shear zone [106]. Machining the lead-free brass alloy (CuZn21Si3P) required higher cutting forces using cemented carbides. As such, the use of coating on carbide tools such as (Ti, V, Zr, Hf, Nb, Ta) N was recommended as a possible solution for reducing the cutting forces [107,108].

Concerning the alloys of the Cu–Sn–Zn system, the lowest values of cutting and feed forces were obtained for Sn contents that led to a dual-phase β–γ microstructure. On the other hand, the presence of Al in alloys of the Cu–Al–Zn system had the opposite effect to that of Sn, rising the cutting force as its content increased [109]. In this ternary system, the highest values of cutting force Fz were always obtained for alloys with a β–γ or pure β structures. Regarding Cu–Fe–Zn alloys, no relationship was found between the Fe content and the machinability parameters. Additionally, no significant correlations were made between the cutting forces and the chemical composition when expressed in terms of effective Cu content. However, a minor correlation between the feed force and the effective

Cu content on conventional alloys was established, although the authors indicated that further studies would be needed to confirm this behaviour [109].

Gaitonde et al. employed Artificial Neural Networks (ANN) to study the effect of cutting speed, feed rate and the required amount of minimum quantity of lubrication (MQL) on two aspects of machinability, namely the specific cutting force and surface roughness in turning of CuZn39Pb3 brass using a K10 carbide tool [112]. In their study, a multilayer feed forward ANN was employed trained by EBPTA (Error-Back Propagation-Training Algorithm). The training patterns required for an input–output database generation were obtained through turning experiments conducted as per Full Factorial Design (FFD). 3D surface plots were generated to analyse the interaction effects of process parameters and the following conclusions were drawn from the present investigation. The minimum specific cutting force existed at high MQL (160–170 mL/h) with medium range of cutting speed (210–220 m/min), Additionally, the specific cutting force was found to be minimal at a higher feed rate (0.20 mm/rev) with a low MQL (100 mL/h). The specific cutting force was highly sensitive to feed rate variations for all values of MQL. The minimum surface roughness resulted in high MQL (160–180 mL/h) with low cutting speed values. Lastly, the surface roughness sharply increased with increasing feed rate irrespective of MQL [94]. They employed the Taguchi method to determine the optimal process parameters for simultaneously minimising the surface roughness and specific cutting force during turning of CuZn39Pb3 brass with a K10 carbide tool [111,112]. In this work, the optimum amount of MQL and the most appropriate cutting speed and feed rate were determined using Analysis of Means (ANOM) and the relative significance of the parameters was identified through ANOVA. The ANOM on multi-response S/N ratio indicated that an optimum MQL of 200 mL/h, cutting speed of 200 m/min and a feed rate of 0.05 mm/rev was necessary to simultaneously minimise surface roughness and specific cutting force. ANOVA illustrated that feed rate was the most dominant parameter followed by MQL and cutting speed in optimising the machinability characteristics. ANOVA also showed that the error contribution was 0.55%, which clearly indicated the absence of the interaction effects of process parameters on optimisation of multiple performance characteristics. Lastly, validation experiments confirmed that the additive model was adequate for determining the optimum quality characteristics at the 95% confidence interval. Utilising ANOVA as means of analysis and optimisation was also explored in the work of Toulfatzis et al. towards the improvement of machinability in CuZn42, CuZn38As and CuZn36 compared to a conventional CuZn39Pb3 brass. To reduce the volume of experiments required for the machinability trials, a DOE Taguchi method in conjunction with ANOVA was used. As such, their statistical analysis converged on the use of a standard L_{16} orthogonal array. Their work further demonstrated the efficiency of ANOVA in identifying the contributions of individual machining parameters towards the optimisation of the cutting force within 90% accuracy and, to a lesser extent, the surface roughness. Furthermore, the S/N response diagrams showed that the depth of cut would be considered the most significant factor in the machining process which was backed up through ANOVA [50]. Plots of S/N response and data means for cutting force and surface roughness are presented in Figures 9 and 10 [50]. As was confirmed by the ANOVA results, the depth of cut is the most influential factor for cutting force (P_p = 64.3992%) while feed rate is the most influential factor (P_p = 55,3085) for surface roughness, see Tables 1 and 2 [50].

Taha et al. investigated the machinability based on surface quality and cutting forces in lead-free silicon brass alloys [104]. In this work, the ease of machinability was derived from the machined surface quality. This was evaluated by measuring the machined average surface roughness (Ra) at selected cutting conditions using H20 and H123 tools. Evidently, for all Si-containing brass alloys investigated, Ra was almost independent of depth of cut. Ra increased as the cutting feed increased and gradually decreased as speed increased. Minimum values of Ra were observed for leaded brass, followed by 1 wt.% Si and 2 wt.% Si Alloys, and finally the leaded brass C37700. Even higher Ra values were observed for 3 wt.% Si and 4 wt.% Si alloys [104]. The Ra was found to increase by 40% as the silicon content increased from 1 wt.% to 4 wt.% Si. For all leaded and unleaded alloys, cutting forces were proportional to the depth of cut.

In both cases, the machining behaviour was attributed to increased resistance of chip formation as a result of the increase in the volume of the material removal. The cutting force increased with increasing cutting speed up to a maximum of 50 m/min. This was attributed to the increase in friction on the tool face caused by the chip flow. Further increase in speed led to the decrease in cutting force, which was attributed to the increase in the cutting temperature at high speeds, which in turn resulted in the decrease in the friction forces and shear strength of the material due to thermally induced softening. Moreover, leaded brass exhibited the lowest cutting force value due to the reduction in the coefficient of friction between the chip and the tool face caused by the lubricating effect of Pb. The maximum cutting force was measured for 1% Si and was attributed to the highest alloy strength [104].

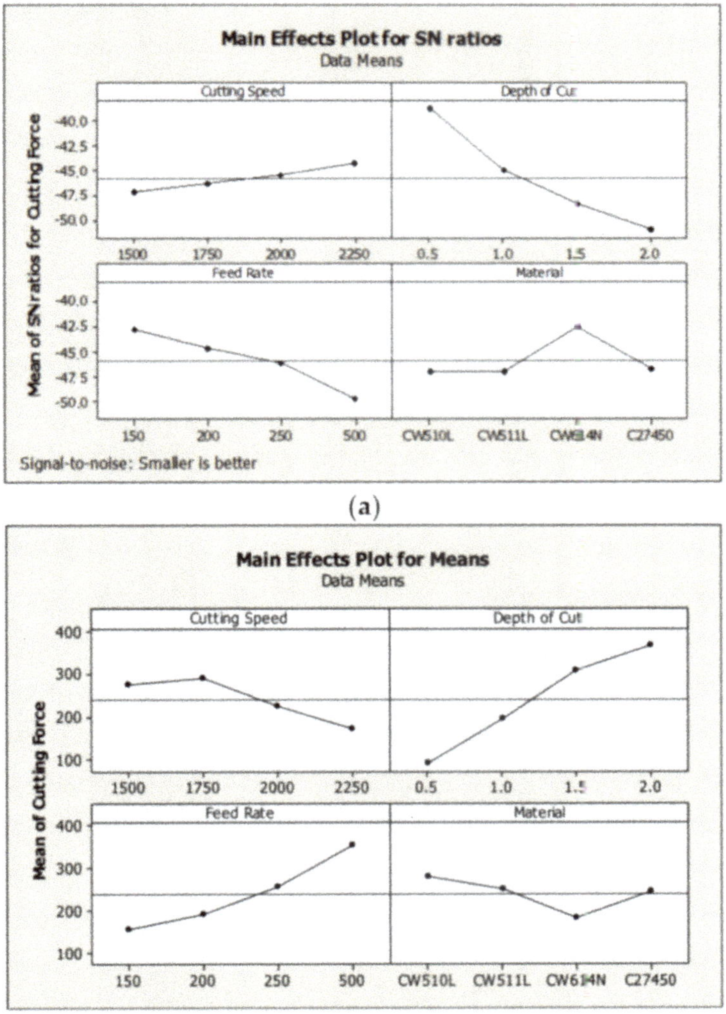

(b)

Figure 9. Diagrams showing the responses of (**a**) S/N ratios and (**b**) data means when evaluated against the cutting parameters which optimise the cutting force [50].

Table 1. Analysis of variance (ANOVA) for the cutting force [50].

Factor	Degrees of Freedom	Sum of Squares (SS)	Corrected Sum of Squares (SS′)	Variance	Percent Contribution P_p (%)	Rank
Cutting Speed	3	19.0425	18.8254	6.3475	3.5728	4
Depth of Cut	3	339.5408	339.3238	113.1803	64.3992	1
Feed Rate	3	108.0831	107.8660	36.0277	20.4715	2
Material	3	58.8660	58.6489	19.6220	11.1308	3
Error	19	1.3746	2.2428	0.0723	0.4257	
Total	31	526.9070	526.9070		100.0000	

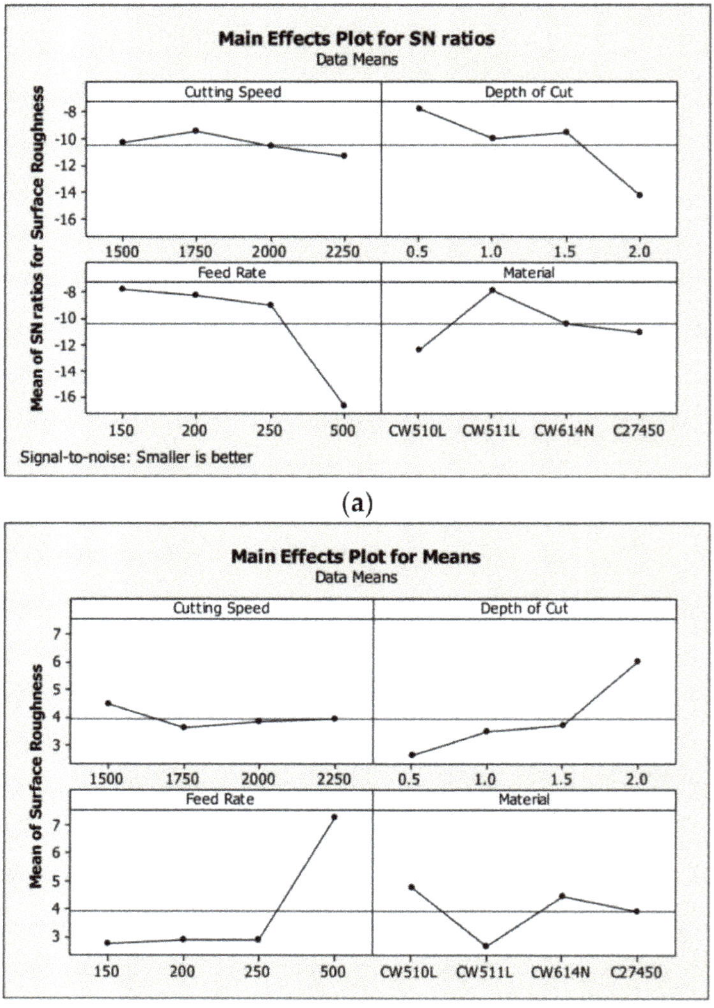

Figure 10. Diagrams showing the responses of (**a**) S/N ratios and (**b**) data means when evaluated against the cutting parameters which optimise the surface roughness [50].

Table 2. Analysis of variance (ANOVA) for the surface roughness [50].

Factor	Degrees of Freedom	Sum of Square (SS)	Corrected Sum of Squares (SS′)	Variance	Percent Contribution P_p (%)	Rank
Cutting Speed	3	7.3498	3.5422	2.4499	0.9460	4
Depth of Cut	3	91.6606	87.8530	30.5535	23.4625	2
Feed Rate	3	210.9051	207.0974	70.3017	55.3085	1
Material	3	40.4101	36.6025	13.4700	9.7753	3
Error	19	24.1149	39.3453	1.2692	10.5078	
Total	31	374.4404	374.4404		100.0000	

4.3. Employment of Vibrational Analysis towards Machinability Evaluation

An emerging trend towards achieving a proper surface finish alongside good economic performance concerns the use of vibrodiagnostics to evaluate the machined components integrity. The term vibrodiagnostics refers to the analysis of the vibrations generated during the machining process. As different materials are characterised by different mechanical and microstructural properties, these variations cause the manifestation of vibrational differences during the cutting process. Specifically, in cases where the workpiece contains for example hard intermetallic phases or inclusions, these irregularities may cause the formation of unwanted vibrations which, in turn can contribute towards an increased oscillation compared to that of a uniform material. Furthermore, tool wear can also be identified through vibrodiagnostics as the geometry of the cutting tool edge is altered during the machining process, generating vibrational differences compared to the initial ones. Thus, by analysing the vibrational characteristics of the cutting process both the workpiece and cutting tool characteristics can be optimised to enable more efficient machinability [113–115]. P. Monka et al. applied such methodologies to optimise the cutting parameters for novel milling tools. In this case study, an S-grade sintered carbide with an AlTiN physical vapor deposition (PVD) coating was used to machine a 16MnCr5 steel. Their analysis pointed at the potential for tool optimisation through modifying the tool geometry [116]. In another study, vibrodiagnostics were used to evaluate the tap failure during internal threads machining. Specifically tailored towards tool behaviour prediction, their analysis showed that cutting tool wear was not affected only by the cutting speed but the angle of the tap as well [117]. Similarly, vibrational studies can be used to unveil the tribological characteristics of the newly formed material surfaces [81].

5. Conclusions and Further Research

The development of eco-friendly brass alloys constitutes a critical research field with significant industrial and environmental impact going forward. Brasses are used in a wide range of applications ranging from plumbing fittings to architectural hardware and decorations. This is due to their superior machinability and corrosion resistance compared to most other industrial materials. The machinability of said alloys has been heavily dependent on the incorporation of Pb, which has traditionally been the key to their success. However, as Pb is currently a heavily regulated substance due to environmental and health concerns its substitution is of paramount importance. Through this review we explored the various ways research has progressed relating to the assessment and resolution of the aforementioned challenge. Two primary approaches for solving this problem were evaluated. The design of novel brass compositions and a compositionally invariant route dependent on a series of heat treatments. Compositional modification aims to replicate the beneficial effects of Pb through the incorporation of second phase particles, while the alternative route is intended to stabilise the β-phase. Overall, in terms of mechanical properties, novel compositions were shown to exhibit much improved mechanical properties compared to the conventional ones. The flexibility pertaining to the inclusion of a wide variety of alloying elements enables various strengthening mechanisms to manifest. Similarly, the β-stabilised heat treated Pb-free conventional compositions showed an increase in hardness and strength at the expense of ductility compared to conventional leaded compositions. Regardless, most of the novel compositions were characterised by mediocre machinability compared to conventional leaded brasses, rarely combining both acceptable mechanical properties and machinability. On the contrary, the heat treated conventional Pb-free compositions showed a degree of machinability similar to the leaded compositions, while retaining a good combination of mechanical properties and fracture resistance. It has thus become clear that both approaches are characterised by complementary benefits. As such, we believe that the optimal way towards the next generation of lead-free brasses lies in the development of alloys that can utilise a combination of both aspects of design. Therefore, we recommend the following pathways for further applied research, aiming at the optimisation of the design and processing routes of lead-free copper alloys to surpass the properties and functionality of the conventional leaded brasses satisfying in parallel the requirements of environmental protection, human health and sustainable development:

(a) The compositional and microstructural design of novel lead-free alloys through computational techniques such as ab initio calculations, CAPHAD (calculation of phase diagrams) approaches alongside general thermodynamic/kinetics and finite element modelling. Such approaches could assess novel compositions in terms of their thermodynamic stability, constituent phases, mechanical properties and corrosion resistance without the expense of physically manufacturing the alloys. Furthermore, these computational approaches could be used alongside experiments to validate and assist in optimization of the thermomechanical processing route of discovered compositions.
(b) Simultaneous "multi-objective" optimisation of the major quality characteristics (chip morphology, power consumption, cutting force and surface roughness) utilising advanced statistical techniques, such as Desirability Functions (DFs) for the machinability of lead-free brass alloys. The optimisation and validation of the cutting conditions in industrial scale for the fabrication of exemplary final brass industrial component by using complex machining operations (such as CNC machining center).
(c) Application of on-line monitoring techniques (e.g., vibration sensors/vibrodiagnostics) might be also employed to forecast the cutting tool durability and service lifetime. In addition, the incurred cutting mechanisms could be further predicted and quantified using appropriate numerical simulation by Finite Element Analysis/FEA software.
(d) Comparative studies concerning the elemental leaching effects between novel lead-free brasses and conventional leaded brasses. This is especially crucial in cases where new alloying additions have been introduced and the resulting alloys have not been investigated in terms of their corrosion performance. This research must be carried out to conduct a risk assessment type study aiming to acknowledge and minimise any impact of these new compositions may have on human health and environment.

Author Contributions: Conceptualization, A.S.P., A.I.T., G.A.P.; methodology, A.I.T. and G.A.P.; literature review A.I.T., G.A.P., P.S.; writing—original draft preparation, A.I.T. and P.S.; writing—review and editing, A.I.T., G.A.P., P.S.; supervision, A.S.P., A.I.T. and G.A.P. All authors have read and agreed to the published version of the manuscript.

Funding: This research received no external funding.

Institutional Review Board Statement: Not applicable.

Informed Consent Statement: Not applicable.

Data Availability Statement: The data supporting the findings of this study are available on request from the corresponding author.

Acknowledgments: The encouragement and support of ELKEME management is highly appreciated.

Conflicts of Interest: The authors declare no conflict of interest.

Appendix A. Chemical Composition of Machinable Brass Alloys

Table A1. Conventional leaded brass compositions.

Conventional Leaded Brasses	
Alloy Composition	**Ref.**
CuZn40Pb	[4,7]
CuZn39Pb3	[8,19,26,27,29,45–50,58,59,64,78,82,104–106]
CuZn40Pb2	[51–55]
CuZn40Pb3	[6,7]
CuZn40Pb4	[7]
CuZn39Pb2	[47,48,57,102]
CuZn36Pb2As	[46,50,59]

Table A2. Conventional lead-free (Pb-free) brass compositions.

Conventional Lead-Free Brasses	
Alloy Composition	**Ref.**
CuZn42	[25,30,79,80,91,104]
CuZn41.5	[31]
CuZn40	[7,23]
CuZn38	[25]
CuZn38As	[24,25,31,50,80–82,92,93,104]
CuZn38Sn	[81]
CuZn37	[108]
CuZn36	[25,50,79]
CuZn32	[64]
CuZn21Si3P	[19,30,65,100,106–109]

Table A3. Novel brass compositions with an increased aluminium (Al) content (alloying elements expressed in wt.%).

Novel Brass Compositions–High Aluminium (Al) Content								
Cu	**Pb**	**Ni**	**Fe**	**Sn**	**Zn**	**Al**	**Si**	**Ref.**
Bal.	-	-	-	-	37.6	3.68	-	[109]
Bal.	2.32	0.24	0.33	0.53	32.67	2.88	0.06	[109]
Bal.	-	-	0.028	-	30.2	1.66	-	[82]
Bal.	-	-	0.03	-	33.7	1.51	-	[82]
Bal.	-	-	0.033	-	34.2	1.49	-	[82]
Bal.	-	-	0.032	-	34.5	1.48	-	[82]

Al addition improves corrosion resistance, mechanical properties and in significant amounts can produce a lightweight effect. Al also affects the stability of $\alpha/\beta/\gamma$ phases.

Table A4. Novel brass compositions with an increased lead (Pb) content (alloying elements expressed in wt.%).

Novel Brass Compositions–High Lead (Pb) Content								
Cu	**Pb**	**Ni**	**Fe**	**Sn**	**Zn**	**Al**	**Si**	**Ref.**
Bal.	4.23	0.24	0.33	1.08	32.28	1.57	0.06	[109]
Bal.	3.5	-	-	-	39.6	-	-	[61]
Bal.	3.04	0.02	0.17	0.06	37.96	0.02	-	[3]
Bal.	2.9	-	-	-	38.8	-	-	[85]
Bal.	2.6	-	0.2	0.2	39	0.02	-	[48]
Bal.	2.46	0.25	0.34	0.55	33.63	0.8	-	[109]
Bal.	2.43	0.24	0.33	0.54	33.15	2.01	0.06	[109]
Bal.	1.65	0.25	0.36	0.54	33.75	0.84	0.1	[109]
Bal.	1.46	-	0.114	0.096	35.02	0.503	-	[33]
Bal.	1.45	-	0.116	0.096	34.89	0.506	-	[33]
Bal.	0.95	-	0.07	-	38.77	1.01	-	[109]
Bal.	0.909	-	0.095	0.104	39.1	0.59	-	[33]
Bal.	0.894	-	0.095	0.104	38.97	0.594	-	[33]

Table A5. Novel brass compositions with an increased tin (Sn) content (alloying elements expressed in wt.%).

Novel Brass Compositions–High Tin (Sn) Content					
Cu	**Pb**	**Fe**	**Sn**	**Zn**	**Ref.**
Bal.	0.02	0.03	17.37	27.49	[103]
Bal.	0.01	0.04	13.89	26.74	[103]
Bal.	-	0.06	11.36	26.72	[103]
Bal.	0.02	0.04	8.02	28.16	[103]
Bal.	0.02	0.05	5.43	28.63	[103]
Bal.	-	-	5.3	39.33	[109]
Bal.	-	-	1.0-4.0	38	[21]
Bal.	-	-	3.84	39.61	[109]
Bal.	-	0.04	3.16	28.57	[103]
Bal.	-	-	1.78	31.53	[109]
Bal.	-	-	1.39	31.9	[109]
Bal.	-	0.03	1.2	28.8	[102]
Bal.	-	-	0.95	34.22	[109]
Bal.	-	-	0.85	32.11	[109]
Bal.	-	-	0.78	39.93	[109]
Bal.	-	-	0.76	38.2	[109]

Sn addition may promote the formation of β' enhancing both the machinability and hardness.

Table A6. Novel brass compositions with an increased bismuth (Bi) content (alloying elements expressed in wt.%).

Novel Brass Compositions–High Bismuth (Bi) Content							
Cu	**Fe**	**Sn**	**Zn**	**Bi**	**Si**	**Cr**	**Ref.**
Bal.	0.22	0.58	40.83	2.85	-	0.22	[75]
Bal.	-	-	40.19	2.20	-	-	[17]
Bal.	0.23	0.6	40.64	2.02	-	0.26	[75]
Bal.	-	2	38	2.00	0.5	-	[105]
Bal.	-	1.5	38	1.50	0.5	-	[105]
Bal.	-	1	38	1.00	0.5	-	[105]
Bal.	0.23	0.60	40.81	0.99	-	0.26	[75]
Bal.	-	0.7	38	0.80	0.5	-	[105]
Bal.	-	0.3	38	0.40	0.5	-	[105]
Bal.	-	-	29.2	0.27	-	-	[77]

Bi addition operates similarly to the Pb addition, improving the machinability through chip segmentation and providing cutting tool lubrication.

Table A7. Novel brass compositions with an increased titanium (Ti) content (alloying elements expressed in wt.%).

Novel Brass Compositions–High Titanium (Ti) Content							
Cu	**Pb**	**Fe**	**Sn**	**Zn**	**Si**	**Ti**	**Ref.**
Bal.	0.02	0.09	0.03	39.6	-	1.03	[65]
Bal.	-	-	0.65	40.1	-	1.03	[67]
Bal.	-	-	-	41.2	-	0.99	[67]
Bal.	0.16	0.58	0.41	38.9	0.13	0.71	[65]
Bal.	-	-	-	39.0	-	0.50	[23]

Ti addition induces Cu_xTi_y type particle precipitation, thus increasing the mechanical properties through precipitation hardening and acting as a grain refiner.

Table A8. Novel brass compositions with an increased silicon (Si) content (alloying elements expressed in wt.%).

Novel Brass Compositions–High Silicon (Si) Content								
Cu	**Pb**	**Ni**	**Fe**	**Sn**	**Zn**	**Al**	**Si**	**Ref.**
Bal.	-	-	-	1.04	34.78	-	3.68	[109]
Bal.	-	-	-	-	31.91	3.37	3.37	[109]
Bal.	2.27	0.25	0.4	0.53	31.77	0.8	2.34	[109]
Bal.	0.01	-	0.01	-	31	-	2.2	[73]
Bal.	-	-	-	-	39.77	-	1.73	[108]
Bal.	0.01	-	0.04	-	29	-	1.65	[73]
Bal.	-	-	-	-	35.98	1.6	1.6	[109]
Bal.	-	-	-	-	40.55	-	1.21	[109]
Bal.	-	-	-	-	39.49	1	1.11	[24]
Bal.	-	-	-	-	39.73	0.7	1.1	[71]
Bal.	-	-	-	-	38.42	0.68	1.1	[24]
Bal.	-	-	-	-	38.44	1.09	1.09	[109]
Bal.	-	-	-	0.5	42	0.5	1	[73]
Bal.	-	-	-	-	38.37	0.34	0.95	[71]
Bal.	-	-	-	-	39.37	0.18	0.95	[24]
Bal.	0.01	-	0.04	-	30	-	0.93	[70]
Bal.	-	-	-	-	40.43	0.38	0.87	[71]
Bal.	-	-	-	-	38.75	0.9	0.85	[24]
Bal.	-	-	-	-	40.9	-	0.83	[102]
Bal.	-	-	-	-	39.13	0.69	0.78	[24]
Bal.	-	-	-	-	38.64	0.51	0.78	[72]
Bal.	-	-	-	-	36.92	0.27	0.78	[71]
Bal.	-	-	-	-	33.68	0.2	0.73	[71]
Bal.	-	-	-	-	38.91	0.5	0.69	[24]
Bal.	-	-	-	-	39.29	0.54	0.54	[109]
Bal.	-	-	-	-	38	-	0.5	[105]
Bal.	-	-	-	-	37.69	0.49	0.49	[109]
Bal.	-	-	-	-	40.67	-	0.34	[102]

Si addition produces fine intermetallic particles, improving machinability and mechanical properties. Si also affects the stability of the $\alpha/\beta/\gamma$ phases in brasses.

Table A9. Novel brass composition with an increased content in antimony (Sb), graphite, magnesium (Mg), manganese (Mn), carbon (C) and cerium (Ce) (alloying elements expressed in wt.%).

Novel Brass Compositions–Uncommon Alloying Element Additions (Sb, Graphite, Mg, Mn, C, Ce)														
Cu	**Fe**	**Sn**	**Zn**	**Al**	**Sb**	**Bi**	**Graphite**	**Si**	**Mg**	**Mn**	**Cr**	**C**	**Ce**	**Ref.**
Bal.	0.095	-	30.1	-	-	0.173	-	-	-	1.26	-	-	-	[77]
Bal.	0.063	-	29.1	-	-	0.2	-	-	-	3.48	-	-	-	[77]
Bal.	0.096	-	29.6	-	-	0.204	-	-	-	5.83	-	-	-	[77]
Bal.	-	-	35	-	0.0–2.34	-	-	-	0.0–1.79	-	-	-	-	[22]
Bal.	-	-	40	-	-	-	0–1.0	-	0.3–1.9	-	-	-	-	[68]
Bal.	-	-	40	-	-	-	-	-	1	-	-	-	-	[59]
Bal.	-	-	39.2	-	-	-	0.0–1.0	-	-	-	-	-	-	[1]
Bal.	4.78	-	38	-	-	-	-	-	-	-	-	0.22	-	[69]
Bal.	-	-	38.6	-	-	-	-	-	-	-	-	-	0.07	[74]
Bal.	-	0.5	41.6	0.5	0.4	-	-	1	-	-	-	-	-	[73]
Bal.	-	0.5	41	0.5	1	-	-	1	-	-	-	-	-	[73]
Bal.	-	0.5	40.6	0.5	1.4	-	-	1	-	-	-	-	-	[73]
Bal.	-	0.5	40	0.5	2	-	-	1	-	-	-	-	-	[73]
Bal.	0.22	0.59	40.86	-	-	-	-	-	-	-	0.34	-	-	[75]

Sb addition improves machinability by promoting the formation of β'. Produces intermetallic particles which compromise the mechanical properties but further improve machinability. Graphite addition acts similarly to Pb, thus increasing machinability and providing cutting tool lubrication. Mg addition improves machinability by promoting the formation of β'. Produces intermetallic particles which compromise the mechanical properties but further improve machinability. Mn addition promotes the formation of β' thus increasing the mechanical properties and machinability. C addition operates similarly to graphite, as in the relevant research [68] it was incorporated in the form of Fe_3C and underwent a graphitisation process. Ce addition reduces the flow stress of the brass.

References

1. Imai, H.; Kosaka, Y.; Kojima, A.; Li, S.; Kondoh, K.; Umeda, J.; Atsumi, H. Characteristics and machinability of lead-free P/M Cu60–Zn40 brass alloys dispersed with graphite. *Powder Technol.* **2010**, *198*, 417–421. [CrossRef]
2. Chakrabarti, A.K.; Spretnak, J.W. The effect of solute (Zn) concentration on the plastic deformation properties of α and $\alpha + \beta$ Cu-Zn alloys. *Mater. Sci. Eng.* **1976**, *24*, 95–100. [CrossRef]
3. Panagopoulos, C.N.; Georgiou, E.P.; Simeonidis, K. Lubricated wear behavior of leaded $\alpha + \beta$ brass. *Tribol. Int.* **2012**, *50*, 1–5. [CrossRef]
4. Achiţei, D.C.; Minciună, M.G.; Vizureanu, P.; Sandu, A.V.; Cimpoeşu, R.; Istrate, B. Study on structure and properties of CuZnPb alloy. *IOP Conf. Ser. Mater. Sci. Eng.* **2019**, *133*, 012015. [CrossRef]
5. La Fontaine, A.; Keast, V.J. Compositional distributions in classical and lead-free brasses. *Mater. Charact.* **2006**, *57*, 424–429. [CrossRef]
6. Wolfenden, A.; Wright, P.K. Role of lead in free-machining brass. *Met. Technol.* **1979**, *6*, 297–302. [CrossRef]
7. Stoddart, C.T.H.; Lea, C.; Dench, W.A.; Green, P.; Pettit, H.R. Relationship between lead content of Cu-40Zn, machinability, and swarf surface composition determined by auger electron spectroscopy. *Met. Technol.* **1979**, *6*, 176–184. [CrossRef]
8. Pantazopoulos, G. Leaded brass rods C 38500 for automatic machining operations: A technical report. *J. Mater. Eng. Perform.* **2002**, *11*, 402–407. [CrossRef]
9. Leaded Brasses, Copper Development Association Inc. Available online: https://www.copper.org/resources/properties/microstructure/lead_brasses.html (accessed on 17 January 2022).
10. Siu, K.W.; Kwok, J.C.M.; Ngan, A.H.W. Thermo-mechanical processing of brass components for potable-water usage increases risks of Pb leaching. *Water Res.* **2020**, *186*, 116414. [CrossRef]
11. Korshin, G.V.; Ferguson, J.F.; Lancaster, A.N. Influence of natural organic matter on the corrosion of leaded brass in potable water. *Corros. Sci.* **2000**, *42*, 53–66. [CrossRef]
12. Triantafyllidou, S.; Parks, J.; Edwards, M. Lead particles in potable water. *Am. Water Work. Assoc.* **2007**, *99*, 107–117. [CrossRef]
13. Tang, M.; Sinsheimer, P.; Sarver, E.; Parks, J.; Edwards, M. Evaluating "lead-free" brass performance in potable water. *Corrosion* **2019**, *75*, 865–875. [CrossRef]
14. Elfland, C.; Scardina, P.; Edwards, M. Lead-contaminated water from brass plumbing devices in new buildings. *Am. Water Work. Assoc.* **2010**, *102*, 66–76. [CrossRef]
15. Triantafyllidou, S.; Nguyen, C.K.; Zhang, Y.; Edwards, M.A. Lead (Pb) quantification in potable water samples: Implications for regulatory compliance and assessment of human exposure. *Environ. Monit. Assess.* **2021**, *185*, 1355–1365. [CrossRef]
16. Estelle, A.A. Drinking water lead regulations: Impact on the brass value chain. *Mater. Sci. Technol.* **2016**, *32*, 1763–1770. [CrossRef]
17. Li, S.; Kondoh, K.; Imai, H.; Atsumi, H. Fabrication and properties of lead-free machinable brass with Ti additive by powder metallurgy. *Powder Technol.* **2011**, *205*, 242–249. [CrossRef]
18. Available online: https://www.epa.gov/sdwa/use-lead-free-pipes-fittings-fixtures-solder-and-flux-drinking-water (accessed on 12 December 2021).
19. Schultheiss, F.; Windmark, C.; Sjöstrand, S.; Rasmusson, M.; Ståhl, J.E. Machinability and manufacturing cost in low-lead brass. *Int. J. Adv. Manuf. Technol.* **2018**, *99*, 2101–2110. [CrossRef]
20. Kuyucak, S.; Sahoo, M. A review of the machinability of copper-base alloys. *Can. Metall. Q.* **1996**, *35*, 1–15. [CrossRef]
21. Suksongkarm, P.; Rojananan, S.; Rojananan, S. Bismuth formation in lead-free Cu-Zn-Si yellow brass with various bismuth-tin alloy additions. *Mater. Trans.* **2018**, *59*, 1747–1752. [CrossRef]
22. Adineh, M.; Doostmohammadi, H. Microstructure, mechanical properties and machinability of Cu-Zn-Mg and Cu-Zn-Sb brass alloys. *Mater. Sci. Technol.* **2019**, *35*, 1504–1514. [CrossRef]
23. Atsumi, H.; Imai, H.; Li, S.; Kondoh, K.; Kousaka, Y.; Kojima, A. The effect of solid solutionizing Ti element on microstructural and mechanical properties of extruded Cu-40Zn-Ti ternary alloy. *Trans. JWRI* **2011**, *40*, 67–71.
24. Yang, C.; Ding, Z.; Tao, Q.C.; Liang, L.; Ding, Y.F.; Zhang, W.W.; Zhu, Q.L. High-strength and free-cutting silicon brasses designed via the zinc equivalent rule. *Mater. Sci. Eng. A* **2018**, *723*, 296–305. [CrossRef]
25. Toulfatzis, A.; Pantazopoulos, G.; David, C.; Sagris, S.; Paipetis, A. Final heat treatment as a possible solution for the improvement of machinability of Pb-free brass alloys. *Metals* **2018**, *8*, 575. [CrossRef]
26. Toulfatzis, A.I.; Pantazopoulos, G.A.; Paipetis, A.S. Microstructure and properties of lead-free brasses using post-processing heat treatment cycles. *Mater. Sci. Technol.* **2016**, *32*, 1771–1781. [CrossRef]
27. Pantazopoulos, G.; Vazdirvanidis, A. Characterization of the microstructural aspects of machinable α–β phase brass. *Microsc. Anal.* **2008**, *22*, 13–16.
28. Vaxevanidis, N.M.; Fountas, N.A.; Koutsomichalis, A.; Kechagias, J.D. Experimental investigation of machinability parameters in turning of CuZn39Pb3 brass alloy. *Procedia Struct. Integr.* **2018**, *10*, 333–341. [CrossRef]
29. Zoghipour, N.; Tascioglou, E.; Atay, G.; Kaynak, Y. Machining-induced surface integrity of holes drilled in lead-free brass alloy. *Procedia CIRP* **2020**, *87*, 148–152. [CrossRef]
30. Nobel, C.; Hofmann, U.; Klocke, F.; Veselovac, D.; Puls, H. Application of a new, severe-condition friction test method to understand the machining characteristics of Cu-Zn alloys using coated cutting tools. *Wear* **2015**, *344*, 58–68. [CrossRef]
31. Naik, D.B.; Dave, A.K. Multipass, multicriterion optimization in turning. *Int. J. Adv. Manuf. Technol.* **1998**, *14*, 544–548. [CrossRef]

32. Besseris, G.J. Non-linear nonparametric quality screening in low sampling testing. *Int. J. Qual. Reliab. Manag.* **2010**, *27*, 893–915. [CrossRef]
33. Amaral, L.; Quinta, R.; Silva, T.E.; Soares, R.M.B.; Castellanos, S.D.; De Jesus, A.M.P. Effect of lead on the machinability of brass alloys using polycrystalline diamond cutting tools. *J. Strain Anal.* **2018**, *53*, 602–615. [CrossRef]
34. Vaxevanidis, N.M.; Kechagias, J.D.; Fountas, N.A.; Manolakos, D.E. Evaluation of machinability in turning of engineering alloys by applying artificial neural networks. *Open Constr. Build. Technol. J.* **2014**, *8*, 389–399. [CrossRef]
35. Abbas, A.; Pimenov, D.; Erdakov, I.; Taha, M.; El Rayes, M.; Soliman, M. Artificial intelligence monitoring of hardening methods and cutting conditions and their effects on surface roughness, performance, and finish turning costs of solid-state recycled aluminum alloy 6061 chips. *Metals* **2018**, *8*, 394. [CrossRef]
36. Hrechuk, A.; Bushlya, V.; M'Saoubi, R.; Ståhl, J.-E. Quantitative analysis of chip segmentation in machining using an automated image processing method. *Procedia CIRP* **2019**, *82*, 314–319. [CrossRef]
37. Besseris, G.J. A methodology for product reliability enhancement via saturated-unreplicated fractional factorial designs. *Reliab. Eng. Syst. Saf.* **2010**, *95*, 742–749. [CrossRef]
38. Fountas, N.; Koutsomichalis, A.; Kechagias, J.D.; Vaxevanidis, N.M. Multi-response optimization of CuZn39Pb3 brass alloy turning by implementing Grey Wolf algorithm. *Frat. Ed Integrità Strutt.* **2019**, *50*, 584–594. [CrossRef]
39. Arvidsson, M.; Gremyr, I. Principles of robust design methodology. *Qual. Reliab. Eng. Int.* **2008**, *24*, 23–35. [CrossRef]
40. Manna, A.; Bhattacharyya, B. Taguchi and Gauss elimination method: A dual response approach for parametric optimization of CNC wire cut EDM of PRAlSiCMMC. *Int. J. Adv. Manuf. Technol.* **2006**, *28*, 67–75. [CrossRef]
41. Ilzarbe, L.; Álvarez, M.J.; Viles, E.; Tanco, M. Practical applications of design of experiments in the field of engineering: A bibliographical review. *Qual. Reliab. Eng. Int.* **2008**, *24*, 417–428. [CrossRef]
42. Viles, E.; Tanco, M.; Ilzarbe, L.; Alvarez, M.J. Planning experiments, the first real task in reaching a goal. *Qual. Eng.* **2009**, *21*, 44–51. [CrossRef]
43. Tanco, M.; Viles, E.; Ilzarbe, L.; Alvarez, M.J. Barriers faced by engineers when applying design of experiments. *TQM J.* **2009**, *21*, 565–575. [CrossRef]
44. Tanco, M.; Viles, E.; Ilzarbe, L.; Alvarez, M.J. Implementation of design of experiments projects in industry. *Appl. Stoch. Models Bus. Ind.* **2009**, *25*, 478–505. [CrossRef]
45. Pantazopoulos, G.; Vazdirvanidis, A. Failure analysis of a fractured leaded-brass (CuZn39Pb3) extruded hexagonal rod. *J. Fail. Anal. Prev.* **2008**, *8*, 218–222. [CrossRef]
46. Toulfatzis, A.I.; Besseris, G.J.; Pantazopoulos, G.A.; Stergiou, C. Characterization and comparative machinability investigation of extruded and drawn copper alloys using non-parametric multi-response optimization and orthogonal arrays. *Int. J. Adv. Manuf. Technol.* **2011**, *57*, 811–826. [CrossRef]
47. Mapelli, C.; Venturini, R. Dependence of the mechanical properties of an α/β brass on the microstructural features induced by hot extrusion. *Scr. Mater.* **2006**, *54*, 1169–1173. [CrossRef]
48. Momeni, A.; Ebrahimi, G.R.; Faridi, H.R. Effect of chemical composition and processing variables on the hot flow behavior of leaded brass alloys. *Mater. Sci. Eng. A* **2015**, *626*, 1–8. [CrossRef]
49. Garcia, P.; Rivera, S.; Palacios, M.; Belzunce, J. Comparative study of the parameters influencing the machinability of leaded brasses. *Eng. Fail. Anal.* **2010**, *17*, 771–776. [CrossRef]
50. Toulfatzis, A.I.; Pantazopoulos, G.A.; David, C.N.; Sagris, D.S.; Paipetis, A.S. Machinability of eco-friendly lead-free brass alloys: Cutting-force and surface-roughness optimization. *Metals* **2018**, *8*, 250. [CrossRef]
51. Holler, K.; Reetz, B.; Müller, K.B.; Pyzalla, A.; Reimers, W. Microstructure and properties of hot extruded brass CuZn40Pb2. *Mater. Sci. Forum* **2003**, *426–432*, 3667–3672. [CrossRef]
52. Loginov, Y.N.; Ovchinnikov, S. Increase in the uniformity of structure and properties of extruded workpieces of alpha + beta lead brasses. *Metallurgist* **2015**, *8*, 342–347. [CrossRef]
53. Kunčická, L.; Jambor, M.; Weiser, A.; Dvořák, J. Structural factors inducing cracking of brass fittings. *Materials* **2021**, *14*, 3255. [CrossRef] [PubMed]
54. Kunčická, L.; Benč, M.; Andreyachshenko, V. Effect of residual stress on cracking of hot-die forged brass fittings. *IOP Conf. Ser. Mater. Sci. Eng.* **2021**, *1190*, 012031. [CrossRef]
55. Kunčická, L.; Kocich, R. Effects of temperature (in)homogeneity during hot stamping on deformation behavior, structure and properties of brass valves. *Adv. Res. Mater.* **2021**, 202101414. [CrossRef]
56. Chandra, T.; Jonas, J.J.; Taplin, D.M.R. Grain-boundary sliding and intergranular cavitation during superplastic deformation of α/β brass. *J. Mater. Sci.* **1978**, *13*, 2380–2384. [CrossRef]
57. Hentati, N.; Makni, A.; Elleuch, R. Study of failure modes affecting a crimped nut related to forging process. *J. Fail. Anal. Prev.* **2012**, *12*, 130–138. [CrossRef]
58. Blaz, L.; Konior, Z.; Majda, T. Structural aspects of α/β transformation in hot deformed CuZn-39Pb3 alloy. *J. Mater. Sci.* **2001**, *36*, 3629–3635. [CrossRef]
59. Pantazopoulos, G.A.; Toulfatzis, A.I. Fracture modes and mechanical characteristics of machinable brass rods. *Metallogr. Microstruct. Anal.* **2012**, *1*, 106–114. [CrossRef]
60. Adineh, M.; Doostmohammadi, H.; Raiszadeh, R. Effect of Si and Al on the microstructure, mechanical properties and machinability of 65Cu-35Zn brass. *Iran. J. Mater. Sci. Eng.* **2019**, *16*, 21–32.

61. Kashyap, B.P.; Verma, S.; Mandlik, P.; Kumar, N.; Toppo, S.P. Effect of test temperature on tensile properties of α/β brass containing lead. *Mater. Sci. Technol.* **2006**, *22*, 363–367. [CrossRef]
62. Laporte, V.; Mortensen, A. Intermediate temperature embrittlement of copper alloys. *Int. Mater. Rev.* **2009**, *54*, 94–116. [CrossRef]
63. Wolley, D.J.; Fox, A.G. The embrittlement of leaded and unleaded α+β (60-40) brasses in the temperature range 300 to 500 °C. *J. Mater. Sci. Lett.* **1988**, *7*, 763–765. [CrossRef]
64. Flegeau, G.; Taleb, L.; Hauet, A.; Vassal, C. Cyclic behavior of an industrial brass Cu Zn32. In *Key Engineering Materials*; Trans Tech Publications Ltd.: Bäch, Switzerland, 2013; Volume 535, pp. 189–192.
65. Johansson, J.; Persson, H.; Ståhl, J.-E.; Zhou, J.-E.; Bushlya, V.; Schultheiss, F. Machinability evaluation of low-lead brass alloys. *Procedia Manuf.* **2019**, *38*, 1723–1730. [CrossRef]
66. Atsumi, H.; Imai, H.; Li, S.; Kondoh, K.; Kousaka, Y.; Kojima, A. Fabrication and properties of high-strength extruded brass using elemental mixture of Cu-40% Zn alloy powder and Mg particle. *Mater. Chem. Phys.* **2012**, *135*, 554–562. [CrossRef]
67. Li, S.; Imai, H.; Atsumi, H.; Kondoh, K.; Kojima, A.; Kousaka, Y.; Yamamoto, K.; Takahashi, M. The effects of Ti and Sn alloying elements on precipitation strengthened Cu40Zn brass using powder metallurgy and hot extrusion. *Mater. Sci. Eng. A* **2012**, *535*, 22–31. [CrossRef]
68. Zhang, X.; Ma, C.; Li, S.; Pan, D.; Zhen, F. Interface design of lead-free free-cutting titanium reinforced graphite brass composites and its effect on mechanical properties and cutting performance. *Mater. Sci. Eng. A* **2020**, *774*, 138909. [CrossRef]
69. Zhuo, H.-O.; Tank, J.-C.; Xue, Y.-Y.; Ye, N. Preparation of lead-free free-cutting graphite brasses by graphitization of cementite. *Trans. Nonferrous Met. Soc. China* **2015**, *25*, 3252–3257. [CrossRef]
70. Alirezaei, M.; Doostmohammadi, H. Microstructure evolution in cast and equilibrium heat-treated CuZn30-(Si) alloys. *Int. J. Cast Met. Res.* **2016**, *29*, 222–227. [CrossRef]
71. Yang, C.; Yang, Y.; Lin, G. Tailoring chip morphology by correlating the microstructure and dynamic yield strength in turning of lead-free silicon brasses. *J. Manuf. Processes* **2020**, *53*, 420–430. [CrossRef]
72. Yang, C.; Ye, W.; Yang, Y.; Wang, F.; Ding, Y.; Zhu, Q.; Zhang, W. Microstructural evolution and mechanical behavior of lead-free silicon brass manufactured by low-pressure die casting. *J. Mater. Eng. Perform.* **2018**, *27*, 5478–5488. [CrossRef]
73. Suksongkarm, D.; Rojananan, S.; Rojananan, S. Microstructure and hardness of Cu-Zn-Si-Al-Sn with antimony addition. *Adv. Mater. Res.* **2013**, *802*, 179–183. [CrossRef]
74. Chandra, T.; Jonas, J.J.; Taplin, D.M.R. The mechanical behavior of cerium-modified alpha-beta brass at high temperatures. *J. Mater. Sci.* **1976**, *11*, 1843–1848. [CrossRef]
75. Atsumi, H.; Imai, H.; Li, S.; Kondoh, K.; Kousaka, Y.; Kojima, A. High-strength, lead-free machinable α–β duplex phase brass Cu–40Zn–Cr–Fe–Sn–Bi alloys. *Mater. Sci. Eng. A* **2011**, *529*, 275–281. [CrossRef]
76. Thomson, J.; Zavadil, R.; Sahoo, M. Development of a lead-free bearing material for aerospace applications. *Int. J. Met.* **2010**, *4*, 19–30. [CrossRef]
77. Basori, I.; Pratiwi, H.I.; Sofyan, B.T. Effects of manganese on the microstructures, mechanical properties and deformation characteristics of Cu-29Zn alloy. *Mater. Sci. Forum* **2017**, *917*, 212–217. [CrossRef]
78. Li, Z.Q.; WeiDong, W.; KaiZhou, L.; GengChun, C.; WeiPing, C. Study on microstructure and properties of brass containing Sb and Mg. *Sci. China Ser. E Technol. Sci.* **2009**, *52*, 2172–2174.
79. Toulfatzis, A.I.; Pantazopoulos, G.A.; Paipetis, A.S. Fracture behavior and characterization of lead-free brass alloys for machining applications. *J. Mater. Eng. Perform.* **2014**, *23*, 3193–3206. [CrossRef]
80. Toulfatzis, A.I.; Pantazopoulos, G.A.; Paipetis, A.S. Fracture mechanics properties and failure mechanisms of environmental-friendly brass alloys under impact, cyclic and monotonic loading conditions. *Eng. Fail. Anal.* **2018**, *90*, 497–517. [CrossRef]
81. Kozana, J.; Garbacz-Klempka, A.; Piękoś, M. Lead-free casting brasses. Investigations of the corrosion resistance and shaping of microstructure and properties. *Arch. Foundry Eng.* **2019**, *19*, 113–118.
82. Iecks, G.; Maiolo, L.M.A.; Bortolozo, A.D.; Osório, W.R. Designing a microstructural array associated with hardness of dual-phase Cu-Zn alloy using investment casting. *Mater. Res.* **2018**, *21*, e20171059. [CrossRef]
83. Lobanov, M.L.; Danilov, S.V.; Pastukhov, V.I. Crystallographic peculiarities of β-α transformation in brass induced by hot extrusion. *Solid State Phenom.* **2019**, *299*, 541–545. [CrossRef]
84. Dhinwal, S.S.; Shukla, A.J.; Biswas, S.; Chouhan, D.K. Evolution of microstructure and crystallographic texture in α–β brass during equal channel angular pressing. *Mater. Characteris.* **2020**, *163*, 110270. [CrossRef]
85. Ibrahim, A.; Rizal, S.; Ali, N.; Huzni, S. The effect of chemical composition on grain size and formability of the free-lead Cu-30Zn alloy: A short review. *IOP Conf. Ser. Mater. Sci. Eng.* **2019**, *536*, 012019. [CrossRef]
86. Pantazopoulos, G. A review of defects and failures in brass rods and related components. *Pract. Fail. Anal.* **2003**, *3*, 14–22. [CrossRef]
87. Gupta, I.; Mishra, R.R.; Nagaraj, M.; Rajesha, S. An overview on modes of failure of brass components. *Int. J. Mech. Prod. Eng.* **2014**, *2*, 33–36.
88. Izumi, O.; Harada, Y. Hot Shortness of α-Brass. *Trans. Jpn. Inst. Met.* **1970**, *11*, 292–299. [CrossRef]
89. Martinez-Hernandez, M.; Juarez-Hernandez, A.; González-Rivera, C.; Hernandez-Rodriguez, M.A.L. Bismuth segregation and crack formation on a free lead yellow brass tap. *Eng. Fail. Anal.* **2013**, *28*, 63–68. [CrossRef]
90. Hsieh, C.-C.; Wang, J.-S.; Wu, P.T.-Y.; Wu, W. Microstructural development of brass alloys with various Bi and Pb additions. *Met. Mater. Int.* **2013**, *19*, 1173–1179. [CrossRef]

91. Chunlei, G.; Dongfu, S.; Kaihong, Z.; Haiyan, W.; Nan, Z. Failure analysis of a lead-free brass tap used in potable water. *Eng. Fail. Anal.* **2016**, *59*, 377–383. [CrossRef]
92. Chunlei, G.; Nan, Z.; Yuehua, K.; Shuncheng, W.; Kaihong, Z. Failure analysis of lead-free brass valve bodies. *Eng. Fail. Anal.* **2019**, *100*, 536–543. [CrossRef]
93. Pantazopoulos, G.; Vazdirvanidis, A. Identification of corrosion and damage mechanisms by using Scanning Electron Microscopy and Energy Dispersive X-ray Microanalysis: Contribution to Failure Analysis Case Histories. *IOP Conf. Ser. Mater. Sci. Eng.* **2014**, *55*, 012015. [CrossRef]
94. Pantazopoulos, G.; Toulfatzis, A.I. Failure analysis of a machinable brass connector in a boiler unit installation. *Case Stud. Eng. Fail. Anal.* **2013**, *1*, 18–23. [CrossRef]
95. Lisenko, N.; Evans, C.D.; Yao, Y.L. Effect of brass composition and phases on stress corrosion mitigation by laser shock peening. *Manuf. Lett.* **2020**, *23*, 5–8. [CrossRef]
96. Choucri, J.; Balbo, A.; Zanotto, F.; Grassi, V.; Touhami, M.E.; Mansouri, I.; Monticelli, C. Corrosion behavior and susceptibility to stress corrosion cracking of leaded and lead-free brasses in simulated drinking water. *Materials* **2021**, *15*, 144. [CrossRef] [PubMed]
97. *ISO 3685*; Tool-Life Testing with Single-Point Turning Tools. International Organization for Standardization (ISO): Geneva, Switzerland, 1993.
98. Gane, N. The effect of lead on the friction and machining of brass. *Philos. Mag. A* **1981**, *43*, 545–566. [CrossRef]
99. Hassan, K.; Kumar, A.; Garg, M.P. Experimental investigation of material removal rate in CNC turning using Taguchi method. *Int. J. Eng. Res. Appl.* **2012**, *2*, 1581–1590.
100. Nobel, C.; Hofmann, U.; Klocke, F.; Veselovac, D. Experimental investigation of chip formation, flow, and breakage in free orthogonal cutting of copper-zinc alloys. *Int. J. Adv. Manuf. Technol.* **2016**, *84*, 1127–1140. [CrossRef]
101. Laakso, S.V.A.; Hokka, M.; Niemi, E.; Kuokkala, V.-T. Investigation of the effect of different cutting parameters on chip formation of low-lead brass with experiments and simulations. *Proc. Inst. Mech. Eng. Part B J. Eng. Manuf.* **2013**, *227*, 1620–1634. [CrossRef]
102. Doostmohammadi, H.; Moridshahi, H. Effect of Si on microstructure, ordering transformation and properties of the $Cu_{60}Zn_{40}$ alloy. *J. Alloys Compd.* **2015**, *640*, 401–407. [CrossRef]
103. Rajabi, Z.; Doostmohammadi, H. Effect of addition of tin on the microstructure and machinability of α-brass. *Mater. Sci. Technol.* **2018**, *34*, 1218–1227. [CrossRef]
104. Taha, M.A.; El-Mahallawy, N.A.; Hammouda, R.R.; Moussa, T.M.; Gheith, M.H. Machinability characteristics of lead free-silicon brass alloys as correlated with microstructure and mechanical properties. *Ain Shams Eng. J.* **2012**, *3*, 383–392. [CrossRef]
105. Suksongkarm, P.; Rojananan, S.; Rojananan, S. Using recycled bismuth-tin solder in novel machinable lead-free brass. *Mater. Trans.* **2017**, *58*, 1754–1760. [CrossRef]
106. Klocke, F.; Nobel, C.; Veselovac, D. Influence of tool coating, tool material, and cutting speed on the machinability of low-leaded brass alloys in turning. *Mater. Manuf. Processes* **2016**, *31*, 1895–1903. [CrossRef]
107. Schultheiss, F.; Johansson, D.; Bushlya, V.; Zhou, J.; Nilsson, K.; Ståhl, J.E. Comparative study on the machinability of lead-free brass. *J. Clean. Prod.* **2017**, *149*, 366–377. [CrossRef]
108. Nobel, C.; Klocke, F.; Lung, D.; Wolf, S. Machinability enhancement of lead-free brass alloys. In Proceedings of the 6th CIRP International Conference on High Performance Cutting, Berkeley, CA, USA, 23–25 June 2014; pp. 95–100.
109. Vilarinho, C.; Davim, J.P.; Soares, D.; Castro, F.; Barbosa, J. Influence of the chemical composition on the machinability of brasses. *J. Mater. Processing Technol.* **2005**, *170*, 441–447. [CrossRef]
110. Tam, P.L.; Schultheiss, F.; Ståhl, J.-E.; Nyborg, L. Residual stress analysis of machined lead-free and lead-containing brasses. *Mater. Sci. Technol.* **2016**, *32*, 1789–1793. [CrossRef]
111. Gaitonde, V.N.; Karnik, S.R.; Davim, J.P. Selection of optimal MQL and cutting conditions for enhancing machinability in turning of brass. *J. Mater. Processing Technol.* **2008**, *204*, 459–464. [CrossRef]
112. Gaitonde, V.N.; Karnik, S.R.; Davim, J.P. Study on the influence of MQL and cutting conditions on machinability of brass using Artificial Neural Network. *Int. J. Mater. Prod. Technol.* **2010**, *3*, 155–172. [CrossRef]
113. Al-Zaharnah, I.T. Suppressing vibrations of machining processes in both feed and radial directions using an optimal control strategy: The case of interrupted cutting. *J. Mater. Processing Technol.* **2006**, *172*, 305–310. [CrossRef]
114. Xiaoliang, J.; Altintas, Y. Chatter stability model of micro-milling with process damping. *J. Manuf. Sci. Eng.* **2013**, *135*, 031011.
115. Kundrak, J.; Szabo, G.; Markopoulos, A.P. Experimental and numerical investigation of the influence of cutting speed and feed rate on forces in turning of steel. *Mater. Sci. Forum* **2016**, *862*, 270–277. [CrossRef]
116. Monka, P.P.; Monkova, K.; Majstrorovic, V.D.; Božić, Ž.; Andrej, A. Optimal cutting parameter specification of newly designed milling tools based on the frequency monitoring. *Int. J. Adv. Manuf. Technol.* **2020**, *115*, 777–794. [CrossRef]
117. Monka, P.; Monkova, K.; Modrak, V.; Hric, S.; Pastucha, P. Study of a tap failure at the internal threads machining. *Eng. Fail. Anal.* **2019**, *100*, 25–36. [CrossRef]

metals

MDPI

Article

Multicomponent Fe-Based Bulk Metallic Glasses with Excellent Corrosion and Wear Resistances

Guan Zhang [1,2], Wenlei Sun [1,*], Lei Xie [3,4], Chengwu Zhang [3,4], Jie Tan [5], Xuan Peng [2], Qiang Li [4,*], Xu Ma [3,4], Dongmei Zhao [6] and Jiangtong Yu [1]

1 School of Mechanical Engineering, Xinjiang University, Urumqi 830046, China; zhangguan0208@163.com (G.Z.); yujtong@163.com (J.Y.)
2 Engineering Training Center, Xinjiang University, Urumqi 830046, China; xjdxpx@126.com
3 Xinjiang Key Laboratory of Solid State Physics and Devices, Xinjiang University, Urumqi 830046, China; guanzxj@163.com (L.X.); zcwxju@163.com (C.Z.); mx2407383624@163.com (X.M.)
4 School of Physics Science and Technology, Xinjiang University, Urumqi 830046, China
5 Northwest Company of China Petroleum and Chemical Corporation, SINOPEC, Urumqi 830011, China; jie26@163.com
6 Laboratory and Equipment Management Office, Xinjiang University, Urumqi 830046, China; zdm1925@163.com
* Correspondence: sunwenxj@163.com (W.S.); qli@xju.edu.cn (Q.L.)

Abstract: In this study, new multicomponent $Fe_{54}M_5Cr_{15}Mo_6Si_2B_4P_{10}C_4$ (M = Fe, Co, and Ni, denoted as Fe_{59}, $Fe_{54}Co_5$, and $Fe_{54}Ni_5$, respectively) bulk metallic glasses (BMGs) with excellent corrosion and wear resistances were synthesized using the J-quenching technique and fluxing treatment. The synthesized Fe-based BMGs possessed a large glass-forming ability, and the maximum diameters of the Fe_{59}, $Fe_{54}Co_5$, and $Fe_{54}Ni_5$ glassy alloy rods reached 5.5, 4.5, and 4.0 mm, respectively. The Fe_{59} BMG had a wide supercooled liquid region of 65 K. Potentiodynamic tests in 3.5 wt.% NaCl solution showed that the corrosion resistances of the synthesized Fe-based BMGs were relatively better than that of the 316L stainless steel. The Fe_{59} BMG had the highest corrosion resistance, with the lowest self-corrosion current density in the order of 10^{-8} $A{\cdot}cm^{-2}$. Wear tests showed that the synthesized Fe-based BMGs exhibited excellent wear resistances, and the wear rate of the Fe_{59} BMG was as low as approximately 1.73×10^{-15} $m^3{\cdot}N^{-1}{\cdot}m^{-1}$. The rare-earth-element-free Fe-based BMGs, especially the Fe_{59} BMG, have a low cost, large glass-forming ability, and excellent wear and corrosion resistance, which make them good candidates for wear-and corrosion-resistant coating materials

Keywords: Fe-based bulk metallic glasses; glass forming ability; corrosion resistance; wear resistance

Citation: Zhang, G.; Sun, W.; Xie, L.; Zhang, C.; Tan, J.; Peng, X.; Li, Q.; Ma, X.; Zhao, D.; Yu, J. Multicomponent Fe-Based Bulk Metallic Glasses with Excellent Corrosion and Wear Resistances. *Metals* **2022**, *12*, 564. https://doi.org/10.3390/met12040564

Academic Editor: George A. Pantazopoulos

Received: 18 February 2022
Accepted: 24 March 2022
Published: 27 March 2022

1. Introduction

Fe-based bulk metallic glasses (BMGs) have gained substantial attention in recent decades owing to their low cost and unique performance, such as high strength and hardness, excellent wear and corrosion resistance, and excellent soft magnetic properties [1–4]. The strength of Fe-based BMGs is generally higher than 2 GPa, which currently exceeds the strength of high-strength steel; in particular, the compressive fracture strength of a $Fe_{33.5}Co_{33.5}Nb_6B_{27}$ BMG reaches 4.84 GPa [5]. Owing to their high strength, the microhardness of the Fe-based BMGs is larger than that of the Zr- and Cu-based BMGs, and their wear resistance is three orders of magnitude higher than that of the Zr- and Cu-based BMGs. Additionally, owing to their unique amorphous structure, Fe-based BMGs exhibit a chemically homogeneous single phase and high atomic reactivity, which results in the formation of an extremely uniform and stable passive film on the surface of the alloys; thus, Fe-based BMGs have higher corrosion resistances compared with their crystalline alloy counterparts. For example, Li et al. reported an $Fe_{59}Cr_{12}Mo_{3.5}Ni_5P_{10}C_4B_4Si_{2.5}$ BMG with excellent corrosion resistance, with I_{corr} and E_{corr} values of 2.47×10^{-7} $A{\cdot}cm^{-2}$ and -0.22 V, respectively [6].

Owing to their low cost, ultra-high strength, and excellent wear and corrosion resistance, Fe-based BMGs have been considered good candidates as wear-and corrosion-resistant coating materials [7]. It has been extensively reported that alloying elements can significantly influence the corrosion and wear properties of Fe-based BMGs. Cr and Mo are the most effective elements for providing a high passivation ability for Fe-based BMGs [8,9]. Cr can form a dense and stable hydrated chromium oxyhydroxide passive film on the surface of alloys to prevent corrosion within the alloys [10]. Generally, the higher the Cr content, the better the corrosion resistance of Fe-based BMGs. However, when the Cr content is greater than 15 at.%, further increasing the Cr content does not significantly improve the corrosion resistance of the Fe-based BMGs but increases the material cost. The effect of Mo on the protective ability of the passivation layer is weaker than that of Cr. However, Mo can promote the enrichment of Cr in the passive film, thereby enhancing the corrosion and pitting resistances of Cr-containing Fe-based BMGs. However, the excess addition of Mo is detrimental to the corrosion resistance of Fe-based BMGs [11]. Recently, it was found that the substitution of Fe with a similar element, such as Ni or Co, can produce positive effects on the glass forming ability (GFA) and the mechanical and magnetic properties of Fe-based BMGs. For example, the substitution of 20 at.% Ni for Fe in $Fe_{80}P_{13}C_7$ BMG resulted in an increase in the critical diameter (D_c) from 2.0 mm to 2.3 mm for full glass formation and a significant increase in room-temperature compressive plastic strain from 1.1% to 11.2% [12]. The room-temperature compressive plasticity of $Fe_{80}P_{13}C_7$ BMG improved from 1.1% to 3.0%, and the saturation magnetization increased from 1.45 T to 1.55 T due to the replacement of 10 at.% Fe with Co [13]. Based on the above considerations, new $Fe_{54}M_5Cr_{15}Mo_6Si_2B_4P_{10}C_4$ (M = Fe, Co, and Ni) BMGs were successfully prepared in this study, and the corrosion and wear properties of the present Fe-based BMGs were systematically investigated.

2. Materials and Methods

$Fe_{54}M_5Cr_{15}Mo_6Si_2B_4P_{10}C_4$ (M = Fe, Ni, and Co) alloy ingots were prepared by torch-melting a mixture of Fe_3P pieces (99.5 mass% purity), Cr powder (99.9 mass% purity), Mo powder (99.9 mass% purity), Fe powder (99.9 mass% purity), Si powder (99.99 mass% purity), boron pieces (99.95 mass% purity), graphite powder (99.95 mass% purity), Co powder (99.9 mass% purity), and Ni powder (99.7 mass% purity) under a high-purity Ar atmosphere. All the as-prepared alloy ingots had a mass of 2 g. Subsequently, the ingots were purified through a fluxing treatment at an elevated temperature of approximately 1450 K for 2–3 h under a vacuum of ~50 Pa, in which a mixture of B_2O_3 and CaO with a mass ratio of 3:1 was used as the fluxing agent. After the fluxing treatment, the alloys were subjected to the J-quenching technique [14–16]. Consequently, $Fe_{54}M_5Cr_{15}Mo_6Si_2B_4P_{10}C_4$ (M = Fe, Ni, and Co) alloy rods with diameters ranging from 1.0–6.0 mm and lengths of several centimeters were prepared.

The amorphous nature of the specimens was examined using an X-ray diffractometer (XRD, Bruker D8 Discover, Bruker Inc., Karlsruhe, Germany) with Co K_α radiation. The thermal behavior of the specimens was examined through differential scanning calorimetry (DSC, NETZSCH DSC 404F1, NETZSCH Inc., Selb, Germany) at a heating rate of 0.33 K/s under an Ar atmosphere. The corrosion behaviors of the specimens, as well as that of 316 stainless steel, which was used for comparison, were evaluated through electrochemical measurements (Zahner Zennium X, Zahner Inc., Kronach, Germany). This was conducted in a three-electrode cell using the glassy rod alloy specimens as the working electrode, a saturated calomel electrode as the reference electrode, and a metal platinum electrode as the counter electrode. The working face of the samples was polished to a mirror face using sandpaper, and the non-working parts were sealed with epoxy resin. The potentiodynamic polarization curves were measured in a 3.5% NaCl solution at room temperature with a potential sweep rate of 1 mV/s after immersing the specimens for approximately 30 min to stabilize the open-circuit potential. Electrochemical impedance spectroscopy (EIS) measurements were carried out in a frequency range of 10 kHz to

10 mHz, using a sinusoidal potential perturbation of ±10 mV relative to the OCP. The fitting of EIS data was operated by using the software ZSimpWin (3.60, AMETEK Inc. Middleboro, MA, USA). The surface morphology of the specimens after immersion was observed using scanning electron microscopy (SEM, SU8010). The chemical states of the surface elements of the specimens after polarization were analyzed using X-ray photoelectron spectroscopy (XPS, ESCALAB 250Xi, Thermo Fisher Scientific Inc., Waltham, MA, USA) with a monochromatic Al K_α X-ray source ($h\nu$ = 1486.6 eV). The binding energies were calibrated using carbon contamination with a C 1s peak value of 284.8 eV. The Vickers hardness (HV) of the specimens was measured at a load of 200 g and dwell time of 15 s using a microhardness test instrument (MT-401MVA, BangYi Co., Shanghai, China). Prior to testing, the specimens were polished to a mirror face using sandpaper and cleaned with alcohol, and the average of ten measurement values was taken as the final hardness value for each specimen. Tribological tests of the specimens were conducted under ambient conditions using a pin-on-disc tribometer (CSM, Graz, Austria) with a pin-on-disk contact geometry; a GCr15 (860HV) plate with a diameter of 60 mm was used as the counter body. A normal load of 7 N and a total sliding distance of 5000 m with a sliding speed of 1.8 m/s were adopted for all the tribological tests. The wear rate was calculated using Equation (1).

$$V = \frac{\Delta m}{PS\rho} = \frac{m_1 - m_2}{PS\rho} \tag{1}$$

where V is the wear rate, Δm is the wear mass loss, m_1 and m_2 are the masses of the sample before and after the test, respectively, S is the sliding distance of the sample, p is the load, and ρ is the density of the test sample, which was measured using density testing equipment (Micromeritics, AccuPyc-1340, Micromeritics Inc., Atlanta, GA, USA). The surface and subsurface morphologies of the worn tracks were observed using scanning electron microscopy (SEM, Hitachi s4800, Hitachi Inc., Tokyo, Japan). The topography and chemical composition of the worn surfaces were examined using energy-dispersive X-ray spectrometry (EDS HORIBA 7593-H, HORIBA Inc., Tokyo, Japan).

3. Results and Discussion

3.1. Material Preparation and Glass Forming Ability (GFA)

The XRD (X-ray diffraction) patterns of the as-prepared $Fe_{54}M_5Cr_{15}Mo_5Si_2B_4P_{10}C_4$ (M = Fe, Ni, and Co) glassy alloy rods with critical diameters (D_{max}) for glass formation are shown in Figure 1a. There was only a broad diffraction peak and no detectable sharp crystalline peaks in the XRD patterns of all the specimens, indicating the fully amorphous structure of the specimens [17]. Among the present Fe-based BMGs, the Fe_{59} alloy had the largest GFA with a D_{max} of 5.5 mm, and the substitution of Fe with 5 at.% Co or Ni degraded the GFA of Fe_{59} alloy; the D_{max} of $Fe_{54}Co_5$ and $Fe_{54}Ni_5$ alloys were 4.5 and 4.0 mm, respectively [17]. Figure 1b shows the DSC curves of $Fe_{54}M_5Cr_{15}Mo_5Si_2B_4P_{10}C_4$ (M = Fe, Ni, and Co) BMGs at a heating rate of 0.33 K/s. The DSC (Differential scanning calorimetry) curves of all specimens exhibited a distinct glass transition, followed by a supercooled liquid region and multistep crystallization peaks. The thermal parameters associated with the glass transition temperature (T_g), crystallization temperature (T_x), and supercooled liquid region ($\Delta T_x = T_x - T_g$) are summarized in Table 1. It can be seen that the substitution of Co and Ni for Fe led to a decrease in the T_g and T_x of the Fe-based BMGs. The T_g of amorphous alloys mainly depends on the atomic bonding strength among the constituent elements, which can be evaluated through the mixing enthalpy between the constituent elements. It is suggested that the dominant short-range order unit for Fe-metalloid amorphous alloys can be characterized as solute-centered atomic clusters, in which the metallic atoms are located on the shell and the metalloid atoms are located in the center [18]. The mixing enthalpies between Fe and the metalloid elements Si, B, P, and C are −35, −26, −39.5, and −50 kJ/mol, respectively; those between Co and the metalloid elements are −38, −24, −35.5, and −42 kJ/mol, respectively; and those between Ni and the metalloid elements are −40, −24, −34.5, and −39 kJ/mol, respectively [19]. Except for Si, the mixing enthalpies between Fe and the metalloid elements were more negative than those between Co/Ni and the metalloid elements, which could be the reason for the decrease in the T_g and T_x of the Fe_{59}, $Fe_{54}Co_5$, and $Fe_{54}Ni_5$ alloys. It was also observed that the ΔT_x of the Fe_{59}, $Fe_{54}Co_5$, and $Fe_{54}Ni_5$ BMGs decreased. Inoue suggested that a wider ΔT_x reflects a better GFA [20], which may account for the compositional dependence of D_{max} of the Fe-based BMGs in this study. Additionally, it can be noted that the ΔT_x of the Fe_{59} BMG reaches 65 K, indicating an extremely high thermal stability.

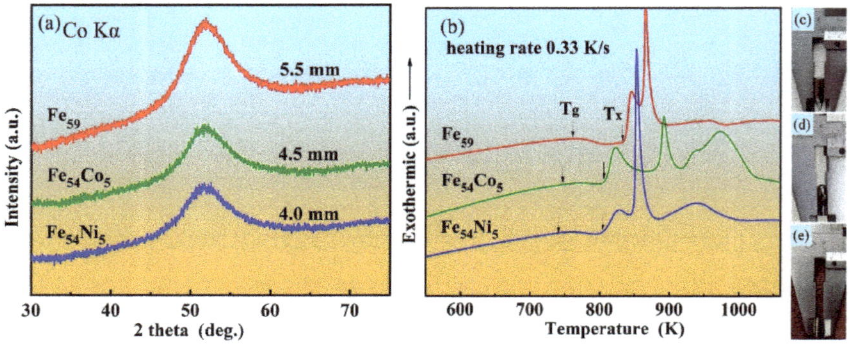

Figure 1. (**a**) XRD patterns and (**b**) DSC curves of as-prepared $Fe_{54}M_5Cr_{15}Mo_6Si_2B_4P_{10}C_4$ (M = Fe, Ni, and Co) glassy alloy rods with the critical diameter (D_{max}) for glass formation. Photos of (**c**) Fe_{59}, (**d**) $Fe_{54}Co_5$, and (**e**) $Fe_{54}Ni_5$ glassy alloy rods with the D_{max} values.

Table 1. Summary of the critical diameters for glass formation (D_{max}), thermal properties, and the GFA indicators of $Fe_{54}M_5Cr_{15}Mo_6Si_2B_4P_{10}C_4$ (M = Fe, Ni, and Co) glassy alloy rods determined from the DSC curves at a heating rate of 0.33 K/s (T_g: glass transition temperature, T_x: onset crystallization temperature, $\Delta T_x = T_x - T_g$).

Alloys	D_{max} (mm)	T_g (K)	T_x (K)	ΔT_x (K)
Fe_{59}	5.5	768	833	65
$Fe_{54}Co_5$	4.5	765	805	40
$Fe_{54}Ni_5$	4.0	762	800	38

3.2. Corrosion Resistance

To evaluate the corrosion resistance, electrochemical measurements of the Fe-based BMGs were performed in a 3.5 wt.% NaCl solution at room temperature. Figure 2 shows representative potentiodynamic polarization curves of the Fe_{59}, $Fe_{54}Co_5$, and $Fe_{54}Ni_5$ glassy alloy rods. The self-corrosion current density (I_{corr}) and self-corrosion potential (E_{corr}) obtained using the Tafel extrapolation method are listed in Table 2.

Table 2. Electrochemical parameters derived from potentiodynamic polarization curves and EIS analysis of the $Fe_{54}M_5Cr_{15}Mo_6Si_2B_4P_{10}C_4$ (M = Fe, Co, and Ni) BMGs and 316L stainless steel (E_{corr}: self-corrosion potential, I_{corr}: self-corrosion current density, R_s: solution resistance, R_t: charge transfer resistance, R_u: charge transfer resistance, Q_t: non-ideal capacitance, Q_u: non-ideal capacitance).

Alloys	E_{corr} (V)	I_{corr} ($\times 10^{-8}$ A·cm^{-2})	R_s (Ω·cm^2)	R_t ($\times 10^6$ Ω·cm^2)	R_u ($\times 10^6$ Ω·cm^2)	Q_t ($\times 10^{-5}$ μF·cm^2)	Q_u ($\times 10^{-5}$ μF·cm^2)
Fe_{59}	0.12	8.4	1.71	0.64	1.8	3.59	9.37
$Fe_{54}Co_5$	−0.19	8.6	1.67	0.76	1.68	3.16	4.12
$Fe_{54}Ni_5$	−0.18	38.3	0.85	3.06	0.82	2.87	2.99
316L	−0.29	98.5	1.16	0.0000013	0.049	2.79	5.28

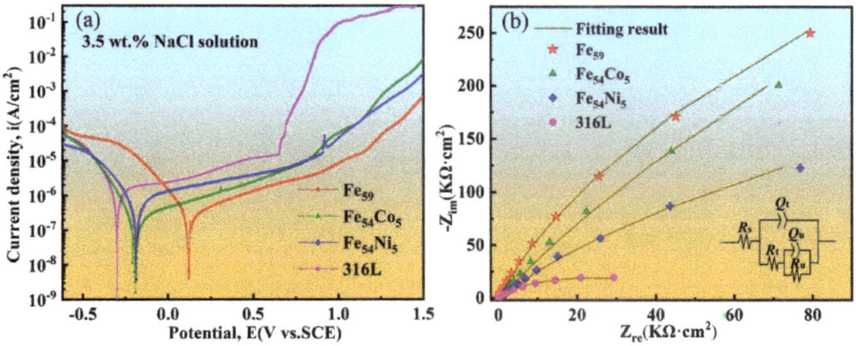

Figure 2. (**a**) Potentiodynamic polarization curves in 3.5 wt% NaCl solution at room temperature for $Fe_{54}M_5Cr_{15}Mo_6Si_2B_4P_{10}C_4$ (M = Fe, Ni, and Co) glassy alloy rods and 316L stainless steel; (**b**) Nyquist plots and the corresponding equivalent circuit (inset) for fitting the impedance spectra of $Fe_{54}M_5Cr_{15}Mo_6Si_2B_4P_{10}C_4$ (M = Fe, Co, and Ni) glassy alloy rods and 316L in 3.5 wt.% NaCl solution at room temperature.

Figure 2a shows the potentiodynamic polarization curves in 3.5 wt.% NaCl solution exposed to air at room temperature for the Fe-based glassy alloy rod specimens as well as the 316L stainless steel (316L) for comparison. The self-corrosion potential (E_{corr}) and self-corrosion current density (I_{corr}) of the specimens listed in Table 2 were obtained from the potentiodynamic polarization curves using the Tafel extrapolation method. Compared with the 316L, the Fe-based BMGs exhibited a more positive E_{corr}, lower I_{corr}, and a more stable and wider passive region, indicating a relatively higher corrosion resistance and superior passive film protection. Among the Fe-based BMGs used in this study, the Fe_{59} BMG exhibited the highest corrosion resistance, with the lowest I_{corr} and I_{pass} in the order of 10^{-8} and 10^{-7} A/cm^2, respectively. Further, the potentiodynamic polarization curve of the Fe_{59} BMG had a smooth shape, as shown in Figure 2a, indicating that a stable passivation layer was formed on the surface and tiny pit corrosion occurred during the entire measurement period. However, few sharp peaks were found in the curves of the $Fe_{54}Co_5$ and $Fe_{54}Ni_5$ BMGs, as shown in Figure 2a, indicating that pit corrosion occurred during the measurement period; evidence of pit corrosion is shown in Figure 3. It is worth noting that the corrosion resistance of the Fe_{59} BMG was better than that of most reported Fe-based BMGs, of which I_{corr} is usually in the order of 10^{-6}~10^{-7} A/cm^2. For example, it has been reported that a $Fe_{36}Cr_{23}Mo_{10}W_8C_{15}B_6Y_2$ BMG possesses good corrosion resistance with an I_{corr} of 6.16×10^{-7} A·cm^{-2} and E_{corr} of -0.275 V [1]. Compared with this Fe-based BMG, the Fe_{59} BMG exhibited a lower I_{corr} but contained less Cr and Mo. This result may be related to the metalloid elements in the alloys, which can significantly influence the corrosion properties of Fe-based BMGs [4]. For example, Si can improve the passivation ability and corrosion resistance of Fe-based BMGs, which can be attributed to the formation of a dense and stable passive film rich in Si and Cr- oxides [21] The $Fe_{43}Cr_{16}Mo_{16}C_{10}B_5P_{10}$ BMGs exhibited a wide passive region, indicating a higher corrosion resistance compared to that of the P-free metallic glass [22].

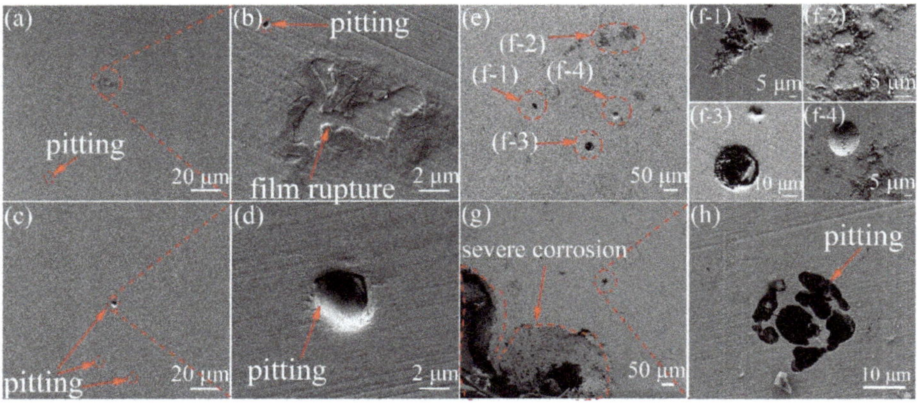

Figure 3. SEM micrographs of the electrochemical corroded surfaces of the Fe_{59} BMG (**a**), $Fe_{54}Co_5$ BMG (**c**), $Fe_{54}Ni_5$ BMG (**e**), and 316L (**g**), respectively, and (**b**,**d**,**f**,**h**) are corresponding partial enlarged views.

Figure 2b shows the Nyquist plots of the Fe-based BMGs and 316L in 3.5 wt.% NaCl solution at room temperature. The Nyquist plots of all samples show a double capacitive loop, implying two time constants in the electrochemical measurements. The semicircle diameter in the Nyquist plots of the Fe_{59}, $Fe_{54}Co_5$, and $Fe_{54}Ni_5$ BMGs and 316L stainless steel decreased, implying that their corrosion resistances also decreased. An appropriate equivalent circuit for fitting the impedance spectra is shown in Figure 2b, where R_s is the solution resistance; R_t and R_u are the charge transfer resistances; and Q_t and Q_u represent the possibility of non-ideal capacitance. These fitting parameters are listed in Table 2. The R_t values of the Fe_{59}, $Fe_{54}Co_5$, $Fe_{54}Ni_5$ BMGs, and 316L stainless steel decreased. It is known that a larger R_t implies better corrosion resistance [23], and thus, the results of the EIS measurements agree well with those of the electrochemical measurements.

Figure 3 shows the SEM micrographs of the electrochemical corroded surfaces of the Fe-based BMGs and 316L specimens. There were a few pitting pits and a film rupture zone on the surface of the Fe_{59} BMG specimen, as shown in Figure 3a,b, revealing the excellent corrosion resistance of the Fe_{59} BMG. It can be seen in Figure 3c,d that there were pitting pits with diameters of approximately 4 μm and of a few nanometers on the surface of the $Fe_{54}Co_5$ BMG specimen. Compared with the Fe_{59} and $Fe_{54}Co_5$ BMGs, there were more and larger pitting pits and more corrosion products on the surface of the $Fe_{54}Ni_5$ BMG specimen, as shown in Figure 3e,f, which could be attributed to the sharp peaks in the potentiodynamic polarization curve of the $Fe_{54}Ni_5$ BMG shown Figure 2a. The surface of the 316L specimen was completely damaged, covered with a thick layer of polarized product, and had huge pitting pits caused by a severe corrosion, as shown in Figure 3g,h. The observed electrochemical corroded surfaces of the specimens confirmed the results of the electrochemical and EIS measurements.

To better understand the effect of the elements on the corrosion mechanism of the Fe-based BMGs, an XPS (X-ray Photoelectron Spectroscopy) analysis was performed to characterize the chemical composition of the oxide films formed on the surfaces of the specimens after electrochemical measurement in 3.5 wt.% NaCl solution. Figure 4 shows the Fe 2p, Cr 2p, Mo 3d, Co 2p, and Ni 2p spectra of the Fe-based BMGs. The Fe 2p spectra of the specimens comprised the peaks of Fe $2p_{1/2}$ and $2p_{3/2}$ corresponding to the peaks of the metallic (Fe) and Fe^{2+}, $Fe_{hy}{}^{3+}$, and $Fe_{ox}{}^{3+}$ oxide (Fe^{ox}) states of Fe [23–27]. The peak intensities of Fe^m in the $Fe_{54}Co_5$ and $Fe_{54}Ni_5$ BMGs were relatively higher than that of the Fe_{59} BMG. The Cr 2p peaks represented the metallic (Cr) and Cr^{3+} and Cr^{6+} oxide (Cr^{ox}) states of Cr [26,28–30]. Cr^{6+} dissolves in water and causes the corrosion current density to increase; however, it has a self-repairing ability and can promote the formation of denser

Cr_2O_3 to prevent chloride ions from penetrating into the internal damage [31,32]. As shown in Figure 4b, the total area of the Cr 2p peaks in the Fe_{59} BMG was the largest, suggesting a higher concentration of Cr on its surface than on the $Fe_{54}Co_5$ and $Fe_{54}Ni_5$ BMGs. The Mo 3d spectrum, consisting of the Mo 3d 3/2 and 3d 5/2 peaks, corresponded to the metallic Mo and Mo^{4+} and Mo^{6+} oxide states [24,27,29,30]. The $Fe_{54}Co_5$ BMG had the highest intensity of the low-valence Mo^{4+} peak on the surface, followed by the Fe_{59} and $Fe_{54}Ni_5$ BMGs. The Fe_{59} BMG had the highest intensity of the high-valence Mo^{6+} peak in the surface film, followed by the $Fe_{54}Co_5$ and $Fe_{54}Ni_5$ BMGs. Mo can promote the enrichment of Cr in the passive film and, consequently, enhance the corrosion resistance of amorphous alloys [33–35]. The Co 2p spectrum consisted of the metallic Co (Co^m), ${Co_{hy}}^{2+}$, and ${Co_{ox}}^{2+}$ peaks, as shown in Figure 4d. The Ni 2p spectrum consisted of the peaks corresponding to the metallic Ni (Ni), Ni^{2+}, and Ni^{3+}, as shown in Figure 4e. It is known that Co and Ni are more effective in providing a high passivation ability for alloys compared with Fe; however, the corrosion resistance of the Fe-based BMGs in this study decreased with the substitution of Co and Ni for Fe. This could be due to the small amount of crystallization in the $Fe_{54}Co_5$ and $Fe_{54}Ni_5$ BMGs owing to the lower GFA, which degraded their corrosion resistance.

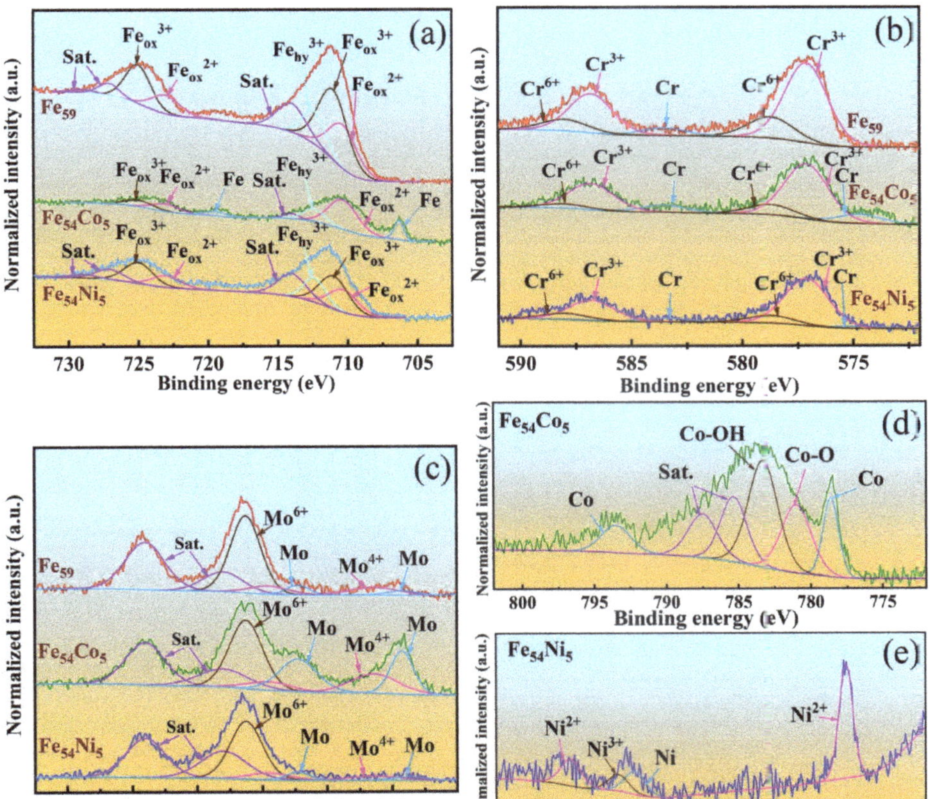

Figure 4. XPS spectra of Fe 2p (**a**), Cr 2p (**b**), Mo 3d (**c**), Co 2p (**d**), and Ni 2p (**e**) in the passive films on the surface of the $Fe_{54}M_5Cr_{15}Mo_6Si_2B_4P_{10}C_4$ (M = Fe, Co, and Ni) glassy alloy rod specimens after electrochemical measurements in a 3.5 wt.% NaCl solution.

The outstanding corrosion resistance of amorphous alloys is primarily influenced by the stability of a uniform passive film. Figure 4b,c show that the Cr^{ox} and Mo^{ox} fractions in the Fe_{59} and $Fe_{54}Co_5$ BMGs in the surface film after the electrochemical test were apparently higher than those in the $Fe_{54}Ni_5$ BMG. The enrichment of Cr^{ox} and Mo^{ox} was beneficial for the modification of the passive film quality of the BMGs and thus improved the corrosion resistance. The XPS results for the Fe_{59}, $Fe_{54}Co_5$, and $Fe_{54}Ni_5$ BMGs were consistent with the SEM morphology of the samples after the electrochemical test, as shown in Figure 3, in which the $Fe_{54}Ni_5$ BMG exhibited a more severely corroded surface than the Fe_{59} and $Fe_{54}Co_5$ BMGs.

3.3. Microhardness and Wear Resistance

The measured microhardness values (HV0.2) of the Fe-based BMG specimens are listed in Table 3 and shown in Figure 5a. The Fe-based BMGs exhibited extremely high microhardnesses, exceeding 1000 HV0.2. The microhardness values of the Fe_{59}, $Fe_{54}Co_5$, and $Fe_{54}Ni_5$ BMGs decreased in turn. Similar to the T_g of amorphous alloys, the hardness of the alloys depends on the atomic bonding strength between the constituent elements; therefore, the microhardness has a similar change trend as the T_g of the synthesized Fe-based BMGs.

Table 3. Density, average microhardness, average COF, and wear rate of the $Fe_{54}M_5Cr_{15}Mo_6Si_2B_4P_{10}C_4$ (M = Fe, Ni, and Co) BMGs.

Alloys	Density (g/cm^3)	Average HV0.2	Average COF	Wear Rate ($\times 10^{-15}$ m^3/N·m)
Fe_{59}	6.62 ± 0.02	1105 ± 31	0.716 ± 0.034	1.73 ± 0.20
$Fe_{54}Co_5$	6.83 ± 0.02	1098 ± 17	0.795 ± 0.059	6.18 ± 0.28
$Fe_{54}Ni_5$	6.79 ± 0.01	1024 ± 22	0.727 ± 0.018	5.05 ± 0.24

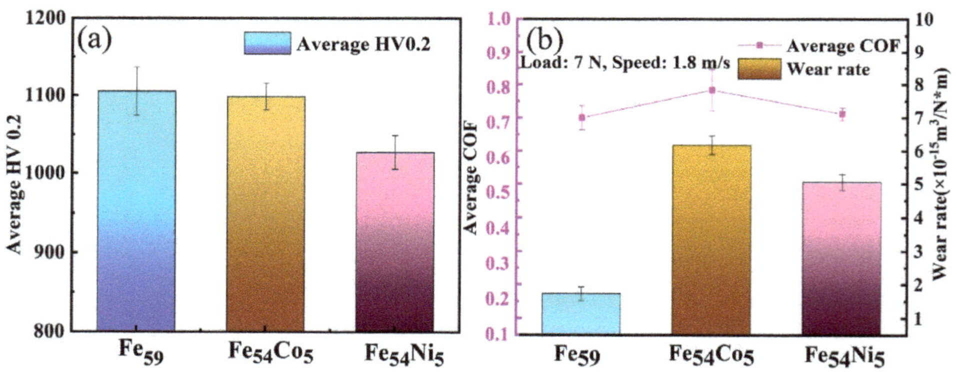

Figure 5. (**a**) Microhardness, (**b**) average coefficient of friction (COF), and wear rate of the $Fe_{54}M_5Cr_{15}Mo_6Si_2B_4P_{10}C_4$ (M = Fe, Ni, and Co) BMGs.

Friction and wear tests of the Fe-based BMGs were performed using pin-on-disk experiments, which were run against a GCr15 plate with a diameter of 60 mm and a hardness of 860 HV under a constant normal load of 7 N, a rolling speed of 1.8 m/s, and a sliding distance of 5000 m. The obtained average steady-state coefficient of friction (COF) and the wear rate, which is calculated by Equation (1), of the Fe-based BMG specimens are listed in Table 3 and shown in Figure 5b. Among the Fe-based BMGs, the Fe_{59} BMG had the smallest wear rate of 1.73×10^{-15} $m^3 \cdot N^{-1} \cdot m^{-1}$, followed by the $Fe_{54}Ni_5$ BMG with a wear rate of 5.05×10^{-15} $m^3 \cdot N^{-1} \cdot m^{-1}$; the $Fe_{54}Co_5$ BMG had the largest wear rate of 6.18×10^{-15} $m^3 \cdot N^{-1} \cdot m^{-1}$. Furthermore, the wear rate of the Fe_{59} BMG was relatively lower than those of the $Fe_{54}Co_5$ and $Fe_{54}Ni_5$ BMGs. The COF values for the Fe_{59}, $Fe_{54}Co_5$, and $Fe_{54}Ni_5$ BMGs under dry friction conditions were 0.716, 0.795, and 0.727, respectively. It was clear that the wear rates of the Fe-based BMG specimens exhibited an obvious positive association with their COF under dry sliding conditions [36]. It should also be noted that the wear rates of the $Fe_{54}Co_5$ and $Fe_{54}Ni_5$ BMGs were negatively correlated, but not inversely proportional, with the microhardness. This indicates that the wear resistance of the Fe-based BMGs cannot be simply determined by their hardness, as predicted by the Archard equation.

Figure 6 shows the SEM (Scanning electron microscope) images of the worn areas of the Fe_{59}, $Fe_{54}Co_5$, and $Fe_{54}Ni_5$ BMGs. An EDS mapping test was performed on typical regions of the worn surfaces; their chemical compositions are summarized in Figure 6c,f,i. The worn surfaces of Fe_{59} exhibited various ploughed grooves parallel to the sliding direction, with different widths and depths, which are typical features corresponding to the abrasive-wear mechanism. In addition, as shown in Figure 6a,b, the worn surface of the Fe_{59} BMG was covered with tribo-patches, in which a high fraction of oxygen was observed through EDS (Energy Disperse Spectroscopy) mapping analysis, as shown in Figure 6c. This result demonstrates the oxidation of the tribo-patches of the Fe_{59} BMG during the sliding process. The worn surface of $Fe_{54}Co_5$, as shown in Figure 6d,e, exhibited a smooth morphology covered with a uniform oxide film (Figure 6f), including some small tribo-patches, indicating an oxidation wear mechanism. As shown in Figure 6g,h, the worn morphology of the $Fe_{54}Ni_5$ BMG exhibited much wear debris with grain-like wear particles, which resulted from a brittle failure characteristic, and many tribo-patches with diverse widths and lengths, indicating an adhesive wear mechanism.

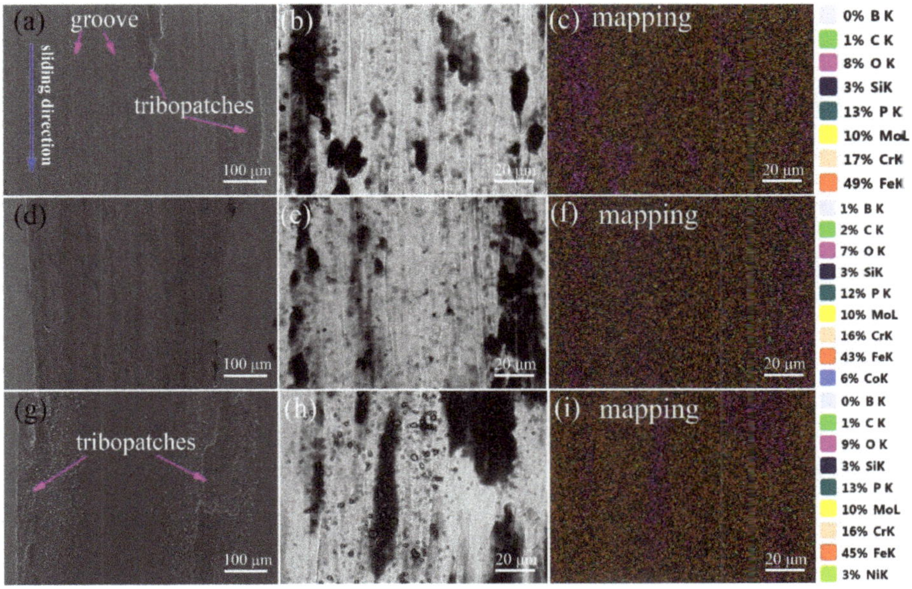

Figure 6. SEM images of wear surface morphology and EDS mapping of the Fe_{59} (**a–c**), $Fe_{54}Co_5$ (**d–f**), and $Fe_{54}Ni_5$ (**g–i**) glassy alloy rod specimens.

4. Conclusions

In this study, a new series of rare-earth-element-free $Fe_{54}M_5Cr_{15}Mo_6Si_2B_4P_{10}C_4$ (M = Fe, Ni, and Co) BMGs were prepared, and the effects of Fe, Co, and Ni contents on the thermal stability, corrosion, and wear resistances of the Fe-based BMGs were investigated. The main results are as follows.

(1) $Fe_{54}M_5Cr_{15}Mo_6Si_2B_4P_{10}C_4$ (M = Fe, Ni and Co) bulk glassy alloy rods were successfully prepared using the J-quenching and fluxing techniques. The D_{max} of the Fe_{59}, $Fe_{54}Co_5$, and $Fe_{54}Ni_5$ alloys reached 5.5, 4.5, and 4.0 mm, respectively.

(2) The Fe-based BMGs exhibited high thermal stability, and the ΔT_x of the Fe_{59} BMG reached 65 K. The T_g, T_x, and ΔT_x of the Fe-based BMGs decreased when Co and Ni substituted for Fe.

(3) Electrochemical and EIS tests showed that, compared with 316L stainless steel, the Fe-based BMGs exhibited excellent corrosion resistance in a 3.5% NaCl solution with very low I_{corr} in the order of 10^{-8}–10^{-7} A/cm^2. Among the Fe-based BMGs, the Fe_{59} BMG exhibited the highest corrosion resistance with the highest positive E_{corr}, the lowest I_{corr}, and the most stable and widest passive region.

(4) Friction and wear tests showed that the COF values of the Fe_{59} (0.716), $Fe_{54}Co_5$ (0.795), and $Fe_{54}Ni_5$ (0.727) BMGs under the dry-friction condition had slight differences. Among the Fe-based BMGs, the Fe_{59} BMG had the lowest wear rate of 1.73×10^{-15} m^3.N^{-1}.m^{-1}, followed by the $Fe_{54}Ni_5$ BMG with a wear rate of 5.05×10^{-15} m^3.N^{-1}.m^{-1} and the $Fe_{54}Co_5$ BMG with a wear rate of 6.18×10^{-15} m^3.N^{-1}.m^{-1}. SEM observation of the wear surface morphology indicated that the wear mechanisms of the Fe_{59}, $Fe_{54}Co_5$, and $Fe_{54}Ni_5$ BMGs were abrasive and oxidation, oxidation, and adhesive and oxidation, respectively.

(5) The rare-earth-element-free Fe-based BMGs, particularly the Fe_{59} BMG, showed potential as a coating material, owing to its combined advantages (low cost, large GFA, and excellent wear and corrosion resistances).

Author Contributions: Conceptualization, G.Z., Q.L. and W.S.; methodology, G.Z. and J.T.; validation, G.Z., Q.L. and W.S.; formal analysis, G.Z.; investigation, G.Z. and L.X.; resources, Q.L. and D.Z.; data curation, Q.L.; writing—original draft preparation, G.Z.; writing—review and editing, G.Z., Q.L., W.S., C.Z., X.P., X.M. and J.Y.; visualization, G.Z.; project administration, Q.L. and W.S.; funding acquisition, G.Z., Q.L. and W.S. All authors have read and agreed to the published version of the manuscript.

Funding: This research was funded by the University Science Research Planning Project of Xinjiang Uygur Autonomous Region, grant number XJEDU2021Y008, the key University Science Research Planning Project of Xinjiang Uygur Autonomous Region, grant number XJEDU2021I003, the Tianshan Innovation Team Program of Xinjiang Uygur Autonomous Region, grant number 2020D14038, the Xinjiang Autonomous Region Key Laboratory Open Fund, grant number 2020520002 and the "Talent Projects" of Urumqi Key Training Object.

Institutional Review Board Statement: Not applicable.

Informed Consent Statement: Not applicable.

Data Availability Statement: Not applicable.

Conflicts of Interest: The authors declare no conflict of interest.

References

1. Zhang, X.; Lai, L.; Xiao, S.; Zhang, H.; Zhang, F.; Li, N.; Guo, S. Effect of W on the Thermal Stability, Mechanical Properties and Corrosion Resistance of Fe-Based Bulk Metallic Glass. *Intermetallics* **2022**, *143*, 107485. [CrossRef]
2. Inoue, A.; Shen, B.L.; Chang, C.T. Super-High Strength of over 4000 MPa for Fe-Based Bulk Glassy Alloys in $[(Fe_{1-x}Co_x)_{0.75}B_{0.2}Si_{0.05}]_{96}Nb_4$ System. *Acta Mater* **2004**, *52*, 4093–4099. [CrossRef]
3. Madge, S.V. Toughness of Bulk Metallic Glasses. *Metals* **2015**, *5*, 1279–1305. [CrossRef]
4. Li, H.X.; Lu, Z.C.; Wang, S.L.; Wu, Y.; Lu, Z.P. Fe-Based Bulk Metallic Glasses Glass Formation, Fabrication, Properties and Applications. *Prog. Mater. Sci.* **2019**, *103*, 235–318. [CrossRef]
5. Zhao, C.; Dun, C.; Man, Q.; Shen, B. Enhancement of Plastic Deformation in FeCoNbB Bulk Metallic Glass with Superhigh Strength. *Intermetallics* **2013**, *32*, 408–412. [CrossRef]
6. Li, J.; Yang, L.; Ma, H.; Jiang, K.; Chang, C.; Wang, J.; Song, Z.; Wang, X.; Li, R. Improved Corrosion Resistance of Novel Fe-Based Amorphous Alloys. *Mater. Design* **2016**, *95*, 225–230. [CrossRef]
7. Ndumia, J.N.; Kang, M.; Gbenontin, B.V.; Lin, J.; Nyambura, S.M. A Review on the Wear, Corrosion and High-Temperature Resistant Properties of Wire Arc-Sprayed Fe-Based Coatings. *Nanomaterials* **2021**, *11*, 2527. [CrossRef]
8. Shi, M.; Pang, S.; Zhang, T. Towards Improved Integrated Properties in FeCrPCB Bulk Metallic Glasses by Cr Addition. *Intermetallics* **2015**, *61*, 16–20. [CrossRef]
9. Suryanarayana, C.; Inoue, A. Iron-Based Bulk Metallic Glasses. *Int. Mater. Rev.* **2013**, *58*, 131–166. [CrossRef]

10. Pang, S.J.; Zhang, T.; Asami, K.; Inoue, A. Bulk Glassy Fe–Cr–Mo–C–B Alloys with High Corrosion Resistance. *Corros. Sci.* **2002**, *44*, 1847–1856. [CrossRef]
11. Li, S.; Wei, Q.; Li, Q.; Jiang, B.; Chen, Y.; Sun, Y. Development of Fe-Based Bulk Metallic Glasses as Potential Biomaterials. *Mater. Sci. Eng. C* **2015**, *52*, 235–241. [CrossRef] [PubMed]
12. Ma, X.; Yang, X.; Li, Q.; Guo, S. Quaternary Magnetic FeNiPC Bulk Metallic Glasses with Large Plasticity. *J. Alloy Compd.* **2013**, *577*, 345–350. [CrossRef]
13. Xu, K.; Ling, H.; Li, Q.; Li, J.; Yao, K.; Guo, S. Effects of Co Substitution for Fe on the Glass Forming Ability and Properties of Fe80P13C7 Bulk Metallic Glasses. *Intermetallics* **2014**, *51*, 53–58. [CrossRef]
14. Qiang, L. Formation of Bulk Ferromagnetic Nanostructured $Fe_{40}Ni_{40}P_{14}B_6$ Alloys by Metastable Liquid Spinodal Decomposition. *Sci. China Ser. E-Tech. Sci.* **2009**, *52*, 1919–1922.
15. Shen, T.D.; Schwarz, R.B. Bulk Ferromagnetic Glasses in the Fe–Ni–P–B System. *Acta Mater.* **2001**, *49*, 837–847. [CrossRef]
16. Zhang, L.; Ma, X.; Li, Q.; Zhang, J.; Dong, Y.; Chang, C. Preparation and Properties of $Fe_{80-x}Ni_xP_{14}B_6$ Bulk Metallic Glasses. *J. Alloy Compd.* **2014**, *608*, 79–84. [CrossRef]
17. Nishiyama, N.; Takenaka, K.; Miura, H.; Saidoh, N.; Zeng, Y.; Inoue, A. The World's Biggest Glassy Alloy Ever Made. *Intermetallics* **2012**, *30*, 19–24. [CrossRef]
18. Cheng, Y.Q.; Ma, E. Atomic-Level Structure and Structure-Property Relationship in Metallic Glasses. *Prog. Mater. Sci.* **2011**, *56*, 379–473. [CrossRef]
19. Takeuchi, A.; Inoue, A. Classification of Bulk Metallic Glasses by Atomic Size Difference, Heat of Mixing and Period of Constituent Elements and Its Application to Characterization of the Main Alloying Element. *Mater Trans* **2005**, *46*, 2817–2829. [CrossRef]
20. Inoue, A. Stabilization of Metallic Supercooled Liquid and Bulk Amorphous Alloys. *Acta Mater.* **2000**, *48*, 279–306. [CrossRef]
21. Zheng, S.; Li, J.; Zhang, J.; Jiang, K.; Liu, X.; Chang, C.; Wang, X. Effect of Si Addition on the Electrochemical Corrosion and Passivation Behavior of Fe-Cr-Mo-C-B-Ni-P Metallic Glasses. *J. Non-Cryst Solids* **2018**, *493*, 33–40. [CrossRef]
22. Pang, S.J.; Zhang, T.; Asami, K.; Inoue, A. Synthesis of Fe–Cr–Mo–C–B–P Bulk Metallic Glasses with High Corrosion Resistance. *Acta Mater.* **2002**, *50*, 489–497. [CrossRef]
23. Yamashita, T.; Hayes, P. Analysis of XPS Spectra of Fe^{2+} and Fe^{3+} Ions in Oxide Materials. *Appl. Surf. Sci.* **2008**, *254*, 2441–2449. [CrossRef]
24. Si, J.; Wu, Y.; Wang, T.; Liu, Y.; Hui, X. Composition-Controlled Active-Passive Transition and Corrosion Behavior of Fe-Cr(Mo)-Zr-B Bulk Amorphous Steels. *Appl. Surf. Sci.* **2018**, *445*, 496–504. [CrossRef]
25. Hua, N.; Hong, X.; Liao, Z.; Zhang, L.; Ye, X.; Wang, Q.; Liaw, P.K. Corrosive Wear Behaviors and Mechanisms of a Biocompatible Fe-Based Bulk Metallic Glass. *J. Non-Cryst Solids* **2020**, *542*, 120088. [CrossRef]
26. Gan, Z.; Zhang, C.; Zhang, Z.-R.; Chen, Z.-J.; Liu, L. Crystallization-Dependent Transition of Corrosion Resistance of an Fe-Based Bulk Metallic Glass under Hydrostatic Pressures. *Corros. Sci.* **2021**, *179*, 109098. [CrossRef]
27. Li, Y.; Zhang, C.; Xing, W.; Guo, S.; Liu, L. Design of Fe-Based Bulk Metallic Glasses with Improved Wear Resistance. *ACS Appl. Mater. Inter.* **2018**, *10*, 43144–43155. [CrossRef]
28. Si, J.J.; Chen, X.H.; Cai, Y.H.; Wu, Y.D.; Wang, T.; Hui, X.H. Corrosion Behavior of Cr-Based Bulk Metallic Glasses in Hydrochloric Acid Solutions. *Corros. Sci.* **2016**, *107*, 123–132. [CrossRef]
29. Liang, D.; Wei, X.; Chang, C.; Li, J.; Wang, Y.; Wang, X.; Shen, J. Effects of W Addition on the Electrochemical Behaviour and Passive Film Properties of Fe-Based Amorphous Alloys in Acetic Acid Solution. *Acta Metall. Sin.-Engl.* **2018**, *31*, 1098–1108. [CrossRef]
30. Li, Y.; Wang, S.; Wang, X.; Yin, M.; Zhang, W. New FeNiCrMo(P, C, B) High-Entropy Bulk Metallic Glasses with Unusual Thermal Stability and Corrosion Resistance. *J. Mater. Sci. Technol.* **2020**, *43*, 32–39. [CrossRef]
31. Wang, Y.; Jiang, S.L.; Zheng, Y.G.; Ke, W.; Sun, W.H.; Wang, J.Q. Electrochemical Behaviour of Fe-Based Metallic Glasses in Acidic and Neutral Solutions. *Corros. Sci.* **2012**, *63*, 159–173. [CrossRef]
32. Xia, H.; Chen, Q.; Wang, C. Evaluating Corrosion Resistances of Fe-Based Amorphous Alloys by YCr/Mo Values. *J. Rare Earth* **2017**, *35*, 406–411. [CrossRef]
33. Tan, M.W.; Akiyama, E.; Habazaki, H.; Kawashima, A.; Asami, K.; Hashimoto, K. The Role of Chromium and Molybdenum in Passivation of Amorphous Fe-Cr-Mo-P-C Alloys in Deaerated 1 M HCl. *Corros. Sci.* **1996**, *38*, 2137–2151. [CrossRef]
34. Tan, M.-W.; Akiyama, E.; Kawashima, A.; Asami, K.; Hashimoto, K. The Effect of Air Exposure on the Corrosion Behavior of Amorphous Fe-8Cr-Mo-13P-7C Alloys in 1 M HCl. *Corros. Sci.* **1995**, *37*, 1289–1301. [CrossRef]
35. Souza, C.A.C.; Ribeiro, D.V.; Kiminami, C.S. Corrosion Resistance of Fe-Cr-Based Amorphous Alloys: An Overview. *J. Non-Cryst. Solids* **2016**, *442*, 56–66. [CrossRef]
36. Hua, N.; Zhang, X.; Liao, Z.; Hong, X.; Guo, Q.; Huang, Y.; Ye, X.; Chen, W.; Zhang, T.; Jin, X.; et al. Dry Wear Behavior and Mechanism of a Fe-Based Bulk Metallic Glass: Description by Hertzian Contact Calculation and FInite-Element Method Simulation. *J. Non-Cryst. Solids* **2020**, *543*, 120065. [CrossRef]

metals

MDPI

Article

Electron Backscatter Diffraction (EBSD) Analysis of Machinable Lead-Free Brass Alloys: Connecting Texture with Fracture

Athanasios Vazdirvanidis, Andreas Rikos, Anagnostis I. Toulfatzis and George A. Pantazopoulos *

ELKEME Hellenic Research Centre for Metals S.A., 61st km Athens—Lamia National Road, 32011 Oinofyta, Greece; avazdirvanidis@elkeme.vionet.gr (A.V.); arikos@elkeme.vionet.gr (A.R.); atoulfatzis@elkeme.vionet.gr (A.I.T.)
* Correspondence: gpantaz@elkeme.vionet.gr; Tel.: +30-2262-60-4463

Abstract: The current paper is related to the study of the microstructure and texture of two machinable lead-free brass alloys, namely CuZn42 (CW510L) and CuZn38As (CW511L), which were evaluated in the as-drawn and post heat treated condition. Electron backscatter diffraction (EBSD) was employed for the examination of the brass rods' crystallographic properties in order to correlate the effect of post processing heat treatment on the evolution of phase structure and texture towards the interpretation of dynamic (impact) fracture properties. It is shown that α- and β-phase volume fractions, mean grain size, and grain boundary misorientation are the most influential factors altering the fracture resistance of single- and dual-phase brass alloy rods. The role of grain boundary engineering, through the formation of coincidence site lattice (CSL) boundaries and their evolution during thermomechanical processing, is of major importance for the design of the mechanical behaviour of new eco-friendly machinable brass alloys.

Keywords: brass; machinability; coincidence site lattice; grain boundary; EBSD; lead-free brass

Citation: Vazdirvanidis, A.; Rikos, A.; Toulfatzis, A.I.; Pantazopoulos, G.A. Electron Backscatter Diffraction (EBSD) Analysis of Machinable Lead-Free Brass Alloys: Connecting Texture with Fracture. *Metals* **2022**, *12*, 569. https://doi.org/10.3390/met12040569

Academic Editor: Shi-Hoon Choi

Received: 1 March 2022
Accepted: 25 March 2022
Published: 28 March 2022

1. Introduction and Literature Review

Brass alloys (Cu–Zn) are broadly used industrial metallic materials in domestic, mechanical, and electrical engineering as a result of their significant formability, corrosion and mechanical resistance, electrical conductivity, and, of course, high machinability [1–3]. Lead (Pb) is a basic alloying element of conventional brass due to its significant positive effect on machinability, resulting in high lubrication activity at the tool/workpiece interface and intense chip-fracturing, affecting the cutting tool service life and process [4–7]. However, the use of brass as a component in drinking water installations has increased the demands for the elimination of lead toxicity, inaugurating a new class of copper alloys, the eco-friendly or lead-free/low-lead brass alloys. A simple version of lead-free/low-lead machinable alloys include (almost) binary Cu–Zn systems, such as CuZn42 (CW510L) and CuZn38As (CW511L) (see [8–10]). New eco-friendly systems containing various (micro)alloying elements, such as Si, Ti, Bi, Sn, Mg, Sb, and Al aim, in their majority, to offer a significant advantage in chip breaking, replacing Pb with other intermetallic or second-phase particle constituents [11–17]. A recent review, summarizing the entire spectrum of such an extended family of copper alloys, highlighting the various technical approaches and machinability improvement strategies, outlines the necessity of the optimization of the manufacturing process towards the fabrication of eco-friendly components, satisfying the requirement of the relevant environmental and health and safety regulations [18].

Evidently, chip rupture constitutes a significant aspect of machinability performance and it is strongly related to the fracture properties of brass alloys under different loading conditions [7,19]. Therefore, the study of the fracture behaviour of leaded and lead-free/low-lead brasses constitutes a significant element, not only for the insightful understanding of the role of microstructure (e.g., lead particle size and distribution, phase structure, grain structure) on chip breaking, but also for the design and optimization of the machinability performance, without compromising the mechanical properties of the fabricated components [20,21].

Failure and embrittlement mechanisms, such as intermediate temperature embrittlement (ITE) and transgranular-to-intergranular fracture, together with properties-processing relationships, have been extensively studied in brass alloys systems (see [22–26]). Among the most typical failure mechanisms, dezincification, stress corrosion cracking (SCC), and liquid metal embrittlement (LME) causing hot-shortness due to localized (intergranular) Pb fusion and grain boundary decohesion are revealed in characteristic case histories studied in the authors' Institute, and representative works are shown in [27–29].

2. Background Information and Research Objectives

The microstructure of common brass alloys typically consists of one (single-phase, α-brasses) or two (dual-phase, α + β brasses) phases, as is shown by the well-known binary Cu–Zn phase diagram [30]. The α-phase is rich in copper, contains less than 37 wt.% zinc, and has a face-centred-cubic (fcc) crystal structure. Single, α-phase alloys have moderate strength and excellent formability at room temperature due to their low stacking-fault energy (SFE < 10 mJ/m^2) (see [31]). Therefore, they easily form twins and they can also form coincidence site lattice (CSL) boundaries.

At a higher zinc content, the β-phase is formed, which possesses a body-centred-cubic (bcc) crystal structure. The copper alloys which include a β-phase have excellent hot workability and machinability. The β-phase is also referred to as an intermetallic compound due to the almost 50–50 atomic stoichiometry of Cu and Zn. It is stable at 800 °C from 39 to 55 wt.% Zn and at 500 °C between 45 and 49 wt.% Zn [30].

Above the temperature of 470 °C, Cu and Zn atoms occupy random locations of the lattice, forming a disordered structure which is known as the β-phase, but below this critical temperature thermal vibrations are decreased and atoms take preferential sites, forming a long-range ordered structure known as the β′-phase. This is constituted by smaller regions of ordered structure called domains. Their boundaries are called antiphase boundaries (APBs) and they can be only observed with transmission electron microscopy (TEM). Quenching from the ordered β-phase region leads to fine domain size and increased hardness. The CuZn or β′-phase belongs to B2 super lattice materials, having a structure similar to that of FeAl [30,32].

Extrusion products with variable volume fractions of α- and β-phases can be produced by altering extrusion temperature. Higher extrusion temperatures favour the formation of a β-phase and result in alloys with a lower yield strength if the α-phase is considerably coarsened [25].

It has been shown that grain boundary engineering (GBE) can be performed in fcc metals of low stacking-fault energy (SFE) through thermo-mechanical processing (TMP) in order to improve their mechanical properties. The proposed mechanism is the formation of coincident site lattice (CSL) boundaries, through the development of annealing twin boundaries with low values of Σ, such as Σ3 boundaries, and its variants, Σ9 and Σ27, which together with low-angle grain boundaries (LAGBs; considered to have 5–15 degrees of misorientation) are resistant to intergranular cracking in comparison to the more brittle behaviour shown by high-angle grain boundaries (HAGBs), which have misorientation values larger than 15 degrees [33].

It has been found that one reason for the more brittle behaviour of HAGBs could be their tendency to attract segregated elements, as opposed to LAGBs and the CSL boundaries. When a crack is formed, propagation is preferential through the network of HAGBs, while the fracture mode can be altered from intergranular to transgranular when the crack meets LAGBs and CSL junctions. The amount of CSL boundaries is also linearly increased with decreasing mean grain size, enabling the manufacturing of fracture-resistant alloy products [34].

In dual-phase brass alloys, the grain boundaries' "character" should be evaluated separately for the two distinct phases, since during deformation and post-processing annealing, the α- and β-phases show different tendencies in HAGBs formation, with the α-phase exhibiting a high volume fraction of CSL boundaries, depending on the thermo-mechanical process scheme and the β-phase random values of HAGBs, which are considered of minor importance in GBE.

The current research is a follow-up work, aiming to shed light, using texture analysis, on the alteration of fracture mechanics and failure mechanisms of environmentally friendly brass alloys under different processing conditions, i.e., as-drawn and heat treated [35].

As presented in [35], apart from the benefits in machinability of CW510L (CuZn42) and CW511L (CuZn38As) alloys provided by heat treatment through the increase and stabilization of β-phase content, controversial results were obtained in impact toughness testing:

- The CuZn42 alloy shows an increasing tendency (from 47 to 52 J).
- The CuZn38As alloy shows a decreasing tendency (from 104 to 84 J)

The fracture surfaces of the respective Charpy specimens have also been studied in detail in [35]. The CuZn42 alloy sample in the as-received condition exhibited a typical, ductile fracture failure mechanism with multiple size dimples, while after the heat treatment it principally showed an intergranular fracture topography, with minute dimples on the grain facets. In the CuZn38As alloy sample, a fully ductile behaviour with multiple size dimples and large, deep voids was observed in the as-received condition as compared to the heat treated condition, where finer and shallower dimples were evident [35].

The present study is exclusively focused on electron backscatter diffraction (EBSD), in order to seek potential fracture–texture relationships that could justify the eminent evolution of fracture behaviour and crack propagation modes in the as-received and heat treated brass alloys. More specifically, further attention was placed on the salient interpretation of the observed "paradox" demonstrated by the CuZn42 sample, described as follows:

> *"The CuZn42 alloy after heat treatment exhibited improved fracture toughness, even though (i) the fully β-phase microstructure, established by the heat treatment, is expected to induce to fracture toughness deterioration and (ii) the obtained impact fracture topography presents an almost complete intergranular pattern which microscopically implies to lower impact energy."*

On the contrary, in the CuZn38As sample, the anticipated decrease in impact energy after the heat treatment was expected considering the attained phase transformations and the observed fracture mechanism demonstrated by the obtained fracture morphology which consists of smaller and shallower dimples.

To the best of the authors' knowledge, there is no similar study concerning the investigation and interpretation of the fracture mechanisms of low-lead/lead-free brass alloys under various thermomechanical processing conditions (as-drawn and heat treated), using detailed texture and grain boundary analysis through electron backscatter diffraction (EBSD) analysis.

3. Materials and Methods

3.1. Brass Alloy Samples

Two distinct lead-free brass alloys were employed for the present study, namely CuZn42 (CW510L) and CuZn38As (CW511L). The chemical composition of the studied alloy samples is shown in Table 1 (see also [35]).

The samples are originated from 35 mm diameter extruded and drawn rods of two types of lead-free brass alloys, namely CuZn42 (CW510L) and CuZn38As (CW511L). The CuZn42 alloy rod was heat treated at 775 °C for 60 min and the CuZn38As at 850 °C for 120 min and they were both water quenched for maximization of the β-phase percentage and suppression of α-phase precipitation with the aim to improve machinability. A simple schematic showing the heat treatment processes for the different brass alloys is illustrated in Figure 1. The respective fracture surfaces after Charpy impact testing are shown in Figure 2.

Table 1. Chemical composition of the brass alloy rods under examination (analysis by optical emission spectroscopy, elemental contents are expressed in wt.%).

Alloy (Spec. Limits)	Cu	Sn	Pb	Fe	Ni	Al	Sb	As	Zn
CuZn42 (CW510L)	57.5	0.006	0.10	0.03	0.003	0.0002	0.003	0.001	42.4
EN 12164	57–59	0.30 max	0.20 max	0.30 max	0.30 max	0.050 max	-	-	Rem.
CuZn38As (CW511L)	62.05	0.004	0.09	0.02	0.001	0.0002	0.003	0.03	37.8
EN 12164	61.5–63.5	0.10 max	0.20 max	0.10 max	0.30 max	0.050 max	-	0.02–0.15	Rem.

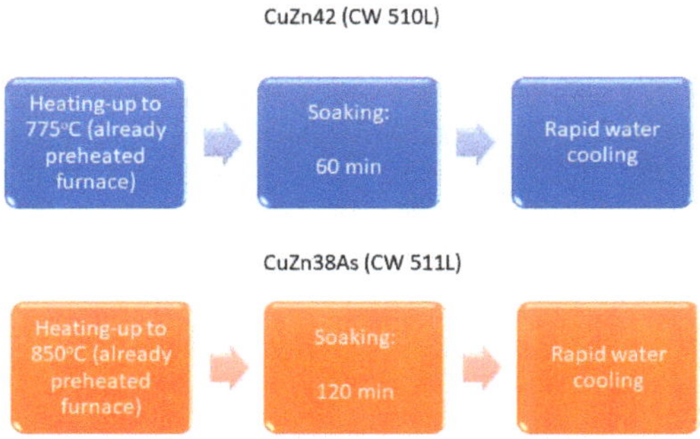

Figure 1. Simple schematic diagram showing the performed heat treatment processes (blue blocks for CuZn42 and orange blocks for CuZn38As alloy samples). Heat treatment conditions—CuZn42: 775 °C for 60 min followed by water quenching; CuZn38As: 850 °C for 120 min followed by water quenching.

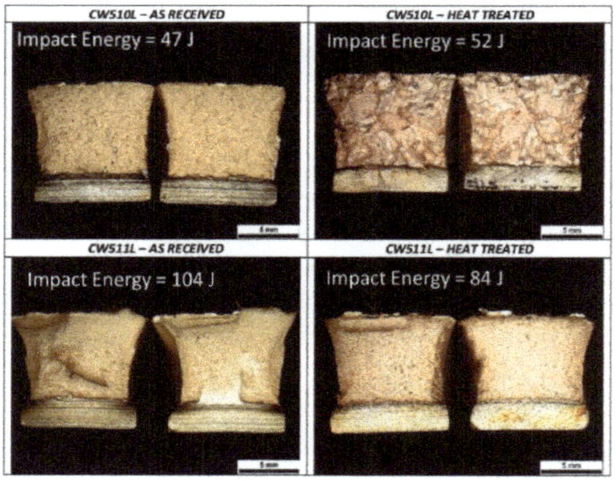

Figure 2. Macroscopic fracture surfaces after impact fracture toughness tests (Charpy) before and after heat treatment of CW510L and CW511L brass alloys. The average values of the induced impact energies are referred too.

3.2. *Electron Backscatter Diffraction (EBSD)*

EBSD analysis was performed in suitably sectioned impact-tested samples using an EDAX Hikari XP camera (EDAX, Mahwah, NJ, USA) mounted on a JEOL IT-800 HL (JEOL Ltd.,Tokyo, Japan) Scanning Electron Microscope (SEM), under 20 kV of accelerating voltage. A tilt of 70 degrees, 4 × 4 binning, and various magnifications and step sizes, depending on the grain size of the samples, prior to and following the heat treatment, were applied to assist in microstructure (phase structure), texture, and grain boundary analysis towards the interpretation of the alterations of fracture mechanisms.

Visualization of the strains caused by the impact tests, near the fracture surfaces, and examination of their distribution on the α- and β-phases, as well as the characterization and calculation of misorientation between adjacent grains, were performed. In the relative statistics, the 15° angle value was deployed as a distinction criterion of LAGBs from HAGBs.

Examination of the microstructure (phase structure) and fracture surface profile was performed in cross-sections parallel to the extrusion direction. The automatic collection and analysis of EBSD patterns was performed. Data mapping and information regarding the orientation, phase distribution, grain size, morphology, grain boundary, and local deformation of crystallographic regions was realized by using Orientation Image Mapping Software (OIM Analysis, version, EDAX, Mahwah, NJ, USA).

Coincident site lattice (CSL) boundaries are special character boundaries. These boundaries are classified in terms of Σ-values. The CSL boundaries are considered as special due to the fact that they possess a given fraction of atoms in the grain boundary plane which are coincident to both lattices which are separated by the grain boundary. The Σ-value denotes the fraction of atoms in coincidence. The detection of CSL boundaries is automatically performed by the OIM Analysis software.

The various EBSD settings used for the presented images/maps are summarized in Table 2.

Table 2. List of settings (scanned area, magnification, step size) used for the collection of the relevant EBSD patterns.

Figure	Scanned Area (μm^2)	Magnification	Step Size (μm)
Figure 3	707 × 693	100×	3
Figure 4	1780 × 2300	50×	7
Figure 5	400 × 500	150×	0.8
Figure 6	400 × 500	150×	2
Figure 7a,b	100 × 240	300×	1
Figure 7c	330 × 710	100×	2
Figure 8	140 × 150	370×	1
Figure 9	30 × 90	500×	0.3
Figure 10a	125 × 170	180×	0.4
Figure 10b	70 × 75	500×	0.3

Two representative EBSD scans per sample from the matrix in the as-drawn and heat treated conditions are shown concerning the phase structure examination. The following maps were acquired from each scan (the results are presented in Section 4.1):

- An inverse pole figure (IPF-Z) map and texture plot for representation of the prevailing orientations. This was achieved after a 90° rotation of the original EBSD data, in order to create an equivalent to the transverse section image. Transverse sections are typically studied for the analysis of extruded and drawn products.
- Phase mapping for calculation of the relative α- and β-phase volume fractions.
- A grain boundary map for characterization of grain boundaries' "character" and misorientation values between neighbouring grains.

One additional scan per sample of the fracture surface profile after Charpy impact testing in the as-drawn and heat treated conditions was performed and the retrieved results were as follows (the results are presented in Section 4.2):

- Phase maps for the correlation of α- and β-phases with surface fracture topography.
- Kernel average misorientation (KAM) maps for representation of the residual strains (deformation) underneath the fracture surface.

4. Results

4.1. Microstructure and Texture Characterization

4.1.1. CuZn42—As-Drawn Condition

The texture and volume fraction percentages of the α- and β-phases in CuZn42 were determined by EBSD (see Figure 3 and Table 3). A (111) fibre structure was exhibited by the α-phase and a single (101) fibre with minor (335) preferred orientations structure was observed by the β-phase (see Figure 3a,b). The results showed a 46 vol.% fraction for the fcc (α-phase) and a 54 vol.% fraction for the bcc (β-phase) (see Figure 3d).

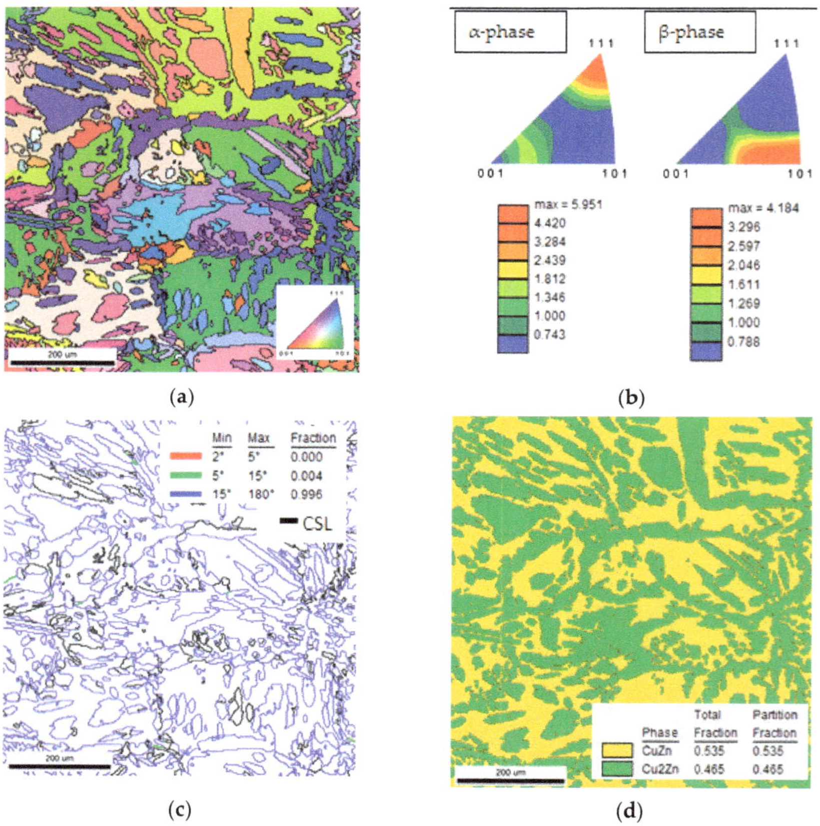

Figure 3. CuZn42 alloy in the as-drawn condition. (**a**) Inverse pole figure (IPF-Z) map and (**b**) texture plots. (c) Misorientation and (**d**) phase maps. The α-phase is precipitated in various morphologies from the β-grains. Note: CuZn (β-phase), Cu2Zn (α-phase). Unit: μm.

The α-phase appeared in various forms, mainly exhibiting an allotriomorphic morphology nucleated on the β-phase grain boundaries, but also as coarse intra-crystalline "islands", typical of a slow cooling rate from a hot working temperature (Figure 3a). The mean grain size of both the α-phase and β-phases was 25 μm. A large fraction (20%) of the α-phase grains exhibited thermal or annealing twins. The microstructure was characterized by the presence of almost exclusively high angle grain boundaries (see Figure 3c and Table 4). No sub-grains (2–5 degrees) were detected, while low angle boundaries (5–15 degrees) were almost null and were located exclusively on α–α interface boundaries (see the green lines in Figure 3c). The high angle boundaries included 19% CSL boundaries, consisting of 10% Σ3 and the remaining 9% distributed to other CSL categories. All CSL high angle boundaries were related to the α-phase, while approximately half of them were located on α–β interphase boundaries, with the other half on α–α boundaries.

Table 3. Results of EBSD scans, % phase volume fractions, mean grain size, % twinned grains, and mechanical testing data.

Alloy	Metallurgical Condition	Phases		Mean Grain Size (μm)		α-Phase Twinned Grains	Charpy Test Results [35]	
		vol.% α	vol.% β	α	β	(%)	Impact Energy (J)	Fracture mode
CuZn42 (CW510L)	As-drawn	46	54	25	25	20	47	Ductile, dimpled fracture
	Heat treated	-	100	-	Macro (several thousand μm, i.e., mm)		52	Intergranular (predominantly)
CuZn38As (CW511L)	As-drawn	98	2	14	3	52	104	Ductile, dimpled fracture (also including large, deep voids)
	Heat treated	86	14	23	6	21	84	Ductile, dimpled fracture (mostly shallow dimples)

Table 4. Results of EBSD scans, distribution of grain boundaries' angles (%).

Alloy	Metallurgical Condition	Sub-Grain Boundaries (2–5°)	Low Angle (5–15°)	High Angle (>15°)		CSL				
				CSL	Rem.	Σ3	Σ5	Σ9	Σ49	Rem.
CuZn42 (CW510L)	As-drawn			19	81	10	2	2	3	2
	Heat treated			20	80	15				5 (Σ7, Σ15, and Σ21)
CuZn38As (CW511L)	As-drawn	1	17	33	49	22	3			8
	Heat treated	22	7	21	50	8	1	2	8	2

4.1.2. CuZn42—Heat Treated Condition

After the heat treatment, the microstructure consisted merely of macroscopic size (several thousand μm, i.e., mm) β-phase grains, while any α-phase precipitation had been suppressed by the rapid water quenching (Figure 4 and Table 3). A double (101) and (113) fibre structure was exhibited by the single β-phase alloy (Figure 4b). Exclusively high angle boundaries were observed (Figure 4c). The β-phase exhibited poor quality in EBSD scans in most of the grains, which rendered difficulties in processing the respective maps. This could be attributed to (i) the disorder–order transformation which led to fine ordered domain size and multiple orientations within a single grain or (ii) due to the minute lattice parameters deviation between the β- and β′-phase, which could not be resolved by the EBSD technique [36]. Interestingly, the examination showed that approximately 20% of the grain boundaries were CSL, with the higher percentage being Σ3 and the remainder being Σ7, Σ15, and Σ21 (Table 4). The coarse grain size is an indication of the exaggerated grain growth which took place during the heat treatment of the brass rod.

The quality of the β-phase regions in the EBSD maps in the as-drawn condition was higher than in the heat treated conditions, since any quenching effects resulting in fine domain sizes were avoided.

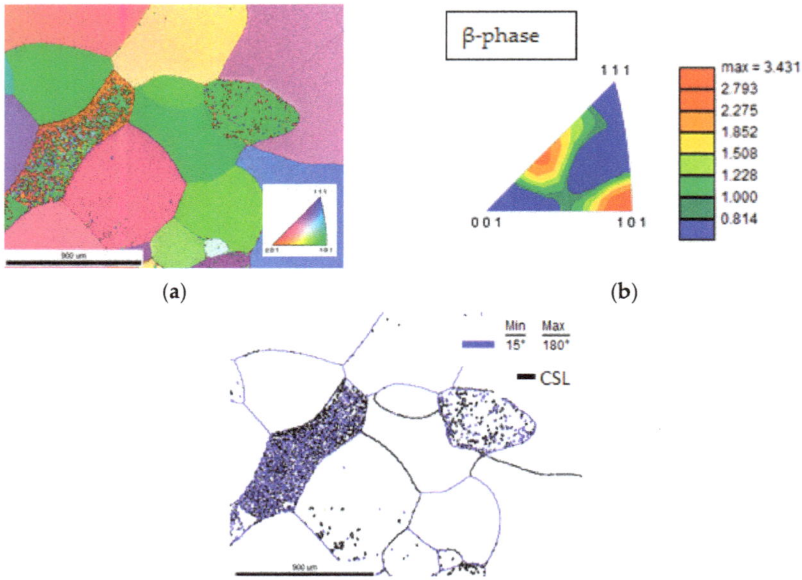

(a) (b) (c)

Figure 4. CuZn42 alloy in the heat treated condition. (**a**) Inverse pole figure (IPF-Z) map and (**b**) texture plot. (**c**) Misorientation map. The poor quality of the map was attributed to fine domain size within the macroscopic β-phase grains. Unit: μm.

4.1.3. CuZn38As—As-Drawn Condition

The texture and microstructure of CuZn38As in the as-drawn condition is shown in Figure 5 and the summary of the results are listed in Tables 3 and 4. The material was strongly textured with the (001) direction oriented parallel to the extrusion/drawing direction (Figure 5a,b). The microstructure consisted of partially recrystallized, equiaxed α-phase grains with a 14 μm mean grain size with a 98 vol.% fraction, while the β-phase was aligned parallel to the extrusion direction (longitudinal sections were performed). The volume fraction of the β-phase was 2% and its mean grain size was 3 μm (Figure 5d). An approximate 52% of α-phase grains exhibited thermal twins. A total of 1% of the grain boundaries constituted subgrain boundaries (2–5 degrees), 17% consisted of LAGBs, and the remaining 82% consisted of HAGBs (Figure 5c). The subgrain and LAGBs were detected within α-phase regions and could be considered as an indication of a fully recovered and partially recrystallized microstructure. The CSL fraction constituted 33% of the total boundaries, which were mainly encountered in the α-phase twins and on α–β interface boundaries in equal amounts. Approximately 22% of the total boundaries were Σ3, 3% were Σ5, and the remaining 8% was distributed to other categories (see Table 4).

4.1.4. CuZn38As—Heat Treated Condition

The texture and microstructure of CuZn38As in the heat treated condition are shown in Figure 6 and the results are summarized in Tables 3 and 4. The α-phase exhibited a (001) and (335) double fibre texture with equal amounts of both directions while the β-phase was strongly textured with prevailing (101) directions and minor (112) (see Figure 6a,b).

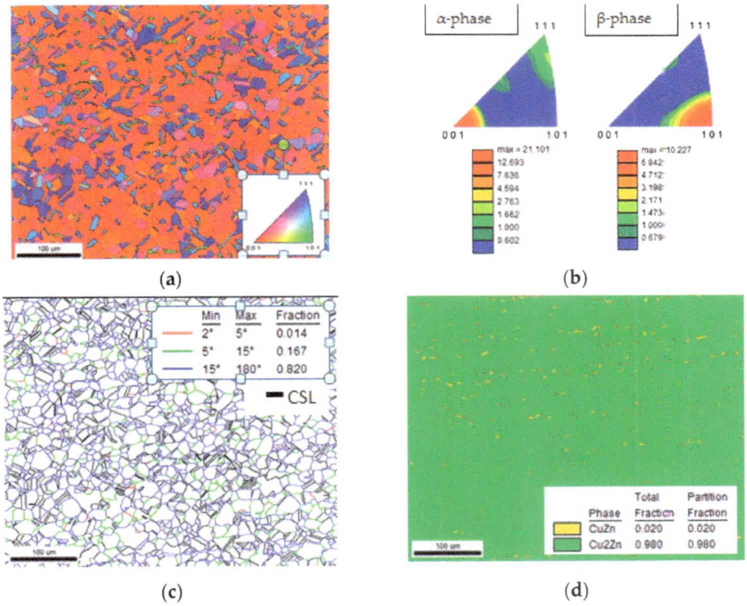

Figure 5. CuZn38As alloy in the as-drawn condition. (**a**) Inverse pole figure (IPF-Z) map and (**b**) texture plot. (**c**) Misorientation and (**d**) phase map. Note: CuZn (β-phase), Cu2Zn (α-phase). Unit: μm.

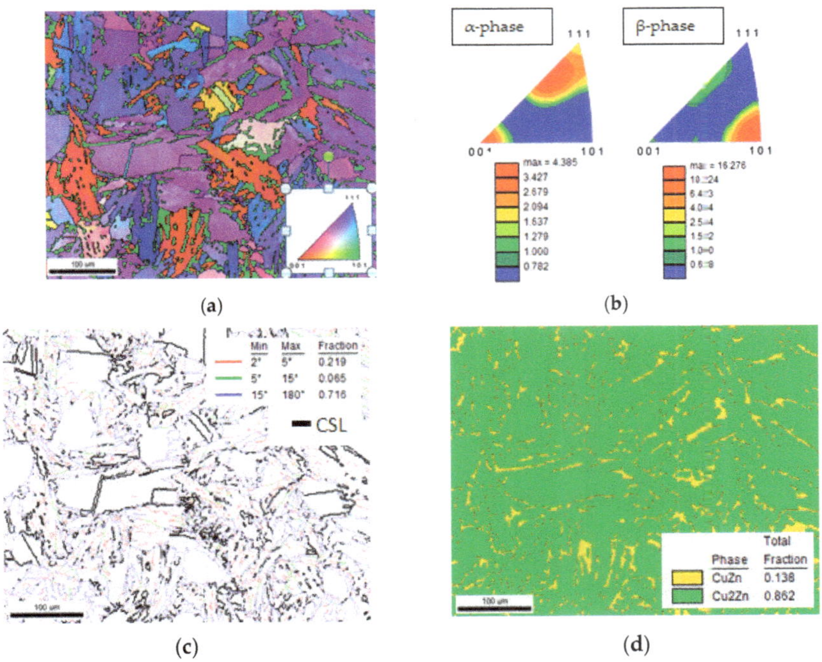

Figure 6. CuZn38As alloy in the heat treated condition. (**a**) Inverse pole figure (IPF-Z) map and (**b**) texture plot. (**c**) Misorientation and (**d**) phases map. Note: CuZn (β-phase), Cu2Zn (α-phase). Unit: μm.

After the heat treatment, the α-phase grains were coarsened, with a 23 μm mean grain size, and the fraction of the twinned grains was reduced to 21% (see Figure 6 and Tables 3 and 4). The β-phase percentage was increased to 14% and the banding morphology was altered since a more uniform distribution of the β-phase was created. The β-phase mean grain size was 6 μm, which was double the size of that prior to the heat treatment condition, and it exhibited a plate-like, elongated morphology (Figure 6d). The grain boundaries' character was also significantly altered by the heat treatment, i.e., 22% of the total boundaries were sub-grain boundaries, 7% were LAGBs, and the remaining 71% were HAGBs. Among them, 21% was characterized as CSL boundaries, with Σ3 and Σ49 occupying 8% of the total boundaries each, Σ5 and Σ9 occupying 3% in total, and the remainder being distributed to other categories (Table 4).

4.2. Microstructure Characterization of the Impact Specimens' Fracture Surface Profile

4.2.1. Sample CuZn42—As-Drawn and Heat Treated Conditions

The profile of the fracture surface of the impact test specimen in the as-drawn condition is shown in Figure 7a. The existence of micro-dimples was apparent as various size protrusions, not exceeding 2 μm. These protrusions did not exhibit the same amount of misorientation values in the KAM map, the latter being dependent on the phase where they were formed. The higher values were found by dimples formed within α-phase regions on the fracture surface or within the β-phase having a short distance from the α-phase, at maximum up to 10 μm. The α-phase network presented high misorientation values in the KAM map at even higher distances from the fracture surface.

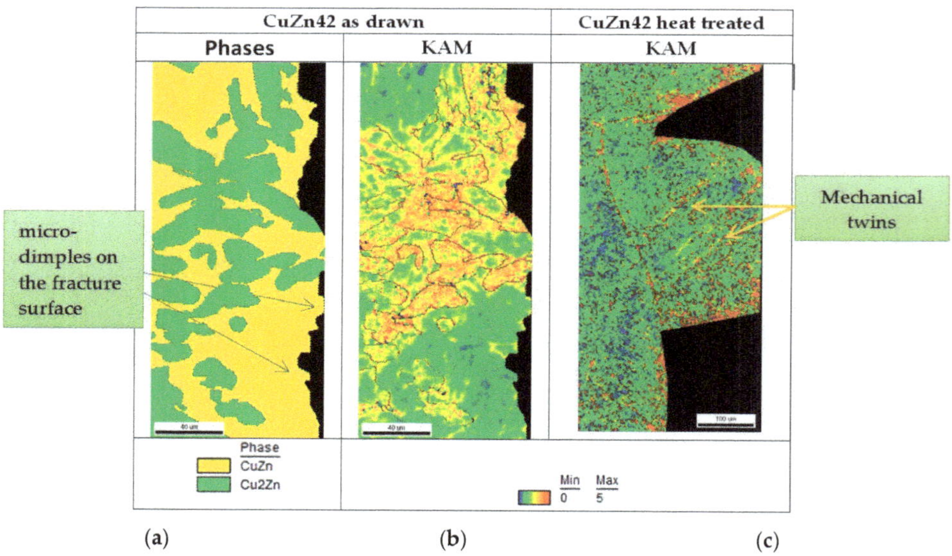

Figure 7. Phase and KAM maps of the fracture surface profile of Charpy impact tested CuZn42 alloy samples: (**a**) phase map, (**b**) KAM map in the as-drawn condition, and (**c**) KAM map in the heat treated condition. Unit: μm.

In Figure 7c, the fracture surface profile of the CuZn42 alloy sample after the heat treatment is shown. It is typical of the occurrence of a brittle, intergranular fracture, with the presence of mechanical twins formed during the impact test. Mechanical twins can be produced in bcc metals under conditions of shock loading in decreased temperatures [31]. The "noise" in the map did not allow a definite recognition of deformed and non-deformed areas, but other than the angle boundaries which were high angle and lead to local high misorientation values, a uniform distribution of lattice rotations was observed, close and far from the fracture surface.

In Figure 8, lens-shaped twins are shown extending across the width of the grain (deformation twins do not extend to adjacent grains). This is the main existing deformation mechanism exhibited by the coarse, single β-phase brass alloy, since the crack tip created a limited plastic zone size compared to the grain size, hindering the activation of multiple sets of slip systems in adjacent grains, and dislocation emission was impeded by the existing microstructural barriers, causing dislocation pile-up to the activated slip systems and grain boundaries [37]. Twinning behaviour in bcc alloys is considered important in orientation changes that enable the activation of new slip systems in order to allow deformation by slip. In the intergranular crack, shown in Figure 8, on both the fracture surface and across the secondary crack, no appreciable deformation has occurred, leading to the hypothesis that mechanical twinning produced by the shock loading was the predominant deformation mechanism, explaining the reason for the uniform misorientation values for the same grain close and far from the fracture surface.

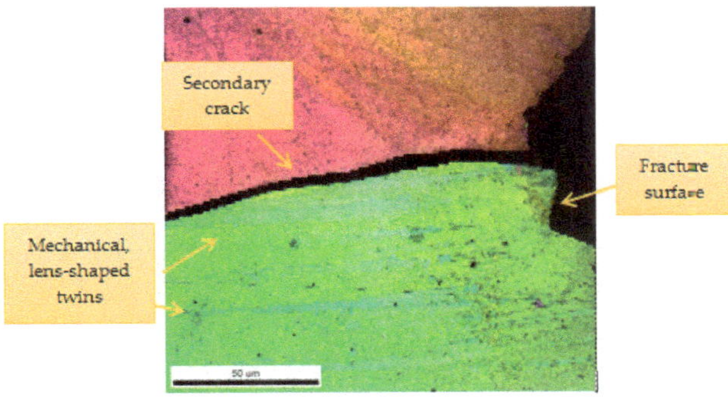

Figure 8. Mechanical twins developed on the fracture surface of β-phase grains after Charpy impact test; CuZn42 alloy sample in heat treated condition, section transverse to fracture surface. Combined image quality and IPF map. Unit: μm.

4.2.2. CuZn38As—As-Drawn and Heat Treated Conditions

The fracture surface profile of the sample in the as-drawn condition is shown in Figure 9a,b. It manifested appreciable ductility and the coarse formation of dimples reaching a size of 30 μm, which is multiple times the size of the dimples of the CuZn42 alloy in the same metallurgical condition. The equiaxed α-phase grains exhibited extensive strain markings close to the fracture surface as a contributing mechanism to the energy absorbance potential (Figure 10a). These constitute very fine deformation twins which are observed as a major deformation mechanism in low stacking-fault energy alloys such as the α-brass (20 mJ·m^{-2}) [38]. The fine grain size of the α-phase grains contributed to improved ductility and the respective high energy values in the Charpy impact test (104 J).

In the heat treated condition, the fracture surface was also deformed, showing the presence of shallower dimples reaching a maximum size of 17 μm, since the more resistant to deformation β-phase did not allow the development of coarser dimples (Figure 9c,d). The α-phase grain structure, which was coarser (23 μm) than in the as-drawn condition, negatively affected the material's deformation capacity during the impact test, leading to a lower energy absorption (84 J). Strain markings in the α-phase were also developed close to the fracture surface and were more readily observed in the coarser α-phase grains, especially in the heat treated condition (see Figure 10).

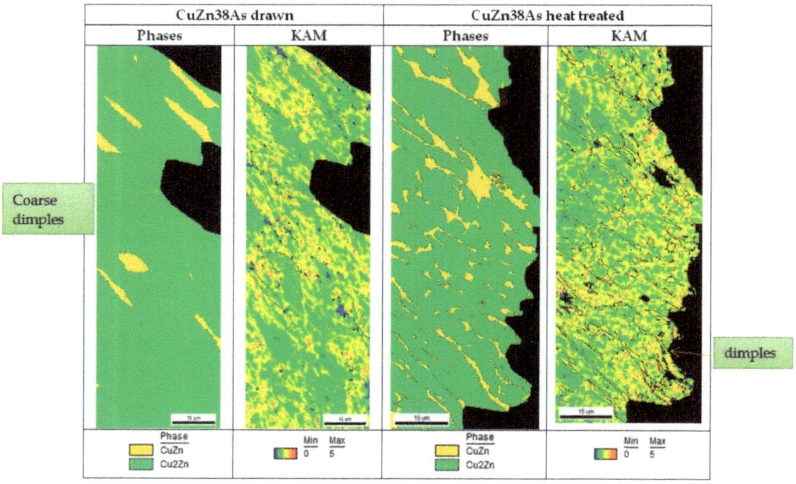

(a) (b) (c) (d)

Figure 9. Phase and KAM maps of the fracture surface profile of Charpy impact tested CuZn38As alloy samples in the (**a**,**b**) as-drawn and (**c**,**d**) heat treated conditions. Unit: μm.

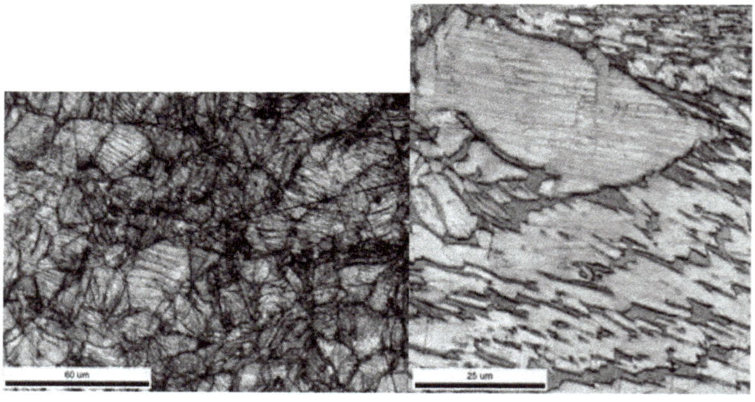

(a) (b)

Figure 10. Strain markings close to the fracture surface of (**a**) CuZn38As as-drawn and (**b**) CuZn38As heat treated alloy sample. EBSD image quality map. Unit: μm.

5. Discussion

The presented results, concerning the fracture behaviour of brass alloys, can be summarized and classified in two main categories: single-phase alloys and dual-phase ($\alpha + \beta$) alloys.

5.1. Single-Phase Alloys: CuZn42, Heat Treated and CuZn38As, As-Drawn Condition

The CuZn38As sample, in the as-drawn condition, possessed a higher amount of α-phase (98 vol.%: it can be considered for simplicity purposes as a "single-phase" alloy) and a finer grain size, rendering a high deformation and energy absorbance capacity. Moreover, it contained the higher fraction of LAGBs and CSL boundaries (~50% in total), which promote ductile, transgranular fracture modes. The high extent of strain markings formation was an additional indication of a strained material with a high energy absorbance capacity.

The opposite condition was observed by the CuZn42 alloy sample in the heat treated condition, which showed a single β-phase structure and macroscopic grain size. These attributes favoured the development of mainly brittle intergranular cracking.

However, the average impact energy value for CuZn42 at the heat treated condition presented a slight increasing tendency (from 47 to 52 J; i.e., approximately 10%), compared with the same alloy at the as-drawn condition. The fact that CSL boundaries were detected (~20%) also led to the appearance of a minor transgranular crack propagation mode in the heat treated CuZn42 alloy. In addition, the occurrence of mechanical twinning constituted the dominant deformation mechanism of this sample with appreciable strain energy absorption during the shock loading. The incremental tendency of impact energy could potentially be explained by the following phenomena:

- The presence of a significant fraction of CSL boundaries a (total of 20%, including a higher percentage of Σ3 boundaries up to 15%).
- The mechanical twinning of the bcc β-phase.

In addition, the higher impact energy can be also supported by the higher hardness of the heat treated CuZn42 (fully β-phase) as compared to the as-drawn one, i.e., approximately 140 HV vs. 130 HV [35].

Grain boundaries are mostly influential on dynamic and static mechanical properties behaviour and can act as preferred locations for crack initiation and propagation. HAGBs hinder dislocation motion as a result of their high degree of misorientation, leading to dislocation pile-up and localized stress concentration that can induce failure [39]. In contrast, the slip systems of neighbour grains at LAGBs possess a higher alignment which favours dislocation mobility and retain slip homogeneity across the grain boundaries. Conclusively, LAGBs resist intergranular fracture, which tends to propagate along HAGBs [39].

5.2. Dual-Phase (α + β) Alloys: CuZn38As, Heat Treated Condition and CuZn42, As-Drawn Condition

Concerning the samples of CuZn42 in the as-drawn condition and CuZn38As in the heat treated condition, which constitute dual-phase alloys, the significant difference in energy absorbance capacity and dimple size, which are indicators of higher plasticity (for CuZn38As), seem to be related to the following factors (see also Tables 3 and 4):

1. The increasing volume fraction of the α-phase (the CuZn42 alloy contains 46 vol.% α-phase and the CuZn38As alloy contains 86 vol.% α-phase) results in the amplification of impact energy from 47 and 84 J, respectively.
2. A decreasing β-phase mean grain size (25 μm for CuZn42 and 6 μm for CuZn38As) leads to a higher toughness (47 J and 84 J, respectively).
3. For the total amount of LAGBs plus CSL boundaries, a higher amount leads to a higher impact toughness (19% for CuZn42 as compared to 28% for CuZn38As), which promotes transgranular ductile fracture.
4. The existence of a subgrain structure (subgrain boundaries were mostly detected in CuZn38As under the heat treated condition, summed up to 22%).

6. Conclusions and Further Research

From the crystallographic examination of the single- and dual-phase brass rods, the following conclusions could be derived:

- There is a strong tendency for energy absorbance in Charpy impact tests to be related to the amount of the sum of subgrains, LAGBs, and CSL boundaries. It is characteristic that the CuZn38As alloy under as-drawn and heat treated conditions, possessing in total 50% of the above interfaces, had the largest energy absorbance values (>80 J).
- Σ3 was the predominant type of CSL boundary occurring in all the studied metallurgical conditions: (i) extruded and drawn and (ii) heat treated in the single-phase region and quenched.
- Post processing heat treatment led to a single β-phase structure for CuZn42 alloy and a 12 vol.% increase for the β-phase in the CuZn38As alloy. The augmentation of the

β-phase volume fraction inherits a negative effect on impact toughness for dual-phase brasses (α + β brasses).

- The size of dimples and the prevalence of transgranular fracture mode in the Charpy impact test specimens' fracture surfaces are strongly related with the amount of the α-phase volume fraction. Strain is also preferentially concentrated in α-phase regions on the fracture surfaces compared with the more brittle β-phase. KAM maps are advantageous in revealing the phases–strains relationships near the fracture surfaces.
- In the absence of an α-phase, as in the case of CuZn42-heat treated condition, the fraction of dimpled fractures is negligible. On the other hand, in samples of CuZn42 in the as-drawn condition, CuZn38As in the heat treated condition, and CuZn38As in the as-drawn condition, as the α-phase volume fraction increases, the size of micro-dimples, viewed transverse to fracture surface sections, respectively increases.
- Interestingly enough, as it was obtained from the systematic impact energy results, the presence of a full β-phase microstructure could have a potential beneficial contribution to fracture toughness, due to the combination of higher hardness with the activation of the mechanical twinning process under shock loading conditions, while the limited size of the plastic zone ahead of the crack tip in combination with the coarse β-phase grains promotes intergranular crack propagation.
- The increase of mean grain size of α- and β-phases offers a negative effect on strength and impact toughness in CuZn38As brass.
- The texture of the α-phase in the as-received condition was different in the two different alloys, a single (111) in CuZn42 and (001) in CuZn38As, revealing a process and/or chemical composition relationship with the resulting texture. The dominant texture of the β-phase was (101) in both alloys and metallurgical conditions.

The fact that CSL boundaries were also observed in the single β-phase CuZn42 heat treated alloy, constituted mostly by Σ3 boundaries, offers a promising ground for further research since the investigation of the influence of texture on single β-phase properties will provide valuable knowledge towards the development of advanced copper alloys with superior mechanical behaviour and machinability.

Author Contributions: Conceptualization, A.V. and G.A.P.; primary materials testing and methodology, A.V., A.R. and A.I.T.; sample preparation, A.R.; writing—original draft preparation, A.V.; writing—review and editing, A.I.T. and G.A.P.; project supervision and management, G.A.P. All authors have read and agreed to the published version of the manuscript.

Funding: This research received no external funding.

Institutional Review Board Statement: Not applicable.

Informed Consent Statement: Not applicable.

Data Availability Statement: The data supporting the findings of this study are available upon request from the corresponding author.

Acknowledgments: The encouragement and support of ELKEME management is highly appreciated.

Conflicts of Interest: The authors declare no conflict of interest.

References

1. Achiţei, D.C.; Minciună, M.G.; Vizureanu, P.; Sandu, A.V.; Cimpoeşu, R.; Istrate, B. Study on structure and properties of CuZnPb alloy. *IOP Conf. Ser. Mater. Sci. Eng.* **2019**, *133*, 012015. [CrossRef]
2. Imai, H.; Kosaka, Y.; Kojima, A.; Li, S.; Kondoh, K.; Umeda, J.; Atsumi, H. Characteristics and machinability of lead-free P/M Cu60–Zn40 brass alloys dispersed with graphite. *Powder Technol.* **2010**, *198*, 417–421. [CrossRef]
3. La Fontaine, A.; Keast, V.J. Compositional distributions in classical and lead-free brasses. *Mater. Charact.* **2006**, *57*, 424–429. [CrossRef]
4. Wolfenden, A.; Wright, P.K. Role of lead in free-machining brass. *Met. Technol.* **1979**, *6*, 297–302. [CrossRef]
5. Stoddart, C.T.H.; Lea, C.; Dench, W.A.; Green, P.; Pettit, H.R. Relationship between lead content of Cu-40Zn, machinability, and swarf surface composition determined by auger electron spectroscopy. *Met. Technol.* **1979**, *6*, 176–184. [CrossRef]

6. Pantazopoulos, G. Leaded brass rods C 38500 for automatic machining operations: A technical report. *J. Mater. Eng. Perform.* **2002**, *11*, 402–407. [CrossRef]
7. Johansson, J.; Alm, P.; M'Saoubi, R.; Malmberg, P.; Ståhl, J.-E.; Bushlya, V. On the Function of Lead (Pb) in Machining Brass Alloys. *Res. Sq.* **2022**, Preprint. [CrossRef]
8. Toulfatzis, A.; Pantazopoulos, G.; David, C.; Sagris, S.; Paipetis, A. Final heat treatment as a possible solution for the improvement of machinability of Pb-free brass alloys. *Metals* **2018**, *8*, 575. [CrossRef]
9. Toulfatzis, A.I.; Pantazopoulos, G.A.; Paipetis, A.S. Microstructure and properties of lead-free brasses using post-processing heat treatment cycles. *Mater. Sci. Technol.* **2016**, *32*, 1771–1781. [CrossRef]
10. Toulfatzis, A.I.; Pantazopoulos, G.A.; David, C.N.; Sagris, D.S.; Paipetis, A.S. Machinability of eco-friendly lead-free brass alloys: Cutting-force and surface-roughness optimization. *Metals* **2018**, *8*, 250. [CrossRef]
11. Schultheiss, F.; Johansson, D.; Bushlya, V.; Zhou, J.; Nilsson, K.; Ståhl, J.E. Comparative study on the machinability of lead-free brass. *J. Clean. Prod.* **2017**, *149*, 366–377. [CrossRef]
12. Nobel, C.; Klocke, F.; Lung, D.; Wolf, S. Machinability enhancement of lead-free brass alloys. In Proceedings of the 6th CIRP International Conference on High Performance Cutting, Berkeley, CA, USA, 23–25 June 2014; pp. 95–100.
13. Atsumi, H.; Imai, H.; Li, S.; Kondoh, K.; Kousaka, Y.; Kojima, A. The effect of solid solutionizing Ti element on microstructural and mechanical properties of extruded Cu-40Zn-Ti ternary alloy. *Trans. JWRI* **2011**, *40*, 67–71.
14. Suksongkarm, P.; Rojananan, S.; Rojananan, S. Bismuth formation in lead-free Cu-Zn-Si yellow brass with various bismuth-tin alloy additions. *Mater. Trans.* **2018**, *59*, 1747–1752. [CrossRef]
15. Adineh, M.; Doostmohammadi, H. Microstructure, mechanical properties and machinability of Cu-Zn-Mg and Cu-Zn-Sb brass alloys. *Mater. Sci. Technol.* **2019**, *35*, 1504–1514. [CrossRef]
16. Vilarinho, C.; Davim, J.P.; Soares, D.; Castro, F.; Barbosa, J. Influence of the chemical composition on the machinability of brasses. *J. Mater. Processing Technol.* **2005**, *170*, 441–447. [CrossRef]
17. Rajabi, Z.; Doostmohammadi, H. Effect of addition of tin on the microstructure and machinability of α-brass. *Mater. Sci. Technol.* **2018**, *34*, 1218–1227. [CrossRef]
18. Stavroulakis, P.; Toulfatzis, A.I.; Pantazopoulos, G.A.; Paipetis, A.S. Machinable leaded and eco-friendly brass alloys for high performance manufacturing processes: A critical review. *Metals* **2022**, *12*, 246. [CrossRef]
19. Toulfatzis, A.I.; Pantazopoulos, G.A.; Paipetis, A.S. Fracture behavior and characterization of lead-free brass alloys for machining applications. *J. Mater. Eng. Perform.* **2014**, *23*, 3193–3206. [CrossRef]
20. Toulfatzis, A.I.; Pantazopoulos, G.A.; Paipetis, A.S. Fracture Analysis of Eco-Friendly Brass Alloys: Comparison Study and Preliminary Assessment. In Proceedings of the 14th International Conference on Fracture (ICF 14), Rhodes, Greece, 18–23 June 2017.
21. Pantazopoulos, G.A.; Toulfatzis, A.I. Fracture modes and mechanical characteristics of machinable brass rods. *Metallogr. Microstruct. Anal.* **2012**, *1*, 106–114. [CrossRef]
22. Laporte, V.; Mortensen, A. Intermediate temperature embrittlement of copper alloys. *Int. Mater. Rev.* **2009**, *54*, 94–116. [CrossRef]
23. Wolley, D.J.; Fox, A.G. The embrittlement of leaded and unleaded α+β (60-40) brasses in the temperature range 300 to 500 °C. *J. Mater. Sci. Lett.* **1988**, *7*, 763–765. [CrossRef]
24. Fefi, A.; Brotzu, A.; Pilone, P. Analysis of the fracture criticality of biphasic brass. *Proc. Struct. Integr.* **2016**, *2*, 2959–2965. [CrossRef]
25. Mapelli, C.; Venturini, R. Dependence of the mechanical properties of an α/β brass on the microstructural features induced by hot extrusion. *Scr. Mater.* **2006**, *54*, 1169–1173. [CrossRef]
26. Pantazopoulos, G. A review of defects and failures in brass rods and related components. *Pract. Fail. Anal.* **2003**, *3*, 14–22. [CrossRef]
27. Pantazopoulos, G.; Vazdirvanidis, A. Identification of corrosion and damage mechanisms by using Scanning Electron Microscopy and Energy Dispersive X-ray Microanalysis: Contribution to Failure Analysis Case Histories. *IOP Conf. Ser. Mater. Sci. Eng.* **2014**, *55*, 012015. [CrossRef]
28. Pantazopoulos, G.; Toulfatzis, A.I. Failure analysis of a machinable brass connector in a boiler unit installation. *Case Stud. Eng. Fail. Anal.* **2013**, *1*, 18–23. [CrossRef]
29. Pantazopoulos, G.; Vazdirvanidis, A. Failure analysis of a fractured leaded-brass (CuZn39Pb3) extruded hexagonal rod. *J. Fail. Anal. Prev.* **2008**, *8*, 218–222. [CrossRef]
30. *Copper and Copper Alloys Specialty Handbook*; ASM International: Materials Park, OH, USA, 2001.
31. Dieter, G.E. *Mechanical Metallurgy*; McGraw-Hill Book Company: New York, NY, USA, 1988.
32. Voort, G.F.V. *ASM Handbook, Vol. 9*; ASM International: Materials Park, OH, USA, 2004.
33. Al-Fadhalah, K.J.; Aleem, R.M.; Nicky, T. Microstructure and texture development in thermomechanically processed leaded brass. *Metals* **2021**, *11*, 998. [CrossRef]
34. Watanabe, T. Structural effects on grain boundary segregation, hardening and fracture. *J. Phys. Colloques* **1985**, *46*, C4-555–C4-566. [CrossRef]
35. Toulfatzis, A.I.; Pantazopoulos, G.A.; Paipetis, A.S. Fracture mechanics properties and failure mechanisms of environmentally-friendly brass alloys under impact, cyclic and monotonic loading conditions. *Eng. Fail. Anal.* **2018**, *90*, 459–517. [CrossRef]
36. Regina, J. Ordered Structures. In *ASM Handbook, Volume 9: Metallography and Microstructures*; ASM International: Materials Park, OH, USA, 2004.

37. Janssen, M.; Zuidema, J.; Wanhill, R. *Fracture Mechanics*, 2nd ed.; SPON Press: Oxford, UK, 2004.
38. Humphrey, F.J.; Hatherly, M. *Recrystallization and Related Annealing Phenomena*; Elsevier: Amsterdam, The Netherlands, 2004.
39. Bond, D.; Zikry, M.A. Microstructural modeling of intergranular fracture in tricrystals with random low- and high-angle grain boundaries. *JOM* **2017**, *69*, 856–862. [CrossRef]

metals MDPI

Article

An Analytical Model for Stress and Curvature Prediction of a Strip Leveling Process

Shih-Kang Kuo [1], Yi-Liang Ou [1] and Dung-An Wang [2,3,*]

1 Steel Research and Development Department, China Steel Corporation, Kaohsiung 81233, Taiwan; 150359@mail.csc.com.tw (S.-K.K.); 165688@mail.csc.com.tw (Y.-L.O.)
2 Graduate Institute of Precision Engineering, National Chung Hsing University, 250 Kuo Kuang Rd., Taichung 40227, Taiwan
3 Faculty of Mechanical Engineering, Industrial University of Ho Chi Minh City, 12 Nguyen Van Bao, Ward 4, Go Vap District, Ho Chi Minh City 800010, Vietnam
* Correspondence: daw@dragon.nchu.edu.tw; Tel.: +886-4-22857207

Abstract: An analytical model of a steel strip under alternate bending/reverse bending during a roller leveling process is developed. A combined isotropic/kinematic hardening model is implemented through a combined hardening parameter. A formulation of the change of the effective stress as a function of the change of the effective strain under cyclic loading is combined with the developed analytical model to predict the stress distributions and residual curvature of a steel strip under roller leveling efficiently and accurately. Dissimilar to the commonly used assumption of one contact point between the stripe and the rolls, an effective radius modelling the wrap-around contact characteristics is proposed. An arc contact of the strip around a roll is described by the contact model. An oscillatory behavior of the residual curvature is observed when a range of roll intermesh setting is considered. The contact model added to the analytical model may enhance the accuracy in predicting the oscillatory behavior of the residual curvatures. A range of the roll intermesh setting can be suggested by the developed model to obtain a flat strip after roller leveling.

Keywords: roller leveling; steel strip; combined hardening; analytical model

Citation: Kuo, S.-K.; Ou, Y.-L.; Wang, D.-A. An Analytical Model for Stress and Curvature Prediction of a Strip Leveling Process. *Metals* **2022**, *12*, 757. https://doi.org/10.3390/met12050757

Academic Editors: George A. Pantazopoulos and Zhengyi Jiang

Received: 14 March 2022
Accepted: 24 April 2022
Published: 28 April 2022

1. Introduction

Steel strips after rolling and annealing may possess defects such as wavy edges, center buckles, cambers, and twists. The flatness defects can be corrected by leveling in steel mills. Amor et al. [1] reported that shape defects of metal strips after rolling process or coiling operation can be removed by leveling in order to meet the quality requirement. Customers may cut the as-received flat stripes in order to achieve end functionalities. Unacceptable shape defects found after cutting the strips are serious issues of concern raised by the customers of steel mills. Li et al. [2] reported that these flatness/shape defects are mainly due to residual stresses generated from rolling, annealing, leveling, coiling, etc. Morris et al. [3] classified the shape defects into two categories—latent and manifest. Shape defects of the latent category appear flat prior to cutting operation, whereas manifest defects, such as wavy edges and center buckles, are visible in the coiled form. Conventionally, tension levelers are used upstream to remove manifest defects. Roller leveling with lower line tension and smaller rolls has been the essential step to flatten metal strips and attenuate the effects of inhomogeneous distribution of residual stress in strips. Roller leveling is a complex forming process involving multiple, alternate bending and reverse bending cycles. In order to achieve more effective leveling operation without resorting to on site trial and error approach, a lot of efforts have been devoted to the development of efficient, accurate simulation tools for roller leveling.

Finite element analyses and analytical modeling are two main approaches adopted to investigate the roller leveling process. Hira et al. [4] developed an analytical model for calculation of curling and residual stress of a strip in longitudinal and width directions after tension leveling. Isotopic material hardening was considered in their model. Doege et al. [5] analyzed a levelling process using the Euler–Bernoulli beam theory and a combined isotropic and kinematic hardening model. Contact points between the strip and the rolls were computed iteratively assuming that only one contact point between the roll and the strip exists. They calculated the range of roll intermeshes for zero strip curvature at the exit roll for a seven-roll leveling system. Details of the verification of their analytical modeling were not described. Behrens et al. [6] developed an analytical model to find suitable settings of a leveler for flat strips. Contact points between the metal strip and the rolls were calculated by assuming that the curve of the strip was composed of arcs and common tangents between two rolls. Lengths of several segments dissected from the strip longitudinally were equalized with different plastic deformations to eliminate wave defects of the strip. Dratz et al. [7] also adopted this common tangent contact condition to develop their roller leveling model. Residual stress and curvature results of the strip after levelling were not reported. The common tangent assumption contradicts with the fact that the curvature between two rolls exhibits a nonlinear distribution. Liu et al. [8] developed an analytical model to analyze the residual stress and curvature of a plate subjected to roller leveling. Isotropic hardening was adopted in their model and a single contact between the plate and each roll was assumed. Chen et al. [9] developed an analytical model of a roller leveler for the strip with transverse and longitudinal wave defects. They divided the strip longitudinally in order to find a suitable bending amount of the rolls to attain equal length of each longitudinal section. Higo et al. [10] presented a theoretical model for a roller leveler based on Euler–Bernoulli beam theory and a single point contact model between a strip and each roll. They examined the influence of the exit roll intermesh on the residual curvature of the strip.

Morris et al. [11] built a finite element model to analyze the deformed shape and surface residual stress of a steel strip after a three-roll leveling process. They concluded that kinematic work-hardening can represent the loading condition experienced during tension levelling more closely than that of isotropic work-hardening by comparing model predictions of the deformed shapes with experiments. Schleinzer and Fischer [12] presented a finite element model for roller leveling of rails. Combined isotropic/kinematic hardening was assumed for the material. A friction coefficient of 0.2 was used between the rail and the rolls. Park and Hwang [13] conducted finite element analyses and experiments to examine the residual curvature and stress distribution of strips after roller leveling. They verified that the strips with various initial curvatures can attain residual curvatures with small deviation given an optimum entrance roll intermesh. Huh et al. [14] built a finite element model for a roller leveler in order to study the effects of process parameters on the curvature solution of roller leveling. They identified the parameters with significant effects on the residual curvature through a design of experiment approach. Roberts et al. [15] carried out a finite element study to compare modeled curvature distributions with experiments of a two-roll stretch-bending process. Their results showed that a reduction in the wrap-around contact length was caused by decreasing intermesh. Jin et al. [16] conducted a three-dimensional finite element analysis of a roller leveler. They presented residual stress distribution in the strip. Kim et al. [17] developed a finite element model for a roller leveling process. They implemented an implicit stress integration procedure in a constitutive material model which can capture the material yield behavior and the Bauschinger effect during reverse bending. Grüber et al. [18] employed a finite element analysis of roller levelling to determine several combinations of roll intermeshes to attain a flat strip after levelling. They identified that the roll intermesh of the center load triangle and the last load triangle of a seven-roll leveler is crucial to reach desired values of flatness and residual stress distribution.

Leveler operators have relied on parameter settings for strips with various yield strengths and through-gauges provided by leveler suppliers. Due to degradation of the leveler through years of usage, empirical methods were used to fine tune the settings for production. Although manifest defects appeared to be invisible after leveling, latent defects may still exist and cause the end products to fail to satisfy customer requirements. Analytical solutions of residual stress and curvature of the roller leveling process are necessary to provide rapid parameter settings to aid in its operation. In this investigation, an analytical model for a nine-roll roller leveler is developed. The nine-roll cold leveler was installed at the plate mill plant of the Chinese Steel Company for leveling heavy gauge steel plates. One of the motivations of this investigation is to provide operational parameters for the leveler to improve the quality of steel plates. Solutions of stress distributions after each bending/reverse bending and residual curvature of the strip are obtained based on the Euler–Bernoulli beam theory within a two-dimensional geometric framework. The material hardening is implemented through a combined isotropic/kinematic hardening parameter. Finite element analyses are carried out to verify the accuracy of the developed model. Results of the stress distributions and residual curvature of a steel strip are compared with the results based on the analyses. Finally, a roll intermesh range to produce a flatness condition of the strip is presented.

2. Analytical Model

Figure 1 is a schematic of a nine-roll leveler. The strip has the thickness t, the length L, and the width W. The rolls have a radius of R and a roll spacing of D. A strip enters the entry roll from the left, undergoes alternate bending and reverse bending, then leaves the exit roll. Flatness defects can be removed by the plastic deformation under the leveler rolls. Partial reduction in the flatness defect is ensued by insufficient plastic deformation. The intermesh settings, usually in a linear, declining trend, provide for the gradual flattening of the shape defects as the strip travels through the leveler. The roll intermeshes p_2, p_4, p_6, and p_8 are indicated in Figure 1. The value of the roll intermesh is positive when the gap between the top roll and the bottom roll is larger than the strip thickness. In this investigation, the amount of the roll intermesh is defined by the inclination angle θ of the upper roll carriage and the roll intermesh at the next-to-last roll p_8 as indicated in Figure 1. The bottom rolls are fixed, and the top rolls are moved downward to specified roll intermeshes. The width of the strip is assumed to be sufficiently large. Therefore, the strain in the width direction is neglected, and the plane strain condition is considered. Since the leveling process can be viewed as a series of three-point bending/reverse bending actions with groups of three rolls, the analytical model of the roller leveling process is conducted based on the framework of Euler–Bernoulli beam theory.

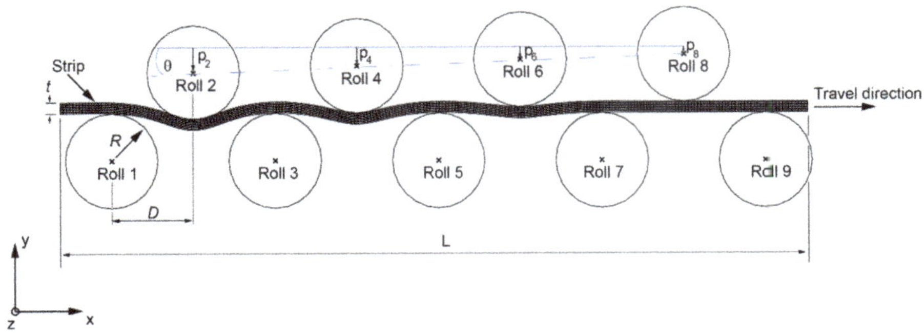

Figure 1. A schematic of a nine-roll leveler.

2.1. Material Model

In development of the analytical model, friction forces at the roll/strip interface, tension forces in the strip, and gravity are neglected. Figure 2 schematically shows a differential element of the strip. The strip has a thickness of t and a width of W. A Cartesian coordinate system is also shown in the figure. x represents the longitudinal direction, and y is along the thickness direction. The neutral axis is assumed to coincide with the mid-surface of the strip. The origin of the y coordinate is at the middle of the beam as shown in the figure. ρ is the radius of the curvature of the differential element under the moment loading M. Longitudinal strain ε_x is expressed as

$$\varepsilon_x = -\kappa y \quad (1)$$

where ε_x and κ [1/m] are the longitudinal strain and the curvature, respectively. In the elastic regime, the longitudinal stress σ_x [N/m^2] is related to the longitudinal strain ε_x by

$$\sigma_x = E\varepsilon_x \quad (2)$$

where E [N/m^2] is the Young's modulus. The material hardening behavior is given by an exponential law

$$\overline{\sigma} = \sigma_s + Q_i\left(1 - e^{-b_i\overline{\varepsilon}}\right) \quad (3)$$

where $\overline{\sigma}$ [N/m^2] and $\overline{\varepsilon}$ are the effective stress and the effective plastic strain, respectively. σ_s [N/m^2] is the initial yield stress. Q_i [N/m^2] and b_i are the isotropic hardening parameters, where Q_i specifies the maximum change in the size of the yield surface, and b_i defines the rate at which the size of the yield surface changes as plastic straining develops. For uniaxial loadings as the strip bending case, $\overline{\sigma}$ and $\overline{\varepsilon}$ can be taken as σ_x and ε_x, respectively. Relatively large plastic deformation is considered for the roller levelling. This simplification of $\overline{\varepsilon} \approx \varepsilon_x$ contributes to a very small fraction of error to the results. Values of (Q_1, b_1) and (Q_2, b_2) are obtained by fitting experimental stress–strain curve with Equation (3) for the first half cycle and the second half cycle of the loading, respectively. Therefore, Q_1 and b_1 are used to model the material hardening behavior in the first half loading cycle. Q_2 and b_2 are adopted for the subsequent loadings. The bending moment M [N·m] of the cross section of the strip is given as

$$M = -2W\int_0^{t/2} \sigma_x y dy \quad (4)$$

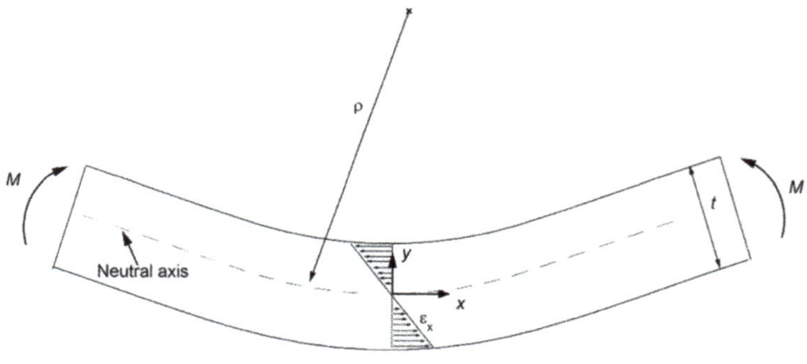

Figure 2. A schematic of a differential element of the strip.

The sign convention for the moment M and the curvature κ is related to the orientation of the coordinate axes.

During the first bend loading, the material of the strip exhibits isotropic hardening behavior. In the subsequent bend loadings, the material hardening of the strip is taken as a combined isotropic/kinematic type. Zhang et al. [19] formulated the change of the effective stress $\Delta\overline{\sigma}$ [N/m^2] as a function of the change of the effective strain $\Delta\overline{\varepsilon}$ under cyclic bend loading as

$$|\Delta\overline{\sigma}| = \begin{cases} E|\Delta\overline{\varepsilon}| & , |\Delta\overline{\varepsilon}| < \frac{|\Delta\overline{\sigma}|_{\text{lim}}}{E} \\ \sigma_s + Q_i\left(1 - e^{-b_i\left(|\Delta\overline{\varepsilon}| - \frac{|\Delta\overline{\sigma}|_{\text{lim}}}{E}\right)}\right) & , |\Delta\overline{\varepsilon}| \geq \frac{|\Delta\overline{\sigma}|_{\text{lim}}}{E} \end{cases} \quad (5)$$

where $|\Delta\overline{\sigma}_{lim}|$ [N/m^2] is the elastic limit of the effective stress change at the stress reversal point and can be written as

$$|\Delta\overline{\sigma}_{lim}| = (1+m)|\overline{\sigma}_r| + (1-m)(2\sigma_s - |\overline{\sigma}_r|) \quad (6)$$

where $|\overline{\sigma}_r|$ [N/m^2] is the magnitude of the effective stress at the point of stress reversal. m is the combined hardening coefficient. $m = 1$ and $m = 0$ correspond to the isotropic hardening case and kinematic hardening case, respectively, and $0 < m < 1$ is for the case of combined hardening. The value of m for each bending during the roller leveling is calibrated by matching the stress–strain curves based on the model of Equations (1)–(6) to the experimental measurements in the corresponding cycles over a strain range expected in the roller leveling process. Kotov et al. [20] experimentally demonstrated the validity of a kinematic hardening model of a steel strip under roller leveling.

2.2. Contact Model

Contact points between the strip and the rolls can be computed iteratively assuming that only one contact point between the roll and the strip exists [5,10]. Behrens et al. [6] and Dratz et al. [7] adopted the common tangent contact condition to develop their roller leveling model. The single point contact and common tangent contact assumptions render a simple means to model the contact between the strip and rolls. A line contact between the strip and each roll is assumed in the initial modeling work here. In the two-dimensional model considered in this investigation, the strip contacts with each roll tangentially as shown in Figure 3. The expanded radius of the roll R_e [m] is given as

$$R_e = R + \frac{t}{2} \quad (7)$$

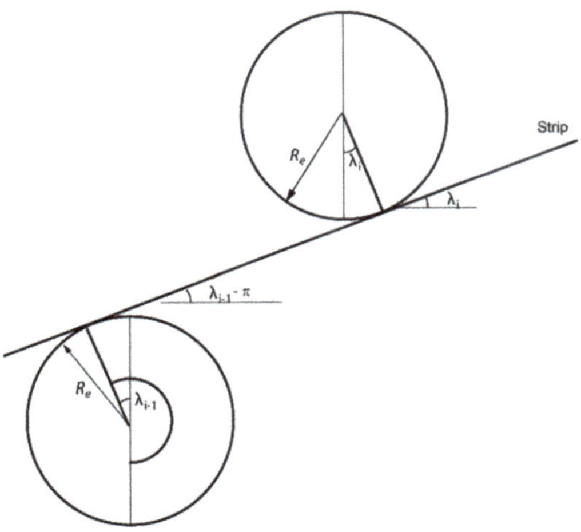

Figure 3. Tangential contact condition of the strip with the rolls.

The tangent of the contact angle λ_i [rad] is the gradient of the longitudinal axis of the strip

$$tan\lambda_i = \left.\frac{dy}{dx}\right|_{x_i} \quad (8)$$

where $y(x)$ is the deflected curve of the longitudinal axis of the strip, (x_i,y_i) is the ith contact point, and λ_i is the ith contact angle. The contact point (x_i, y_i) [m] is

$$\begin{aligned} x_i &= x_c + R_e sin\lambda_i \\ y_i &= y_c - R_e cos\lambda_i \end{aligned} \quad (9)$$

where (x_c, y_c) [m] is the coordinates of the center of the roll. Müller et al. [21] also used the expanded radius of the roll to calculate the contact point between a strip and a roll in a leveler.

A recursive scheme is adopted to calculate the curvature, bending moment, deflected curve of the strip, and the contact points. Initially, the location of the contact points and the curvature κ_i of the strip curve at the ith contact point are assumed. The bending moment M_i at the ith contact point is computed using Equations (1)–(6). Based on the assumption of a linear distribution of the moment between the contact points, the curvature of the deflected strip curve is estimated. Using the curvature distribution $\kappa(x)$, the deflected strip curve y [m] is calculated by

$$y(x) = \kappa(x)dx + C_1 x + C_2 \tag{10}$$

where C_1 and C_2 are the integration constants, and can be determined by the known contact points. A Newton–Raphson algorithm is used to obtain converged solutions of the deflected strip curve, moment distribution, curvature distribution, and contact points that satisfy both the Euler–Bernoulli beam theory and the geometrical constraints. Guan et al. [22] adopted a similar curvature integration approach to find the contact points for a roller leveler. They verified their model by comparing the deflection curves calculated by their model with those from experiments.

2.3. Residual Curvature

The flatness of the strip after exiting the leveler is related to the residual curvature. The bending and reverse bending cycle is repeated until the material reaches the last roll in the leveler. The strip is free to rotate at the last roll where the bending moment applied to the strip can be considered as zero. The internal stress distribution of the cross section of the strip at the last roll should result in a zero-bending moment. This condition can be enforced by imposing an artificial bending moment of the same magnitude but opposite sign to the moment at the last roll before unloading. Assuming no reverse yielding during unloading, the residual curvature κ' [1/m] of the strip at the last roll can be expressed as

$$\kappa' = \kappa - \frac{M}{EI} \tag{11}$$

where κ and M are the curvature and moment, respectively, before unloading, and I is the second moment of inertia of the cross section of the strip. Hosford and Caddell [23] and Guan et al. [24] also adopted this approach to calculate the residual curvature for beam bending problems.

3. Finite Element Analysis

In order to examine the accuracy of the developed analytical model, a two-dimensional finite element analysis of a roller leveler is carried out. A quasi-static condition is assumed in the finite element analyses. Due to the sufficiently large width, the strain in the width direction can be neglected, and the plane strain condition is considered in the finite element analysis. Both the upper and lower rolls are modeled as rigid bodies. A strip and a nine-roll leveler are shown in Figure 1. A Cartesian coordinate system is also shown in the figure. The direction of the z axis is given by the right-hand rule. In this investigation, the strip is moved forward by a displacement boundary condition and the bottom rolls are assumed to be fixed in the x and y directions and free to rotate with respect to the z axis. The displacements in the $-y$ direction of the upper rolls are specified to represent the amount of roll intermeshes, while their displacements in the x direction are constrained. As shown in Figure 1, a uniform displacement is applied in the $+x$ direction to the right edge surface of the strip, and the displacement in the y direction for the right edge surface is constrained to represent the strip travel horizontally along the $+x$ direction.

The commercial software, Abaqus (6.14, Dassault Systemes, Walthem, MA, USA), is adopted to compute the stress and deformation of the strip during the leveling process. Mises yield surface is used with the nonlinear isotropic/kinematic hardening model in Abaqus. The nonlinear kinematic hardening component is modeled through the back stress to describe the translation of the yield surface. The isotropic hardening component is modeled through the equivalent stress as a function of the equivalent plastic strain to define the size of the yield surface. The hardening law for the back stress α [N/m^2] is

$$\dot{\alpha} = \frac{C}{\sigma_t}(\sigma_x - \alpha)\dot{\varepsilon}^{pl} - \gamma\alpha\dot{\varepsilon}^{pl} \quad (12)$$

where C [N/m^2] and γ are kinematic hardening parameters that are calibrated from symmetric strain, cyclic test data. σ_t and $\dot{\varepsilon}^{pl}$ are the current size of the yield surface and the plastic strain rate, respectively. Figure 4 schematically shows a stabilized cycle. The plastic strain ε^{pl} is determined by

$$\varepsilon^{pl} = \varepsilon_x - \frac{\sigma_x}{E} - \varepsilon_0 \quad (13)$$

where ε_0 is the strain value of the intercept of the left half of the stabilized cycle with the strain axis as shown in Figure 4. The value of α [N/m^2] is given as

$$\alpha = \sigma_x - \sigma_t \quad (14)$$

where $\sigma_t = (\sigma_1 + \sigma_n)/2$ is the current size of the yield surface. The stresses σ_1 and σ_n are marked in Figure 4. To solve the back stress hardening law of Equation (12) over this stabilized cycle, with the first data point of $\sigma_x = \sigma_1$ and $\varepsilon^{pi} = 0$, the expression of α [N/m^2] is

$$\alpha = \frac{C}{\gamma}\left(1 - e^{-\gamma\varepsilon^{pl}}\right) + (\sigma_1 - \sigma_t)\varepsilon^{-\gamma pl} \quad (15)$$

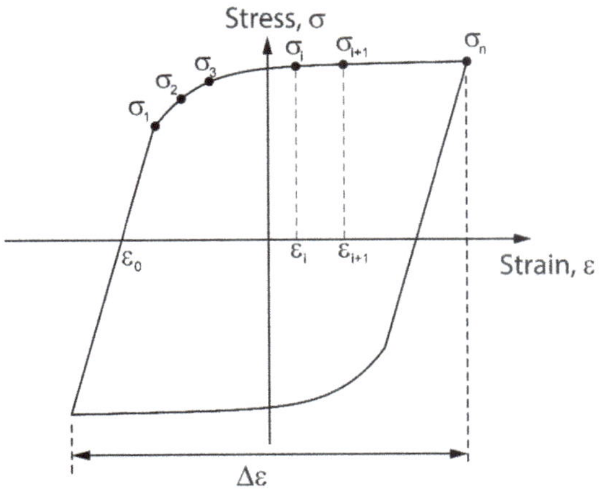

Figure 4. A schematic of a stabilized cycle.

Data pairs of $\left(\alpha, \varepsilon^{pl}\right)$ extracted from the stabilized stress–strain curve are used to calibrate values of the kinematic hardening parameters C and γ.

The isotropic hardening model of the Abaqus is defined as

$$\sigma_t = \sigma_s + Q_\infty\left(1 - e^{-b\bar{\varepsilon}^{pl}}\right) \quad (16)$$

where σ_t [N/m^2] and σ_s [N/m^2] are the current size and the initial size of the yield surface, respectively, and $\bar{\varepsilon}^{pl}$ is the equivalent plastic strain. Q_∞ [N/m^2] and b_i are the isotropic hardening parameters of the finite element model. The isotropic hardening parameters are calibrated from the data of the current size of the yield surface σ_t-equivalent plastic strain $\bar{\varepsilon}^{pl}$ relation based on a symmetric strain-controlled cyclic experiment with strain range $\Delta\varepsilon$ as shown schematically in Figure 4. For the ith cycle, the current size of the yield surface σ_t [N/m^2] is calculated by

$$\sigma_t = \sigma_i^p - \alpha_i \tag{17}$$

where σ_i^p [N/m^2] and α_i [N/m^2] are the peak tensile stress and the back stress, respectively, in the ith cycle. α_i [N/m^2] is calculated by

$$\alpha_i = \frac{\sigma_i^p + \sigma_i^n}{2} \tag{18}$$

where σ_i^n [N/m^2] is the compressive stress with the same ε^{pl} value as the peak tensile stress σ_i^p in the ith cycle. Since the value of the back stress α_i in each cycle at a particular strain level is nearly the same based on the model, α_i is approximated by the values of σ_1^p and σ_1^n. The equivalent plastic strain $\bar{\varepsilon}^{pl}$ corresponding to the ith cycle is

$$\bar{\varepsilon}^{pl} = \frac{4i-3}{2}\Delta\varepsilon^{pl} \tag{19}$$

where $\Delta\varepsilon^{pl}$ can be approximated by $\Delta\varepsilon - 2\sigma_1^p/E$.

Figure 5a is a mesh of the finite element model. Two-dimensional plane strain 4-noded CPE4R element is employed in the model. The total number of elements is 8188. Figure 5b is a close-up view of the mesh near a work roll. Ten elements are used in the thickness direction of the strip in order to obtain an accurate solution of the stress–strain distribution. The "analytical rigid surface" in Abaqus is used to model the rolls. Frictionless contact is assumed at the interface between the strip and the rolls.

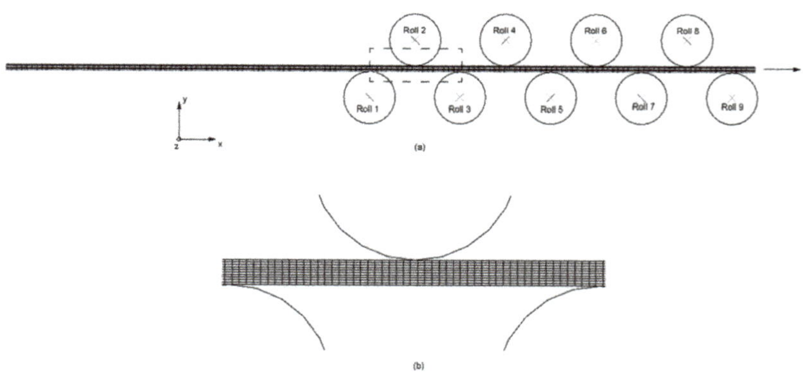

Figure 5. (**a**) A finite element model. (**b**) A close-up view of the mesh near a roll.

4. Analyses, Results, and Discussions

4.1. Analysis and Results of the Analytical Model

Figure 6 shows the stress–strain curves of a mild steel under a uniaxial, symmetric strain-controlled, cyclic test. The strain range $\Delta\varepsilon$ is 0.015. Based on the first half of the stress–strain curve of the first cycle, the calibrated values of the Young's modulus E and the initial yield stress σ_s are 219.8 GPa and 329.7 MPa, respectively, and Q_1 and b_1 have values of 66.3 MPa and 742.7, respectively. The values of Q_2 and b_2 are calibrated as 128.3 MPa and 604.7, respectively, based on the second half of the stress–strain curve of the first cycle.

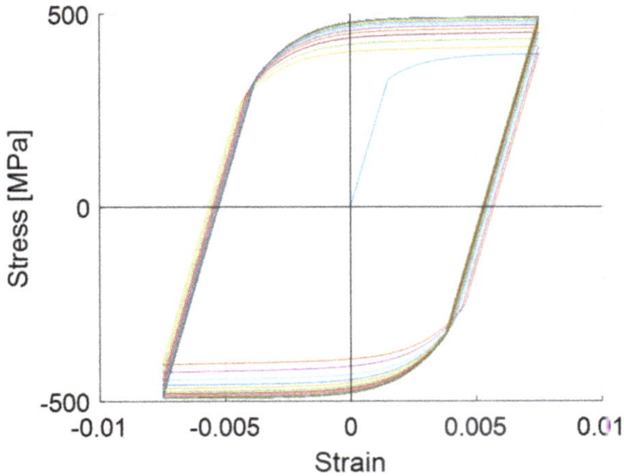

Figure 6. Stress–strain curves of a uniaxial, symmetric strain-controlled, cyclic test.

By specifying the combined hardening coefficient m, computational stress–strain curves based on the analytical model of Equations (1)–(11) with the strain range $\Delta\varepsilon$ of 0.0015 can be obtained. For a nine-roll roller leveler, the total number of bends equals seven. The values of the combined hardening coefficient m for the seven bends can be calibrated by comparing the computational stress–strain curves with the stress–strain curves of the mild steel shown in Figure 6 for the first seven bend loadings. Figure 7 shows the computational stress–strain curves and the stress–strain curves of Figure 6. The computational curves agree with those of the mild steel. The calibrated m values for the second to the seventh bending are 0.0784, 0.2, 0.3, 0.4, 0.45, and 0.5, respectively. The hardening behavior of the material for the first bending is taken as isotropic.

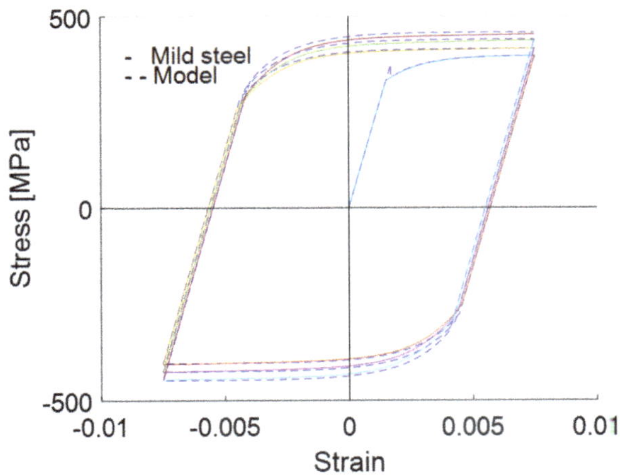

Figure 7. Stress–strain curves for calibration of the combined hardening coefficients.

Consider the case where the strip has the thickness t (=20 mm), the length L (=2480 mm), and the width W (=100 mm). The radius R of the rolls is 85 mm. The roll spacing D is 150 mm. The strip is taken as flat and free of residual stress. The upper roll carriage has an inclination angle θ of 0.1° and the roll intermesh at the next to the last roll p_8 ranges from −1.2 mm to 0.8 mm. Note that the value of the roll intermesh is positive when the gap between the top roll and the bottom roll is larger than the strip thickness. Therefore, a negative value of the intermesh means the work roll plunges into the strip.

With $p_8 = -0.70$ mm and the upper roll carriage inclined at the angle θ of 0.1°, p_2, p_4, and p_6 have values of −2.32 mm, −1.78 mm, and −1.24 mm, respectively. Sixty divisions are taken along the thickness direction of the strip for calculation of the stress and strain distributions. Figure 8 shows the deformed strip between roll 1 and roll 9 based on the analytical model. Figure 9a–c shows the distribution of the bending moment per unit width, the curvature, and the deformed center line of the strip, respectively, between roll 1 and roll 9. The contact points between the strip and the rolls are marked by circles in the figure. One hundred nodes are used between the contact points in the computations. Figure 9a shows the linear distribution of the bending moment between the contact points. At the entry roll and the exit roll, the values of the bending moment are zero. Figure 9b shows the curvature distribution. Values of the curvatures are nearly two orders smaller than the roll curvature 1.176×10^{-2} mm^{-1}. The curvature of the strip at the exit roll has a value of 3.054×10^{-5} mm^{-1}. The positive value of the curvature indicates that the strip is bent upward. Higo et al. [10] pointed out that the abrupt increase in magnitude of the curvature near all contact points except the final one is due to the nonlinear material hardening behavior. This is evidenced by the nonlinear sections of the moment–curvature curves during the leveling process plotted in Figure 10. The moment and the curvature are normalized by the maximum bending moment M_0 and the maximum curvature C_0 for which elastic conditions hold, respectively.

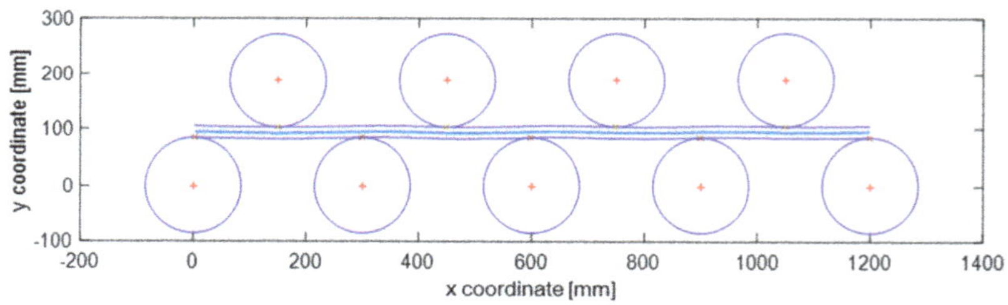

Figure 8. The deformed strip between roll 1 and roll 9.

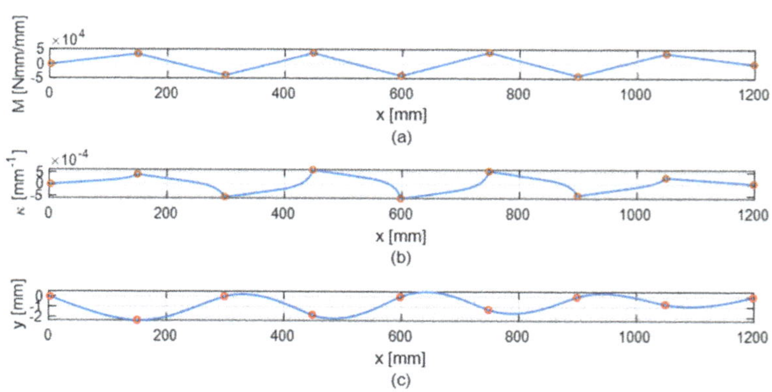

Figure 9. Distributions of (**a**) bending moment per unit width, (**b**) curvature, and (**c**) deformed center line of the strip between roll 1 and roll 9.

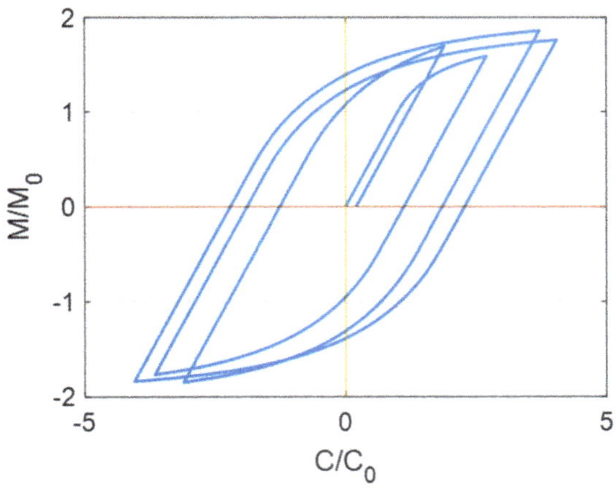

Figure 10. Normalized moment as a function of normalized curvature during leveling.

Figure 11a–c shows the stress distributions in the strip thickness direction when the strip travels through the leveler. The results computed by the analytical model are displayed by lines in the figure. Significant plastic deformations appear in the first three bends as seen in Figure 11a. The fractions of the plastic deformation are nearly 40%, 50%, and 60% for the first, the second, and the third bend, respectively. As the strip traverses through the fifth roll and the sixth roll, the region of plastic deformation stays at 60% as seen in Figure 11b. Figure 11c shows the stress distributions in the strip thickness direction at the eighth roll and the exit roll. The stress distribution at the exit roll can be taken as the residual stress distribution since the total moment applied to the strip at the exit roll is nearly zero. The stress at the ith roll is calculated based on a linear superposition assumption. The remnant stress of the cross section of the strip at the $(i-1)$th roll and the loading stress at the ith roll are superimposed to obtain the stress distribution at the ith roll. Guan et al. [25] also adopted this stress inheritance law in their roller leveling model. Yonetani [26] reported that the stress of a microscopic segment at the cross section in a uniaxial stress state satisfies the linear superposition assumption. The uniaxial stress loading condition is also assumed in the model considered in this investigation.

4.2. Analysis and Results of the Finite Element Model

Finite element analyses are carried out to verify the accuracy of the analytical model. The material properties for the mild steel used in the finite element analyses are listed in Table 1. The isotropic hardening parameters, Q_∞ and b, and the kinematic hardening parameters, C and γ, are calibrated from the uniaxial, symmetric strain-controlled, cyclic test with the strain range $\Delta\varepsilon$ of 0.015 as shown in Figure 6. The calibration procedure is described in Section 3.

Table 1. Material properties employed in the finite element analyses

Property	E (Gpa)	σ_s (MPa)	Poisson's Ratio	C (Mpa)	γ	Q_∞ (Mpa)	b
Value	219.8	329.7	0.3	44,600.6	673.1	96.9	12.7

The strip is taken as flat and free of residual stress in the beginning of the leveling process. Smith [27] reported that the initially flat condition is useful in the leveling analysis since the incoming flatness defects of a strip may vary from one location to another, and a strip generally has some initially flat area. Figure 12a shows the initial configuration of the finite element model. A Cartesian coordinate system is also shown in the figure. A displacement of 1200 mm in the $+x$ direction is given to the right edge surface of the strip. The displacement equals the distance from the center of roll 1 to the center of roll 9. Therefore, the cross-section A (marked in Figure 12a) travels from roll 1 at the beginning of the analysis to roll 9 at the end of the analysis. The y-displacement of the nodes at the right edge surface is constrained to represent the horizontal motion of the strip. Figure 12b shows the deformed shape of the mesh when the cross-section A reaches roll 9. Then, the constraints at the right edge surface of the strip are released for the strip to spring back. Figure 12c shows the deformed shape of the mesh after spring-back.

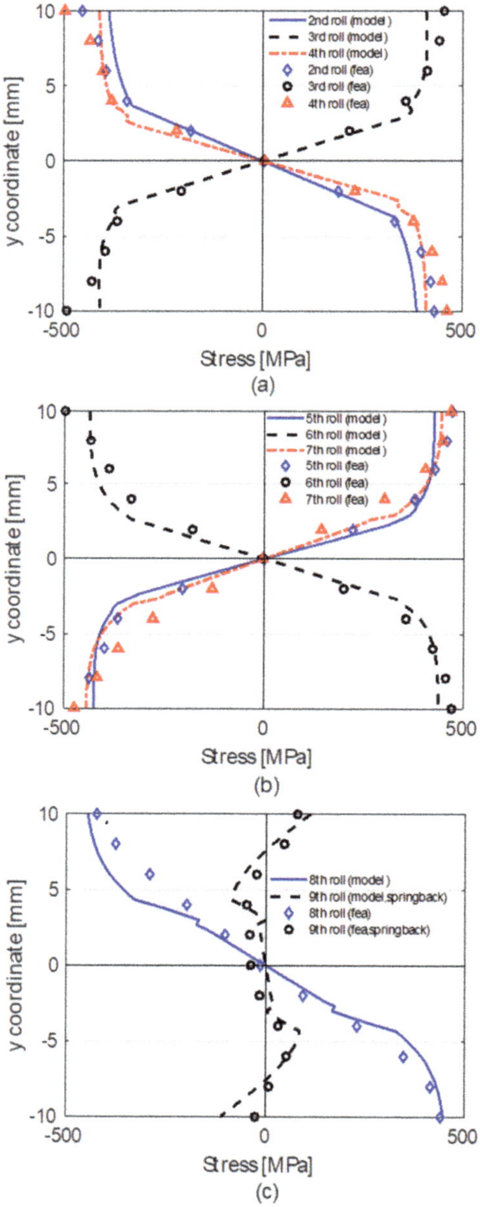

Figure 11. Stress distributions of the strip at (**a**) rolls 2, 3, and 4; (**b**) rolls 5, 6, and 7; (**c**) rolls 8 and 9.

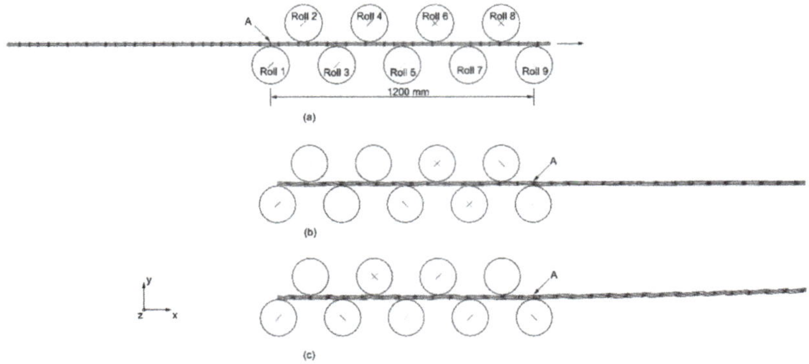

Figure 12. (**a**) The initial configuration, (**b**) a deformed configuration before spring-back, and (**c**) a deformed configuration after spring-back of the finite element model.

The stress distributions of the cross-section A at rolls 2–4, rolls 5–7, and rolls 8–9 based on the finite element analyses are shown in Figure 11a–c, respectively. The results based on the finite element analyses (FEA) are plotted with the markers in the figure. As shown in the figure, the analytical results agree with the results of finite element analyses. The discrepancy can be attributed to different contact conditions between the analytical model and the finite element model. Multiple contact points between the strip and each roll are observed in the finite element model. However, a single contact point between the strip and each roll is assumed in the analytical model.

4.3. Residual Curvature

Flatness is an important factor to evaluate the strip quality after roller leveling processes. The residual curvature of the strip at the exit roll can be used as a metric to evaluate the strip flatness after leveling. Figure 13 shows the residual curvatures κ' of the strip as a function of the averaged intermesh based on the analytical model and the finite element analyses. The averaged intermesh is the averaged value of the roll intermeshes, p_2, p_4, p_6, and p_8, where p_8 ranges from -1.2 mm to 0.8 mm. The residual curvature κ' is an indicator of the deviation of the strip from an initially flat surface at the entry roll. The curve with the dash-dot line and the markers represent the results based on the analytical model and the finite element analyses, respectively. The analytical predictions generally agree with the finite element analyses for the averaged intermeshes ranging from -1.71 mm to 0. As the averaged intermesh decreases from zero, the values of κ' decreases. κ' appears to exhibit an oscillatory behavior when the averaged intermesh is less than -0.11 mm based on the analytical model. As the averaged intermesh decreases further, the amplitude of the oscillation of κ' grows. When the averaged intermesh is less than -1.41 mm (analytical predictions), κ' has positive values of increasing magnitude. Five crossover points are observed at the averaged intermesh of -0.36 mm, -0.61 mm, -0.86 mm, -1.11 mm, and -1.41 mm (analytical predictions). Smith [27] reported that the several crossover points with zero residual curvature underlie the reason why successful leveling can be achieved by the series roll leveling process in practice.

In the analytical model, a point contact is assumed between the strip and each roll, which means the strip does not wind around the work rolls. This contradicts with the fact that multi-point contact between the strip and the roll predicted in the finite element analyses. Morris et al. [3] reported that the wrap angle near the exit roll has a significant influence on the flatness of the strip. Wrap-around contact length between the strip and the roll may depend on the intermesh and roll spacing. In describing the arc of contact of the strip around a roll, an effective radius can be assumed to model the wrap-around contact characteristics. The concept of the effective radius is illustrated schematically in Figure 14. Figure 14a shows the original contact model, where the strip contacts with roll i

tangentially. R_e is the expanded radius of the roll, which is defined by Equation (7). (x_i,y_i) is the ith contact point, and λ_i is the ith contact angle. Figure 14b shows that the strip makes contact elastically with roll i when subjected to an external force F_i. The circumferences of the deformed roll and the original roll are represented by the solid line and the dashed line, respectively, in Figure 14b. (x'_i,y'_i) is the ith contact point, and λ'_i is the ith contact angle for the Hertz contact model. A local deformation is ensued to cause a reduction in the local radius of roll i. The effective radius of the roll R_s is given as

$$R_s = R_e - sF_i \tag{20}$$

where s is the contact compliance of roll i. The external force F_i is computed from the moment distribution in the analytical model. Based on the Hertz contact model, the contact point (x'_i, y'_i) is

$$\begin{aligned} x'_i &= x_c + R_s sin\lambda'_i \\ y'_i &= y_c - R_s cos\lambda'_i \end{aligned} \tag{21}$$

where (x_c, y_c) is the coordinates of the center of the roll. As shown in Figure 14b, the deformed center line of the strip based on the analytical model with the Hertz contact compliance may be thought to have an arc segment bounded by the two virtual contact points, U and V, wrapped around the circumference of the original roll i, shown as a dashed circle in Figure 14b. When the contact angle λ'_i is very small which is the case in the roller leveler, the line segment and the arc segment bounded by the two end points U and V are approximately equal. The contact compliance of the rolls is taken as a fitting parameter in the analytical model with the Hertz contact compliance to fit the analytical predictions with the finite element computations. Yi et al. [28] considered the wrap angle on a roll during a roller leveling process by fitting an arc curve around a roll. A parameter determined by experiments is needed for curve fitting.

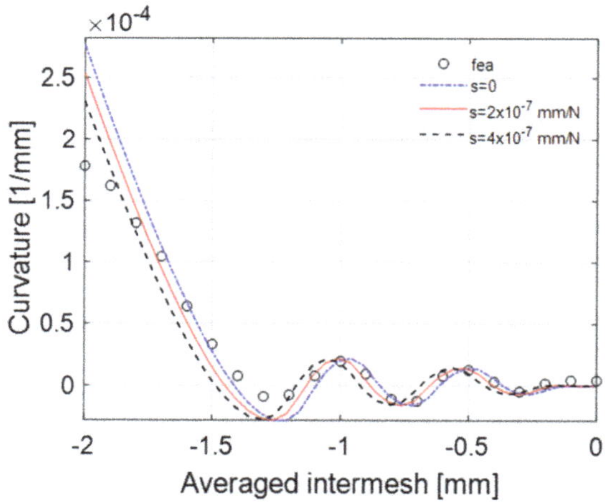

Figure 13. Residual curvature as a function of the averaged intermesh.

The residual curvatures κ' of the strip calculated by the analytical model with the Hertz contact compliance are plotted in Figure 13. The curves with the compliance $s = 0$ and 4×10^{-7} mm/N appear to envelop the results based on the finite element analyses. The curve with the compliance $s = 2 \times 10^{-7}$ mm/N may be able to predict the general trend of the results based on the finite element analyses for the averaged intermesh ranging from 0 mm to -2 mm. The wrap-around contact condition between the strip and the rolls can be manifested by the analytical model with the Hertz contact compliance within an acceptable accuracy, compared with the results of the finite element analyses.

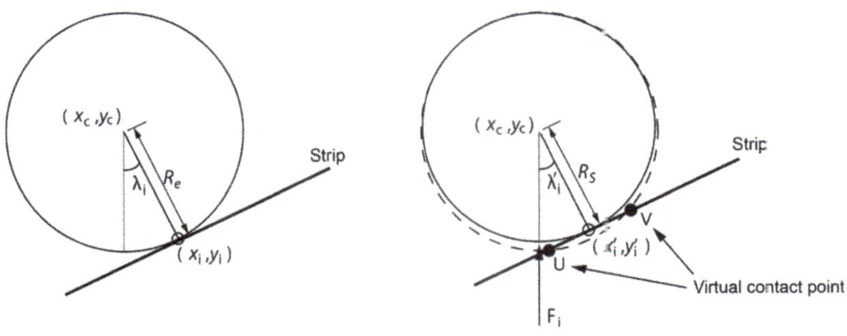

(a) (b)

Figure 14. Illustration of (**a**) the original contact model; (**b**) the Hertz contact model.

A series of simulations was run to calculate the residual curvatures κ' of the strip with an initial curvature κ_0 based on the analytical model with the Hertz contact compliance $s = 2 \times 10^{-7}$ mm/N. The initial stress in the strip is neglected without losing generality of the residual curvature predictions. Mathieu et al. [29] considered initial flatness defects in their finite element analyses of a leveling process. They introduced the flatness defects in the strip which was free of stress. Figure 15 shows the residual curvature as a function of the averaged intermesh for the strip with the initial curvature κ_0 varying between -1×10^{-4} mm^{-1} to 1×10^{-4} mm^{-1}. As shown in Figure 15, at low levels of roll intermesh (averaged intermeshes greater than −0.1), the residual curvatures κ' for the three cases of κ_0 = -1×10^{-4} mm^{-1}, 0, and 1×10^{-4} mm^{-1} deviate from each other significantly. For the averaged intermesh in this level, the curve for the case of κ_0 = -1×10^{-4} mm^{-1} oscillates mostly in the positive-curvature region ($\kappa' > 0$), in contrast to the cases of κ_0 = 0 and 1×10^{-4} mm^{-1}, which oscillate between the positive κ' region and the negative κ' region. The initially bowed-down defect for the strip with κ_0 =-1×10^{-4} mm^{-1} exits the leveler with the residual curvature in the same direction for the averaged intermesh greater than −0.71. As the extent of the averaged intermesh increases, the residual curvatures for the three cases of κ_0 = -1×10^{-4} mm^{-1}, 0, and 1×10^{-4} mm^{-1} gradually converge to the same values. For the values of the averaged intermesh less than −1.1, the curves for the three cases appear identical, where two cross over points with the interpolated averaged intermesh values of −1.46 and −1.13 were found. For the values of the averaged intermesh less than −1.46, positive residual curvatures were produced. For the averaged intermesh values within the interval of −1.46, −1.13, the minimum of the residual curvature is -0.27×10^4 mm^{-1}. In this region of the leveler settings, the residual curvatures of the strip seem to be insensitive to its initial curvatures. This result can serve to the advantage of leveler operators to obtain nearly zero residual curvature for strips with various initial curvatures. Grüber et al. [18] also demonstrated robustness of the roll intermesh settings for a roller leveler regarding a change in the initial curvature. In this investigation, the plane strain condition is considered in the finite element analyses. The width-to-thickness ratio of the strip considered in the analyses is 5. Carvalho et al. [30] reported that, in order to develop near plain strain conditions, it is important to maintain a ratio of width-to-thickness greater than 5.

A combined isotropic/kinematic hardening is implemented to describe the material hardening of the strip in this investigation. Doege et al. [5] also adopted combined isotropic/kinematic hardening for their leveling model, where mathematical formulations of their hardening model were not presented. Detailed formulations of the hardening model are provided in this investigation. Doege et al. [5] presented analysis results of their model. Results of the stress distributions and residual curvature of a steel strip based on our analytical model are verified by the finite element analyses. Doege et al. [5] computed the contact points between the strip and the rolls by assuming only one contact point between the strip and each roll. An effective radius modelling the wrap-around contact characteristics by the Hertz contact compliance is proposed to describe the arc of contact of the strip around a roll. A roll inter-mesh range to produce a flatness condition of the strip is presented based on the analytical model with the Hertz contact compliance.

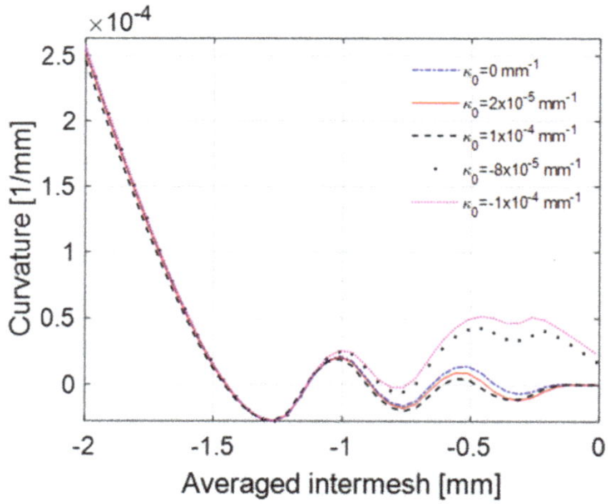

Figure 15. Residual curvature as a function of the averaged intermesh for the strip with an initial curvature.

Indeed, the initial curvature considered in the model is the longitudinal wave defect of a strip. Behrens et al. [6] sectioned a strip longitudinally and showed that the length of all sections after leveling should be the same in order to remove a transverse wave defect. Therefore, bendable rolls, as practiced in the industry, can be applied into a leveler to cause various degree of plastic deformation in each longitudinal section to achieve equal length. Chen et al. [9] developed an analytical model of a roller leveler with consideration to the bending of the rollers to eliminate transverse wave defects. This approach can be implemented in the analytical model to extend its applicability. Park and Hwang [13] slit a strip longitudinally to calculate the initial curvature of each longitudinal section. Given proper roll intermesh settings, the longitudinal sections with various initial curvatures can reach similar values of residual curvature after leveling based on finite element analyses and experiments. The results shown in Figure 15 based on the developed model also provide evidence that the residual curvatures of the strips with different initial curvatures can converge to the same value given enough amount of roll intermesh.

Given the multiple forming processes and complex machine settings involved in the roller leveling, tradeoffs between a simple, efficient model and an elaborate, detailed model should be balanced. Baumgart et al. [31] described that the effects of parts of the leveler, such as support rolls, frames, posts, and adjustment screws, should also be considered in order to obtain a more accurate leveler model. Wang and Li [32] reported that stiffness of roll cassettes and leveler housing are important factors of the leveling process. In this study, a relatively simple analytical model was developed based on the assumptions of two-dimensional geometry, pure bending of the strip, and the uniaxial loading condition. Compared to previously reported models, the Hertz contact compliance implemented in the model relaxed the single point contact condition between the strip and the roll, and a relatively accurate prediction of the residual curvature can be attained. The model could serve as a guide in the development of strategies for effective adjustment of roller levelers.

5. Conclusions

An analytical model for analyzing the residual stress distribution and residual curvatures of a strip during roller leveling is developed. The model verified by the finite element analyses is based on a simplified, two-dimensional geometric framework. The combined isotropic and kinematic material hardening model is implemented through a combined hardening parameter. The stress distributions after each bending/reverse bending and the residual curvature at the exit roll subjected to various leveler settings can be obtained by the developed model. The model was further enhanced by considering a Hertz contact model to eliminate the discrepancies between the model predictions and the finite element analyses. Based on the analytical model with the Hertz contact compliance, a range of the leveler settings can be determined to robustly obtain a strip with a nearly zero curvature at the exit roll.

Analytical models with combined isotropic/kinematic material hardening for roller leveling have been reported previously. A formulation of the change of the effective stress as a function of the change of the effective strain under cyclic bend loading is adopted in the developed analytical model with the combined hardening. The model is efficient and accurate in predicting the stress distributions and residual curvature of a steel strip. Unlike the assumption of a single contact point between the strip and the rolls, an effective radius modelling the wrap-around contact characteristics by the Hertz contact compliance is proposed in this investigation. An arc contact of the strip around a roll is described by the Hertz contact model. A roll intermesh range to produce a flatness condition of the strip can be obtained by the analytical model with the Hertz contact compliance.

Author Contributions: Conceptualization, S.-K.K. and D.-A.W.; methodology, D.-A.W.; software, Y.-L.O.; validation, D.-A.W.; formal analysis, D.-A.W.; resources, S.-K.K.; data curation, D.-A.W.; writing—original draft preparation, D.-A.W.; writing—review and editing, D.-A.W.; funding acquisition, S.-K.K. All authors have read and agreed to the published version of the manuscript.

Funding: This research was funded by China Steel Corporation, grant number 106T7D-RE18 and the Ministry of Science and Technology, Taiwan, grant number MOST 106-2221-E-005-038.

Data Availability Statement: Data presented in this article are available at request from the corresponding author.

Acknowledgments: The authors would like to express their appreciation to the National Center for High-Performance Computing (NCHC), Taiwan for their assistance.

Conflicts of Interest: The authors declare no conflict of interest.

References

1. Amor, A.; Rachik, M.; Sfar, H. Combination of finite-element and semi-analytical models for sheet metal leveling simulation. *Key Eng. Mater.* **2011**, *473*, 182–189. [CrossRef]
2. Li, S.-Z.; Yin, Y.-D.; Xu, J.; Hou, J.-M.; Yoon, J. Numerical simulation of continuous tension leveling process of thin strip steel and its application. *J. Iron Steel Res. Int.* **2007**, *14*, 8–13. [CrossRef]
3. Morris, J.W.; Hardy, S.J.; Lees, A.W.; Thomas, J.T. Some fundamental considerations for the control of residual flatness in tension levelling. *J. Mater. Processing Technol.* **2002**, *120*, 385–396. [CrossRef]
4. Hira, T.; Abe, H.; Azuma, S. Analysis of sheet metal bending deformation behavior in processing lines and its effectiveness. *Kawasaki Steel Eng. Rep.* **1988**, *19*, 54–62.
5. Doege, E.; Menz, R.; Huinink, S. Analysis of the levelling process based upon an analytic forming model. *CIRP Ann. Manuf. Technol.* **2002**, *51*, 191–194. [CrossRef]
6. Behrens, B.-A.; El Nadi, T.; Krimm, R. Development of an analytical 3D-simulation model of the levelling process. *J. Mater. Processing Technol.* **2011**, *211*, 1060–1068. [CrossRef]
7. Dratz, B.; Nalewajk, V.; Bikard, J.; Chaste, Y. Testing and modelling the behavior of steel sheets for roll levelling applications. *Int. J. Mater. Form.* **2009**, *2*, 519. [CrossRef]
8. Liu, Z.; Wang, Y.; Yan, X. A new model for the plate leveling process based on curvature integration method. *Int. J. Mech. Sci.* **2012**, *54*, 213–224. [CrossRef]
9. Chen, W.-H.; Liu, J.; Cui, Z.-S.; Wang, Y.-J.; Wang, Y.-R. A 2.5-dimensional analytical model of cold leveling for plates with transverse wave defects. *J. Iron Steel Res. Int.* **2015**, *22*, 664–671. [CrossRef]
10. Higo, T.; Matsumoto, H.; Ogawa, S. Influence of delivery-side roll position of roller leveler to plate flatness. In Proceedings of the 2016 AISTech Conference, Pittsburgh, PA, USA, 16–19 May 2016; pp. 2129–2137.
11. Morris, J.W.; Hardy, S.J.; Lees, A.W.; Thomas, J.T. Formation of residual stresses owing to tension levelling of cold rolled strip. *Ironmak. Steelmak.* **2001**, *28*, 44–52. [CrossRef]

12. Schleinzer, G.; Fischer, F.D. Residual stress formation during the roller straightening of railway rails. *Int. J. Mech. Sci.* **2001**, *43*, 2281–2295. [CrossRef]
13. Park, K.-C.; Hwang, S.-M. Development of a finite element analysis program for roller leveling and application or removing blanking bow defects of thin steel sheet. *ISIJ Int.* **2002**, *42*, 990–999. [CrossRef]
14. Huh, H.; Heo, J.H.; Lee, H.W. Optimization of a roller levelling process for Al7001T9 pipes with finite element analysis and Taguchi method. *Int. J. Mach. Tools Manuf.* **2003**, *43*, 345–350. [CrossRef]
15. Roberts, I.; Wang, C.; Mynors, D.; Adams, P.; Lane, K.; Unwin, P. Numerical analysis of strip-roll conformity in tension levelling. In Proceedings of the 10th International Conference of Technology of Plasticity (ICTP 2011), Aachen, Germany, 25–30 September 2011; pp. 315–319.
16. Jin, H.-R.; Yi, Y.-L.; Han, X.-Y.; Liang, Y. Roller straightening process and FEM simulation for stainless steel clad plate. *Open Mech. Eng. J.* **2014**, *8*, 557–561.
17. Kim, J.; Park, K.-C.; Kim, D.-N. Investigating the fluting defect in v-bending due to the yield-point phenomenon and its reduction via roller-leveling process. *J. Mater. Processing Technol.* **2019**, *270*, 59–81. [CrossRef]
18. Grüber, M.; Kümmelb, L.; Hirt, G. Control of residual stresses by roller leveling with regard to process stability and one-sided surface removal. *J. Mater. Processing Technol.* **2020**, *280*, 116600. [CrossRef]
19. Zhang, D.; Cui, Z.; Ruan, X.; Li, Y. An analytical model for predicting springback and side wall curl of sheet after U-bending. *Comput. Mater. Sci.* **2007**, *38*, 707–715. [CrossRef]
20. Kotov, K.A.; Bolobanova, N.L.; Nushtaev, D.V. Modeling the stress state of a steel strip with a roller leveling machine under cyclic alternating deformations. *Steel Transl.* **2020**, *50*, 750–755. [CrossRef]
21. Müller, U.; Krambeer, H.; Wolff, A.; Viella, A.E.; Richardson, A.D.; Perä, J.-O.; Luoto, P.; Weberm, W. *Optimisation of Final Plate Flatness by Set-Up Coordination for Subsequent Manufacturing Process (Final Plate Flatness)*; Final Report, EUR 25852 EN; Europe Commission: Brussels, Belgium, 2013.
22. Guan, B.; Zhang, C.; Zang, Y.; Wang, Y. Model for the whole roller leveling process of plates with random curvature distribution based on the curvature integration method. *Chin. J. Mech. Eng.* **2019**, *32*, 47. [CrossRef]
23. Hosford, W.F.; Caddell, R.M. *Metal Forming–Mechanics and Metallurgy*; Prentice Hall: Hoboken, NJ, USA, 1993.
24. Guan, B.; Zang, Y.; Wu, D.; Qin, Q. Study on mechanical behavior of thin-walled member during precision straightening process. *Sens. Transducers* **2014**, *179*, 36–42.
25. Guan, B.; Zang, Y.; Wu, D.; Qin, Q. Stress-inheriting behavior of H-beam during roller straightening process. *J. Mater. Processing Technol.* **2017**, *244*, 253–272. [CrossRef]
26. Yonetani, S. *The Engender Theory and Countermeasure of Residual Stress*; China Machine Press: Beijing, China, 1983.
27. Smith, R.P. The effect of the number of leveling rolls on the straightening process. *Iron Steel Technol.* **2007**, *4*, 57–68.
28. Yi, G.; Wang, Z.; Hu, Z. A novel modeling method in metal strip leveling based on a roll-strip unit. *Math. Probl. Eng.* **2020**, *2020*, 1486864. [CrossRef]
29. Mathieu, N.; Dimitriou, R.; Parrico, A.; Potier-Ferry, M.; Zahrouni, H. Flatness defects after bridle rolls: A numerical analysis of leveling. *Int. J. Mater. Form.* **2013**, *6*, 255–266. [CrossRef]
30. Carvalho, A.P.; Reis, L.M.; Pinheiro, R.P.R.P.; Pereira, P.H.R.; Langdon, T.G.; Figueiredo, R.B. Using Plane Strain Compression Test to Evaluate the Mechanical Behavior of Magnesium Processed by HPT. *Metals* **2022**, *12*, 125. [CrossRef]
31. Baumgart, M.; Steinboeck, A.; Kiefer, T.; Kugi, A. Modelling and experimental validation of the deflection of a leveler for hot heavy plates. *Math. Comput. Model. Dyn. Syst.* **2015**, *21*, 202–227. [CrossRef]
32. Wang, X.; Li, X. Research on force simulation of main leveler housing and roll cassettes in medium and heavy plate leveling. *J. Converg. Inf. Technol.* **2012**, *7*, 153–161.

metals

MDPI

Article

Research of Resistance of Selected Materials to Abrasive Wear to Increase the Ploughshare Lifetime

Monika Vargova [1,*], Miroslava Tavodova [1], Katarina Monkova [2,3,*] and Miroslav Dzupon [4]

[1] Faculty of Technology, Technical University in Zvolen, Studentska 26, 960 01 Zvolen, Slovakia; tavodova@tuzvo.sk
[2] Faculty of Manufacturing Technologies with a Seat in Presov, Technical University of Kosice, Sturova 31, 080 01 Presov, Slovakia
[3] Faculty of Technology, Tomas Bata University in Zlin, Vavreckova 275, 760 01 Zlin, Czech Republic
[4] Institute of Materials Research, Slovak Academy of Sciences, Watsonova 47, 040 01 Kosice, Slovakia; mdzupon@saske.sk
* Correspondence: monika.vargova@tuzvo.sk (M.V.); katarina.monkova@tuke.sk (K.M.)

Abstract: Road maintenance and cleaning in winter are performed with ploughshares. Due to the fact that the layer of snow and ice that is removed from the road surface contains various hard impurities, ploughshares are exposed to high intensity abrasive wear. This article deals with the resistance to abrasive wear of originally used ploughshare materials and the materials that were designed as a suitable modification of the ploughshare to increase its service life. The chemical composition of materials used to manufacture ploughshare components is unknown. For this reason, they were analyzed with an ARL 4460 spectrometer, which was used to analyze the element content. The main part of the research was focused on the abrasion resistance test, which was performed according to the GOST 23.208-79 standard. Based on the chemical analysis, it was found that the basic body of the ploughshare was made of S355J2G3 steel, and the raking blade material was made of 37MnSi5 steel. The original material (steel S355J2G3) of the ploughshare body as a reference standard was compared to steel HARDOX 450. Furthermore, a sample made of the original material of the raking blade (steel 37MnSi5) was used as a reference standard, the properties of which were compared to the newly designed OK 84.58 and UTP 690 hardfacing materials. The parametric test method of statistical hypotheses was also used to process and evaluate the weight losses of the selected materials.

Keywords: abrasive wear; wear resistance; ploughshare; hardfacing materials; hypothesis testing

Citation: Vargova, M.; Tavodova, M.; Monkova, K.; Dzupon, M. Research of Resistance of Selected Materials to Abrasive Wear to Increase the Ploughshare Lifetime. *Metals* **2022**, *12*, 940. https://doi.org/10.3390/met12060940

Academic Editor: George A. Pantazopoulos

Received: 27 April 2022
Accepted: 26 May 2022
Published: 30 May 2022

Publisher's Note: MDPI stays neutral with regard to jurisdictional claims in published maps and institutional affiliations.

1. Introduction

Forest roads are important in terms of making forests accessible for heavy machinery and timber transportation. Therefore, it is necessary to maintain these roads both in summer and winter. It is necessary to remove a layer of snow or ice from the road surface in the winter time. Snow is removed from the road using a snow ploughshare. Since the removed layer of snow contains various hard impurities, the ploughshares are exposed to high—and especially abrasive—wear and their replacement is relatively economical. Therefore, it is necessary to look for ways to increase their service life and examine their resistance to abrasive wear.

Wear is one of the main factors affecting the performance and lifetime of the components of various mechanisms [1]. Abrasive wear is one of the dominant and common tribological wear processes [2–4] leading to expensive costs for repairing worn mechanisms and replacing spare parts [1]. This wear is characterized by a reduction in the volume of the material and a change in its shape, causing a decrease in work efficiency [5]. Wear caused by shocks and abrasion by hard abrasive particles is a major problem in many industries, especially in agriculture, forestry, mining, mineral processing, etc. [6,7].

Abrasive wear occurs when a hard rough surface slides across a softer surface; in this case, wear is defined as damage to a solid surface that generally involves the progressive loss of material and is due to relative motion between that surface and a contacting substance or substances.

The material from grooves is usually removed in the form of loose particles [8–10]. The material resistance against abrasive wear is higher with higher hardness [5,11].

This type of wear is caused by sharp particles that slide or flow over the metal surface at different speeds and pressures, grinding the material from the surface like small cutting tools. The harder the particles and the sharper their shapes, then the more intense the abrasion is. Typical cases of abrasive wear are encountered in earthworks, material transportation and agricultural equipment [12,13].

During abrasive wear, the material is removed from the surface when the stress in it reaches a critical value. This local fracture stress can be achieved by one or more abrasive particles coming into contact with the tool surface. Material that is not removed in this way can be deformed due to abrasion as well as impact [14,15].

One of the very effective measures to increase wear resistance is to protect functional surfaces with a suitable coating material [16]. Hardfacing is a commonly used method to improve the surface properties of agricultural tools, mining components and soil preparation equipment, among others [17]. The alloy is applied to the surface of the base material (usually on low-carbon or medium-carbon steels) by hardfacing in order to increase the hardness and wear resistance without a significant loss in the ductility and toughness of the base material [16,17].

Important factors that determine the resistance to abrasive wear include the hardness, size, shape and intensity of the particles; as well as the hardness, shape, size and number of hard phases and their distribution in the parent metal. Wear resistance increases with the increasing hardness of hard structural components (carbides, borides, etc.) and with their increasing share in the structure [9,12].

The main components of Fe-based alloys are chromium, molybdenum and boron, which make Fe-based alloys resistant to wear. Wang et al. studied the microstructure, hardness and wear resistance of Fe-based alloys by adding the elements ferrotitanium (Fe–Ti), ferromolybdenum (Fe–Mo), ferrovanadium (Fe–V) and graphite, which were applied by the arc welding process. The results showed that the hardness and resistance of the hardfacing metal layer increase with an increasing proportion of Fe-Ti, Fe-V, Fe-Mo and graphite. Their amounts must be controlled in the range of 8–10% graphite, 12–15% Fe-Ti, 10–12% Fe-V and 2–4% Fe-Mo. However, cracks begin to form in the layer of hardfacing metal if the amount of Fe-Mo is more than 5% [18,19].

Xu et al. studied the effect of the carbide type on the resistance to abrasive wear at different sizes and abrasive particle loads. The results showed that M_6C carbides showed higher resistance to wear by fine abrasive particles compared to M_2C [20].

During impact wear, the material is exposed to impact and high pressure, due to which it deforms or locally breaks—it can even crack. Due to the relatively high working speeds of ploughshare vehicles, it is necessary to protect the ploughshare from the effects of the impact of the ploughshare hitting with a solid obstacle, e.g., ice and abrasion by mineral particles, gravel and stones [9,12].

For this reason, it is necessary to pay attention to the possibilities of increasing the life of the blades and examine their resistance to abrasive wear, which was also the main goal of the authors' long-term research.

Based on the experience and study of already published outputs in the field, the authors tried to bring new solutions to the issue [16,21]. To the best of the authors' knowledge, no relevant studies have yet been published that compare the resistance of a selected combination of materials (37MnSi5, S355JG3 and HARDOX 450 steels, as well as OK 84.58 and UTP 690 hardfacing materials) to abrasive wear. The achieved results will be contributing and applicable not only to ploughshares, construction machines and forest manipulators, but can also be used in the mechanical engineering, civil and mining industries.

2. Materials and Methods

The ploughshare (Figure 1), the resistance of which to abrasive wear is the subject of our research, is a commonly used tool for modifying forest roads during wintertime. The ploughshare works in a heterogeneous environment, coming into contact with the road and thus with snow, stones, mineral particles, gravel, sand, etc. First, the raking blade comes into contact with the road. It is stressed by the impact of the first contact with the road and with corrosion and abrasion during the raking. In addition, the rest part of the ploughshare is also stressed because the accumulated snow, together with sand and gravel, creates a force on the ploughshare body from the point of pick-up until it is pushed to the side of the road. As a result, wear occurs due to abrasion and corrosion not only on the raking blade but on the entire working part of the ploughshare (weldment).

Figure 1. The arrow ploughshare [22].

The basic dimensions of the currently used steel raking blades and their geometry, symmetrically placed in the shape of an arrow on the front part of the tractors, are shown in Figure 2. The values of these basic dimensions, declared by the manufacturer [23], are listed in Table 1.

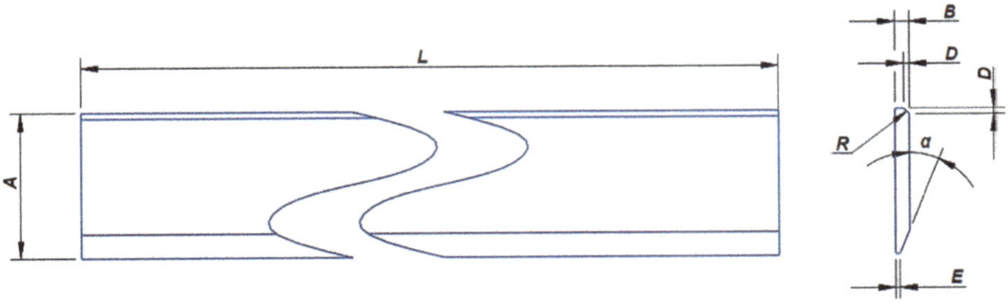

Figure 2. Basic dimension of steel raking blade.

Table 1. Basic dimension of steel raking blade [23].

A (mm)	*B* (mm)	α (°)	*R*/*D* (mm)	*E* (mm)	*L* (mm)
80–400	12–50	20.2–24	R3/5–12	3–28	6000–6300

2.1. Materials of Research Samples

The raking blades of the ploughshares can be attached to the ploughshare body (to the base material of the ploughshare) in two ways:

- Threaded joint;
- Welded joint.

Due to the rapid wear of the ploughshare and the need for its frequent replacement, the operator of such ploughshares initiated research that would have increased their lifetime and ensured greater efficiency in the use of funds spent on their purchase and replacement.

For the ploughshare that was the subject of the research, the raking blade was attached by a weld, so it can be said that the ploughshare consisted of 3 main parts, namely the basic ploughshare body, the raking blade and the weld (Figure 3a). There is a detail of the raking blade which is that its significant damage due to the heterogeneity of the working environment is visible in Figure 3b.

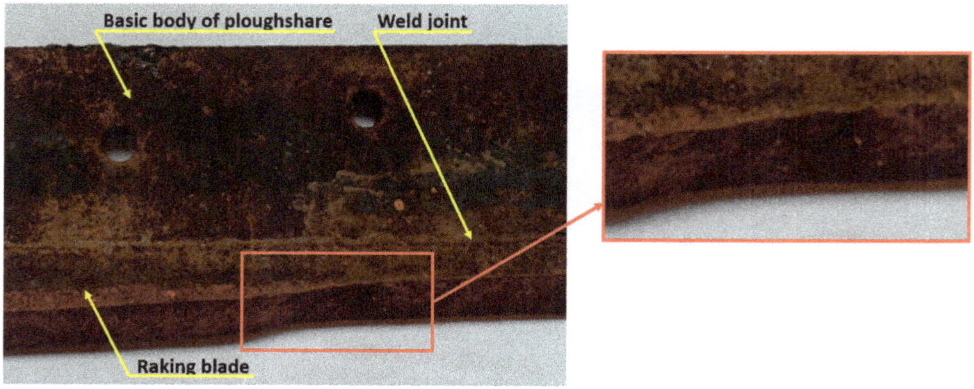

(a) (b)

Figure 3. Part of the discarded ploughshare: (**a**)—main parts of the ploughshare; and (**b**)—detail of a raking blade).

The chemical composition of materials used to manufacture ploughshare components is unknown. From this concern, their elemental analyses were carried out using ARL 4460 spectrometer. The results of the chemical analysis of the individual parts of the originally used ploughshare are shown in Table 2.

Table 2. Chemical composition of the individual parts of the original ploughshare.

Element	C	Mn	Si	Cr	Ni	Cu	Mo	P	S	Fe
Basic body of the ploughshare (wt.%)	0.2	1.4	0.3	0.04	0.055	0.16	<0.01	<0.005	<0.15	balance
Base weld metal (wt.%)	0.06	1.3	0.55	-	-	-	-	-	-	balance
Raking blade (wt.%)	0.33	1.24	1.36	-	-	0.04	0.178	0.017	<0.15	balance

Based on the performed chemical analysis, it is possible to state that:

- The basic body of the ploughshare was made of S355J2G3 steel. It is the steel of the usual quality for low temperatures with guaranteed weldability. It is used in parts of equipment operating at temperatures up to −50 °C made of sheet metal with a guaranteed value of notch toughness up to −50 °C. The yield strength is Rm = 355 MPa and the ultimate tensile strength is Re = 470–610 MPa [24]. The chemical composition of the tested steel, according to the data from the material sheet, is shown in Table 3;
- The base weld metal—an electrode for general use, suitable for welding carbon and low-alloy steels—was used for welding the basic ploughshare body with a raking blade. The weld metal of the electrode is tough and resistant to cracking [14];
- The raking blade was made of 37MnSi5 material (Wr. Nr. 1.5122). It is manganese-silicon steel suitable for tempering. It is used on medium-stressed machine parts and particularly wear-resistant road vehicle parts, e.g., shafts, axles, connecting rods,

levers, bolts, etc. The ultimate tensile strength of 37MnSi5 steel is Rm = max. The 760 MPa and hardness 217 HBW [25]. The chemical composition of the tested steel, according to the data from the material sheet, is shown in Table 3.

The ploughshare that was the subject of the research worked for 6–8 weeks. After this time, the raking blade of the ploughshare was worn and unusable. The frequent replacement of the raking blades caused economic and technical problems. Due to the fact that the currently used ploughshare was functionally usable only for a short time and did not meet the requirements for the lifetime of this type of product, it was necessary to design a new solution that improved the situation. Based on the authors' own research and practical experience, as well as on the results described in the professional and scientific literature presented in the Introduction, the following options are suggested to increase the lifetime of the ploughshare and thus increase its resistance to abrasive wear (Figure 4):

(a) Change of ploughshare material for HARDOX 450 material;
(b) Application of the OK 84.58 hardfacing material to the exposed surfaces of the functional parts of the ploughshare;
(c) Application of the UTP 690 hardfacing material to exposed areas of functional parts of the ploughshare.

The first alternative solution was the idea of replacing the original ploughshare with one compact body made of HARDOX material, replacing all three parts of the ploughshare: the ploughshare body, the weld metal, and the raking blade. HARDOX steels are produced in several designs. HARDOX sheets are considered a world-wide standard in the field of abrasion-resistant steel due to their unique combination of hardness and toughness. The hardness of this steel ranges from 400 HBW to 600 HBW. It is available in the form of sheet metal in thicknesses from 0.7 mm to 160 mm but also in the form of tubes and rods [9,26]. HARDOX 450 material is an abrasion-resistant steel with excellent construction properties. The hardness of this steel is 450 HBW, and the yield strength is Re = 1100–1300 MPa. HARDOX 450 material is a well flexible material that has guaranteed weldability. It provides good abrasion resistance and thus a longer lifetime. It has better wear resistance [27], higher load capacity, and a longer lifetime compared to ordinary steel with the same hardness. It is used in various components and constructions that are subject to wear [9,27,28]. The chemical composition of HARDOX 450 material is shown in Table 3.

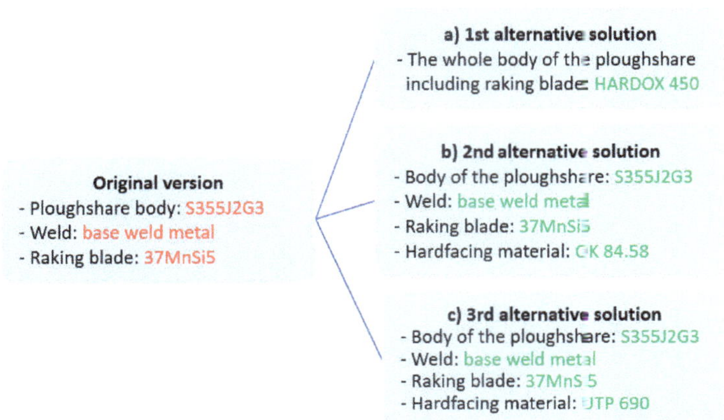

Figure 4. Suggested research solutions regarding the current state.

The second alternative solution was to apply a new layer in the form of a hardfacing material to the original 37MnSi5 steel of the raking blade part (see Figure 4). Such a sample with OK 84.58 hardfacing material (Figure 5a) was prepared in the company ŽOS Zvolen, a.s. in a single layer by certified welders. The electrode was dried for 2 h at 200 °C without preheating the sample. The applied hardfacing voltage was U = 21.5 V and the hardfacing current was I = 140 A. The stated parameters were chosen according to the experience of experts from the company.

(**a**) (**b**)

Figure 5. Sample with hardfacing materials; (**a**) OK 84.58; and (**b**) UTP 690.

The OK 84.58 electrode is a high-yield electrode for hardfacing functional surfaces resistant to wear under simultaneous shock stress with the required partial corrosion resistance, used, e.g., for parts of agricultural and forestry machines, mixers, transport equipment, etc. It forms a martensitic structure after hardfacing. It is possible to treat hardfacing materials by grinding [29,30]. The chemical composition of the OK 84.58 electrode is shown in Table 3.

The sample with hardfacing material UTP 690 (Figure 5b), prepared in the company ŽOS Zvolen, a.s. by certified welders, was used as the third alternative solution (see Figure 4). The hardfacing layer was applied to the original 37MnSi5 material of the raking blade part. The electrode was dried for 2 h at 300 °C with the preheating of the sample to 230 °C. The applied hardfacing voltage was U = 21.5 V, and the hardfacing current was I = 105 A. In this case, the parameters were also designed according to the experience of the experts from the company.

The UTP 690 electrode is used for the repair and production of cutting tools, especially for the restoration of cutting edges and work surfaces. The coating is highly resistant to friction, compression and impact even at elevated temperatures up to 550 °C [31]. Using this electrode, it is also possible to produce new tools by hardfacing on carbon and low-alloyed base steels [31]. This forms a martensitic structure after hardfacing [16]. The chemical composition of the UTP 690 electrode is shown in Table 3.

Table 3. Chemical composition of S355J2G3 [24], 37MnSi5 [25], HARDOX 450 [27] steels and OK 84.58 [29] and UTP 690 [31] hardfacing materials according to the datasheets.

Material	Element (wt.%)													
	C	Mn	Si	Cr	Ni	Cu	Cr+Ni+Cu	Ti	P	Mo	B	S	V	W	Fe
S355J2G3	max. 0.20	max. 1.40	max. 0.55	max. 0.30	max. 0.30	max. 0.30	max. 0.70	max. 0.20	max. 0.03	-	-	max. 0.03	-	-	balance
37MnSi5	0.33–0.41	1.10–1.40	1.10–1.40	-	-	-	-	-	max. 0.035	-	-	max. 0.035	-	-	balance
HARDOX 450	max. 0.23	max. 1.60	max. 0.5	max. 1.20	max. 0.25	-	-	-	max. 0.025	max. 0.25	max. 0.005	max. 0.010	-	-	balance
OK 84.58	0.67	0.70	0.7	10.40	-	-	-	-	-	-	-	-	-	-	balance
UTP 690	0.90	0.50	0.80	4.50	-	-	-	-	-	-	-	-	1.20	2.00	balance

All samples, both standard (S355J2G3 and 37MnSi5) and reference ones (HARDOX, OK 84.58 and UTP 690), were prepared according to the GOST 23.208-79 standard by abrasive water jet cutting technology (AWJM), machined by milling and ground by a magnetic surface grinder for obtaining dimensions 30 mm × 30 mm × 10 mm and surface roughness

Ra = 0.4 μm (Figure 6). OTTAWA SiO_2 silica sand with a grain size of 0.1–0.3 mm was used as an abrasive material [31]. The hardness corresponds to the hardness value according to Vickers 500 HV and Rockwell 54 HRC.

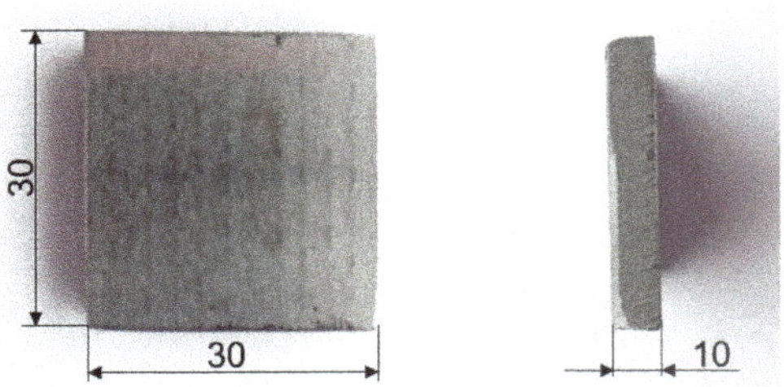

Figure 6. Sample for abrasion resistance test with marked dimensions.

2.2. Testing Methods

There are several methods for testing the wear resistance of materials. Based on the ploughshare operating environment, the method simulating its load in the laboratory conditions at the Technical Faculty of CULS in Prague was chosen. The abrasion resistance test was performed according to standard GOST 23.208-79—Ensuring wear resistance of products. The wear resistance testing of materials by friction against loosely fixed abrasive particles [32].

The principle of the method is to compare the loss of test material and the loss of standard material under the same test conditions. The abrasive material used is electrocorundum with a grain size of 100–250 μm [33] and with a maximum relative mass moisture content of 0.15%. The standard further states that when assessing wear resistance under specific wear conditions, it is permitted to use abrasive material corresponding to the material used during operation but with a maximum grain size of 1.0 mm [32].

Each test sample (standard, test sample) is weighed before the test and placed in the test equipment. Subsequently, the abrasive supply is started, and the rubber disc is pressed against the test sample. A schematic diagram of test equipment for testing the abrasive resistance of materials with a sample ready for testing is shown in Figure 7.

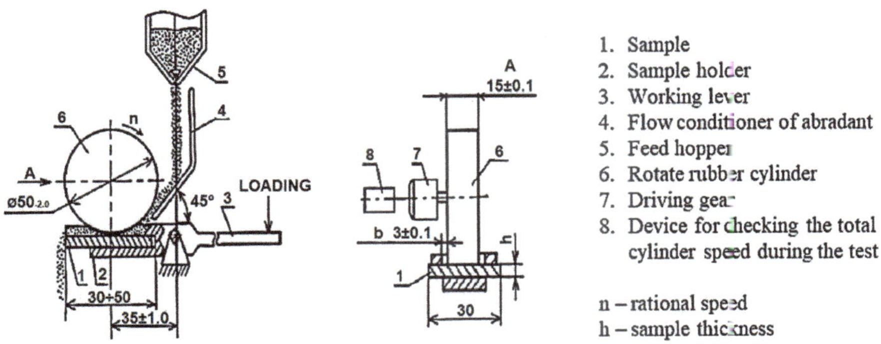

Figure 7. Scheme of the test equipment with a test sample [32].

The test equipment Tester T-07, ITC PIB Institution of Technology Radom and the BT-16 Controller (Instytut Technologii Eksploatacji, Radom, PL) used in the presented research are shown in Figure 8b,c, together with the scales (Figure 8a). Three samples were used for each material. After each cycle, the sample was weighed three times on Kern ABS analytical scales with a maximum weight of 120 g and an accuracy of g = 1.5 mg. The parameters of the test equipment and samples are shown in Table 4.

(a) (b) (c)

Figure 8. Abrasion resistance and weight measuring equipment (scales); (**a**)—scales; (**b**)—abrasion resistance equipment; (**c**)—detail of a driving gear with abrasive particles.

Table 4. Parameters of the test equipment.

Length of the Friction Path in One Cycle R (m)	153.6
Diameter of rubber cylinder D (mm)	48.9
Compression force F (N)	15.48
Cylinder revolutions in one cycle (revolutions)	1000
Abrasive material	OTTAWA silica sand

The sample is taken and weighed again after passing the track. Weight loss was calculated as a mean value based on the measurements of each sample.

The hardness coefficient K_T (-) is calculated from Equation (1) [32] as follows:

$$K_T = \frac{H}{H_a}\ (-) \tag{1}$$

where H—standard material hardness, which is the reference sample (HRC), and H_a—abrasive hardness (HRC).

The relative abrasion resistance is calculated from Equation (2) [12,32] as follows:

$$\Psi_h = \frac{W_{hE}}{W_{hPV}}\ (-) \tag{2}$$

where W_{hE}—weight loss standard (g); W_{hPV}—a weight loss of the reference sample (g).

2.3. Evaluation Methods

TESCAN VEGA3 electron microscope (Tescan Orsay Holding, a. s., Brno, CZ) was used for microscopy. The samples were etched in a solution of HNO_3 in ethyl alcohol—2% Nital and in etchant Cor (120 mL CH_3COOH, 20 mL HCl, 3 g picric acid, 144 mL CH_3OH).

The parametric test method of statistical hypotheses was chosen to verify the established hypothesis of the weight loss of materials 37MnSi5, S355JG3, HARDOX 450 and OK 84.58 and UTP 690 hardfacing materials. STATISTICA 12 software (Version number 1.0-0, StatSoft, Inc., Tulsa, OK, USA) was used to test statistical hypotheses.

The significance level of the test is chosen to be $\alpha = 0.05$. It is necessary to test the equality of the variances before testing the difference between the two means. Additionally, therefore [34,35]:

$$H_0 : \sigma_1^2 = \sigma_2^2 \tag{3}$$

against

$$H_1 : \sigma_1^2 \neq \sigma_2^2 \quad (4)$$

where σ_1^2—variance of material weight loss values of standards S355J2G3, 37MnSi5; σ_2^2—variance of material weight loss values of reference samples HARDOX 450 material, OK 84.58 and UTP 690 hardfacing materials.

A test of the two means difference was then performed The software calculates the appropriate p-level. In order to verify the hypothesis on the basis of the graph, that the average value of the weight losses of the standard material is greater or smaller than the material of the reference sample. In this case, the software-calculated p-level must be divided by 2, as the software works with two-way hypothesis testing.

If the value of the p-level is greater than α, then H_0 is not rejected, and thus the equality of the variances is confirmed. Conversely, if the p-level value is less than α, then H_0 is rejected in favor of H_1.

3. Results and Discussion

3.1. Preliminary Research

Individual materials were analyzed in terms of microstructure during the initial phase of the research. Figure 9a shows the microstructure of the standard material of 37MnSi5. The samples were etched in a solution of HNO_3 in ethyl alcohol—2% Nital. The microstructure of the material was predominantly sorbite, referring to the fact that the raking blade was heat treated-hardened and then tempered at a high temperature.

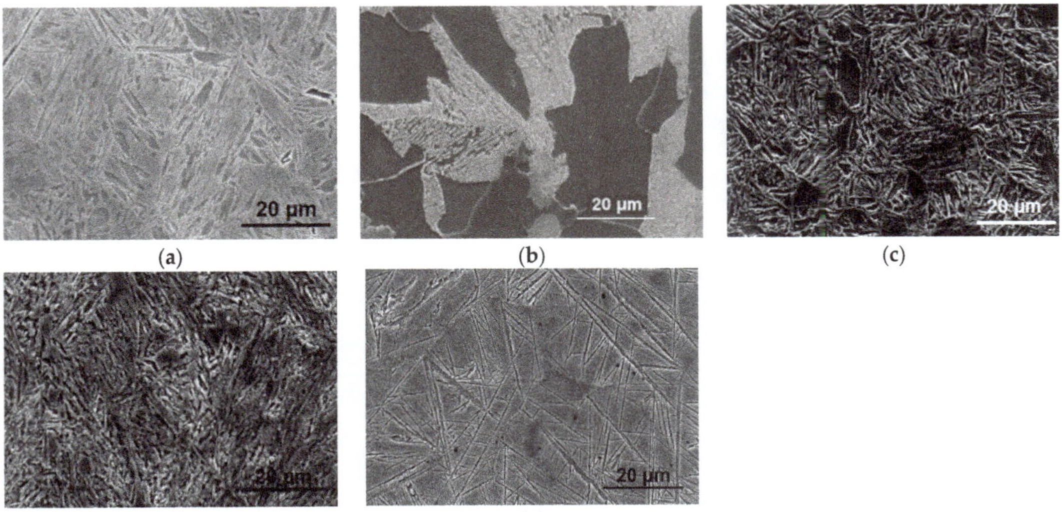

(a) (b) (c) (d) (e)

Figure 9. Microstructure of the materials; (**a**) 37MnSi5; (**b**) S355J2G3; (**c**) HARDOX 450; (**d**) hardfacing material OK 84.58; (**e**) hardfacing material UTP 690.

Figure 9b shows the microstructure of the S35J2G3 material. A ferritic-pearlitic structure was observed, with a predominance of ferrite, corresponding to the type of microstructure of structural low carbon steels. We used 2% Nital as an etchant. This part of the ploughshare is not directly exposed to abrasion in contact with the road. It is a supporting part for the raking blade connected to the base material by a weld. However, with a larger amount of material to be cleaned, this part also comes into contact with the abrasive particles but with a lower degree of attack on its surface and is thus exposed to an abrasive load. The ferritic-pearlitic structure is characterized by a low degree of abrasive resistance.

This was the reason for the idea of replacing such a weldment with one unit made of abrasion-resistant material.

Figure 9c shows the microstructure of HARDOX 450 material. The sample was etched with Cor etchant (120 mL CH_3COOH, 20 mL HCl, 3 g picric acid, 144 mL CH_3OH). The fine-grained structure corresponds to the state after heat treatment-hardening and tempering, which is stated and declared by the manufacturer in his technical sheet [27].

Figure 9d shows the microstructure of OK 84.58 hardfacing coating. The samples were also etched with Cor etchant. The microstructure corresponds to the chemical composition with a high Cr and medium C content. These elements create a precondition for the formation of carbides and intermediate phases, hardening the softer ferritic matrix. This increases the hardness and also partly the abrasion resistance while maintaining the good toughness of the material.

Figure 9e shows the microstructure of the UTP 690 hardfacing coating. Samples were also etched with Cor etchant. The microstructure of UTP 690 material is very similar to the microstructure of OK 84.58 material. The hardfacing material UTP 690, with its chemical composition represented by Cr and W, increases wear resistance and hardness. The elements V and W refine the grain, which also promotes the abrasion resistance of the material where V increases the toughness needed to balance the increased hardness with carbides C, Cr and W. Compared to OK 84.58 hardfacing material, we can observe a needle-like structure in the UTP 690 hardfacing material, which could better transfer the loads that act on it and is thus a premise for better resistance to abrasive wear.

Microstructural analyses confirmed that the choice of hardfacing materials, which due to their chemical composition, create different and more suitable structures than the analyzed structures of the original ploughshare materials, was correct. This creates the premise of better abrasion resistance and impact resistance on the ploughshare.

The hardness was measured after verifying the microstructure of the materials. Hardness is the ability of a metallic material to resist the ingress of a foreign body. Therefore, the hardness of the material is an important value providing information on the resistance of the material to abrasive wear. At the same time, hardness values are needed to calculate K_T hardness coefficients. The hardness of steel 37MnSi5, S355J2G3, HARDOX 450 and OK 84.58 and UTP 690 hardfacing materials was measured according to Vickers and Rockwell. Vickers hardness was measured according to the procedure stated in ISO 6507-1: 2018 [36] with a Vickers 432SVD hardness tester. The duration of the load was t = 15 s, and the loading force was F = 98.07 N. The Rockwell hardness was measured according to the procedure stated in ISO 6508-1: 2016 [37], with a UH250 hardness tester, and the loading force was F = 1471 N. Vickers hardness was used because it is more sensitive to small differences in material hardness. Rockwell hardness was used to better compare harder hardfacing materials. The average values of measured hardnesses obtained from ten measurements are shown in Table 5.

The measured hardness of 37MNsi5 steel (raking blade) was 538 HV0.5 and 50 HRC. The hardness of S355J2G3 steel (raking blade body) was 168 HV0.5 and 18 HRC. During its work, the ploughshare is stressed by abrasive wear not only on the raking blade but also on its supporting part. The idea of modifying the ploughshare within the meaning of changing the weldment into a solid piece of HARDOX 450 steel was based not only on removing the connecting part acting as a notch under stress, but also increasing the hardness of the new material compared to S355J2G3 material. The hardness of the HARDOX 450 material is higher compared to the S355J2G3 material, its value according to Vickers is 519HV0.5 and Rockwell is 45HRC. The aim of hardfacing additional materials to the raking blade was to increase the lifetime of the raking blade (37MnSi5) of the ploughshare. The OK 84.58 hardfacing material was the same hardness as the 37MnSi5 material, and its value was 546HV0.5 and 50 HRC. Significantly higher hardness compared to the 37MnSi5 material was achieved only by hardfacing material UTP 690, with which the hardness value was 677HV0.5 and 62 HRC. The results of the material hardness measurements predicted that the UTP 690 hardfacing material could be a suitable solution for ploughshare modification.

Table 5. Measured hardness values.

Hardness Type	37MnSi5	S355JG3	HARDOX 450	OK 84.58	UTP 690
Vickers hardness (HV 0.5)	538 ± 40	168 ± 20	519 ± 30	546 ± 40	677 ± 40
Rockwell hardness (HRC)	50 ± 2	18 ± 2	45 ± 2	50 ± 2	62 ± 2

3.2. Abrasive Wear Resistance Analysis

The abrasion resistance test was firstly performed on standard samples of 37MnSi5 and S355J2G3 materials. Subsequently, the test was also performed on reference samples made of HARDOX 450 steel and with OK 84.58 and UTP 690 hardfacing materials. In Figure 10, samples after an abrasive wear test with traces of a rubber disc are presented.

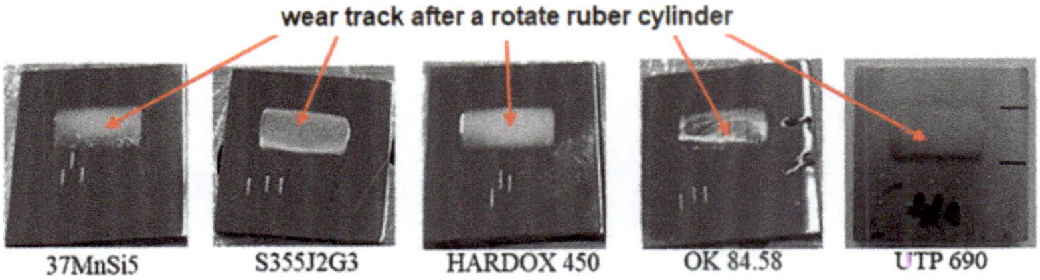

Figure 10. Samples after abrasive wear test.

For each material, three samples were used for the abrasion test. After each cycle, the sample was weighted three times on Kern ABS analytical scales, while the average values of the weight losses that were used for the next processing are shown in Table 6.

Table 6. Average weight losses of 37MnSi5, S355J2G3, HARDOX 450 steels and OK 84.58 and UTP 690 hardfacing materials.

Tracked Distance R (m)	Weight Loss (g)				
	37MnSi5	S355J2G3	HARDOX 450	OK 84.58	UTP 690
153.6	0.0062	0.0158	0.0133	0.0062	0.0022
307.2	0.0088	0.0223	0.0108	0.0075	0.0009
460.8	0.0067	0.0208	0.0091	0.0078	0.0013
614.4	0.0072	0.0145	0.0077	0.0062	0.001
768.0	0.0072	0.0126	0.0066	0.0078	0.0011
921.6	0.0043	0.0197	0.0069	0.0072	0.0010
1075.2	0.0042	0.0133	0.0081	0.0072	0.0008
1228.8	0.0098	0.0141	0.0076	0.0070	0.0009
1382.4	0.0053	0.0131	0.0067	0.0050	0.0012
1536.0	0.0056	0.0118	0.0068	0.0062	0.0012
1689.6	0.0061	0.0141	0.0072	0.0053	0.0038
Average weight loss W_{hi}	0.0064	0.0157	0.0082	0.0066	0.0014

Equation (1) [16] was used to calculate the hardness coefficients K_{TE} and K_{TV} of the standard materials 37MnSi5 and S355J2G3 and reference samples HARDOX 450, OK 84.58 and UTP 690 materials. The resulting values are shown in Table 7.

Table 7. Hardness coefficient values K_T.

Sample Material	37MnSi5	S355J2G3	HARDOX 450	OK 84.58	UTP 690
Hardness coefficient K_T	0.93	0.33	0.83	0.93	1.15

If this coefficient is higher than 1, it is assumed that the material and the hardfacing material can withstand abrasive particles. It is clear from the graph that the largest value of the K_T hardness coefficient was reached by the UTP 690 hardfacing material, namely 1.15. Other steels (37MnSi5, S355J2G3, HARDOX 450) and hardfacing material (OK 84.58) reached a value of K_T inferior to 1. The S355J2G3 material of the ploughshare body reached the lowest value of K_{TE2} = 0.33 in comparison with all examined materials and the standard. Based on this result, it would also be appropriate to consider changing the base material to a HARDOX one.

The relative abrasion resistance Ψ_h was then calculated according to Equation (2) from the weight losses of W_h. The relative resistance to abrasive wear was also calculated for the S355J2G3 basic material of the ploughshare, as it is also exposed to abrasive wear during raking [8]. The resulting values are shown in Table 8.

Table 8. Relative abrasion resistance Ψ_h.

Sample Material	37MnSi5	S355J2G3	HARDOX 450	OK 84.58	UTP 690
Relative abrasion resistance Ψ_h	1	1	1.91	0.97	4.57

Materials 37MnSi5 and S355J2G3 have the value Ψ_h = 1 because they are standard materials. The relative resistance to abrasive wear is the ratio of the weight loss of the standard material to the weight loss of the material, respectively, of the reference sample. If this coefficient is higher than 1, it is assumed that the material and the hardfacing material of the reference sample can withstand abrasive wear better than the material of the standard sample. Based on the test of resistance to abrasive wear, we can state that the sample from HARDOX 450 material achieved almost twice as good results as the standard sample from S355J2G3 steel.

As the abrasion resistance increases in direct proportion to the hardness of the material [5,9] by using higher grade HARDOX abrasion resistant steel with higher hardness, we can achieve even better abrasion resistance. With OK 84.58 hardfacing material, we achieved almost the same resistance to abrasive wear as the standard sample made of 37MnSi5 material. The highest value of Ψ_h was reached by the UTP 690 hardfacing material, up to 4.5 times higher than the 37MnSi5 standard material. Based on these results, we can state that UTP 690 hardfacing material could be the most suitable for modifying the ploughshare.

Figure 11 shows the box plots with 95% confidence intervals. In the first case (Figure 11a), the weight loss values of S355J2G3 and HARDOX 450 materials were compared. It can be stated from the graph that the average values of weight loss of S355J2G3 and HARDOX 450 materials are not the same. Based on these results, it can be assumed that HARDOX 450 material will have better resistance to abrasive wear.

In the second case (Figure 11b), the values of the weight loss of 37MnSi5 material and OK 84.58 hardfacing material were compared. Based on the graph, it can be stated that the average values of the weight loss of 37MnSi5 material and OK 84.58 hardfacing material are almost the same, and thus hardfacing with such a hardfacing material on the exposed parts of the ploughshare will not significantly increase the lifetime.

In the third case, the values of mass losses of the 37MnSi5 material and UTP 690 hardfacing material were compared (Figure 11c). A comparison of the graphs clearly shows that the average values of weight loss of 37MnSi5 material and UTP 690 hardfacing material that UTP 690 hardfacing material will have better resistance to abrasive wear.

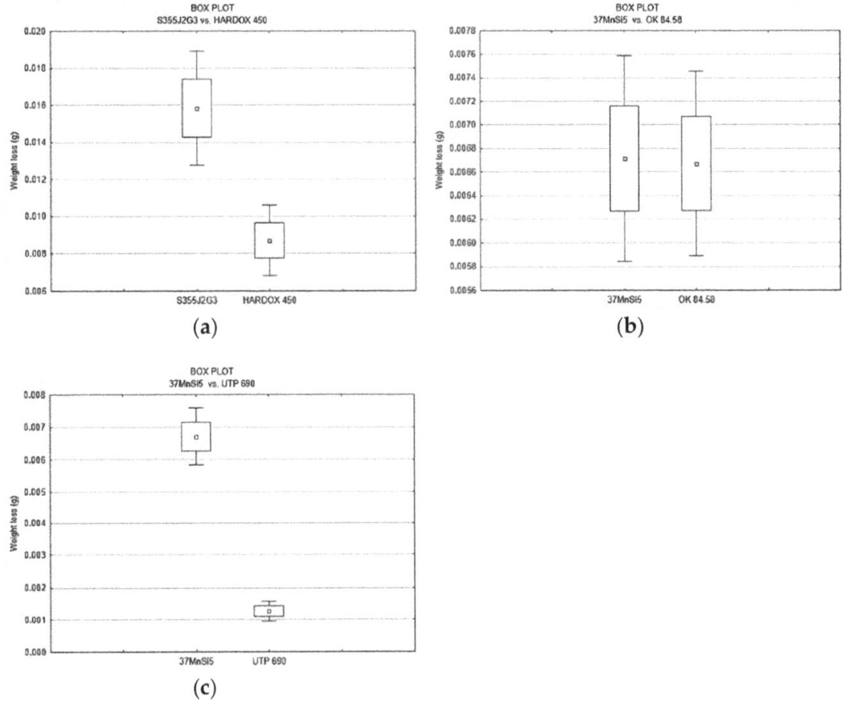

Figure 11. Box charts: (**a**) weight loss of S355J2G3 and HARDOX 450 materials; (**b**) weight loss of 37MnSi5 material and OK 84.58 hardfacing material; and (**c**) weight loss of 37MnSi5 material and UTP 690 hardfacing material.

3.3. Discussion

(a) Change of the S355J2G3 material of the supporting part of the ploughshare for the HARDOX 450 material

Based on the results, we can state that HARDOX 450 material has significantly better results compared to the S355J2G3 material of the basic ploughshare body. The hardness of HARDOX 450 material was 45 HRC, and the hardness coefficient of K_{TV1} reached the value of 0.83. The hardness of the basic body S355J2G3 material was 18 HRC, and the hardness coefficient of K_{TE2} reached the value of 0.33. These two basic data provide us with the initial information that HARDOX 450 material will better resist the abrasive particles present during the work cycle than the S355J2G3 material of the basic ploughshare body. The relative abrasion resistance Ψ_{h1} reached the value of 1.91. Based on this datum, we can say that HARDOX 450 material achieved almost twice as good results as the standard sample No. 2 made of S355J2G3 steel. In this case, it would be appropriate to verify the suitability of the higher class HARDOX material, e.g., HARDOX 550, which has higher hardness.

(b) Hardfacing on the exposed part of the raking blade with the OK 84.58 hardfacing material

Based on the results, we can state that the OK 84.58 hardfacing material achieved almost the same results in comparison with the 37MnSi5 material of the raking blade. The hardness of the raking blade 37MnSi5 material was 50 HRC, and the hardness coefficient of K_{TE1} reached the value of 0.93. The hardness of the OK 84.58 hardfacing material was 50 HRC, and the hardness coefficient of K_{TV2} reached the value of 0.93. These two basic data provide us with the initial information that the OK 84.58 hardfacing material will be worse as well as resist the abrasive particles present during the work cycle as a 37MnSi5 material. The relative resistance to abrasive wear Ψ_{h2} reached the value of 0.97. Based on this datum, we can say that the OK 84.58 hardfacing material achieved worse results by only 3% compared to the 37MnSi5 standard material of the raking blade. These elements create a precondition for the formation of carbides and intermedial phases hardening the softer ferritic matrix. This increases the hardness as well as partly the abrasion resistance while maintaining the good toughness of the material.

(c) Hardfacing on the exposed part of the raking blade with the UTP 690 hardfacing material

Based on the results, it can be stated that UTP 690 hardfacing material achieved better results compared to the 37MnSi5 material of the raking blade. The hardness of the UTP 690 hardfacing material was 62 HRC, and the hardness coefficient K_{TV3} reached the value of 1.15. These two basic data provide us with initial information that the UTP 690 hardfacing material will better withstand the abrasive particles present during the working cycle. The relative resistance to abrasive wear Ψ_{h3} reached the value of 4.57. Based on this data, we can say that the UTP 690 hardfacing material achieved more than 4.5 times better results compared to the 37MnSi5 standard material of the raking blade. In this case, we can state that the UTP 690 hardfacing material could significantly affect the lifetime of the ploughshare. The hardfacing material of the UTP 690 coating with its chemical composition represented by Cr and W increases wear resistance and hardness. The elements V and W soften the grain promoting the abrasion resistance of the material, where V increases the toughness needed to balance the increased hardness with carbides C, Cr and W. Compared to the OK 84.58 hardfacing material to transfer the loads that act on it and is, therefore, an assumption for better resistance to abrasive wear.

Based on the results, we can state that the hardness of the material strongly correlates with the resistance to abrasive wear. The hardness of HARDOX 450 was almost 2.5 times higher than S355J2G3. The relative abrasion resistance was almost twice as high for HARDOX 450 as for S355J2G3. The hardness and relative abrasion resistance of the 37MnSi5 material and the OK 84.58 hardfacing material were almost the same. Compared to the 37MnSi5 material, the UTP 690 material achieved higher hardness and higher resistance to abrasive wear.

4. Conclusions

Ploughshares are exposed to high wear, especially of the abrasive type, and their replacement is relatively economical. For this reason, it is necessary to look for ways to increase their lifetime and examine their resistance to abrasive wear.

When working with a ploughshare, it is not only the raking blade that wears rapidly but also its supporting part which the raking blade welded to. The article analyzes three ways to modify ploughshares. Based on the results, it is possible to state the following:

(a) The HARDOX 450 material achieved significantly better results compared to the basic S355J2G3 material, not only in terms of hardness and hardness coefficient K_T but also in terms of relative resistance to abrasive wear, where the coefficient Ψ_{h1} for HARDOX 450 material achieved almost twice as good results compared to the standard sample from S355J2G steel;

(b) The OK 84.58 hardfacing material achieved almost the same results compared to the 37MnSi5 original material of the raking blade and therefore does not significantly affect the lifetime of the raking blade on the ploughshares;

(c) The UTP 690 hardfacing material achieved more than 4.5 times better results than the 37MnSi5 standard material of the raking blade, so it could significantly affect the lifetime of the ploughshare. Compared to the OK 84.58 hardfacing material, we can observe a needle-like structure in the UTP 690 hardfacing material, which could better transfer the loads that act on it, and thus it is an assumption for the better resistance to abrasive wear.

Based on the results, the manufacturer of ploughshares was recommended to use HARDOX 450 material as a basic material and UTP 690 coating material hardfacing on exposed parts of functional surfaces as a suitable means of increasing the lifetime of ploughshares.

In further research, it would be appropriate to also perform a test of resistance to abrasive wear for abrasion-resistant HARDOX sheet but of a higher class, i.e.. with higher hardness, e.g., HARDOX 550 material, as well as for OK 84.58 hardfacing material which would be hardfaced in several layers. Then, the most suitable adjustments would be selected and tested under real operating conditions.

Author Contributions: Conceptualization, M.V., M.T. and K.M.; methodology, M.V. and M.T.; software, M.V., M.T. and K.M.; validation, M.V., M.T., K.M. and M.D.; formal analysis, M.V., investigation, M.V. and M.D.; resources, M.T. and M.V.; data curation, M.T., K.M. and M.D.; writing—original draft preparation, M.V., M.T. and K.M.; writing—review and editing, M.V. and K.M.; visualization, M.V. and K.M.; supervision, M.T.; project administration, M.T. and K.M.; funding acquisition, K.M. All authors have read and agreed to the published version of the manuscript.

Funding: This research was supported by the Ministry of Education, Science, Research, and Sport of the Slovak Republic through project APVV-19-0550.

Institutional Review Board Statement: Not applicable.

Informed Consent Statement: Not applicable.

Data Availability Statement: Not applicable.

Acknowledgments: The article was supported by the APVV-16-0194 "Research on Impact of Innovation in Production Processes on the Life of Tooling and Components of Forest Mechanisms".

Conflicts of Interest: The authors declare no conflict of interest.

References

1. Jankauskas, V.; Katinas, E.; Pusvaškis, M.; Leišys, R. A Study of the Durability of Hardened Plough Point. *J. Frict. Wear* **2020**, *41*, 78–84. [CrossRef]
2. Stawicki, T.; Kostencki, P.; Białobrzeska, B. Roughness of Ploughshare Working Surface and Mechanisms of Wear during Operation in Various Soils. *Metals* **2018**, *8*, 1042. [CrossRef]
3. Singh, J.; Chatha, S.S.; Sidhu, B.S. Abrasive wear behavior of newly developed weld overlaid tillage tools in laboratory and in actual field conditions. *J. Manuf. Process.* **2020**, *55*, 143–152. [CrossRef]
4. Doluk, A.; Rudawska, A.; Stančeková, D.; Mrázik, J. Influence of surface treatment on the strength of adhesive joints. *Manuf. Technol.* **2021**, *21*, 585–591. [CrossRef]
5. Müller, M.; Novák, P.; Chotěborský, R.; Hrabě, P. Reduction of Ploughshare Wear by Means of Carbide Overlay. *Manuf. Technol.* **2018**, *18*, 72–78. [CrossRef]
6. Kováč, J.; Krilek, J.; Štefánek, M.; Kuvik, T. Analysis of the selected technical parameters of skidders. *MM Sci. J.* **2016**, *2016*, 1273–1276. [CrossRef]
7. Zdravecká, E.; Tkáčová, J.; Ondáč, M. Effect of microstructure factors on abrasion resistance of high-strength steels. *Res. Agric. Eng.* **2014**, *60*, 115–120. [CrossRef]
8. Javaheri, V.; Porter, D.; Kuokkala, V.T. Slurry erosion of steel–Review of tests, mechanisms and materials. *Wear* **2018**, *408*, 248–273. [CrossRef]
9. Ťavodová, M.; Vargová, M.; Falat, L. Possibilities of modification of ploughshares used for winter maintenance of forest roads. *Manuf. Technol.* **2020**, *20*, 834–844. [CrossRef]
10. Petru, J.; Zlámal, T.; Čep, R.; Stančeková, D.; Pagač, M.; Vortel, O. Mechanism of cutting insert wear and their influence on cutting ability of the tool during machining of special alloys. In Proceedings of the 3rd International Conference on Manufacturing Engineering and Technology for Manufacturing Growth (METMG 2015), Vancouver, BC, Canada, 1–2 August 2015. [CrossRef]
11. Hnilica, R.; Kotus, M.; Jankovský, M.; Hnilicová, M.; Dado, M. Qualitative classification of mulchers. *Agron. Res.* **2017**, *15*, 1890–1896. [CrossRef]
12. Balla, J. *Náuka o Materiáloch*; SPU: Nitra, Slovakia, 2003; pp. 48–54.
13. Natsis, A.; Petropoulos, G.; Pandazaras, C. Influence of Local Soil Conditions on Mouldboard Ploughshare Abrasive Wear. *Tribol. Int.* **2008**, *1*, 151–157. [CrossRef]
14. ESAB. Opravy a Údržba. Available online: http://products.esab.com/ESABImages/prirucka%206%20vydani%202011.pdf (accessed on 1 April 2022).

15. Chintha, A.R.; Valtonen, K.; Kuokkala, V.T.; Kundu, S.; Peet, M.J.; Bhadeshia, H.K.D.H. Role of fracture toughness in impact-abrasion wear. *Wear* **2019**, *428*, 430–437. [CrossRef] [PubMed]
16. Müller, M.; Hrabě, P. Overlay materials used for increasing lifetime of machine parts working under conditions of intensive abrasion. *Res. Agric. Eng.* **2013**, *59*, 16–22. [CrossRef]
17. Buchely, M.F.; Gutierrez, J.C.; Leon, L.M.; Toro, A. The effect of microstructure on abrasive wear of hardfacing alloys. *Wear* **2005**, *259*, 52–61. [CrossRef]
18. Wang, X.; Han, F.; Liu, X.; Qu, S.; Zou, Z. Microstructure and wear properties of the Fe–Ti–V–Mo–C hardfacing alloy. *Wear* **2008**, *265*, 583–589. [CrossRef]
19. Wang, X.H.; Han, F.; Liu, X.M.; Qu, S.Y.; Zou, Z.D. Effect of molybdenum on the microstructure and wear resistance of Fe-based hardfacing coatings. *Mater. Sci. Eng. A* **2008**, *489*, 193–200. [CrossRef]
20. Xu, L.; Wei, S.; Xiao, F.; Zhou, H.; Zhang, G.; Li, J. Effects of carbides on abrasive wear properties and failure behaviours of high speed steels with different alloy element content. *Wear* **2017**, *376*, 968–974. [CrossRef]
21. Konat, Ł.; Zemlik, M.; Jasiński, R.; Grygier, D. Austenite Grain Growth Analysis in a Welded Joint of High-Strength Martensitic Abrasion-Resistant Steel Hardox 450. *Materials* **2021**, *14*, 2850. [CrossRef] [PubMed]
22. Šípová Snežná Radlica. Available online: http://www.agrowest.cz/produkty/agrometall-celni-snehove-radlice/sipova-radlice-pracovni-sirka-3000-mm (accessed on 8 February 2022).
23. Zhrňovací Nôž. Available online: https://detvaservis.sk/brity-na-radlice/ocelove-brity-s-jednostrannym-ukosom/ (accessed on 1 April 2022).
24. Steel S355J2G3. Lexikon kovov 2.7, Verlag Dashöfer, Praha, 4/2019. Available online: https://www.dashofer.cz/ (accessed on 6 March 2022).
25. Steel 37MnSi5. Lexikon Kovov 2.7, Verlag Dashöfer, Praha, 4/2019. Available online: https://www.dashofer.cz/ (accessed on 6 March 2022).
26. HARDOX. Available online: https://www.gamaocel.cz/4861/hardox/ (accessed on 6 March 2022).
27. HARDOX Wear Plate. Available online: https://www.ssab.com/products/brands/hardox (accessed on 6 February 2022).
28. Ulewicz, R.; Mazur, M.; Bokuvka, O. Structure and Mechanical Properties Offine-grained Steels. *Transp. Eng.* **2013**, *41*, 111–115. [CrossRef]
29. OK 84.58. Available online: https://www.esab-slovakia.sk/sk/sk/products/filler-metals/repair-and-maintenance/hardfacing-alloys/ok-weartrode-55-hd.cfm (accessed on 7 February 2022).
30. Vargová, M.; Ťavodová, M. Proposal for modification of a snow ploughshare by hard surfacing to increase its service life. *Acta Fac. Tech. Sci. J. Fac. Technol. Zvolen* **2020**, *2*, 59–70.
31. UTP 690. Available online: https://www.zvarcentrum.sk/obchod_homedir/data/2616/prilohy/UTP%20690.pdf (accessed on 7 February 2022).
32. GOST 23.208-79 Ensuring of Wear Resistance of Products. Wear Resistance Testing of Materials by Friction against Loosely Fixed Abrasive Particles. Available online: http://docs.cntd.ru/document/gost-23-208-79 (accessed on 10 February 2022).
33. GOST 3647-80 Abrasives. Grain Sizing. Graininess and Fractions. Test Methods. Available online: https://docs.cntd.ru/document/1200016841 (accessed on 10 February 2022).
34. Schmidtová, J.; Vacek, V. *Applied Statistics*; Technical University in Zvolen: Zvolen, Slovakia, 2013; pp. 85–102.
35. Markechová, D.; Tirbáková, A.; Stehlíková, B. *Základy Štatistiky pre Pedagógov*; Fakulta Prírodných vied UKF: Nitra, Slovakia, 2011; pp. 115–154.
36. *ISO 6507-1:2018*; Metallic Materials—Vickers Hardness Test—Part 1: Test Method. International Organization for Standardization: London, UK, 2018.
37. *ISO 6508-1:2016*; Metallic Materials—Rockwell Hardness Test—Part 1: Test Method. International Organization for Standardization: London, UK, 2018.

metals MDPI

Article

Effect of Machining Conditions on Temperature and Vickers Microhardness of Chips during Planing

Peter Pavol Monka [1,2], Katarina Monkova [1,2,*], Martin Vasina [2,3], Milena Kubisova [2], Martin Korol [1] and Adriana Sekerakova [1]

[1] Faculty of Manufacturing Technologies with the Seat in Presov, Technical University of Kosice, Sturova 31, 080 01 Presov, Slovakia
[2] Faculty of Technology, UTB Tomas Bata University in Zlin, Vavreckova 5669, 760 01 Zlin, Czech Republic
[3] Faculty of Mechanical Engineering, VŠB-Technical University of Ostrava, 17. Listopadu 15/2172, 708 33 Ostrava-Poruba, Czech Republic
* Correspondence: katarina.monkova@tuke.sk; Tel.: +421-55-602-6370

Abstract: For the machining of long and narrow surfaces and when processing multiple pieces, planing technology is used, the productivity of which can be higher than that of milling, although it is relatively slow machining. The article aims to study the degree of influence of the geometry of the tool (the angle of cutting-edge inclination and the angle of the tool-orthogonal rake), as well as the cutting conditions (cutting depth and cutting speed) on the chip characteristics (temperature and microhardness) in orthogonal and oblique slow-rate machining of steel 1.0503 (EN C45). The experiments were carried out on specially prepared workpieces designed for immediate stopping of machining. The results of the experiments were statistically processed, and behavioural models were created for temperature and Vickers microhardness of chips for individual combinations of factors. The obtained dependencies revealed how the geometry of the cutting tool and the cutting conditions affect the temperature and microhardness in the cutting area and at the same time allowed the best conditions for both orthogonal and oblique machining to be set up.

Keywords: planing; chip; Vickers microhardness; temperature; cutting conditions

Citation: Monka, P.P.; Monkova, K.; Vasina, M.; Kubisova, M.; Korol, M.; Sekerakova, A. Effect of Machining Conditions on Temperature and Vickers Microhardness of Chips during Planing. *Metals* **2022**, *12*, 1605. https://doi.org/10.3390/met12101605

Academic Editor: Sergey Konovalov

Received: 27 August 2022
Accepted: 21 September 2022
Published: 26 September 2022

1. Introduction

Cutting operations with relatively low cutting velocity include planing and broaching technologies. Planing is a cutting process in which the formed chip is visible to the eye, whereas in broaching, the chip generation process is hidden and invisible to the naked eye. Therefore, it is possible to monitor the development of the temperature field in the area of chip generation during planing with a temperature measuring device by laser aiming and thus providing an approximate idea of the temperature generated not only during cutting but also during broaching [1,2].

The planing process is one of the oldest among the so-called single-point machining processes, used in the production of long cuts. This type of machining technology uses a linear motion with a single-pointed cutting tool to create a flat surface. The cutting tool is clamped in the tilting head, which prevents damage to the cutting edge when returning the tool. Although the process seems to be easy, setting the cutting conditions is difficult [3–6].

During the planing, the workpiece moves linearly and is usually machined with one tool. Because the large, heavy workpiece and table are in motion at relatively low cutting speeds, planers come with several tool heads and multiple tools in one head [7]. In addition, many planers are constructed with tool heads to make cuts in both directions of table movement [8]. Nevertheless, as mentioned above, since only single-pointed tools and low cutting speeds are used, planing is classified as a low-productivity cutting process compared to other cutting methods. However, when machining long and narrow surfaces (such as guide rails, long grooves, etc.) and when cutting multiple pieces with a planer, the efficiency of this type of workpiece processing can be much higher compared to milling [9,10]. The planing accuracy can reach IT9~IT8 and the surface roughness can be in the range of Ra is 3.2~1.6 μm [11–13].

In spite of planing being considered an obsolete technology, for performing and observing experimental tests for machining at relatively low cutting speeds, planing appears to be a very attractive machining process [14]. At slow-rate machining, the chip-formation process can be observed in both cutting methods, orthogonal and oblique.

For orthogonal cutting (Figure 1a), it is characteristic that the cutting edge is perpendicular to the direction of movement of the tool and wider than the width of the cut. The shear force acts on a smaller area, so shear force per unit area is higher and the tool life is shorter than obtained in oblique cutting [15,16]. The forces that occur in this type of machining can be drawn in a plane, which is why it is also known as 2D machining [17]. If the angle between the cutting edge and the movement of the tool is not "right" (90°), it is an oblique cutting, which is also called 3D machining, because the cutting force is not planar, but spatial and is represented by 3 components [18]. All these facts affect the cutting conditions, tool life, durability as well as the quality of machined surface.

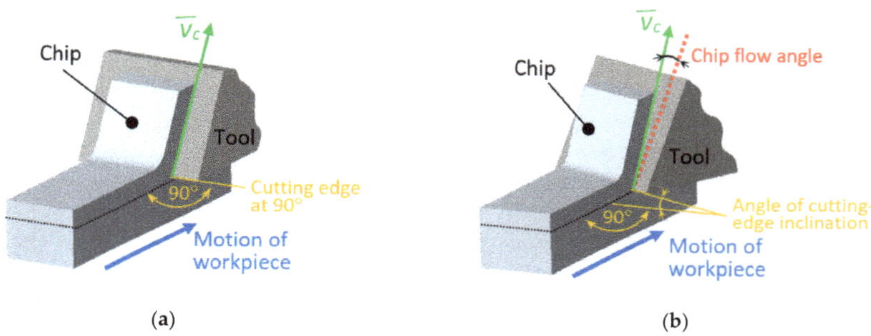

Figure 1. Principle of (**a**) orthogonal and (**b**) oblique machining.

In both types of machining, the chip resulting from the interaction of the tool with the workpiece plays a crucial role in determining the properties of the machined surface and tool life. Properties that reflect its behaviour in the shear plane (acceleration of shear deformation, shear stress and shear angle) can also be represented by temperature and microhardness [19,20].

Jain et al. [21] verified the degree of influence of shear strain accelerations on the microhardness of chips during machining under the same cutting conditions. The basis of the hypothesis was the fact that the acceleration of shear deformation controls machining parameters such as tool-chip interface temperature, shear angle, tool wear, etc., and the hypothesis was confirmed.

The process of chip development in high-speed machining of hardened steel was studied by Wang et al. [22]. Research has confirmed that the formation of chips, when cutting materials of different hardness, can be controlled by setting a suitable combination of technological parameters. The hardness of the material as well as the local temperature increases with increasing cutting speed. When the parameters are in equilibrium, the deformation is concentrated in the shear zone and adiabatic shear occurs.

The effect of cutting conditions on the microhardness of the material in the cutting area along with chip thickness was investigated by Alrabii and Zumot [23] in experimental tests. They reported that microhardness increases with increasing cutting parameters (depth of cut, speed and feed), but only to a certain extent. Above this level, an increase in cutting parameters causes a decrease in microhardness.

The influence of changes in the microstructure of AISI 1045 steel on diffusion wear during turning in a zone of chip formation as a function of variable cutting speed was investigated Pu et al. [24]. The results of the research showed that with the cutting speed increasing, the grain size of the various phases also increases.

Response surface methodology was used in the study of Senussi [25] to specify the root cause of an effect of the relationship between chip microhardness and input control variables influencing the response as a 2D or 3D hypersurface. He stated that the combined effect of cutting speed at its lower level, feed rate and cutting depth at their higher values could result in increasing the microhardness of the chip.

Various approaches have been used by many researchers to study the effect of cutting parameters on chip properties and behaviour. The issues of microhardness, temperature, geometry and chip shape within analytical and numerical models have been developed in several studies [26–30], however, the complexity of the chip generation process forced the authors to simplify the simulation conditions, which, however, affected the accuracy of the obtained results, which did not correspond to the real state.

It can be said that many studies were focused on the influence of cutting parameters on microhardness and temperature evaluation in the past, but mostly the object of their investigation was the machined material or the cutting tool. However, only a few studies have addressed the microhardness and temperature of the chips generated during machining, but no study has compared the degree of influence of both the tool geometry (angle of tool-cutting-edge inclination and angle of the tool-orthogonal rake) and the technological conditions (depth of cut, cutting speed) on the chip characteristics (temperature and microhardness) in orthogonal and oblique slow-rate machining of 1.0503 (EN C45) steel that is included in the presented research. Based on the measured data, behavioural models were created for the temperature and Vickers microhardness of chips at individual combinations of the factors. The experiments were carried out on specially prepared workpieces designed for immediate stopping of machining.

On the basis of the above, it can be said that the novelty of the research lies not only in its complexity and the combination of included research factors (from the type of machining and machined material, through the geometry of the tool to the technological conditions), which according to the authors' best knowledge have not yet been studied, but also in the newly proposed methodology for obtaining chips in an attempt to bring the observed chip samples as close as possible to the state in which they were created so that the texture of such a chip corresponds to the current cutting speed.

2. Materials and Methods

Planing technology, in which the workpiece performs the main movement, was selected for this study. In the experiments carried out as part of the presented research, a planing machine HJ8A type (KOVOSVIT inc., Holoubkov; Czech Republic) shown in Figure 2a was used. The process of planing of EN C45 steel is captured in Figure 2b. The machine is designed for the production of medium-heavy workpieces in piece and series production. The worktable has three levels of working and retraction speeds. The hydraulic drive is equipped with two gear pumps with separate electric motors, electromagnetic, hydraulic distribution and a combination of pump devices to change the speed levels of the table. The crossbar with support is vertically adjustable on the stand. The planer has two supports on the crossbar and one on the stands. Feeds and rapid feeds are controlled by the slide box at the end of the crossbar, either hydraulically or electrically. The table feed and the fast feeds of the crossbar, as well as the carriage, are controlled by the buttons from the hanging control box.

No cooling medium was used during machining, as the tool path was relatively short (about 200 mm), so there was no overheating of the tool and the material being processed.

Steel 1.0503 (EN C45), which is usually used in experiments as a standard [31], was chosen as the machined material within the experiments. It is a medium carbon-unalloyed structural steel that provides moderate tensile strength (in the range of 570–700 MPa), Brinell hardness of 170–210 and a good wear resistance. The material can be processed with hardening by means of quenching and tempering on focused and restricted areas. C45 can also be instigated with induction hardening up to the hardness level of HRC 55. This grade is in majority situations delivered in an unattended heat treatment state i.e., normalized condition, although it can be furnished also in numerous heat treatment variations. Machinability of C45 is equivalent to that of mild steel for example CR1 grade; on the other hand weld ability is exhibited lower than that of mild steel [32]. Steel is suitable for producing shafts of turbochargers, pumps, traction machines, electric motors, more giant gears, screws, automobile crankshafts, connecting rods, steering levers, spring hinges or pins. The chemical composition of machined 1.0503 (EN C45) steel is in Table 1.

Figure 2. Experimental device (**a**) HJ8A planer, (**b**) experimental processing of a workpiece made of EN C45 steel.

Table 1. Chemical composition of the 1.0503 (EN C45) steel [33].

Steel	C (%)	Mn (%)	Si (%)	Cr (%)	Ni (%)	Cu (%)	P (%)	S (%)
EN C45	0.42–0.50	0.50–0.80	0.17–0.37	max. 0.25	max. 0.30	max. 0.30	max. 0.040	max. 0.040

The values of the cutting parameters (cutting speed v_c, cutting depth a_p) for experimental testing were chosen based on the requirements of technical practice, whereas the lowest cutting speed was given by the planer and corresponded to 60% of the machine's power. Values for cutting depth a_p were set based on machine limitations for maximum chip cross-section.

The planing necking tool type 32×20 ON 36550 HSS00 was used at orthogonal machining, whereas the straight roughing tool 32×20 ON 36500 HSS00 was employed at oblique machining. The brazed high-speed steel tips with three different values of the angle of cutting-edge inclination λ_s = 0°, 10° and 20° were applied within experimental testing [33].

The cutting tools used in the experiments are shown in Figure 3, and the next tools' characteristics are given in Table 2.

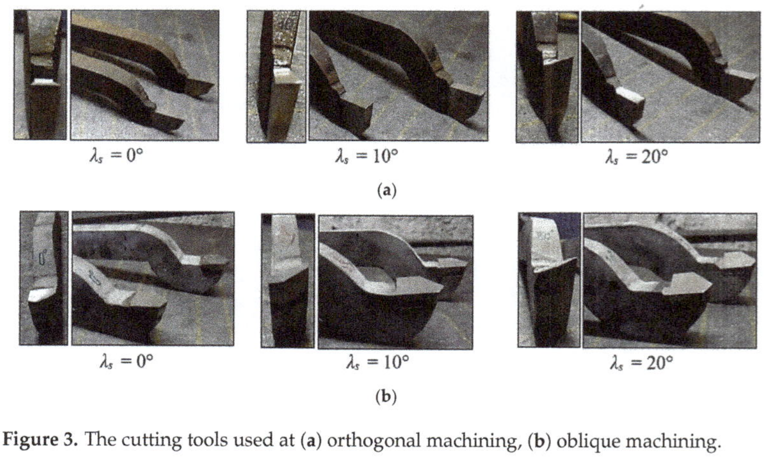

Figure 3. The cutting tools used at (**a**) orthogonal machining, (**b**) oblique machining.

Table 2. Values of tool angles [33].

Type of Tool Angle	Planing Necking Tool	Straight Roughing Tool
Tool-cutting-edge angle κ_r	0°	60°
Tool minor (end)-cutting-edge angle κ_r'	-	20°
Tool-included angle ε_r	-	100°
The angle of the tool-orthogonal rake γ_o	8°/12°/16°	3°/7°/11°
The angle of the tool-orthogonal clearance α_o	15°/11°/7°	15°/11°/7°

The geometrical characteristics of the cutting tool were chosen in accordance with the machined material and the characteristics of the planing process. This method of machining requires the use of a more robust tool compared to turning. Increased robustness is required from the point of view of a significantly more dynamic way of first contact between the tool face and the workpiece [34]. The following requirements were also taken into account when choosing the tool geometry:

- Compliance with the condition of the tool's physical resistance to dynamic shocks when entering the cut;
- The range of variable values ensuring the statistical significance of the experiment;
- The difference in the characteristics of the chip formation process for orthogonal and oblique cutting.

According to the tool manufacturer's recommendations, the rake angle when planing type C45 steels is chosen in the range of 8°–14°, and the angle of tool-orthogonal clearance α_o for this type of steel is chosen in the range of 7°–11° [35].

Considering all the demands and recommendations, the angle of tool-orthogonal clearance α_o, adopted the strategy of using the smallest value (7°) in terms of production recommendations and the other two values are in a linear series with an increment of 4° for the tools within the presented experiments.

Similarly, in the research, the angles of the tool-orthogonal rake γ_o were set up so they reflected the requirement to ensure the lowest possible specific resistance, whereas the maximum value was related to the bending stiffness of the cutting wedge. Specifically, for the planing necking tool, an interval of three values graduated by 4° from the smallest recommended value (8°) was chosen for the angle of the tool-orthogonal rake γ_o. For the straight roughing tool, the manufacturer's recommended values of the angle of tool-orthogonal rake γ_o were reduced due to an effort to increase the strength properties of the tool tip and were used in gradations of 3°/7°/11°. For this tool, for the same reason of increasing the strength of the tip, the tool included angle ε_r in the value of 100° was used.

In order to observe changes in the chip formation zone, it was necessary to stop the machining process and break the contact of the tool with the workpiece. For this purpose, a methodology of immediate machining stop was developed based on the observation of chip behaviour during face milling, in which the end of the chip breaks off when leaving the cut. Observations have shown that the texture of such a chip corresponds to the actual speed at the moment of interruption of the cut. This know-how led the authors to use it in the presented research, and so they have developed a new methodology of obtaining chip roots based on the special modifications of the machined sample. The principle of the methodology is based on the designing of a specific workpiece (shown in Figure 4), and it is described below.

After milling the grooves of 10 mm × 10.5 mm into a semi-finished product with dimensions of 240 mm × 87.5 mm × 30 mm in the longitudinal direction, 11 protrusions with a rectangular cross-section of 10.5 mm × 2.5 mm were created, and each protrusion was used for a separate experiment.

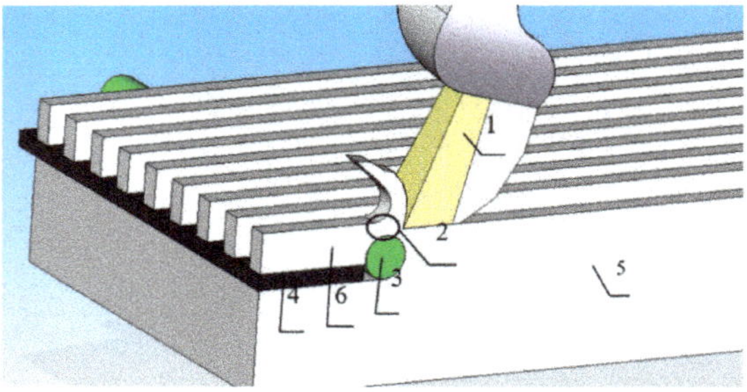

Figure 4. The principle of the methodology to immediately break the contact of the tool and the workpiece for orthogonal cutting (1—cutting wedge of the tool, 2—the place of rupture of the sample, 3—a drilled hole reinforced with a metal rod due to the narrowing of the cross-section, 4—supporting plate preventing the deformation of the final sample, 5—workpiece, 6—the final sample used for observation after tearing away from the workpiece).

At the end of the sample, where the tool comes out of the grip, the protrusions were undercut, and a plate was inserted into the resulting gap, which had the task of preventing the deformation of the final sample. The most crucial role in the cutting interruption process was played by a transverse hole with a diameter of 8 mm, which was drilled in front of the reinforcing plate (from the point of view of the cutting direction) and which was subsequently also reinforced with a metal rod. When the tool passed over the hole, the cross-section narrowed, and the material was torn, and the end part of the protrusion was thrown sharply in the direction of the tool's movement at speed greater than the cutting speed. The separated end parts together with the resulting chip served as samples that captured the state of plastic deformation corresponding to the actual cutting speed.

An infrared thermometer UNI-T UT305C (Uni-trend Technology Co., Ltd., Dongguan City, Guangdong Province, China) was employed to measure the temperature of various surfaces by measuring the infrared radiation emitted from a focused surface and whose specifications were sufficient for the experiments carried out in this research. Its basic characteristics were: Temperature range −50~1550 °C; Accuracy ±1.8%; Repeatability ±0.5%; Resolution 0.1; Response time 250 ms (95% of reading); Emissivity 0.1~1.0 adjustable; Laser power <1 mW; Laser wavelength 630~670 nm; Spectral response 8~14 μm.

Since the temperature measurement was performed only on a cutting length of 200 mm, it was assumed that the increase in temperature in the cutting area was directly proportional to the increase in the size of the cutting path of the tool.

The obtained samples were cast in technical dentacryl, which has good insulating properties, high mechanical strength and perfect thermal insulation. The cast samples were ground using water, which had a cooling effect and thus prevented the sample from being affected by heat. In Figure 5a, the ground and polished sample are shown. After polishing, the samples were etched with Nital, which is a 2% solution of nitric acid HNO_3 in ethyl alcohol. After being etched on the surface, a sample structure of the material was observed with the Platinum USB digital microscope UM019 (Shenzhen Handsome Technology Co., Ltd., Shenzhen, China) with a magnification of 25–220×. An example of an etched sample is shown in Figure 5b.

Microhardness was measured with a Micro–Vickers hardness tester CV–403DAT (Figure 6a) according to STN EN ISO 6507-1 standard. The measuring device involves a microscopic and a static method expressed on Vickers or Knoop scales, with magnifications of 200× and 600×. The microhardness measuring according to the Vickers method is determined by optical magnification by pressing a diamond regular quadrilateral needle with a peak wall angle of 136° into the tested material and then measuring the diagonals perpendicular to each other after relieving the load. The loading force acts in a perpendicular direction on the surface of the test specimen.

Figure 5. Investigated sample of a chip, (a) ground and polished, (b) etched, magnified 200×.

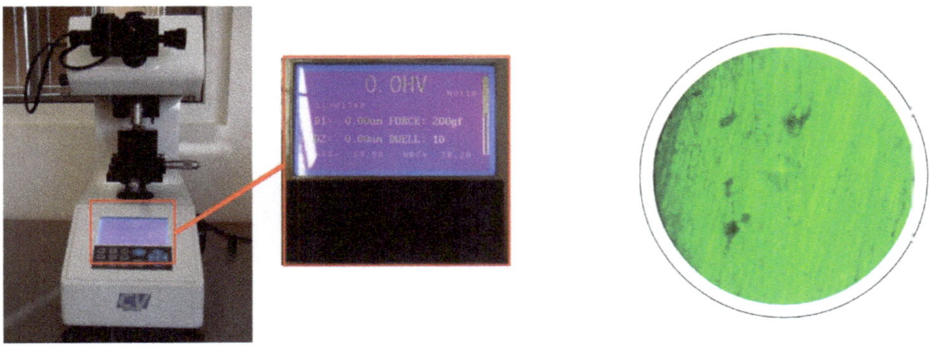

Figure 6. (a) Micro-Vickers hardness tester CV-403DAT; (b) tested samples—imprint, magnification 600×.

Hardness measurements were performed in the area of the shear angle at a load of 200 gf (gram-force) for a dwell of 10 s. An example of the tested sample with an imprint under 600× magnifications is shown in Figure 6b.

3. Results and Discussion

The experiments were carried out within the so-called "planned experiment" that was designed at three levels (lower, basic, and upper). The temperatures were measured in a cutting zone for orthogonal and oblique machining, and the micro-hardness of the material was measured in the area where the shear angle was formed. Measured data were evaluated through the statistical method using a regression function. Outliers from each group of measurements were specified using Grubbs test criteria and were excluded from further statistical processing of the results. Regression coefficients were calculated by means of Matlab software, and their significance was examined according to Student's test criterion. The adequacy of the regression function was evaluated according to the Fisher-Snedecor test criterion $F < F_{0.05}$ (f_1, f_2), where the degrees of freedom $f_1 = Nq$ (N is a number of measures, q is a number of significant coefficients) and $f_2 = N$ $(m - 1)$, where m is a number of assessed measurements within the Grubbs' test [36–39].

The variables listed in Table 3 were considered within the experimental study, and the codes for individual variables' specific values are referred in Table 4.

Table 3. The variables and the ranges of individual values [33].

Variables		Symbols	−1	−α	0	+α	+1
Cutting speed (m·min^{-1})	v_c	x_1	6	8.25	10.5	12.75	15
Cutting depth (mm)	a_p	x_2	0.2	0.25	0.3	0.35	0.4
The angle of the tool-orthogonal rake—*orthogonal* cutting (°)	γ_o	x_3	8	10	12	14	16
The angle of the tool-orthogonal rake—*oblique* cutting (°)	γ_o	(x_3)	3	5	7	9	11
The angle of the tool-cutting-edge inclination (°)	λ_s	x_4	0	5	10	15	20

Table 4. The codes for specific values of individual variables.

Code	Specification of Variables
−1	x_{min}
−α	$[(x_{max} + x_{min})/2] - [(x_{max} - x_{min})/2\alpha^2]$
0	$(x_{max} + x_{min})/2$
+α	$[(x_{max} + x_{min})/2] + [(x_{max} - x_{min})/2\alpha^2]$
+1	x_{max}

Although the evaluation of tool wear was not the main goal of the research, it can be said that the dominant mechanism of tool wear during planing in the experiments was abrasive wear, which was manifested in the change of the geometry of the tool on the flank surface.

3.1. Temperature Evaluation

To evaluate the behaviour of the temperature and the influence of the monitored parameters on it, the regression dependencies for orthogonal machining of EN C45 steel were expressed by Equation (1), and for oblique machining by Equation (2), and the meaning of parameters $x_1 \div x_4$ is explained in Tables 3 and 4 above. The calculated Fisher–Snedecor criterion [40] confirmed that the equations were functionally dependent and the coefficients of determination of the relations were R^2 = 0.90 and 0.91, respectively:

$$\begin{gathered} y = 3.4146 - 5.7289x_1 + 6.0652x_2 + 6.6299x_3 + 0.1732x_4 - 0.3957x_1x_2 + 0.3796x_1x_3 + \\ +0.0634x_1x_4 - 0.3435x_2x_3 + 0.0202x_2x_4 + 0.0559x_3x_4 + 2.6854x_1^2 + 5.0529x_2^2 - 4.7508x_3^2 + \\ +0.0934x_4^2 \end{gathered} \quad (1)$$

$$\begin{gathered} y = -2.415 + 5.9456x_1 - 3.438x_2 + 0.5381x_3 + 0.1701x_4 - 0.0292x_1x_3 + 0.0163x_1x_4 \\ +0.2491x_2x_3 + 0.0355x_2x_4 - 0.0181x_3x_4 - 2.9253x_1^2 - 3.0763x_2^2 - 0.3104x_3^2 + 0.0592x_4^2 \end{gathered} \quad (2)$$

Using Equations (1) and (2), the temperature dependences on a combination of two parameters were plotted out of four monitored (depth of cut, cutting speed, face angle and inclination of the cutting edge of the tool) so that the behaviour of the temperature field could be evaluated.

When orthogonal cutting of carbon steel EN C45 in the interaction of cutting speed and cutting depth, the cutting depth had a more significant effect (Figure 7a). The cutting speed had a much milder effect, and the highest temperature values were achieved at the minimum cutting speed and the maximum values of the cutting depth. Conversely, in the oblique cutting of carbon steel (Figure 7b), the maximum temperatures were reached at the minimum cutting depth and the maximum cutting speed.

The influence of the angle of the tool-orthogonal rake and the cutting speed in orthogonal cutting is almost equally important/intense (Figure 8a), whereas in oblique cutting, the angle of orthogonal rake has an almost imperceptible effect compared to the impact of the cutting speed (Figure 8b). In both cases, the maximum temperature values were reached at the mean values of the angle of the tool-orthogonal rake, but as already mentioned, the temperature reacts in the opposite way to a change in cutting speed. The maximum temperature in orthogonal cutting is at the minimum cutting speed, and in oblique cutting, it is at the maximum cutting speed.

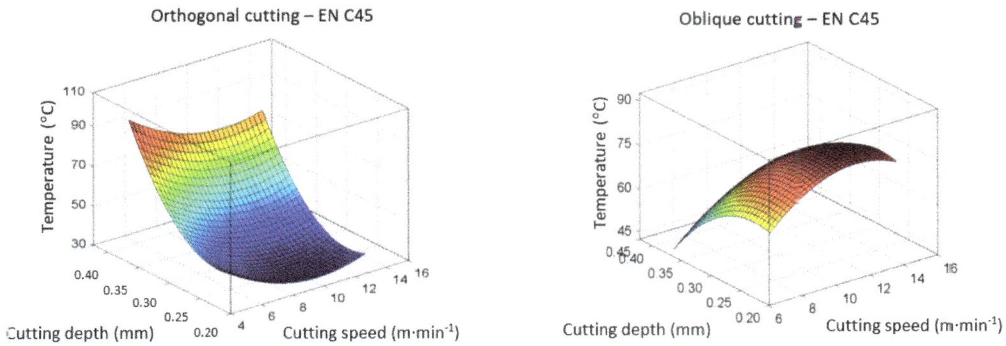

(**a**) (**b**)

Figure 7. The graphs of the influence of cutting speed v_c and cutting depth a_p on temperature change T, (**a**) orthogonal cutting; (**b**) oblique cutting.

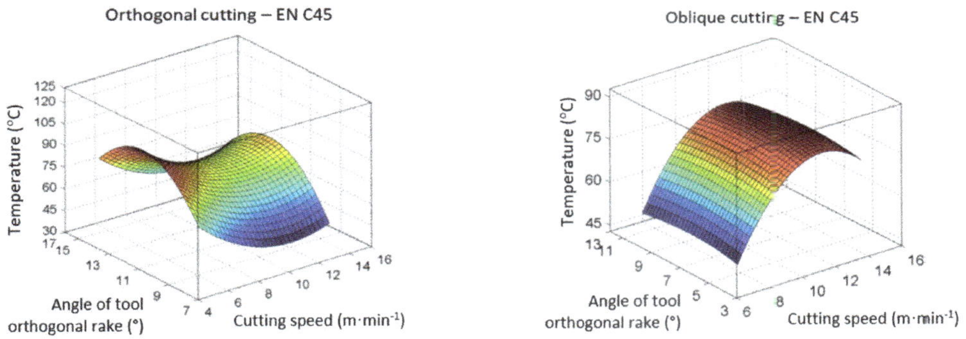

(**a**) (**b**)

Figure 8. The graphs of the influence of cutting speed v_c and the angle of the tool-orthogonal rake γ_o on temperature change T, (**a**) orthogonal cutting; (**b**) oblique cutting.

The influence of the angle of tool-cutting-edge inclination λ_s is almost undetectable; the more significant impact on temperature has the cutting speed (Figure 9).

The influence of the angle of the tool-orthogonal rake on the temperature is more evident in the orthogonal cutting of carbon steel (Figure 10a) than in oblique cutting (Figure 10b), and the maximum temperature values are reached at their mean values. The cutting depth has a more pronounced effect, and it is precisely in the opposite way. At orthogonal cutting, the maximum temperature values are reached at the maximum cutting depth values, and at oblique cutting, the maximum temperature is reached at the minimum cutting depth. In both types of cutting, the angle of tool-cutting-edge inclination λs has no noticeable effect on the change in cutting temperature compared to the cutting depth (Figure 11).

The angle of tool-cutting-edge inclination at the orthogonal cutting of carbon steel (Figure 12a) has an almost unnoticeable effect compared to the angle of the tool-orthogonal rake. However, it is different at oblique cutting (Figure 12b), where maximum temperature values are reached at the maximum angle of tool-cutting-edge inclination. The angle of the tool-orthogonal rake had a more prominent effect in orthogonal cutting, but the maximum temperature was reached at its mean values in both cutting methods.

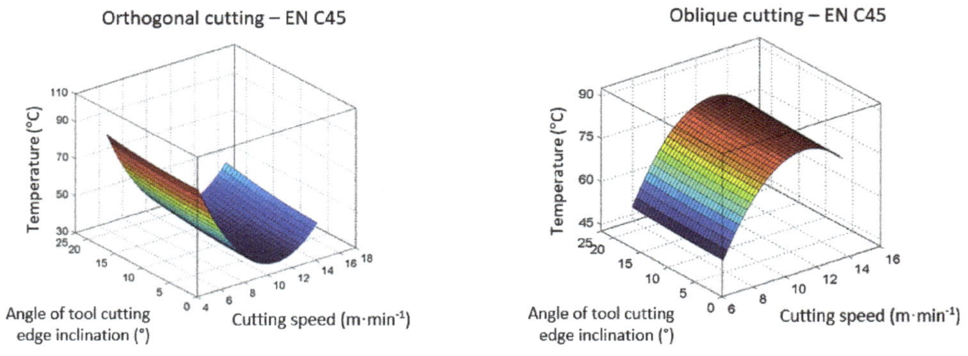

(a) (b)

Figure 9. The graphs of the influence of cutting speed v_c and the angle of tool-cutting-edge inclination λ_s on temperature change T, (**a**) orthogonal cutting; (**b**) oblique cutting.

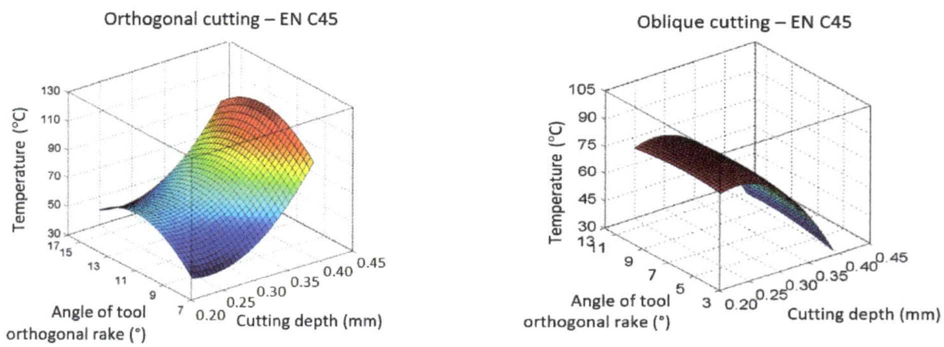

(a) (b)

Figure 10. The graphs of the influence of the tool-orthogonal rake γ_o and cutting depth a_p on temperature change T, (**a**) orthogonal cutting; (**b**) oblique cutting.

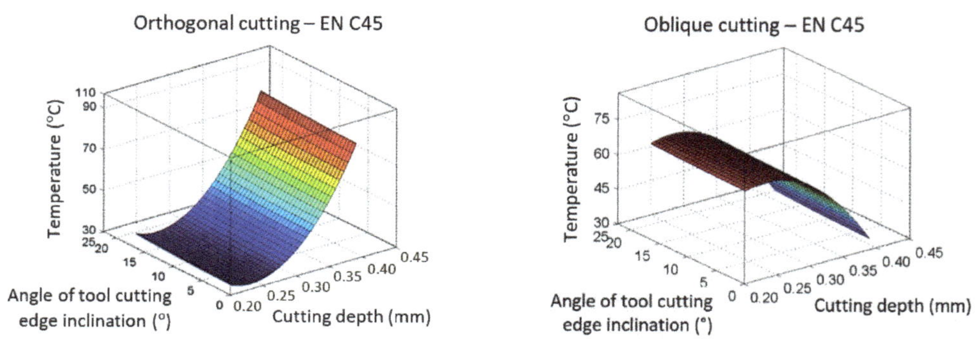

(a) (b)

Figure 11. The graphs of the influence of cutting depth a_p and the angle of tool-cutting-edge inclination λ_s on temperature change T, (**a**) orthogonal cutting; (**b**) oblique cutting.

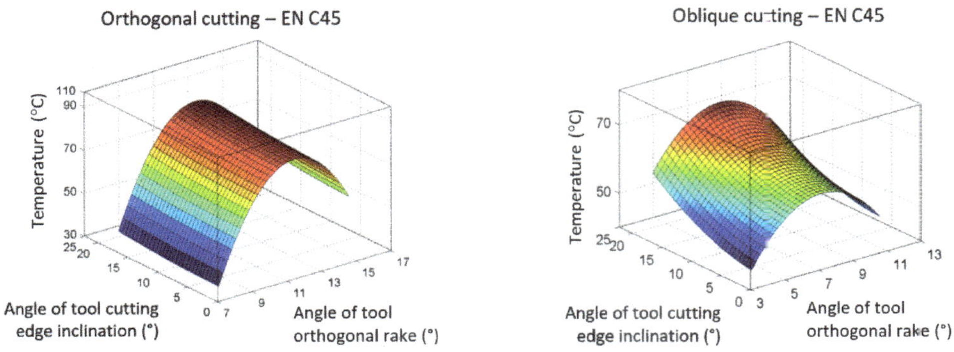

(a) (b)

Figure 12. The graphs of the influence of the tool-orthogonal rake γ_o and the angle of tool-cutting-edge inclination λ_s on temperature change T, (**a**) orthogonal cutting; (**b**) oblique cutting.

On the basis of experiments aimed at determining the influence of the investigated factors on the temperature of the chip, it is possible to evaluate achieved results from three points of view:

- From the point of view of the cutting speed, the results of the research are characterized for orthogonal cutting by a slight decrease in the temperature of the chip as the cutting speed increases. From the point of view of theoretical and practical knowledge, such a trend occurs at very high cutting speeds, when the chip removal speed from the cutting site is significantly higher than the heat conduction speed in the machined material, which is used in high-speed machining. In oblique cutting, the maximum temperature of the chip is reached in roughly 2/3 of the range of cutting speeds, which decreases slightly with the further increase in the cutting speed values. However, according to theoretical knowledge, the trend should be rising. The deviation between the research results and the theoretical starting point is probably caused by the influence of the complexity of the experimental system, as well as the factors of the experimental conditions that are uncontrollable.
- From the point of view of the change in cutting depth, significantly opposite tendencies for orthogonal and oblique cutting occurred. With this phenomenon, it should be noted that the temperature gradient in the graphical interpretation of the results is about 45 °C for orthogonal cutting and about 15 °C for oblique cutting. The influence of the cutting depth on the cutting temperature in orthogonal cutting represents the current theoretical and practical knowledge very well—with increasing depth of cut, the degree of deformation of the machined material increases, which outwardly manifests as an increase in temperature in a geometric trend. In oblique cutting, this experiment showed a trend of a slight increase in chip temperature while reducing its thickness. This phenomenon is in contrast to the simplified theoretical model of cutting. This opposite trend in the behavior of the acquired dependence can be explained by the interaction of two factors causing a slight decrease in chip temperature when increasing its thickness (influence i).
- From the perspective of the angle of the tool-orthogonal rake γ_o, the research results indicate for orthogonal cutting that the maximum chip temperature is reached in roughly half of the range of angles examined. This trend is most likely caused by the interaction of influences of both used methodologies of experimental measurement and data evaluation. In oblique cutting, the effect of the dependence of chip temperature on the angle of the tool-orthogonal rake γ_o is insignificant, which is in accordance with current knowledge.

- The influence of the angle of tool-cutting-edge inclination λ_s on chip temperature is almost undetectable in accordance with theoretical knowledge.

The influence of parameters on the maximum measured temperature during orthogonal and oblique EN C45 machining is clearly shown in Table 5.

Table 5. Effect of parameters on the maximum measured temperature during orthogonal and oblique machining EN C45.

Machining Type	Tool-Cutting-Edge Angle (°)	Cutting Parameters				Temperature
		v_c (m·min^{-1})	a_p (mm)	γ (°)	λ_s (°)	T (°C)
orthogonal	$\kappa_r = 0$	6 [1]	0.4 [2]	12 [3]	- [4]	111.4
oblique	$\kappa_r = 60$	15 [1]	0.2 [2]	7 [3]	20 [4]	103.5

[1] significant influence; [2] the most significant influence; [3] little significant influence; [4] insignificant effect.

3.2. Vickers Microhardness Evaluation

Similar to the measured temperature results, as well as for the obtained HV microhardness values, regression analysis was used to determine the dependencies on the investigated parameters, which are expressed by Equations (3) and (4), for orthogonal and oblique cutting, respectively. The reliability of the relations is $R^2 = 0.87$ and 0.90, respectively. Plotted dependencies of microhardness on different combinations of two parameters are shown in Figures 13–18.

$$y = -0.0771 + 4.5677x_1 - 0.484x_2 - 0.2907x_3 - 0.1027x_4 + 1.2326x_1x_2 + 0.0893x_1x_3 + 0.0603x_1x_4 \\ -0.0892x_2x_4 + 0.0476x_3x_4 - 1.8966x_1^2 + 0.8444x_2^2 + 0.1503x_3^2 + 0.0208x_4^2 \quad (3)$$

$$y = -5.9254 - 2.2198x_1 + 12.092x_2 + 3.0554x_3 - 0.2568x_4 + 0.9296x_1x_2 - 0.0616x_1x_3 - \\ -0.0356x_1x_4 - 0.6367x_2x_3 + 0.0233x_3x_4 + 1.4725x_1^2 + 11.3708x_2^2 - 2.1329x_3^2 - 0.096x_4^2 \quad (4)$$

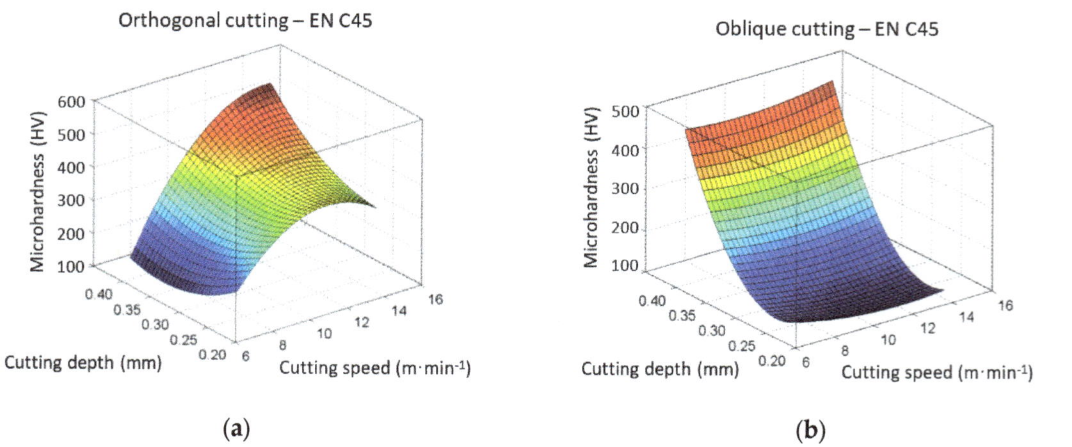

Figure 13. The graphs of the influence of cutting speed v_c and cutting depth a_p on change of microhardness according to Vickers HV, (**a**) orthogonal cutting; (**b**) oblique cutting.

The microhardness in orthogonal machining (Figure 13a) is much more influenced by the cutting speed than the cutting depth. In oblique cutting, it is precisely the opposite (Figure 13b), and the cutting depth has a much more pronounced effect, and the greatest micro-hardness is achieved at its maximum values.

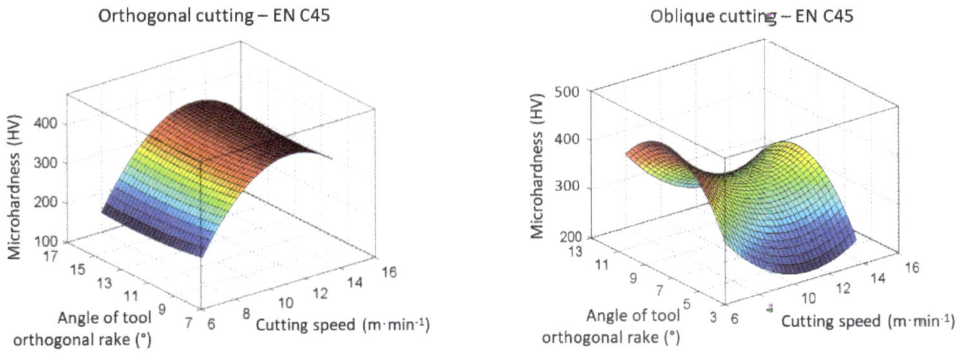

(**a**) (**b**)

Figure 14. The graphs of the influence of cutting speed v_c and the angle of the tool-orthogonal rake γ_o on change of microhardness according to Vickers HV; (**a**) orthogonal cutting; (**b**) oblique cutting.

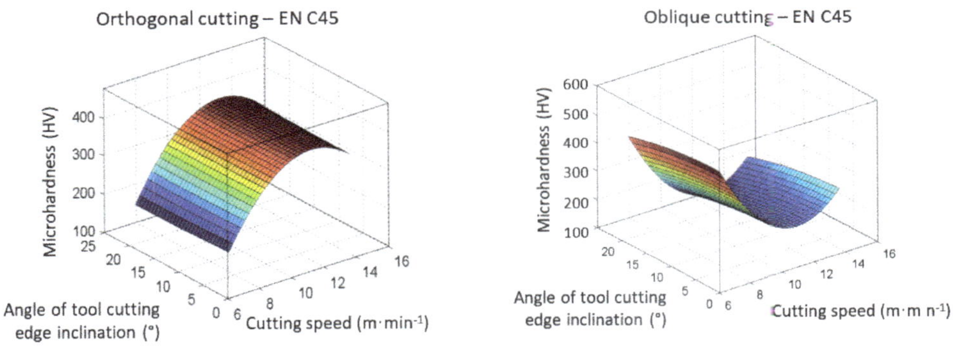

(**a**) (**b**)

Figure 15. The graphs of the influence of cutting speed v_c and the angle of tool-cutting-edge inclination λ_s on change of microhardness according to Vickers HV; (**a**) orthogonal cutting; (**b**) oblique cutting.

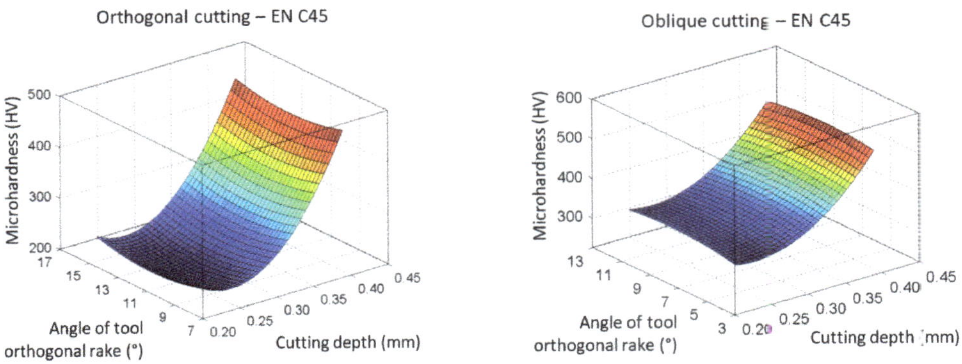

(**a**) (**b**)

Figure 16. The graphs of the influence of the angle of the tool-orthogonal rake γ_o and cutting depth a_p on the change of microhardness according to Vickers HV; (**a**) orthogonal cutting; (**b**) oblique cutting.

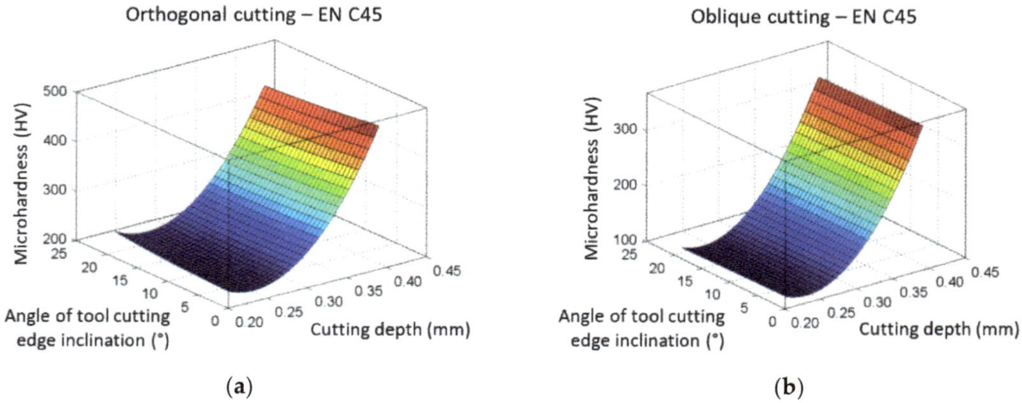

Figure 17. The graphs of the influence of the cutting depth a_p and the angle of inclination of the main cutting edge λ_s on change of microhardness according to Vickers HV; (**a**) orthogonal machining; (**b**) oblique machining.

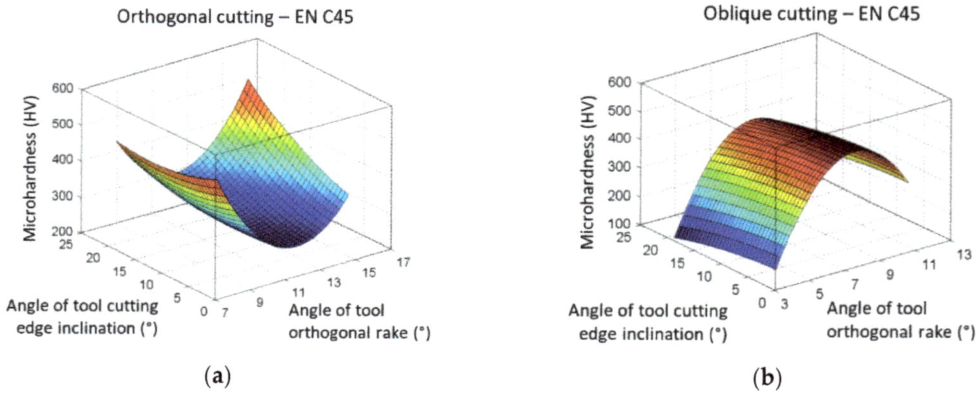

Figure 18. The graphs of the influence of the tool-orthogonal rake angle γ_o and the angle of inclination of the main cutting edge λs on change of microhardness according to Vickers HV; (**a**) orthogonal machining; (**b**) oblique machining.

The dependencies on Figure 14 show that the angle of the tool-orthogonal rake during orthogonal cutting (Figure 14a) affects the microhardness only imperceptibly compared to the effect of the cutting speed, whereas during oblique cutting (Figure 14b) the intensity due to the influence of the angle of the tool-orthogonal rake and the cutting speed is almost the same. In orthogonal machining, the maximum values of microhardness were achieved at the maximum values of the cutting speed and marginal values of the angle of the tool-orthogonal rake. During oblique cutting, the maximum microhardness was achieved at the minimum cutting speed and medium values of the angle of the tool-orthogonal rake.

The angle of inclination of the main cutting edge in both carbon steel cutting methods (Figure 15) does not significantly affect the change in microhardness HV compared to the cutting speed.

The effect of the thickness of the cut layer in interaction with the face angle on the change of microhardness appears to be the same in orthogonal (Figure 16a) and in oblique cutting (Figure 16b). In this case, the influence of the face angle is interesting, where in orthogonal cutting the microhardness maxima are reached at its extreme values, in oblique cutting at its mean values.

The influence of the angle of inclination of the main cutting edge (Figure 17) is not noticeable in both cutting methods due to the influence of the thickness of the cut layer, which is very significant.

In the interaction of the tool-orthogonal rake angle and the angle of inclination of the main cutting edge (Figure 18), the angle of inclination has a non-significant effect on the microhardness change HV in both machining methods when compared to the effect of the angle of the tool-orthogonal rake.

The results of research aimed at determining the dependence of chip microhardness on selected factors of the cutting process pointed out that:

- The growing depth of the cut causes an increase in the rate of deformation in the primary deformation zone, which leads to an increase in strengthening and therefore of hardness. The increased rate of deformation, however, also tends to generate a significantly greater amount of heat, which reduces the manifestations of strengthening and thus microhardness. The experimentally obtained results are in accordance with theoretical knowledge for both orthogonal cutting and oblique cutting.
- From the point of view of the cutting speed, the values of the theoretical microhardness are achieved by a combination of two basic opposing phenomena—an increasing cutting speed participates in a higher strengthening of the removed material, but at the same time generates a greater amount of heat, which reduces the strengthening. Then, the mutual combination of these two opposing events with material and process properties participates in the final trend of microhardness dependence. For orthogonal cutting, the dependence shows a maximum in about 2/3 of the cutting speed range with a gradient of roughly 200 HV. In oblique cutting, the minimum microhardness of the chip with a gradient of approx. 125 HV is achieved in approximately 1/2 of the range of cutting speeds.
- From the perspective of the angle of the tool-orthogonal rake γ_o, the research results point to insignificant changes in microhardness with respect to this angle for orthogonal cutting. For oblique cutting, the trend of microhardness dependence on this angle shows a maximum in roughly half of the range of investigated values with a gradient of approximately 150 HV.
- The influence of the angle of tool-cutting-edge inclination λ_s on HV microhardness is almost undetectable in accordance with theoretical knowledge.

The overview of the influence of selected parameters on Vickers microhardness in the area of occurrence of the shear angle on samples made of carbon steel EN C45 is shown in Table 6.

Table 6. Effect of parameters on the maximum measured temperature during orthogonal and oblique machining EN C45.

Machining Type	Tool-Cutting-Edge Angle (°)	Cutting Parameters				Microhardness
		v_c (m·min^{-1})	a_p (mm)	γ (°)	λ_s (°)	(HV)
orthogonal	$\kappa_r = 0$	15 [1]	0.4 [2]	8/16 [3]	20 [4]	550.6
oblique	$\kappa_r = 60$	6 [2]	0.4 [1]	7 [3]	0 [4]	472.1

[1] the most significant influence; [2] significant influence; [3] little significant influence; [4] insignificant effect.

4. Conclusions

Planing is one of the slow-speed machining types, and although it is a technology that is rarely used today, its effectiveness for specific types of workpieces can be much higher than that of milling. However, the phenomena taking place at slow-speed machining are similar to, for example, broaching, but in this case the chip remains between the teeth of the broaching mandrel and inside the workpiece, and therefore the process of chip formation cannot be observed or investigated (or only under very complicated conditions).

The presented research aimed to analyse the influence of selected parameters on the temperature during cutting operations performed at relatively low cutting speeds and to compare the results at orthogonal and oblique machining. The new methodology for obtaining the chip root was designed so that the chip samples correspond as much as possible to the conditions of the ongoing event and to the current speed at the given moment of machining,

The experiment was designed by the orthogonal experimental composition plan of the investigation through the "star points". The effect of four parameters (cutting speed v_c, the angle of inclination of the main cutting edge λ_s, the depth of cut a_p and the angle of the tool-orthogonal rake γ) on temperature T (°) and Vickers microhardness HV was investigated in the area where the shear angle was formed.

Carbon steel EN C45, which is usually used in the experiments as a standard, was tested in two types of cutting, i.e., orthogonal cutting with a planing necking tool and oblique cutting with a straight roughing tool without cooling. Based on the measured values, the statistical regression dependencies were compiled, where various test criteria were used to verify the measured results, e.g., Grubbs criterion for detecting outlying measured results or gross errors, Cochran's criterion for homogeneity of variance, Student's test criterion for determining the significance of regression coefficients, Fisher-Snedecor test criterion for confirming the adequacy of the statistical model. Subsequently, graphical dependencies were constructed and evaluated.

Based on the dependencies of the temperature on the selected parameters on samples made of carbon steel EN C45 in the area of occurrence of the shear angle, it can be concluded:

- The most significant influence has the cutting depth in both orthogonal and oblique cutting, followed by the cutting speed, and the angle of the tool-orthogonal rake.
- The observations have shown that the influence of the angle of inclination of the main cutting edge is inconspicuous (or almost non-existent).
- The maximum temperature was reached at the maximum cutting depth at orthogonal cutting, in contrast to oblique cutting, in which the maximum temperature was measured at the minimum cutting depth. In both cases of cutting at the maximum temperature, the mean values of the angle of the tool-orthogonal rake played a role.

When evaluating the influence of selected parameters on Vickers microhardness, it can be stated:

- The significance of the parameters for the change of the shear angle was manifested differently. In orthogonal cutting, the highest microhardness was measured at the highest cutting speed. In this case, the highest values of the cutting depth and the angle of cutting-edge inclination were also acquired.
- During oblique cutting, the most significant influence on the change of microhardness HV had the cutting depth.

Information on the temperature development in the chip and microhardness values achieved under certain conditions indicate the suitability (or unsuitability) of choosing the selected combination of technological parameters, and together with the geometry of the tool, it is possible to subsequently increase/regulate the quality of the machined surface and the service life of the tool.

Although in the presented experimental research, the effort was to capture as much as possible the real state of the chip at the moment of separation from the machined surface and to prepare samples not only for measuring microhardness, but in the future also for observing microstructure changes, the authors realize that the machining process is so complex that obtaining samples that would perfectly reflect the state of chip formation at a given moment is almost impossible.

Author Contributions: Conceptualization, P.P.M. and A.S.; methodology, P.P.M. and K.M.; software, P.P.M. and M.K. (Martin Korol); validation, M.V. and M.K. (Milena Kubisova); investigation, P.P.M. and K.M.; resources, M.K. (Martin Korol) and K.M.; data curation, P.P.M. and M.K. (Milena Kubisova); writing—original draft preparation, K.M.; writing—review and editing, K.M.; visualization, M.V. and K.M.; supervision, P.P.M.; project administration, K.M.; funding acquisition, K.M. and P.P.M. All authors have read and agreed to the published version of the manuscript.

Funding: This research was funded by the grants APVV-19-0550, SK-CN-21-0046, KEGA 005TUKE-4/2021 and KEGA 032TUKE-4/2022.

Institutional Review Board Statement: Not applicable.

Informed Consent Statement: Not applicable.

Data Availability Statement: Not applicable.

Acknowledgments: The article was prepared thanks to support of the Ministry of Education of the Slovak Republic through the grants APVV-19-0550, SK-CN-21-0046, KEGA 005TUKE-4/2021 and KEGA 032TUKE-4/2022.

Conflicts of Interest: The authors declare no conflict of interest.

References

1. Varga, G.; Sovilj, B.; Jakubowicz, M.; Babič, M. Experimental Examination of Surface Roughness in Low-Environmental-Load Machining of External Cylindrical Workpieces. In *Advances in Manufacturing II*; Lecture Notes in Mechanical Engineering; Springer: Cham, Switzerland, 2019; Volume 2, pp. 307–321. [CrossRef]
2. Prasetyo, L.; Tauviqirrahman, M. Study of chip formation feedrates of various steels in low-speed milling process. *IOP Conf. Ser. Mater. Sci. Eng.* **2017**, *202*, 012097. [CrossRef]
3. Pastucha, P.; Majstorovic, V.; Kucera, M.; Beno, P.; Krile, S. Study of Cutting Tool Durability at a Short-Term Discontinuous Turning Test. In *Advances in Manufacturing Engineering and Materials*; Lecture Notes in Mechanical Engineering; Springer: Cham, Switzerland, 2019; pp. 493–501. [CrossRef]
4. Karkalos, N.E.; Markopoulos, A.P.; Manolakos, D.E. Cutting Speed in Nano-Cutting as MD Modelling Parameter. *Int. J. Manuf. Mater. Mech. Eng.* **2016**, *6*, 1–13. [CrossRef]
5. Toulfatzis, A.I.; Pantazopoulos, G.A.; Besseris, G.J.; Paipetis, A.S. Machinability evaluation and screening of leaded and lead-free brasses using a non-linear robust multifactorial profiler. *Int. J. Adv. Manuf. Technol.* **2016**, *86*, 3241–3254. [CrossRef]
6. Neslušan, M.; Mrkvica, I.; Čep, R.; Kozak, D.; Konderla, R. Deformations after heat treatment and their influence on cutting process. *Teh. Vjesn.* **2011**, *18*, 601–608.
7. Tomczewski, L. The application of planing tools during turning. *Adv. Sci. Technol. Res. J.* **2015**, *9*, 83–86. [CrossRef]
8. Chukarin, A.; Meskhi, B.; Shoniya, D. Theoretical analysis on regularities of the process of noise generation of planing, slotting and planing-milling machines. *Akustika* **2021**, *41*, 173–177. [CrossRef]
9. Zetek, M.; Zetkova, I. Increasing of the Cutting Tool Efficiency from Tool Steel by Using Fluidization Method. *Procedia Eng.* **2015**, *100*, 912–917. [CrossRef]
10. Kundrák, J.; Markopoulos, A.P.; Karkalos, N.E.; Makkai, T. The Examination of Cutting Force as Function of Depth of Cut in Cases with Constant and Changing Chip Cross Section. In *Advances in Manufacturing II, Volume 4—Mechanical Engineering*; Springer: Cham, Switzerland, 2019; pp. 405–415.
11. Saglam, H.; Yaldiz, S.; Unsacar, F. The effect of tool geometry and cutting speed on main cutting force and tool tip temperature. *Mater. Des.* **2007**, *28*, 101–111. [CrossRef]
12. Jurko, J.; Panda, A.; Behun, M. Prediction of a new form of the cutting tool according to achieve the desired surface quality. *Appl. Mech. Mater.* **2013**, *268–270*, 473–476. [CrossRef]
13. Filippov, V.; Filippova, E.O. Determination of Cutting Forces in Oblique Cutting. *Appl. Mech. Mater.* **2015**, *756*, 659–664. [CrossRef]
14. Bagci, E. 3-D numerical analysis of orthogonal cutting process via mesh-free method. *Int. J. Phys. Sci.* **2011**, *6*, 1267–1282.
15. Filice, L.; Micari, F.; Rizzuti, S.; Umbrello, D. A critical analysis on the friction modeling in orthogonal machining. *Int. J. Mach. Tools Manuf.* **2007**, *47*, 709–714. [CrossRef]
16. Aydin, M.; Koklu, U. A study of ball-end milling forces by finite element model with Lagrangian boundary of orthogonal cutting operation. *J. Fac. Eng. Archit. Gazi Univ.* **2018**, *33*, 507–516.
17. Seah, K.H.W.; Rahman, M.; Li, X.P.; Zhang, X.D. A three-dimensional model of chip flow, chip curl and chip breaking for oblique cutting. *Int. J. Mach. Tools Manuf.* **1996**, *36*, 1385–1400. [CrossRef]
18. Saglam, H.; Unsacar, F.; Yaldiz, S. Investigation of the effect of rake angle and approaching angle on main cut-ting force and tool tip temperature. *Int. J. Mach. Tools Manuf.* **2006**, *46*, 132–141. [CrossRef]
19. Dewangan, S.; Chattopadhyaya, S.; Hloch, S. Wear Assessment of Conical Pick used in Coal Cutting Operation. *Rock Mech. Rock Eng.* **2015**, *48*, 2129–2139. [CrossRef]
20. Vukelic, G.; Vizentin, G.; Ivosevic, S.; Bozic, Z. Analysis of prolonged marine exposure on properties of AH36 steel. *Eng. Fail. Anal.* **2022**, *135*, 106132. [CrossRef]
21. Jain, V.K.; Kumar, S.; Lal, G.K. Effects of Machining Parameters on the Microhardness of Chips. *J. Eng. Ind.* **1989**, *111*, 220. [CrossRef]
22. Wang, C.; Xie, Y.; Zheng, L.; Qin, Z.; Tang, D.; Song, Y. Research on the Chip Formation Mechanism during the high-speed milling of hardened steel. *Int. J. Mach. Tools Manuf.* **2014**, *79*, 31–48. [CrossRef]

23. Alrabii, S.A.; Zumot, L.Y. Chip Thickness and Microhardness Prediction Models during Turning of Medium Carbon Steel. *J. Appl. Math.* **2007**, *2017*, 051905. [CrossRef]
24. Pu, C.L.; Zhu, G.; Yang, S.B.; Yue, E.B.; Subramanian, S.V. Effect of Dynamic Recrystallization at Tool-Chip Interface on Accelerating Tool Wear During High-Speed Cutting of AISI1045 Steel. *Int. J. Mach. Tools Manuf.* **2016**, *100*, 72–80. [CrossRef]
25. Senussi, G.H. Interaction Effect of Feed Rate and Cutting Speed in CNC-Turning on Chip Micro-Hardness of 304-Austenitic Stainless Steel. *World Acad. Sci. Eng. Technol.* **2007**, *28*, 121–126.
26. Mabrouki, T.; Courbon, C.; Zhang, Y.; Rech, J.; Nélias, D.; Asad, M.; Hamdi, H.; Belhadi, S.; Salvatore, F. Some insights on the modelling of chip formation and its morphology during metal cutting operations. *C. R. Mécanique* **2016**, *344*, 335–350. [CrossRef]
27. Umer, U.; Qudeiri, J.A.; Ashfaq, M.; Al-Ahmari, A. Chip morphology predictions while machining hardened tool steel using finite element and smoothed particles hydrodynamics methods. *J. Zhejiang Univ. Sci. A* **2016**, *17*, 873–885. [CrossRef]
28. Adekunle, A.S.; Adedayo, S.M.; Ohijeagbon, I.O.; Olusegun, H.D. Chip morphology and behaviour of tool temperature during turning of AISI 301 using different biodegradable oils. *J. Prod. Eng.* **2015**, *18*, 18–22.
29. Bai, W.; Sun, R.; Roy, A.; Silberschmidt, V.V. Improved analytical prediction of chip formation in orthogo-nal cutting of titanium alloy Ti6Al4V. *Int. J. Mech. Sci.* **2017**, *133*, 357–367. [CrossRef]
30. Karmiris-Obratański, P.; Papazoglou, E.L.; Leszczyńska-Madej, B.; Karkalos, N.E.; Markopoulos, A.P. An Optimalization Study on the Surface Texture and Machining Parameters of 60CrMoV18–5 Steel by EDM. *Materials* **2022**, *15*, 3559. [CrossRef]
31. Saez-de-Buruaga, M.; Soler, D.; Aristimuño, P.X.; Esnaola, J.A.; Arrazola, P.J. Determining tool/chip temperatures from thermography measurements in metal cutting. *Appl. Therm. Eng.* **2018**, *145*, 305–314. [CrossRef]
32. Kraišnik, M.; Čep, R.; Kouřil, K.; Baloš, S.; Antić, A.; Milutinović, M. Characterization of Microstructural Damage and Failure Mechanisms in C45E Structural Steel under Compressive Load. *Crystals* **2022**, *12*, 426. [CrossRef]
33. Monkova, K.; Monka, P.P.; Sekerakova, A.; Tkac, J.; Bednarik, M.; Kovac, J.; Jahnatek, A. Research on Chip Shear Angle and Built-Up Edge of Slow-Rate Machining EN C45 and EN 16MnCr5 Steels. *Metals* **2019**, *9*, 956. [CrossRef]
34. Abdallah, F.; Abdelwahab, S.A.; Aly, W.I.; Ahmed, I. Influence of cutting factor on the cutting tool temperature and surface roughness of steel C45 during Turning process. *Int. J. Eng. Res. Technol.* **2019**, *7*, 1065–1072.
35. Puneeth Kumar, N.; Srikantappa, A.S. Temperature and Tool Wear Analysis on Machining of Two Different Composition of C45 Steel. *Int. J. Eng. Res. Technol.* **2022**, *11*, 208–211.
36. D'Amato, R.; Amato, G.; Ruggiero, A. Adaptive Noise Cancellation-Based Tracking Control for a Flexible Rotor in Lubricated Journal Bearings. In Proceedings of the 23rd International Conference on Mechatronics Technology (ICMT), Fisciano, Italy, 23–26 October 2019.
37. Obeng, D.P.; Morrell, S.; Napier-Munn, T.J. Application of central composite rotatable design to modeling the effect of some operating variables on the performance of the three-product cyclone. *Int. J. Miner. Process.* **2005**, *76*, 181–192. [CrossRef]
38. Davim, J.P.; Silvia, J.; Baptista, A.M. Experimental cutting model of metal matrix composites (MMCs). *J. Mater. Process. Technol.* **2007**, *183*, 358–362. [CrossRef]
39. Arioli, M.; Gratton, S. Linear regression models, least-squares problems, normal equations, and stopping criteria for the conjugate gradient method. *Comput. Phys. Commun.* **2012**, *183*, 2322–2336. [CrossRef]
40. Paris, A.S.; Tarcolea, C.; Croitoru, S.M.; Majstorović, V.D. Statistical Study of Parameters in the Process of Orthogonal Cutting Surface Hardness. In Proceedings of the 4th International Conference on the Industry 4.0 Model for Advanced Manufacturing, Industry 4.0 and Internet of Things for Manufacturing (AMP), Belgrade, Serbia, 3–6 June 2019; pp. 68–77.

metals

Article

Tribological Behaviors of Inconel 718–Tungsten Carbide Friction Pair with Sulfur Additive Lubrication

Ye Yang [1,*], Hao Luan [1], Songshan Guo [1], Fengbin Liu [1], Yuanjing Dai [2], Chenhui Zhang [3], Duzhou Zhang [4] and Gang Zhou [4]

1 School of Mechanical and Material Engineering, North China University of Technology, Beijing 100144, China
2 Tianjin Research Institute for Advanced Equipment, Tsinghua University, Tianjin 300308, China
3 State Key Laboratory of Tribology, Tsinghua University, Beijing 100084, China
4 Beijing Key Laboratory of Long-Life Technology of Precise Rotation and Transmission Mechanisms, Beijing Institute of Control Engineering, Beijing 100094, China
* Correspondence: yangye@ncut.edu.cn

Abstract: This work investigated the lubricating and anti-wear properties of several sulfur additives for a nickel-based superalloy–tungsten carbide friction pair. Compared with PAO40 without any active chemical compounds, the three kinds of sulfur additives could decrease the friction coefficient from 0.2 to 0.1 and the wear volume by 90%. Sulfurized fatty acid ester had the best performance under high temperature and heavy load with COF below 0.1 and the smallest wear volume. Furthermore, the lubricating mechanism was investigated by XPS. The physical adsorptive film and the tribochemical film together enhanced the friction-reducing and anti-wear performances of the lubricants. This effective lubricant for Inconel 718 can be applied to the machining of nickel-based alloy.

Keywords: Inconel 718; sulfur additives; boundary lubrication; wear mechanism

Citation: Yang, Y.; Luan, H.; Guo, S.; Liu, F.; Dai, Y.; Zhang, C.; Zhang, D.; Zhou, G. Tribological Behaviors of Inconel 718–Tungsten Carbide Friction Pair with Sulfur Additive Lubrication. *Metals* **2022**, *12*, 1841. https://doi.org/10.3390/met12111841

Academic Editor: George A. Pantazopoulos

Received: 7 October 2022
Accepted: 24 October 2022
Published: 28 October 2022

1. Introduction

Nickel-based superalloys such as Inconel 718 are widely used in aerospace, gas turbine, nuclear, and automotive industries because of their excellent high-temperature mechanical strength and corrosion resistance [1]. At the same time, nickel-based superalloys are recognized as difficult-to-machine materials. These metals exhibit serious problems during machining such as high cutting force, rapid tool wear, short tool life, and poor surface quality of the machined surface due to the physical, chemical, and thermal properties of metals [2–5]. Tungsten carbide tools are the first choice for their good thermal conductivity, high strength, and poor affinity with nickel [6]. It is of great importance to understand the relationship between nickel-based alloy and carbide material, especially the tribological behaviors, to reduce tool wear and improve tool life. The use of suitable cutting fluids can effectively improve the machining conditions of nickel-based superalloys [7–9]. The cutting fluid acts as a lubricant to reduce friction and as a coolant to cool the temperature at the cutting zone. The lubricating ability of a cutting fluid greatly influences the quality of the machined surface, as well as the tool life [10,11].

The chemical additives in the cutting fluid can act with the metal surface at high temperature and pressure; thus, a lubricant film is formed to reduce friction between the rake face and chips, the flank face, and machined surfaces. Sulfur additives are well known for their extreme-pressure performance and anti-wear characteristics [12–14]. The sulfur compounds, under extreme-pressure conditions, undergo chemical decomposition causing sulfur release (rupture of the R–S bond) and their reaction with the metallic surface that promotes the formation of an inorganic iron sulfide layer [12,15]. Most of the cutting fluids are designed for ferrous metals [11,16–18]. However, there is a matching problem between the additives and the workpiece materials [19,20]. Whether these additives have the same reaction with nickel-based superalloys is not clear. Few scholars have studied the lubricants for nickel-based alloy–tungsten carbide contacts. Moreover, metal working fluid which consists of several chemically active additives has a very complex chemical composition. The individual actions of each component are not easily identifiable. However, it is necessary for new metals such as nickel-based superalloys whose characteristics are very different from traditional ferrous metals. Therefore, it is important to separate contributions from different types of additives to the lubrication so as to select the most efficient lubricant molecules and optimize the components of the cutting fluid. This work focuses on the lubricating effect of different sulfur additives for the nickel-based superalloy (Inconel 718)–tungsten carbide (YG8) tribopair, and the lubricating mechanism is further investigated.

2. Experimental Details

Inconel 718 is the most widely used superalloy. YG8 (WC-Co) tungsten carbide is the optimal tool material for nickel-based superalloy machining. The specimen used in this paper was bought from Hengshihui company, Jiangsu, China. Table 1 shows the chemical composition of Inconel 718 superalloy. The main component is Ni (60.01 wt.%) followed by Cr (17.28 wt.%) and Fe (15.58 wt.%). The Table 2 displays the mechanical parameters of Inconel 718.

Table 1. Chemical compositions of Inconel 718 superalloy (wt.%).

Ni	C	O	Al	Ti	Cr	Mo	Nb	Fe
60.01	2.31	0.62	0.29	0.51	17.28	1.25	2.15	15.58

Table 2. The mechanical parameters of Inconel 718.

Young's Modulus (GPa at 20 °C)	Poisson's Ratio	Hardness (HRC)	Yield Strength (MPa)
199.9	0.3	26.4	1035

The commercial sulfur additives were bought from Symarin company, Shanghai, China. Each additive contained a different amount of sulfur in the molecule (by weight), and the sulfur bonding mechanism was different in each additive. According to the manufacturer's data, the sulfur content of sulfurized olefin was as much as 40%, more than that of sulfurized fatty acid ester (17%) and sulfurized lard (10%). The kinematic viscosity of sulfurized olefin was much smaller than that of the other two kinds of additives for the smallest molecular weight. Table 3 shows the available information about the lubricants. Furthermore, PAO40, the viscosity of which is 396 mm^2/s, was used as pure oil without any active elements for comparison.

Table 3. The basic parameters of sulfur additives.

Name	Sulfur Content	Kinematic Viscosity mm^2/s (at 40 °C)	Appearance	Flash Point (°C)
Sulfurized olefin	40%	45	pale yellow	150
Sulfurized fatty acid ester	17%	582	tan	210
Sulfurized lard	10%	900–1300	dark brown	>160

The frictional tests were carried out utilizing a ball-on-disc apparatus SRV-IV (Optimol, Munich, Germany) under different lubricating conditions. The schematic diagram of the tester is displayed in Figure 1. The discs were Inconel 718 with a hardness of HRC 35. All of the specimens were polished before frictional tests by an automatic polishing/grinding machine, and a surface roughness (Sa) less than 40 nm was obtained. The counter specimen was tungsten carbide YG8 ball with a diameter of 10 mm and surface roughness (Sa) of 25 nm. The hardness of the carbide ball was 89HRA. The samples were ultrasonically cleaned using acetone and ethanol and then ultrapure water successively, each for 10 min, before tests. The upper ball slid reciprocally against the stationary disc with an amplitude of 2 mm and frequency of 20 Hz for 5 min. Before the frictional test, plenty of lubricant was dropped onto the surface of the disc, and the ball returned to yield a normal load of 100 N.

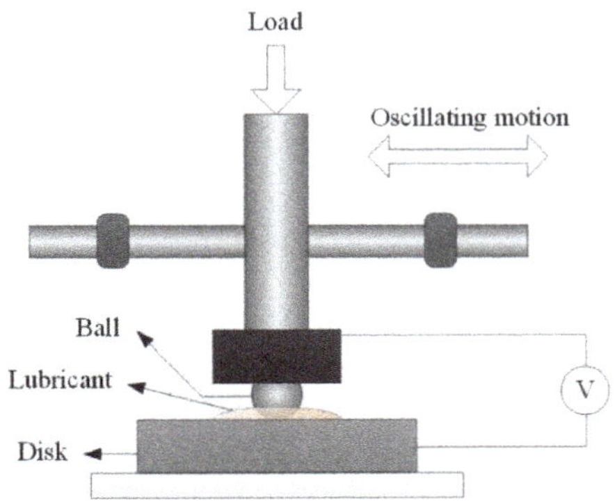

Figure 1. The schematic diagram of SRV-IV tribo-tester.

After the frictional tests, the morphology of rubbing surfaces was observed using a LEXT™OLS5100 laser scanning confocal microscope (Olympus, Tokyo, Japan), and wear volume is calculated. Each test was repeated three times, and the average values were used. The relative errors were on the order of ±5%. A Quanta200 scanning electron microscope (SEM) (FEI, Hillsboro, OR, USA) combined with energy dispersion spectrometry (EDS) (FEI, Hillsboro, OR, USA) was used for the surface analysis of the investigated materials. The chemical compositions of the worn surfaces were characterized using a PHI Quantera SXM X-ray photoelectron spectrometer (ULVAC-PHI, Chigasaki, Japan).

3. Results and Discussions

3.1. Friction and Wear Properties Lubricated with Sulfur Additives

The curves of the friction coefficient and wear volume of the lower samples lubricated by the four kinds of lubricants are shown in Figure 2. For PAO40, the friction coefficient increased to 0.7 rapidly in the running-in period. Then, it decreased gradually and stabilized at about 0.2 after 200 s. When using sulfurized olefin as lubricant, the initial friction coefficient fluctuated to 0.17, before remaining stable at about 0.14 after 100 s. For sulfurized fatty acid ester and sulfurized lard, the friction coefficients remained stable at 0.1 and 0.11, respectively, until the end of the experiment. Compared with PAO40, the three kinds of sulfur-containing additives could reduce the friction coefficient significantly for the nickel-based superalloy. The friction coefficient of sulfurized fatty acid ester was the smallest, while that of sulfurized olefin was 27% larger.

Considering that tungsten carbide is much harder than nickel-based superalloys, wear occurs on the surface of the disc apparently while no wear is observed on the ball. After the frictional tests, the remaining solution and debris on the wear track are washed away by water and alcohol. From Figure 2b, it can be seen that the wear volume of the disc lubricated by PAO40 was 2.4×10^7 μm^3, about ten times that lubricated by the sulfur additives. Thus, basic oil without any active chemical elements failed to lubricate nickel-based superalloy under boundary lubrication. Compared with PAO40, the sulfur additives showed good lubricating and anti-wear performance for the Inconel 718–tungsten carbide friction pair. It is easy to generate inorganic protective films in the tribochemical reaction for an excellent anti-wear effect. For sulfurized fatty acid ester and sulfurized lard, the wear volumes were 2.6×10^6 μm^3 and 2.8×10^6 μm^3, respectively. With sulfurized olefin lubrication, the wear volume was 4.1×10^6 μm^3. The relatively big wear volume is related to the long running-in period during the friction test. A lubricating film was formed, but it was not stable enough to prevent direct contact between Inconel 718 and tungsten carbide completely during the running-in period. It can be concluded that sulfurized fatty acid had the best lubrication for Inconel 718 with the smallest COF and wear volume.

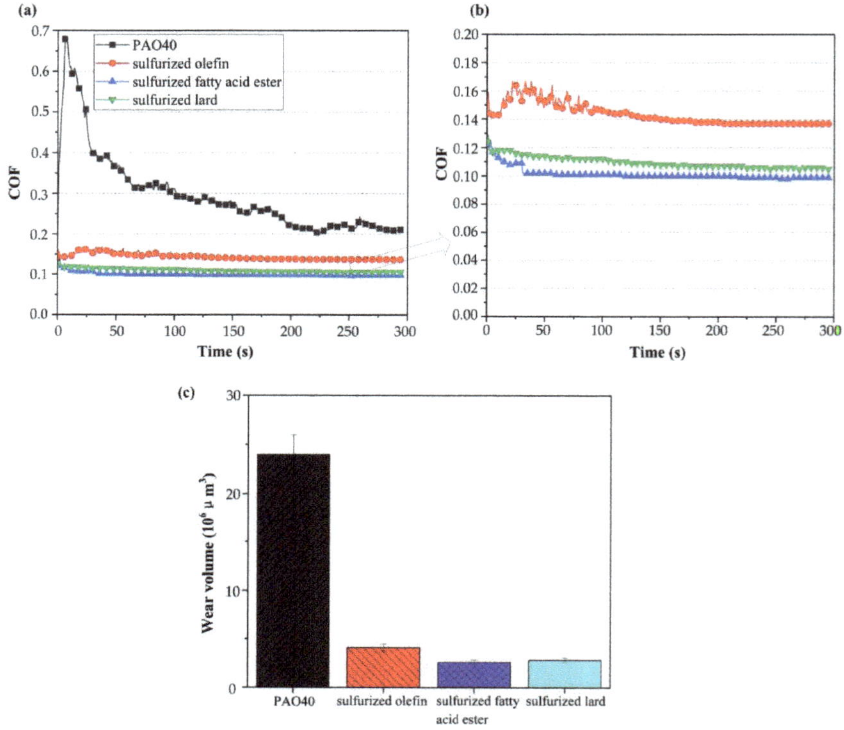

Figure 2. (**a**) Friction coefficients with (**b**) enlarged view, and (**c**) wear volume of the discs lubricated by different lubricants ($F = 100$ N, $f = 20$ Hz, $L = 2$ mm).

The micro-morphology of the friction pairs was obtained by SEM as shown in Figure 3. It can be seen that there were furrows on the surface of the superalloy discs and irregular block materials on the surface of the balls lubricated by the three kinds of additives. From Figure 3(a1), delaminated scars and abrasive particles can be seen on the surface of the disc lubricated by sulfurized olefin. There were obvious furrows on the worn surface lubricated by sulfurized lard (Figure 3(c1)). Furthermore, large dark blocky materials were detected on the ball surface (Figure 3(c2)). The worn surface was compared to the smoothest with no adhesive scar when lubricated by sulfurized fatty acid ester (Figure 3(b1)). However, there were still patches on the surface of the ball (Figure 3(b2)). To ascertain the ingredients of the materials on the surface (the red frame in Figure 3), EDS analysis was conducted. The results show that the nickel-based superalloy on the ball surface was transferred from the discs and adhered to the balls (Figure 3d). The materials on the balls were the same; thus, the EDS result is listed once. The main wear mechanism of the superalloy was significant adhesion, delamination, and ploughing. From the EDS results of the worn surface of the discs, the sulfur element was detected, which did not belong to the alloy. This demonstrates that sulfides formed on the metal surface during the friction process. However, the sulfur contents on the surface of the discs lubricated by the additives were different, seen from Table 4. The sulfur content of the worn surface lubricated by sulfurized olefin was the highest (12.37%), while the adhesion on the ball was relatively mild. It should be considered that the active sulfur content of sulfurized olefin was also the highest. The sulfur content of the worn surface lubricated by sulfurized lard was the lowest (1.34%) with the most adhesive blocky material on the ball. For the sulfurized fatty acid ester, the adhesion on the ball was significantly decreased compared with the sulfurized lard. This indicates that the sulfide generated on the metal surface during the friction process played an important role in reducing the adhesion of nickel-based superalloy to tungsten carbide ball during friction. Sulfurized fatty acid ester contains active sulfur that can react with the metal surface at a lower temperature; hence, the surface quality of the disc and the adhesion of the ball were better than those lubricated by sulfurized lard. Meanwhile, the sulfurized fatty acid ester molecule has a long carbon chain which helps to separate the two surfaces; thus, direct contact was avoided compared with sulfurized olefin.

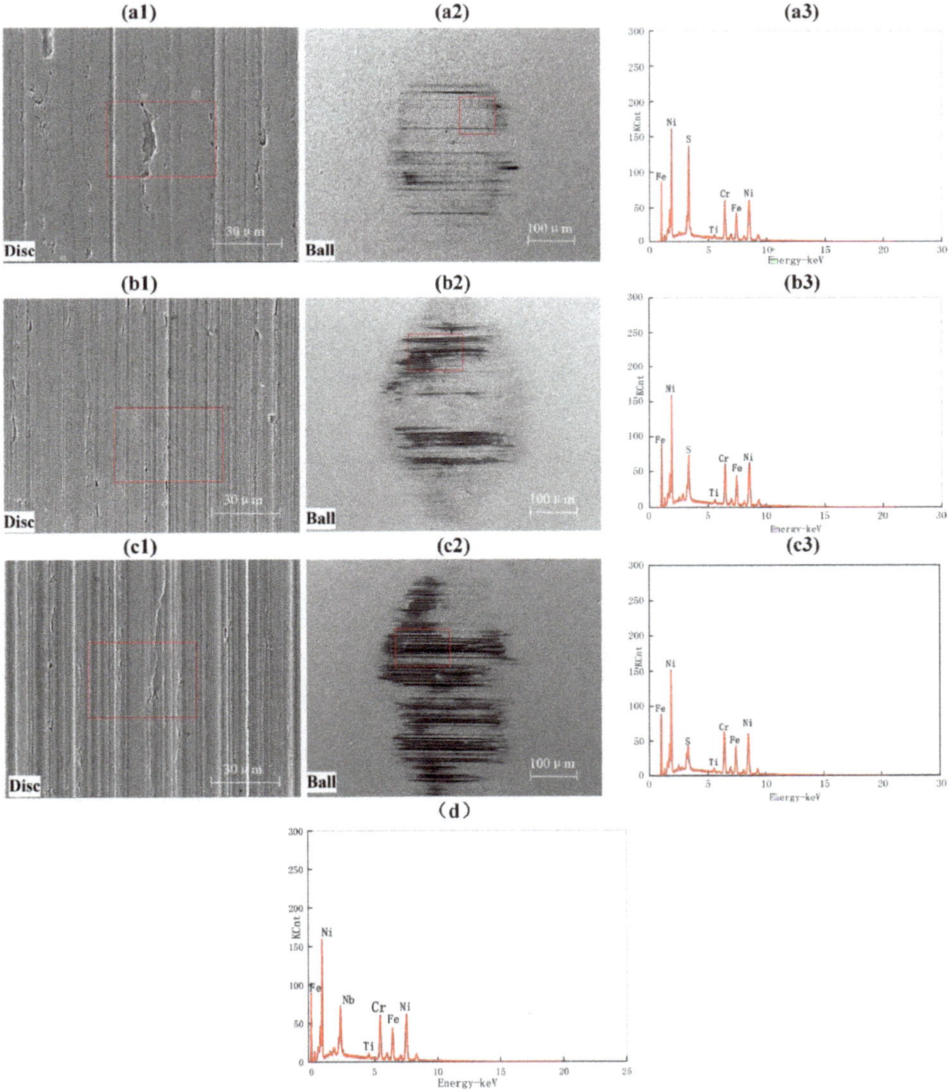

Figure 3. SEM morphology of the tracks on Inconel 718 and tungsten carbide ball lubricated by (**a1**,**a2**) sulfurized olefin, (**b1**,**b2**) sulfurized fatty acid ester, and (**c1**,**c2**) sulfurized lard; (**a3**,**b3**,**c3**) EDS spectrum of the red box in (**a1**,**b1**,**c1**); (**d**) EDS spectrum of the red box in (**a2**).

Table 4. The elements content of the above EDS analysis.

Element (wt.%)	Ni	Fe	Cr	S	C	Al	Si	Ti	Nb
(a3)	45.23	18.2	16.8	12.37	6.09	0.36	0.16	0.96	–
(b3)	43.71	17.67	16.75	2.71	6.75	0.37	0.97	0.87	10.19
(c3)	46.44	18.93	18.36	1.34	4.54	0.52	0.14	1.35	8.38
Element (wt.%)	**Ni**	**Fe**	**Cr**	**Nb**	**Ti**	**C**	**W**	**O**	–
(d)	33.92	16.54	16.01	8.54	1.13	7.31	9.04	1.5	–

3.2. Temperature Effect

Temperature is very important for the chemical reaction of additives with tribo-surface. The influence of temperature on the friction and wear behavior of Inconel 718 sliding against WC-Co under the lubrication of sulfur additives was investigated. The experimental temperature was increased to 150 °C with the other experimental parameters the same as those in Section 3.1. Figure 4a shows the friction coefficient curve at 150 °C. For sulfurized olefin, the fluctuation of the friction coefficient was reduced at the beginning of the test compared with that at 30 °C. Furthermore, the friction coefficient stabilized at about 0.18 after 20 s. For sulfurized fatty acid ester and sulfurized lard, the friction coefficient remained at 0.1 from the beginning to the end of the test without running-in time at 150 °C. This indicates that the lubricant film formed rapidly and remained more stable under high temperature as expected. The wear volumes of the Inconel 718 discs are shown in Figure 4b. Compared with that at 30 °C, the wear volume of the discs with different sulfur lubricants did not change much. The wear volume lubricated by sulfurized fatty acid ester was slightly smaller compared to the other two kinds of additives. It can be seen that the sulfur lubricants had very good temperature stability for nickel-based superalloys.

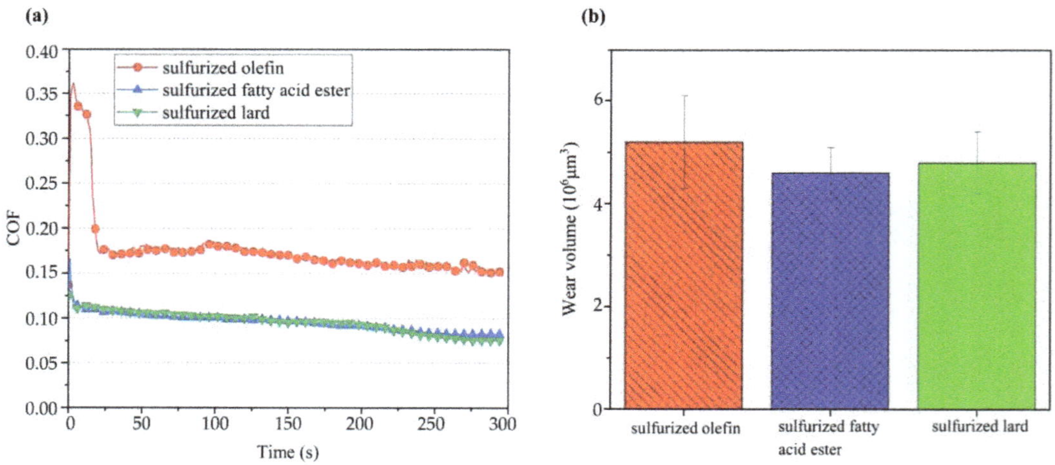

Figure 4. (**a**) Friction coefficients and (**b**) wear volume when lubricated by sulfur additives at 150 °C (F = 100 N, f = 20 Hz, L = 2 mm).

The morphology of the lower disc and the upper tungsten carbide ball after frictional tests at 150 °C was observed by SEM, and the pictures are shown in Figure 5. Compared with 30 °C, the furrows on the surface of the discs decreased, especially for sulfurized fatty acid ester and sulfurized lard (Figure 5(b1,c1)). This indicates that high temperature promoted the formation of a lubricant film, thus increasing the surface quality of the nickel-based superalloy. With respect to the surface of the balls, the adhesive materials were also decreased when lubricated by sulfurized fatty acid ester and sulfurized lard (Figure 5(b2,c2)). However, the block material on the surface of the ball lubricated by sulfurized olefin increased significantly, indicating that adhesion was aggravated, as seen in Figure 5(a2). EDS analysis was performed on the wear scar of the nickel-based superalloy and the adhesion area on the tungsten carbide (the red box in Figure 5). The dark materials on the balls were the same; thus, the EDS results are listed once (Figure 5d). A nickel-based superalloy was transferred from the disc and adhered to the ball. It can be speculated that, under high temperature, the main wear mechanism of the superalloy was adhesion. The elements content of the EDS analysis is listed in Table 5. From the EDS spectrum of the surface lubricated by sulfurized fatty acid ester (Figure 5(b3)), it can be seen that the peak of sulfur in the wear scar increased significantly and the sulfur content increased from 2.71% (30 °C) to 16.33% (150 °C). The sulfide acted as an effective lubricant film, which significantly decreased the adhesion on tungsten carbide balls and the ploughing on nickel-based superalloy discs. For sulfurized lard, the sulfur content increased from 1.34% (30 °C) to 5.43% (Figure 5(c3)), and the surface quality of the frictional pairs improved. As for sulfurized olefin, the sulfur peak did not change very much, and the content of sulfur was 11.63%. This indicates that high temperature did not promote the tribochemical reaction of sulfurized olefin with nickel-based superalloy. Considering that the flash point of sulfurized olefin is 150 °C, when the experimental temperature increased to 150 °C, the sulfurized olefin began to volatilize continuously, and the material left in the friction area was unstable. This led to the aggravation of adhesion at 150 °C lubricated by sulfurized olefin. The flash points of the other two sulfur lubricants were both higher than 150 °C, and the molecules remained stable with temperature increasing. Moreover, the tribochemical reaction was promoted, leading to better surface quality and less adhesion.

Table 5. The elements content of the above EDS analysis.

Element (wt.%)	Ni	Fe	S	C	Ti	Cr	Nb	–
(a3)	41.8	16.12	11.63	6.22	0.72	16.26	3.58	–
(b3)	43.71	17.67	16.33	6.45	0.71	15.4	10.19	–
(c3)	47.56	17.7	5.43	4.35	0.81	17.27	4.16	–
Element (wt.%)	**Ni**	**Fe**	**Nb**	**Cr**	**Ti**	**C**	**W**	**O**
(d)	45.29	14.14	3.73	13.7	0.7	6.22	10	3.51

3.3. Load Capacity

The extreme-pressure performance of the lubricants is also an important index used to measure the lubricating property. To illustrate the performance under extreme pressure, the load slope test results of the three kinds of sulfur additives are shown in Figure 6a. The test load increased from 100 N to1000 N in steps of 100 N. The friction coefficient of sulfurized olefin fluctuated significantly with each load increase. For sulfurized fatty acid ester and sulfurized lard, the friction coefficients remained stable at about 0.1 before the load increased to 600 N. When the load was more than 600 N, the friction coefficients had a little fluctuation. Thus, the load-bearing capacity of sulfurized fatty acid ester and sulfurized lard was better than that of sulfurized olefin.

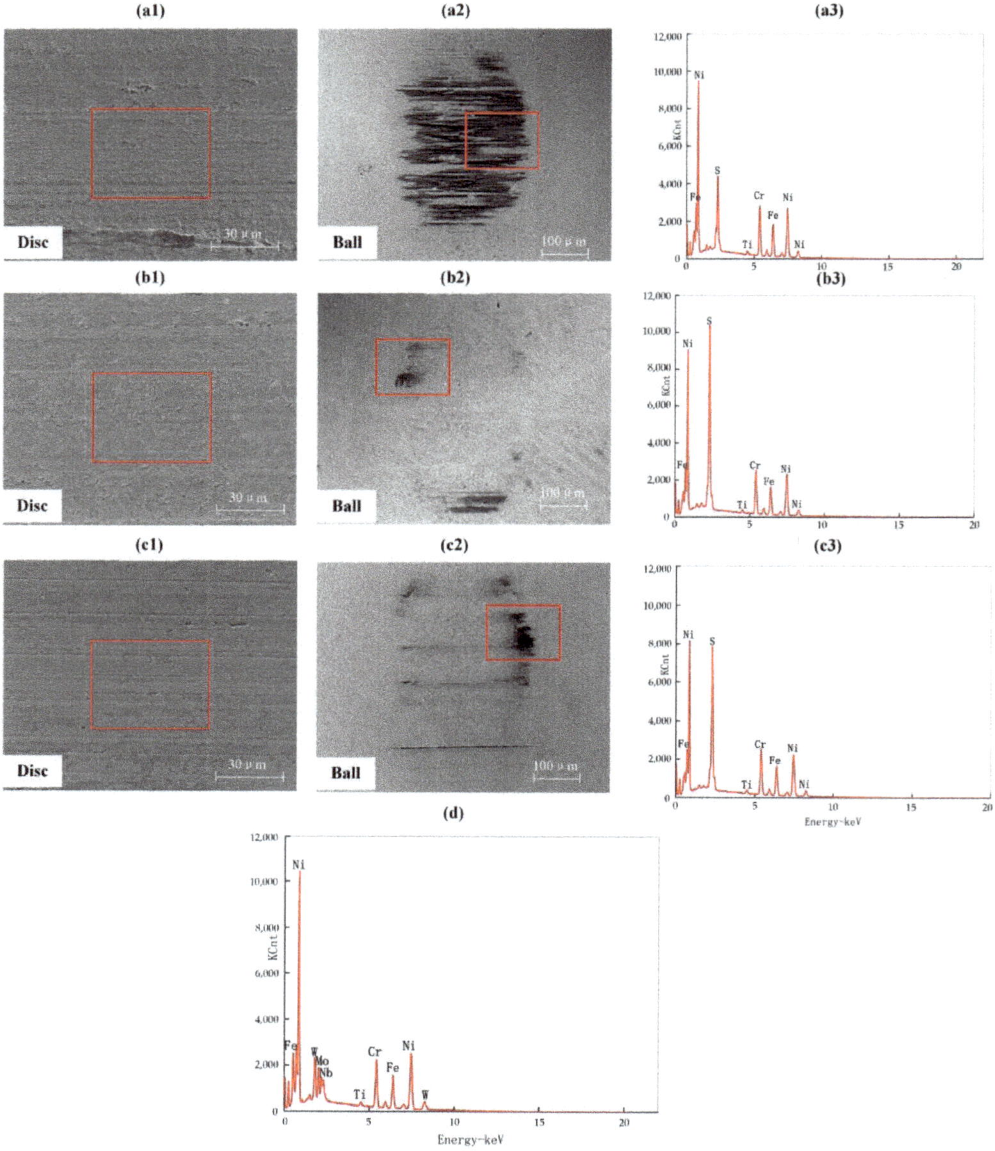

Figure 5. SEM morphology of the tracks on Inconel 718 and tungsten carbide ball lubricated under 150 °C lubricated by (**a1**,**a2**) sulfurized olefin, (**b1**,**b2**) sulfurized fatty acid ester, and (**c1**,**c2**) sulfurized lard; (**a3**,**b3**,**c3**) EDS spectrum of the red box in (**a1**,**b1**,**c1**); (**d**) EDS spectrum of the red box in (**a2**).

The wear volumes of sulfurized fatty acid ester and sulfurized lard on nickel-based superalloy under different loads were further tested. The experimental load was set to 300 N, 500 N, 700 N, and 900 N, and the results are shown in Figure 6b. When sulfurized fatty acid ester lubricated the nickel-based superalloy, the wear volume increased steadily with the increase in load. As for sulfurized lard, the wear volume had a larger increase compared to sulfurized fatty acid when the load increased to 700 N. The big wear volume was consistent with the large friction coefficient, indicating unstable lubricant condition.

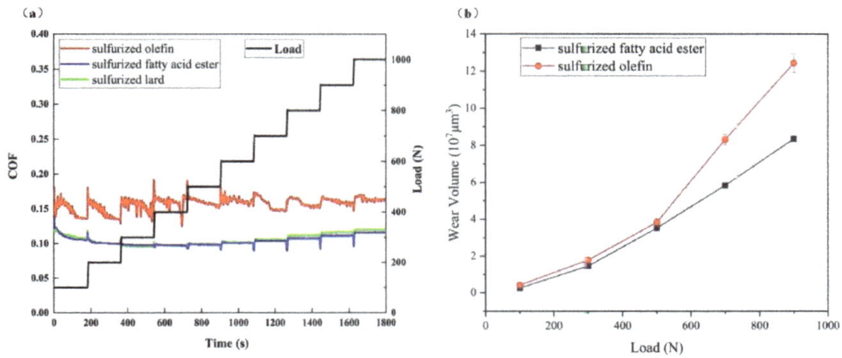

Figure 6. (**a**) Friction coefficient vs. time during a load slope test from 100 N to 1000 N and (**b**) wear volume under different loads (f = 20 Hz, L = 2 mm, t = 25 °C).

3.4. Exploration of the Adsorption Characteristics

XPS is a practical method to clarify the chemical states of elements within the adsorption film on the surface of tribopair. To further explain the lubrication mechanism of sulfur additives for nickel-based superalloy, the worn surfaces were tested by XPS. Figures 7 and 8 show the spectra of the several elements lubricated by sulfurized fatty acid ester and sulfurized olefin. It can be observed that the peak shapes and binding energies of the corresponding elements were similar. Therefore, the tribochemical reaction processes were the same when sulfurized fatty acid ester and sulfurized lard were used as lubricants for the nickel-based superalloy. Figure 7a shows typical XPS survey scans inside and outside the wear track over a binding energy at the range of 0–1400 eV with the lubrication of sulfurized fatty acid ester. The values were shifted 400,000 upward from the second line to show a clear contrast. The peak intensities of Ni2p and Fe2p inside the wear track are lower than those outside the wear track. Furthermore, a sulfur peak at 168 eV appeared at the position inside the wear track, while no sulfur was detected outside the track. This demonstrates that sulfide compounds remained after the frictional test when lubricated by sulfurized fatty acid ester. Ni2p inside the wear track was apparently lower than the substrate, which further demonstrates that some film existed on the wear track. To further investigate the way in which the sulfide compounds acted with nickel-based superalloy, detailed high-resolution XPS scans of Ni2p, Fe2p, S2p, and O1s were recorded, and the results are shown in Figure 7b–e. The peak at 852.8 eV is Ni–S, and the peak at 855.2 eV is Ni–SO_4 (Figure 7b). The peaks at 710.11 eV and 723.4 eV correspond to Fe–S and $FeSO_4$, respectively (Figure 7c) [11]. The S2p spectrum is shown in Figure 7d. The peak at 161 eV–162 eV corresponds to Fe–S, the peak at 162.8 eV corresponds to Ni–S, and the peak at 169.7 eV is the metal sulfate [21]. The O1s spectrum is shown in Figure 7e. The peak at 531.7 eV corresponds to S–O in –SO_4, and the peak at 530.2 is metallic oxide. Combining the Ni2p, S2p, and O1s data, it can be inferred that $NiSO_4$ may have existed on the surface of the wear track. Compared with the S2p spectra on the surface lubricated by sulfurized fatty acid ester, the peak intensities of Ni2p and Fe2p inside the wear track were higher when lubricated by sulfurized olefin (Figure 8a). Moreover, the peak of metal sulfates was much lower than that lubricated by sulfurized fatty acid ester (Figure 8c). It can be speculated that the tribochemical reaction film on Inconel 718 surface lubricated by sulfurized olefin was thinner than that lubricated by sulfurized fatty acid ester, leading to a higher COF and bigger wear volume.

According to the XPS results, the lubricating mechanism of the sulfur additives for Inconel 718–tungsten carbide contact can be summarized. In the frictional process, the molecules adsorb on the metal surface to form a physical adsorption protective film. The active element S reacts with the metal matrix Inconel 718. A tribochemical protective film composed of high-toughness inorganic salts such as nickel sulfide and nickel sulfate is formed, playing a role in lubrication. The physical adsorptive film and the tribochemical film together enhance the friction-reducing and anti-wear performances of the lubricants.

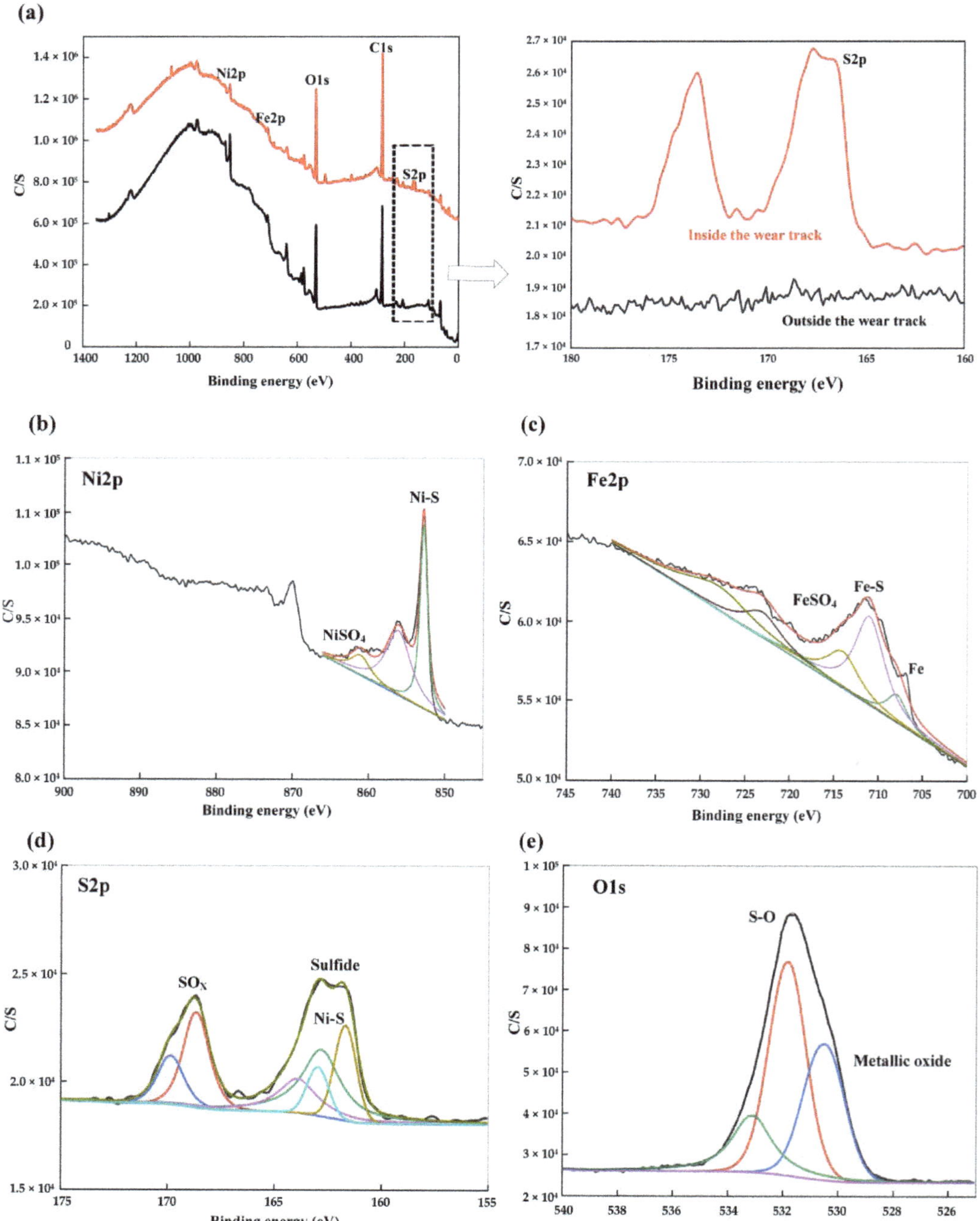

Figure 7. (**a**) XPS survey scans of Inconel 718 surface after tribological test: (**b**) Ni2p, (**c**) Fe2p, (**d**) S2p, and (**e**) O1s inside the wear track lubricated by sulfurized fatty acid ester.

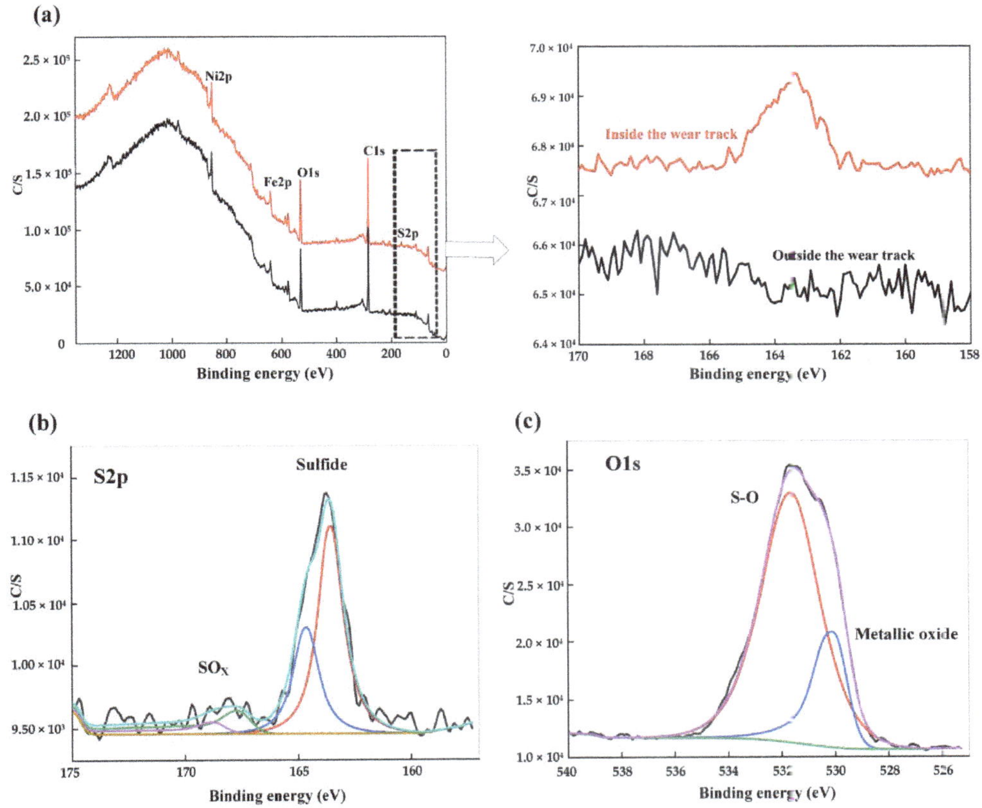

Figure 8. (**a**) XPS survey scans of Inconel 718 surface after tribological test, (**b**) S2p, and (**c**) O1s inside the wear track lubricated by sulfurized olefin.

3.5. The Improvement of the Cutting Fluid

The experimental results show that, for the nickel-based superalloy–tungsten carbide friction pair, sulfurized fatty acid ester had the best lubricating and anti-wear performance among the three kinds of extreme-pressure agents. The effect of sulfurized fatty acid ester on the lubricating performance of some kind of cutting fluid without sulfur-containing additives was tested. The cutting fluid was diluted to 5 wt.% with different content of sulfur additives. The frictional experiments were completed, and the friction coefficients are shown in Figure 9. The results show that the friction coefficient dropped to 0.133 with the concentration of sulfurized fatty acid ester at 1 wt.%. When the concentration of sulfurized fatty acid ester increased to 2%, the friction coefficient further dropped to 0.127. The ester concentration was further increased to 3%, and the friction coefficient did not continue to decrease and remained stable at 0.127. It can be seen that 2 wt.% sulfurized fatty acid ester could successfully decrease the friction coefficient of the cutting fluid.

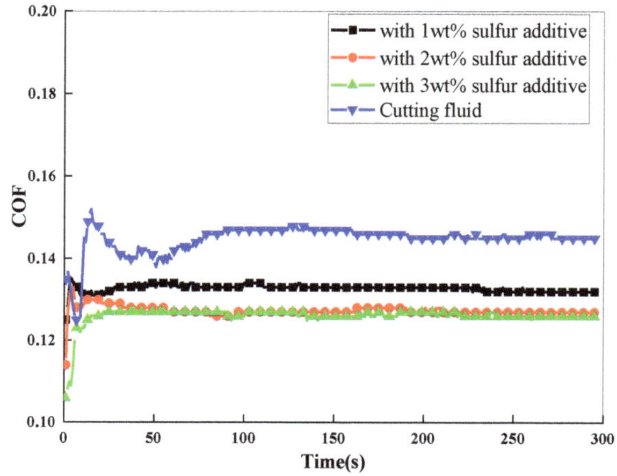

Figure 9. The COF of some cutting fluids with different contents of sulfurized fatty acid ester (F = 100 N, f = 20 Hz, L = 2 mm, t = 25 °C).

4. Conclusions

In the present work, the tribological performance of three kinds of sulfur additives for the Inconel 718–tungsten carbide friction pair was investigated. The friction experiment results showed that sulfurized fatty acid ester possessed excellent antifriction (COF 0.1) and anti-wear performance (wear volume 90% smaller than that lubricated by PAO 40), particularly at the high temperature of 150 °C and at heavy load. The lubrication mechanism of the sulfur additives for the Inconel 718–tungsten carbide friction pair was investigated using XPS. The physical adsorptive film and the tribochemical film together enhanced the friction-reducing and anti-wear performance of the lubricants. This effective lubricant for Inconel 718 can be applied to the machining of nickel-based superalloy.

Author Contributions: Conceptualization, Y.Y. and C.Z.; methodology, Y.Y. and Y.D.; validation, F.L.; formal analysis, F.L.; investigation, H.L., S.G. and Y.Y.; resources, C.Z., Y.Y. and G.Z.; data curation, H.L. and S.G.; writing—original draft preparation, H.L.; writing—review and editing, Y.Y.; visualization, Y.Y.; supervision, F.L.; project administration, F.L. and D.Z.; funding acquisition, C.Z., Y.Y. and G.Z. All authors have read and agreed to the published version of the manuscript.

Funding: This research was funded by National Key Research and Development Program of China grant number 2018YFB2002204 and National Natural Science Foundation of China grant number 52005010 and U1837602.

Data Availability Statement: The data presented in this study are available on request from the corresponding author.

Conflicts of Interest: The authors declare no conflict of interest.

References

1. Mohsan, A.U.; Liu, Z.; Padhy, G.K. A review on the progress towards improvement in surface integrity of Inconel 718 under high pressure and flood cooling conditions. *Int. J. Adv. Manuf. Technol.* **2017**, *91*, 107–125. [CrossRef]
2. Devillez, A.; Schneider, F.; Dominiak, S.; Dudzinski, D.; Larrouquere, D. Cutting forces and wear in dry machining of Inconel 718 with coated carbide tools. *Wear* **2007**, *262*, 931–942. [CrossRef]
3. Ezugwu, E.O. Key improvements in the machining of difficult-to-cut aerospace superalloys. *Int. J. Mach. Tools Manuf.* **2005**, *45*, 1353–1367. [CrossRef]
4. Pusavec, F.; Deshpande, A.; Yang, S.; M'Saoubi, R.; Kopac, J.; Dillon, O.W., Jr.; Jawahir, I.S. Sustainable machining of high temperature Nickel alloy–Inconel 718: Part 1—Predictive performance models. *J. Clean. Prod.* **2014**, *81*, 255–269. [CrossRef]
5. Pusavec, F.; Deshpande, A.; Yang, S.; M'Saoubi, R.; Kopac, J.; Dillon, O.W., Jr.; Jawahir, I.S. Sustainable machining of high temperature Nickel alloy–Inconel 718: Part 2—Chip breakability and optimization. *J. Clean. Prod.* **2015**, *87*, 941–952. [CrossRef]

6. Pervaiz, S.; Rashid, A.; Deiab, I.; Nicolescu, M. Influence of tool materials on machinability of titanium and nickel-based alloys: A review. *Mater. Manuf. Process.* **2014**, *29*, 219–252. [CrossRef]
7. Debnath, S.; Reddy, M.M.; Yi, Q.S. Environmental friendly cutting fluids and cooling techniques in machining: A review. *J. Clean. Prod.* **2014**, *83*, 33–47. [CrossRef]
8. Busch, K.; Hochmuth, C.; Pause, B.; Stoll, A.; Wertheim, R. Investigation of cooling and lubrication strategies for machining high-temperature alloys. *Procedia CIRP* **2016**, *41*, 835–840. [CrossRef]
9. Eskandari, B.; Bhowmick, S.; Alpas, A.T. Flooded drilling of Inconel 718 using graphene incorporating cutting fluid. *Int. J. Adv. Manuf. Technol.* **2021**, *112*, 1–14. [CrossRef]
10. Brinksmeier, E.; Meyer, D.; Huesmann-Cordes, A.G.; Herrmann, C. Metalworking fluids-mechanisms and performance. *CIRP Ann.-Manuf. Technol.* **2015**, *64*, 605–628. [CrossRef]
11. Bierla, A.; Fromentin, G.; Minfray, C.; Martin, J.M.; Le Mogne, T.; Genet, N. Mechanical and physico-chemical study of sulfur additives effect in milling of high strength steel. *Wear* **2012**, *286–287*, 116–123. [CrossRef]
12. Li, Y.R.; Pereira, G.; Lachenwitzer, A.; Kasrai, M.; Norton, P.R. X-ray Absorption spectroscopy and morphology study on antiwear films derived from ZDDP under different sliding frequencies. *Tribol. Lett.* **2007**, *27*, 245–253. [CrossRef]
13. Li, Y.R.; Pereira, G.; Kasrai, M.; Norton, P.R. Studies on ZDDP anti-wwear films formed under different conditions by XANES spectroscopy, atomic force microscopy and 31P NMR. *Tribol. Lett.* **2007**, *28*, 319–328. [CrossRef]
14. Najman, M.N.; Kasrai, M.; Bancroft, G.M. X-ray Absorption spectroscopy and atomic force microscopy of films generated from organosulfur extreme-pressure (EP) oil additives. *Tribol. Lett.* **2003**, *14*, 225–235. [CrossRef]
15. Alves, S.M.; Oliveira, J.F.; Klocke, F.; Maier, B. Effects of sulfur additive EP in ester coolants during friction with CBN. *Tribol. Trans.* **2008**, *51*, 278–284. [CrossRef]
16. Chan, C.H.; Tang, S.W.; Mohd, N.K.; Lim, W.H.; Yeong, S.K.; Idris, Z. Tribological behavior of biolubricant base stocks and additives. *Renew. Sustain. Energy Rev.* **2018**, *93*, 145–157. [CrossRef]
17. Yu, Q.; Zhang, C.; Dong, R.; Shi, Y.; Wang, Y.; Bai, Y.; Zhang, J.; Cai, M.; Zhou, F. Novel N, P-containing oil-soluble ionic liquids with excellent tribological and anti-corrosion performance. *Tribol. Int.* **2019**, *132*, 118–129. [CrossRef]
18. Pettersson, A.; Elisabet, K.; Minami, I. Additives for environmentally adapted lubricants-friction and wear protection. *Tribol. Online* **2008**, *3*, 163–167. [CrossRef]
19. Jayal, A.D.; Balaji, A.K. Effects of cutting fluid application on tool wear in machining: Interactions with tool-coatings and tool surface features. *Wear* **2009**, *267*, 1723–1730. [CrossRef]
20. Xavior, M.A.; Adithan, M. Determining the influence of cutting fluids on tool wear and surface roughness during turning of AISI 304 austenitic stainless steel. *J. Mater. Process. Technol.* **2009**, *209*, 900–909. [CrossRef]
21. Guo, B.; Li, Y.; Zheng, J.; Li, F.; Li, X.; Du, X.; Yuan, L. Tribological properties of a halogen-free ionic liquid for Inconel 690–tungsten carbide contact. *Tribol. Int.* **2021**, *163*, 107153. [CrossRef]

metals

MDPI

Article

Assessment of the Effect of Thermal-Assisted Machining on the Machinability of SKD11 Alloy Steel

Thi-Bich Mac [1], The-Thanh Luyen [1] and Duc-Toan Nguyen [2,*]

[1] Faculty of Mechanical Engineering, Hungyen University of Technology and Education, Hungyen 100000, Vietnam
[2] School of Mechanical Engineering, Hanoi University of Science and Technology, 1A-Dai Co Viet Street, Hai Ba Trung District, Hanoi City 100000, Vietnam
* Correspondence: toan.nguyenduc@hust.edu.vn; Tel.: +84-43-869-2007

Abstract: This study aimed to investigate the effects of Thermal-Assisted Machining (TAM) on SKD11 alloy steel using titanium-coated hard-alloy insert cutting tools. The microstructure, material hardness, chip color, cutting force, chip shrinkage coefficient, roughness, and vibration during TAM were evaluated under uniform cutting conditions. The machining process was monitored using advanced equipment. The results indicated that thermal-assisted processing up to 400 °C did not alter the microstructure and hardness of the SKD11 alloy steel. However, a significant variation in chip color was observed, indicating improved heat transfer through TAM. The cutting force, vibration amplitude of the workpiece, and surface roughness all decreased with increasing TAM. Conversely, the chip shrinkage coefficient of the machined chips tended to increase due to the high temperatures.

Keywords: SKD11 alloy steel; machinability; thermal-assisted machining (TAM)

Citation: Mac, T.-B.; Luyen, T.-T.; Nguyen, D.-T. Assessment of the Effect of Thermal-Assisted Machining on the Machinability of SKD11 Alloy Steel. *Metals* **2023**, *13*, 699. https://doi.org/10.3390/met13040699

Academic Editors: George A. Pantazopoulos and Badis Haddag

Received: 17 February 2023
Revised: 27 March 2023
Accepted: 31 March 2023
Published: 3 April 2023

1. Introduction

The cutting mode and tool geometry parameters play a critical role in machining techniques, especially when working with materials that are challenging to machine due to their high hardness. These parameters affect various output parameters of the machining process, such as cutting force, cutting heat, cutting tool wear, cutting process vibration, surface roughness, and chip geometry. To improve productivity, part quality, and cost-effectiveness, researchers are constantly seeking new technological solutions to support the machining process, including the use of smooth cold techniques, new cutting tool materials, vibration-assisted cutting, and thermal-assisted machining. Difficult-to-machine materials, characterized by high hardness, good wear resistance, and little change in mechanical properties at high temperatures, are widely used in various industries, such as mechanical, automotive, aerospace, aviation, defense, medical, and electrical-electronics-automation. Surveys have shown that over 30% of milling, turning, and drilling operations are performed on such materials [1].

Thermal-assisted machining (TAM) is a machining process that involves heating the workpiece before machining it using traditional or CNC machines. This technique has been extensively used in industrial production since its introduction in 1945 [1,2]. Compared to traditional machining methods, TAM offers several advantages, such as increased tool life, reduced cutting force, decreased tool wear, and improved surface quality, resulting in enhanced productivity [3–5]. TAM is suitable for both cutting processes (e.g., turning, milling, drilling) and deformation processes (e.g., forging, stamping, drawing). Several heating methods can be employed for TAM, including electrical current heating, laser beam heating (LAM), plasma heating (PEM), furnace heating (FAM), and induction heating (IAM). While each method has its own benefits and drawbacks, induction heating (IAM) is particularly effective due to its high heating capacity, ease of use, and affordability, making it a great choice for metalworking operations [6].

Many studies have suggested eco-friendly dry and minimum quantity lubrication or optimization to improve machining productivity, increase tool life, and enhance product quality [7–9]. However, for challenging-to-machine materials, like those commonly used in heavy-duty industries such as automotive, marine, and aviation, advanced machining methods such as diamond grinding or discharge machining are often used but limited by low material removal rates, expensive tools, and rapid wear [2]. Among these materials, SKD11 Tool Steel is frequently utilized in mold and automotive sectors due to its hardness, strength, and ductility [10]. Although traditional techniques like diamond grinding or discharge machining are limited by high costs and specialized technologies, TAM has been introduced as an innovative solution to the challenges of machining SKD11. Studies have indicated that machining with TAM can reduce cutting forces by up to 40%, increase tool life, enhance surface roughness by up to 50%, and increase material removal rate when compared to machining at room temperature [11–13].

Recent research has shown that using thermal-assisted machining (TAM) can offer several advantages, including improved tool life, enhanced productivity, and better working conditions between the tool and workpiece [14–16]. Studies have revealed that workpiece temperature is directly proportional to tool life and inversely proportional to cutting speed. Induction heating has been investigated for milling titanium alloy Ti-6Al-4V, with heat treatment significantly improving tool life and material removal rate, by up to 169.4% at a temperature of 650 °C. However, heating above 640 °C may negatively impact the workpiece's mechanical properties, reducing machining efficiency [5,17]. Studies have also shown that cutting and shear forces are inversely proportional to temperature, with a 13% reduction in cutting force observed when machining at 500 °C compared to room temperature. A novel approach to machining Inconel 718 was presented by Wang et al. [18], which combines traditional turning, cryogenic enhanced machining, and plasma heating. Cold working techniques can reduce the heat generated during cutting, enhancing the cutting tool's longevity. The implementation of this method has been shown to result in a substantial enhancement of the machining process, with the surface gloss of the machined material increased by a factor of 2.5, the required machining force decreased by half, and the cutting tool's lifespan prolonged by a factor of 1.7 compared to traditional machining at room temperature.

Efficiently machining hard alloys has long been a challenge, but thermally assisted machining (TAM) offers a promising solution. TAM involves cutting the material and then softening it with an external heat source, which has been shown to reduce the hardness and tensile strength of the workpiece material, thereby improving machinability [19]. Research indicates that TAM is linked to higher material removal rates, better control of machining time, and improved surface finishes. TAM is particularly advantageous for machining bio-implant titanium alloys, which require high precision. Hard steels can also be successfully machined using TAM, resulting in lower cutting forces, increased tool life, and higher material removal rates [20–24]. TAM can reduce cutting force amplitudes and chip morphology changes, resulting in less vibration and better surface integrity. However, machining certain hardenable steels, such as 1090 steel, can lead to increased cutting forces due to phase transformation hardening when the laser-preheated part enters the cutting zone. Softening the workpiece by reaching surface temperatures of 300–400°C for an uncut chip thickness of 0.05 mm, on the other hand, can reduce the magnitude and amplitude variation of cutting forces, and limit the evolution of tool wear [25]. TAM also induces a change in chip morphology from sawtooth to continuous, improving the surface finish. While TAM can lead to higher cutting forces in some hardenable steels, it can also enhance surface integrity [26].

TAM techniques have been extensively researched and used in production. However, little is known about the use of magnetic induction heating for mold steel, particularly when milling difficult materials like SKD11 steel, which is commonly used in industry. This study aims to evaluate the influence of TAM on milling SKD11 steel and its effect on the material's machinability by analyzing cutting force, vibration amplitude of the workpiece, and surface roughness. This analysis provides a basis for selecting appropriate TAM process parameters. A significant challenge when using electromagnetic induction heating in machining is its application to large parts with varying sizes or complex shapes. This study proposes a potential technological solution for processing difficult materials using electromagnetic induction heating. The results demonstrate that the heating process effectively reduces cutting force, cutting heat, and vibration during cutting, while improving the surface quality of the workpieces. The findings of this study have practical implications and are applicable to the manufacturing industry.

2. Experimental Setup

The ease or difficulty of processing a material is known as its machinability, which can be evaluated using parameters such as tool life, MRR, cutting force, cutting process vibration, and surface gloss of the machined material. The machinability of a material is greatly affected by its microstructure, which can be further influenced by the cutting mode. Therefore, it is essential to assess the machinability of materials during heating processing and compare it to conventional machining to determine the impact of heating on the machinability of SKD11 steel materials.

2.1. Schematic of Experimental System

This study establishes an experimental research model that is based on the research objectives and the available experimental equipment, as depicted in Figure 2.

2.2. Materials and Testing Equipment

SKD11 steel is a frequently used alloy steel in mold processing, as defined by JIS-G4404. This steel is notable for its hardness, strength, and ductility, and can maintain its hardness even at high temperatures for an extended period, making it suitable for use in the production of molds for extrusion, plastic injection, pressure, and components that require specific performance characteristics. Table 1 presents a breakdown of the SKD11 steel components, expressed as a weight percentage (wt%).

Table 1. The chemical composition of SKD11 steel (wt%) [27].

C	Cr	Mo	Si	Mn	Ni	V
1.4–1.6	11–13	0.7–1.2	$\leq$0.6	$\leq$0.6	-	0.15–0.3

In this study, a machining experiment was conducted on SKD11 steel test specimens using a Taiwanese MC500 milling machine. The billet was rough-worked and had dimensions of 70 $\times$ 31 $\times$ 80 mm^3, and a chamfer of 7 mm $\times$ 7 mm was applied for uniform contact with the induction coil. For the milling process, a 40 mm face milling cutter was used, equipped with a titanium-coated hard alloy insert manufactured by PRAMET, a tool manufacturer from the Czech Republic. Specifically, the APKT 1604PDR–GM hard alloy piece was utilized, with cutting edge parameters such as the rake angle, clearance angle, and cutting-edge radius set at 24°, 11°, and 0.8 mm, respectively. Other geometric parameters, including l = 16 mm, d = 9.44 mm, s = 5.67 mm, and d_1 = 4.6 mm, were also specified. No coolant was applied during the machining process. One significant challenge in this research was the application of electromagnetic induction heating in machining, as depicted in Figure 1d.

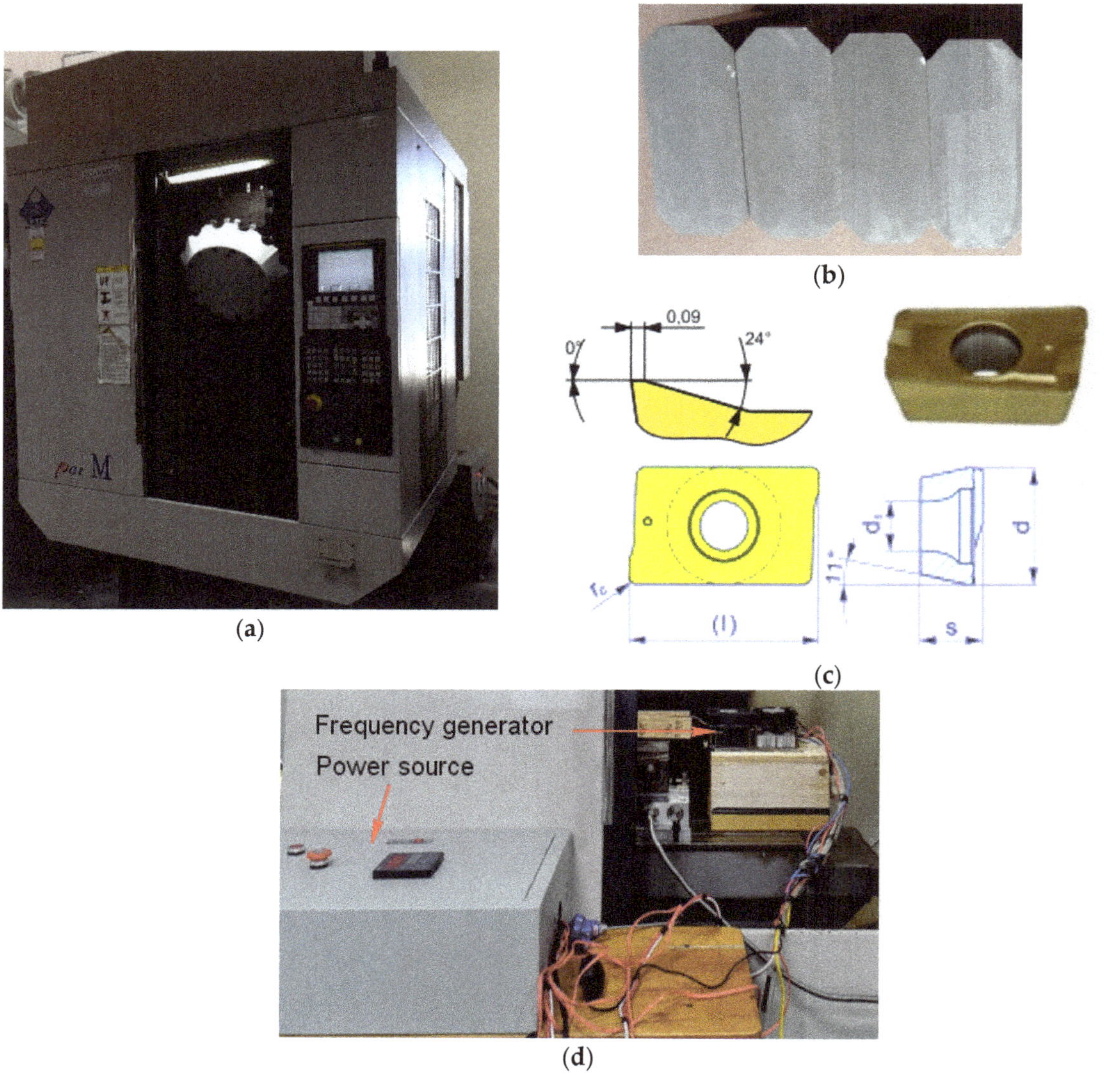

Figure 1. Components of the experimental system. (**a**) MC500 Milling Machine, (**b**) Sample Workpices, (**c**) Hard Alloy Insert, and (**d**) Power Source-Frequency Generator.

The study utilized the Axiovert 25 CA optical microscope (Figure 3a) to examine the microstructure of the material after heating at different temperatures and compare it with the original sample. This microscope, specifically designed for materials research, is used in conjunction with Image-Pro Plus image analysis software to analyze the material phase. The study also employed the Brinell hardness tester (Figure 3b) to test the initial and post-heating samples for hardness.

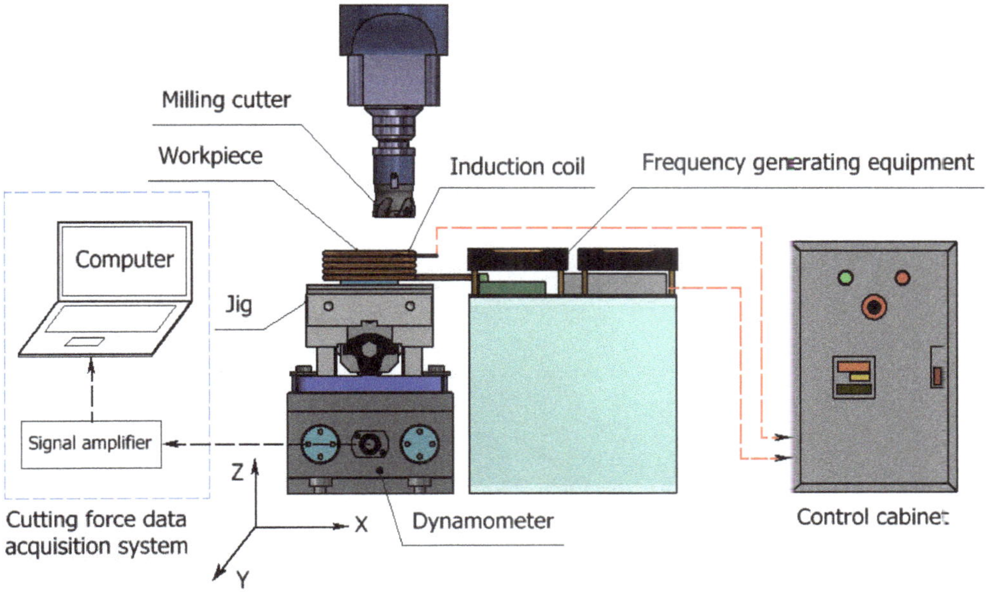

(**a**)

(**b**)

Figure 2. Schematic of Experimental System. Schematic diagram (**a**) and experiment photograph in working zone (**b**).

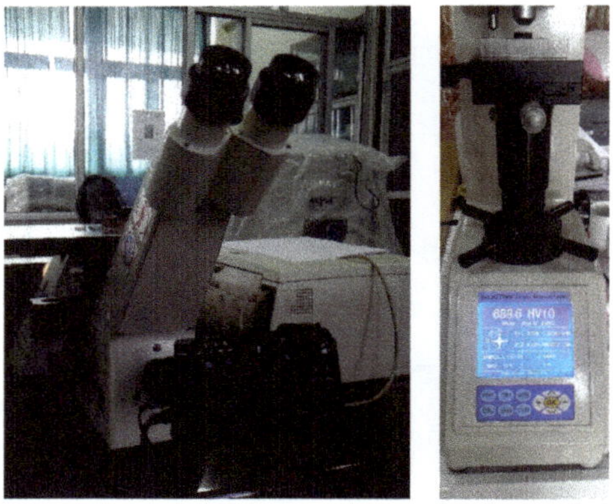

(a) (b)

Figure 3. Equipment used in the tests: (**a**) Axiovert 25 CA Optical Microscope, (**b**) Brinell Hardness Tester.

3. SKD11 Steel Material Machinability Results

3.1. Effect of TAM on Microstructure of SKD11 Steel

To conduct an accurate examination of the microscopic structure of materials, proper sample preparation is crucial. The process of preparing metal samples involves multiple stages, including cutting, grinding, polishing, and impregnation. Specimens are initially machined following established standards to achieve a size that can be impregnated. Figure 4 displays the experimental samples and their cross-sections for SEM examination.

Figure 4 illustrates the microstructural analysis of SKD11 alloy steel samples before and after thermal-assisted machining (TAM) at different temperatures, using optical microscopy. Figure 4a presents an image of the microstructure of the original material sample, while Figure 4b–d show images of the microstructure after TAM at different temperatures. The analysis reveals that the microstructure of the samples after TAM at room temperature and 400 °C is similar to that of the original material. Specifically, the microstructure of the four samples includes white plates and bright round particles of chromium Cr7C3 carbides, spherical dark dots of Cementite, and a light background of Pearlite. These observations suggest that the microstructure of the specimen remains unchanged even when the temperature is increased to 400 °C, which is below the phase transition temperature of 700 °C for the SKD11 alloy steel [28].

3.2. Effect of TAM on Hardness

To evaluate the impact of elevated temperatures on workpiece hardness after heating, the workpiece was exposed to the designated temperature and allowed to cool naturally via exposure to ambient air. Surface hardness measurements necessitated meticulous sample preparation, including cutting, grinding, polishing, and impregnation. In order to prevent the influence of cutting heat, wire cutters were employed during sample cutting. The grinding process aimed to minimize undulation caused by the varying hardness of structural elements, while polishing removed any coarse grinding marks and scratches. Finally, deformations from the cutting, grinding, and polishing stages were either eliminated or leveled to a size sufficient for removal with an impregnating agent. Three hardness measurements were taken on each sample using a Brinell hardness tester at distinct locations, as illustrated in Figure 5. The results, presented in Figure 5, showed that increasing temperatures resulted in decreased specimen hardness to 2–3 HB at 200 °C and 300 °C. However, at 400 °C the sample demonstrated increased hardness compared to the other temperatures, although this increase was not significant within the chosen experimental temperature range.

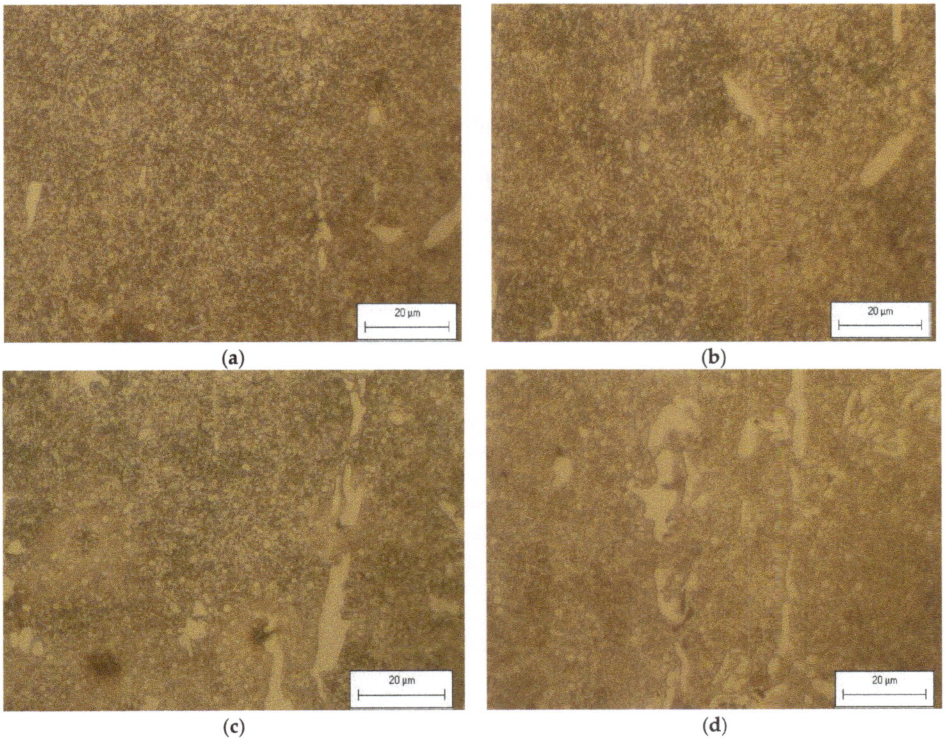

Figure 4. Microstructure of Material Samples at Different Temperatures: (**a**) Room Temperature, (**b**) 200 °C, (**c**) 300 °C, and (**d**) 400 °C.

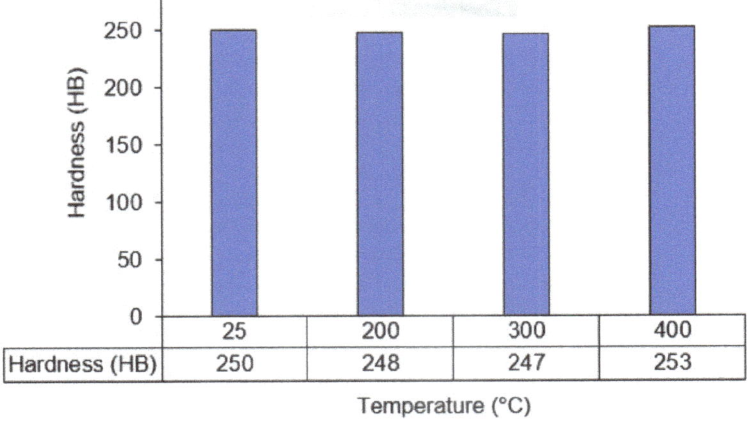

Figure 5. Workpiece hardness after heating at different temperatures and natural air-cooling.

3.3. Effect of TAM on Chip Color

The geometry and morphology of chips play a crucial role in assessing the machinability of materials. Chip geometry affects cutting force, cutting heat, and tool wear, while chip morphology provides valuable insights for milling tool design. Therefore, when analyzing the impact of heating on the machinability of SKD11, examining chip formation is crucial. Figure 6 shows chip images obtained during the milling of SKD11 steel under different cutting modes and heating conditions. Figure 6a–d display chip formations during machining at various temperatures. The formation of wire chips is due to the workpiece material's ductile properties. The chip color varies significantly, with the normal machining chip appearing purple-black in color due to excessive cutting heat generated during machining at room temperature (Figure 6a). In contrast, chips appear bright white (Figure 6b) and yellow (Figure 6c) when machined at 200 °C and 300 °C, respectively. The chip appears darker yellow when machined at 400 °C (Figure 6d). This variation can be attributed to the uniform transfer of heat between the cutting tool, workpiece, and chip during heating processing. The high temperature reduces the material's tensile strength, mechanical strength, and yield stress, while increasing its deformation, improving heat transfer conditions, and decreasing the bonding force between metal molecules [19,20]. As a result, chip removal becomes easier, cutting heat is reduced, and chips appear light-colored during hot machining. The results suggest that temperatures between 200–300 °C preserve the chip color compared to the original substrate material.

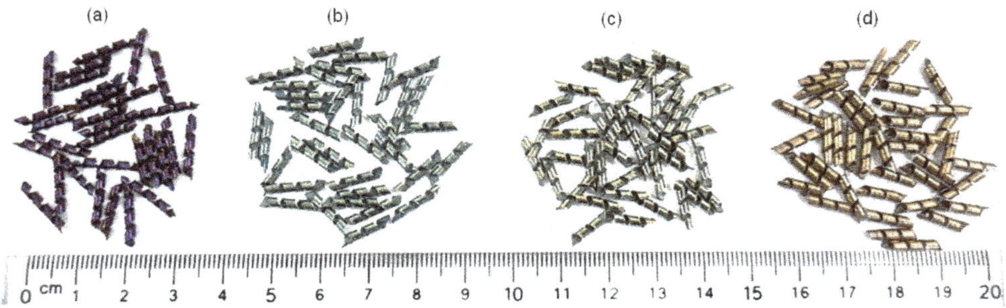

Figure 6. Chip Formation at Different Temperatures: (**a**) Room Temperature, (**b**) 200 °C, (**c**) 300 °C and (**d**) 400 °C.

3.4. Effect of TAM on Cutting Force

An experimental study was conducted to investigate the effect of elevated temperature on cutting forces (F) during milling with identical cutting parameters at room temperature and elevated temperature. The cutting force is a dynamic phenomenon and changes throughout the machining cycle [29]. Figure 7 displays the cutting force transformation with respect to tool distance during milling at room temperature (25 °C) and elevated temperature (200 °C). The cutting parameters used were a cutting speed of 235 m/min, feed rate of 305 mm/min, and cutting depth of 1.5 mm. The average cutting force (F) was calculated from the component cutting forces (F_f, F_p, F_c) using Equation (1). Table 2 shows the component cutting force values and the average cutting force under conventional machining and elevated temperature conditions. The results indicate that the F_p component cutting force (direct force) has the largest value, while the radial force (F_f) and axial force (F_c) have relatively smaller values. The average cutting force reduction at elevated temperature compared to conventional machining is 37.5%, as determined by Equation (2)).

$$F = \sqrt{{F_f}^2 + {F_p}^2 + {F_c}^2} \quad (1)$$

$$\Delta F(\%) = \frac{F_T - F_R}{F_R} \times 100\% \quad (2)$$

where F_T represents the cutting force during heating and F_R represents the cutting force when machining at room temperature (T = 25 °C).

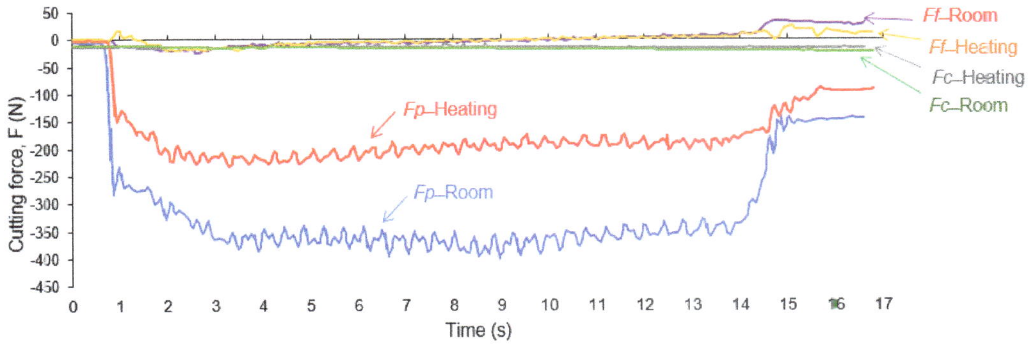

Figure 7. Cutting force during machining at room temperature and at elevated temperature (200 °C).

Table 2. Average cutting force at room and elevated temperature.

Temperature	F_f (N)	F_p (N)	F_c (N)	F (N)
Room	4.317	360.031	9.028	360.170
200 °C	8.958	214.583	9.071	214.962

The experimental results (shown in Figure 8) indicate a significant reduction in cutting force (F) during machining at 200 °C compared to conventional machining. As the heating temperature increased to 300–400 °C, the cutting force decreased at a slower rate. The maximum reduction in cutting force (ΔF) was 65.1% during milling at 400 °C. This reduction can be attributed to the weakening of the strength and bonding between metal molecules due to the heating process, making the cutting process easier. Moreover, in the second strain region, the compressive stress was found to decrease during heating, which led to a decrease in cutting force.

3.5. Effect of TAM on Surface Roughness (Ra)

The surface quality of a workpiece is critical and influenced by various factors, including the machining method, cutting tool geometry, and machining environment. Surface roughness, in particular, is essential in determining the workability and performance of the final product. This study used the average deviation criterion *Ra* to evaluate surface roughness. According to ISO standards, Ra is the arithmetic average of the absolute values of the profile over the reference length range (L = 250 µm) for machined surfaces with a smoothness level of 8–11. The research investigated the effect of elevated temperatures on Ra during Thermal Assisted Machining (TAM) while maintaining the same cutting parameters, including speed, feed rate, and cutting depth, under various temperature conditions. The method of surface roughness measurement is shown in Figure 9a, where the measuring head moves perpendicularly to the machining trace, and each sample is measured at three locations (1, 2, 3) to ensure reliability, within the reference length range of 250 µm, as shown in Figure 9b,c. Results indicated that heating the workpiece before machining significantly reduced surface roughness compared to conventional machining.

This is because the thermal softening of the material results in increased smoothness and stability during the cutting process.

$$\Delta \mathrm{Ra}(\%) = \frac{\mathrm{Ra_R} - \mathrm{Ra_T}}{\mathrm{Ra_R}} \times 100\% \tag{3}$$

where $\mathrm{Ra_R}$ and $\mathrm{Ra_T}$ represent the surface quality (Ra) at room and high temperature, respectively.

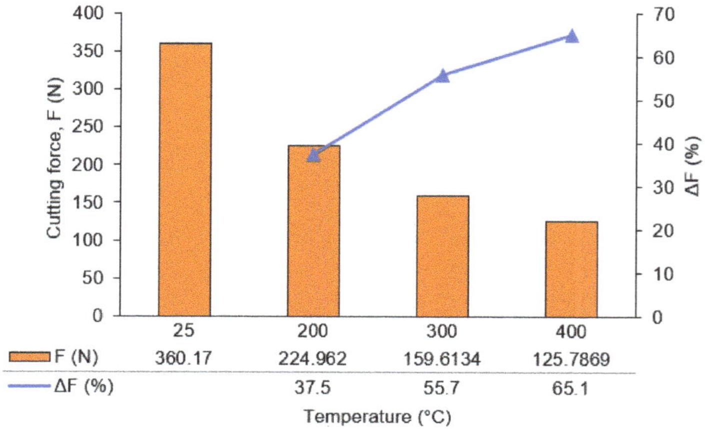

Figure 8. The mean cutting forces and the corresponding reductions observed during machining at various temperatures.

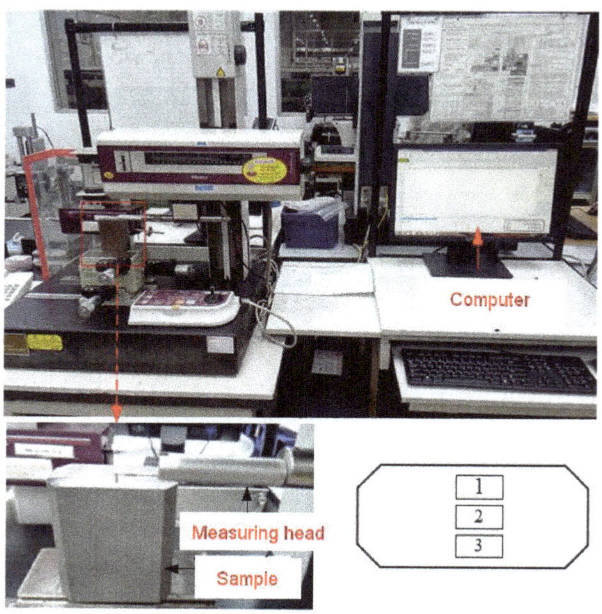

(a)

Figure 9. *Cont.*

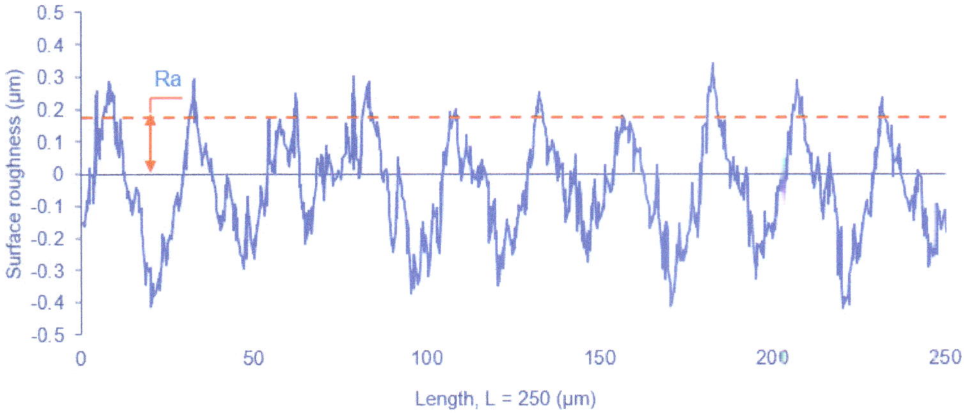

(b)

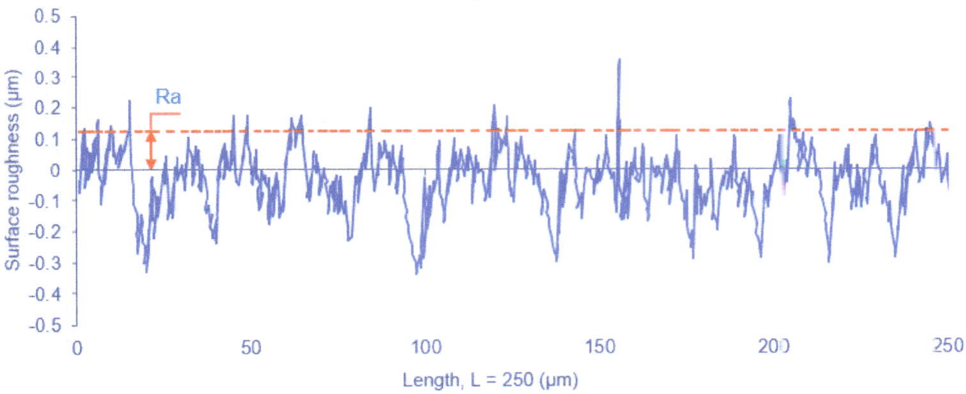

(c)

Figure 9. Measurement of Surface Roughness with Equipment and Method (**a**) and Comparison of Surface Roughness Profiles for Machining at Room Temperature (**b**) and 200 °C (**c**).

In Figure 10, the average Ra values for machining at various temperatures are plotted. The decrease in Ra was calculated using Equation (3). The results indicate that there is an inverse relationship between Ra and workpiece temperature. The greatest reduction in roughness, 47.1%, was observed at 400 °C.

3.6. Effect of TAM on Cutting Vibration

To investigate the impact of temperature-assisted machining (TAM) on vibration, experiments were conducted using the same cutting parameters of V, f, and t at 235 m/min, 305 mm/min, and 1.5 mm, respectively, but at various temperature conditions. The vibration amplitudes in directions of (X, Y) as (A_X and A_Y) were summed to obtain the A_{XY} using Equation (4).

$$A_{XY} = \sqrt{A_X{}^2 + A_Y{}^2} \tag{4}$$

Figure 11 displays the vibration amplitudes at room temperature (a) and at 200 °C (b) in directions of (X, Y). The analysis of the no-load and individual vibration data did not indicate any resonance. The results showed that TAM reduced the vibration amplitude compared to conventional machining. The reduction in cutting vibration amplitude (ΔA_{XY})

during TAM, as compared to milling at room temperature, is calculated using Equation (5).

$$\Delta A_{XY}(\%) = \frac{A_{XY-R} - A_{XY-T}}{A_{XY-R}} \times 100\% \quad (5)$$

where $A_{XY\text{-}R}$, $A_{XY\text{-}T}$ represent the vibration amplitudes in directions of (X, Y) during machining at room and high temperature, respectively.

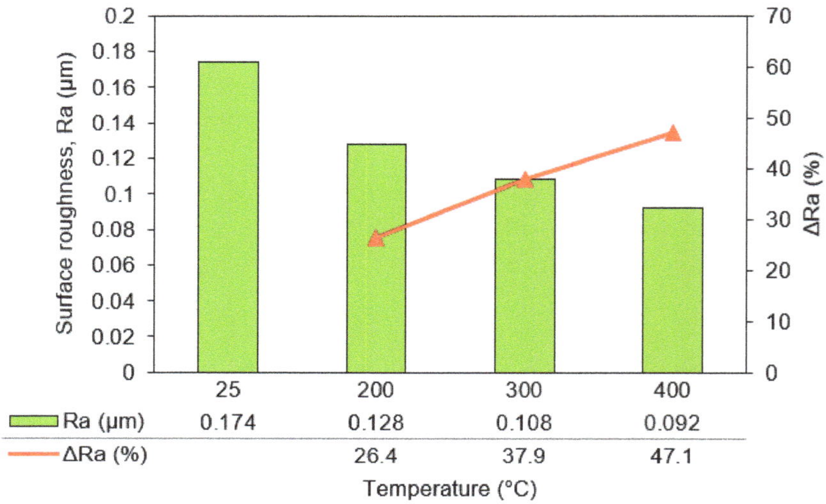

Figure 10. Surface Roughness and Reduction at Elevated Machining Temperatures.

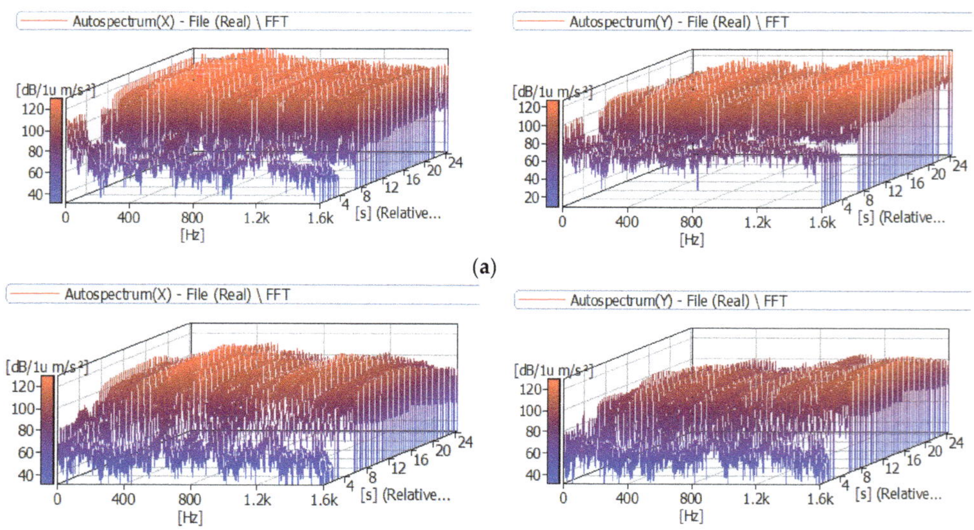

(a)

(b)

Figure 11. The vibration results at room (**a**) and heating at 200 °C (**b**) in the X and Y directions.

Figure 12 illustrates the vibration amplitude and reduction in vibration amplitude during SKD11 steel machining at various temperatures, as compared to milling at room temperature. The results indicated that A_{XY} decreased by 13.2% at 200 °C, 14.9% at

300 °C, and 15.8% at 400 °C. These findings indicate increased stability during SKD11 steel processing when conducted at higher temperatures, due to the reduction in metallic bond strength resulting in an easier cutting process. However, the reduction in vibration amplitude did not exhibit a significant change when high temperature was used to facilitate the cutting process.

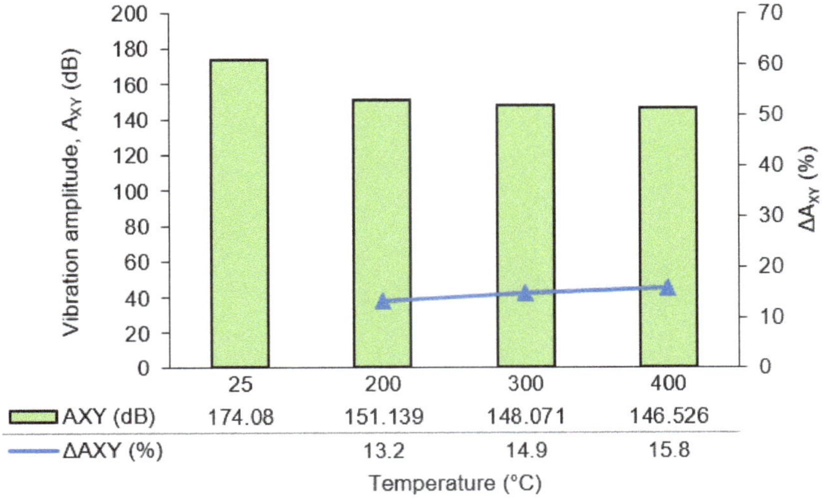

Figure 12. Vibration amplitude and vibration amplitude reduction during machining at various temperature conditions.

3.7. Effect of TAM on Chip Shrinkage Coefficient (K)

K is a vital parameter used to assess the plastic strain of a material during machining, and it has a significant influence on the dimensional accuracy of the machined surface. Several factors affect *K*, including the mechanical properties of the workpiece material, cutting tool geometry, cutting mode, and other cutting conditions. This study aims to analyze the impact of elevated temperatures on *K* and compare it to conventional machining under the same cutting mode. The objective is to assess the material's softening and formability at high-temperature conditions. The chip shrinkage coefficient, computed using Equation (6), is utilized to determine *K*.

$$K = \frac{1000 \cdot Q}{\rho \cdot L_f \cdot f \cdot t} \tag{6}$$

where Q and ρ represent the mass of chip (g) and the material density (g/cm^3), respectively, L_f is the chip cutting length (mm), and f and t are the feed rate (mm/rev) and cutting depth (mm), respectively.

Figure 13 displays the chip shrinkage coefficient at different temperatures for the machining process conducted at cutting parameters of V = 235 m/min, f = 305 mm/min, and t = 1.5 mm. The results demonstrate that the chip shrinkage coefficient increases with heating compared to conventional machining at room temperature. The increase in *K*, denoted as Δ*K*, is calculated using Formula (7). The chip shrinkage coefficient experiences a 27.6% increase when the workpiece is heated to 200 °C, and the maximum increase in *K* of 46.5% is observed at a temperature of 400 °C. This trend suggests that an increase in temperature enhances the cutting process by improving the chip shrinkage coefficient. The observed phenomenon can be attributed to the material softening under high temperatures, resulting in weakened atomic bonds and increased metal deformation, leading to a higher chip shrinkage coefficient.

$$\Delta K(\%) = \frac{K_T - K_R}{K_R} \times 100\% \tag{7}$$

where K_T and K_R represent the chip shrinkage coefficient during heating and at room temperature (T = 25 °C), respectively.

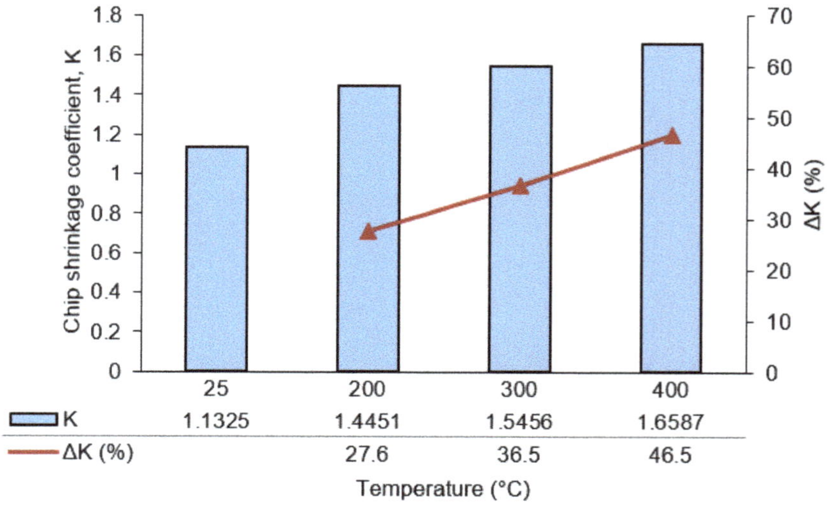

Figure 13. Chip shrinkage coefficient and chip shrinkage increase during machining at various temperatures.

4. Conclusions

The aim of this study was to investigate and compare the effectiveness of using electromagnetic induction heating versus conventional methods for machining SKD11 steel, a material that is difficult to cut. The study analyzed various output parameters, including chip geometry, chip shrinkage coefficient, vibration amplitude, surface roughness, and cutting force. Experimental results indicated that the TAM process did not alter the material's microscopic structure in the temperature range of 200 °C to 400 °C, and the machined workpiece, cooled in air, maintained its original hardness. Additionally, the chip geometry changed and cutting force significantly decreased, with a maximum reduction of 65.1%. The K value increased by 31.7%, while surface roughness decreased notably, with a maximum reduction of 47.1% at 400 °C. Lower vibration amplitude also indicated a more stable machining process compared to traditional methods. These findings demonstrate the potential of this method to enhance machining performance and contribute to the development of high-precision and high-efficiency machining processes for difficult-to-cut materials.

Author Contributions: Conceptualization, D.-T.N.; Data curation, T.-B.M.; Formal analysis, D.-T.N.; Funding acquisition, T.-T.L.; Investigation, T.-T.L.; Methodology, D.-T.N.; Project administration, T.-B.M.; Resources, T.-B.M.; Software, T.-T.L.; Supervision, D.-T.N.; Validation, D.-T.N.; Visualization, D.-T.N.; Writing—original draft, T.-B.M.; Writing—review & editing, D.-T.N., T.-B.M. All authors have read and agreed to the published version of the manuscript.

Funding: This work was supported by Vietnam Ministry of Education and Training (MOET) under grant number B2022-BKA-08.

Acknowledgments: This work was supported by the Vietnam Ministry of Education and Training (MOET) under grant number B2022-BKA-08.

Conflicts of Interest: The authors declare no conflict of interest.

References

1. Irving, A.D.; Clarance, L.M.; Russell, F.H. *High Temperature Machining Methods*; Armed Services Techinical Information Agency, Arlington Hall Station: Arlington, VA, USA, 1963.
2. Rajput, M. CFD Modelling of Hot Machining Operation. Ph.D. Thesis, National Institute of Technology Rourkela, Rourkela, Orissa, India, 2010.
3. Makwana, R.; Prajapati, H. Experimental Investigation on Effect of Machining Parameters on Surface Roughness in Thermally Assisted Turning of mild steel. In Proceedings of the International Conference on Multidisciplinary Research & Practice, Ahmedabad, India, 30 November 2014; Volume 1, pp. 488–489.
4. Harpreet, E.M.G.S.; Sharma, E. Analysis of Surface Roughness and Material Removal Rate in Dry and Thermal Assisted Machining of EN8 Steel. *Int. J. Eng. Sci. Res. Technol.* **2015**, *4*, 577–583.
5. Baili, M.; Wagner, V.; Dessein, G.; Sallaberry, J.; Lallement, D. An Experimental Investigation of Hot Machining with Induction to Improve Ti-5553 Machinability. *Appl. Mech. Mater.* **2011**, *62*, 67–76. [CrossRef]
6. Rahim, E.A.; Warap, N.; Mohid, Z. Thermal-assisted machining of nickel-based alloy. In *Superalloys*; IntechOpen: London, UK, 2015; pp. 3–29.
7. Pimenov, D.Y.; Mia, M.; Gupta, M.K.; Machado, R.A.; Pintaude, G.; Unune, D.R.; Khanna, N.; Khan, A.M.; Tomaz, I.; Wojciechowski, S.; et al. Resource saving by optimization and machining environments for sustainable manufacturing: A review and future prospects. *Renew. Sustain. Energy Rev.* **2022**, *166*, 112660. [CrossRef]
8. Barewar, S.D.; Kotwani, A.; Chougule, S.S.; Unune, D.R. Investigating a novel Ag/ZnO based hybrid nanofluid for sustainable machining of inconel 718 under nanofluid based minimum quantity lubrication. *J. Manuf. Process.* **2021**, *66*, 313–324. [CrossRef]
9. Unune, D.R.; Mali, H.S. Current status and applications of hybrid micro-machining processes: A review. *Proc. Inst. Mech. Eng. Part B J. Eng. Manuf.* **2014**, *229*, 1681–1693. [CrossRef]
10. Xavierarockiaraj, S.; Kuppan, P. Investigation of cutting forces, surface roughness and tool wear during Laser assisted machining of SKD11Tool steel. *Procedia Eng.* **2014**, *97*, 1657–1666. [CrossRef]
11. Thi-Bich, M.; Van-Chien, D.; Tien-Long, B.; Duc-Toan, N. Cutting Force Model for Thermal-Assisted Machining of Tool Steel Based on the Taguchi Method. *Metals* **2018**, *8*, 992. [CrossRef]
12. Mac, T.-B.; Banh, T.-L.; Nguyen, D.-T. Study on cutting force and chip shrinkage coefficient during thermal-Assisted machining by induction heating of SKD11 steel. *J. Korean Soc. Precis. Eng.* **2019**, *36*, 803–811. [CrossRef]
13. Mac, T.-B.; Luyen, T.-T.; Nguyen, D.-T. A Study for Improved Prediction of the Cutting Force and Chip Shrinkage Coefficient during the SKD11 Alloy Steel Milling. *Machines* **2022**, *2022*, 229. [CrossRef]
14. Thanh-Huan, N.; Duc-Toan, N. Experimental Researches of Turning Hardened 9CrSi Alloy Tool Steel with Laser-Assisted Machining. *Arab. J. Sci. Eng.* **2021**, *46*, 11725–11738. [CrossRef]
15. Dong, T.P.; Toan, N.D. A study on the investigation of the microstructure of SKD61 steel after selected quenching and tempering processes. *Mod. Phys. Lett. B* **2023**, 2340022. [CrossRef]
16. Ginta, T.L.; Amin, A.K.M.N. Thermally-Assisted End Milling of Titanium Alloy Ti-6Al-4V Using Induction Heating. *Int. J. Mach. Mach. Mater.* **2013**, *14*, 194–212. [CrossRef]
17. Wang, Y.; Ding, C.; Tang, F.; Zheng, D.; Li, L.; Xie, S. Modeling and simulation of the high-speed milling of hardened steel SKD11 (62 HRC) based on SHPB technology. *Int. J. Mach. Tools Manuf.* **2016**, *108*, 13–26. [CrossRef]
18. Wang, Z.Y.; Rajurkar, K.P.; Fan, J.; Lei, S.; Shin, Y.C.; Petrescu, G. Hybrid machining of Inconel 718. *Int. J. Mach. Tools Manuf.* **2003**, *43*, 1391–1396. [CrossRef]
19. Dong, T.; Toan, N.; Dung, N. Influence of heat treatment process on the hardness and material structure of SKD61 tool steel. *Mod. Phys. Lett. B* **2023**, 2340014. [CrossRef]
20. Mruthunjaya, M.; Yogesha, K.B. A review on conventional and thermal assisted machining of titanium based alloy. *Mater. Today Proc.* **2021**, *46*, 8466–8472. [CrossRef]
21. Sun, S.; Brandt, M.; Dargusch, M.S. Thermally enhanced machining of hard-to-Machine materials—A review. *Int. J. Mach. Tools Manuf.* **2010**, *50*, 663–680. [CrossRef]
22. Maitya, K.; Swain, P. An experimental investigation of hot-machining to predict tool life. *J. Mater. Process. Technol.* **2008**, *198*, 344–349. [CrossRef]
23. Germain, G.; Lebrun, J.-L.; Robert, P.; Santo, P.; Poitou, A. Experimental and numerical approaches of laser assisted turning. *Int. J. Form. Process.* **2005**, *8*, 347–361.
24. Jeon, Y.; Pfefferkorn, F. Effect of laser preheating the workpiece on micro end milling of metals. *J. Manuf. Sci. Eng. Trans. ASME* **2008**, *130*, 1–9. [CrossRef]
25. Dumitrescu, P.; Koshy, P.; Stenekes, J.; Elbestawi, M. High-power diode laser assisted hard turning of AISI D2 tool steel. *Int. J. Mach. Tools. Manuf.* **2006**, *46*, 2009–2016. [CrossRef]
26. Kannatey-Asibu, E. *Principles of Laser Materials Processing*; John Wiley & Sons, Inc.: Hoboken, NJ, USA, 2009.
27. Wang, C.; Xie, Y.; Zheng, L.; Qin, Z.; Tang, D.; Song, Y. Research on the Chip Formation Mechanism during the high-speed milling of hardened steel. *Int. J. Mach. Tools Manuf.* **2014**, *79*, 31–48. [CrossRef]
28. Rudnev, V.; Love30less, D.; Cook, R.; Black, M. *Handbook of Induction Heating*; Marcel Dekker: New York, NY, USA, 2003.

29. Thi-Hoa, P.; Thi-Bich, M.; Van-Canh, T.; Tien-Long, B.; Duc-Toan, N. A study on the cutting force and chip shrinkage coefficient in high-speed milling of A6061 aluminum alloy. *Int. J. Adv. Manuf. Technol.* **2017**, *98*, 177–188. [CrossRef]

metals

Article

Effect of Wear on Vibration Amplitude and Chip Shape Characteristics during Machining of Eco-Friendly and Leaded Brass Alloys

Peter Pavol Monka [1,2], Katarina Monkova [1,2,*], George A. Pantazopoulos [3,*] and Anagnostis I. Toulfatzis [3]

1 Faculty of Manufacturing Technologies, Technical University in Kosice, Sturova 31, 080 01 Presov, Slovakia; peter.pavol.monka@tuke.sk
2 Faculty of Technology, Tomas Bata University in Zlin, Nam. T.G. Masaryka 275, 760 01 Zlin, Czech Republic
3 ELKEME Hellenic Research Centre for Metals S.A., 61st km Athens—Lamia National Road, 32011 Oinofyta, Greece; atoulfatzis@elkeme.vionet.gr
* Correspondence: katarina.monkova@tuke.sk (K.M.); gpantaz@elkeme.vionet.gr (G.A.P.)

Abstract: The dynamic stability of the machining set and the entire cutting process, together with the appropriate form of chips generated during machining under the given conditions, are the basic prerequisites for autonomous machining in accordance with the Industry 4.0 trend. The research, based on a newly designed method, aims to study the frequency response of the machining system to different values of tool wear and cutting speed, which cause the worsening of the machined parts' quality and the instability of the whole cutting process. The new idea is based on the inverse principle, in which the wear with various values of VB was artificially prepared in advance before machining. Consequently, the effect of artificial wear and cutting speed on vibration and chip shape characteristics were studied. Three types of brass alloys were used within the experiments as the machined materials. Measured data were statistically processed and the desired dependencies were plotted. Chips were collected for each combination of machining conditions, while the article presents a database of the obtained chip shapes at individual cutting speeds so that they can be compared and classified. The results showed that brass alloys CW510L and CW614N exhibit an average of three times lower vibration damping compared to the CW724R alloy, while relatively good chip formation was noted in the evaluated machining conditions even without the use of a chip breaker. The problematic chip shape occurred only in some cases at the machining of CW510L and CW724R, which cannot be generalized.

Keywords: artificial wear; vibration response; brass alloys; machining process; chip form

Citation: Monka, P.P.; Monkova, K.; Pantazopoulos, G.A.; Toulfatzis, A.I. Effect of Wear on Vibration Amplitude and Chip Shape Characteristics during Machining of Eco-Friendly and Leaded Brass Alloys. *Metals* **2023**, *13*, 828. https://doi.org/10.3390/met13050828

Academic Editor: Yadir Torres Hernández

Received: 31 March 2023
Revised: 18 April 2023
Accepted: 20 April 2023
Published: 23 April 2023

1. Introduction

Material removal has been a significant set of operations in production since the beginning of the Industrial Revolution. Its importance is highlighted by the significant contribution of machining to the development of the industry by enabling the development of new, more precise and efficient technological solutions. This field includes concentrated knowledge in a wide range of expertise—physical laws, chemistry, mathematics, material sciences, technical sciences, economic disciplines, environmental research, and many others.

Cutting off the material rate (measured by the machined volume over time) and durability of the cutting tool (represented by the duration of its work) are considered the most important results of the cutting process [1–3].

The results of the cutting process are directly influenced by machining environment variation, while the most important factors are [1]: Cutting tool; Workpiece; Machine Tool; Technological Process; and Technological Settings.

Following the rules of economic efficiency, ecological reform demands, societal regulations, and high demands for constant refinement of production (micro and nano dimensions), the complexities of the above factors unite in a significant way. Therefore, the integrity of modern machining systems requires a very precise and demanding design and formulation with optimization for the given production conditions [4,5].

In the finally determined production system (given mainly by the product, machine tool, cutting tool, and process), only cutting factors can be used for the optimization of efficiency and productivity (durability of cutting tool and material removal rate). Generally, the basic cutting factors are depth of cut, feed, and cutting speed. However, the depth of cut can be a factor related to machining efficiency that increases only in the case of small series and piece production if it is not possible to provide material with dimensions and shapes close to the manufactured product. In that case, it is an operation with an inefficient use of input material, and such a production method would need to be optimized for the application of more economical technologies. Therefore, it is not assumed that this factor will be used for optimization. The general effect of depth of cut a_p on durability T is shown in Figure 1a [1,6].

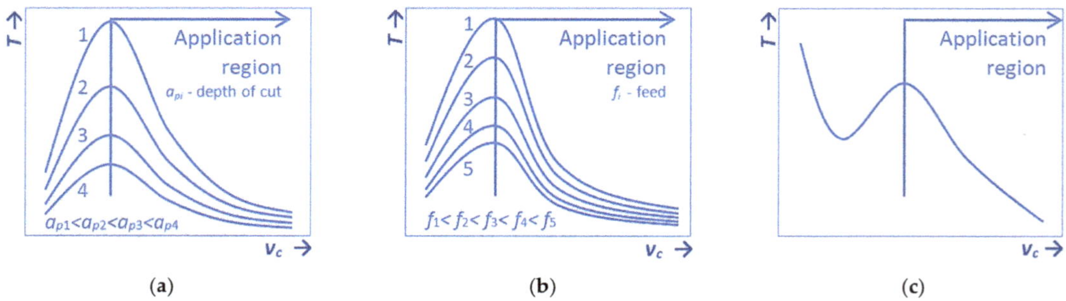

Figure 1. Influence of cutting parameters on tool life: (**a**) a_p—depth of cut, (**b**) f—feed, (**c**) v_c —cutting speed.

Feed rate linearly affects productivity. However, its speed is limited by the required quality parameters of the final surfaces, and therefore the value of this speed cannot be increased without special solutions, e.g., smoothing geometry of the cutting tool, etc. The general effect of feed f on durability T is shown in Figure 1b [1].

Cutting speed affects machining productivity in a linear relationship, but tool life, expressed by cutting ability, is degraded exponentially. The relationship between these two physical regularities of the cutting tool is determined by the kinematics of the machine tool and the events (both physical and chemical) in contact between the tool and the machined material. The general effect of cutting speed v_c on durability T is documented in Figure 1c. This dependence shows the region appropriate for the good relationship of cutting speed and durability of the tool [1,7,8].

The next important parameter that enters the machining process is temperature. The general regularities of the influence of the temperature reached at the point of the cut on the nonlinearity of the course of the investigated factors are shown in Figure 2. The temperature mainly depends on the type of machined material, the rate of deformation in shear volume, the friction conditions on the front and back plane of the tool, and the cooling characteristics of the process. All these physical factors are also influenced by variations in other factors of the machining environment, as mentioned above.

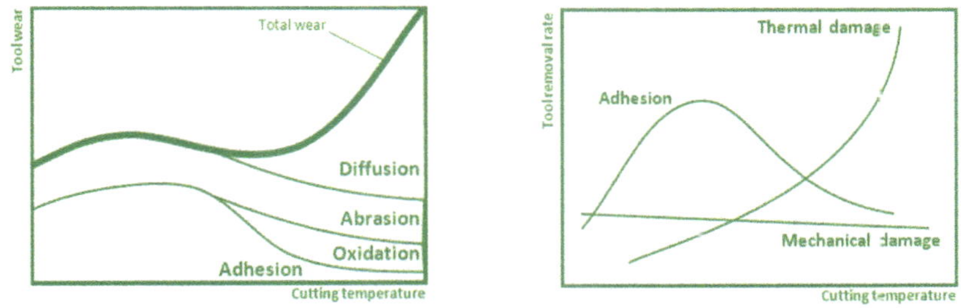

(a) (b)

Figure 2. Regularities of the (**a**) cutting tool degradation and (**b**) removal rate—depending on the temperature at the point of the cut.

The dependence of the achieved wear limit on the cutting speed is also a very important regulator of machining. The wear limit is the maximum value of admissible degradation of the cutting edge, at which the admissible quality characteristic of the workpiece surface by machining can still be achieved [9,10]. In the given conditions, wear of the flank or face on the cutting tool depends not only on the type of machined material but also on the proportion between feed and cutting speed. The dimensions of worn parts of the tool are usually specified by [11]:

- VB—width of the wear area on the flank;
- KT—depth of wear groove on the face;
- KVs—wear of the tool corner by wear.

An effective machining process requires controlling and checking the tool's wear propagation. The best way of the wear propagation is uniform flank wear of the cutting insert as shown in Figure 3.

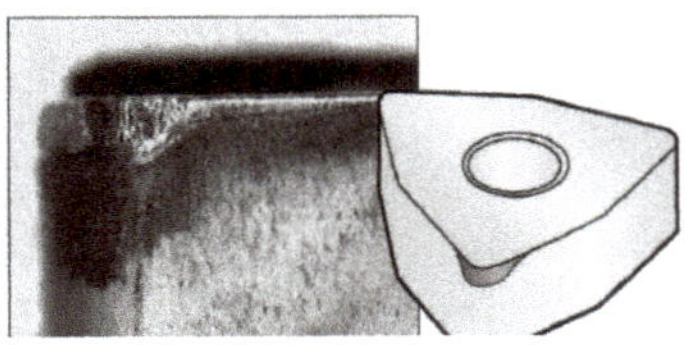

Figure 3. Optimal wear of the cutting tool—uniform flank wear.

Tool wear is closely related to tool durability since it directly affects the tool's durability and life. Figure 4 presents the dependence of the decrease in durability on the cutting speed for the determined wear limit and the area of degradation effects as a function of the overall cutting time T on the cutting speed v_c in logarithmic coordinates [12,13].

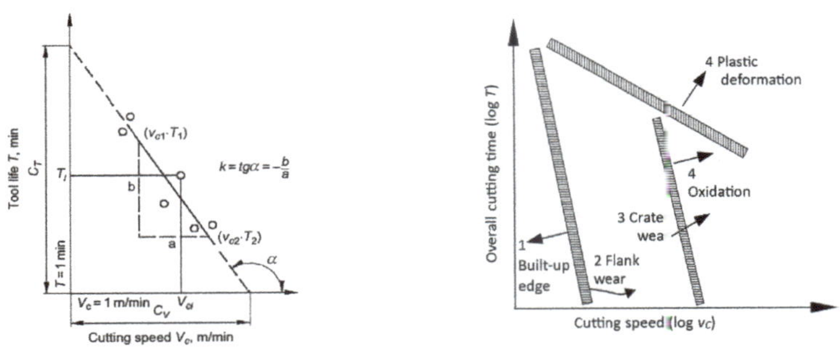

(a) (b)

Figure 4. (**a**) Taylor tool-life curve, (log-log scale); (**b**) The wear and failure criteria for carbide inserts.

Following the above it can be said that the joint action of a group of factors in the production system (tools and tool materials, tool geometry, tool coating, and procedural media), as well as groups of process factors (cutting speed, depth of cut, and feed) creates the resulting heat generation and degradation indices of the tool. The intensity of abrasion appears to be a relationship between a combination of processed material and cutting tool, and in terms of the dependence of the tool wear on the achieved cutting temperature, it shows quasi-static characteristics. The adhesion intensity slowly increases as the cutting temperature increases and gradually decreases when the temperature of the diffusion and oxidation processes is reached. The oxidation intensity begins to develop after critical cutting temperatures and its magnitude stabilizes when the maximum saturation is reached. The intensity of the diffusion processes is imperceptible at low cutting temperatures and increases very progressively when the critical cutting temperature is reached.

Several studies have already dealt with machining process monitoring in which the influence of cutting conditions on the machining set and system stability was investigated. Dimla [14] identified that time domain features were deemed to be more sensitive to cutting conditions than tool wear, whereas frequency-based features correlated well with the tool wear and measured wear values correlated well with certain resonant peak frequencies. Chuangwen [15] obtained results for 022Cr17Ni12Mo2 stainless steel where the root-mean-square value of vibration acceleration signals increased with wear progression in general. Bouhalais [16] confirmed a good match between high-frequency vibration components' energy values and tool wear monitoring. Sridhar [17] found ten times larger vibration amplitudes of a worn drill compared to a sharp one. Babu [18] found that the correlation of experimentally measured tool wear with tool acceleration is practically feasible for real-time flank wear prediction at any time. Antic [19] noted that most research studies on the development of tool wear monitoring systems focus on the type of machining and the machining process to which they are applicable. The types and types of input signals, methods and techniques of signal processing and feature extraction are not primarily considered. This approach limits the scope of the investigation. Data processing of the instrument monitoring system proposed by Anticom includes these steps: signal acquisition and processing; data classification by grouping into an input function; and definition of tool wear status [19].

The study of the effect of conditions on the machining process and the durability of the tool can generally be carried out by tests categorized into two groups: short-term and long-term tests. Short-term durability tests are more commonly used in practice because of shorter time and less amount of processed material, but also for efficiency and economy. This testing is employed usually as the first step in checking the material of a workpiece or as a quick comparison of the durability of the examined cutting tools with a standardized etalon [20]. The principle of the short-term test, in which the material is removed by facing, consists in machining a workpiece with a relatively large diameter (D_{max} is about 300 mm). The material is removed in the radial direction, i.e., from a pre-drilled coaxial hole in the center of the workpiece towards the edge of the disc. This means that the cutting speed gradually increases, and, at a specific D_n diameter, the cutting speed reaches the maximum v_n value at which the ability to cut the material is lost. Pastucha et al. [21] have proposed a prospective variant to a short-term test, which is based on discontinuous cutting thus leading to higher intensity of tool wear to make this method more effective. In this method, the workpiece is radially placed on a large-diameter disk, while the cutting tool removes the material from the workpiece in circular segments.

However, the short-term tests described above also have their weaknesses. Except for certain short-term tests where the cutting speed is not constant, making it difficult to evaluate the effect of cutting speed on tool wear, such types of tests are not applicable when the workpieces with a larger diameter are for various reasons unavailable, or the turning machine does not allow workpieces with such a large diameter to be clamped.

These disadvantages were taken into account in the present study, and in order to eliminate them, the authors proposed a new methodology for investigating the appearance of tool wear that can be considered a novelty and a contribution of the authors to the scientific field. Since the dynamic stability of the machining set and the entire cutting process, together with the appropriate form of chips generated during machining under the given conditions, are the basic prerequisites for autonomous machining in accordance with the Industry 4.0 trend, it is, therefore, important to deal with this issue.

Manufacturing within Industry 4.0 is planned and programmed to perform tasks with minimal human intervention, with autonomous manufacturing equipment varying greatly in size and function, from drone inventory scanning to autonomous mobile robots for pick and place operations. Equipped with cutting-edge software, artificial intelligence, sensors and machine vision, the devices are capable of performing demanding and delicate tasks. They should be able to recognize, analyze, and act on the information they receive from their surroundings. It is very important to ensure that the manufacturing machines work smoothly and that the impact of external conditions causing problems was eliminated where possible. A number of such problems can be caused by inappropriate chip shaping. Too short chips cause intermittent machining, which leads to micro-cracks on the cutting edge and shortens the tool life. In terms of life, long chips are slightly more favorable. Long, smooth-shaped chips show less micro-vibration when machining, which contributes to improving the quality of the machined surface. For the cutting process itself, however, long chips are also not suitable. Long chips can cause the cutting tool to become tangled and lose its ability to cut. They may damage the machine tool, workpiece, and cutting tool, and induce dangerous working conditions for operators. They can also cause problems with jamming in the conveyor and outages in operation. Thus, inappropriate chip problems can cause the cutting process not to be autonomous and will therefore not be suitable for inclusion in the automated process.

Based on these trends in the automation of production processes and their requirements, the present research focused on knowing the frequency response of the system for various tool wear values, resulting in deterioration of the quality of the processed parts and unsteadiness of the whole cutting operation. Understanding and knowledge of the behavior trend of vibration characteristics allow not only setting the most suitable technological conditions for machining the given material and thus increasing the tool life but also comparing the durability of the tool when machining selected brass alloys. The new idea is based on the inverse principle, at which the wear with various values of VB was artificially prepared before machining. Consequently, the effect of artificial wear and cutting speed on vibration and chip shape characteristics were studied. The combination of the studied parameters and the designed methodology that removes the disadvantages of the traditional short-term tests is unique, so this approach can be considered a novelty and the authors' contribution to the research area.

2. Materials and Methods

2.1. Machined Materials

Two types of eco-friendly (CW510L and CW724R) and one conventional standard (CW614N) brass alloy were selected as the material to be machined in this study.

Brass alloys are generally characterized by high thermal and electrical properties and conductivity as well as excellent antibacterial properties. Therefore, they are widely used in various industries such as electrical and electronic, automotive, and sanitary. Due to a large number of cutting operations in the production of brass components, various alloying elements improving machinability are usually added to brass. The most important element in this context is lead, which improves machinability due to excellent chip breaking, low tool wear and high usable cutting parameters. These aspects can be explained by two basic phenomena. First, the solubility of lead in brass is very low, and consequently, lead segregation occurs throughout the microstructure, especially at the grain boundaries. This will significantly reduce shear strength, resulting in very good chip breaking. Second, lead exhibits a low melting point T_m = 327.5 °C. During cutting, lead decreases the chip ductility while a thin, semi-liquid lead film reduces friction and thus cutting forces and tool wear [22–25].

However, since Pb is currently a strongly regulated substance due to environmental and medical concerns, its substitution is of primary importance. Stavroulakis [26] and his team have proposed four directions for further applied research to optimize the design and processing of alloys without lead so as to overcome the characteristics and functionality of conventional lead brass, which meet the requirements of environmental protection, human health and sustainable development.

Brass alloy CW614N is the standard leaded alloy for free cutting purposes, while CW510L and CW724R are lead-free brasses with good machinability, belonging to the group of so-called eco-friendly brass alloys since, for both environmental and health reasons, the reduction of lead in brass parts has great importance [27,28]. The basic characteristics of the brass alloys used in the study are given in Table 1 [29–31] and chemical compositions are given in Table 2 [32,33].

Table 1. Basic characteristics of machined brass alloys data from [29–31].

Characteristic	Unit	CW510L (Material 1)	CW614N (Material 2)	CW724R (Material 3)
Tensile strength R_m	(MPa)	478–484 [25] 220–500 [28] *	456 [25] 360–500 [28] *	654 [25] 500–670 [28] *
Yield strength $R_{p0.2}$	(MPa)	310–315 [25]	324 [25]	400 [25]
Hardness	HB	134–157 [25] 90–160 [28] *	154 [25] 90–160 [10] *	210 [25] 130–220 [28] *
Elongation to break A	(%)	25–29 [25] 5–20 [28] *	26 [25] 5–20 [28] *	21 [25] 10–15 [28] *
Thermal conductivity λ	(W/mK)	113–139 [25] 139 [28]	123 [25] 113 [28]	35 [25] 35 [28]

* Depending on the method of production.

Table 2. Indicative chemical compositions of machined brass alloys (wt%) data from [32,33].

	Cu	Zn	Pb	Si	As	P
CW510L (Material 1)	57.38	42	0.07	-	-	-
CW614N (Material 2)	57.61	39	3.32	-	-	-
CW724R (Material 3)	75.86	21	0.02	3.4	-	0.05

The presence of lead (Pb) in conventional leaded brasses (typical composition range ~2.5–3.5 wt%) favors machinability and machining operations, since it results in chip fracturing, enhances lubrication and reduces cutting force minimizing tool wear, see for instance [23,26,32]. It can therefore be said that when replacing the original (leaded) brass with environmentally friendly brass alloys, the conditions of machining deteriorate (due to a decreasing amount of lead), which can virtually directly reduce the productivity of the machining process by reducing technological factors (especially cutting speed and feed) to ensure the required qualitative factors (surface roughness, dimensions) and the efficiency of the process (durability of the cutting tool).

The workpiece was provided in the form of rods with a diameter of 35 mm and a length of 400 mm (Figure 5a). The bar was clamped with a maximum overhang of 200 mm to maintain the required rigidity and before measurements, it was finely machined to a diameter of 34.6 mm. The machined bar after experimental testing is shown in Figure 5b.

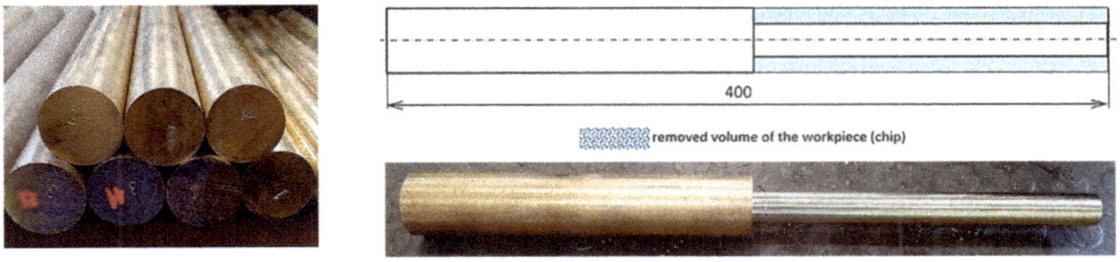

(a) (b)

Figure 5. Workpiece (**a**) rods provided by the supplier; (**b**) machined part.

2.2. Machining Process Using the Newly Designed Artificial Wear Methodology

A newly designed methodology was used for the machining of brass alloys based on artificial wear to find out the response of the cutting tool and the whole system to machining under given conditions. The approach was selected because of the existing settings, which were mainly connected to the small diameter of the supplied bars. A diameter of 35 mm is not suitable for long-term testing as it does not enable reaching the desired cutting speed, and when the diameter of the workpiece is becoming smaller, the stiffness of the bar reduces very quickly [33–35].

A total of 80 pieces of cutting inserts were specially modified by the experienced supplier, while the artificial tool wear VB_{artif} was created only on the major flank of the tool. A tool wear VB was chosen as the most important factor affecting the quality of the machined surface and the dimensional accuracy of the machined component The values of artificial wear within the experiments were selected in a geometric series so that they could capture critical values and which are reflected in vibrations. From the point of view of the machined surface, VB = 0.4 mm wear is considered critical during finishing. In the experiment, this range was doubled to better understand the behavior of the machining system and its responses to machining with a worn tool.

At every new test, a cutting tip with a new cutting wedge was used, so one cutting insert was active for four experiments (with different corners). The geometry of created artificial wear VB_{artif}, as well as the values of the VB_{artif} (mm) and grinding depth H (mm) for the individual cutting inserts, are shown in Figure 6a, while prepared cutting inserts are shown in Figure 6b.

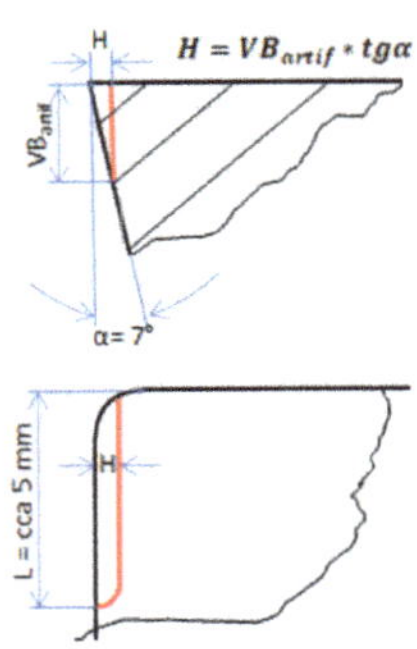

Cutting insert designation	VB_{artif} (mm)	H (mm)
T01...T10	0.000	0.000
T11...T20	0.050	0.006
T21...T30	0.080	0.010
T31...T40	0.128	0.016
T41...T50	0.205	0.025
T51...T60	0.328	0.040
T61...T70	0.524	0.064
T71...T80	0.839	0.103

(a)

(b)

Figure 6. (**a**) Geometrical and dimensional characteristics of artificial wear; (**b**) Cutting inserts with prepared artificial wear.

Durability evaluation, with a constant cutting depth $a_p = 1$ mm, constant feed $f = 0.06$ mm, and cutting speeds v_c = 100, 150, 225, and 337.5 m/min, was realized according to the relation of diameters to the artificial wears listed in Table 3.

Table 3. Relation of artificial wear and the workpiece diameters used in machining.

Range of the Workpiece Diameters (mm)	VB_{artif} (mm)
34.6–32.6	0.84
32.6–30.6	0.52
30.6–28.6	0.33
28.6–26.6	0.21
26.6–24.6	0.13
24.6–22.6	0.08
22.6–20.6	0.05
20.6–18.6	0.00

As is seen from Table 3, the tool with the highest wear (0.84 mm) was used for turning the workpiece with the largest diameter (34.6 mm) and the tool with a sharp wedge was used for machining the smallest considered diameter (20.6 mm) to ensure the best possible degree of stability.

ISO 3685 recommendations were used for the maximum extent of geometric and dimensional characteristics of the cutting tool, including working conditions as the basis for settings, given that a similar standard for turning brass alloys with single-point turning tools is not available. The cutting tool consists of a cutting insert (type SCGW 12T304 K10) and a tool holder (type SSBCR 2525 M 12-M-A); the basic characteristics are shown in Table 4. On every machined surface, a new corner of a cutting insert was used.

Table 4. Geometrical and dimensional characteristics of the cutting tool.

Geometrical and Dimensional Characteristic	Value
Cutting insert	
Rake angle γ	0°
Clearance angle α	7°
Cutting edge inclination angle λ_s	0°
Tool cutting edge angle κ_r	75°
Tool included angle ε_r	90°
Corner radius r_ε	0.4 mm
Tool holder	
Tool cross-section	25 × 25 mm
Insert shape	Square
Insert size	12.7 mm
Insert thickness	3.18 mm

The universal CNC turning machine DMG MORI ecoTurn 450 was used for experiments. A three-jaw chuck and a rotational tailstock were used for the workpiece clamping and supporting as shown in Figure 7.

Figure 7. Positions of the individual items of the machining set.

2.3. Measurement and Evaluation of Vibration Intensity

A very important accompanying factor of machining is the generation of vibrations. Machining vibrations are forced and self-excited in general. For the safety of the process and the quality of its outputs, it is necessary to monitor self-excited vibrations [36,37].

The self-excitation mechanism in the chip generation during cutting contributes significantly to the characteristic vibrational spectrum of the entire process. Chipping can cause the edge fracture of the machining tool, impairing cutting-edge sharpness and leading to tool blunting. This type of tool failure increased cutting forces, resulting therefore in further degradation of the machining performance (longer chips, higher fluctuations and chattering, part distortion), causing serious defects in the machined surface quality [38–40]. The friction on surfaces between the tool and machined material generated a dynamically discontinuous process. Tool wear significantly affects these friction-induced nonlinear vibrations and can be observed through the vibration characteristics of new and worn cutting tools. The evolution of the vibration characteristics can be used for the tool wear progress monitoring, too [41–43].

The advantage of installing a system for collecting and evaluating the vibration characteristics of the machining system for monitoring the tool wear process is that it only minimally interferes with the construction of the machine tool and the cutting tool, the monitoring takes place in real time, and this system can also be used to identify the condition of the machine tool.

The assessment of experiments focused on the vibration response analysis was performed in the complex system (Machine tool—Cutting tool—Fixture—Workpiece), therefore the vibration characteristics of the process were examined over a wide range of frequencies (0–25 kHz).

The acceleration sensor Wilcoxon Research WR-712F-M4 was fixed on the bottom of the tool holder, on the opposite side as a cutting insert was attached. The location of the sensor was selected as close as possible to the cutting edge and at the same time in such a way that it was protected from being attacked by chips. A National Instruments PXI-4462 measurement card and LabView software were used to collect and analyze vibration data [44].

An example of a vibro-diagnostics signal in the time domain obtained for one minute of a running machining process is shown in Figure 8. Signals obtained by all measurements were analyzed by the Fast Fourier Transformation (FFT) method. Figure 9 shows the same signal as in Figure 8 but is processed using Fast Fourier Transform (FFT) to observe the measured amplitudes in the frequency domain.

The frequencies at which the peaks occurred are the so-called natural frequencies (indicated in Figure 9 with red crosses). These correspond to the resonant frequencies of the individual elements of the complex machining set, which can seriously influence the overall oscillating properties of the system.

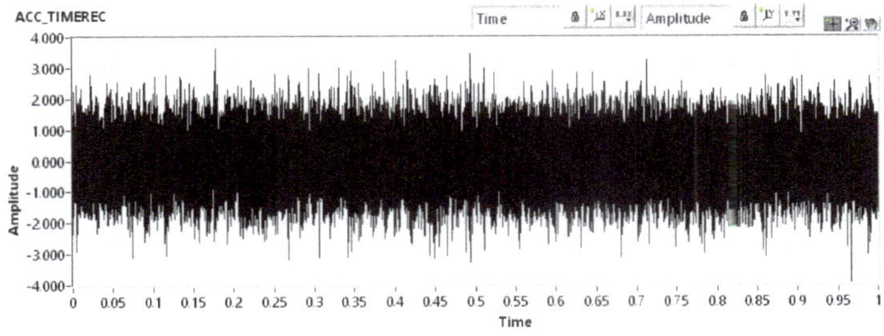

Figure 8. Example of signal collected for the evaluation process (time domain).

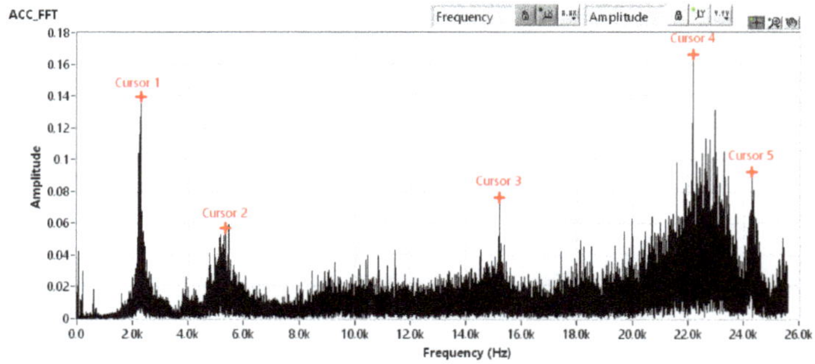

Figure 9. Example of the signal in the frequency domain processed using Fast Fourier Transformation.

To specify the effect of individual "components" of the machining set on the frequency characteristics and their rigidity, a numerical modal analysis of individual components of the set was performed using the software Ansys 2022 R2. Figure 10 shows a modal analysis of a cutting tool with VB_{artif} = 0.05 mm. (Note: The numerical analysis is not the goal of the presented research, so the authors do not go into detail with the boundary conditions).

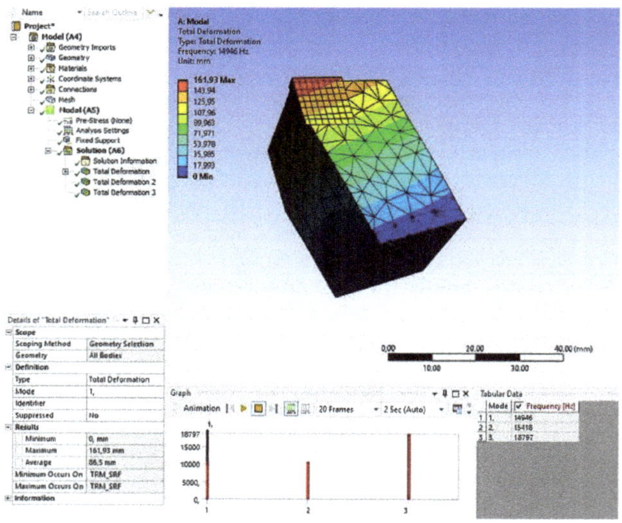

Figure 10. Numerical modal analysis of the cutting tool.

Based on the results, it was specified that, e.g., the peak highlighted in Figure 9 as the "Cursor 3" corresponds to the cutting tool, since the first natural frequency of the cutting tool identified by numerical analysis is close to 15,000 Hz (i.e., it was precisely 14,946 Hz for the presented numerical model consisting of the cutting insert and the tool holder, the back planar surface of which was fixed).

A mean value of amplitude as an output of measurement of accelerations was taken into account to evaluate the vibration intensity at the given conditions (specified by the alternate combination of cutting speed and tool wear, and constant parameters cutting depth a_p = 1 mm, constant feed f = 0.06 mm).

The measured data were statistically processed using multiple linear regression methodology in this sequence of steps [45,46]:

- Exclusion of outliers through the Grubbs test;
- Computing the regression coefficients b_i for a linear model with independent variables VB_{artif} and v_c and their mutual interaction;

- Diagnosis of outliers using residuals;
- Determination of significance of linear regression relationships;
- Generation of graphical representation of statistically processed data.

An example of the datasheet with processed statistical evaluation of measured data realized in MATLAB software is shown in Figure 11.

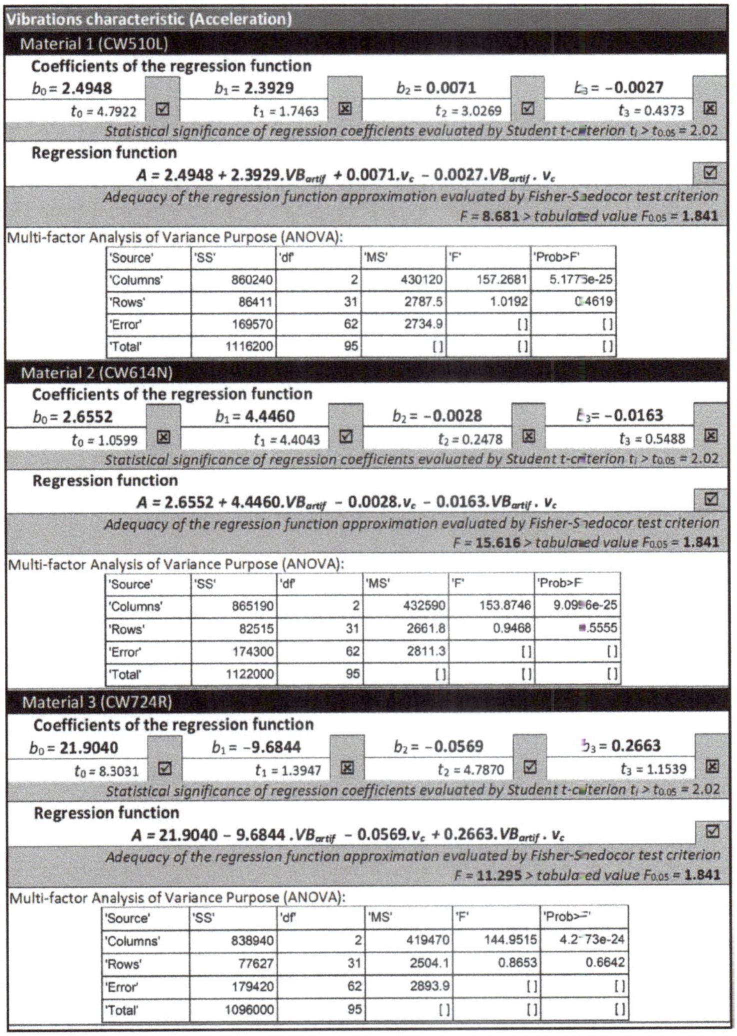

Vibrations characteristic (Acceleration)

Material 1 (CW510L)

Coefficients of the regression function

$b_0 = 2.4948$	$b_1 = 2.3929$	$b_2 = 0.0071$	$b_3 = -0.0027$
$t_0 = 4.7922$ ☑	$t_1 = 1.7463$ ☒	$t_2 = 3.0269$ ☑	$t_3 = 0.4373$ ☒

Statistical significance of regression coefficients evaluated by Student t-criterion $t_i > t_{0.05} = 2.02$

Regression function

$$A = 2.4948 + 2.3929.VB_{artif} + 0.0071.v_c - 0.0027.VB_{artif}.v_c$$ ☑

Adequacy of the regression function approximation evaluated by Fisher-Snedecor test criterion $F = 8.681 >$ tabulated value $F_{0.05} = 1.841$

Multi-factor Analysis of Variance Purpose (ANOVA):

'Source'	'SS'	'df'	'MS'	'F'	'Prob>F'
'Columns'	860240	2	430120	157.2681	5.1775e-25
'Rows'	86411	31	2787.5	1.0192	0.4619
'Error'	169570	62	2734.9	[]	[]
'Total'	1116200	95	[]	[]	[]

Material 2 (CW614N)

Coefficients of the regression function

$b_0 = 2.6552$	$b_1 = 4.4460$	$b_2 = -0.0028$	$b_3 = -0.0163$
$t_0 = 1.0599$ ☒	$t_1 = 4.4043$ ☑	$t_2 = 0.2478$ ☒	$t_3 = 0.5488$ ☒

Statistical significance of regression coefficients evaluated by Student t-criterion $t_i > t_{0.05} = 2.02$

Regression function

$$A = 2.6552 + 4.4460.VB_{artif} - 0.0028.v_c - 0.0163.VB_{artif}.v_c$$ ☑

Adequacy of the regression function approximation evaluated by Fisher-Snedecor test criterion $F = 15.616 >$ tabulated value $F_{0.05} = 1.841$

Multi-factor Analysis of Variance Purpose (ANOVA):

'Source'	'SS'	'df'	'MS'	'F'	'Prob>F'
'Columns'	865190	2	432590	153.8746	9.0996e-25
'Rows'	82515	31	2661.8	0.9468	0.5555
'Error'	174300	62	2811.3	[]	[]
'Total'	1122000	95	[]	[]	[]

Material 3 (CW724R)

Coefficients of the regression function

$b_0 = 21.9040$	$b_1 = -9.6844$	$b_2 = -0.0569$	$b_3 = 0.2663$
$t_0 = 8.3031$ ☑	$t_1 = 1.3947$ ☒	$t_2 = 4.7870$ ☑	$t_3 = 1.1539$ ☒

Statistical significance of regression coefficients evaluated by Student t-criterion $t_i > t_{0.05} = 2.02$

Regression function

$$A = 21.9040 - 9.6844.VB_{artif} - 0.0569.v_c + 0.2663.VB_{artif}.v_c$$ ☑

Adequacy of the regression function approximation evaluated by Fisher-Snedecor test criterion $F = 11.295 >$ tabulated value $F_{0.05} = 1.841$

Multi-factor Analysis of Variance Purpose (ANOVA):

'Source'	'SS'	'df'	'MS'	'F'	'Prob>F'
'Columns'	838940	2	419470	144.9515	4.2173e-24
'Rows'	77627	31	2504.1	0.8653	0.6642
'Error'	179420	62	2893.9	[]	[]
'Total'	1096000	95	[]	[]	[]

Figure 11. The example of a datasheet with statistical evaluation of measured data.

2.4. Chip Form Evaluation

The form of the chip created in the cutting process is an essential feature of the process evaluation. The chip form influences mainly [47,48]:

- The quality of the process—e.g., long-length chips can attack finished surfaces;
- The stability of the process—e.g., long-length chips do not leave the machine, fill the working space of the machine, and can be wound on a tool or workpiece, and they harm the cutting process itself;

- The environmental effect of the production—e.g., small, broken chips are far easier to handle, store, transport and recycle.

A very good system of chip form evaluation is given in ISO 3685. The standard classifies the chip form from the following points of view [49]:

1. The first aspect of chip classification consists of the basic characteristics of the chip form:
   (a) Shape characteristic (represented by the first number in the classification, e.g., 1—Ribbon, 6—Arc, 7—Elementary);
   (b) Chip form extended characteristics (represented by the second number in the classification, e.g., 1—Long, or 2—Short, etc.);
2. The second aspect is the direction of chip movement (represented by the third number within the classification; these can be numbers from 1 to 4, e.g., 1—from the workpiece in the feed direction; etc.).
3. The third aspect is the area on which the chip breaks (represented by the third number within the classification; there can be used numbers from 5 to 8; e.g., 7—broken against the workpiece surface).

Following the above chip form evaluation characteristics, it is possible to state the criteria for the chip form suitability, while several priorities are considered in the criteria statement:

- Secure required quality characteristics of machined surface;
- Safety automatic chip removal from the machine tool working area;
- Efficient chip handling, storage and removal, etc.

The chip form suitability classification scale can be tailored to specific requirements, e.g., Good/Acceptable/Unacceptable, etc.

In order to simplify the above-mentioned chip form evaluation approach by using a binary scale for basic and extended chip form characteristics, for the sake of clarity and a better summary of the characteristics of the created chip, the authors chose to divide the chip form in the final statement into two basic groups:

- Favorable
- Unfavorable.

No chip breaker was used for the cutting tool used for evaluation.

3. Results and Discussion

3.1. The Influence of Tool Wear and Cutting Speed on the Intensity of Tool Vibrations

During the vibration measurements, many data were recorded that needed to be processed. Therefore, they had to be studied using several practices and methods related to the phenomena of wear with respect to the chosen scope of cutting speed [50]. The most important analysis focused on increasing the credibility of durability assessment was the determination of the "general intensity of the vibration intensity of the system". Specification of "intensity" had to take into account the vibro-diagnostic changes of the experimental system to determine the needed machining factors, since any change in spindle rotation (due to the required cutting speeds) considerably affects the frequency range that is interconnected with the tool wear rate and manifested in the machining characteristics [51].

Data recorded during measurements for individual brass alloys are shown in Figure 12.

At first sight, it is clear from the dependencies in Figure 12 that the vibration intensity of the tool is different for each of the three tested materials. The mean value of the vibration intensity under the given conditions was further considered when generating the dependence of the amplitude on wear and cutting speed. A graphical representation of statistically processed data (mesh color surface) is shown in Figure 13.

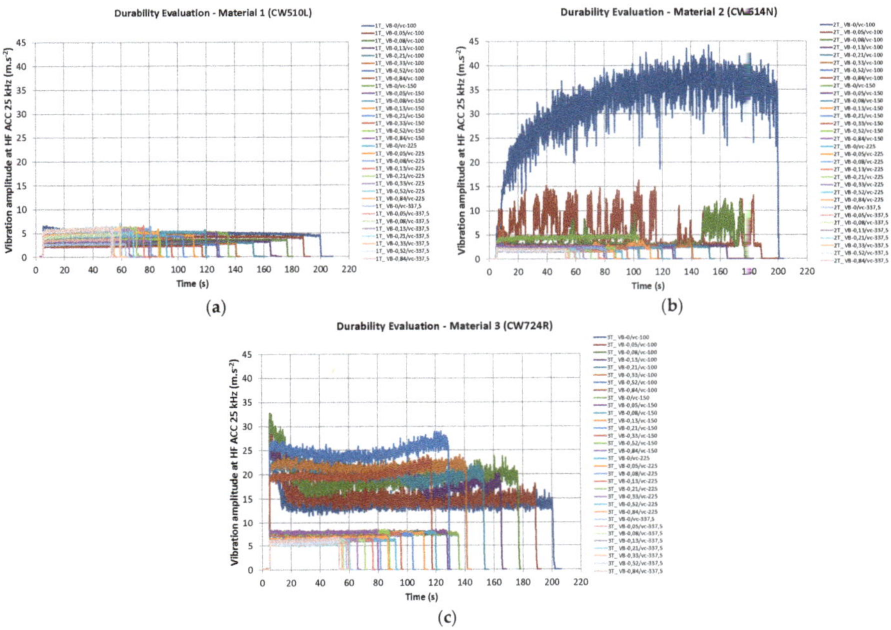

Figure 12. Measured vibrations in a time domain for individual materials; (**a**) CW510L; (**b**) CW614N; (**c**) CW724R.

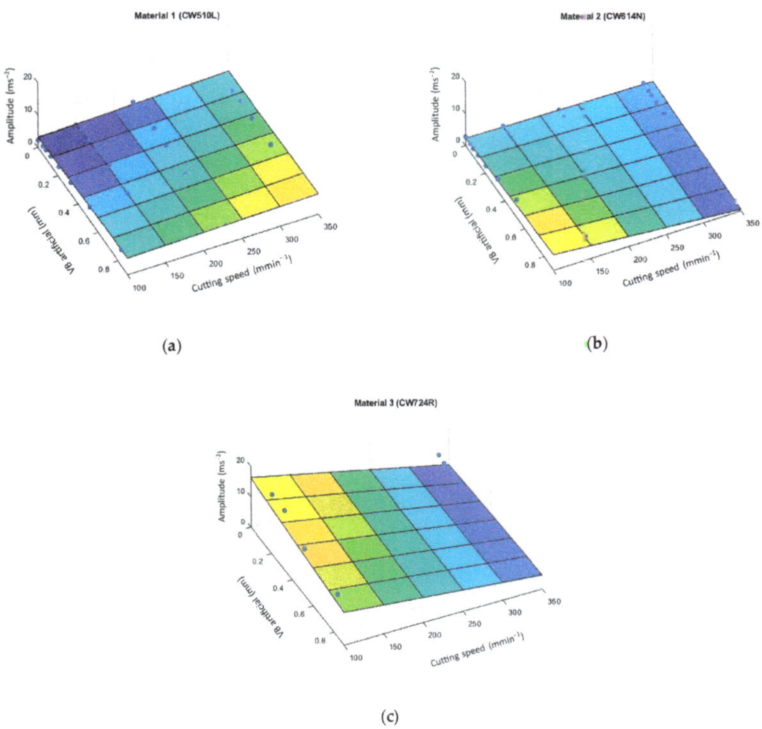

Figure 13. Graphical interpretation of vibration relationship on cutting speed and artificial tool wear for individual materials; (**a**) CW510L; (**b**) CW614N; (**c**) CW724R.

The results of the vibration amplitude divided the materials into two groups:

I. Material 1 (CW510L) and Material 2 (CW614N) show relatively low sensitivity for generating vibrations through tool wear;

II. Material 3 (CW724R) shows the higher vibration intensity generated by tool wear at lower cutting speeds.

To analyze the behavior of vibrating responses of the machining tool during the turning of individual brass alloys, the following statements can be made.

When machining CW510L brass alloy, the vibration amplitude value increased with wear and the cutting speed increased. This logically expected fact may be attributed to the heterogeneous structure of the material, made of a mixture of alpha and beta phases of the alloy, and the resulting very good value ductility, which is about twice as high as the remaining two compared materials.

From the strength characteristics ($R_m/R_{p0.2}$) point of view, this material is at the lowest amplitude values of the materials studied, which certainly contributes to a more suitable force ratio in the chip root and the creation of the lowest cutting resistance. In this case, mechanical properties are very well correlated with the lowest hardness values of all compared materials and the resulting effect on the nature of chip formation and vibration generation. Excellent thermal conductivity is certainly included in the comprehensive action of the external manifestations during machining—which is comparable to the thermal conductivity of the CW614N material but is three times better compared to the CW724R. This gives a prerequisite for better heat distribution to the chips and machined material.

Monitoring of the vibrations during the machining of the CW614N alloy showed that the value of the vibration amplitude increased uniformly with increasing wear on the tool flank area, but decreased with increasing cutting speed. This was probably caused by the duplex microstructure of the material containing insoluble lead particles, which take over the function of reducing the friction of the contact surfaces of the tool and at the same time serve in the place of the cutting plane of the chip as non-cohesive particles, ensuring very good conditions for breaking the chip. The CW614N alloy showed half the ductility value compared to the CW510L material, which indicated its lower plasticity and lower ability to dampen vibrations. From the point of view of strength characteristics ($R_m/R_{p0.2}$), the CW614N alloy is at the middle values of the studied materials, which also indicates the middle values of the strength ratios at the root of the chip and the degree of cutting resistance. Even in this case, the mechanical properties were well correlated with the average hardness values of the compared materials and the resulting effect on the character of chip formation and vibration generation. Excellent thermal conductivity, which is comparable to the material CW510L but is three times better compared to the material CW724R, certainly entered into the complex action of external manifestations during machining. This gives the assumption of a better distribution of heat to the chip and the machined material.

From the dependence of vibration intensity of the CW724R brass alloy on the wear and cutting speed it can be seen that the value of the amplitude decreased uniformly with increasing wear on the back surface of the tool and with increasing cutting speed. This resulted from the material's complex heterogeneous structure in which, compared to other materials, the content of silicon is high. The influence of silicon is manifested mainly by increasing the strength characteristics of brasses, which causes this material to have the highest. In the case of machining, the influence of silicon in the structure is also significant, with considerable grinding action on the tool flank surfaces, and increasing the width of the worn surface as a resulting effect of increasing the damping properties of this frictional contact. The lowest thermal conductivity of the compared materials probably caused a significant increase in temperature in the cutting plane of the chip, thus a substantial increase in local plasticity, and considerably different vibrational manifestations of the machining system.

3.2. Evaluation of Chip Form

Waste material from machining (chips, residues) was collected separately for each evaluated group of materials so that it could be recycled and thus ensure a higher degree of environmental protection and non-contamination.

Results of chip evaluation within durability tests are summarized in Figure 14.

$VB_{artificial}$ (mm)	Material 1 (CW510L)				Material 2 (CW614N)				Material 3 (CW724R)			
	v_c (m.min^{-1})				v_c (m.min^{-1})				v_c (m.min^{-1})			
	100	150	225	337	100	150	225	337	100	150	225	337
0.00	2.2 / 6.2	2.2	2.2	2.2	7	2.2 / 6.2	6.2	6.2	6.2	4.2 / 6.2	6.2	1.2
0.05	2.2 / 6.2	2.2	2.2	2.2	7	6.2	6.2	6.2	6.2	6.2	6.2	5.2
0.08	2.2 / 6.2	2.2	2.2	2.2	7	6.2	6.2	6.2	6.2	6.2	6.2	2.2 / 2.3
0.01	6.2 / 1.1	2.2	2.2 / 1.1	2.2	7	6.2	6.2	6.2	6.2	6.2	6.2	6.2
0.21	1.3	2.2	2.2 / 1.1		7	6.2	6.2	6.2	6.2	6.2	6.2	2.3
0.33	6.2	2.2	2.2 / 1.1	2.2	7	6.2	6.2	6.2	6.2	6.2	6.2	2.3
0.52	2.2 / 6.2	2.2	2.2	2.2	7	6.2	6.2	6.2	6.2	6.2	6.2	1.3
0.84	2.2 / 6.2	2.2	2.2	2.2	7	7	6.2	6.2	6.2	6.2	6.2	1.3

Unfavourable chip form

Favourable chip form

Figure 14. Evaluation of chip form produced during experimental verification.

It can be stated that the machined brass alloys have shown relatively good chip formation. The problematic chip form occurred only in some cases of Material 1 (CW510L) and Material 3 (CW724R) that is marked in Figure 14 in yellow with the statement "Unfavourable chip form".

For each combination of machining conditions (given by the type of brass alloy, cutting speed and artificial wear), a representative sample of the removed material was documented in the form of a chip, which was registered and stored in the authors' internal database. Figures 15–18 document the comparison of the resulting forms of chips arising at individual cutting speeds.

From the practical point of view, based on the results related to the chip form originating at the machining of selected brass alloys, it can be said that there are good production conditions in terms of chip shape since it was affected very little by the type of alloy. Undesirable areas of unsuitable chip forms can be easily adjusted by changing the cutting conditions (cutting speed and feed) or—the best practical solution—by using a cutting tool with a chip breaker.

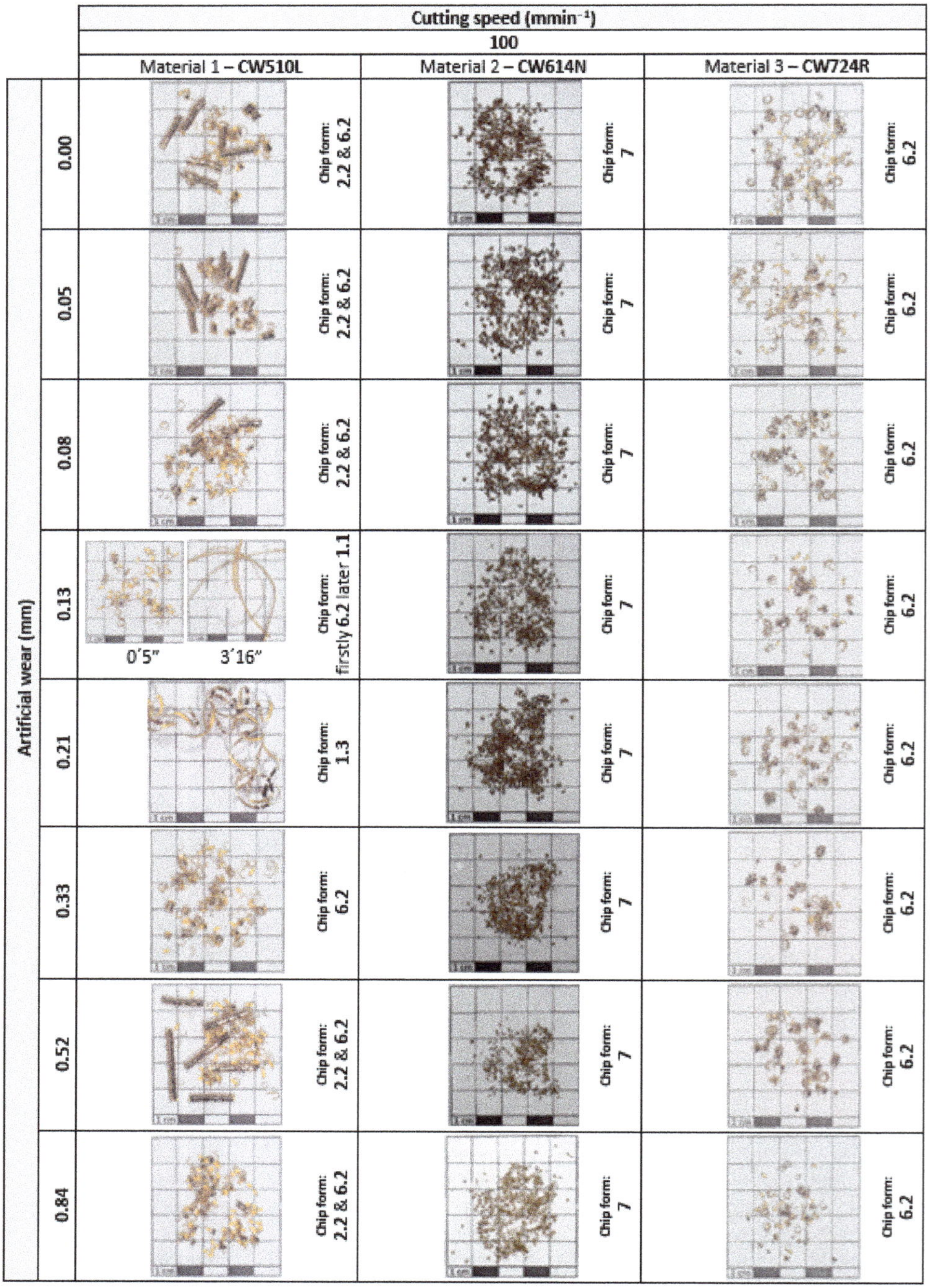

Figure 15. Comparison of chips originated at cutting speed 100 m/min.

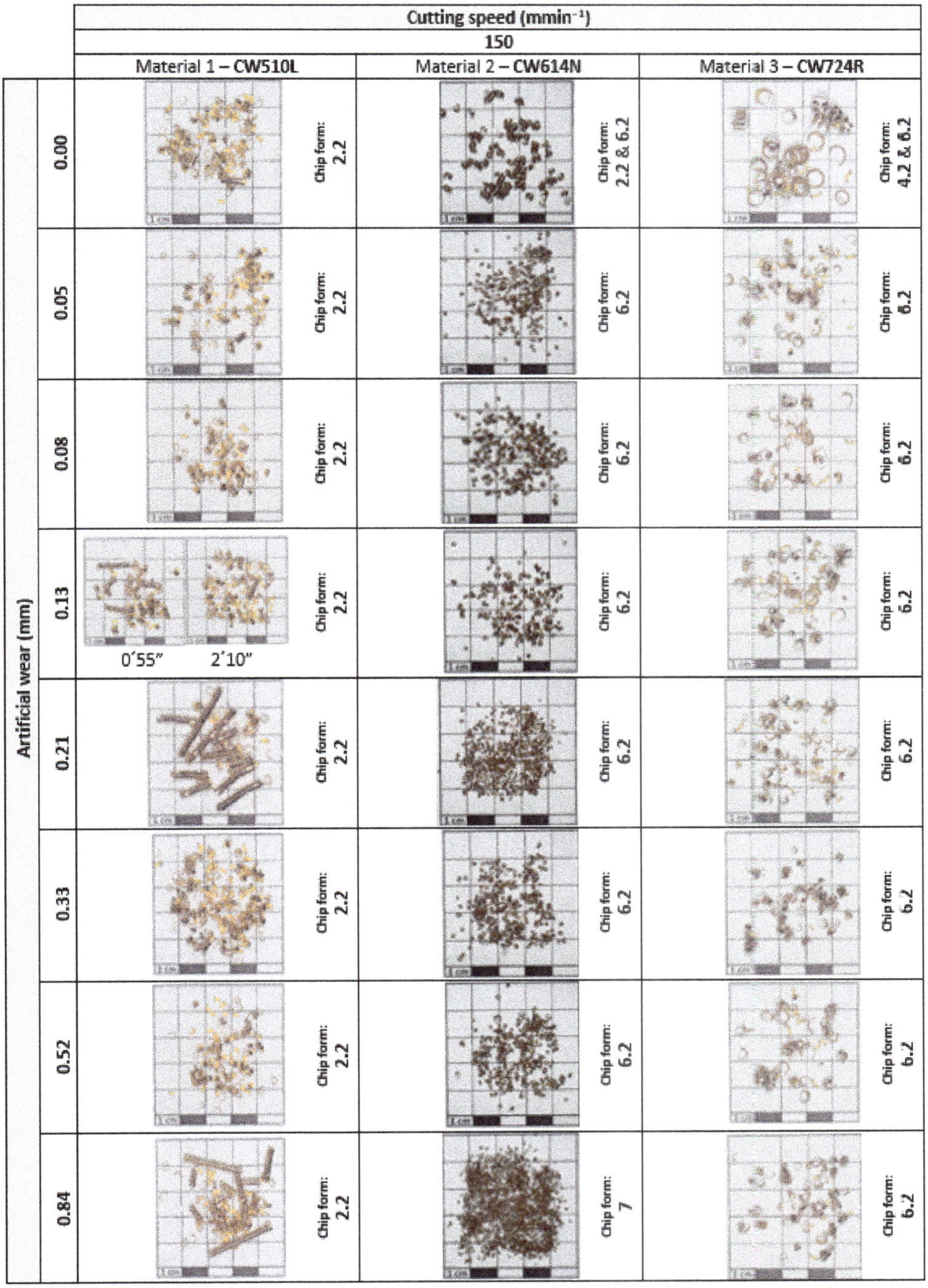

Figure 16. Comparison of chips originated at cutting speed 150 m/min.

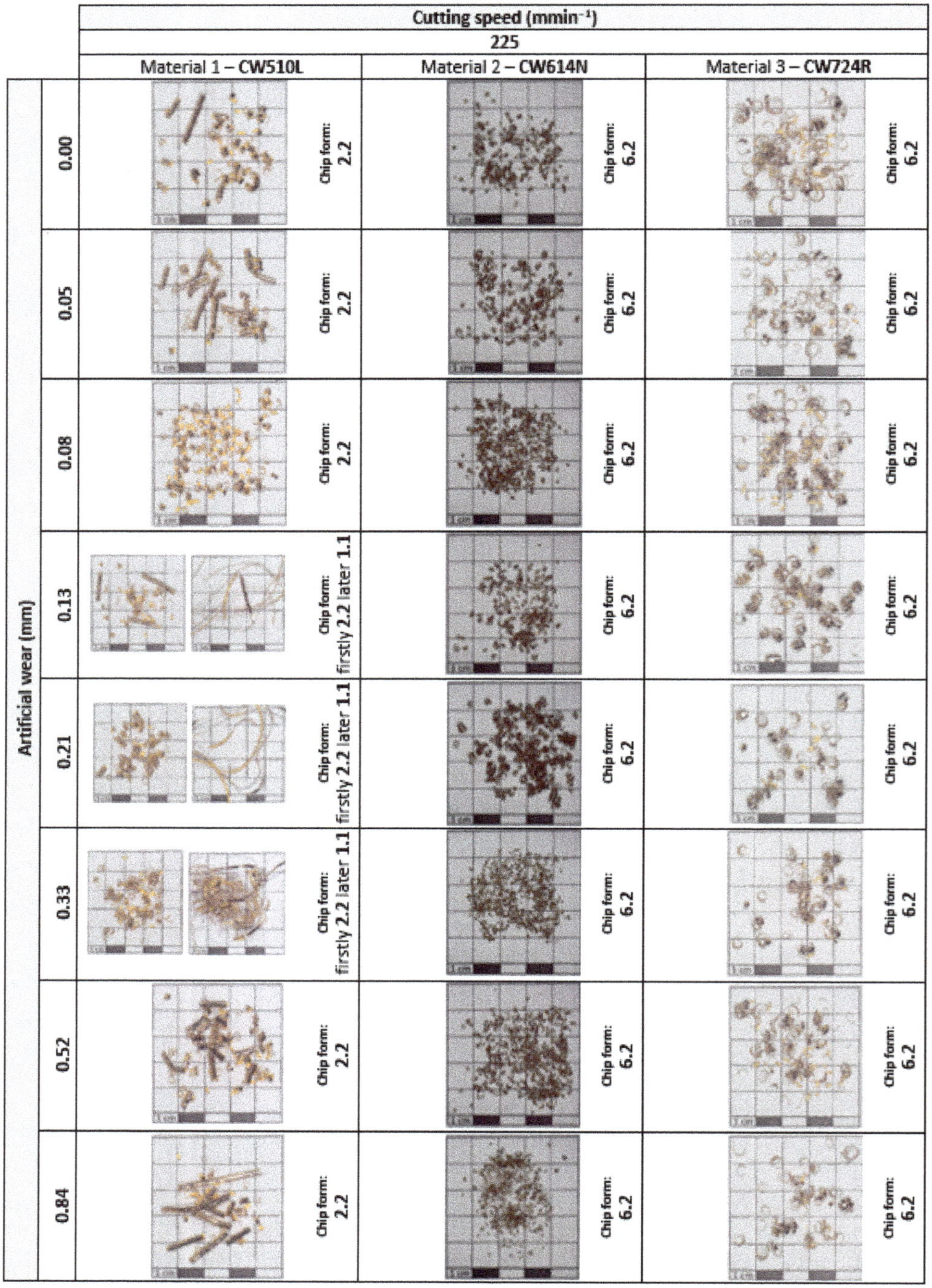

Figure 17. Comparison of chips originated at cutting speed 225 m/min.

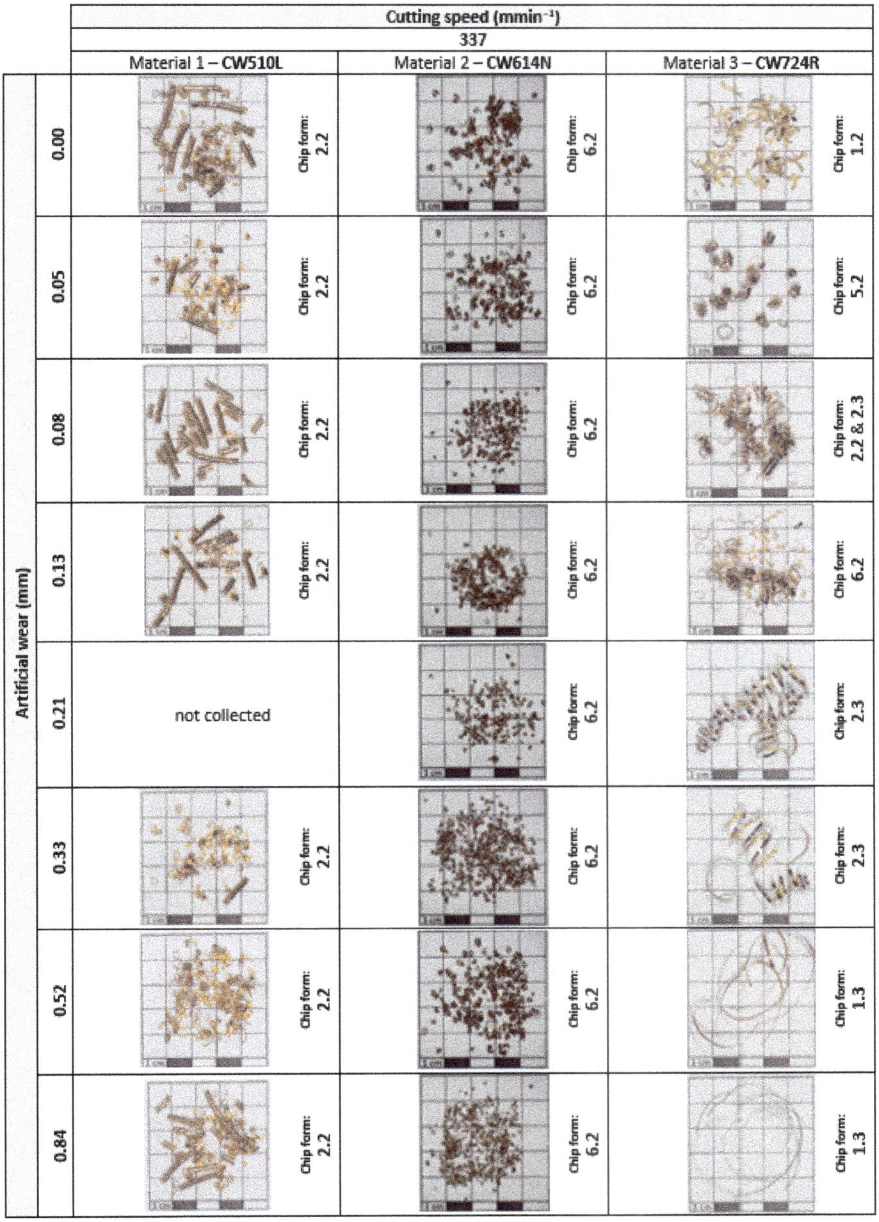

Figure 18. Comparison of chips originated at cutting speed 337.5 m/min.

4. Conclusions

Machining is a complex process involving many variables and parameters. In many cases, the theoretically derived dependencies of the basic factors of the cutting process optimization cannot be directly applied in specific production settings, because the boundary conditions (properties of the processed material, product characteristics, machine tool characteristics, process conditions, etc.) in real practice significantly change the expected results. It follows that for each individual system (machine tool, cutting tool, and product), lengthy experimental verification and evaluation would be necessary, which is not possible in practice due to the required high productivity and production efficiency.

A huge disadvantage of the classic approach when creating optimization models is that, despite very precise preparation and processing of experimental results, it is not possible to exclude the occurrence of random events that will significantly shorten the life of some cutting tools and thus cause complications in production. Therefore, in real production conditions, monitoring the response of the system to real stimuli appears to be the most suitable way of monitoring the actual state of the cutting tool, to which the presented research contributes to a large extent.

The primary objective of this research was to determine the level of vibration intensity of machining in response to a particular tool wear while turning brass alloys, and thus determine the level (borderline) at which the process is stable in the newly proposed methodology with an inverse approach that is based on artificially generated tool wear (prepared in advance in advance on the major flank). A supplementary part of the research was the study of chip formation when machining brass alloys in the given machining conditions, as this is directly connected to the wear of the tool. Specifically, the research results can be summarized as follows:

- The collection and processing of the experimentally obtained data were carried out using the standard steps of the multiple linear regression methodology to evaluate the dependence parameters (vibration characteristic and chip shape characteristic) on the experimental factors (cutting speed and artificial wear VB_{artif}).
- Based on the measured data, the statistical dependences of the acceleration amplitude (A) as a vibration characteristic on the experimental factors (v_c, VB_{artif}) were determined—all at the required level of statistical reliability.
- CW510L and CW614N brass alloys were found to exhibit an average of three times lower vibration damping compared to CW724R alloy.
- The CW724R brass alloy showed significantly steeper dependences of vibration generation in terms of changes in cutting speed and VB_{artif} wear compared to the other two materials.
- Differences in susceptibility to the generation of vibration manifestations are probably caused by the structure of the evaluated materials.
- The evaluated material production conditions significantly affect the vibro-diagnostic characteristics of the machining process, while for CW510L the slope of the dependence of the amplitude on the cutting speed is opposite to that for the other two materials (with increasing speed, the amplitude grew).

Regarding the evaluation of chip shape properties, it can be concluded:

- The experiments showed relatively good chip formation in the evaluated machining conditions even without the use of a chip breaker.
- The problematic chip shape occurred only in some cases with material 1 (CW510L) and material 3 (CW724R), which cannot be generally determined.
- From the point of view of suitable chip formation, the production conditions were very little affected by the type of alloy.
- Undesirable areas of inappropriate chip shape can be easily adjusted in practical applications by changing the feed, cutting speed, or—the best practical solution—by using a cutting tool with a chip stiffener optimized for the given process.

In the near future, the authors would like to continue the research with the measurement of cutting forces for the same or for a reduced experimental design.

Author Contributions: Conceptualization, P.P.M., K.M., G.A.P. and A.I.T.; methodology, P.P.M. and K.M.; software, P.P.M. and K.M.; validation, P.P.M., K.M. and G.A.P.; formal analysis, A.I.T.; investigation, P.P.M. and K.M.; resources, P.P.M., K.M. and A.I.T.; data curation, K.M., P.P.M. and A.I.T.; writing—original draft preparation, K.M.; writing—review and editing, P.P.M. and G.A.P.; visualization, K.M., P.P.M. and G.A.P.; supervision, P.P.M. and G.A.P.; project administration, K.M.

and G.A.P.; funding acquisition, K.M., P.P.M. and G.A.P. All authors have read and agreed to the published version of the manuscript.

Funding: This research was funded by the Ministry of Education, Science, Research and Sport of the Slovak Republic by grants APVV-19-0550, KEGA 005TUKE-4/2021, KEGA 032TUKE-4/2022, and SK-CN-21-0046 as well as by ELKEME S.A.

Data Availability Statement: Not applicable.

Acknowledgments: The support and encouragement of ELKEME S.A. management and ELVAL-HALCOR Brass Rods and Tubes Plant are greatly appreciated. The article was prepared thanks to the support of the Ministry of Education, Science, Research and Sport of the Slovak Republic through the grants APVV-19-0550, KEGA 005TUKE-4/2021, KEGA 032TUKE-4/2022 and SK-CN-21-0046.

Conflicts of Interest: The authors declare no conflict of interest.

References

1. Grzesik, W. *Advanced Machining Processes of Metallic Materials—Theory, Modelling and Applications*; Elsevier: Amsterdam, The Netherlands, 2017; ISBN 978-0-444-63711-6.
2. Trent, E.M.; Wright, P.K. *Metal Cutting: Theory and Application*; Butterwort Heinemann: Auckland, New Zealand, 2000; ISBN 0-7506-7069-X.
3. Jurko, J.; Panda, A.; Behun, M. Prediction of a new form of the cutting tool according to achieve the desired surface quality. *Appl. Mech. Mater.* **2013**, *268–270*, 473–476. [CrossRef]
4. Suh, C.S.; Liu, M.-K. *Control of Cutting Vibration and Machining Instability: A Time-Frequency Approach for Precision, Micro and Nano Machining*; John Wiley & Sons: Chichester, UK, 2013; ISBN 978-1-118-37182-4.
5. Kuyucak, S.; Sahoo, M. A review of the machinability of copper-base alloys. *Can. Metall. Q.* **1996**, *35*, 1–15. [CrossRef]
6. Jerzy, J.; Kuric, I.; Grozav, S.; Ceclan, V. Diagnostics of CNC machine tool with R-Test system. *Acad. J. Manuf. Eng.* **2014**, *12*, 56–60.
7. Pitel, J.; Matiskova, D.; Marasova, D. A new approach to evaluation of the material cutting using the artificial neural networks. *TEM J.* **2019**, *8*, 325–332.
8. Markopoulos, A.P.; Pressas, I.S.; Papantoniou, I.G.; Karkalos, N.E.; Davim, J.P. Machining and machining modeling of metal matrix composites—A review. In *Modern Manufacturing Engineering*; Springer: Cham, Switzerland, 2015; pp. 99–141.
9. Twardowski, P.; Legutko, S.; Krolczyk, G.M.; Hloch, S. Investigation of wear and tool life of coated carbide and cubic boron nitride cutting tools in high speed milling. *Adv. Mech. Eng.* **2015**, *7*, 1687814015590216. [CrossRef]
10. Petru, J.; Schiffner, J.; Zlamal, T.; Cep, R.; Kratochvil, J.; Stancekova, D. Wear progress of exchangeable cutting inserts during Ti (6) Al (4) V alloy machining. In Proceedings of the 24th International Conference on Metallurgy and Materials, Brno, Czech Republic, 3–5 June 2015; pp. 1147–1155.
11. Stavropoulos, P.; Papacharalampopoulos, A.; Vasiliadis, E.; Chryssolouris, G. Tool wear predictability estimation in milling based on multi-sensorial data. *Int. J. Adv. Manuf. Technol.* **2016**, *82*, 509–521. [CrossRef]
12. Zetek, M.; Zetkova, I. Increasing of the cutting tool efficiency from tool steel by using fluidization method. *Procedia Eng.* **2015**, *100*, 912–917. [CrossRef]
13. Bakša, T.; Kroupa, T.; Hanzl, P.; Zetek, M. Durability of cutting tools during machining of very hard and solid materials. *Procedia Eng.* **2015**, *100*, 1414–1423. [CrossRef]
14. Dimla, D.E. The Correlation of Vibration Signal Features to Cutting Tool Wear in a Metal Turning Operation. *Int. J. Adv. Manuf. Technol.* **2002**, *19*, 705–713. [CrossRef]
15. Xu, C.; Jianming, D.; Yuzhen, C.; Huaiyuan, L.; Zhicheng, S.; Jing, X. The relationships between cutting parameters, tool wear, cutting force and vibration. *Adv. Mech. Eng.* **2018**, *10*, 1687814017750434. [CrossRef]
16. Bouhalais, M.L.; Nouioua, M. The analysis of tool vibration signals by spectral kurtosis and ICEEMDAN modes energy for insert wear monitoring in turning operation. *Int. J. Adv. Manuf. Technol.* **2021**, *115*, 2989–3001. [CrossRef]
17. Sridhar, A.V.; Prasad, B.S.; Mouli, K.V. Evaluation of tool performance and wear through vibration signature analysis in drilling of IS3048 steel. *J. Eng. Appl. Sci.* **2021**, *68*, 27. [CrossRef]
18. Babu, M.S.; Rao, T.B. Real-time cutting tool condition assessment and stochastic tool life predictive models for tool reliability estimation by in-process cutting tool vibration monitoring. *Int. J. Interact. Des. Manuf.* **2022**, 1–17. [CrossRef]
19. Antić, A.; Šimunović, G.; Šarić, T.; Milošević, M.; Ficko, M. A model of tool wear monitoring system for turning. *Tech. Gaz.* **2013**, *20*, 247–254.
20. Childs, T.; Maekawa, K.; Obikawa, T. *Metal Machining: Theory and Applications*; Butterworth-Heinemann: London, UK, 2000; ISBN 0-340-69159-X.
21. Pastucha, P.; Majstorovic, V.; Kučera, M.; Beno, P.; Krile, S. Study of Cutting Tool Durability at a Short-Term Discontinuous Turning Test. In Proceedings of the International Conference on Manufacturing Engineering and Materials (ICMEM 2018), Novy Smokovec, Slovakia, 18–22 June 2018; pp. 493–501. [CrossRef]
22. Toulfatzis, A.I.; Pantazopoulos, G.A.; David, C.N.; Sagris, D.S.; Paipetis, A.S. Machinability of eco-friendly lead-free brass alloys: Cutting-force and surface-roughness optimization. *Metals* **2018**, *8*, 250. [CrossRef]

23. Pantazopoulos, G. Leaded brass rods C38500 for automatic machining operations. *J. Mater. Eng. Perform.* **2002**, *11*, 402–407. [CrossRef]
24. Gane, N. The effect of lead on the friction and machining of brass. *Philos. Mag.* **1981**, *43*, 545–566. [CrossRef]
25. Nobel, C.; Klocke, F.; Lung, D.; Wolf, S. Machinability Enhancement of Lead-free Brass Alloys. *Procedia CIRP* **2014**, *14*, 95–100. [CrossRef]
26. Stavroulakis, P.; Toulfatzis, A.I.; Pantazopoulos, G.A.; Paipetis, A.S. Machinable Leaded and Eco-Friendly Brass Alloys for High Performance Manufacturing Processes: A Critical Review. *Metals* **2022**, *12*, 246. [CrossRef]
27. Toulfatzis, A.; Pantazopoulos, G.; Paipetis, A. Fracture behavior and characterization of lead-free brass alloys for machining applications. *J. Mater. Eng. Perform.* **2014**, *23*, 3193–3206. [CrossRef]
28. Johansson, J.; Alm, P.; M'Saoubi, R.; Malmberg, P.; Ståhl, J.-E.; Bushlya, V. On the Function of Lead (Pb) in Machining Brass Alloys. *Res. Sq.* **2022**, *120*, 7263–7275. [CrossRef]
29. Company Sarbak Materials Data Sheets. Available online: www.sarbak.com.tr (accessed on 19 June 2022).
30. *EN 1982;* Copper and Copper Alloys—Ingots and Castings. CEN: Brussels, Belgium, 2017.
31. *EN 12164;* Copper and Copper Alloys—Rod for Free Machining Purposes. CEN: Brussels, Belgium, 2016.
32. Pantazopoulos, G.; Vazdirvanidis, A. Characterization of the microstructural aspects of machinable? *Phase brass. Microsc. Anal.* **2008**, *22*, 13–16.
33. Toulfatzis, A.I.; Pantazopoulos, G.A.; Paipetis, A.S. Fracture mechanics properties and failure mechanisms of environmental-friendly brass alloys under impact, cyclic and monotonic loading conditions. *Eng. Fail. Anal.* **2018**, *90*, 497–517. [CrossRef]
34. Garcia, P.; Rivera, S.; Palacios, M.; Belzunce, J. Comparative study of the parameters influencing the machinability of leaded brasses. *Eng. Fail. Anal.* **2010**, *17*, 771–776. [CrossRef]
35. Filippov, A.V.; Filippova, E.O. Determination of cutting forces in oblique cutting. *Appl. Mech. Mater.* **2015**, *756*, 659–664. [CrossRef]
36. Altintas, Y. Manufacturing automation: Metal cutting mechanics, machine tool vibrations, and CNC design. *Appl. Mech. Rev.* **2012**, *54*, B84.
37. Vazdirvanidis, A.; Rikos, A.; Toulfatzis, A.I.; Pantazopoulos, G.A. Electron Backscatter Diffraction (EBSD) Analysis of Machinable Lead-Free Brass Alloys: Connecting Texture with Fracture. *Metals* **2022**, *12*, 569. [CrossRef]
38. Hagarová, M.; Peterka, P.; Mantič, M.; Vojtko, M.; Baranová, G.; Matvija, M. Failure analysis of leaded brass bolt. *Eng. Fail. Anal.* **2023**, *143*, 106899. [CrossRef]
39. Toulfatzis, A.I.; Besseris, G.J.; Pantazopoulos, G.A.; Stergiou, C. Characterization and comparative machinability investigation of extruded and drawn copper alloys using non-parametric multi-response optimization and orthogonal arrays. *Int. J. Adv. Manuf. Technol.* **2011**, *57*, 811–826. [CrossRef]
40. Yaqoob, K.; Hashmi, F.; Hassan Tanvir, W. Failure Analysis of Cartridge Brass Shell. *Eng. Fail. Anal.* **2022**, *138*, 106325. [CrossRef]
41. Baron, P.; Dobransky, J.; Pollak, M.; Kocisko, M.; Cmorej, T. The parameter correlation of acoustic emission and high-frequency vibrations in the assessment process of the operating state of the technical system. *Acta Mech. Autom.* **2016**, *10*, 112–116. [CrossRef]
42. Filippov, A.V.; Nikonov, A.Y.; Rubtsov, V.E.; Dmitriev, A.I.; Tarasov, S.Y. Vibration and acoustic emission monitoring the stability of peakless tool turning: Experiment and modeling. *J. Mater. Process. Technol.* **2017**, *246*, 224–234. [CrossRef]
43. Chen, J.; Feng, J.; Wang, F.; Peng, Q.; Lan, G.; Zhao, L.; Wu, L. Cracking Analysis of a Brass Clamp Mounted on the Main Transformer in the Power Grid System. *Energies* **2023**, *16*, 3460. [CrossRef]
44. Monka, P.P.; Monkova, K.; Pantazopoulos, G.; Toulfatzis, A.I. Effect of Wear on the Vibrating Behaviour of the Tool at Turning CW724R Alloy. In Proceedings of the 13th International Conference on Mechanical and Aerospace Engineering (ICMAE), Bratislava, Slovakia, 20–22 July 2022; pp. 51–55. [CrossRef]
45. Panda, A.; Dobránsky, J.; Jančík, M.; Pandova, I.; Kačalová, M. Advantages and effectiveness of the powder metallurgy in manufacturing technologies. *Metalurgija* **2018**, *57*, 353–356.
46. Obeng, D.P.; Morrell, S.; Napier-Munn, T.J. Application of central composite rotatable design to modeling the effect of some operating variables on the performance of the three-product cyclone. *Int. J. Miner. Process.* **2005**, *76*, 181–192. [CrossRef]
47. Malotova, S.; Cep, R.; Cepova, L.; Petru, J.; Stancekova, D.; Kyncl, L.; Hatala, M. Roughness Evaluation of the Machined Surface at Interrupted Cutting Process. *Manuf. Technol.* **2016**, *16*, 168–173. [CrossRef]
48. Kouadri, S.; Necib, K.; Atlati, S.; Haddag, B.; Nouari, M. Quantification of the chip segmentation in metal machining: Application to machining the aeronautical aluminum alloy AA2024-T351 with cemented carbide tools WC-Co. *Int. J. Mach. Tools Manuf.* **2013**, *64*, 102–113. [CrossRef]
49. *ISO 3685: 1993(E);* International Standard, Tool Testing with Single Point Turning Tools. ISO: Geneva, Switzerland, 1993.
50. Sahraoui, Z.; Mehdi, K.; Jaber, M.B. Analytical and experimental stability analysis of AU4G1 thin-walled tubular workpieces in turning process. *Proc. Inst. Mech. Eng. Part B J. Eng. Manuf.* **2020**, *234*, 1007–1018. [CrossRef]
51. Glaa, N.; Mehdi, K.; Moussaoui, K.; Zitoune, R. Numerical and experimental study of the drilling of multi-stacks made of titanium alloy Ti-6Al-4V: Interface and burr behaviour. *Int. J. Adv. Manuf. Technol.* **2020**, *107*, 1153–1162. [CrossRef]

metals

MDPI

Article

Inverse Thermal Analysis as a Tool for Optimizing Concentrated Solar Energy Elaboration of Wear Resistant Surface Layers

Anna D. Zervaki [1], Samuel G. Lambrakos [2], Athanasios G. Mourlas [3], Ioannis G. Papantoniou [4,5], José Rodríguez [6] and Pandora P. Psyllaki [3,*]

1 Shipbuilding Technology Laboratory, School of Naval Architecture and Marine Engineering, National Technical University of Athens, 9 Heroon Polytechniou Ave., 157 73 Athens, Greece
2 U.S. Naval Research Laboratory, Washington, DC 20375, USA
3 Surface Engineering Laboratory, Department of Mechanical Engineering, University of West Attica, 250 Thivon & P. Ralli Str., Ancient Olive Grove Campus, 122 41 Egaleo, Greece
4 Laboratory of Manufacturing Technology, School of Mechanical Engineering, National Technical University of Athens, Heroon Polytechniou 9, 157 80 Athens, Greece
5 Department of Naval Architecture, School of Engineering, University of West Attica, Ag. Spyridonos Str., Egaleo Park Campus, 122 43 Egaleo, Greece
6 CIEMAT-Plataforma Solar de Almería, Aptdo, 22, 04200 Tabernas, Spain
* Correspondence: psyllaki@uniwa.gr; Tel.: +30-210-5381292

Abstract: Concentrated Solar Energy (CSE) processing is considered a promising renewable energy source technique for elaborating thick, wear-resistant claddings onto metallic surfaces of large dimensions that are expected to operate in heavy duty applications, such as excavator shovels, mineral crushers, etc. However, the prediction of surface processing effects on the microstructure and the properties of the main construction base metal are of crucial importance, as they are commonly required in all surface modification techniques. Thus, the present study is focused on the inverse thermal analysis and parametric modeling of heat deposition associated with CSE surface processing of metals. In this preliminary attempt, experimental findings that concern the elaboration of TiC- and chromium carbide-reinforced clads onto common steel base metals were used to quantify the evaluation of the temperature histories within the volume of workpieces undergoing solar heating, where direct temperature measurements contain uncertainties and/or are not even possible. Results of prototype inverse thermal analyses of heat transfer in processed layer-substrate systems are presented, demonstrating the general aspects of a parametric model for thermal analysis and simulation.

Keywords: inverse thermal analysis; parametric modeling; concentrated solar energy processing; cladding; wear-resistant surface layers

Citation: Zervaki, A.D.; Lambrakos, S.G.; Mourlas, A.G.; Papantoniou, I.G.; Rodríguez, J.; Psyllaki, P.P. Inverse Thermal Analysis as a Tool for Optimizing Concentrated Solar Energy Elaboration of Wear Resistant Surface Layers. *Metals* **2023**, *13*, 942. https://doi.org/10.3390/met13050942

Academic Editor: Robert B. Heimann

Received: 30 March 2023
Revised: 9 May 2023
Accepted: 10 May 2023
Published: 12 May 2023

1. Introduction

Concentrated Solar Energy (CSE), that is, the production of medium- to high-temperature heat by means of movable mirrors that track the sun and concentrate its radiation on a focal point/area has been proposed, among others, as a "green" technique for implementing surface modifications of metallic substrates [1], leading to surface layers of superior performance during operation via surface hardening [2,3], gas nitrating [4], cladding [5–8] and surface alloying [7,9]. Almost twenty-five years ago, Flamant et al. [3] introduced solar processing to perform such high temperature material treatments, documenting on technical and economic data, when comparing concentrated solar beams to other nonconventional ones, namely laser and plasma. Although the capital cost of such solar systems is higher than those of laser and plasma sources, the higher overall energy efficiency and the feasibility of treating surfaces of much larger dimensions are attractive, important features for considering concentrated solar energy for applications in the specific field of surface modification of solids. Despite its main disadvantage related to the sun availability [10], CSE remains a promising alternative to conventional heat sources [11] for specific applications, such as cladding of large dimension metallic surfaces.

The feasibility of CSE treatment in the field of materials science and metallurgy has been proven for various processes and several materials. However, the heat effects during the whole processing history of the treated material are still a matter of concern for the relevant scientific community, since the metallurgical transformations take place not only during solar exposure, but they last during cooling from above or near melting to ambient temperature, controlling in this way the microstructure and the properties of both the directly solar-treated layer, as well as those of the underlying material.

In the case of solar cladding process of AISI 316 stainless steel over AISI 1026 common steel [12], the satisfactory agreement between experimental observations and numerical simulation results carved out a solid base for further optimization of the solar cladding parameters. More precisely, stainless steel powder was pre-deposited on the base metal surface and the system was solar irradiated, whilst the temperature evolution of the base metal was recorded in real-time by the aid of K-type thermocouples welded at three different locations on the opposite side of the irradiated surface. These results, together with the thermophysical properties of the material system, were used to develop and validate a finite element model (FEM) that could reliably predict the level and extent of the thermal fields that occurred during solar processing. Microscopic observations revealed, mainly, the dendritic structure of the obtained clads and verified further the good agreement to the numerical thermal model's predictions.

In order to improve the homogeneity inside the test chamber for materials processing [13], a numerical thermal model was developed, taking as main assumptions that heat transfer occurs via: (i) conduction through the receiver plate; (ii) radiation between the up-plate and the down-plate; (iii) natural convection between the plates and the surrounding medium. For the validation of the numerical model, two materials of vastly different physical properties -thermal conductivity, specific heat capacity and emissivity- were used as receivers (up-plates) under three set-up configurations and the temperature evolution was recorded in real-time by the aid of a set of five K-type thermocouples placed at specific locations of the tested materials system. The combined evaluation of the numerical simulation and experimental results revealed that the thermal fields developed during solar irradiation are controlled by the thermal properties of both the down–plate and the surrounding medium; thus, the optimization of a required solar processing is strongly dependent on the proper selection of the set-up configuration and not exclusively on the solar irradiation parameters.

In the case where solar processing incorporates melting of the irradiated materials, as in welding [14], the thermal fields developed under the constraints imposed by the particular experimental configuration and solar exposure parameters determine also the microstructure obtained during materials re-solidification, as well as that of the heat affected zone (H.A.Z.). In the case of solar sintering of Al_2O_3-based ceramics [15], high levels of densification of the green bodies were achieved, which cannot be obtained by conventional sintering in electric furnaces without applying pressure. The relevant dense microstructure obtained was attributed to the high heating rates and short sintering duration during solar treatment that restrained grain growth, whilst the use of a $95N_2{:}5H_2$ mixture as surrounding medium was found to result in pores elimination. Similar observations were reported in the case of solar sintering of Ni-Zn ferrites [16]; compared to sintering in conventional electric furnace, the solar-based technique led to ferrites of higher densification that influenced both their mechanical and magnetic properties.

Solar sintering of bulk carbide-reinforced metallic-matrix composites [17] was achieved by pre-mixing vanadium carbide and AISI M2 high speed steel powders, and successive solar exposure using (a) Fresnel lens and (b) solar furnace, for various reinforcing particles percentages and treatment parameters. The experimental results revealed the high microstructural quality of the treated material and are significantly encouraging for performing further such metallurgical processes via concentrated solar energy.

Finally, the particular thermal cycles during solar processing, that is, high heating rate and controlled dwell time at a temperature level, as well as the thermal field implied in the receiver's volume allowed foaming of Al-Si precursors and led to stable porous metallic structures, overcoming the common drawback of pore collapse when thermal treatment takes place in electric furnaces [18].

Inverse analysis of volumetric heat-deposition processes [19–24], for example, welding [25,26], poses a specific problem, which follows from the unique characteristics and relative complexity of such processes. These processes are typically characterized by regions of volumetric heat deposition having complex shapes and heat-source-workpiece interactions. Modeling and inverse analysis of thermal processes within these regions of volumetric deposition, for example, inverse thermal analysis, remains an open problem, distinct from other problems of inverse analysis. Heat-deposition fields, for example, temperature fields, associated with these processes are a function of the nature of the energy source interaction (or coupling) with the ambient medium. The general characteristics of interactions between sources and ambient media, especially for the case of energy-deposition processes such as welding, motivates a relatively well-defined posing of the problem for inverse analysis of deposition processes. This posing establishes the need for parametric models of deposition processes whose formulations are optimal for inverse analysis of complex source-ambient-medium interactions. For example, inverse thermal analysis of welding processes still remains an open problem, because of the complexity of energy-source-workpiece coupling related to the wide range of different welding processes [27].

The present study continues upon experimental research previously published by members of the authors' team [7,8], where the feasibility of using CSE technology for the elaboration of wear-resistant surface layers has been proven. Nevertheless, the evaluation of the temperature field within the workpiece is the crucial factor that controls the microstructure obtained in both the solar-treated surface layers and the underlying metal and, thus, their mechanical properties that finally affect the machinability of the so-treated workpiece using conventional tools, as well as the in-service performance of the surface-treated mechanical components. For this purpose, the particular work—part of which has been presented at the Advances in Welding & Additive Manufacturing Research Conference [28]—is targeted to the inverse thermal analysis and parametric modeling of heat deposition associated with the specific heat-deposition taking place during CSE surface processing, in order to describe temperature histories within the volume of the solar-treated workpieces, where direct temperature measurements contain uncertainties and/or are not even possible.

2. Materials and Methods

The inverse thermal analysis developed is based on microscopic observations of two CSE-treated materials systems, namely TiC and Cr_3C_2 powders pre-deposited onto similar carbon steel substrates, at the installations of Plataforma Solar de Almería, PSA (Spain), in particular at the horizontal solar furnace SF40 [29]. As mentioned above [7,8], the general aspects of the surface solar processing have been described in detail in previous works. Carbon steel (DIN 17100 St 37-2) specimens with dimensions 50 × 70 × 15 mm were used as base metals on the polished surface on which the carbide powders were deposited, under preparation conditions such that no interactions between the powder and the substrate material could take place. More precisely, 3.37 g of TiC powder with a particle size distribution −149 + 74 µm and 5.71 g Cr_3C_2 powder with a particle size distribution of −106 + 45 µm had been manually deposited in the form of disks (∅35, thickness: 1.5 mm), prior to solar processing (Figure 1a). The so-prepared specimens were placed at the center of the processing chamber (Figure 1b), onto a thermal insulating Al_2O_3 support and exposed to concentrated solar energy (Figure 1c) under continuous flow of Ar inert gas.

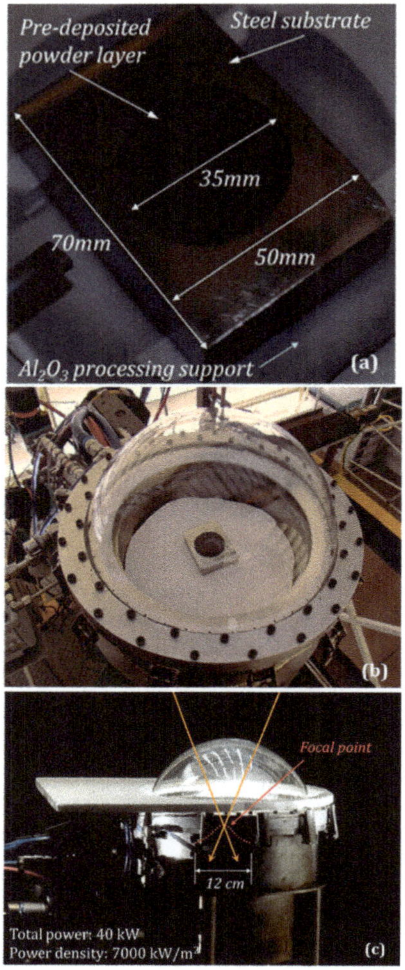

Figure 1. Experimental details for the CSE processing aiming to surface cladding: (**a**) material system to be solar-treated; (**b**) positioning of the prepared specimen in the processing chamber; (**c**) experimental arrangement for solar processing.

For both material systems, the microscopic observations provide quantitative results on both the extent and geometry of the surface layer that has been led to melting, and the underlying Heat Affected Zone (H.A.Z.), as well as on the metallurgical reactions taking place in the melt and its microstructure obtained during solidification.

3. Experimental Findings and Physical Model

The two material systems considered for prototyping solar-heating analysis in this study exhibited extremely different behavior [7]. In the case of TiC powder (Figure 2), the ceramic particles were dispersed in the liquid metallic matrix, without any metallurgical reaction with it, leading to a "clear" surface composite layer after cooling to ambient temperature (Figure 3). On the contrary, after sufficient solar exposure (Figure 4) the Cr_3C_2 particles were totally incorporated in the molten pool and diluted in it, leading to a new surface alloy rich in Cr rather than to a surface composite layer (Figure 5).

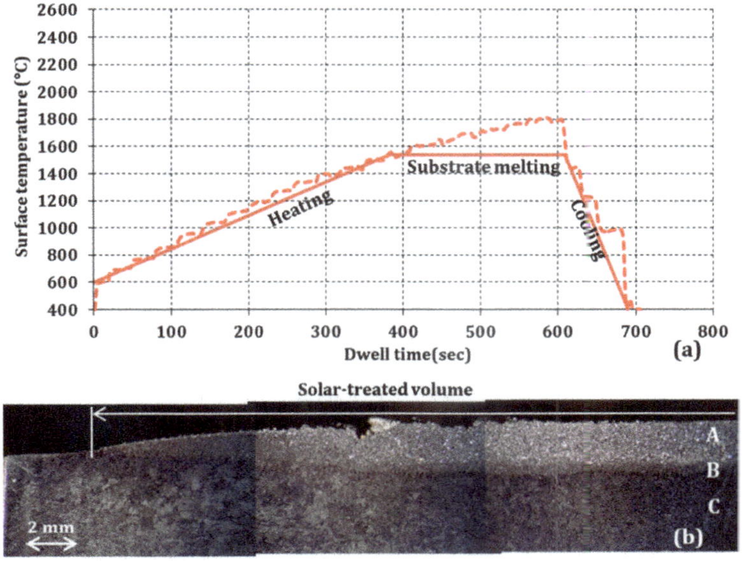

Figure 2. Solar processing of TiC/carbon steel system: (**a**) successive stages during solar processing (dashed line: experimental measurements, reproduced from [7], license number: 5545980711788); (**b**) half area of the solar-treated specimen obtained (stereoscope cross-section).

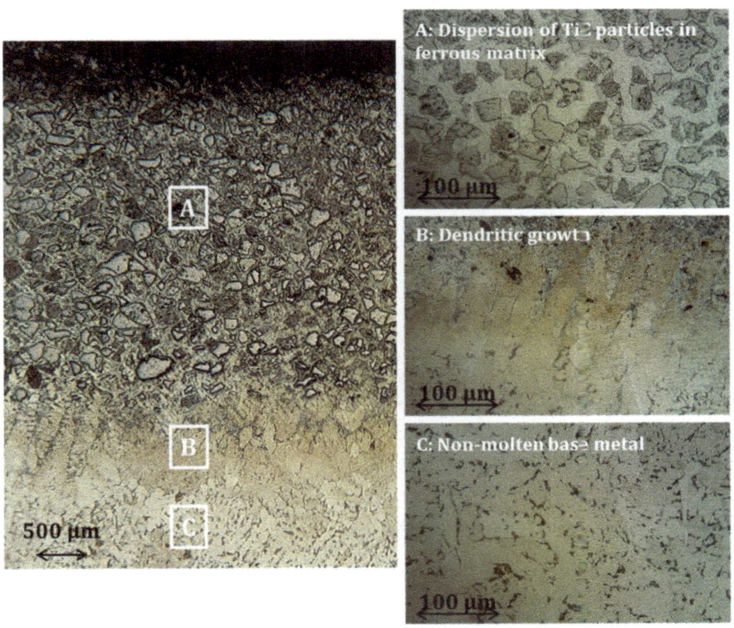

Figure 3. Optical micrograph of the solar-treated TiC/carbon steel system cross-section, where areas of different microstructure can be clearly identified (middle of the CSE-affected area). Areas marked as A, B and C on the left image, are presented in higher magnification at the right column.

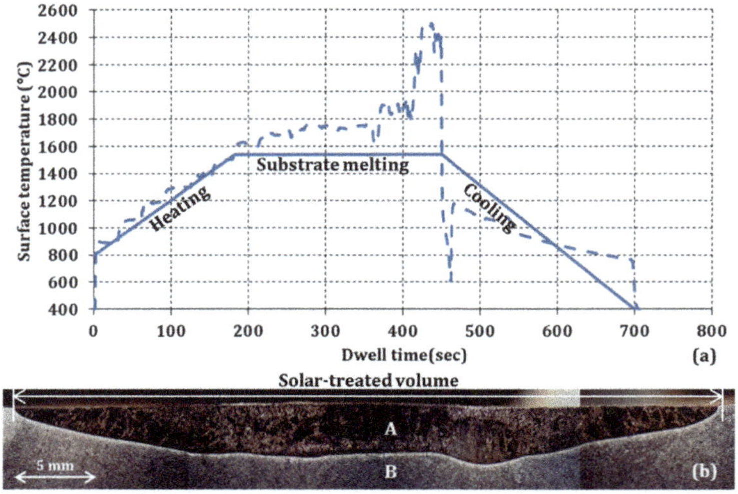

Figure 4. Solar processing of Cr_3C_2/carbon steel system: (**a**) successive stages during solar processing (dashed line: experimental measurements, reproduced from [7], license number: 5545980711788); (**b**) overall cross section of the solar-treated specimen obtained (stereoscope cross-section).

Figure 5. Optical micrograph of the solar-treated Cr_3C_2/carbon steel system cross-section, where areas of different microstructure can be identified (middle of the CSE-affected area).

During solar treatment, the surface temperatures of irradiated specimens were monitored in real-time via an optical pyrometer. It should be noted that with respect to inverse thermal analysis, it is assumed that the measured temperatures are estimates, owing to uncertainties due to liquid-surface movement during process-evolution and ambient vapor in the sealed testing bell-jar chamber.

Based on the above experimental findings, a generic physical model describing phenomena that take place during successive stages can be proposed (Figure 6):

- During heating at surface temperature below the melting temperature of the substrate (T_{max}: 1540 °C), the concentrated solar energy is absorbed only by the area covered by the black-colored ceramic particles, while it is reflected by the noncovered polished surface of the base metal (Figure 6a).

- During heating at surface temperature above the melting temperature of the substrate, the solar energy absorbed results in progressive melting of the underlying metal and immersion of the pre-deposited powder in the melting pool (Figure 6b).
- Lasting the solar exposure of the workpiece, the depth of the melting pool increases and the circulation currents in the ferrous melt lead to the homogeneous distribution of the carbide particles in the liquid metal (Figure 6c). Depending on the nature of the incorporated carbides, they either remain non-attacked (TiC case), or are diluted in the molten steel (Cr_3C_2 case).
- During cooling to ambient temperature, the solidification of the melts follows the thermodynamic path imposed by their chemical composition, whilst solid state metallurgical transformations complete the final microstructure of the obtained solar-treated surface, as well as that of the heat-affected area, H.A.Z., (Figure 6d). In both material systems, the areas of dendritic growth (Figures 3 and 5)—at the bottom of the re-solidified pool, in contact to the nonmolten base metal—are indicative of the solidification starting points and the heat dissipation directions.

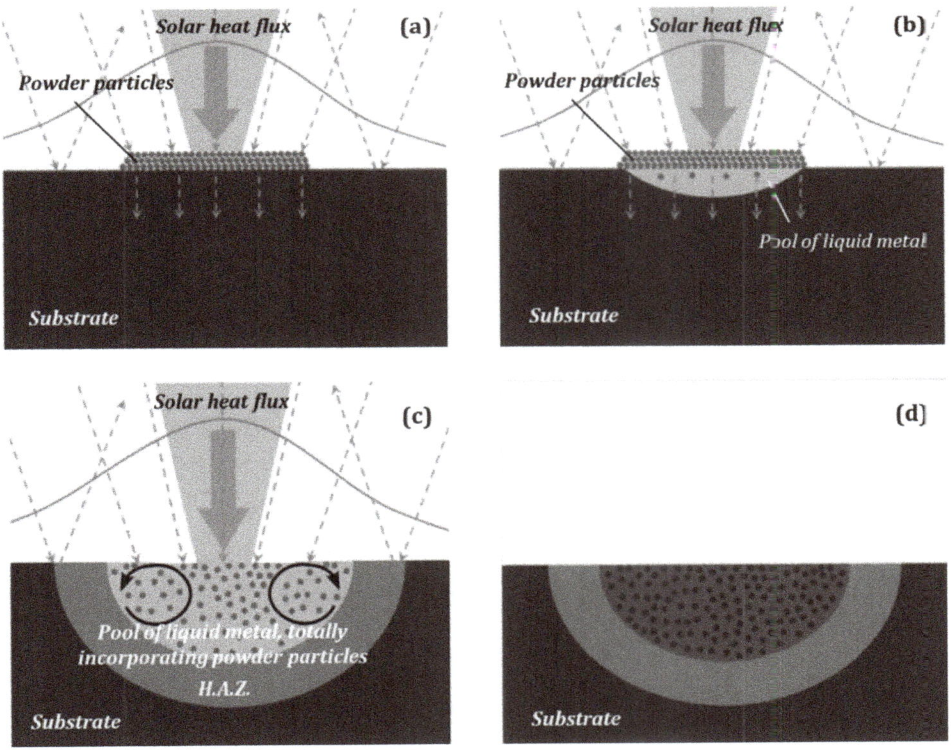

Figure 6. Simplified schematic representation of the successive stages that can be distinguished during surface solar processing: (**a**) absorption of concentrated solar energy by the area covered by the black-colored powder particles; (**b**) heat transfer to the metallic substrate and surface liquidation of the volume wherein temperature exceeded its characteristic melting value; (**c**) immersion and dispersion of the pre-deposited particles in the molten metallic pool; (**d**) re-solidified surface volume, consisting of dispersion (case of TiC powder) or total dilution (case of Cr_3C_2 powder) of the ceramic particles within the substrate metallic matrix and cooled to ambient temperature.

This physical model (Figure 6) is applied to describe thermal phenomena taking place during the particular solar processing. As depicted, the heat transfer associated with this process is time-dependent and rather complex, as it is governed by the:

- nature of coupling between the surface-heating source and the workpiece that consists of a powder layer onto a metallic substrate. This coupling is not purely diffusive, but in principle, is a combination of Beer—Lambert-type volumetric energy deposition and advection-diffusion of heat;
- thermal contact properties at the pre-deposited powder layer-metallic substrate interface,
- complexity of the metallurgical transformations taking place between the powder layer and substrate materials;
- changes in volume and shape of the molten region that are produced by the dispersion of the powder particles in the solar-affected region of the substrate materials.

It should be noted that the schematic description presented in Figure 6 does not include the process phases of dilatation of the metallic substrate during liquidation of the metal and shrinkage of the composite volume, which is the sum of layer and substrate material volumes during re-solidification and cooling to ambient temperature.

4. Inverse Thermal Analysis

The posing of the associated inverse thermal analysis problem can be derived from the above experimental observations, the surface temperature measurements and the generic physical model described. Given the known qualitative characteristics of the solar-heating process, thermal properties of the constitute materials, shapes and geometries of the powder-layer-substrate system at onset of heating and the re-solidified region after termination of process, and estimates of surface temperature as a function of time, the first step is to construct a parametric model structured for input of this information for prediction of temperature histories within the volume of processed workpieces. Then, the inverse problem is solved accordingly, constrained by the resulting molten volume of mixed material at the surface and estimated surface-temperature history, for the purpose of controlling treatment depth as a function of irradiation time. In principle, treatment depth would be defined by metallurgical transformations taking place within the volume of mixed material during processing. Temperature histories calculated by inverse thermal analysis would be used as input to transformation models and associated software for prediction of metallurgical transformations [26]. The inverse problem, as defined by the solar-heating process, consists of two separate parts, namely estimation of time-dependent temperature fields during heating and cooling, each one associated with the respective stage of the process. As revealed in what follows, more detailed information tends to be available for inverse thermal analysis during the cooling phase, owing to its accessibility by convenient measurement and post-process metallographic analysis.

Presented in this section is a parametric model for inverse thermal analysis of heat transfer through a thin layer of surface material, and a volume of an underlying substrate material, which is characteristic of the process considered here. This model, which uses the heat-kernel solution of the heat-conduction equation as ansatz [27], is given by:

$$T(x,t) = C(t)G(x,x_k,t) \tag{1}$$

and

$$T(x = x_k,t) = T_{Surface}(t) \tag{2}$$

for $t < t_h$, the heating period, and for $t > t_h$, the cooling period,

$$T(x,t) = C(t_h)G(x,x_k,t) \tag{3}$$

where:

$$G(x,x_k,t) = \frac{1}{\sqrt{4\pi\kappa_u t}}\exp\left[-\frac{(x-x_k)^2}{4\kappa_u t}\right] + \frac{W_0}{\sqrt{4\pi\kappa_u t}}\exp\left[-\frac{(x_{imag})^2}{4\kappa_u t}\right] \tag{4}$$

for $x \geq x_l$:

$$G(x,x_k,t) = \frac{W_1}{\sqrt{\pi\kappa_d t}}\exp\left[-\frac{1}{4t}\left(\frac{x-x_l}{\sqrt{\kappa_d}} + \frac{x_l - x_k}{\sqrt{\kappa_u}}\right)^2\right] \tag{5}$$

and for $x \geq x_s$:

$$G(x,x_k,t) = \frac{W_1}{\sqrt{\pi\kappa_d t}}\exp\left[-\frac{1}{4t}\left(\frac{x-x_s}{\sqrt{\kappa_s}} + \frac{x_s - x_l}{\sqrt{\kappa_d}} + \frac{x_l - x_k}{\sqrt{\kappa_u}}\right)^2\right] \tag{6}$$

where the weight coefficients W_0 and W_1 are given by:

$$W_0 = \frac{k_u\sqrt{\kappa_d} - k_d\sqrt{\kappa_u}}{k_u\sqrt{\kappa_d} + k_d\sqrt{\kappa_u}} \tag{7}$$

and

$$W_1 = \frac{2k_u\kappa_d}{\sqrt{\kappa_u}(k_u\sqrt{\kappa_d} + k_d\sqrt{\kappa_u})} \tag{8}$$

The distance within the layered material from the surface (at x_k) is $(x - x_k)$, and the location of the interface between the thin-layer and the substrate is $(x_s - x_k)$. The quantities k_u, k_u are the estimated thermal conductivity and diffusivity of the thin-layer, and *ks* is the thermal diffusivity of substrate materials, respectively. The quantities x_l, k_d and k_d are adjustable phenomenological parameters for modeling thermal coupling of layer to substrate at their interface. Accordingly, the distance parameter x_l is not associated with an actual physical location within the layer system. For $x < x_1$, $x_{imag} = x + x_k - 2x_l$, which is also a phenomenological parameter, is not associated with the image-distance of a physical location. Similarly, although k_d and k_d are formally conductivity and diffusivity, respectively, they do not represent bulk material properties, relevant schema is presented in Figure 7. The function *C(t)* is specified by the boundary and initial conditions defined by Equations (2) and (3), respectively. Specifically, the model assigns two conditions during heating, namely the measured surface temperature and the time period of heating, and one after termination of heating, the time period of cooling. Equations (4)–(8) are the heat-kernel solution to the heat-conduction equation for a one-dimensional system consisting of a layer of a material of finite thickness, on a substrate of essentially infinite extent, where the layer and substrate materials have different thermal properties [25], and there exists complex coupling between the surface-heating source and workpiece, that can be a combination of:

- Beer—Lambert-type volumetric energy-deposition and advection diffusion of heat;
- differentiation of the thermal properties of the surface layer-substrate at the contact interface;
- nonuniform mixing of layer and substrate materials;
- temporal changes of the volume and the exact shape of the molten pool.

This complex coupling is modeled via the adjustable phenomenological parameters x_l, k_d and κ_d; whilst for the TiC/steel system, the estimated thermal properties and model parameters are: k_u = 5.64 W/(kg °C), κ_u= 1.306 × 10^{-6} m^2/s [30], κ_s = 1.483 × 10^{-5} m^2/s and x_s = 1.5 mm, and those for the Cr_3C_2/ steel system are: k_u = 15 W/(kg °C), κ_u= 1.612 × 10^{-5} m^2/s [31–34], κ_s = 1.483 × 10^{-5} m^2/s and z_s = 3.0 mm. For the analyses whose results are shown in Figures 8–13, the quantity g is an adjustable coefficient for assigning values to the phenomenological parameters k_d and *kd*, which is completed by scaling of measured thermal properties of material layers.

The results of inverse thermal analyses for the solar processing of the TiC/common steel system are presented in Figures 8–10, whilst those of the Cr_3C_2/common steel system are presented in Figures 11–13. In all cases, the values of the adjustable phenomenological parameters x_l, k_d and κ_d are provided in the figure captions.

Shown in Figure 8 are different volumetric temperature distributions at completion of surface heating of the TiC/common steel system, which is on a timescale that is small relative to that of heat conduction in the layered material (see inset of Figure 9b). As indicated in Figure 8, the shapes of these temperature distributions are a function of the phenomenological parameters x_l, k_d and k_d. Referring to Figure 8, the temperature distributions for x_l = (70/100)1.5 mm and x_l = (10/100)1.5 mm correspond, respectively, to total and partial volumetric energy deposition.

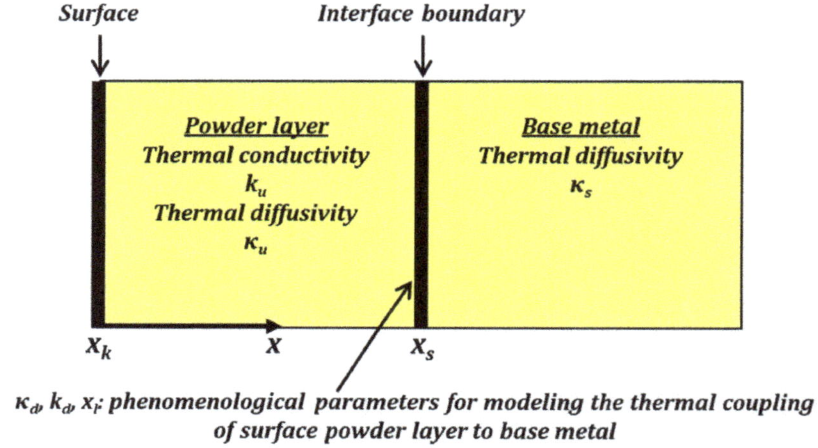

Figure 7. 2D-Schematic representation of regions of the layered-system, and associated material properties and phenomenological parameters for the 1D model defined by Equations (1)–(7).

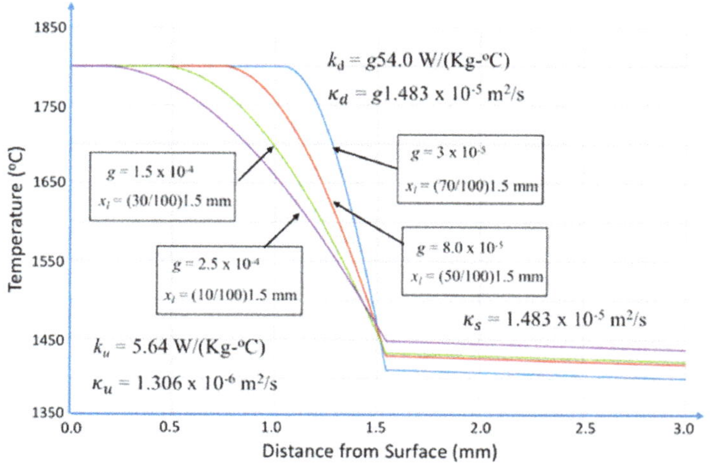

Figure 8. Temperature field at end of heating period as a function of adjustable phenomenological parameters x_l, k_d and κ_d (system TiC/common steel).

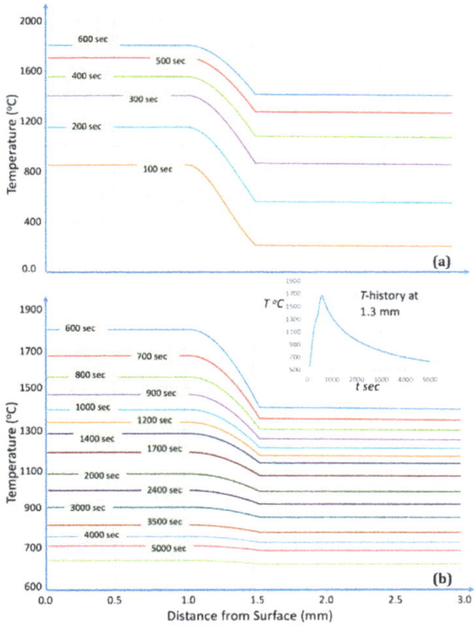

Figure 9. Simulations for the system TiC/common steel at x_l = (70/100) 1.5 mm: (**a**) heating period; (**b**) cooling period [$k_d = g54.0$ W/(kg. °C), $\kappa_d = g1.483 \times 10^{-5}$ m^2/s, where $g = 3 \times 10^{-5}$].

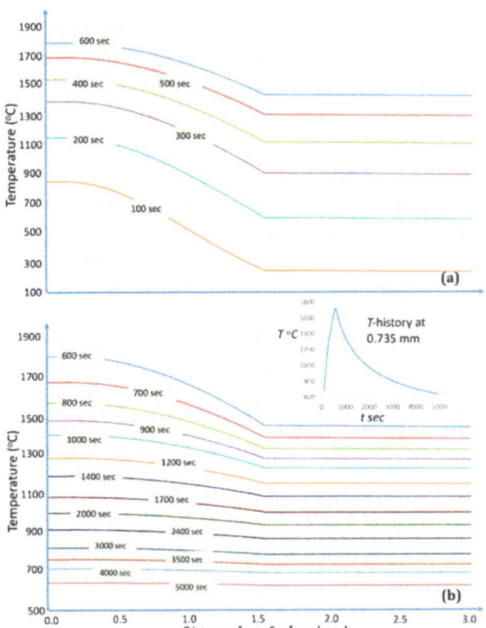

Figure 10. Simulations for the system TiC/common steel at x_l = (10/100) 1.5 mm: (**a**) heating period; (**b**) cooling period [$k_d = g54.0$ W/(kg. °C), $\kappa_d = g1.483 \times 10^{-5}$ m^2/s, where $g = 2.5 \times 10^{-4}$].

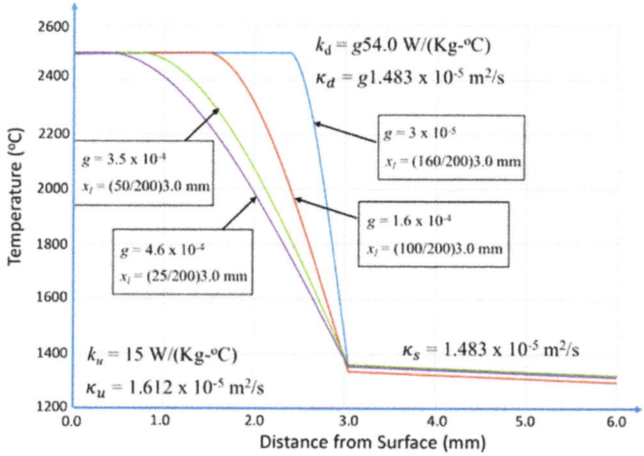

Figure 11. Temperature field at end of heating period as a function of adjustable phenomenological parameters x_l, k_d and κ_d (system Cr_3C_2/common steel).

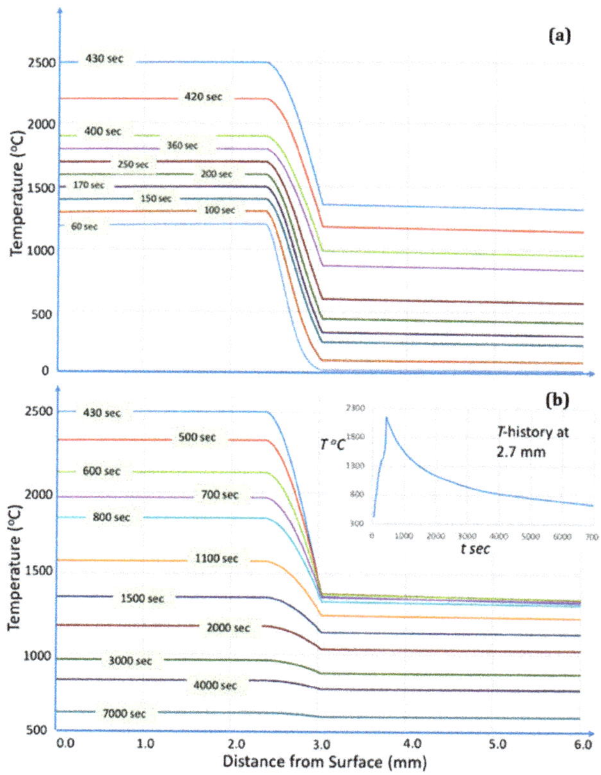

Figure 12. Simulations for the system Cr_3C_2/common steel at x_l = (160/200) 3.0 mm: (**a**) heating period; (**b**) cooling period [$k_d = g54.0$ W/(kg−°C), $\kappa_d = g1.483 \times 10^{-5}$ m^2/s, where $g = 3 \times 10^{-5}$].

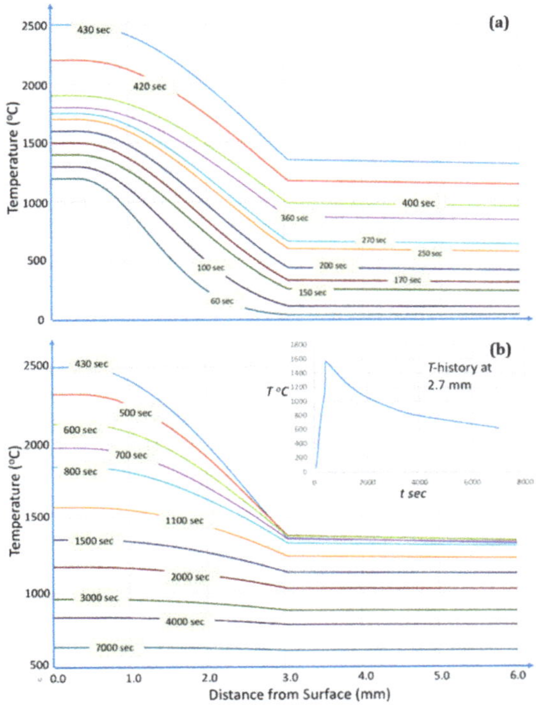

Figure 13. Simulations for the system Cr_3C_2/common steel at x_l = (25/200)3.0 mm: (**a**) heating period; (**b**) cooling period [$k_d = g54.0$ W/(kg−°C), $\kappa_d = g4.6 \times 10^{-4}$ m^2/s, where $g = 3 \times 10^{-5}$].

Shown in Figures 9 and 10 are heating and cooling periods for the TiC/common steel system, corresponding, respectively, to total and partial energy deposition during surface heating. Comparison of temperature histories shown in Figures 9 and 10 indicates a relative insensitivity of temperature histories in terms of trend during cooling to the range of volumetric energy deposition from partial to total. This result is significant in that estimating the temperature distribution at completion of surface heating, because the complexity of the solar heating process, consisting of multiple coupled physical processes, poses a more difficult problem than estimating the temperature history in the layered system, which is based on heat conduction alone. In particular, of significance is that the temperature distribution at completion of surface heating provides a boundary condition, in terms of the heat conduction equation [27], for prediction of temperature histories in the layer system. There is a direct correlation between maximum temperature at layer-substrate interface and the extent of energy deposition at the surface (see insets of Figures 9 and 10).

From a practical point of view, the feasibility of a quantitative analysis of temperature fields developed during solar processing could be used as a useful tool for the prediction of the microstructure of the solar-treated surface layers (e.g., top layer reinforced by a dispersion of TiC particles, dendritic growth at the bottom melt pool area in contact with the unmolten metal), the extent and the microstructure of the underlying heat affected zone, as well as the solid state modifications of the base metal microstructure (e.g., stress relief, grains growth).

The analysis whose results are shown in Figures 11–13, for the Cr_3C_2/common steel system, follows the same procedure as above for the TiC/common steel system. Shown in Figure 11 are different volumetric temperature distributions at completion of surface heating of the Cr_3C_2/common steel system, ranging from partial to total volumetric energy deposition. Referring to Figure 11, the temperature distributions for x_l = (160/200) 3.0 mm and x_l = (25/200) 3.0 mm correspond, respectively, to total and partial volumetric energy deposition. Shown in Figures 12 and 13 are heating and cooling periods for the Cr_3C_2/common steel system, corresponding, respectively, to total and partial energy deposition during surface heating. As for the previous analysis, comparison of temperature histories shown in Figures 12 and 13 indicates a relative insensitivity of temperature histories with respect to trend during cooling to the range of volumetric energy deposition from partial to total, and direct correlation between maximum temperature at layer-substrate interface and the extent of energy deposition at the surface (see insets of Figures 12 and 13).

From a practical point of view, the verification of the quantitative description potency in the case of Cr_3C_2 powder that is totally dissolved in the liquid ferrous matrix constitutes strong evidence for the development of a generic model, valid for the quantitative description of solar surface processing of other pre-deposited carbide powder/base metal systems, whatever the metallurgical reactions could be. Thus, such an inverse thermal analysis could be applied as a prediction tool for optimising the solar processing parameters, in order to control the microstructure and, consequently, the mechanical properties and in-service performance of solar-treated material systems.

Overall, the energy deposition process is multiscale with respect to heat transfer, owing to a region of rapid heating (top, directly irradiated layer) being coupled to one of purely diffusive heat transfer (underlying workpiece volume), thus posing a specific problem for inverse thermal analysis, as discussed above (see Figures 6 and 7). The model, Equations (1)–(8), and simulations using this model, whose results are shown in Figures 8–13, examines the nature of the coupling between two regions where heat transfer occurs on different timescales, and is controlled by characteristics of their interface coupling. The timescale and rate for heat transfer at the top layer can be estimated by in-situ surface-temperature measurements and ex-situ optical microscopy observations of the volume of the workpiece, while rate for heat transfer within the workpiece is predictable given the thermal diffusivity and observed microstructural changes within the workpiece, see Figures 2–5. The unknown characteristic of the layered system, to be estimated by inverse thermal analysis, is the coupling between these regions.

What is known a priori concerning the thermal physics of interfaces between two materials having different thermal properties is represented formally by the model Equations (1)–(8). Considered in this model is that even for the ideal case, where thermal contact resistance is not present, differences in thermal diffusivity are sufficient to induce reflected thermal transport during time-dependent heat transfer across an interface [27]. For the nonideal case, which should be assumed for the process considered here, contact resistance at the interface will contribute to reflected heat transfer within the top layer. The characteristics of contact resistance with respect to reflected heat transfer is by nature extremely complex, having multivariable dependence, and thus modelled using phenomenological parameters. The results of simulations show that a range of reasonable estimated interface-coupling strengths, defined by phenomenological adjustable parameters (see Figures 8 and 11), can be determined by inverse thermal analysis, given the measured boundary constraints imposed by the top-layer and workpiece regions. With respect to process control, these simulations show that characteristics of the interface coupling are a significant controlling factor for heat transfer, as demonstrated by comparison of Figures 9 and 12 to Figures 10 and 13, respectively, and thus may suggest methods of process control in the future. Overall, the solar heating process is defined by rate of solar-energy coupling to the top layer, and subsequently, coupling of this energy to the workpiece, which is a function of interface heat-transport characteristics.

Recently, such an inverse analysis approach has been successfully adopted for the evaluation of the solidification path of stainless steel melts [35] and microstructure modifications of Inconel 718 [36], both taking place during additive manufacturing via Selective Laser Melting.

5. Conclusions

The purpose of this study was to describe a generic parametric model for the estimation of temperature histories within bi-material workpieces subjected to surface treatment applying Concentrated Solar Energy. During solar processing, details on the exact amount of the volumetric energy deposition within the workpiece are difficult to be determined. Thus, reverse thermal analysis constitutes a powerful tool for describing quantitatively the temporal and spatial temperature evolution, based on post-processing metallographic observations of the microstructures obtained.

In summary, the CSE surface treatment of two carbide powder/common steel systems resulted in reinforced surface clads that have been previously proven to exhibit acceptable machinability using conventional cutting tools and superior in-service wear resistance. The present first attempt revealed the potential of applying reverse thermal analysis and parametric modeling for the prediction of the temperature history within the CSE-affected area and to correlate, by extension, the solar-layers performance to the solar processing parameters.

Author Contributions: Conceptualization, P.P.P.; thermal analysis, S.G.L. and A.D.Z.; experimental validation, A.G.M., I.G.P. and J.R.; metallographic observation, I.G.P.; solar processing, A.G.M. and J.R.; initial manuscript preparation, S.G.L. and A.D.Z.; final manuscript preparation and editing, P.P.P. All authors have read and agreed to the published version of the manuscript.

Funding: The experimental work received partial funding from EU-DG RTDs Programme "Solar Facilities for the European Research Area—Second Phase –SFERA II" (projects: CarbiSol P1404300058, P1503060136 and P1602050194). The work of S.G.L. was supported by a Naval Research Laboratory (NRL) internal energy-deposition processes modeling core program.

Institutional Review Board Statement: Not applicable.

Informed Consent Statement: Not applicable.

Data Availability Statement: Data presented in this article are available at request from the corresponding author.

Conflicts of Interest: The authors declare no conflict of interest.

References

1. Pitts, J.R.; Tracy, E.; Shinton, Y.; Fields, C.L. Applications of solar energy to surface modification processes. *Crit. Rev. Surf. Chem.* **1993**, *2*, 247–269.
2. Rodríguez, G.P.; de Damborenea, J.J.; Vázquez, A.J. Surface hardening of steel in a solar furnace. *Surf. Coat. Technol.* **1997**, *92*, 165–170. [CrossRef]
3. Flamant, G.; Ferriere, A.; Laplaze, D.; Monty, C. Solar processing of materials: Opportunities and new frontiers. *Sol. Energy* **1999**, *66*, 117–132. [CrossRef]
4. Rodríguez, G.P.; Herranz, G.; Romero, A. Solar gas nitriding of Ti6Al4V alloy. *Appl. Surf. Sci.* **2013**, *283*, 445–452. [CrossRef]
5. Pantelis, D.I.; Psyllaki, P.; Sarafoglou, C. Surface alloying on cast iron using concentrated solar energy. *Fonderie Fondeur Aujourd'hui* **2002**, *211*, 23–33.
6. Mourlas, A.; Psyllaki, P.; Pantelis, D. Anti-wear TiC-based surface layers using concentrated solar energy. *Key Eng. Mater.* **2016**, *674*, 296–301. [CrossRef]
7. Mourlas, A.; Pavlidou, E.; Vourlias, G.; Rodríguez, J.; Psyllaki, P. Concentrated solar energy for in-situ elaboration of wear-resistant composite layers. Part I: TiC and chromium carbide surface enrichment of common steels. *Surf. Coat. Technol.* **2019**, *377*, 124882. [CrossRef]
8. Mourlas, A.; Pavlidou, E.; Vourlias, G.; Rodríguez, J.; Psyllaki, P. Concentrated solar energy for in-situ elaboration of wear-resistant composite layers. Part II: Tungsten carbide surface enrichment of common steels. *Surf. Coat. Technol.* **2019**. *375*, 739–751. [CrossRef]
9. Pantelis, D.I.; Griniari, A.; Sarafoglou, C. Surface alloying of pre-deposited molybdenum-based powder on 304L stainless steel using concentrated solar energy. *Sol. Energy Mat. Sol. Cells* **2005**, *89*, 1–11. [CrossRef]
10. Fernández-González, D.; Ruiz-Bustinza, Í.; González-Gasca, C.; Piñuela-Noval, J.; Mochón-Castaños, J.; Sancho-Gorostiaga, J.; Verdeja, L.F. Concerntrated solar energy applications in materials science and metallurgy. *Sol. Energy* **2018**, *170*, 520–540. [CrossRef]
11. International Energy Agency. Renewable Energy for Industry: From Green Energy to Green Materials and Fuels. 2017. Available online: https://www.iea.org/publications/freepublications/publication/ (accessed on 25 March 2023).

12. Sánchez Bautista, C.; Rodríguez, G.P.; Ferriere, A. Numerical modelling of the solar cladding process. *Surf. Coat. Technol.* **2008**, *202*, 1594–1605. [CrossRef]
13. Li, B.; Oliveira, F.A.C.; Rodríguez, J.; Fernandes, J.C.; Rosa, L.G. Numerical and experimental study on improving temperature uniformity of solar furnaces for materials processing. *Sol. Energy* **2015**, *115*, 95–108. [CrossRef]
14. Romero, A.; García, I.; Arenas, M.A.; López, V.; Vázquez, A. High melting point materials welding by concentrated solar energy. *Sol. Energy* **2013**, *95*, 131–143. [CrossRef]
15. Román, R.; Cañadas, I.; Rodríguez, J.; Hernández, M.T.; González, M. Solar sintering of alumina ceramics: Microstructural development. *Sol. Energy* **2008**, *82*, 893–902. [CrossRef]
16. Gutiérrez-López, J.; Levenfeld, B.; Várez, A.; Pastor, J.Y.; Cañadas, I.; Rodríguez, J. Study of the densification, mechanical and magnetic properties of Ni-Zn ferrites sintered in a solar furnace. *Ceram. Int.* **2015**, *41*, 6534–6541. [CrossRef]
17. Herranz, G.; Romero, A.; DeCastro, V.; Rodríguez, G.P. Processing of AISI M2 high speed steel Reinforced with vanadium carbide by solar sintering. *Mater. Des.* **2014**, *54*, 934–946. [CrossRef]
18. Cambronero, L.E.G.; Cañadas, I.; Martínez, D.; Ruiz-Román, J.M. Foaming of aluminium-silicon alloy using concentrated solar energy. *Sol. Energy* **2010**, *84*, 879–889. [CrossRef]
19. Ozisik, M.N.; Orlande, H.R.B. *Inverse Heat Transfer, Fundamentals and Applications*; Taylor & Francis: New York, NY, USA, 2000.
20. Kurpisz, K.; Nowak, A.J. *Inverse Thermal Problems—Computational Engineering*; WIT Press: Boston, MA, USA, 1995.
21. Alifanov, O.M. *Inverse Heat Transfer Problems*; Springer: Berlin/Heidelberg, Germany, 1994.
22. Beck, J.V.; Blackwell, B.; St. Clair, C.R., Jr. *Inverse Heat Conduction: Ill-Posed Problems*; John Wiley & Sons Inc.: New York, NY, USA, 1985.
23. Beck, J.V. Inverse Problems in Heat Transfer with Application to Solidification and Welding. In Proceedings of the 5th Conference on Modeling of Casting, Welding and Advanced Solidification Processes, Davos, Switzerland, 16–21 September 1990.
24. Beck, J.V. Inverse Problems in Heat Transfer. In *Mathematics of Heat Transfer*; Tupholme, G.E., Wood, A.S., Eds.; Oxford University Press: New York, NY, USA, 1998; pp. 13–24.
25. Lambrakos, S.G. Parametric Modeling of welding processes using numerical-analytical basis functions and equivalent source distributions. *J. Mater. Eng. Perform.* **2016**, *25*, 1360–1375. [CrossRef]
26. Prosgolitis, C.G.; Lambrakos, S.G.; Zervaki, A. Phase-Field Modeling of Nugget Zone for a AZ31-Mg-Alloy Friction Stir Weld. *J. Mater. Eng. Perform.* **2018**, *27*, 5102–5113. [CrossRef]
27. Carslaw, H.S.; Jaegar, J.C. *Conduction of Heat in Solids*, 2nd ed.; Oxford University Press: New York, NY, USA, 1986; p. 374.
28. Zervaki, A.D.; Mourlas, A.G.; Psyllaki, P.P.; Lambrakos, S.G. Parametric Model of Concentrated Solar Energy Surface Processing. In Proceedings of the Advances in Welding & Additive Manufacturing Research Conference 2022, Miami, FL, USA, 13–16 June 2022.
29. Rodríguez, J.; Cañadas, I.; Zarza, E. New PSA high concentration solar furnace SF40. *AIP Conf. Proc.* **2016**, *1734*, 070028.
30. Shackelford, J.F.; Han, Y.-H.; Kim, S.; Kwon, S.-H. *CRC Materials Science and Engineering Handbook*, 4th ed.; Taylor & Francis: Boca Raton, FL, USA, 2015; pp. 50, 263, 279.
31. Pankratz, L.B. *Thermodynamic Properties of Carbides, Nitrides, and Other Selected Substances*; US Department of the Interior, Bureau of Mines: Washington, DC, USA, 1994; 696, p. 223.
32. Desorbo, W. Heat Capacity of Chromium Carbide (Cr_3C_2) from 13 to 300 °K. *J. Am. Chem. Soc.* **1953**, *75*, 1825–1827. [CrossRef]
33. Oriani, R.A.; Murphy, W.K. The Heat Capacity of Chromium Carbide (Cr_3C_2). *J. Am. Chem. Soc.* **1954**, *76*, 343–345. [CrossRef]
34. Kelley, K.K.; Boericke, F.S.; Moore, G.E.; Huffman, E.H.; Bangert, W.M. *Thermodynamic Properties of the Carbides of Chromium*; US Department of the Interior, Bureau of Mines: Washington, DC, USA, 1944; 662, pp. 7–15.
35. Abolhasani, D.; Hossein Seyedkashi, S.M.; Kang, N.; Kim, Y.J.; Woo, Y.Y.; Moon, Y.H. Analysis of Melt-Pool Behaviors during Selective Laser Melting of AISI 304 Stainless-Steel Composites. *Metals* **2019**, *9*, 876. [CrossRef]
36. Tabaie, S.; Rézaï-Aria, F.; Jahazi, M. Microstructure Evolution of Selective Laser Melted Inconel 718: Influence of High Heating Rates. *Metals* **2020**, *10*, 587. [CrossRef]

metals MDPI

Article

Evaluating Surface Quality of Inconel 617 by Employing Deep Cryogenically Treated Electrodes in Surfactant-Added Dielectrics of Transformer Oil

Kashif Ishfaq [1], Muhammad Sana [1], Muhammad Arif Mahmood [2,*], Saqib Anwar [3] and Muhammad Umair Waseem [1]

1 Department of Industrial and Manufacturing Engineering, University of Engineering and Technology, Lahore 54890, Pakistan
2 Intelligent Systems Center, Missouri University of Science and Technology, Rolla, MO 65409, USA
3 Industrial Engineering Department, College of Engineering, King Saud University, P.O. Box 800, Riyadh 11421, Saudi Arabia
* Correspondence: mmahmood@mst.edu; Tel.: +1-573-341-4908

Abstract: Over the past few decades, better surface quality has remained of great interest to researchers. It deteriorates the fatigue life of the workpiece. The criticality arises when a material of greater strength is selected to work in high-temperature areas such as nickel (Ni)-based superalloys, categorically Inconel 617. Conventional machining operations are not the best choice for the machining of this alloy because of its low density and greater strength. Therefore, electric discharge machining (EDM) is generally engaged. Still, there is a great necessity to make a more reliable surface using EDM, which performs better even in harsh working areas. Therefore, this study examined the potential of deep-cryogenically treated electrodes under the modified dielectrics of transformer oil in the said context, which has not been discussed so far. A set of 30 experiments was performed, designed using the full factorial technique. Deep-cryogenically treated electrodes provided better surface quality in comparison to the non-treated electrodes. Amongst the deep-cryogenically treated electrodes, brass performed outstandingly and provided the lowest value of surface roughness (SR), 6.65 μm, in the modified dielectric of transformer oil with Span 80. The surface finish of deep-cryogenically treated brass is 28.72% better compared to the average value of the overall deep-cryogenically treated electrodes. The lowest value of SR (8.35 μm) was gained by engaging a non-cryogenically treated Cu electrode with a T-20–transformer oil-modified dielectric. The said value of SR is 17.7% better than the highest value of SR achieved in the case of S-80–transformer oil with a non-cryogenically treated Cu electrode.

Keywords: surface quality; cryogenic treatment; transformer oil; Inconel 617; surfactant

Citation: Ishfaq, K.; Sana, M.; Mahmood, M.A.; Anwar, S.; Waseem, M.U. Evaluating Surface Quality of Inconel 617 by Employing Deep Cryogenically Treated Electrodes in Surfactant-Added Dielectrics of Transformer Oil. *Metals* **2023**, *13*, 1092. https://doi.org/10.3390/met13061092

Academic Editor: George A. Pantazopoulos

Received: 3 May 2023
Revised: 2 June 2023
Accepted: 4 June 2023
Published: 9 June 2023

1. Introduction

Surface quality is an essential aspect of all machining processes, i.e., traditional, and nontraditional. The better surface quality of difficult-to-machine alloys increased the fatigue life of work in harsh cutting areas [1]. The material removal phenomena cause overlapping craters in nontraditional machining operations, resulting in a rougher surface [2]. There is a functional relationship between the roughness and the performance/functional properties of the machined surface. The roughness of a machined surface has an important role in determining the interaction between the surface and its functional environment. A rough surface usually demonstrates a high wear rate and higher coefficient of friction than a smooth surface. It influences the mechanical and chemical properties of the machined component [3]. An increase in surface roughness (SR) upsurges the stress concentration factor, leading to a decrease in fatigue strength [4]. Moreover, a high roughness value promotes grain slip and crack initiation at the surface of the material [5]. The corrosion rate is also affected by a variation in this parameter. It has been reported that the corrosion rate significantly improves as the arithmetic mean roughness (R_a)—one of the measures of surface roughness—is increased [6].

During the spark-off time, eroded material resolidified on the workpiece's surface, forming a recast layer that increased the material's SR. The SR parameter 'Ra' is termed as the arithmetic average of the irregularities produced on the surface of the machined specimen after the machining process. The recast layer showed different characteristics compared to the original material. Properties such as brittleness and hardness of the material after the recast layer are changed from the previous material [7]. Therefore, the recast layer in the non-conventional machining process must be avoided so that the fatigue life of work can be increased. Hard-to-cut Ni-based superalloys such as Inconel 617 are widely used in aerospace, gas turbines, super boilers, aircraft, and combustion cans [8]. Ni-based superalloys are employed in the above fields due to their high stability under vigorous cutting, strength, and low density [9]. Considering its unique properties, the increased strength, low thermal conductivity, and rapid strain hardening make Inconel 617 difficult to machine using typical methods such as milling, drilling, and lathe [10]. A non-traditional setup was built to process the Ni-alloy for this purpose. Electric discharge machining (EDM), wire-EDM, electric discharge drilling, and milling are examples of nonconventional processes [11,12]. Because of its new qualities, the EDM die sinker was chosen over all other choices. One of the great benefits of EDM is that it can be used for machining with a thin tool, resulting in precise geometric perfection as well as an excellent surface quality that cannot be achieved using other traditional machining procedures such as grinding [13]. As illustrated, regardless of mechanical properties, EDM can process difficult-to-cut materials such as Inconel 617 [14,15].

EDM, or spark erosion, is preferred over older methods because it can precisely build complicated morphologies, dies, and molds regardless of material type [16,17]. Researchers say EDM can cut strong metals, alloys, and composites without respect to metallurgical properties [18]. EDM uses repeated electric sparks between the workpiece–electrode gap to melt and vaporize the material. Discharges create a plasma channel in the gap by ionizing the dielectric fluid surrounding the electrode [12,19]. The tool's surface is opposite the object's needed shape [20,21]. EDM requires dielectric fluid [22,23]. It cools the material during pulse off-time and flushes debris/chips from the machined surface [24]. EDM uses several dielectrics based on need. Dielectric fluids include mineral oils, hydrocarbons, mineral seal oils, water-based dielectrics, water with and without additives, gaseous dielectrics, and powder-added dielectrics. Dielectrics differ in material removal rate (MRR), surface quality, wear ratios, and accuracy [25,26]. In EDM, kerosene oil is commonly utilized as a traditional dielectric liquid. However, its high flammability, minimum cutting rates, and poor performance as a result of material removal rate, tool wear rate, and surface polish make it unsuitable for operating Inconel 617 using EDM [27]. As a result, researchers have shifted their focus away from kerosene and onto another hydrocarbon-based dielectrics, i.e., transformer oil, with a higher flash point, fire point, lower viscosity, and higher thermal conductivity. Furthermore, transformer oil produces better results than kerosene oil, but the MRR, tool wear rate (TWR), and SR obtained using it as EDM oil are insignificant [28]. To overcome the said flaws, certain additives have been added to increase the dielectric properties and the machine reactions.

To address the restrictions of poor surface quality, an electrically conductive metal powder, such as alumina, copper, and silicon, is added to the dielectric fluid, reducing its insulating property [29]. These conductive microparticles/nanoparticles operate as a bridge when an appropriate voltage is supplied across the tool and workpiece material. Because these powder particles are scattered, they form chains in several locations beneath the sparking area, increasing the spark gap [30]. Fast sparking occurs when the number of sparks per unit time increases, causing the workpiece material to be removed quickly [29,31]. By expanding the plasma channel, added powder enhances spark density. As a result, regular sparking occurs on the material. This uniform sparking contributes to the workpiece's smooth surface finish [32]. However, the introduction of micro/nano-powder to the dielectric medium experienced a serious difficulty of agglomeration, which curtailed sparking through the dielectric medium and hence limited EDM performance [33,34]. The creation of agglomeration during EDM reduces sparking, thereby affecting output results. Surfactants lower dielectric medium surface tension and facilitate additive mixing and mobility [35]. Surfactants are hydrophilic and lipophilic surface-active substances [36]. Emulsifiers, which are classified as a subgroup of the substance class known as surfactants, are also employed in dielectrics to increase surface smoothness [37].

Poor surface finish during material EDM is the key concern. Surfactants and metallic particles in dielectric media as well as electrodes treatment are used to solve the problem. Deep-cryogenically treated (CT) electrodes during EDM have solved this problem in previous investigations. Cryogenic treatment of electrodes increased wear characteristics by eliminating poor surface finish [38]. Ozdemir [39] performed shallow CT high Cr-Fe and low carbon cast steel at −84 °C and discovered improved wear characteristics. Shallow cryogenic treatment improved microstructure, the study found. Senthilkumar and Rajendran [40] found that cryogenic treatment on En 19 steel increased its wear characteristics by 114% after shallow cold working and 214% after deep cold working. Jamadar and Kavade [32] measured the SR of AISI D3 steel using a copper electrode in EDM and aluminum powder particles in IPOL oil as a dielectric. At low peak current, higher pulse on time, and six grams per liter of aluminum powder in the dielectric medium, the SR of the steel improved significantly. Jafarian [20] used Inconel 718 alloy as base material and EDM against a square copper electrode in kerosene oil as a dielectric. By increasing the pulse on time from 120 to 240 μs, the machined specimen's SR improved significantly, according to the author. Kumar et al. [30] examined Inconel 718 alloy workpiece and copper electrode EDM output responses. Stirrers suspended aluminum powder in EDM oil. Suspended aluminum powder increased the specimen's SR. The right current input parameter improved MRR and TWR output responses.

Kolli and Kumar [18] used the Taguchi design of the experiment to determine how the surfactant and graphite powder dispersed in dielectric medium-affected titanium alloy EDM against the copper electrode. The authors found that using the least surfactant, graphite powder, and discharge current resulted in the best surface finish. Abdul Razak et al. [41] investigated surfactant(s) addition during EDM of reaction-bonded silicon carbide against a copper electrode using EDM oil as the dielectric medium. MRR, SR, and TWR were better with the surfactant Span 20 than Span 80. Adding 0.4 weight % to the dielectric medium during EDM improved the surface finish. Li et al. [42] evaluated the machining proficiency of EDM by taking Inconel 718 alloy as the workpiece and two different electrodes, namely simple copper and a modified electrode, i.e., Cu-SiC, during the machining operation. The authors concluded that the modified electrode provided high surface quality compared to the simple one when the machined specimen was testified. The supremacy of EDM was examined by Wu et al. [43] by using SKD61 steel as the workpiece material and copper as the electrode, and in addition, surfactant was added to enhance the conductivity of the dielectric medium. The authors found that due to the addition of the surfactant in the dielectric medium, the output response, i.e., MRR, increased by 40%, but when the optimal settings were engaged, the said response was enhanced by 80%. The authors also claimed that the SR did not significantly deteriorate when MRR was of greater value. Kumar et al. [38] used the cryogenic treatment on different electrodes to determine tool life and study its effects on electrode materials. The cryogenic treatment on the electrodes increased tool life by 92% and improved tribological characteristics (wear and friction coefficient), hardness, and toughness. AbdulKareem et al. [44] examined the effect of cryogenic treatment on a copper electrode during EDM of titanium alloy. By cryogenically treating the copper electrode, the electrical and thermal conductivities increased, improving output responses.

Srivastava and Pandey [45] examined how cryogenic treatment affected a copper electrode against high-speed steel during EDM of M2-grade steel. The authors found that cryogenic treatment on the copper electrode lowered EWR by 20% and improved surface quality by 27% compared to standard machining. Ram et al. [46] examined the influence of cryogenic treatment on a graphite electrode during the EDM of EN31 as the base material. The authors found that the CT graphite electrode enhanced the surface finish of EN31 when the machined specimen was testified, and statistical data approaches were engaged. Jafferson and Hariharan [47] investigated the wear properties of different electrode materials, i.e., tungsten, brass, and copper, against the workpiece material AISI304 during EDM. The authors found that a CT tungsten electrode provided better microhardness, less TWR, and better electrical conductivity compared to other electrodes.

Bartkowiak et al. [48] explored the surface topography of steel using graphite electrodes during the EDM process and by utilizing motif and multiscale analyses. The authors found that with the change in the machining parameters the surface topography, i.e., craters, changed. Along with that, the authors also claimed that the discharge energy was the most influential effect, which results in the production of large amounts of craters. Gogolewski [49] used spherical objects to assess the surface morphology of Ti alloy-based powder prepared using additive technology. The surface morphology was measured using a multiscale approach. The author claimed that the findings aid in the advancement of cutting-edge diagnostic systems that can examine surface topography swiftly and thoroughly across several phases of data collection and processing.

Based on the research discussed above, the effectiveness of surfactant-based dielectrics using deep cryogenically treated (CT) electrodes is still to be determined. Additionally, the surfactant(s) capacity for Inconel 617 EDM utilizing CT electrodes has not been thoroughly discovered using transformer oil. The SR of a Ni-based superalloy using CT and non-deep cryogenically treated (NT) electrodes in various dielectrics examined in this work will be a great contribution to impeding the recast layer and obtaining a better surface finish during the EDM process. The experiment used CT and NT electrodes with distinct dielectrics with the addition of surfactants in transformer oil. Energy-dispersive X-ray spectroscopy (EDS) analysis was also explored to investigate the deposition of materials on the workpiece surface after EDM. The optimal dielectric and electrode are recommended and tested for this project.

2. Materials and Methods

Using five tailored dielectric mediums, deep cryogenically treated (CT) and non-deep cryogenically treated (NT) electrodes were tested against a Ni-based superalloy during machining. Surface roughness (SR) was used to evaluate the most important process yield. Optical microscopy was used to determine Inconel 617's composition. SR was measured in terms of arithmetic mean (R_a), which is a commonly used roughness parameter. The selection of this parameter was based on the rationale that it provides the mean value of surface asperities distributed about the mean line within the defined evaluation length. So, this parameter shows a holistic picture of the surface's quality. For each of the machined cavities, roughness measurements were made at five different points. Considering that the roughness values are closely related (standard deviation is small), an average of these five points was used for further analysis. Table 1 lists the chemical composition of the selected workpiece material and details the parameters kept constant in this study. These parameters were kept constant, as the focus of the current research is to evaluate the impact of nontreated and cryogenically treated electrodes on surface quality in EDM of Inconel-617 (IN617) under surfactant-based transformer oil dielectrics. Preliminary trials were carried out to determine the values of these parameters. Those parametric levels were identified for final experimentation, which ensures the availability of complete machining expression on the workpiece surface. Minimum chance of arcing was another important consideration that was kept in mind while finalizing the parametric values. A rectangular workpiece with dimensions 60 mm × 60 mm × 5 mm was used for experimental investigation.

Table 1. Chemical attributes of nickel-based superalloy and constant factor [50].

Element	%Wt.	Element	%Wt.	Constant Factors (Units)	Values
Ni	44.5	B	0.006	Surfactant concentration (vol %)	6
Cr	20.0–24.0	S	0.015	Spark voltage (volts)	4
Co	10.0–15.0	Cu	0.5	Current (amperes)	10
Mo	8.0–10.0	Ti	0.6	Pulse off time (μsec)	26
Fe	3.0	Mn	1.0	Pulse on time (μsec)	100
Al	0.8–1.5	Si	1.0	-	-
C	0.05–0.15	-	-	-	-

This investigation used 9 mm diameter copper, brass, and graphite electrodes. Electrode materials' efficiency was tested before and after cryogenic treatment. The electrodes were inserted in the nitrogen container and exposed to liquid nitrogen at −185 °C to cryogenically operate. Electrodes were held at this temperature for around 24 h to finish the cryogenic process. Five dielectrics—transformer oil, Span 20+ transformer oil, Span 80+ transformer oil, Tween 20+ transformer oil, and Tween 80+ transformer oil—were tested against three electrodes. Table 2 shows the surfactants' main traits.

Table 2. Chemical and physical properties of surfactants [27,51,52].

Properties	Chemical Formula	Molecular Weight (g/mol)	Density (g/cm^3 at 25 °C)	Flashpoint (°C)	Hydrophilic-Lipophilic Balance (HLB) Value
T-80	$C_{64}H_{124}O_{26}$	1309	1.08	148	15
T-20	$C_{58}H_{114}O_{26}$	1227.54	1.095	>110	16.7
S-80	$C_{24}H_{44}O_6$	428.60	1.068	186.2	4.6
S-20	$C_{18}H_{34}O_6$	346.46	1.032	>110	8.6

Introductory tests determined how parametric settings affected the process. The technique was developed during the first experiments and used for final experimentation to match the workpiece's complete machining impression. These traits also reduce tool and workpiece burns. Table 3 lists the machine settings based on the experiments, Table 4 compiles the properties of electrodes [1,53] and Table 5 shows the properties of transformer oil dielectric. Determining transformer oil dielectric surfactant concentration was crucial. Preliminary experiments against the given selection criterion determined the final choice; however, a literature introduction guideline was used. Early results suggest a surfactant concentration of 6% for the criteria. A motorized stirrer tank was designed to mix surfactant and transformer oil. The stirrer mechanism mixed dielectric and surfactant during the experiment. Figure 1 shows an EDM machine (model: RJ230) used in this experiment, while Figure 2 exhibits the cross-sectional views of the electrodes.

Table 3. Input parameters levels.

Input Variable	Level 1	Level 2	Level 3	Level 4	Level 5
Treatment type	NT	CT	-	-	-
Electrode type	Copper	Brass	Graphite	-	-
Type of dielectric	Pure transformer oil	S-20–transformer oil	S-80–transformer oil	T-20–transformer oil	T-80–transformer oil

Table 4. Physical and electrical properties of the electrodes [1,53].

Properties	Copper	Brass	Graphite
Density (g/mm^3)	8.904×10^{-3}	8.55×10^{-3}	1.77×10^{-3}
Electrical resistivity (Ω.m)	1.96×10^{-8}	4.7×10^{-7}	60×10^{-5}
Melting point (°C)	1083	990	3350
Specific heat capacity (J/g °C)	0.835	0.380	7.1
Electrical conductivity (S/m)	59×10^6	16×10^6	0.3×10^6

Table 5. Properties of transformer oil and physical attributes of superalloy [8,54].

Properties	Values	Properties	Values
Breakdown voltage (kV)	56.8	Density (kg/m^3)	8360
Flash point (°C)	>140	Melting range (°C)	1332–1380
Pour point (°C)	−40	Specific heat (J/kg C)	419
Viscosity (Pa-s)	7.9×10^{-3}	Electrical resistivity (μ Ωm)	1.22
Density (kg/dm^3) at 20 °C	0.81	-	-

The investigation was carried out using a full factorial experimental approach. Table 3 summarizes the input parameters with levels. Each test was machined 0.3 mm and 30 experiments were conducted. The first fifteen experiments used NT electrodes, and the last fifteen used CT electrodes. Once the experiment had been conducted satisfactorily, the output parameter was evaluated. The SR of the machined specimen was measured using a surface roughness tester (Surtronic-128) manufactured by Taylor Hobson, as shown in Figure 3. Before taking measurements, the instrument was calibrated using a standard specimen provided by the manufacturer. Afterward, the machined specimen was placed below the stylus probe of the equipment. It was ensured that the surface of the machined specimen was parallel to the stylus. A cutoff length of 0.8 mm and an evaluation length of 4 mm were used in this study to measure the roughness value. ISO-4287 roughness standard was used for measurements. As in the current scenario, the designed diameter of the machined cavity is 9 mm; therefore, the aforementioned cutoff length and evaluation length were used, which is the closest match as per the ISO standard for surface roughness measurement. The collected data were then analyzed using bar charts. Three-dimensional surface profilometry graphs can been seen in the Section 3 to determine surface attributes. Considering Inconel 617 EDM's physical phenomena, results are described. The best electrode material and most efficient dielectric were provided for the desired response characteristics.

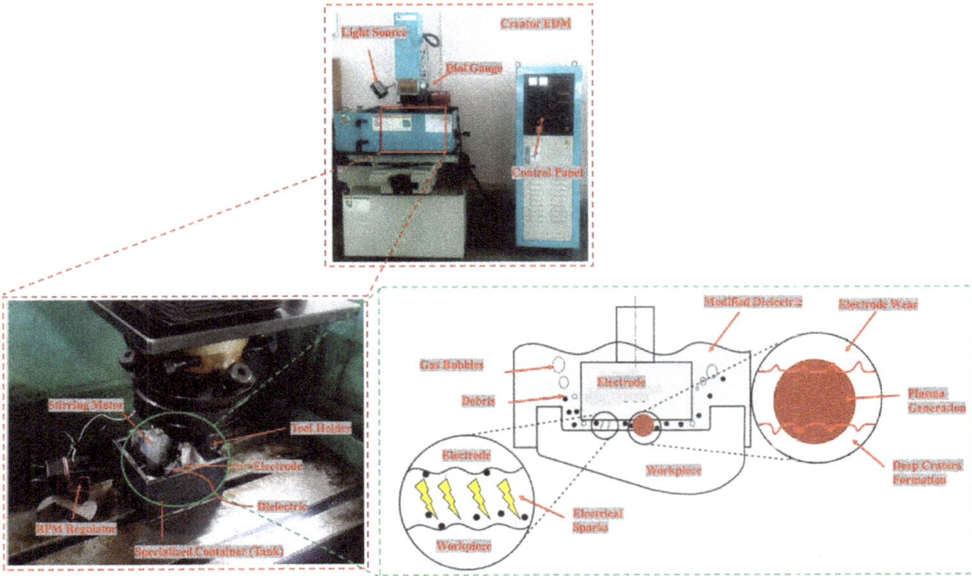

Figure 1. Experimental setup of electric discharge machining.

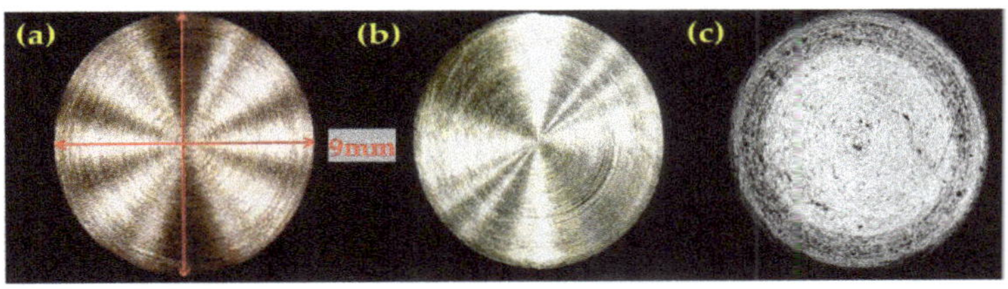

Figure 2. Profile of used deep-cryogenically treated (**a**) copper, (**b**) brass, and (**c**) graphite electrodes.

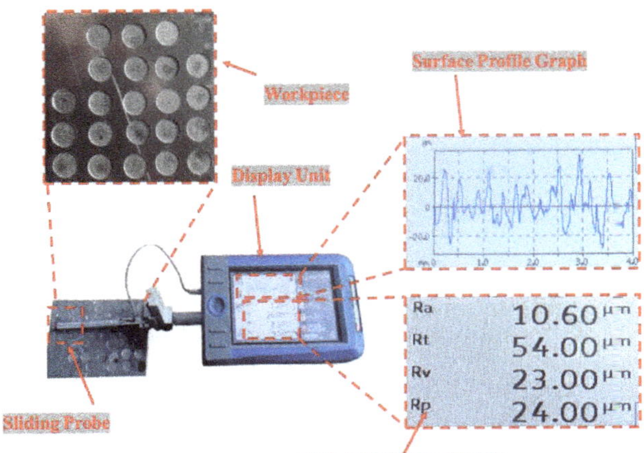

Figure 3. Surface roughness meter.

3. Results and Discussion

The machining performance of EDM of Inconel 617 in terms of surface roughness (SR) is evaluated using deep-cryogenically treated (CT) and non-deep cryogenically treated (NT) electrodes under different types of modified dielectric media to determine the effect of machining on the Ni-based superalloy. The results pertaining to SR, along with results due to NT electrodes, are presented in the form of bar charts, as shown in Figure 4.

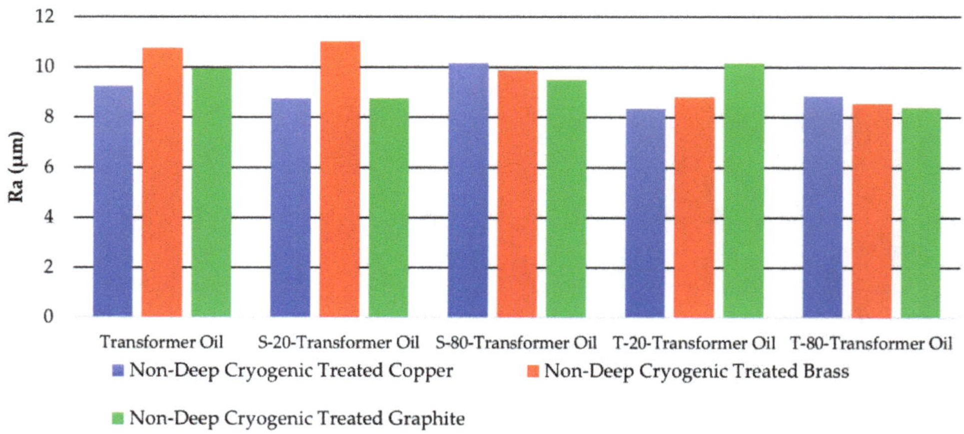

Figure 4. Surface roughness comparison of distinct non-deep cryogenic-treated electrodes.

EDM was used to study the effect of regime change—dielectric medium and electrode material in terms of surface roughness (SR)—on Inconel 617's surface. In pure transformer oil, non-deep cryogenically treated (NT) Copper (Cu)-electrodes have the lowest SR, 9.25 µm. The high flash point (140 °C) of transformer oil and the high breakdown voltage (56.8 kV) of the dielectric utilized in this investigation explain the low SR in pure transformer oil due to the NT Cu-electrode. The transformer oil's high flash point prevents erratic sparking and lowers the machined specimen's SR. Figure 5a shows EDM-formed shallow craters on an Ni-based superalloy. Additionally, the Taylor Hobson roughness meter graph showed the machined profile's low SR. The machining capability of EDM was investigated using an NT graphite electrode in pure transformer oil. It was found that the NT graphite electrode ranked second in achieving the minimum SR (9.95 µm) during the cutting phenomenon of Ni-based superalloy. The reason for the second-highest SR is due to the high pulse on time (100 µs), which leads to the generation of irregular and nonuniform sparking. This irregular sparking produced deep craters on the surface of Inconel 617 during EDM. The microscopic image shown in Figure 5c illustrates the high SR and depicts the deep crater formed on the machined profile. The graph attached below the microscopic image in Figure 5c demonstrates the peak values, which are an indication of high SR. In pure transformer oil, the NT brass electrode had the highest SR, 10.75 µm. Pure transformer oil had the greatest SR value because brass has the lowest melting point (990 °C) of the electrodes utilized in this investigation. The spark density melts and solidifies Ni-based superalloy due to its low melting point. The phenomenon increased the SR of the specimen due to the brass electrode, and the microscopic image shown in Figure 5b indicates the presence of shallow and wide craters.

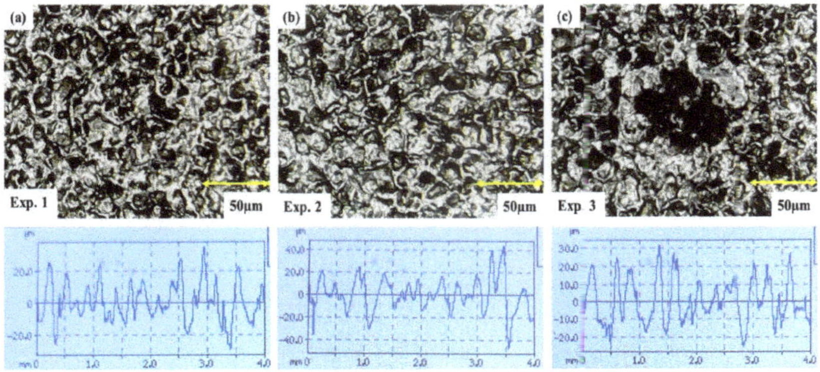

Figure 5. Microscopic surface roughness of non-deep cryogenically treated (**a**) copper, (**b**) brass, and (**c**) graphite electrodes in transformer oil.

S-20, a span representative, was studied for EDM using several NT electrodes. When non-deep cryogenically treated (NT) copper (Cu) and graphite electrodes are machined against Inconel 617 in S-20–transformer oil, they offer the same lowest value of surface roughness (SR) 8.75 μm, which is 5.4% and 12.1% lower than in pure transformer oil. S-20 disperses dielectric molecules, which directly did not agglomerate the degraded particles of the Ni-based superalloy, reducing the SR of NT Cu and graphite electrodes. The Cu and graphite electrode and S-20 in transformer oil after EDM provide Inconel 617 with a low SR. The SEM analysis is shown in Figure 6a, representing shallow craters that illustrate better surface finish. EDM was used to measure SR using an NT brass in S-20–transformer oil-modified dielectric. The NT brass provided the maximum SR, i.e., 11.0 μm, compared to other electrodes, and an increment of 2.2% was found using S-20–transformer oil relative to pure transformer oil. The reason for the increment in the SR is due to the NT brass electrode employed possessing the low electrical conductivity of brass (16×10^6 S/m). Due to the low electrical conductivity and low melting temperature (990 °C) of the brass, eroded material melted and resolidified on the machined profile of the said superalloy. To justify this, the specimen machined by the NT brass electrode was testified using SEM. The SEM image shown in Figure 6b depicts the presence of the recast layer on the surface of Inconel 617. This recast layer is the primary reason for the increment in SR of the nickel (Ni)-based superalloy even after the addition of surfactant. Figure 7 shows a 3D surface profilometry of superalloy in S-20–transformer oil with NT brass, showing heightened peaks and valleys and a higher average Ra.

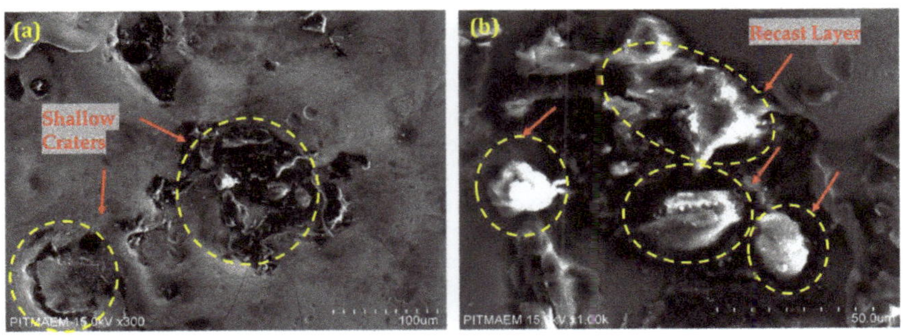

Figure 6. SEM of nickel-based superalloy using non-deep cryogenically treated (**a**) copper, and (**b**) brass electrodes.

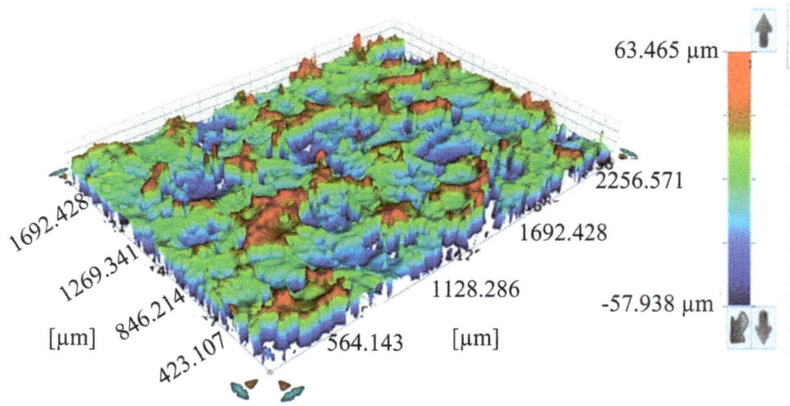

Figure 7. Three-dimensional profilometry in S-20–transformer oil using non-deep cryogenically treated brass.

During Inconel 617 machining using S-80-transformer oil, the non-deep cryogenically treated (NT) graphite electrode had the lowest surface roughness (SR), 9.5 µm. The rough surface in S-80-transformer oil was 9.7% more uneven than the profile acquired in S-20–transformer oil. NT graphite's porousness and lower electrical conductivity (0.3×10^6 S/m) explain why S-80's modified dielectric SR was higher than S-20's. Graphite's characteristics caused erratic sparking. Thus, irregular and nonuniform sparking at a high pulse on time (100 µs) melts more material, but in pulse off time, this eroded material resolidifies on the surface of Ni-based superalloy, resulting in poor surface quality. NT brass electrodes were used to study EDM machining efficiency. The S-80–transformer oil-modified dielectric had the second-lowest SR (9.85 µm). Compared to S-20–transformer oil, S-80–transformer oil had 10.5% more designed surface. The reason for the greater surface finish in S-80 is linked with its high flash point (186.2 °C) and that of the dielectric medium (140 °C); because of the high flash points, the burning phenomenon did not take place and provided a greater surface finish in the modified dielectric of S-80 surfactant. The NT copper (Cu) electrode produced the highest SR (10.15 µm) in Inconel 617 machining using the S-80–transformer oil dielectric. S-80–transformer oil's SR was 13.8% higher than that of S-20–transformer oil due to the NT Cu electrode's rougher surface. Cu's high electrical conductivity (59×10^6 S/m) caused the NT Cu electrode's high SR. The S-80's poor surface quality was caused by the NT Cu electrode's increased electrical conductivity eroding more material from Inconel 617.

The representative of tweens, T-20, with varied NT electrodes machining nickel (Ni)-based superalloy, was used to determine EDM's proficiency. Figure 4 shows that the NT Cu electrode had the lowest SR (8.35 µm) in the T-20–transformer oil-modified dielectric, 13.8% better than S-80. T-20 had the highest HLB (16.7), which lowered SR and improved value compared to S-80 surfactant-based modified dielectric. Therefore, the higher the HLB was, the lower the tension of the dielectric was, which resulted in the impeding of the phenomenon of eroded debris agglomeration. SEM was performed on the machined profile of Ni-based superalloy gained after engaging with the NT Cu and T-20 surfactant, as shown in Figure 8a. The SEM image of the machined specimen shown in Figure 8. depicts the shallow craters formed, which is an indication of good surface finish. Figure 9 shows a 3D surface profilometry in T-20–transformer oil dielectric with NT Cu, showing short-heightened peaks and troughs. The NT brass electrode had the second-lowest SR (8.8 µm) in T-20–transformer oil during Inconel 617 machining. Compared to the machined surface in S-80–transformer oil, a refined machined surface was achieved in T-20–transformer oil, which was a 10.7% improved surface compared to the machined surface gained in the case of S-80 with the same electrode. The reason for the low SR is due to the NT brass electrode being linked with the said findings in the case of the NT Cu electrode and T-20 surfactant. The machining performance of EDM due to use of an NT graphite electrode was investigated against the work part and in the presence of the T-20–transformer oil dielectric medium. The NT graphite provided the highest SR (10.15 μm), which was 6.4% higher than the value achieved using S-80–transformer oil. The porous nature of NT graphite is the primary reason for the highest SR even with the highest HLB of the T-20. This porosity of the graphite electrode causes an agglomeration phenomenon that leads to the poor surface finish in the modified dielectric of T-20. Moreover, the SEM image shown in Figure 8b depicts the melts redeposited and deep craters that increased the SR of the base material.

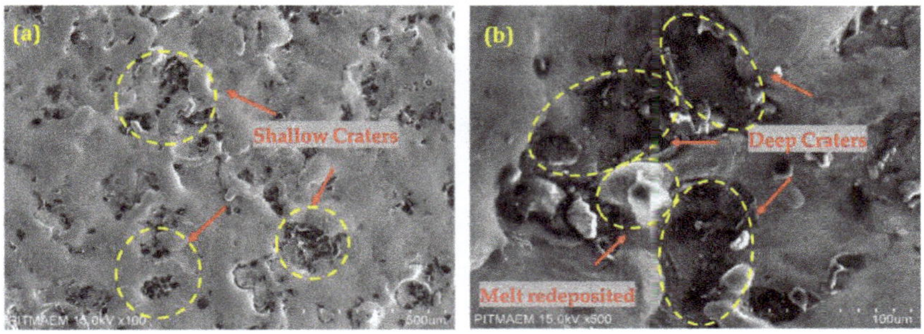

Figure 8. SEM images of nickel-based superalloy using non-deep cryogenically treated (**a**) copper, and (**b**) graphite electrodes in T-20 transformer oil.

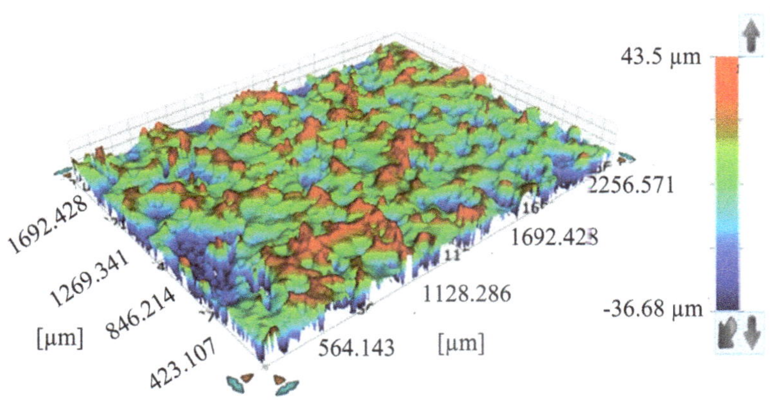

Figure 9. Three-dimensional profilometry in T-20–transformer oil using non-deep cryogenically treated copper.

The non-deep cryogenically treated (NT) graphite electrode had the lowest surface roughness (SR) (8.4 μm) with Inconel 617 and T-80–transformer oil-modified dielectric metal. T-80–transformer oil with NT Graphite had 17.2% lower machining performance than S-80. Graphite had the greatest melting point (3350 °C) of all electrodes, which lowered SR in T-80–transformer oil-modified dielectric. This high melting point prevented electrode melting and vaporization, resulting in a high surface finish. Figure 10c shows small, shallow craters on the machined specimen, indicating a higher surface quality. In T-80–transformer oil, the NT brass had the second highest SR, 8.55 μm. The T-80–transformer oil with NT brass yielded 2.8% less than the T-20–transformer oil dielectric medium. The fact that greater surface finish was achieved due to NT brass electrode is due to brass having lower electrical conductivity (16×10^6 S/m), and this lower electrical conductivity incorporated with the second highest HLB value (15) of T-80 to flush away the eroded material during the pulse-off time. In this way, solidification of worn debris is avoided, which helps in achieving a better surface finish. Figure 10b shows a microscopic view of the nickel (Ni)-based superalloy's machined profile with shallow craters. The potentiality of EDM was tested in the presence of T-80–transformer oil dielectric medium and NT copper (Cu). SR, 8.85 μm, was 5.60% larger than that achieved during machining using NT Cu and T-20–transformer oil. The NT Cu electrode had high SR because Cu has high electrical conductivity (59×10^6 S/m) compared to other electrodes tested in this investigation. Cu sparks irregularly and erodes material due to its strong electrical conductivity. Figure 10a shows deep and small superalloy craters.

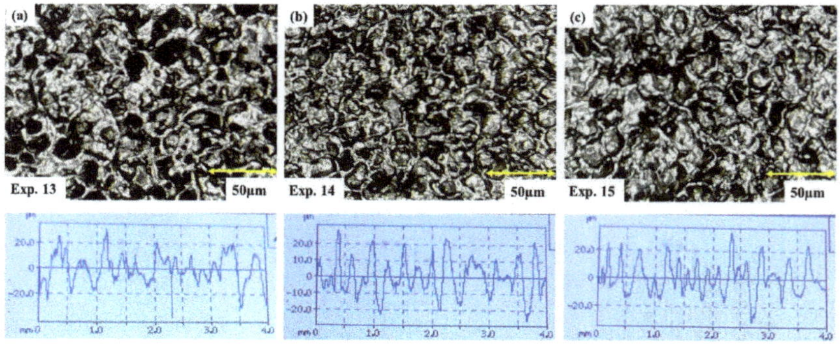

Figure 10. Microscopic surface roughness of non-deep cryogenically treated (**a**) copper, (**b**) brass, and (**c**) graphite electrodes in T-80–transformer oil.

The following is the overall summary of non-deep cryogenically treated (NT) electrode performance during the EDM of nickel (Ni)-based superalloy. The NT copper (Cu) electrode provided the lowest surface roughness (SR) (8.35 μm) in the T-20–transformer oil-modified dielectric medium. The machined surface obtained in the presence of modified dielectric medium (T-20–transformer oil) was 9.7% more improved and refined than the machined profile achieved without the use of surfactant. The NT graphite electrode was observed as ranking second in achieving the lowest SR (8.4 μm) in the presence of T-80–transformer oil during the machining of hard-to-cut superalloy. An enhancement of 15.6% in the machined profile was observed when the Inconel 617 was cut in the presence of modified dielectric (T-80–transformer oil). The NT brass electrode achieved the lowest SR (8.55 μm) in the same modified dielectric, i.e., T-80–transformer oil. It was found that if T-80–transformer oil dielectric is used, then an improvement of 20.5% can be achieved in the machined surface other than if the pure transformer oil dielectric is considered. The graph shown in Figure 11. elaborates on the proficiency of NT electrodes in achieving the maximum and minimum values of SR in different modified dielectrics.

The machining ability of EDM of Inconel 617 in terms of surface roughness (SR) was investigated using deep cryogenically treated (CT) electrodes under different types of modified dielectric media. The results pertaining to SR, along with results due to CT electrodes are presented in the form of a bar chart, shown in Figure 12.

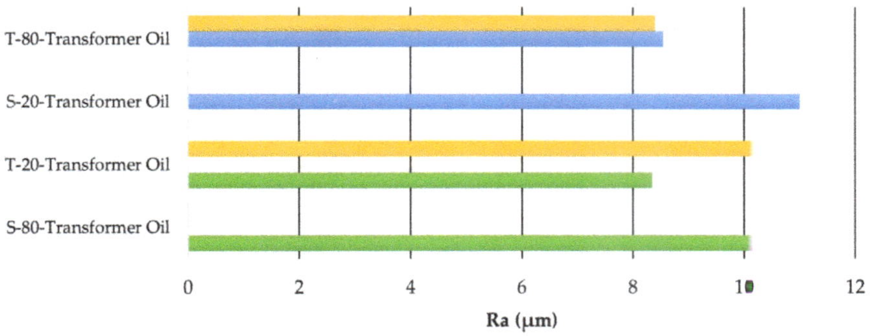

Figure 11. Comparison of different non-deep cryogenically treated electrodes in various modified dielectrics for maximum and minimum average surface roughness (R_a).

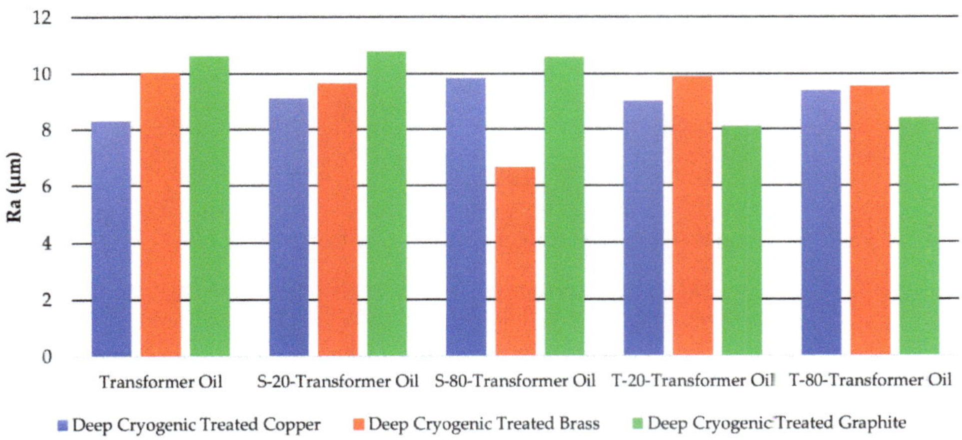

Figure 12. Surface roughness comparison of distinct deep cryogenically treated electrodes.

Different deep cryogenically treated (CT) electrodes and modified dielectrics were used to investigate Inconel 617 surface roughness (SR) in EDM. Figure 12 shows that the CT copper (Cu) electrode had the lowest SR of 8.3 μm in pure transformer oil compared to other cryogenic electrodes. The SR value was 10.3% better than the NT Cu electrode with the same dielectric medium. Cryogenic treatment refines the grain structure, preventing erratic sparking and lowering SR, resulting in higher surface quality in pure transformer oil. In pure transformer oil CT, copper achieved less SR, i.e., a better surface finish than the non-deep cryogenically treated (NT) electrode. This is quite understandable considering the phenomenon of crater size. Craters were larger and deeper on the surface machined using the NT Cu electrode as compared to CT electrodes, as shown in Figure 13a. It is well known that smaller crater dimensions result in a better surface finish [55]. Yildiz et al. [56] also reported that the use of CT electrodes results in lower SR compared to NT electrodes. The proficiency of EDM was evaluated using a CT brass electrode against work part 617 in pure transformer oil. It was found that the CT brass stood second in achieving the lowest SR (10.05 μm) in the pure transformer oil. The machined surface achieved in the case of CT brass and pure transformer oil dielectric showed a 6.5% improved machined surface compared to the NT brass electrode. The surface asperities achieved due to the NT brass electrode were rougher and more crowded than the surface achieved in the case of the CT brass electrode. Figure 13b depicts those asperities as more random and wider in the NT scenario than the profiles achieved after the cryogenic treatment on the brass electrode. CT graphite achieved the highest SR (10.65 µm) in pure transformer oil compared to other CT electrodes. The reason for the high SR value due to the CT graphite electrode is that the low electrical conductivity (0.3×10^6 S/m) is assisted by cryogenic treatment. Moreover, the porous nature of the graphite electrode also improved due to said treatment, and these two properties were enhanced and incorporated to uplift the outcome response. The microscopic image shown in Figure 13c illustrates that even the CT graphite electrode achieved a higher SR compared to the NT graphite electrode, but the burn marks were reduced, which were produced by irregular sparking due to cryogenic treatment on the graphite electrode.

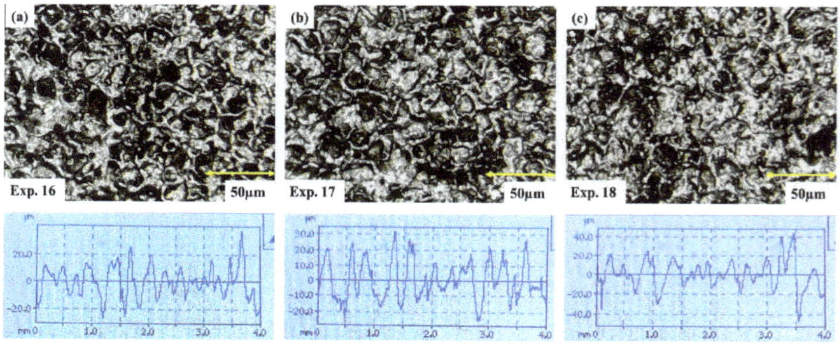

Figure 13. Microscopic surface roughness of deep cryogenically treated (**a**) copper, (**b**) brass, and (**c**) graphite electrodes in transformer oil.

S-20 with transformer oil as dielectric and various deep cryogenically treated (CT) electrodes against an Inconel 617 work part were used to measure EDM potentiality. In the modified dielectric, S-20–transformer oil, the CT copper (Cu) electrode had the lowest surface roughness (SR), 9.1 µm. SR was 8.8% greater than CT Cu–pure transformer oil. Cryogenic treatment helps Cu's electrical conductivity (59×10^6 S/m) increase the SR value. Due to the assistance of cryogenic treatment in conductivity, higher spark density was produced during the pulse on time, and greater melting and vaporization occurred, which led to a poor surface finish. The microscopic image shown in Figure 14a describes the smaller but shallow crater formation on the surface of a Ni-based superalloy, which is an indication of poor surface finish. SEM of the specimen, machined using a CT Cu electrode in S-20–transformer oil dielectric, was performed, as shown in Figure 15a. The SEM image also shows a little melt redeposited on the surface of the Inconel 617, which is an indication of a greater surface finish. Figure 12 depicts that the CT brass electrode achieved the second lowest SR with a value of 9.65 µm in the presence of S-20–transformer oil and, compared to the values with pure transformer oil in CT brass and S-20–transformer oil in non-deep cryogenically treated (NT) brass, improvements of 4.0% and 12.3% in SR value were observed using the same electrode. The reason for achieving a better surface finish is due to the CT brass electrode linked with the addition of S-20 in transformer oil. The surfactant addition in the dielectric medium lowers the surface tension of the dielectric, which incorporates flushing away the debris from the workpiece material. Due to this recast layer, a better surface finish is achieved. Figure 14b depicts the more refined surface of the specimen due to S-20 addition, and the cryogenic treatment on the electrode. The performance of EDM was evaluated using CT graphite against the machining of the said work part in S-20–transformer oil. CT graphite electrode achieved the highest SR value in S-20–transformer oil, and there was an increase in the value of SR by 1.40% compared to the value gained in pure transformer oil. The reason for the increment in the SR due to the CT graphite electrode is that the porous nature of graphite is improved by the treatment and the conductivity of graphite is therefore assisted by the said treatment. These characteristics are responsible for providing a poor surface finish. The microscopic image shown in Figure 14c, describes the poor surface finish achieved in S-20–transformer oil due to the CT graphite electrode. Taylor Hobson's SR graph also depicts higher average surface roughness (R_a) peaks, which is an indication of poor surface finish. The SEM shown in Figure 15b illustrates the high deposition of melts redeposited that increased the SR of the specimen. Figure 16 shows a 3D surface profilometry of Inconel 617 in S-20–transformer oil dielectric with CT graphite, showing heightened peaks and valleys and a higher average R_a value.

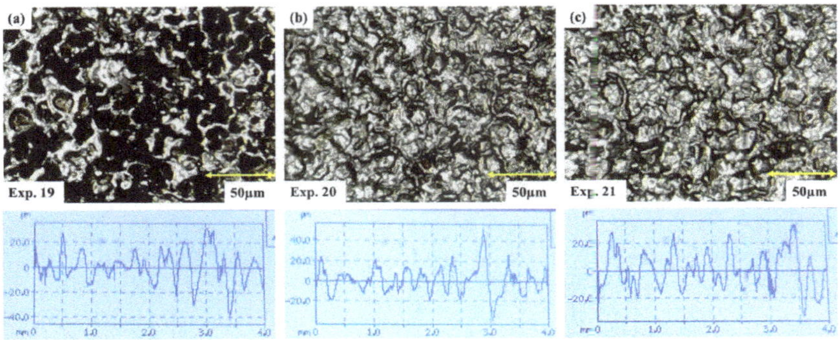

Figure 14. Microscopic surface roughness of deep cryogenically treated (**a**) copper, (**b**) brass, and (**c**) graphite electrodes in S-20–transformer oil.

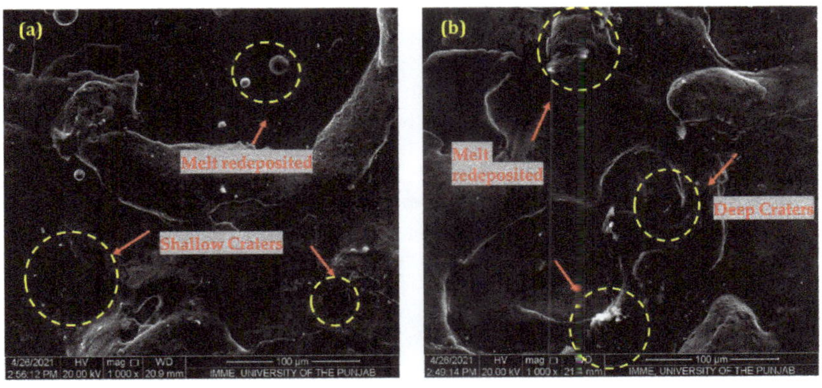

Figure 15. SEM images of nickel-based superalloy in S-20–transformer oil after the EDM using deep cryogenically treated electrodes (**a**) copper, and (**b**) graphite electrodes.

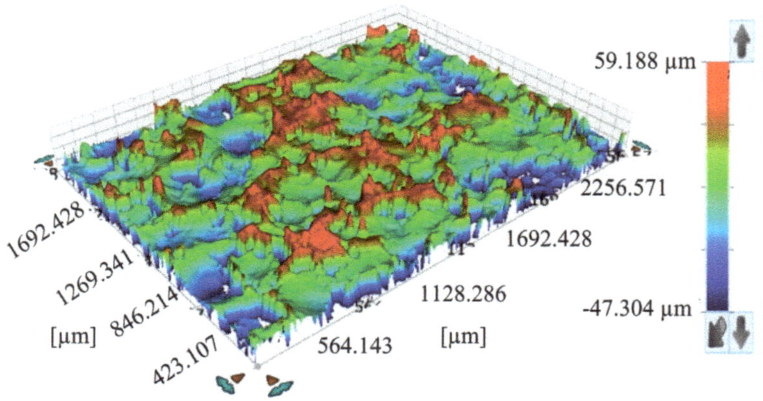

Figure 16. Three-dimensional profilometry in S-20–transformer oil using deep cryogenically treated graphite.

It was revealed that the deep cryogenically treated (CT) brass electrode stood first in achieving the lowest surface roughness (SR) value, and the minimum value of SR was 6.65 µm compared to other CT electrodes in the S-80–transformer oil. The reduction in SR value was 31.1% and 32.5% compared to the value obtained using the S-20–transformer oil in CT brass and S-80–transformer oil in non-deep cryogenically treated (NT) brass, respectively. The reason for the improvement in the surface finish is linked to the high flash point of S-80 (186.2 °C) and the high flash point of the transformer oil (144 °C). The high flash point of the dielectric and surfactant impedes the flammability of the dielectric during the pulse on time, and the surfactant's addition curtails the surface tension of the dielectric. These two characteristics of the dielectric and the surfactant incorporated uplifting the surface finish of the CT brass electrode in the modified dielectric of S-80–transformer oil during the EDM of Inconel 617. SEM was performed on the nickel (Ni)-based superalloy after EDM due to the CT brass electrode in the modified dielectric T-80–transformer oil, as shown in Figure 17a. The SEM result pertains that shallow craters were formed on the surface of the machined specimen, which is an indication of an excellent surface finish. Figure 18 shows a 3D surface profilometry in S-80–transformer oil with CT brass, showing minor peaks and valleys. The CT copper (Cu) electrode, with the second lowest SR (9.85 µm), determined the EDM performance. S-80–transformer oil had 7.60% more SR than S-20. Cryogenic treatment helps Cu's electrical conductivity (59 × 106 S/m) increase the SR value. Due to the assistance of cryogenic treatment in conductivity, greater spark density was produced during the pulse on time, and greater melting and vaporization occurred, which led to a poor surface finish. The SEM image shown in Figure 17b depicts a high volume of melts redeposited, which increased the SR and the basic reason for the high SR of the machined specimen. The supremacy of EDM was evaluated using a CT graphite electrode and workpiece, i.e., Inconel 617 in the presence of S-80–transformer oil dielectric. It was found that the CT graphite electrode ranked third in maximum SR (10.60 µm), and this value was 1.90% less than the value achieved in the S-20–transformer oil-modified dielectric medium. The reason for the reduction in SR in the presence of S-80–transformer oil-modified dielectric is that graphite has the highest melting point (3350 °C) compared to other electrodes. This high melting point impeded the high melting and vaporization of the electrode due to which a high surface finish was achieved. Moreover, the high breakdown voltage of transformer oil (56.8 kV) was the other reason for the reduction in SR in the S-80–transformer oil.

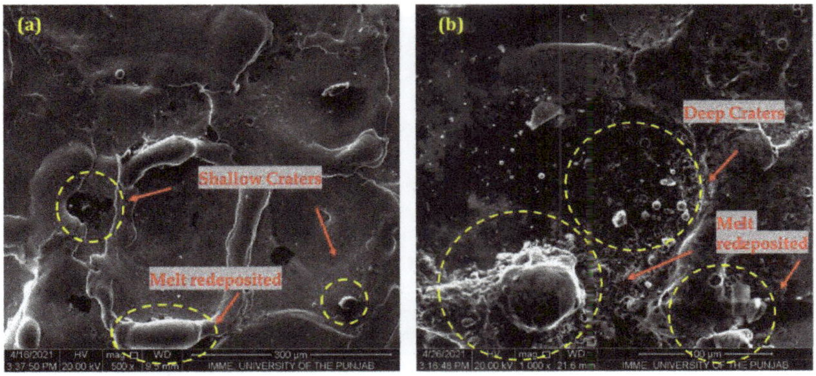

Figure 17. SEM images of nickel—based superalloy in S-80–transformer oil after the EDM using deep cryo-genically treated (**a**) brass, and (**b**) copper electrodes.

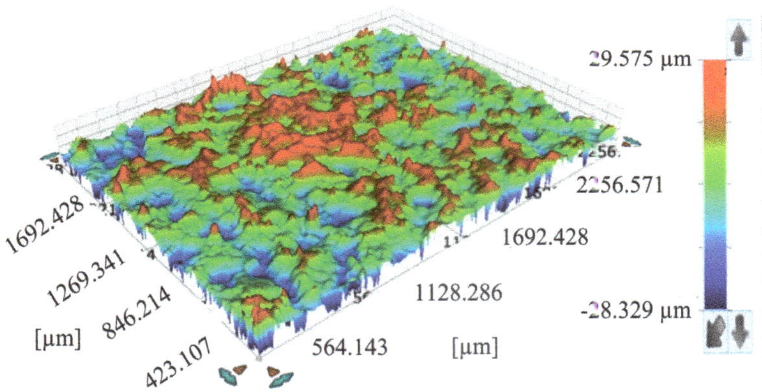

Figure 18. Three-dimensional profilometry in S-80–transformer oil using deep cryogenically treated brass.

Machining proficiency of EDM was determined in tweens, i.e., T-20 with different types of deep cryogenically treated (CT) electrodes against the machining of nickel (Ni)-based superalloy. From Figure 12, the CT graphite electrode stood first in achieving the lowest surface roughness (SR) (8.1 µm) in the presence of T-20–transformer oil-modified dielectric, a 23.6% improved surface compared to the surface gained in S-80-Transformer oil. There are possibly two reasons for the improvement in surface finish due to the CT graphite in the presence of T-20–transformer oil. The first and foremost reason for the reduction in SR of the Ni-based superalloy is linked with the highest HLB (16.2) of T-20. The high HLB value of the surfactant resulted in a reduction in surface tension of the dielectric, and this reduction in surface tension yielded a better mix of surfactant with the dielectric and greater flushing of the eroded debris from the workpiece. The second reason for the improvement in the surface finish was employed with the enhancement in the porosity of the graphite electrode. These two properties integrated to uplift the surface finish of the Ni-based superalloy. The microscopic image shown in Figure 19c describes the very small number of shallow craters present on the machined surface of the Inconel 617, which is in favor of a greater surface finish. The CT Cu electrode stood second in achieving the minimal value of SR (9.0 µm) in the presence of T-20–transformer oil during the machining of Inconel 617. Compared to the machined surface in S-80–transformer oil, an 8.62% refined and better-machined surface was achieved in T-20–transformer oil. The reason for the better surface finish in the CT Cu electrode is that cryogenic treatment enhanced the grain structure due to which a uniform sparking falls on the workpiece irrespective of the nonuniform sparking, which occurred in the noncryogenic treatment. This phenomenon helps in achieving a better surface finish due to cryogenic treatment in the modified dielectric medium. The microscopic image shown in Figure 19a illustrates the small and shallow crater formation on the surface of the said specimen. The machining performance of EDM due to the CT brass electrode was investigated against the work part and in the presence of T-20–transformer oil dielectric medium. The CT brass electrode achieved the highest value of SR (9.90 μm), which was 32.8% higher than the value achieved in S-80–transformer oil. The reason for the higher SR due to the CT brass electrode is elaborated in earlier findings, and Figure 19b shows the random, rougher surface obtained in the T-20–transformer oil scenario.

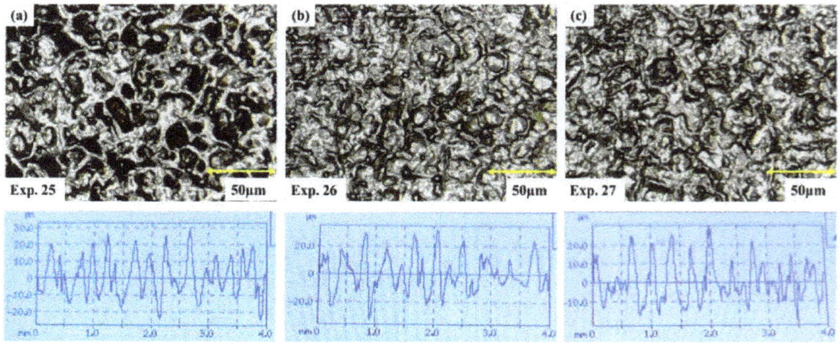

Figure 19. Microscopic surface roughness of deep cryogenically treated (**a**) copper, (**b**) brass, and (**c**) graphite electrodes in T-20–transformer oil.

With a deep cryogenically treated (CT) graphite electrode and T-80–transformer fluid, EDM outperformed Inconel 617 machining. CT graphite electrodes in T-80–transformer oil had the lowest surface roughness (SR) value of 8.40 μm. T-20–transformer oil had 3.6% less SR. The reason for the increment in the SR value is linked with the second highest value of HLB (15) of T-80. The lower the HLB, the higher the surface tension and vice versa. T-20 has the highest HLB, so it achieved the lowest SR due to the CT graphite electrode. T-80 has a slightly lower HLB than T-20, so it achieved a slightly higher SR compared to T-20. Figure 20c shows a rougher surface than T-20–transformer oil. The CT copper (Cu) electrode had the second-lowest SR (9.40 μm) in T-80–transformer oil. Inconel 617 machined against the CT Cu electrode increased SR by 4.3% compared to T-20–transformer oil. Cryogenic treatment helps Cu's electrical conductivity (59×10^6 S/m) increase the SR value. Due to the assistance of cryogenic treatment in conductivity, higher spark density was produced during the pulse on time, and greater melting and vaporization occurred, which led to a poor surface finish. Figure 20a describes the smaller but deeper crater formation on the surface of nickel (Ni)-based superalloy, which is an indication of poor surface finish. The potentiality of EDM was observed in T-80–transformer oil using a CT brass electrode, which ranked third in achieved the lowest SR with a value of 9.9 μm. T-20–transformer oil lowered the SR by 3.5%. High HLB (15) and transformer oil flash point (144 °C) improve surface finish. The pulse on time dielectric's high flash point and surfactant reduces surface tension. These two said characteristics of the dielectric and the surfactant facilitated uplifting the surface finish of the CT brass electrode in the modified dielectric of T-80–transformer oil during the EDM of Inconel 617. The microscopic image shown in Figure 20b depicts the wider and shallow craters present on the machined specimen.

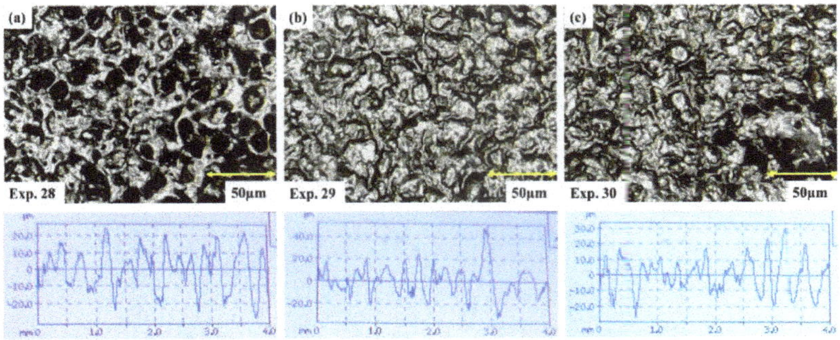

Figure 20. Microscopic surface roughness of deep cryogenically treated (**a**) copper, (**b**) brass, and (**c**) graphite electrodes in T-80–transformer oil.

The deep cryogenically treated (CT) electrode's surface roughness (SR) results can be interpreted as follows: The CT brass electrode achieved the lowest value of SR, i.e., 6.65 µm in the S-80–transformer oil, against the machining of Inconel 617. This achieved value of SR in the modified dielectric was a 33.8% improvement over the SR gained in pure transformer oil dielectric. The second lowest value of SR (8.3 µm) was found against the CT copper (Cu) electrode and in the presence of pure transformer oil during the cutting of Ni-based superalloy. The surface achieved in pure transformer oil using a CT Cu electrode had a 10.3% more polished surface than the machined surface achieved in pure transformer oil using an non-deep cryogenically treated (NT) Cu electrode. The CT graphite electrode was last in achieving the lowest SR (8.4 µm) value in the T-80–transformer oil. The said value achieved a more polished surface, i.e., 21.1%, than the machined surface did in pure transformer oil. The graph shown in Figure 21 elaborates on the proficiency of CT electrodes in achieving the maximum and minimum values of SR in different modified dielectrics.

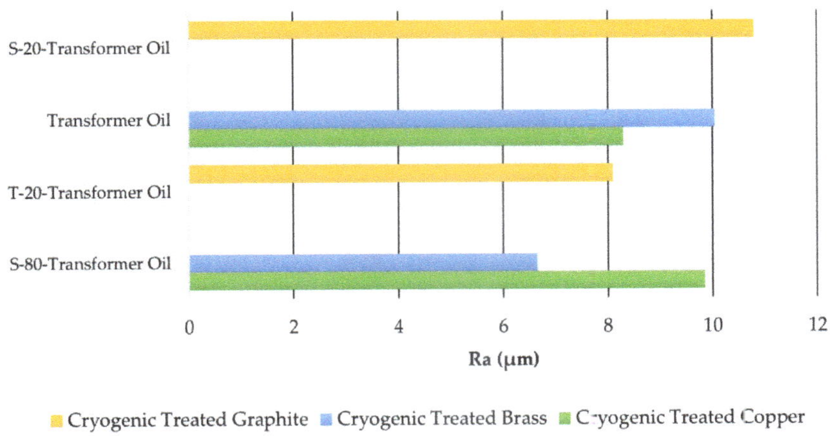

Figure 21. Comparison of different deep cryogenically treated electrodes in various modified dielectrics for maximum and minimum average surface roughness (R_a).

Energy-dispersive X-ray spectroscopy (EDS) analysis of the work part was performed to investigate the overall presence of the various metals in the base alloy. By the continuous dispersion of the dielectric fluid in the customized tank, the debris is flushed away from the workpiece material. During the discharge gap, deposition of the eroded material on the machined specimen occurs due to the incomplete flushing of debris from the workpiece material. Thereof, a little amount of the material is solidified on the surface of the work part from an electrode to the base material that is emersed in the dielectric medium. The

EDS examination provided a detailed composition of the Inconel 617 before and after the experimentation. Figure 22a,b depicts the EDS examination in terms of improved chemical composition, quality, and quantity as well as the investigation of the recast layer of electrode material on the surface of the Ni-based superalloy. When any surfactant is added to the dielectric medium, and EDM is performed using the CT Cu electrode, then some of the debris moves from electrode material to base material.

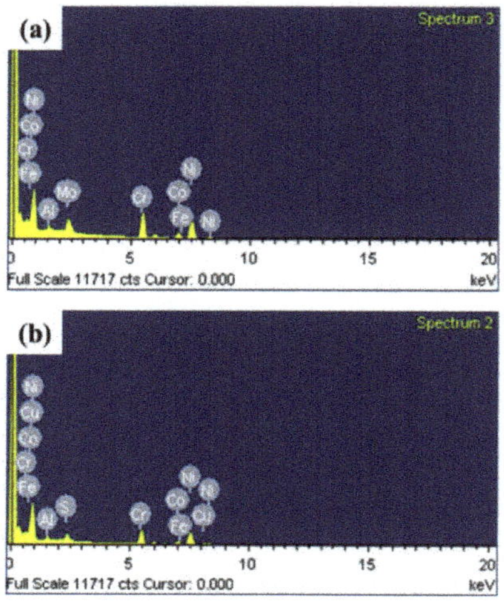

(a)

Element	Weight%	Atomic%
Al	1.11	2.41
Cr	23.61	26.54
Fe	1.57	1.65
Co	12.34	12.24
Ni	51.17	50.95
Mo	10.20	6.22
Totals	100.00	

(b)

Element	Weight%	Atomic%
Al	1.94	3.97
S	2.40	4.14
Cr	20.83	22.17
Fe	1.65	1.63
Co	10.99	10.32
Ni	50.34	47.45
Cu	11.86	10.33
Totals	100.00	

Figure 22. Energy-dispersive X-ray spectroscopy analysis of nickel-based superalloy (**a**) before EDM, and (**b**) after EDM.

Figure 22a shows the EDS analysis of the Inconel 617 before EDM was performed against the CT Cu electrode. The EDS examination discloses that before the machining, Ni, Cr, and Mo were the major peaks of the other metals of the Inconel 617. However, when the CT Cu electrode was engaged for the machining of Inconel 617, debris of the Cu from the electrode to the workpiece material were observed. This is because, during the pulse on time, high spark density is generated, which melts and vaporizes the minimum quantity of the electrode material, and this eroded material solidifies on the surface of the base material. Then, EDS analysis was performed after the EDM, and Figure 22b shows that Cu was the peak constituent of the base material after Ni.

Statistical analysis for the evaluation of average surface roughness (R_a) was performed and is shown in Figure 23. A residual plot indicates the fitness of the data either using regression or analysis of variance. Studying the residual plots allows to see if the regular least-squares assumptions were met. From the normal probability plot, there is a close relationship between the observed data points of the response measure and the predicted line, i.e., a linear relationship exists, which is the indication of less error in the observed magnitudes of the response measure (surface roughness, SR). However, versus fits also presents less randomness in values of magnitudes. Moreover, the histogram favors less negative residual data.

If a comparison is developed between the current research findings and other research conducted by various researchers, then it can be concluded that the surface finish achieved by this study is 23.57% better than the highest magnitude of SR achieved using EDM of die steel by Jamadar and Kavade [32]. Now, if the comparison is made between the highest value of SR achieved in this study using the CT electrode, then it can be found that the surface finish is still better by about 30.56% than the highest SR gained using EDM of Inconel 718 by Jafarian [20].

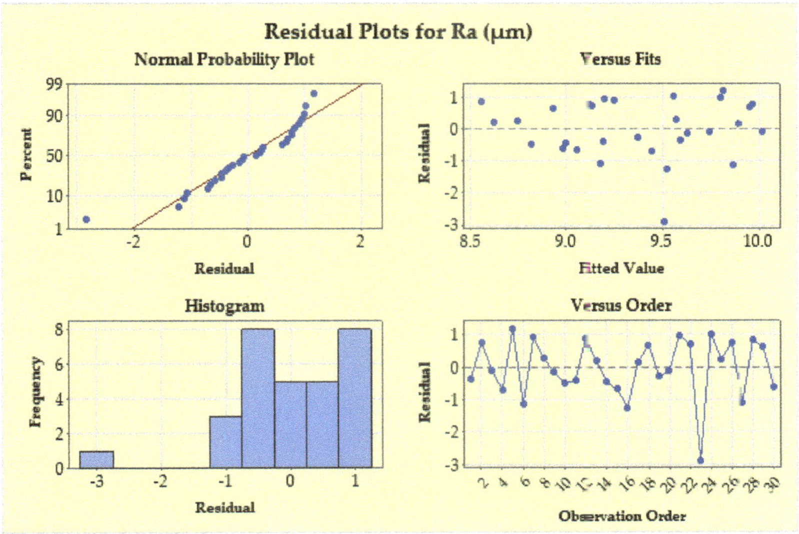

Figure 23. Statistical analysis of surface roughness.

4. Conclusions

The machining proficiency of electric discharge machining (EDM) of nickel (Ni)-based superalloy in terms of surface roughness was examined using various types of modified dielectrics of transformer oil along with different kinds of non-deep cryogenically treated (NT) and deep cryogenically treated (CT) electrodes. The results were explained with the help of process physics, the possible primary reasons, and Energy-dispersive X-ray spectroscopy (EDS) analysis of the workpiece material. After studying the comprehensive investigation of the effect of NT and CT electrodes under various modified dielectrics of transformer oil, the following conclusions can be made:

- Brass excelled among CT electrodes, delivering the lowest surface roughness (SR) (6.65 μm) in the modified dielectric of transformer oil (S-80). CT brass provides a surface polish that is 28.72% higher than the average value provided by all CT electrodes.
- Across a range of changed dielectric media, CT electrodes outperformed compared to NT electrodes in terms of machining performance, with an average of 1.2% higher surface polish.
- The performance of CT electrodes was also examined using span and tween in transformer oil. It was found that tweens had 4.1% better surface polish or less SR than spans.
- CT copper (Cu) electrode in pure transformer oil achieves the lowest SR (8.3 μm), which is a 10.3% better-machined surface compared to pure transformer oil in the NT case. Similarly, average surface roughness (R_a) of 8.1 μm was achieved by using a CT graphite electrode in T-20–transformer oil, which was a 23.9% improved surface finish compared to the value obtained in pure transformer oil.
- In this investigation modified dielectrics, employed against NT electrodes in T-80–transformer oil, yielded the lowest values of SR. R_a magnitudes obtained in T-80–transformer oil during the EDM were lower by 4.32% for NT Cu, 20.5% for NT brass, and 15.6% for NT graphite compared to pure transformer oil.
- NT Cu electrode and T-20–transformer oil dielectric yielded the lowest SR (8.35 μm). The greatest SR for S-80–transformer oil with NT Cu electrode was 17.7% higher than the T-20 value.
- In CT electrodes, span-based modified dielectrics performed 4.13% worse than tween-based ones. Tweens perform better because their high flash points reduce flammability.

- Similarly, the proficiency of tween-based modified dielectrics of transformer oil in NT electrodes was witnessed as superior by 8.41% compared to the span-added dielectrics
- NT Cu, graphite, and brass were machined to the lowest SR of 8.35 µm, 8.40 µm, and 8.55 µm in T-20-, T-80-, and T-80-modified dielectrics, respectively. SR values were 9.72%, 15.6%, and 20.5% higher than pure transformer oil.
- If a comparison is made between the current research findings with the other research conducted by various researchers, then it can be concluded that the surface finish achieved in this study is 23.57% better than the highest magnitude of SR achieved by EDM of die steel. Now, if the comparison is made between the highest value of SR achieved in this study using a CT electrode, then it is found that the surface finish is still better by about 30.56% than the highest of SR gained by EDM of Inconel 718.

In this work, EDM of Inconel 617 was carried out using five dielectric media under three different types of NT and CT electrodes. This work can be extended considering the influence of added powder and machining responses for further improvement of the process. Evaluation of a blend of dielectrics instead of single dielectrics is another valuable possible future direction. Furthermore, a comparative assessment using a multiscale approach is another area that can be explored. The measurement and analysis of other roughness-related parameters such as R_z, R_p, R_{SM}, etc., will be comprehensively explored in future investigations.

Author Contributions: Conceptualization, K.I., M.S. and M.U.W.; methodology, K.I., M.S. and M.U.W.; validation, K.I., M.S. and M.U.W.; formal analysis, K.I., M.S. and M.U.W.; investigation, K.I., M.S., S.A. and M.U.W.; resources, K.I., M.S., M.A.M., S.A. and M.U.W.; data curation, K.I., M.S., M.A.M., S.A. and M.U.W.; writing—original draft preparation, K.I., M.S., M.A.M., S.A. and M.U.W.; writing—review and editing, K.I., M.S., M.A.M., S.A. and M.U.W.; supervision, K.I., M.A.M. and S.A.; project administration, K.I., M.A.M. and S.A.; funding acquisition, K.I., M.A.M. and S.A. All authors have read and agreed to the published version of the manuscript.

Data Availability Statement: Not applicable.

Acknowledgments: The authors appreciate the support from Researchers Supporting Project number (RSPD2023R702), King Saud University, Riyadh, Saudi Arabia.

Conflicts of Interest: The authors declare no conflict of interest.

References

1. Khan, A.R.; Rahman, M.; Kadirgama, K. An experimental investigation on surface finish in die-sinking EDM of Ti-5Al-2.5Sn. *Int. J. Adv. Manuf. Technol.* **2015**, *77*, 1727–1740. [CrossRef]
2. Abbas, N.M.; Solomon, D.G.; Bahari, F. A review on current research trends in electrical discharge machining (EDM). *Int. J. Mach. Tools Manuf.* **2007**, *47*, 1214–1228. [CrossRef]
3. Lee, G.M.; Lee, J.U.; Park, S.H. Effects of surface roughness on bending properties of rolled AZ31 alloy. *J. Magnes. Alloy.* **2022**, *11*, 1224–1235. [CrossRef]
4. Arola, D.; Williams, C.L. Estimating the fatigue stress concentration factor of machined surfaces. *Int. J. Fatigue* **2002**, *24*, 923–930. [CrossRef]
5. Zhao, B.; Song, J.; Xie, L.; Hu, Z.; Chen, J. Surface roughness effect on fatigue strength of aluminum alloy using revised stress field intensity approach. *Sci. Rep.* **2021**, *11*, 19279. [CrossRef] [PubMed]
6. Ye, H.; Sun, X.; Liu, Y.; Rao, X.-X.; Gu, Q. Effect of ultrasonic surface rolling process on mechanical properties and corrosion resistance of AZ31B Mg alloy. *Surf. Coat. Technol.* **2019**, *372*, 288–298. [CrossRef]
7. Azam, M.; Jahanzaib, M.; Abbasi, J.A.; Abbas, M.; Wasim, A.; Hussain, S. Parametric analysis of recast layer formation in wire-cut EDM of HSLA steel. *Int. J. Adv. Manuf. Technol.* **2016**, *87*, 713–722. [CrossRef]
8. Ishfaq, K.; Waseem, M.U.; Sana, M. Investigating cryogenically treated electrodes' performance under modified dielectric(s) for EDM of Inconel(617). *Mater. Manuf. Process.* **2022**, *37*, 1902–1911. [CrossRef]
9. Li, W.; Guo, Y.B. A Two-Parameter Method to Monitor and Characterize Tool Wear in End Milling Inconel 718. In Proceedings of the ASME 2012 International Manufacturing Science and Engineering Conference, Notre Dame, IN, USA, 4 June 2012; pp. 799–804.
10. Guo, Y.B.; Li, W.; Jawahir, I.S. Surface integrity characterization and prediction in machining of hardened and difficult-to-machine alloys: A state-of-art research review and analysis. *Mach. Sci. Technol.* **2009**, *13*, 437–470. [CrossRef]

11. Voigt, O.; Peuker, U.A. Suitability of Eroded Particles from Die-Sink Electro Discharge Machining for Additive Manufacturing—Review, Characterization and Processing. *Metals* **2022**, *12*, 1447. [CrossRef]
12. Abu Qudeiri, J.E.; Zaiout, A.; Mourad, A.-H.I.; Abidi, M.H.; Elkaseer, A. Principles and Characteristics of Different EDM Processes in Machining Tool and Die Steels. *Appl. Sci.* **2020**, *10*, 2082. [CrossRef]
13. Shastri, R.K.; Mohanty, C.P.; Dash, S.; Gopal, K.M.P.; Annamalai, A.R.; Jen, C.-P. Reviewing Performance Measures of the Die-Sinking Electrical Discharge Machining Process: Challenges and Future Scopes. *Nanomaterials* **2022**, *12*, 384. [CrossRef] [PubMed]
14. Alhodaib, A.; Shandilya, P.; Rouniyar, A.K.; Bisaria, H. Experimental Investigation on Silicon Powder Mixed-EDM of Nimonic-90 Superalloy. *Metals* **2021**, *11*, 1673. [CrossRef]
15. Masuzawa, T. State of the Art of Micromachining. *CIRP Ann.* **2000**, *49*, 473–488. [CrossRef]
16. Singh, S.; Maheshwari, S.; Pandey, P.C. Some investigations into the electric discharge machining of hardened tool steel using different electrode materials. *J. Mater. Process. Technol.* **2004**, *149*, 272–277. [CrossRef]
17. Gangil, M.; Pradhan, M.K. Modeling and optimization of electrical discharge machining process using RSM: A review. *Mater. Today Proc.* **2017**, *4*, 1752–1761. [CrossRef]
18. Kolli, M.; Kumar, A. Effect of dielectric fluid with surfactant and graphite powder on Electrical Discharge Machining of titanium alloy using Taguchi method. *Eng. Sci. Technol. Int. J.* **2015**, *18*, 524–535. [CrossRef]
19. Razak, M.A.; Abdul-Rani, A.M.; Nanimina, A.M. Improving EDM Efficiency with Silicon Carbide Powder-Mixed Dielectric Fluid. *Int. J. Mater. Mech. Manuf.* **2015**, *3*, 40–43. [CrossRef]
20. Jafarian, F. Electro discharge machining of Inconel 718 alloy and process optimization. *Mater. Manuf. Process.* **2020**, *35*, 95–103. [CrossRef]
21. Tiwary, A.P.; Pradhan, B.B.; Bhattacharyya, B. Application of multi-criteria decision making methods for selection of micro-EDM process parameters. *Adv. Manuf.* **2014**, *2*, 251–258. [CrossRef]
22. Saodaen, R.; Janmanee, P.; Rodchanarowan, A. Characteristics of Ternary Metal (Cu-Ni-TiN) Electrodes Used in an Electrical Discharge Machining Process. *Metals* **2021**, *11*, 694. [CrossRef]
23. Singh, K.; Agarwal, A.K.; Yadav, R. Effect of Dielectric Fluids Used on EDM Performance: A Review. *Int. J. Emerg. Technol. Eng. Res.* **2017**, *5*, 7.
24. Bhattacharyya, B.; Doloi, B. Machining Processes Utilizing Thermal Energy. In *Modern Machining Technology*; Elsevier Amsterdam, The Netherlands, 2020; pp. 161–363. ISBN 978-0-12-812894-7.
25. Kumar, D.; Kumar, K.; Payal, D.H.S. A State-of-the-Art Review on Dielectric Fluid in Electric Discharge Machining: Uses and Its Effects. *Int. Res. J. Eng. Technol.* **2017**, *4*, 6.
26. Joshi, A.Y.; Joshi, A.Y. A systematic review on powder mixed electrical discharge machining. *Heliyon* **2019**, *5*, e02963. [CrossRef]
27. Razak, M.R.A.; Liew, P.J.; Hussein, N.I.S.; Ahsan, Q.; Yan, J. Effect of Surfactants and Additives on Electrical Discharge Machining of Reaction Bonded Silicon Carbide. *ARPN J. Eng. Appl. Sci.* **2017**, *12*, 4334–4339.
28. Tanteh, D.N.; Al, S.Y.; Ssekasiko, D. *Properties of Transformer Oil that Affect Efficiency*; Department of Electrical Engineering, Blekinge Institute of Technology: Karlskrona, Sweden, 2014.
29. Zhao, W.; Meng, Q.; Wang, Z. The application of research on powder mixed EDM in rough machining. *J. Mater. Process. Technol.* **2002**, *129*, 30–33. [CrossRef]
30. Kumar, A.; Maheshwari, S.; Sharma, C.; Beri, N. Analysis of Machining Characteristics in Additive Mixed Electric Discharge Machining of Nickel-Based Super Alloy Inconel 718. *Mater. Manuf. Process.* **2011**, *26*, 1011–1018. [CrossRef]
31. Kumar, A.; Maheshwari, S.; Sharma, C.; Beri, N. Research Developments in Additives Mixed Electrical Discharge Machining (AEDM): A State of Art Review. *Mater. Manuf. Process.* **2010**, *25*, 1166–1180. [CrossRef]
32. Jamadar, M.U.M.; Kavade, M.V. Effect of aluminium powder mixed edm on machining characteristics of die steel (AISI D3). *IJMPE* **2014**, *2*, 120–123.
33. Dewan, P.R.; Kundu, P.K.; Phipon, R. Powder Mixed Electric Discharge Machining—A Review. In Proceedings of the AIP Conference Proceedings 2273, Seoul, Republic of Korea, 20 January 2020; Volume 2, p. 050075.
34. Amin, A.; Abdul-Rani, A.M.; Rana, M.; Hastuty, S.; Danish, M.; Rubaiee, S.; bin Mahfouz, A. Evaluation of modified 316L surface properties through HAp suspended EDM process for biomedical application. *Surf. Interfaces* **2022**, *28*, 101600. [CrossRef]
35. Qazi, M.J.; Schlegel, S.J.; Backus, E.H.; Bonn, M.; Bonn, D.; Shahidzadeh, N. Dynamic Surface Tension of Surfactants in the Presence of High Salt Concentrations. *Langmuir* **2020**, *36*, 7956–7964. [CrossRef] [PubMed]
36. Dave, N.; Joshi, T. A Concise Review on Surfactants and Its Significance. *Int. J. Appl. Chem.* **2017**, *13*, 663–672. [CrossRef]
37. Bart, J.C.J.; Gucciardi, E.; Cavallaro, S. Formulating Lubricating Oils. In *Biolubricants*; Elsevier: Amsterdam, The Netherlands, 2013; pp. 351–395. ISBN 978-0-85709-263-2.
38. Kumar, S.; Khedkar, N.K.; Jagtap, B.; Singh, T.P. The Effects of Cryogenic Treatment on Cutting Tools. *IOP Conf. Series Mater. Sci. Eng.* **2017**, *225*, 012104. [CrossRef]
39. Özdemir, Z. Shallow Cryogenic Treatment (SCT) Effects on the Mechanical Properties of High Cr Cast Iron: Low-Carbon Cast Steel Bimetallic Casting. *Int. J. Met.* **2021**, *15*, 952–961. [CrossRef]
40. Senthilkumar, D.; Rajendran, I. Influence of Shallow and Deep Cryogenic Treatment on Tribological Behavior of En 19 Steel. *J. Iron Steel Res. Int.* **2011**, *18*, 53–59. [CrossRef]

41. Razak, M.R.A.; Liew, P.J.; Hussein, N.I.S.; Ahsan, Q. Effect of Surfactant on EDM of Low Conductivity Reaction-Bonded Silicon Carbide. *Key Eng. Mater.* **2016**, *701*, 107–111. [CrossRef]
42. Li, L.; Li, Z.Y.; Wei, X.T.; Cheng, X. Machining Characteristics of Inconel 718 by Sinking-EDM and Wire-EDM. *Mater. Manuf. Process.* **2015**, *30*, 968–973. [CrossRef]
43. Wu, K.L.; Yan, B.H.; Lee, J.-W.; Ding, C.G. Study on the characteristics of electrical discharge machining using dielectric with surfactant. *J. Mater. Process. Technol.* **2009**, *209*, 3783–3789. [CrossRef]
44. Abdulkareem, S.; Khan, A.A.; Konneh, M. Reducing electrode wear ratio using cryogenic cooling during electrical discharge machining. *Int. J. Adv. Manuf. Technol.* **2009**, *45*, 1146–1151. [CrossRef]
45. Srivastava, V.; Pandey, P.M. Performance Evaluation of Electrical Discharge Machining (EDM) Process Using Cryogenically Cooled Electrode. *Mater. Manuf. Process.* **2012**, *27*, 683–688. [CrossRef]
46. Ram, N.R.; Rao, K.V.; Kanth, C.L.; Sri, M.N.S. Parametric Analysis on the Effect of Cryogenic Treatment on the Work Piece Material of EDM Process. *Int. J. Eng. Resea. Technol.* **2014**, *3*, 8.
47. Jafferson, J.M.; Hariharan, P. Machining Performance of Cryogenically Treated Electrodes in Microelectric Discharge Machining: A Comparative Experimental Study. *Mater. Manuf. Process.* **2013**, *28*, 397–402. [CrossRef]
48. Bartkowiak, T.; Mendak, M.; Mrozek, K.; Wieczorowski, M. Analysis of Surface Microgeometry Created by Electric Discharge Machining. *Materials* **2020**, *13*, 3830. [CrossRef] [PubMed]
49. Gogolewski, D. Multiscale Data Treatment in Additive Manufacturing. *Materials* **2023**, *16*, 3168. [CrossRef]
50. Ishfaq, K.; Sana, M.; Waseem, M.U.; Anwar, S.; Alfaify, A.Y.; Zia, A.W. Surface quality investigation in surfactant-based EDM of Inconel 617 using deep cryogenically treated electrodes. *Int. J. Adv. Manuf. Technol.* **2023**. [CrossRef]
51. Reddy, V.V.; Kumar, A.; Valli, P.M.; Reddy, C.S. Influence of surfactant and graphite powder concentration on electrical discharge machining of PH17-4 stainless steel. *J. Braz. Soc. Mech. Sci. Eng.* **2015**, *37*, 641–655. [CrossRef]
52. Kassem, M.G.; Ahmed, A.-M.M.; Abdel-Rahman, H.H.; Moustafa, A.H. Use of Span 80 and Tween 80 for blending gasoline and alcohol in spark ignition engines. *Energy Rep.* **2019**, *5*, 221–230. [CrossRef]
53. Khan, A.A. Electrode wear and material removal rate during EDM of aluminum and mild steel using copper and brass electrodes. *Int. J. Adv. Manuf. Technol.* **2008**, *39*, 482–487. [CrossRef]
54. Tola, O.J.; Zungeru, A.M.; Usifo, F.; Garba, A.J. Experimental Study of the Characteristics of Transformer Oil and Some Selected Vegetable Oils. *Int. J. Eng. Res. Afr.* **2016**, *23*, 13–23. [CrossRef]
55. Cetin, A.; Cakir, G.; Aslantas, K.; Ucak, N.; Cicek, A. Performance of cryogenically treated Cu and CuCrZr electrodes in an EDM process. *Met. Mater.* **2018**, *55*, 431–440. [CrossRef]
56. Yildiz, Y.; Sundaram, M.M.; Rajurkar, K.P.; Nalbant, M. The Effects of Cold and Cryogenic Treatments on the Machinability of Beryllium-Copper Alloy in Electro Discharge Machining. In Proceedings of the 44th CIRP Conference on Manufacturing Systems, Madison, WI, USA, 31 May–3 June 2011.

MDPI
St. Alban-Anlage 66
4052 Basel
Switzerland
Tel. +41 61 683 77 34
Fax +41 61 302 89 18
www.mdpi.com

Metals Editorial Office
E-mail: metals@mdpi.com
www.mdpi.com/journal/metals

www.ingramcontent.com/pod-product-compliance
Lightning Source LLC
Chambersburg PA
CBHW061825210926
51850CB00006B/810
* 9 7 8 3 0 3 6 5 8 2 8 0 1 *